Index to Illustrations of Living Things
Outside North America

Index to Illustrations
of Living Things
Outside North America

Where to Find Pictures of Flora and Fauna

Lucile Thompson Munz
and
Nedra G. Slauson

A Companion Volume to John W. Thompson's
Index to Illustrations of the Natural World:
Where to Find Pictures of the Living Things
of North America

Archon Books
1981

Library of Congress Cataloging in Publication Data

Munz, Lucile Thompson.
   Index to illustrations of living things outside North
America.

   "A companion volume to Thompson's Index to illustrations of the natural world (North America)."
   Includes index.
   Summary: Indexes photographs, paintings and sketches of
more than 9,000 species of animals and plants found all
over the world.
   1. Zoology—Pictorial works—Indexes. 2. Botany—
Pictorial works—Indexes. 3. Zoology—Indexes.
4. Botany—Indexes. [1. Zoology—Pictorial works—
Indexes. 2. Botany—Pictorial works—Indexes] I. Slauson, Nedra G. II. Thompson, John W., 1891-

Index to illustrations of the natural world. III. Title.
Z7998.N67M85 [QL46]   016.574022'2   81-8037
ISBN 0-208-01857-3   AACR2

First published 1981 as an Archon Book
an imprint of
The
Shoe String Press, Inc.
995 Sherman Avenue
Hamden, CT 06514

Printed in the United States of America

Contents

Biographical Data

Lucile Thompson Munz

Mrs. Munz accompanied her father, distinguished botanist John W. Thompson, on field trips to photograph plants and wildlife for more than 25 years. They travelled widely, taking thousands of photographs that are the basis of the J. W. Thompson Company's color slides, which are used by science teachers across the U.S., in Canada and around the world.

Since Mr. Thompson's death in 1978, Mrs. Munz has continued to operate the Thompson scientific materials business. Slides and sets of biology diagrams for all life forms are furnished to hundreds of colleges and schools each year. Commuting between the business in Seattle and her home in the foothills of the Cascade mountains means a lot of travelling, but Mrs. Munz finds time to increase her scientific knowledge and enhance her photographic skills. With it all, she allows herself time to look for additions to her favorite collections of early American household items such as trivets, glassware and miniatures associated with the "Pennsylvania Dutch" culture of the 19th century.

Nedra G. Slauson

Mrs. Slauson is a free-lance publications specialist and librarian, associated with a Seattle communications/publications firm. A journalism graduate of the University of Washington, she also has a master's degree in librarianship from that institution. She has more than 20 years' experience as a writer and editor, community college instructor and librarian. She served as the editor of Thompson's North American index.

An enthusiastic sportswoman and outdoors adventurer, Nedra Slauson enjoys backpacking, hiking, skiing, cycling and canoeing with her three grown children and/or fellow members of The Mountaineers, a Seattle-based outdoor recreation/conservation organization. She is a member also of the Seattle Chamber Singers, a 35-voice group which is planning its first European concert tour for the summer of 1981.

Introduction

The person who needs a picture of a particular plant, fish, bird, mammal, reptile or other living thing often faces a formidable search, especially if the item is uncommon. This index will simplify locating good-quality illustrations to more than 9,000 species of animals and plants found all over the world. The user will find that the 206 books covered by the index include the work of many of the world's outstanding biologists, botanists and other natural scientists, plus the photographs, paintings and drawings of scores of leading artists and photographers.

Books chosen for the index share at least three attributes: authoritativeness in their field, the fine quality of their illustrations, and their availability in medium-size and larger libraries. Virtually all have been published since 1963; many bear publication dates in the 1970s. The few titles published prior to 1963 have been included because of their excellent illustrations and/or coverage.

Every book indexed has at least some full-color illustrations; most use color illustrations exclusively. All art forms are represented—photographs, paintings and sketches.

Every animal and plant included has at least two citation; most have from six to twelve citations given, so the user should be able to find several of the books cited for any item in libraries or library media centers close at hand, and more if needed through the cooperative information network.

The index is world-wide in scope except for animals and plants found only in North America, which are covered by John W. Thompson's **Index to Illustrations of the Natural World,** first published in 1977. Some plants and animals are truly world-wide, however, and are included in this volume even though they also were cited in Thompson's index. Living things only are cited; those generally accepted as extinct are excluded.

How the Index is Organized

Entries are listed by common name in **boldface** type, followed by the scientific name (in parentheses). An item with more than one widely-used common name is cited under the name used by a majority of the books indexed; "see" references guide the reader from other names. If the same name applies to two or more categories of living things, each category is named (in boldface type) after the common name. For example, **Locust, Clammy (tree)** and **Locust, Desert (insect).**

The **Scientific Name Index** (back of the book) lists each item's scientific name followed by its common name as used in this index. Sometimes several species are grouped under the genus name: i.e., **Moth, Oak Silk** (Antherea sp.).

When several species of the same genus, or several genera, share the same common name, their citations may be grouped in one entry. More than one scientific name may be given for items which have been reclassified or whose classification is in dispute.

Each book title is identified by a three-letter code. For example, **The Birds of Britain and Europe** by Richard Fitter, illustrated by Herman Heinzel, published by Lippincott, Philadelphia, in 1972, has title code FBE. The **Hamlyn Encyclopedia of Plants** by J. Triska, London, Hamlyn Publishing Group, 1975, is coded TEP.

In citations, each title code is followed by page number and/or plate number of the picture(s) in that book. Most pictures are in color; those that are not are labeled (bw). Book titles are identified by consulting the **Bibliography by Code Letters** (following this Introduction) which lists, in alphabetical order by title code, all books indexed. Title codes in each citation are in alphabetical order to simplify the decoding process.

Here is an analysis of a typical entry in the index:

Monkey, Diana (Ceropithecus diana) ALE 10:393 (cp 8), DDM 64 (cp 9), DSM 81, HMΛ 71, HMW 24, SMW 85 (cp 45).

The illustration in ALE (decoded: Grzimek's Animal Life Encyclopedia. Bernard Grzimek, editor-in-chief. 13 volumes. New York, Van Nostrand Reinhold, 1974.) is in Volume 10, color plate 8 on page 393.

DDM (A Field Guide to the Larger Mammals of Africa, by Jean Dorst, illustrated by Pierre Dandelot. Boston, Houghton Mifflin, 1970.) pictures this monkey on color plate 9 on page 64.

DSM (World Guide to Mammals, by Nicole Duplaix and Noel Simon, illustrated by Peter Barrett. New York, Crown, 1876.) has a picture on page 81. Since (bw) is not indicated, the picture is in color.

HMA (Mammals, by Donald F. Hoffmeister. New York, Golden Press, 1963.) illustrates this monkey on page 71.

HMW (Mammals of the World, by Hans Hvass, illustrated by Wilhelm Eigener. London, Methuen, 1961.) has an illustration on page 24.

SMW (Living Mammals of the World, by Ivan T. Sanderson. Garden City, New York, Doubleday, 1961.) illustrates the diana monkey on color plate 45 on page 85.

Abreviations are: cp—color plate; cps—color plates; pl—black/white plate; pls -black/white plates; sp.—species; var.—variety. Note that cps 6, 7 means two color plates while cp 6, 7 means color plate 6 plus another illustration on page 7. Some books use an identifying number (in parentheses) following the page or plate number. When a genus name is used a second time in the same entry, it is abbreviated by its initial letter: i.e., **Lark, Short-toed** (Calandrealla cinerea and C. brachydactyla).

Entries are arranged alphabetically word by word. So **Spider** (without a modifying adjective) precedes **Spider, Barrel** and **Sparrow, Tree** precedes **Sparrow-hawk** and **Partridge, Tree** precedes **Partridge-berry.** A hyphenated word is considered one word, so **Flame Vine, Mexican** precedes **Flame-of-the-forest.** Names starting with an abbreviation are placed as if the abbreviation were spelled out: i.e., **St. Augustine Grass** is located as if it were **Saint Augustine Grass.** The category of item and punctuation are disregarded in alphabetizing, thus **Monkey, Proboscis** (a mammal) is followed by **Monkey Puzzle Tree** (a plant) and then by **Monkey, Red-eared Nose-spotted** (a mammal).

This index makes no claims to infallibility—although enormous effort has been made to be accurate—nor to universality—there are far too many thousands of animals and plants for that. The editors hope that users of this index will agree that the 9,000-plus items covered are an acceptable and useful cross-section of the life found in our vast and wonderful world.

Nedra G. Slauson
Lucile Thompson Munz

Seattle, Washington
May 1981

Bibliography by Code Letters

ABW Birds of the World. Oliver L. Austin, Jr. Illustrated by Arthur Singer. New York, Golden Press, 1961.

ALE Grzimek's Animal Life Encyclopedia. Bernard Grzimek, editor-in-chief. 13 vol. New York, Van Nostrand Reinhold, 1974.

AMP Health Plants of the World. Francesco Bianchini and Francesco Corbetta. Illustrated by Marilena Pistoia. New York, Newsweek Books, 1977.

ART The World of Amphibians and Reptiles. Milli Ubertazzi Tanara. Translated by Simon Pleasance. New York, Abbeville Press, 1978.

BBB Book of British Birds. Richard Fitter, editor. London, Drive Publications for Reader's Digest Association, 1973.

BBE Birds of Europe. Bertel Bruun. Illustrated by Arthur Singer. New York, McGraw-Hill, 1970.

BBT Birds of the Tropics. John A. Burton, New York, Crown, 1973.

BFB Handbook of British Flowering Plants. Edward A. Melderis and E. B. Bangerter. New York, Abelard-Schuman, 1959.

BIN The Amazing World of Insects. Arend T. Bandsma and Robin T. Brandt. New York, Macmillan, 1963.

BIP The ABC of Indoor Plants. Jocelyn Baines and Katherine Key. New York, Alfred A. Knopf, 1977.

BKT The Glory of the Tree. B.K. Boom and H. Kleijn. Garden City, N.Y., Doubleday, 1966.

BMA Encyclopedia of Mammals. Maurice and Robert Burton. London, Octopus Books, 1975.

BNZ Buller's Birds of New Zealand. Edited and revised by E.G. Turbott. Honolulu, East-West Center, 1967.

BOW Owls of the World. John A. Burton, editor. Illustrated by John Rignall. New York, E.P. Dutton, 1973.

BWG The Wild Garden. Lys de Bray. London, Weidenfeld and Nicolson, 1978.

CAA Animals of the World: Africa. Allan Cooper et al. London, Paul Hamlyn, 1968.

CAF Africa. Leslie Brown. (The Continents We Live On series.) New York, Random House, 1965.

CAS Asia. Pierre Pfeffer. (The Continents We Live On series.) New York, Random House, 1968.

CAU Australia and the Pacific Islands. Allen Keast. (The Continents We Live On series.) New York, Random House, 1966.

CAW Living Amphibians of the World. Doris M. Cochran. (World of Nature series.) Garden City, N.Y., Doubleday, 1961.

CDB Dictionary of Birds in Color. Bruce Campbell. New York, Viking Press, 1974.

CEU Europe. Kai Curry-Lindahl. (The Continents We Live On series.) New York, Random House, 1964.

CGF A Field Guide to the Ferns and Their Related Families. Boughton Cobb. Illustrated by Laura Louise Foster. (Peterson Field Guide series.) Boston, Houghton Mifflin, 1963.

CIB A Field Guide to the Insects of Britian and Northern Europe. Michael Chinery. (International series.) Boston, Houghton Mifflin, 1974.

COG Australian Reptiles in Colour. Harold Cogger. Honolulu, East-West Center, 1967.

CSA South America and Central America. Jean Dorst. (The Continents We Live On series.) New York, Random House, 1967.

CSF Collins Guide to the Sea Fishes of Britain and
     Northwestern Europe. Bent J. Muus. Illustrated
     by Preben Dahlstrom. London, Collins, 1974.

CSS Collins Pocket Guide to the Seashore. John
     Barrett and C.M. Yonge. Illustrated by Elspeth
     Yonge et al. London, Collins, 1970.

CTB Color Treasury of Butterflies and Moths.
     Introduction by Michael Tweedie. London,
     Crescent, 1972.

CTF Color Treasury of Aquarium Fish. Elso Lodi.
     London, Crescent, 1972.

CTH Color Treasury of Herbs and Other Medicinal
     Plants. Introduction by Jerry Cowhig. London,
     Crescent, 1972.

CTI Color Treasury of Insects: World of Miniature
     Beauty. Umberto Parenti. London, Crescent 1971.

CTM Color Treasury of Mushrooms and Toadstools.
     Introduction by Uberto Tosco and Annalaura
     Fanelli. London, Crescent, 1972.

CTS Color Treasury of Sea Shells. Serio Angeletti.
     Introduction by Michael Tweedie. London,
     Crescent, 1973.

DAA Animals of Asia. Robert Wolff. Illustrated by
     Robert Dallet. New York, Lion Press, 1969.

DAE Animals of Europe. Robert Wolff. Illustrated by
     Robert Dallet. New York, Lion Press, 1969.

DDM A Field Guide to the Larger Mammals of Africa.
     Jean Dorst. Illustrated by Pierre Dendelot.
     Boston, Houghton Mifflin, 1970.

DEW Plants of the World. H.C.D. deWitt. 3 vols. New
     York, E.P. Dutton, 1966-1969.

DFP Color Dictionary of Flowers and Plants for Home
     and Garden. Roy Hay and Patrick M. Synge. New
     York, Crown Publishers, 1969.

DLM The Encyclopedia of Mushrooms. Colin
     Dickinson and John Lucas. New York, G.P.
     Putnam's Sons, 1979.

DPB Philippine Birds. John Eleuthere duPont.
     Illustrated by George Sandstrom and John R.
     Peirce. Greenville, Del., Delaware Museum of
     Natural History, 1971.

DSM World Guide to Mammals. Nicole Duplaix and
     Noel Simon. Illustrated by Peter Barrett. New
     York, Crown, 1976.

DTB Portraits of Tropical Birds. John S. Dunning.
     Wynnewood, Pa., Livingston, 1970.

ECS Cacti and Succulents. Philip Perl. (Time-Life
     Encyclopedia of Gardening.) Alexandria, Va.,
     Time-Life Books, 1978.

EDP Decorating with Plants. Oliver E. Allen.
     (Time-Life Encyclopedia of Gardening.)
     Alexandria, Va., Time-Life Books, 1978.

EEG Easy Gardens. Donald Wyman and Curtis
     Prendergast. (Time-Life Encyclopedia of
     Gardening). Alexandria, Va., Time-Life Books,
     1978.

EET The Illustrated Encyclopedia of Trees. Herbert
     Edlin et al. New York, Harmony, 1978.

EFC Flowers in Color. W. Rytz and Herbert Edlin.
     Illustrated by Hans Schwarzenbach. New York,
     Viking Press, 1960.

EFP Foliage House Plants. James Underwood
     Crockett. (Time-Life Encyclopedia of Gardening.)
     New York, Time-Life Books, 1972.

EFS Flowering Shrubs. James Underwood Crockett.
     Illustrated by Allianora Rosse. (Time-Life
     Encyclopedia of Gardening.) New York,
     Time-Life Books, 1972.

EGA Annuals. James Underwood Crockett.
     Illustrated by Allianora Rosse. (Time-Life
     Encyclopedia of Gardening.) New York,
     Time-Life Books, 1971.

EGB Bulbs. James Underwood Crockett. Illustrated by
     Allianora Rosse. (Time-Life Encyclopedia of
     Gardening.) New York, Time-Life Books, 1971.

13

EGC   Lawns and Ground Covers. James Underwood Crockett. Illustrated by Allianora Rosse. (Time-Life Encyclopedia of Gardening.) Alexandria, Va., Time-Life Books, 1971.

EGE   Evergreens. James Underwood Crockett. Illustrated by R.A. Merrilees and John Murphy. (Time-Life Encyclopedia of Gardening.) New York, Time-Life Books, 1971.

EGF   Ferns. Philip Perl. Illustrated by Richard Crist. (Time-Life Encyclopedia of Gardening.) Alexandria, Va., Time-Life Books, 1977.

EGG   Greenhouse Gardening. James Underwood Crockett. (Time-Life Encyclopedia of Gardening.) Alexandria. Va., Time-Life Books, 1977.

EGH   Herbs. James Underwood Crockett and Ogden Tanner. Illustrated by Richard Crist. (Time-Life Encyclopedia of Gardening.) Alexandria, Va., Time-Life Books, 1977.

EGL   Gardening Under Lights. Wendy B. Murphy. (Time-Life Encyclopedia of Gardening.) Alexandria. Va., Time-Life Books, 1978.

EGO   Orchids. Alice Skelsey. (Time-Life Encyclopedia of Gardening.) Alexandria, Va., Time-Life Books, 1978.

EGP   Perennials. James Underwood Crockett. Illustrated by Allianora Rosse. (Time-Life Encyclopedia of Gardening.) New York, Time-Life Books, 1972.

EGT   Trees. James Underwood Crockett. (Time-Life Encyclopedia of Gardening.) New York, Time-Life Books, 1979.

EGV   Vines. Richard H. Cravens. (Time-Life Encyclopedia of Gardening.) Alexandria, Va., Time-Life Books, 1979.

EGW   Winter Gardens. Oliver E. Allen (Time-Life Encyclopedia of Gardening.) Alexandria, Va., Time-Life Books, 1979.

EHP   Hallucinogenic Plants. Richard Evans Schultes. Illustrated by Elmer W. Smith. (Golden Guide series.) New York, Golden Press, 1976.

EJG   Japanese Gardens. Wendy B. Murphy. (Time-Life Encyclopedia of Gardening.) Alexandria, Va., Time-Life Books, 1979.

ELG   Landscape Gardening. James Underwood Crockett. Illustrated by R.A. Merrilees and B. Wolff. (Time-Life Encyclopedia of Gardening.) Alexandria, Va., Time-Life Books, 1971.

EMB   Miniatures and Bonsai. Philip Perl. (Time-Life Encyclopedia of Gardening.) Alexandria, Va., Time-Life Books, 1979.

EPF   Pictorial Encyclopedia of Plants and Flowers. F.A. Novak. New York, Crown Publishers, 1966.

EPG   Pruning and Grafting. Oliver E. Allen. (Time-Life Encyclopedia of Gardening.) Alexandria, Va., Time-Life Books, 1978.

ERA   Encyclopedia of Reptiles, Amphibians and other Cold-blooded Animals. Introduction by Maurice Burton. London, Octopus Books, 1975.

ERW   Rock and Water Gardens. Ogden Tanner. (Time-Life Encyclopedia of Gardening.) Alexandria, Va., Time-Life Books, 1979.

ESG   Shade Gardens. Oliver E. Allen (Time-Life Encyclopedia of Gardening.) Alexandria, Va., Time-Life Books, 1979.

EVF   Vegetables and Fruits. James Underwood Crockett. Illustrated by Richard Crist. (Time-Life Encyclopedia of Gardening.) New York, Time-Life Books, 1972.

EWF   Wild Flowers of the World. Brian D. Morley. Illustrated by Barbara Everard. New York, G.P. Putnam's Sons, 1970.

EWG   Wildflower Gardening James Underwood Crockett and Oliver E. Allen. Illustrated by Richard Crist. (Time-Life Encyclopedia of Gardening.) Alexandria, Va., Time-Life Books, 1977.

FBE The Birds of Britain and Europe. Richard Fitter. Illustrated by Herman Heinzel. Philadelphia. Lippincott, 1972.

FBW Britain's Wildlife; Rarities and Introductions. Richard Fitter. Illustrated by John Leigh-Pemberton. London, Nicholas Kaye, 1966.

FEF The Pictorial Encyclopedia of Fishes. S. Frank. London, Hamlyn, 1971.

FHP Flowering House Plants. James Underwood Crockett. Illustrated by Allianora Rosse. (Time-Life Encyclopedia of Gardening.) New York, Time-Life Books, 1971.

FNZ A Field Guide to the Birds of New Zealand and Outlying Islands. R.A. Falla et al. Illustrated by Chloe Talbot-Kelly. Boston, Houghton Mifflin, 1967.

FSA A Field Guide to the Snakes of Southern Africa. V.F.M. Fitz Simons. London, Collins, 1970.

FWF Freshwater Fishes. Juraj Holcik and Jozef Mihalik. Feltham, England, Hamlyn, 1968.

GAL Animal Life of Europe. Jakob Graf. New York, Frederick Warne, 1968.

GBP Birds of Prey of the World. Mary Louise Grossman and John Hamlet. Photos by Shelly Grossman. New York, Potter, 1964.

GDS The Color Dictionary of Shrubs. S. Millar Gault. New York, Crown, 1976.

GFI Flowers of the Islands in the Sun. Graham Gooding. Illustrated by Clarence E. Hall. New York, Barnes, 1966.

GPB The Great Book of Birds. John Gooders. Foreword by Roger Tory Peterson. New York, Dial Press, 1975.

GSP Spiders and their Kin. Herbert W. and Lorna R. Levi. (Golden Nature Guide series.) New York, Golden Press, 1968.

GSS Sea Shells of the World. R. Tucker Abbott. Illustrated by George and Marita Sandstrom. (Golden Nature Guide series.) New York, Golden Press, 1962.

HAB Australian Birds. Robin Hill. New York, Funk and Wagnall, 1967.

HBG A Field Guide of the Birds of Galapagos. Michael Harris. Illustrated by Barry Kent McKay. London, Collins, 1974.

HFI Fishes of the World in Color. Hans Hvass. Illustrated by Wilhelm Eigener. New York, E.P. Dutton, 1965.

HFW Living Fishes of the World. Earl S. Herald. Garden City, N.Y., Doubleday, 1961.

HMA Mammals. Donald F. Hoffmeister. (Golden Bookshelf of Natural History series.) New York, Golden Press, 1963.

HMW Mammals of the World. Hans Hvass. Illustrated by Wilhelm Eigener. London, Methuen, 1961.

HPW Flowering Plants of the World. V. H. Heywood, editor. Illustrated by Victoria Goaman et al. Oxford, (England) University Press, 1978.

HRA Reptiles and Amphibians of the World. Hans Hvass. Illustrated by Wilhelm Eigener. London, Methuen, 1964.

HSC Shrubs in Colour. A.G.L. Hellyer. Garden City, N.Y., Doubleday, 1966.

HWR The World of Reptiles and Amphibians. John Honders, editor. New York, Peebles Press, 1975.

JAW Animal Atlas of the World. E.L. Jordan. Maplewood, N.J., Hammond, 1969.

KBA Field Guide to the Birds of Southeast Asia. Ben F. King and Edward C. Dickinson. (International series.) Boston, Houghton Mifflin, 1975.

KBM The World of Butterflies and Moths. Alexander B. Klots. New York, McGraw-Hill, (no date).

KIW Living Insects of the World. Alexander B. and Elsie B. Klots. (World of Nature series.) Garden City, N.Y., Doubleday, 1965 (?).

KMF Mushrooms and other Fungi. H. Kleijn. Garden City, N.Y., Doubleday, 1962.

KVP The Beauty of the Wild Plant. H. Kleijn and P. Vermeulen. London, George G. Harrap, 1964.

LAW Larousse Encyclopedia of the Animal World. Introduction by Desmond Morris. New York, Larousse, 1975.

LBR The Book of Reptiles. R.A. Lanworn. London, Hamlyn, 1972.

LCS Pocket Encyclopedia of Cacti and other Succulents in Color. Edgar and Brian Lamb. New York, Macmillan, 1970.

LEA Larousse Encyclopedia of Animal Life. New York, McGraw-Hill, 1967.

LFT Wild Flowers of the Transvaal. Cythna Letty. Johannesburg, South Africa, Hortors, 1962.

LFW Flowers of the World (in Full Color). Robert S. Lemmon and Charles L. Sherman. Garden City, N.Y., Hanover House, 1958.

LHM Collins Guide to Mushrooms and Toadstools. Morten Lange and F. Bayard Hora. London, Collins, 1963.

LHP The Instant Guide to Successful House Plants. David Longman. New York, New York Times Books, 1979.

LMF Marine Game Fishes of the World. Francesca laMonte. Illustrated by Janet Roemhild. Garden City, N.Y., Doubleday, 1952.

LVA The World's Vanishing Animals: The Mammals. Cyril Littlewood. Illustrated by D.W. Ovenden. New York, Arco, 1970.

LVB The World's Vanishing Birds. Cyril Littlewood. Illustrated by D.W. Ovenden. New York, Arco, 1972.

LVS Vanishing Species. Introduction by Romain Gary. New York, Time-Life Books, 1974.

MAR The World of Amphibians and Reptiles. Robert Mertens. New York, McGraw-Hill, 1960.

MBA Birds of Australia. Michael Morcombe. New York, Scribner, 1971.

MBF The Concise British Flora in Colour. W. Keble Martin. New York, Holt, Rinehart, Winston, 1965.

MCW Common Weeds of Canada. Gerald A. Mulligan. (no place), McClelland and Stewart, Ltd., 1976.

MEA An Illustrated Encyclopedia of Australian Wildlife. Michael Morcombe. South Melbourne, Australia, MacMillan, 1974.

MEP Exotic Plants. Julia F. Morton. Illustrated by Richard E. Younger. (Golden Nature Guide series.) New York, Golden Press, 1971.

MFF Collins Pocket Guide to Wild Flowers. David McClintock and R.S.R. Fitter. Illustrated by Dorothy Fitchew et al. London, Collins, 1971.

MLP Living Plants of the World. Lorus and Margery Milne. 2nd ed., rev. New York, Random House, 1975.

MLS The Mammals: A Guide to the Living Species. Desmond Morris. New York, Harper and Row, 1965.

MOL Ocean Life in Color. Norman Marshall. Illustrated by Olga Marshall. (Macmillan Color series.) New York, Macmillan, 1971.

MTB Field Guide to the Trees of Britain and Northern Europe. Alan Mitchell. Boston, Houghton Mifflin, 1974.

MWA Wild Australia. Michael K. Morcombe. New York, Taplinger, 1972.

MWF What Flower is That? Stirling Macoboy. New York, Crown, 1971.

NBA Garden Birds of South Africa. Kenneth Newman. New York, Elsevier, 1961.

NBB The Beauty of Birds. Cyril Newberry. New York, Hanover, 1958.

NFP Non-flowering Plants. Floyd S. Shuttleworth and Herbert S. Zim. (Golden Nature Guide series.) New York, Golden Press, 1967.

NHE Natural History of Europe. Harry Garms. Illustrated by Wilhelm Eigener. London, Paul Hamlyn, 1967.

OBB Oxford Book of Birds. Bruce Campbell. Illustrated by Donald Watson. Oxford (England) University Press, 1964.

OBI Oxford Book of Insects. John Burton et al. Illustrated by Joyce Bee et al. Oxford (England) University Press, 1968.

OBT Oxford Book of Trees. A.R. Clapham. Illustrated by B.E. Nicholson. Oxford (England) University Press, 1975.

OBV Oxford Book of Vertebrates. Marion Nixon. Illustrated by Derek Whiteley. Oxford (England) University Press, 1972.

OFP Oxford Book of Food Plants. S. G. Harrison et al. Illustrated by B.E. Nicholson. Oxford (England) University Press, 1969.

OGF Oxford Book of Garden Flowers. E.B. Anderson et al. Illustrated by B.E. Nicholson. Oxford (England) University Press, 1963.

OIB Oxford Book of Invertebrates. David Nichols with John A.L. Cooke. Illustrated by Derek Whiteley. Oxford (England) University Press, 1971.

ONP Oxford Book of Flowerless Plants. Frank H. Brightman. Illustrated by B.E. Nicholson. Oxford (England) University Press, 1966.

OWF Oxford Book of Wild Flowers. S. Ary and M. Gregory. Illustrated by B.E. Nicholson. Oxford (England) University Press, 1960.

PBE Field Guide to the Birds of Britain and Europe. Roger Tory Peterson (Peterson Field Guide series.) Boston, Houghton Mifflin, 1967.

PEI Pictorial Encyclopedia of Insects. V.J. Stanek. London, Paul Hamlyn, 1969.

PFE Flowers of Europe. Oleg Polunin. London, Oxford University Press, 1969.

PFF Fieldbook of Natural History. E. Laurence Palmer and H. Seymour Fowler. 2nd ed. New York, McGraw-Hill, 1975.

PFM Flowers of the Mediterranean. Oleg Polunin and Anthony Huxley. Boston, Houghton Mifflin, 1966.

PFW Flowers of the World. Frances Perry. Illustrated by Leslie Greenwood. New York, Crown, 1972.

PMU Mushrooms. Albert Pilat. Illustrated by Otto Usak. London, Spring Books, (no date).

PSA A Field Guide to the Birds of Southern Africa. O.P.M. Prozesky. Illustrated by Dick Findlay. London, Collins, 1970.

PUM Mushrooms and other Fungi. Albert Pilat. Illustrated by Otto Usak. London, Peter Nevill, 1961.

PWC Wildlife Crisis. H.R.H. Prince Philip, Duke of Edinburgh, and James Fisher. New York, Cowles, 1970.

PWF Wild Flowers of Britain. Roger Phillips. New York, Quick Fox, 1977.

PWI The World of Insects. Paul Pesson. New York, McGraw-Hill, 1959.

RBE Birds of Europe. A. Rutgers. Illustrated by John Gould. London, Methuen, 1966.

RBI Birds of Asia. A. Rutgers. Illustrated by John Gould. New York, Taplinger, 1969.

RBT Beetles. Ewald Reitter. New York, G.P. Putnam's Sons, 1961.

RBU Birds of Australia. A. Rutgers. Illustrated by John Gould. London, Methuen, 1967.

RDP Success with House Plants. Anthony Huxley, editor. Pleasantville, N.Y., Reader's Digest Association, 1979.

REM Exotic Mushrooms. Henri Romagnesi. New York, Sterling, 1971.

RTM The Complete Book of Mushrooms. Augusto Rinaldi and Vassili Tyndale. New York, Crown, 1974.

RWF The World of Flowers. Herbert Reisigl, editor. New York, Viking, 1964.

SAB A Field Guide to Australian Birds (Passerines). Peter Slater. Wynnewood, Pa., Livingston, 1974.

SAO Birds of the Atlantic Ocean. Ted Stokes. Illustrated by Keith Shackleton. New York, Macmillan, 1968.

SAR Animals of the Antarctic. Bernard Stonehouse. New York, Holt, Rinehart, Winston, 1972.

SBB Strange and Beautiful Birds. Josef Seget. London, Spring Books, 1965.

SBM Illustrated Encyclopedia of Butterflies and Moths. V.J. Stanek. Edited by Brian Turner. London, Octopus Books, 1977.

SCS Cacti and Succulents; A Concise Guide in Colour. Rudolf Subik. London, Hamlyn, 1968.

SFW Freshwater Fishes of the World. Gunther Sterba. Translated and revised by Denys W. Tucker. New York, Viking, 1963.

SHU Hummingbirds. Walter Scheithauer. New York, Crowell, 1967.

SIR Living Reptiles of the World. Karl P. Schmidt and Robert F. Inger. (World of Nature series.) Garden City, N.Y., Doubleday, 1957.

SLP The Last Paradises: On the Track of Rare Animals. Eugen Schuhmacher. Garden City, N.Y., Doubleday, 1967.

SLS Last Survivors. Noel Simon and Paul Geroudet. Illustrated by Helmut Diller and Paul Barruel. New York, World, 1970.

SMF Mushrooms and Fungi. Moira Savonius. New York, Crown, 1973.

SMG The Mushroom Hunter's Field Guide. Alexander H. Smith. Rev. ed. Ann Arbor, University of Michigan Press, 1963.

SMR The Mammals of Rhodesia, Zambia and Malawi. Reay H.N. Smithers. Illustrated by E.J. Bierly. London, Collins, 1966.

SMW Living Mammals of the World. Ivan T. Sanderson. (World of Nature series.) Garden City, N.Y., Doubleday, 1961.

SNB A Field Guide to Australian Birds (Non-passerines). Peter Slater. Wynnewood, Pa., Livingston, 1972.

SNZ New Zealand Flowers and Plants in Colour. J.T. Salmon. Wellington, New Zealand, Reed, 1963.

SPL Complete Guide to Plants and Flowers. Frances Perry, ed. New York, Simon and Schuster, 1974.

SSA A Guide to the Birds of South America. Rodolphe Meyer deSchauensee. Wynnewood, Pa., Livingston, 1970.

SST Guide to Trees. Stanly Schuler, editor. (Fireside series.) New York, Simon and Schuster, 1977.

SSW Snakes of the World. John Stidworthy. Illustrated by Dougal MacDougal. New York, Bantam, 1972.

TEP The Hamlyn Encyclopedia of Plants. J. Triska. London, Hamlyn, 1975.

TGF Guide to Garden Flowers. Norman Taylor.
Illustrated by Eduardo Salgado. Boston,
Houghton Mifflin, 1958.

TGS Guide to Garden Shrubs and Trees. Norman
Taylor. Illustrated by Eduardo Salgado. Boston,
Houghton Mifflin, 1965.

TVG A Field Guide in Color to Plants. Jan Tykac and
Vlastimil Vanek. London, Octopus Books, 1978.

TWA Wild Animals of the World. T.L.C. Tomkins.
Illustrated by Rein Stuurman. (no place),
Pitman, 1962.

VRA Reptiles and Amphibians. Zdenek Vogel.
Illustrated by P. Pospisil and M. Rada. London,
Paul Hamlyn, 1966.

VWA Vanishing Wild Animals of the World. Richard
Fitter. Illustrated by John Leigh-Pemberton.
London, Midland Bank, 1968.

VWS Strange Wonders of the Sea. H. Gwynne Vevers.
Garden City, N.Y., Hanover House, 1957.

WAB World Atlas of Birds. London, Mitchell Beazley,
1974.

WAN Birds of the Antarctic. Edward Wilson. Brian
Roberts, editor. New York, Humanities Press,
1968.

WAW Australian Wildlife. Eric Worrell. London,
Angus and Robertson, 1966.

WBA A Field Guide to the Birds of East and Central
Africa. J.G. Williams. Illustrated by R.
Fennessy. Boston, Houghton Mifflin, 1964.

WBV Birds of South Vietnam. Philip Wildash.
Rutland, Vt., Charles E. Tuttle, 1968.

WDB Dictionary of Butterflies and Moths in Color.
Allan Watson and Paul E.S. Whalley. New York,
McGraw-Hill, 1975.

WEA The World Encyclopedia of Animals. Maurice
Burton, editor. New York, World Publishing,
1972.

WFB The Wild Flowers of Britain and Northern
Europe. Richard and Alastair Fitter. Illustrated
by Marjorie Blamey. New York, Scribner, 1974.

WFI The Birds of the Falkland Islands. Robin W.
Woods. New York, Anthony Nelson, 1975.

WFW Fishes of the World: An Illustrated Dictionary.
Illustrated by Alwyne Wheeler and Peter
Stebbing. New York, Macmillan, 1975.

WPW A Guide to the Pheasants of the World. Philip
Wayre. Illustrated by J.C. Harrison. London,
Country Life, 1969.

WUS Common Weeds of the United States. United
States Department of Agriculture. New York,
Dover, 1971.

WWT The World of Wildflowers and Trees. Uberto
Tosco. New York, Crown, 1973.

ZWI The World of Insects. Adriano Zanetti. Translated
by Catherine Atthill. New York, Abbeville Press,
1978.

Index to Illustrations of the Natural World

A

Accentor, Siberian (Prunella montanella)
BBE 221; FBE 221.

Achimenes, Trumpet (Achimenes longiflora)
MEP 141; MWF 29.

Acidanthera (Acidanthera bicolor murielae)
DEW 2:223 (cp 140); DFP 83 (cp 657); EGB 90;
OGF 117; TGF 44; TVG 181.

Aconite, Winter (Eranthis hyemalis) AMP
53; DEW 1: 59 (cp 26); DFP 92 (cp 729); EFC
cp 47; EGB 112; EGW 78, 109; EPF pl 202;
ESG 110; HPW 47; LFW 50 (cp 117); MBF cp
4; MFF cp 38 (5); NHE 178; PFE cp 18 (202);
PFW 250; PWF cp 12a; TGF 61; TVG 193;
WFB 67 (4); (E. tubergeniana) DFP 92 (cp
730); OGF 7.

Acouchi (Myoprocta pratti) MLS 230 (bw);
SMW 142 (cp 93); (M. acouchi) ALE 11:438.

Adam-and-Eve-in-a-Bower—See **Dead-
nettle, White.**

Addax (Addax nasomaculatus) ALE 13:402;
CAF 26; DDM 188 (cp 31) DSM 259; HMW 73;
LVA 13; LVS 131; PWC 96-97; SLS 207; SMW
262 (bw).

Adder—See also **Viper.**

Adder, Burrowing—See **Viper, Mole.**

Adder, Death (Acanthophis antarcticus) COG
cp 42; HRA 50; MAR 174 (pl 70); SIR 235 (bw);
SSW 106; WAW 86 (bw).

Adder, Desert (Bitis peringueyi) FSA 177 (pl
30); MAR 185 (pl 75).

Adder, European—See **Viper, Northern.**

Adder, Gaboon or West African (Bitis gabonica) CAA 113 (bw), FSA 173 (cp 28); MAR
186 (pl 76), SIR 249 (cps 130, 131); SSW 129.

Adder, Horned (Bitis caudalis) FSA 173 (cp
28), 193 (pl 32); MAR 189 (cp XV); (B.
cornuta) FSA 192 (pl 31).

Adder, Night (Causus rhombeatus) FSA 173
(cp 28), 176 (pl 29); LBR 37; SSW 128.

Adder, Puff (Bitis arietans) CAA 109; ERA
234; FSA 173 (cp 28); HRA 44; LAW 432; LBR
112; LEA 328 (bw); SIR 248 (cp 129); SSW
127.

Adder's tongue—See **Fern, Adder's-tongue.**

Adenostyles, Glabrous (Adenostyles glabra)
EFC cp 31; NHE 285.

Adenostyles, Gray (Adenostyles alliariae)
KVP cp 119; NHE 286; PFE cp 150 (1445).

Adonis—See **Pheasant's Eye.**

Addra—See **Gazelle, Dama.**

Adjutant—See **Marabou.**

Aechmea (Aechmea caudata) MWF 32; (A.
mariae-reginae) LFW 237 (cp 528); (A.
nudicaulis) HPW 295; (A. tillandsioides)
PFW 52 (A. chantinii) MEP 13; RDP 72; (A.
miniata) DEW 2:254; EPF pl 1033.

Aeonium (Aeonium arboreum) ECS 93; LCS
cps 196-197; PFE cp 38 (386); RDP 10, 73; (A.
glutinosum): MEP 43.

Agama—See **Lizard.**

Agapanthus, Pink or Nerine (Nerine
bowdenii) DFP 109 (cps 865, 866); EWF cp
86a; MWF 212; OGF 175; SPL cp 433.

Agaric, Broad-gilled (Collybia platyphylla)
NFP 66; PMU cp 85; REM 136 (cp 104); RTM
87.

Agaric, Dingy (Tricholoma portentosum)
DLM 261; LHM 87; PMU cp 74b; REM 121 (cp
89); RTM 101; TEP 221.

Agaric, Fragrant (Lepiota clypeolaria) DLM
216; LHM 127; NFP 59; PMU cp 109; REM 78
(cp 46).

Agaric, Gray (Tricholoma terreum) CTM 30;
DLM 263; LHM 89; NFP 61; PMU cp 74a;
REM 117 (cp 85); RTM 99.

Agaric, Red-haired (Tricholoma rutilans)
CTM 35; NFP 60; REM 119 (cp 87).

Agaric, Spindle-stem (Collybia fusipes) DLM
181; LHM 101; ONP 123; PMU cp 87; REM
138; RTM 86.

Agaric, Spring (Agrocybe praecox) DLM 165;
LHM 153; REM 85 (cp 53); RTM 64.

Agaric, Stout (Amanita spissa) PMU cp 114;
REM 60 (cp 28); RTM 20.

Agaric, Wood (Collybia dryophila) DLM 180;
LHM 101; ONP 127; PMU cp 88; RTM 89;
SMG 164.

Agave (Agave attenuata) EWF cp 168 b; MWF
33; RDP 75; (A. univittata) ECS 94.

Agave, Victoria-American (Agave victor-
iae-americanae) BIP cp 9; SPL cp 126.

Agouti, Orange-rumped (Dasyprocta aguti)
ALE 11:438; DSM 133; HMA 23; HMW 177;
LEA 523 (bw); MLS 229 (bw); SMW 151 (bw);
TWA 140.

Agrimony, Common (Agrimonia eupatoria)
AMP 171; BFB cp 9; BWG 36; EGH 93; KVP
cp 84; cp 60 (363); NHE 211; OWF 17; PWF
99E; TEP 109; WFP 107 (3); (A. odorata)
NHE 27; (A. pilosa) NHE 211.

Agrimony, Hemp (Eupatorium cannabinum) AMP 35; BFB cp 19; EPF pl 527; MBF cp 44; MFF cp 21 (822); NHE 100; OWF 155; PFE cp 141 (1357); PFW 131i; TEP 257; WFB 231.

Agrion—See **Damselfly, Agrion.**

Aichryson (Aichryson sp.) BIP 21 (cp 7); ECS 95; EGL 84; ROP 78.

Akeake (Dodonaea viscosa) SNZ 52 (cps 123-125); (D. bursarifolia) HPW 193.

Akebia, Five-leaf (Akebia quinata) DEW 1:95; EGC 121; EGV 91; EJG 99; EWF cp 90 d; HPW 50; HSC 11, 12 (bw); MWF 35; OGF 63; PFW 156; TGS 71.

Albacore (Thunnus alalunga) ALE 5:191; CSF 145; OBV 27; (T. germo) LMF 49.

Albacore, False (Euthynnus alletteratus) CSF 145; LMF 45 (bw).

Albatross, Black-browed (Diomedea melanophris) BBE 25; CDB 36 (cp 24); FBE 25; HAB 9; PSA 65 (pl 6); RBU 100; SAO 29; SAR 107, 108; SSA pl 20; WAN 81-84; WEA 42; WFI 83, 84.

Albatross, Black-footed (Diomedea nigripes) ABW 30; ALE 7:138; CAU 222 (bw); CDB 37 (cp 25).

Albatross, Gray-headed (Diomedea chrysostoma) BBT 118; CSA 283, 284 (bw); HAB 9; LEA 363 (bw); SAO 29; SAR 107, 108; WAB 222; WAN 85, 86, 90.

Albatross, Laysan (Diomedea immutabilis) ABW 30; CAU 228-229 (bw); WEA 42.

Albatross, Light-mantled Sooty (Phoebetria palpebrata) ALE 7:138; CAU 62 (bw); CDB 37 (cp 26); HAB 9; SAR 107, 109; WAN 87-89.

Albatross, Royal (Diomedea epomophora) SAO 29, SLP 150; WAN 77.

Albatross, Sooty (Phoebetria fusca) HAB 8; SAO 29; WAN 88.

Albatross, Wandering (Diomedea exulans) ABW 30; ALE 7:138; BNZ 205; CDB 36 (cp 22); GPB 44; HAB 8-9; LAW 447; PSA 65 (pl 6); SAO 24, 29; SAR 85, 106, 107; WAB 223; WAN 77, 80; WEA 42.

Albatross, Waved (Diomedea irrorata) ALE 7:138; CDB 36 (cp 23); CSA 252 (bw); GPB 45; HBG 52 (bw), 64 (pl 1); LAW 464; LVS 194-195.

Albatross, Yellow-nosed (Diomedea chlororhynchos) HAB 9; PSA 65 (pl 6).

Alderfly—See **Fly, Alder.**

Alder, Black (Ilex sp.)—See **Winterberry.**

Alder, Black or Buckthorn or Berry-bearing (Frangula sp.)—See **Buckthorn, Alder.**

Alder, European or Common (Alnus glutinosa) BKT cp 62; DEW 1:108 (cp 63), 187 (bw); EET 142; EPF pl 329; HPW 59; MBF cp 77; MFF pl 71 (560); NHE 23; OBT 13, 17; OWF 201 (bw); PFE cp 4 (40); PWF 13 n, 159; SST cp 80; TEP 47; TGS 87; WFB 31 (8).

Alder, Gray or Speckled (Alnus incana) DEW 1:186; EET 143; MTB 193 (cp 18); NHE 23; OBT 153; PFE cp 4 (41); PFF 139; PWF 20f, 159p; SST cp 81.

Alder, Green (Alnus viridis) EPF pl 331; MTB 193 (cp 18); NHE 23.

Alder, Italian (Alnus cordata) BKT cp 64; DFP 181 (cp 1443); EET 142-143; EGT 98; MTB 193 (cp 18); OBT 153.

Alder, White (Alnus rhombifolia) EGT 98; EPG 104.

Alexanders (Smyrnium olusatrum) BFB cp 11 (9); BWG 40; MBF cp 36; MFF cp 51 (466); OWF 47; PWF 42g; WFB 167 (2); (S. perfoliatum) PFE cp 84 (866); PFM cp 118.

Alfalfa (Medicago sativa) BFB cp 7; EPF pl 465; MBF cp 21; MFF cp 5 (278); NHE 212; OWF 135; PFE cp 58 (582, 583); PFF 214; PWF 84 d; TEP 165; WFB 127 (9).

Alfileria—See **Storksbill, Common.**

Alga, Green (Volvox sp.) DEW 3: cp 7; NFP 22; PFF 50.

Alga, Yellow (Botrydium granulatum) DEW 3: cp 3; NFP 20; PFF 59.

Alison, Hoary (Berteroa incana) EPF pl 247; MFF pl 92 (80); NHE 217; PWF 123 g; WFB 97 (4).

Alison, Small (Alyssum alyssoides) MFF cp 44 (78); NHE 217; WFB 85 (3).

Alkanet (Alkanna sp.) EWF cp 37d; PFE cp 103 (1061); PFM cp 152, pl 14 (388)

Alkanet, Green (Pentaglottis sempervirens) EWF cp 20a; MBF cp 60; MFF cp 13 (653); OWF 171; PFE cp 102 (1058); PWF cp 24b; WFB 193 (2).

Alkanet, Large Blue (Anchusa azurea) DFP 121 (cp 963); EGP 95; LFW 10 (cp 21); OGF 85; PFE cp 102 (1056); PFM cp 144; PFW 49; TGF 172; WFB 192.

Alkanet, True (Anchusa officinalis) EGH 97; EPF pl 796; HPW 236; NHE 229; PFE cp 103 (1055); PWF 125 l; TEP 117; WFB 193 (4).

Allamanda, Purple (Allamanda violacea) EWF cp 180a; MEP 108; MWF 36.

Allamanda, Yellow (Allamanda neriifolia) DEW 2:102 (cp 55); MWF 36.

Allgood—See **Good King Henry.**

All-heal—See **Self-heal** (Prunella).

Alligator, Caiman—See **Caiman.**

Alligator, Chinese (Alligator sinensis) ALE 6:126; HRA 78; LVS 168.

Alligatorweed—See **Alternanthera.**

Allium, Giant (Allium albopilosum) EGB 57, 92; OGF 71.

Allseed (Radiola linoides) MBF cp 19; MFF pl 105 (232); NHE 95; OWF 83; PFE 220; PWF 146e; WFB 133 (10).

Allseed, Four-leaved (Polycarpon tetraphyllum) MBF cp 16; MFF pl 105 (193); WFB 57(9).

Allspice or Pimento (Pimenta dioica) OFP 133; SST cp 291.

Alluaudia (Alluaudia sp.) ECS 96; HPW 74.

Almond (Prunus dulcis) EET 173, 174; EVF 134; OBT 180; OFP 27; SPL cp 318.

Almond, Flowering (Prunus triloba) DFP 221 (cp 1767); EPF pl 427; SPL cp 320; TGS 198; TVG 237.

Aloe (Aloe speciosa) AMP 29; MLP 253; MWF 37; WWT 102.

Aloe, American—See **Century Plant, American.**

Aloe, Candelabra or Tree (Aloe arborescens) MWF 37; RDP 80.

Aloe, Lace (Aloe aristata) EPF pl 889; RDP 80.

Aloe, Partridge-breasted or Tiger (Aloe variegata) BIP 23 (cp 13); DFP 51 (cp 407); EDP 144; EFP 88; RDP 79; SCS 69.

Aloe, True (Aloe vera) EGH 96; EGL 84.

Alpaca (Lama pacos) ALE 13:133; CSA 208, 209; HMA 65; HMW 57; JAW 171; MLS 378 (bw); SMW 237 (cp 150); WEA 44.

Alpencress—See **Hutchinsia, Alpine.**

Alpenrose (Rhododendron ferrugineum) DFP 223 (cp 1781); EFC cp 10; NHE 266; PFE cp 87 (919); RWF 30; SPL cp 494; TEP 97.

Alpenrose, Hairy (Rhododendron hirsutum) EFC cp 10; EPF pl 478; NHE 266; PFE cp 87 (919).

Alpine Bells (Cortusa matthioli) NHE 279; PFE cp 91 (955).

Alternanthera (Alternanthera amoena) EGA 92; MWF 38; (A. philoxeroides) WUS 141.

Aluminum Plant (Pilea cadieri) BIP 145 (cp 218); DFP 79 (cp 627); EFP 134; EMB 128; LHP 150, 151; MWF 233; PFW 302; RDP 317; SPL cp 29.

Alyssum, Golden—See **Basket of Gold.**

Amadina—See **Finch.**

Amaryllis (Hippeastrum hybrids) BIP 99 (cp 149); DFP 69 (cps 550, 551); EDP 119; EGB 119; EGG 115; EPF pl 928; FHP 125; PFW 24; RDP 239, 240; SPL cps 289, 291.

Amaryllis, Hardy or Garden (Lycoris squamigera) BIP 115 (cp 176), EGB 128; LFW 67 (cp 156); TGF 13.

Amazon—See **Parrot.**

Amazon Vine—See **Golden Vine** (Stigmaphyllon).

Amblydoras, Hancock's (Amblydoras hancocki) FEF 251 (bw); HFI 110.

Amoeba (Amoeba sp.) ALE 1:107, 305; (Arcella sp.) ALE 1:305; 2:97; LEA 16; OIB 1; (Difflugia sp.) ALE 2:97; OIB 1.

Amorphophallus (Amorphophallus sp.) DEW 2:311; EPF pl 1003; EWF cp 67a; MEP 11.

Ampelopsis (Ampelopsis brevipedunculata) EGV 49, 92; MWF 40; PFW 307; TGS 71.

Amphioxus (Branchiostoma sp.) ALE 3:428; HFI 152; LAW 277; LEA 204 (bw); OIB 177; WEA 218.

Anacampseros (Anacampseros sp.) ECS 12, 98; LCS cp 216; LFT 134 (cp 66); WWT 85.

Anaconda (Eunectes murinus) ALE 6:376; CSA 61; ERA 205, 206; HRA 67; LBR 10, 11; LEA 320 (bw); SIR 154 (cp 74); SSW 51; VRA 213; (E. notaeus) ALE 6:376; VRA 213.

Anaptychia (Anaptychia sp.) DEW 3:225;
ONP 165.

Anchovy (Engraulis encrasicholus) CSF 71;
HFI 118; (Stolephorus heterolobus) WFW cp
29.

Andromeda, Japanese (Pieris japonica)
DFP 218 (cp 1738); EGE 140; EGW 132;
EJG 136; ELG 142; EPF pls 482, 484; ESG
136; GDS cps 319-321; LFW 113 (cp 250);
MWF 233; PFW 107; TGS 406.

Andryala (Andryala integrifolia) PFE cp 159
(1533); PFM cp 27.

Anemone, Alpine (Pulsatilla alpina) DFP 21
(cp 167); NHE 269; PFE cp 23 (220); (P.
montana) EFC cp 43; NHE 269.

Anemone, Blue Wood (Anemone apennina)
NHE 199; PFE cp 22 (214); WFB 76.

Anemone, Daffodil (Anemone narcissiflora)
DFP 121 (cp 966); EPF pl 216; NHE 269;
PFE cp 22 (218); WFB 76.

Anemone, Great Peacock of Greece
(Anemone pavonina) DFP 85 (cps 674, 675);
OGF 31; PFE cp 21 (216).

Anemone, Greek (Anemone blanda) DFP 84
(cp 671) EGB 95; EGW 78, 89; EMB 98; EWF
cp 8 d; OGF 9; PFE cp 21 (214); PFM cp 21;
PFW 252.

Anemone, Japanese (Anemone japonica or
A. hupehensis) DFP 121 (cp 964); EGP 95;
EJG 100; OGF 179; PFW 252; SPL cp 188;
TGF 61; TVG 79.

Anemone, Poppy or Crown (Anemone
coronaria) DFP 84 (cp 672); EGB 54, 55,
95; EGG 87; EWF cp 27 f; HPW 48; KVP cp
164; LFW 11 (cps 22, 23); MLP 53; NHE 199;
PFE cp 21 (216); PFM cps 24, 26, 27; SPL cp
187; TVG 181.

Anemone, Sea—See **Sea Anemone.**

Anemone, Spring or Pale (Pulsatilla
vernalis) DFP 22 (cp 170); EPF pl 217; NHE
71; PFE cp 22 (221); RWF 24, 25; WFB 77 (2);
(P. pallida) EPF pl 219.

Anemone, Star (Anemone hortensis) NHE
199; PFE cp 21 (216); PFM cps 22, 23, 25;
SPL cp 470.

Anemone, Wood (Anemone nemorosa) AMP
115; DFP 85 (673); EFC cp 44; EPF pl 213;
ERW 95; MBF cp 1; MFF pl 84 (8); NHE 26;
OWF 67; PFE cp 22 (214); PWF cp 19 n; TEP
233; WFB 77 (5).

Anemone, Yellow (Anemone ranunculoides)
AMP 115; NHE 26; OGF 7; TEP 233; WFB 77
(6).

Anemone-fish (Amphiprion sp.) ALE 5:133,
134, 137; CTF 62 (cp 85); FEF 458, 459 (bw),
463; HFI 37; HFW 162 (cp 64) LAW 341;
MOL 115; VWS 5 (bw); WFW cps 385-387.

Angel Wings—See **Caladium.**

Angelfish —See also **Discus** (Symphysodon).

Angelfish (Pomacanthus sp. or Pomacanth-
odes sp.) ALE 5:95, 96, 115; CTF 62 (cp 84);
FEF 336, 353, 354, 371, 372, 387; HFI 33;
HFW 94 (cps 55, 56), 95 (cp 57), 161 (cp 61);
LAW 340, 342; VWS 15, 17 (bw), 29; WFW
cps 342-349.

Angelfish (Euxiphipops asfur) WFW cp 332;
(E. xanthometopon) FEF 390 (bw); (Holacanthus sp.) ALE 5:95; WFW cp 339.

Angelfish or **Bullfish** (Heniochus sp.) ALE
5:115; CTF 61 (cp 82); FEF 381, 392 (bw);
HFI 34; WFW cps 336-338.

Angelfish, Freshwater (Pterophyllum sp.)
ALE 5:121; CTF 53 (cps 68, 69); FEF 409,
431 (bw); HFI 35; HFW 10 (bw); LEA 209
(bw); SFW 588 (pl 159), 589 (pl 160); 601 (cp
166) WFW cp 377.

Angelfish, Royal (Pygoplites diacanthus)
HFI 34, HFW 95 (cp 58); WFW cp 351.

Angelica, Common or Eurasian (Angelica
archangelica) AMP 41; EGH 98; MLP 196;
NHE 95; OFP 149; PFE cp 85 (895); PWF
63g.

Angelica Tree, Japanese (Aralia elata) DFP
181 (cp 1446); EPF pl 754; GDS cp 9; SST cp
82.

Angelica, Wild (Angelica sylvestris) BFB cp
11 (10); MBF cp 40; MFF pl 97 (503); NHE
182; OWF 87; PWF 148c; TEP 247; WFB 161
(2).

Angel's Trumpet (Datura candida) EHP 146,
147; MEP 128; (D. inoxia) EHP 144; LFW
244 (cp 543).

Angel's Trumpet (Datura metel) EGA 113;
EHP 53; MEP 129; PFE cp 118 (1186); PFM
cp 168; TGF 208.

Angel's Trumpet (Datura suaveolens) BIP 67
(cp 95); DFP 62 (cp 495); EHP 147; LFW 245
(cp 544) MWF 106; PFW 283; SPL cp 80;
TVG 223; (D. arborea) EHP 146; RWF 96.

Angler, Ceratiid or Holboell's (Ceratias holboelli) ALE 4:421; HFI 11; MOL 99.

Angler, Deep-sea (Linophryne sp.) HFI 11; MOL 100; (Oneirodes carlsbergi) WFW cp 185;(Melanocetus sp.) ALE 4:421; HFI 11; HFW 285 (bw); LAW 330; MOL 100.

Angler, European (Lophius piscatorius) CSF 195; FEF 535 (bw); HFI 10; MOL 106; OBV 41; WFW cp 181.

Angwantibo—See **Potto, Golden.**

Anhinga (Anhinga rufa) ALE 7:168; DPB 9; FBE 33; LAW 502; MBA 51; PSA 32 (cp 1); SNB 26 (cp 12).

Ani, Greater (Crotophaga major) ALE 8:370; CDB 112 (cp 424).

Ani, Smooth-billed (Crotophaga ani) CDB 112 (cp 423); GPB 184; WAB 99 (bw).

Anise or Aniseed (Pimpinella anisum) AMP 43; EGH 134; EVF 139; OFP 137; PFF 267.

Anoa, Lowland (Anoa depressicornis) DAA 114; DSM 250; HMW 89; LVA 53; MLS 400 (bw); PWC 246 (bw); SMW 258 (bw); (Bubalus depressicornis) ALE 13:302.

Annatto Tree (Bixa orellana) HPW 106; MEP 87; MWF 57; SST cp 269.

Anole (Anolis sp.) ALE 6:193, 194; MAR pl 47; (Tropidurus grayi) ALE 6:194.

Anostomus, Striped (Anostomus anostomus) ALE 4:291; FEF 117 (bw); HFI 89; SFW 176 (cp 39); WFW cp 96.

Anostomus, Three-spot (Anostomus trimaculatus) FEF 118 (bw); SFW 161 (pl 35); WFW cp 98.

Ant (Anergates atratulus) OBI 151; (Myrmecia sp.) BIN pls 113, 114; (Paltothyreus tarsatus) LAW 224.

Ant, Army (Eciton sp.) ALE 2:523; KIW 266 (cp 141); PEI 166 (bw).

Ant, Carpenter (Camponotus sp.) CTI 60 (cp 83); KBM 165 (pl 57); KIW 290 (bw); PWI 123 (bw).

Ant, Harvester (Messor barbarus) CIB 260 (cp 39); LEA 123 (bw); PWI 108 (bw), 159.

Ant, Large Red—See **Ant, Wood.**

Ant, Leaf-cutter (Atta sp.) KIW 266 (cp 140); PEI 165 (bw).

Ant, Leaf-tying or Weaving (Oecophylla sp.) CAS 250 (bw); LEA 123 (bw); PWI 119 (bw); ZWI 208.

Ant, Meadow (Lasius sp.) EPD 86; LAW 225; OBI 151; PWI 119 (bw).

Ant, Velvet—See **Wasp, Velvet-ant.**

Ant, Wood (Formica rufa) ALE 2:51; CIB 260 (cp 39); CTI 61 (cp 84); LEA 113 (bw), 148 (bw); OBI 153; PEI 162-164 (bw); PWI 108 (bw); WEA 51; ZWI 95, 117, 206; (F. fusca) PWI 111 (bw).

Antbear—See **Aardvark.**

Antbird (Drymophila sp.) DTB 45 (cp 22); SSA cp 10; (Gymnopithys sp.) SSA cp 10; WAB 84; (Gymnocichla nudiceps) ALE 9:126.

Antbird, Chestnut-backed or -tailed (Myrmeciza exsul) ABW 200; ALE 9:126; (M. hemimelaena) SSA pl 26.

Antbird, Ocellated (Phaenostictus mcleannani) DTB 51 (cp 25); SSA cp 36.

Antbird, Spectacled (Phlegopsis nigromaculata) ABW 200; ALE 9:126; CDB 142 (cp 569); DTB 53 (cp 26).

Antbird, Spotted (Hylophylax naevioides) ABW 200; DTB 49 (cp 24); SSA pl 47; (H. poecilonata) ALE 9:126.

Antbird, White-faced or -plumed (Pithys albifrons) ABW 200; ALE 9:126; CDB 142 (cp 570); DTB 47 (cp 23); GPB 234; SSA cp 36.

Anteater, Banded or Marsupial (Myrmecodius fasciatus) ALE 10:89; CAU 40 (bw); DSM 29; HMW 200; LVA 57; LVS 32; MEA 92; MLS 55 (bw); MWA 81 (bw); TWA 198; VWA cp 33.

Anteater, Giant or Great (Myrmecophaga tridactyla or M. jubata) ALE 11:173, 190; DSM 100; HMA 24; HMW 180; JAW 65; LAW 578; LEA 514 (bw); LVS 65; MLS 167 (bw); SMW 90 (cp 55); TWA 142.

Anteater, Lesser or Three-toed (Tamandua tetradactyla or T. tridactyla) ALE 8:359; 11:173, 190; DSM 101; HMA 24; HMW 181; LAW 577; LEA 515 (bw); MLS 168 (bw); SMW 106 (bw).

Anteater, Scaly—See **Pangolin.**

Anteater, Spiny—See **Echidna.**

Anteater, Two-toed or Dwarf (Cyclopes didactylus) ALE 11:190; HMA 24; MLS 169 (bw).

Antelope, Bates Pygmy (Neotragus batesi) DDM 257 (cp 42); JAW 196.

Antelope, Desert—See **Addax.**

Antelope Ears—See **Fern, Stag's-horn.**

Antelope, Four-horned (Tetracerus quadricornis) ALE 13:302; DAA 89.

Antelope, Hunter's—See **Hartebeest, Hunter's.**

Antelope, Indian—See **Blackbuck.**

Antelope, Roan (Hippotragus equinus) ALE 13:402; DDM 188 (cp 31); DSM 257; HMW 72; SMR 97 (cp 8) TWA 25.

Antelope, Royal (Neotragus pygmaeus) ALE 13:426; DDM 257 (cp 42); DSM 263; HMW 76; MLS 414 (bw).

Antelope, Sable (Hippotragus niger) ALE 13:402; CAA 74 (bw); DDM 188 (cp 31); DSM 257; HMA 25; HMW 72; LVA 17; LVS 140-141; MLS 408 (bw); SLP 235; SMR 97 (cp 8); SMW 282 (cp 176); TWA 24; VWA cp 10.

Antelope, Saiga—See **Saiga.**

Antelope, Tibetan (Pantholops hodgsoni) ALE 13:438; DAA 40; HMW 79.

Anthias—See **Seaperch.**

Antholyza (Antholyza sp.) EWF cp 87a; TGF 44.

Anthurium, Crystal (Anthurium crystallinum) EPF pl 998; LHP 24; RDP 20, 82.

Anthurus (Anthurus sp.) DEW 3: cp 71; DLM 6, 7; RTM 224, 244.

Ant-lion (Ascalaphus sp.) ALE 2:291; CIB 145 (cp 14); LAW 241; PEI 301 (bw); PWI 33; (Creoleon lugdunense) LAW 211; (Eidoleon bistrigatus) BIN pls 93, 94; (Formicaleon tetragrammicus) LEA 129; (Myrmeleon sp.) ALE 2:301; PEI 298 (bw); ZWI 119.

Antpitta (Grallaricula ferrugineipectus) CDB 142 (cp 568); CSA 55; (G. flavirostris) SSA cp 10; (G. nana) ALE 9:126; (Pittasoma sp.) DTB 55 (cp 27); SSA cp 36.

Antpitta, Streak-chested (Grallaria perspicillata) ABW 200; (G. squamigera) ALE 9:126; (Hylopezus perspicillatus) SSA cp 36.

Antshrike, Barred (Thamnophilus doliatus) ABW 200; ALE 9:126; DTB 41 (cp 20); SSA pl 47; (T. multistriatus) SSA cp 36.

Antshrike, Fasciated (Cymbilaimus lineatus) CDB 141 (cp 566); SSA pl 47.

Antshrike, Great (Taraba major) ABW 200; ALE 9:126; DTB 39 (cp 19); SSA cp 36, pl 45.

Ant-tanager (Saltator sp.)—See **Saltator.**

Ant-tanager, Red-crowned (Habia rubica) ABW 291; ALE 9:361.

Ant-thrush (Formicarius analis) ALE 9:126; CDB 141 (cp 567); (F. rufipectus) SSA cp 36.

Antvireo, Plain (Dysithamnus mentalis) GPB 234; SSA pl 47.

Antwren (Myrmotherula sp.) ABW 200; ALE 9:126; SSA pl 47; (Formicivora grisea) ALE 9:126; (Terenura callinota) SSA cp 10.

Antwren, Dot-winged (Microrhopias quixensis) DTB 43 (cp 21); SSA cp 36, pl 47.

Anu—See **Nasturium, Tuberous** (Tropaeolum).

Aoudad—See **Sheep, Barbary or Maned.**

Ape, Barbary (Macaca sylvana) ALE 10:387 (cp 2); DAE 10; HMA 71; HMW 21; JAW 49; MLS 143 (bw); (Simia sylvanus) SLP 39.

Ape, Black or Celebes (Cynopithecus niger) ALE 10:387 (cp 2); DSM 82; LEA 508 (bw); MLS 144 (bw).

Ape, Night (Aotus trivirgatus) ALE 10:318; DSM 77; HMW 27; MLS 127 (bw); SMW 83 (cp 41).

Aphelandra (Aphelandra sp.) DEW 2:181; MEP 142.

Aphid, Bean (Aphis fabae) CIB 144 (cp 13); OBI 35.

Aphid, Green Peach (Myzus persicae) CIB 144 (cp 13); EPD 88.

Aphid, Spruce Pineapple Gall (Adelges abietis) CIB 144 (cp 13); OBI 35.

Aphid, Woolly Apple (Eriosoma lanigerum) CIB 144 (cp 13); OBI 35; PEI 137.

Aphyllanthes (Aphyllanthes monspeliensis) PFE cp 162 (1585); PFM cp 252.

Apostle-bird (Struthidea cinerea) ALE 9:469; HAB 152; RBU 295; SAB 18 (cp 8).

Apple (Malus sp.) BKT cps 130-132; DFP 213 (cp 1701-1704); EDP 143; EET 171; EGT 41, 45, 127; ELG 107; GDS cp 295; MBT 289 cp 26; MWF 195; OBT 181; OFP 47-61; SST cp 255; TGS 201; TVG 233.

Apple, Common (Malus pumila) DEW 1:276 (cp 156); EVF 46-48, 111, 112; GDS cp 294; LFW 94 (cps 215, 216); MWF 195; OFP 47.

Apple, Crab—See **Crabapple.**

Apple-of-Peru (Nicandra physaloides) PFE cp 119 (1179); WFB 207 (6).

Apostle Plant —See **Marica** (Neomarica).

Apteryx —See **Kiwi.**

Arabis, Garden (Arabis caucasica) EPF pl 250; LFW 14 (cp 28); MFF pl 90 (94).

Aracari (Pteroglossus sp.) ABW 189; ALE 9:81, 107; CDB 135 (cp 536); GPB 224; SSA cp 31.

Aralia, Balfour (Polyscias balfouriana) BIP 147 (cp 223); EFP 138; RDP 326; (P. guilfoylei) RDP 327.

Aralia, Castor (Kalopanax pictus) EEG 117; EGT 46, 120.

Aralia, False (Dizygotheca elegantissima) BIP 71 (cp 101); DFP 65 (cp 514); EDP 128; EFP 107; LHP 28; RDP 27; SPL cp 16; (D. veitchii) RDP 178.

Aralia, Five-leaved (Acanthopanax sieboldianus) EEG 92; EFS 94; TGS 294; (A. henryi) HPW 218-3.

Aralia, Japanese —See **Fatsia, Japanese.**

Arapaima (Arapaima gigas) ALE 4:207; HFI 129; HFW 99 (bw); WEA 52.

Arawana (Osteoglossum bicirrhosum) ALE 4:207, 223; FEF 69 (bw); HFI 128; HFW 100 (bw); SFW 60 (pl 7); WFW cp 30.

Arborvitae, False or Hiba (Thujopsis dolobrata) EPF pl 155; MTB 65 (cp 4); OBT 104, 128; SST cp 44; TGS 23.

Arborvitae, Oriental (Thuja orientalis) DFP 256 (cps 2044, 2045); EGE 109; EJG 148; EPF pl 157; GDS cp 480; OBT 128; SST cp 43; TGS 24; TVG 255.

Archangel, Balm-leaved (Lamium orvala) EWF cp 20e; PFE cp 111 (1131).

Archangel, Yellow (Lamium galeobdolon or Galeobdolon luteum) BFB cp 12; DFP 154 (cp 1230); EEG 144; EFC cp 53 (3); KVP cp 60; MBF cp 70; MFF cp 60 (769); NHE 231; OWF 29; PFE cp 112 (1132); TEP 253; WFB 201 (9).

Archerfish (Toxotes jaculator) ALE 5:111; HFI 31; HFW 194 (bw); LAW 345; SFW 608 (cp 167); WFW cp 311; (T. chatareus) FEF 366 (bw); WFW cp 310.

Ardisia (Ardisia sp.) BIP 27 (cp 19); DFP 52 (cp 409); EPF pl 255; FHP 105; HPW 136; MEP 104; PFW 190.

Argali (Ovis ammon ammon) DAA 33; HMW 82; MLS 431 (bw); TWA 177 (bw).

Argentine (Argentina sp.) CSF 81; OBV 23.

Argus, Great —See **Pheasant, Great Argus.**

Argus-fish (Scatophagus argus) ALE 5:111; CTF 50 (cp 63), 51 (cp 64); FEF 370 (bw); HFI 39; SFW 601 (cp 166); (Cephalopholis argus) HFW 86 (cp 42).

Ark Shell —See **Noah's Ark.**

Armadillo, Giant (Priodontes giganteus) ALE 11:189; JAW 67; LVA 29; MLS 171 (bw); SMW 108 (bw).

Armadillo, Hairy (Euphractus villosus) ALE 11:173, 189; TWA 143; (Chaetophractus villosus) DSM 102; (Zaedus pichyi) CSA 168 (bw).

Armadillo, Nine-banded (Dasypus novem-cinctus) ALE 11:173, 189; CSA 129; DSM 103; FWA 135; HMA 26; LAW 517; LEA 544; MLS 172 (bw); PFF 738; SMW 109 (bw).

Armadillo, Six-banded (Euphractus sex-cinctus) ALE 11:189; HMW 180.

Armadillo, Three-banded (Tolypeutes tricinctus) ALE 11:189; LEA 517 (bw); LVS 66; (T. matacus) JAW 68.

Arna —See **Buffalo, Indian Water.**

Arnica, Mountain (Arnica montana) AMP 169; CTH 55 (cp 86); EPF pl 566; KVP cp 145; NHE 189; PFE cp 151 (1446); SPL cp 383; WFB 237 (7).

Arrow-grass, Marsh (Triglochin palustris) MBF cp 79; MFF pl 106 (970); NHE 102; PFE 480; PWF 126d; WFB 221 (6).

Arrow-grass, Sea (Triglochin maritima) MBF cp 79; MLP 268; NHE 137; OWF 103; PWF 53k; WFB 221 (7).

Arrowhead (Sagittaria sagittifolia) BFB cp 22 (8); DFP 170 (cp 1357); EPF pls 1020, 1021; ERW 148; HPW 270; KVP cp 5; MBF cp 79; MFF pl 69 (962); MLP 266; NHE 102; OWF 103; PWF 133f; SPL cp 511; TEP 33; TVG 133; WFB 259 (6).

Arrowhead (Sagittaria sp.) DEW 2:226; EWF cp 166e; PFF 338; WUS 27.

Arrowhead Vine or Plant (Syngonium sp.) EFP 144; EGV 139; LHP 180, 181; RDP 380.

Arrowroot (Maranta arundinacea) EWF cp 189d; HPW 300; OFP 181.

Arrow-worm —See **Worm, Arrow.**

Artemis, Rayed (Dosinia exoleta) CSS pl XIX; OIB 75.

Artichoke, Globe (Cynara scolymus) AMP
31; DEW 2:205; EPF pl 586; EWF cp 40a;
MLP 249; MWF 101; OFP 165; PFF 331;
PFW 85.

Artichoke, Jerusalem (Helianthus
tuberous) OFP 179; PFF 321; SPL cp 431.

Artichoke, Wild (Cynara cardunculus) CTH
63 (cp 102); DEW 2:205; EGH 111.

Artillery or Gunpowder Plant —See
Panamiga (Pilea sp.).

Arum, Bog or Water (Calla palustris) DEW
2:286 (cp 169); EPF pl 1002; ERW 139; KVP
cp 26; MLP 291; NHE 107; PFE cp 183
(1817); PFF 365; PFW 33; TEP 38.

Arum, Dragon (Dracunculus vulgaris) DFP
91 (726); NHE 246; PFE cp 183 (1819); PFM
cp 223; PFW 32.

Arum, Italian (Arum italicum) DFP 85 (cp
676); MBF cp 88; MWF 47; NHE 246; OGF
13; PFE cp 183 (1818); PFM cp 220; PWF
77g; SPL cp 385.

Arum, Ivy —See **Ivy, Devil's** (Scindapsus).

Asarabacca —See **Ginger, European Wild.**

Ascidian —See **Sea Squirt.**

Ash, European or Common (Fraxinus
excelsior) BKT cp 189, 190; DEW 2:116, 119;
EET 204; EPF pls 672-674; MBF cp 58; MFF
pl 70 (629); MTB 369 (cp 38); MWF 135; NHE
24; OBT 20, 24; OWF 197, 202; PWF 26d,
175f; SST cp 118; TEP 289; WFB 36, 37 (4).

Ash, Flowering or Manna (Fraxinus ornus)
AMP 59; BKT cps 185-187; EET 204, 205;
EGT 115; MTB 369 (cp 38); NHE 24; OBT
196; PFE cp 94 (979); SPL cp 89; SST cp 120;
TGS 327.

Ash, Moraine (Fraxinus holotricha) EEG
116; EGT 115.

Ash, Weeping (Fraxinus excelsior pendula)
DFP 202 (cp 1615); EET 205; OBT 196.

Ashanti Blood —See **Red-flag Bush.**

Asity, Velvet (Philepitta castanea) ABW 212;
ALE 9:150; 10:283; GPB 237; WAB 159.

Asp (fish) (Aspius aspius) ALE 4:342; FEF
147; FWF 71; SFW 212 (pl 49); (A. rapax)
HFI 99.

Asp (snake) —See **Viper, Asp.**

Asparagus (Asparagus officinalis) EVF 80;
MFF pl 65 (993); NHE 192; OFP 163; PFF
377; PWF 152b; WFB 269 (7).

Asparagus, Bath —See **Star of Bethlehem,
Spiked.**

Asparagus, Sharp-leaved (Asparagus
acutifolius) NHE 246; PFE cp 168 (1650);
PFM pl 23 (431).

Asparagus-fern (Asparagus plumosus or A.
setaceus) BIP 29 (cp 22); EGL 87; EGV 79;
LHP 33; PFF 377.

Asparagus-fern (Asparagus sprengeri or A.
densiflora) BIP 29 (cp 22); EGC 123; EGV
79, 94; LHP 32; MWF 48; RDP 27, 88; SPL cp
3.

Asparagus-fern, Foxtail (Asparagus
myersii) EDP 140; EFP 90; LHP 32; RDP 88.

Aspen, European (Populus tremula) BFB cp
14 (3); DFP 218 (cp 1744); MBF cp 78; MFF
pl 70 (566); MTB 176 (cp 15); NHE 24; OBT
13, 17; OWF 199; PFE cp 3 (32); PWF 13f,
161g; TEP 281; WFB 31 (2).

Asphodel (Asphodel sp.) DEW 2:212 (cp 213);
NHE 42; PFM cp 233.

Asphodel, Bog (Narthecium ossifragum)
BFB cp 23 (7); EWF cp 24a; KVP cp 70; MBF
cp 86; MFF pl 63 (987); NHE 77; OWF 15;
PWF 15g; WFB 263 (2).

Asphodel, German (Tofieldia calyculata)
NHE 77; PFE cp 162 (1583); WFB 261 (8).

Asphodel, Hollow-stemmed (Asphodelus
fistulosus) PFE cp 163 (1592); PFM cp 238.

Asphodel, Scottish (Tofieldia pusilla) EWF
cp 7c; MBF cp 86; MFF pl 83 (986); NHE 290;
WFB 261 (7).

Asphodel, White (Asphodelus albus) DEW
2:237; KVP cps 159, 159a; PFM cp 232; (A.
aestivus) PFE cp 163 (1591).

Asphodel, Yellow —See **King's Spear.**

Ass, African Wild (Equus asinus) DDM 141
(cp 24); DSM 211; HMW 40; LAW 556; LVS
117; WEA 53; (E. africanus) ALE 12:551.

Ass, Asian or Indian Wild (Equus hemionus
var.) ALE 12:554, 573; CAS 76, 164 (bw);
DAA 28; DSM 210; HMW 42, 43; JAW 157,
158; LEA 584 (bw); LVA 45; LVS 116; MLS
349 (bw); PWC 241, 242 (bw); SLP 187; SLS
141; SMW 223 (bw); TWA 160 (bw); VWA cp
12.

Astarte (Astarte sulcata) OIB 83; (A.
borealis) ALE 3:142.

Astelia (Astelia sp.) EWF cp 7j; SNZ 73 (cps
202-203), 162 (cp 514).

Aster, Alpine (Aster alpinus) DFP 3 (cp 17);
EPF pl 533, cp XXb; ERW 101; KVP cp 130;
NHE 285; OGF 35; PFE cp 142 (1363); SPL
cp 441; TEP 95; TVG 143.

Aster, China (Callistephus chinensis) DFP
32 (cps 251-254); EGA 101; EPF pl 537; EWF
cp 106c; LWF 27 (cps 55, 56); MWF 68; OGF
127; PFF 314; SPL cp 417; TGF 253; TVG 37.

Aster, Daisy-star (Aster bellidiastrum) EPF
pl 534; NHE 285.

Aster, Heath (Aster ericoides) MWF 48; OGF
179.

Aster, Mexican —See Cosmos.

Aster, New England (Aster nova-angliae)
DFP 124 (cp 985); EEG 131; EWG 89.

Aster, New York —See Daisy, Michaelmas.

Aster, Sea (Aster tripolium) BFB cp 19 (3);
KVP cp 98; MBF cp 44; MFF cp 9 (875); NHE
136; OWF 137; PFE cp 142 (1365); PWF
153j; TEP 63; WFB 233 (4).

Aster, Willow-leaved (Aster salignus) EPF
pl 535; NHE 100.

Asterophora (Asterophora lycoperdioides)
DEW 3:166; LHM 81; PUM cp 82b; (A.
parasitica) LHM 81; PUM cp 82a; SMF 55.

Astilbe (Astilbe sp.) DFP 125 (cps 994-996);
EDP 141; EEG 131; EJG 101; EPF pl 389;
ESG 69, 95; LFW 17 (cp 33); OGF 75; PFW
272; TGF 112; TVG 81.

Astrapia —See Bird of Paradise.

Atelopus —See Frog.

Aubergine —See Egg Plant.

Auger Shell (Terebra sp.) CTS 55 (cp 100);
GSS 120-123; MOL 67.

Auk, Little —See Dovekie.

Auk, Razorbill (Alca torda) ABW 138; ALE
7:61; 8:208; BBB 258; BBE 161; CDB 95 (cp
343); DAE 31; FBE 165; GPB 170; LEA 417
(bw); OBB 101; PBE 125 (pl 36); RBE 179;
SAO 143; WAB 72.

Aulacomnium (Aulacomnium androgynum)
DEW 3:247; ONP 175.

Auroch —See Cattle (Bos taurus).

Australian Nut —See Queensland Nut.

Avadavat, Red (Amandava amandava) ALE
9:440; DPB 413; KBA 433 (cp 64); (Estrilda
amandava) ABW 302; WBV 221.

Avahi, Woolly —See Lemur, Avahi

Avens, Boris (Geum borsii) DFP 144 (cp
1151); EPF cp Xa; MWF 142.

Avens, Common Wood (Geum urbanum)
BWG 36; MBF cp 29; MFF cp 42 (360); NHE
179; OWF 17; PWF 37j; SNZ 31 (cp 49); TEP
243; WFB 111 (6).

Avens, Creeping (Geum reptans) ERW 119;
NHE 272.

Avens, Mountain (Dryas octopetala) BFB cp
9 (3); DFP 8 (cp 60); EPF pl 409; ERW 112;
EWF cp 2b; KVP cp 156; MBF cp 29; MFF pl
77 (362); NHE 266; OWF 81; PFE cp 45
(441); PWF 52c, 149f; TEP 97; TGS 122;
WFB 111 (7).

Avens, Mountain (Geum montanum) DEW
1:298; NHE 272; (G. uniflorum) SNZ 155 (cp
487).

Avens, Scarlet (Geum chiloense or G.
coccineum) DFP 144 (cp 1152); EWF cp 28a;
LFW 201 (cp 454); MWF 142; OGF 107; SPL
311; TGF 108; TVG 95; (G. quellyon) PWF
263.

Avens, Water (Geum rivale) BFB cp 9 (8);
EWF cp 16d; MBF cp 29; MFF cp 25 (361);
NHE 179; OWF 117; PFE cp 45 (443); PWF
33k; TEP 43; WFB 111 (5).

Avocado Pear (Persea americana) DEW
1:90; OFP 115; SST cp 212; (P. gratissima)
BIP 143 (cp 217); HPW 37.

Avocet, Australian or Red-necked
(Recurvirostra novaehollandiae) BNZ 135;
HAB 73; LEA 413 (bw); MEA 39; SNB 84 (cp
41).

Avocet, European (Recurvirostra avosetta)
ALE 8:187; BBB 222; BBE 135; CDB 89 (cp
302); FBE 119; FBW 65; GPB 151; KBA 146
(bw); LEA 333 (bw); NBB 80 (bw); OBB 81;
PBE 92 (pl 29), 101 (cp 32); PSA 145 (pl 16);
RBE 151.

Awlwort (Subularia aquatica) MBF cp 9;
MFF pl 66 (76); NHE 94; WFB 291 (1).

Aye-aye (Daubentonia madagascariensis)
ALE 10:247 (cp 1), 251 (cp 5); 283; HMA 64;
HMW 32; LVA 25; LVS 63; MLS 120 (bw);
PWC 108, 174 (bw); SLS 187; SMW 66 (bw);
VWA cp 1.

Azalea (Rhododendron luteum) GDS cp 352;
HSC 102; OGF 57; PWF 39k; WFB 170

Azalea Indica (Rhododendron simsii) LHP
38, 39; PFW 109; RDP 338; SPL cp 53.
Azalea, Wild or Alpine or Creeping
(Loiseleuria procumbens) EWF cp 6b; MBF
cp 56; MFF cp 24 (583); NHE 266; PFE cp 87
(921); TGS 39; WFB 173 (6).
Azara (Azara sp.) DFP 182 (cp 1454); HPW
100; HSC 15, 16; PFW 118.
Azarole (Crataegus azarolus) OFP 63; PFE 171.

B

Babbler, Arrow-marked (Turdoides
jardineii) CDB 169 (cp 716); GPB 276; PSA
240 (cp) 27.
Babbler, Chestnut-crowned(Pomatostomus ruficeps) HAB 171; SAB 18 (cp 8).
Babbler, Common Tit (Parisoma subcaeruleum) CDB 169 (cp 713); PSA 273 (cp 32).
Babbler, Gray-crowned Scimitar (Pomato-
stomus temporalis) ABW 240; ALE 9:222;
HAB 171; RBU 228; SAB 18 (cp 8); WAW 29
(bw).
Babbler, Ground (Trichastoma sp.) DPB
271; KBA 289 (cp 46); (Ptilocichla sp.) DPB
271.
Babbler, Hall's (Pomatostomus halli) HAB
171; SAB 18 (cp 8).
Babbler, Jungle (Turdoides sp.) ALE 9:222;
FBE 263; KBA 284 (pl 43); PSA 240 (cp 27);
WAB 168.
Babbler, Large Scimitar (Pomatorhinus
hypoleucos) KBA 284 (pl 43); WBV 199.
Babbler, Nun (Alcippe sp.) ALE 9:221; KBA
304 (cp 47); WBV 205.
Babbler, Rail (Eupetes macrocercus) ALE
9:222; KBA 284 (pl 43).
Babbler, Red-capped (Timalia pileata)
ABW 240; ALE 9:221; KBA 284 (pl 43).
Babbler, Rusty-cheeked Scimitar
(Pomatorhinus erythrogenys) ALE 9:222;
CDB 169 (cp 714); KBA 284 (pl 43); RBI 163.
Babbler, Shrike (Pteruthius sp.) ABW 240;
KBA 288 (cp 45); RBI 159.
Babbler, Striped Jungle (Pellorneum rufi-
ceps) ABW 240; ALE 9:221; KBA 284 (pl 43).
Babbler, Tit (Macronous sp.) DPB 275; KBA
305 (cp 48).

Babbler, Tree (Stachyris sp.) ALE 9:221;
DPB 273, 275; KBA 305 (cp 48); (Malacop-
teron sp.) ALE 9:221; DPB 271; KBA 289 (cp
46).
Babbler, White-browed Scimitar (Pomatostomus superciliosus) CDB 169 (cp 715);
HAB 171; SAB 18 (cp 8); (Pomatorhinus
schisticeps) KBA 284 (pl 43); WBV 199.
Babbler, White-hooded (Gampsorhynchus
rufulus) KBA 285 (pl 44); WBV 205.
Babbler, Wren (Napothera sp.) DPB 271;
KBA 304 (cp 47); WBV 199; (Kenopia
striata) KBA 305 (cp 48).
Babirusa (Babirussa babyrussa) ALE 13:92;
DAA 126; DSM 221; HMA 61; HMW 51; MLS
372 (bw); SMW 245 (bw).
Baboon, Chacma (Papio ursinus) ALE 10:31,
388 (cp 3); CAA 100; DDM 53 (cp 6); HMA
27; LEA 510 (bw); SMR 16 (cp 1).
Baboon, Dog-faced or Olive (Papio anubis)
ALE 10:388 (cp 3), 411; CAA 93 (bw); CAF
99 (bw), 159 (bw); DDM 53 (cp 6); DSM
84-85; SMW 86 (cp 48); WEA 58.
Baboon Flower (Babiana stricta) EGB 96·
MEP 23; MWF 51.
Baboon, Gelada (Papio gelada or Theropith-
ecus gelada) ALE 10:389 (cp 4); CAF 51;
DDM 53 (cp 6); DSM 87; MLS 149 (bw); WEA
163.
**Baboon, Hamadryas or Sacred—See
Hamadryas.**
Baboon, Western or Guinea (Papio papio)
ALE 10:388 (cp 3); DDM 53 (cp 6).
Baboon, Yellow (Papio cynocephalus) ALE
10:388 (cp 3); DDM 53 (cp 6); SMR 16 (cp 1);
TWA 18; (P. richei) HMW 19.

Baby's Tears (Hypoestes sp.) BIP 103 cp 156; EWF cp 64c; LFT cp 158; LHP 114; RDP 245.

Baby's-breath (Gypsophila paniculata) DFP 145 (cps 1153, 1154); EDP 128; LFW 54 (cp 124); PFF 164; TGF 71; TVG 95; (G. fastigiata) NHE 226; PFE 90.

Baby's-breath, Annual (Gypsophila elegans) DFP 38 (cp 303); EGA 122; EGG 114; MWF 145; OGF 131.

Baby's-breath, Creeping (Gypsophila repens) DFP 11 cp 81; EMB 139; ERW 119; NHE 277; PFE cp 16 (178).

Baby's-breath, Wall (Gypsophila muralis) NHE 73; WFB 63 (5).

Bachelor's Button (Centaurea) —See **Cornflower.**

Bachelor's Buttons—See **Buttercup, Common Meadow.**

Bachelor's Buttons—See **Feverfew** (Chrysanthemum).

Bachelor's Buttons or **Billy Buttons** (Craspedia) —See **Woollyhead.**

Backswimmer (Notonecta sp.) ALE 2:188; CIB 128 (cp 11); GAL 300 (bw); LAW 229; OBI 31; PEI 113 (bw); PWI 44 (bw); WEA 59.

Backswimmer, Lesser (Corixa punctata) CIB 128 (cp 11); OBI 31; PEI 114 (bw).

Bacon and Eggs—See **Trefoil, Common Bird's-foot.**

Badger, Eurasian (Meles meles) ALE 42, 83; 13:163; CEU 193 (bw), 222; DAE 35; DSM 173; GAL 26 (bw); HMW 138; LEA 564 (bw); MLS 291 (bw); OBV 165; TWA 88; WEA 59.

Badger, Ferret (Melogale sp.) ALE 12:84; DAA 80; DSM 172; MLS 293 (bw).

Badger, Hog or Sand (Arctonyx collaris) ALE 12:83; DAA 81; DSM 172; MLS 292 (bw).

Badger, Honey —See **Honey-badger.**

Badis (Badis badis) ALE 5:116; FEF 395 (bw); HFI 34; SFW 540 (pl 145); WFW cp 460.

Baikal —See **Teal, Baikal or Formosa.**

Balanophora (Balanophora sp.) EWF cp 117a; HPW 176.

Baldmoney —See **Spignel** (Meum).

Balicassiao —See **Drongo.**

Balloon Flower (Platycodon grandiflora) DFP 18 (cp 138); EEG 137; EJG 138; EPF pl 524; LFW 112 (cp 248); MWF 236; OGF 109; PFW 63; SPL cp 243; TGF 200; TVG 117.

Balloon-fish —See **Porcupine-fish.**

Balm, Bastard (Melittis melissophyllum) AMP 111; BFB cp 17; DEW 2:176 (cp 110); EFC cp 52; MBF cp 69; MFF pl 112 (760); OWF 97; PFE cp 109 (1121); TEP 253; WFB 199 (7).

Balm, Lemon (Melissa officinalis) AMP 19; EGH 123; EVF 139; MBF cp 68; MFF pl 112 (757); OFP 143; PWF 125f; WFB 203 (4).

Balsam (Impatiens sp.) DEW 2: cp 74; EGG 65; EPF pl 741; ESG 80, 81; LFT cp 101; MWF 163; PFW 39.

Balsam, Garden (Impatiens balsamina) DFP 39 (cp 311); EGA 127; EPF pl 739, 740; ESG 120; HPW 211; LFW 168 (cp 377); MWF 163; SPL cp 337; TGF 141.

Balsam, Orange —See **Jewel-weed.**

Balsam, Small (Impatiens parviflora) DEW 2:74; MBF cp 20; MFF pl 60 (259); NHE 31; PFE cp 70 (715); PWF 121f; WFB 139 (3).

Balsam, Touch-me-not (Impatiens nolitangere) BFB cp 7 (7); EFC cp 3; EPF pl 738; KVP cp 58; MBF cp 20; MFF pl 62 (257); NHE 31; PFE cp 70 (714); TEP 245; WFB 139 (2).

Balsam-pear (Momordica charantia) DEW 1:166 (cp 84), 168 (cps 89, 90), 169 (cp 92); EGA 138; OFP 121; SPL cp 176; (M. clematidea) LFT (cp 165).

Bamboo (Arundinaria sp.) EMB 122; GDS 12; HPW 287; HSC 13; PFF 340; (Bambusa sp.) BIP 29 (cp 24); MLP 310; OFP 163; (Phyllostachys sp.) EJG 72, 73, 136; MWF 232.

Bamboo, Giant (Dendrocalamus giganteus) EPF pl 1068; MLP 311; WAW 107; WWT 27.

Bamboo, Heavenly or Sacred (Nandina domestica) EEG 109; EGE 136; EGW 69, 129; EJG 67, 130; ELG 140; MWF 207; TGS 294.

Bamboofish —See **Saupe.**

Bamburanta (Ctenanthe sp.) MWF 98; RDP 163.

Banana (Musa paradisiaca or M. sapientum) CAS 227; DEW 2:292; EPF pls 1036, 1037; HPW 296; MLP 319; MWF 204; OFP 109; PFF 388; SST cp 210; WWT 116, 117.

Banana (Musa sp.) BIP 123 (cp 187); EGG 127; EWF cp 123a; LFW 160 (cp 354); LHP 128, 129; MEP 24.

Banana Shrub (Michelia sp.) EWF cp 110b; MWF 200; PFW 181.

Bananaquit (Coereba flaveola) ABW 286; ALE 9:371; CDB 203 (cp 886) GPB 307; SSA cp 39; WAB 99.

Bandicoot, Long-nosed (Perameles nasuta) ALE 10:72; DSM 30; LEA 485 (bw); MLS 56 (bw); SMW 22 (bw).

Bandicoot, Rabbit (Macrotis lagotis) ALE 10:72; LVS 28; (Thalacomys lagotis) HMW 201; TWA 191.

Bandicoot, Short-nosed (Isoodon sp.) DSM 30; MWA 42.

Bandicoot, Tasmanian Barred (Perameles gunni) ALE 10:72; WAW 54 (bw).

Bandy-bandy (Vermicella annulata) ALE 6:406; 10:89; COG cp 47; HWR 20; LEA 326; SSW 113; (Rhynchoelaps bertholdi) MAR 179 (cp XIV).

Baneberry, Common (Actaea spicata) MBF cp 4; MFF pl 80 (7); NHE 26; OWF 67; TEP 235; WFB 73 (5); (A. alba) EWG 84.

Banksia, Coast (Banksia integrifolia) DEW 2:95; MWF 52.

Banksia, Heath (Banksia ericifolia) MEP 33; MWF 52.

Banksia, Saw (Banksia serrata) EET 243; EWF cp 134d.

Banksia, Scarlet (Banksia coccinea) MEP 33; MLP 76; MWF 52; RWF 76.

Banteng (Bos banteng) DAA 113; DSM 251; HMW 91; JAW 188; MLS 402 (bw); TWA 159 (bw); (Bibos sondaicus) CAS 241, 272; SLP 194; (Bos javanicus lowi) ALE 13:342.

Banyan Tree (Ficus bengalensis) DEW 1:178; HPW 97; LHP 94; MLP 66, 67, 71; RDP 211, 212; SST cp 112; WWT 38.

Baobob (Adansonia digitata) CAF 172(bw), 191 (bw); CAU 149; DEW 1:217 (cp 127); EET 231; EPF pl 608; EWF cp 76b; LAF 159; LFT 215 (cp 107), 216 (fig. 10); MEP 84; MLP 149; PFW 48 (bw); SST cp 78; WWT 28.

Barasingha —See **Deer, Swamp.**

Barb (Barbus barbus) ALE 4:341; FEF 152 (bw); FWF 81; GAL 242 (bw); HFI 99; OBV 99; SFW 220 (pl 51); (B. callepterus) WFW cp 108; (B. arulius) SFW 237 (cp 56).

Barb, Black-spot (Barbus filamentosus) FEF 192 (bw); SFW 260 (pl 63); WFW cp 110.

Barb, Blind Cave (Caecobarbus geertsi) ALE 4:417; HFI 102; HFW 117 (bw); SFW 257 (pl 62).

Barb, Cherry (Barbus titteya) CTF 29 (cp 24); FEF 202 (bw); SFW 237 (cp 56); WFW cp 115.

Barb, Clown (Barbus everetti) FEF 192 (bw); HFI 101; SFW 261 (pl 64).

Barb, Five-banded (Barbus pentazona) FEF 171, 194 (bw); SFW 236 (cp 55).

Barb, Flying (Esomus danricus) ALE 4:318; HFI 101.

Barb, Malayan Flying (Esomus malayensis) ALE 4:318; FEF 185 (bw); HFI 101; SFW 269 (pl 68).

Barb, Red or Rosy (Barbus conchonius) FEF 191 (bw); SFW 236 (cp 55); WFW cp 109.

Barb, Schwanenfeld's or Tinfoil (Barbus schwanenfeldi) CTF 29 (cp 24); FEF 200 (bw); HFI 101; SFW 261 (pl 64).

Barb, Tiger (Barbus tetrazona) CAA 12-13 (bw), 14; CTF 28 (cps 22, 23); FEF 196, 197 (bw); HFI 101; SFW 228 (cp 53); WFW cp 114; (Puntius sp.) ALE 4:300, 351, 352; HFW 39 (cp 16), 40 (cp 18), 41 (cp 20).

Barb, Two-spot (Barbus ticto) FEF 201 (bw); HFI 101; SFW 256 (pl 61).

Barbados Cherry (Malpighia glabra) BIP 117 (cp 178); MLP 114.

Barbados Gooseberry (Pereskia aculeata) ECS 14, 137; EWF cp 175c; LCS cp 158; PFF 255.

Barbary Nut (Iris sisyrinchium) NHE 42; PFE cp 175 (1684); PFM cp 269.

Barbel —See **Barb.**

Barberry (Berberis sp.) DFP 183 (cp 1457); GDS cps 14, 16, 17, 18, 19; HSC 16, 17, 18; TGS 226, 230.

Barberry, Common or European (Berberis vulgaris) AMP 145; BFB cp 4 (10); EPF pls 195-197; MBF cp 4; MFF cp 64 (30); MWF 56; NHE 267; OFP 190; OWF 191; PFE cp 27 (261); PFF 178; PWF 60b; TEP 127; WFB 78 (1).

Barberry, Darwin's (Berberis darwinii) DFP 182 (cps 1455, 1456); GDS cp 15; HPW 45; HSC 16, 17; MLP 54; MWF 56; OGF 23; PWF 42.

Barberry, Japanese (Berberis thunbergii) DEW 1:99 (cps 42, 43); DFP 183 (cp 1459); EEG 92; EFS 32, 96; EGW 92; EJG 103; EPG 106; GDS cp 21; PFF 178; TGS 230; TVG 217.

Barberry, Rosemary (Berberis stenophylla) DFP 183 (cp 1458); GDS cp 20; SPL cp 65.

Barberry, Wintergreen (Berberis julianae) EGE 113; TGS 230.

Barbet (Capito sp.) DTB 26-27 (cp 13); SSA pl 45.

Barbet, Black-browed (Megalaima oorti) KBA 225 (cp 30); RBI 128.

Barbet, Blue-cheeked or -throated (Megalaima asiatica) ALE 9:97; KBA 225 (cp 30).

Barbet, Collared (Lybius torquatus) ABW 185; ALE 9:97; CDB 133 (cp 527); NBA 51; PSA 225 (cp 26).

Barbet, Coppersmith or Crimson-breasted (Megalaima haemacephala) DPB 209; KBA 225 (cp 30); WBV 135.

Barbet, Crested (Trachyphonus vaillantii) CDB 134 (cp 532); NBA 49; PSA 225 (cp 26).

Barbet, D'Arnaud's (Trachyphonus darnaudii) CDB 134 (cp 530); GPB 222.

Barbet, Golden-throated (Megalaima franklinii) ABW 185; KBA 225 (cp 30).

Barbet, Great Hill (Megalaima virens) ALE 9:97; KBA 240 (cp 31).

Barbet, Pied (Lybius leucomelas) CDB 133 (cp 526); PSA 225 (cp 26).

Barbet, Red-and-Yellow (Trachyphonus erythrocephalus) ALE 9:97; CDB 134 (cp 531).

Barbet, Red-headed (Eubucco bourcierii) CDB 133 (cp 525); DTB 29 (cp 14); SSA cp 34.

Barbet, Tinker —See **Tinkerbird.**

Barbet, Toucan (Semnornis ramphastinus) ABW 185; ALE 9:97; SSA cp 34.

Barclaya (Barclaya sp.) DEW 1:128; HPW 44.

Bare-eye, Black-spotted —See **Antbird, Spectacled.**

Barilius, Goldlip (Barilius christyi) ALE 4:352; HFI 102.

Barley, Foxtail —See **Grass, Squirrel-tail.**

Barley, Meadow (Hordeum secalinum) MBF pl 99; MFF pl 124 (1212); NHE 194.

Barley, Sea (Hordeum marinum) MBF pl 99; MFF 124 (1214); NHE 194.

Barley, Six-rowed (Hordeum vulgare) EPF pl 1048; MLP 305; OFP 5; PFF 356.

Barley, Two-rowed (Hordeum distichon) EPF pls 1043, 1047; OFP 5; TEP 170.

Barley, Wall (Hordeum murinum) BWG 177; DEW 2:267; EPF pl 1049; MBF pl 99; MFF pl 124 (1213); NHE 242; PFE cp 178 (1735); TEP 206.

Barley, Wood (Hordelymus europaeus) MFF pl 124 (1215); NHE 40.

Barnacle (Elminius modestus) CSS pl VI; LAW 148.

Barnacle, Acorn (Balanus sp.) CSS pl VI; LEA 165 (bw); OIB 129; VWS 64 (bw); (Verruca stroemia) CSS cp VI; OIB 129.

Barnacle, Goose (Lepas sp.) CSS pl VI; LEA 165 (bw); MOL 85; OIB 129; VWS 70 (bw).

Barnacle, Stalked (Scalpellum sp.) MOL 85; OIB 129.

Barracuda, Great (Sphyraena barracuda) ALE 5:138, 150; FEF 325 (bw); HFI 62; HFW 217 (cp 105); LEA 225 (bw); LMF 35 (bw); WEA 62.

Barrenwort (Epimedium macranthum or E. grandiflorum) EEG 142; EGC 131; EJG 115; EPF pl 198; ERW 113; ESG 62, 110.

Barrenwort (Epimedium sp.) HPW 45; MBF cp 4; NHE 268; OGF 13, 49; TGF 91.

Barringtonia (Barringtonia sp.) DEW 2:17; EWF cp 148a; HPW 99.

Bartsia, Alpine (Bartsia alpina) MBF cp 65; MFF cp 3 (731); NHE 282; PFE cp 126 (1236); WFB 213 (8).

Bartsia, Red (Odontites verna) BWG 120; MBF cp 65; MFF cp 28 (729); NHE 74; OWF 141; PFE cp 126 (1241); PWF 144b; WFB 213 (6).

Bartsia, Yellow (Parentucellia viscosa) MBF cp 65; MFF cp 61 (730); PFE cp 126 (1238); PWF 111f; WFB 213 (9); (P. latifolia) PFE cp 126 (1239); (Odontites lutea) NHE 74; PFE 384; WFB 213 (7).

Basil, Sweet (Ocimum basilicum) DEW 2:191; EDP 41; EGA 140; EGH 82, 128, 129; EGL 128; EVF 139; OFP 143; SPL cp 398.

Basil, Wild (Clinopodium vulgare) BFB cp 16; MBF cp 68; MFF cp 29 (756); NHE 35; OWF 147; PFE cp 115 (1158); PWF 103h, 105j; WFB 205 (7).

Basilisk (Basiliscus basiliscus) ALE 6:180; ERA 159; HRA 29; LAW 410; MAR 116 (pl 48); (B. vittatus) SIR 113 (bw).

Basilisk, Double-crested or Plumed (Basiliscus plumbifrons) ART 111; MAR 116 (pl 48); SIR 114 (bw); VRA 183; WEA 62.

Basil-thyme (Acinos arvensis) MBF cp 68; MFF cp 10 (755); NHE 231; OWF 143; PWF 46c; WFB 205 (9).

Basket Grass (Oplismenus hirtellus) BIP 133 (cp 199); RDP 283.

Basket of Gold (Alyssum saxatile) DFP 1 (cp 3); EPF pl 252; ERW 93; MWF 39; NHE 217; OGF 41; PFM cp 40; SPL cp 440; TEP 89; TGF 77; TVG 137.

Basket Shell (Nassarius sp.) CSS pl XVI; OIB 43; (Corbula gibba) CSS cp 23; OIB 81.

Basket-star or **Brittle-star** (Gorgonocephalus sp.) ALE 3:389; MOL 88.

Basketvine or **Basket Plant** —See **Lipstick Plant.**

Bass (Dicentrarchus labrax) CSF 123; OBV 59; WEA 63.

Bass, Sea —See **Sea-perch, Banded.**

Basslet, Fairy (Gramma loreto) ALE 5:94; WFW cp 271.

Bat, Barbastelle (Barbastella barbastellus) ALE 2:363; 11:135; HMW 186; OBV 161.

Bat, Bechstein's (Myotis bechsteini) ALE 11:134; HMW 187; LAW 604; OBV 159.

Bat, Blood-sucking —See **Bat, Vampire.**

Bat, Daubenton's or Water (Myotis daubentoni) ALE 11:105; DAE 66; HMW 187; OBV 159.

Bat, Egyptian Fruit (Rousettus aegyptiacus) ALE 11:87; WEA 65.

Bat, False Vampire (Vampyrus spectrum) HMW 184; (Cartioderma cor) WEA 64; (Macroderma gigas) WAW 62, 63 (bw).

Bat Flower (Tacca sp.) DEW 2:249, 278 (cp 156); EPF pl 936; EWF cp 125b; MEP 21; PFW 290.

Bat, Gray-headed Flying Fox (Pteropus polio cephalus) ALE 11:87; MWA 40 (bw); WAW 56.

Bat, Greater Horse-shoe (Rhinolophus ferrumequinum) ALE 11:105, 125; DSM 56; HMW 186; LAW 602, 603; LEA 497 (bw); MLS 102 (bw); OBV 155.

Bat, Greater Indian (Pteropus giganteus) DAA 130; DSM 63; LEA 496 (bw); MLS 98 (bw); TWA 176.

Bat, Hammerhead (Hypsignathus monstrosus) ALE 11:87; DSM 59; WEA 64.

Bat, Hoary (Lasiurus cinereous) DSM 59; HMA 29; LEA 498; PFF 681; SAA 136; SMW 45 (cp 31).

Bat, Indian Fruit (Cynopterus sp.) ALE 11:85, 87; CAS 215.

Bat, Lesser Horse-shoe (Rhinolophus hipposideros) ALE 11:125; CAS 93; DAE 66; HMW 186; LAW 604; OBV 155.

Bat, Long-eared (Plecotus auritus) ALE 11:133, 135; DAE 67; DSM 56; HMW 186; LAW 605; LEA 494 (bw); OBV 161; SMW 56 (bw).

Bat, Malay Fruit (Pteropus vampyrus) HMW 184; JAW 37; SMW 44 (cp 25).

Bat, Mouse-eared (Myotis myotis) ALE 11:88, 134; HMW 187; OBV 157.

Bat, Naked (Cheiromeles torquatus) DSM 62; HMW 185; MLS 104 (bw).

Bat, Natterer's (Myotis nattereri) HMW 187; OBV 157.

Bat, Noctule or Nocturnal (Nyctalus noctula) ALE 10:223, 11:136; DSM 57; HMW 186; LEA 498 (bw); OBV 163.

Bat, Particolored (Vespertilio sp.) ALE 2:135; HMW 187; OBV 161.

Bat, Pipistrelle (Pipistrellus pipistrellus) ALE 2:363; 11:135; DAE 67; DSM 57; HMW 187; OBV 163; TWA 100 (bw); (P. nathusii) OBV 159.

Bat, Rat-tailed or Tomb (Rhinopoma microphyllum) ALE 11:123; HMW 185; MLS 101 (bw).

Bat, Serotine (Eptesicus serotinus) ALE 11:135; HMW 186; LEA 499 (bw); OBV 163; SMW 44 (cp 28); (E. pumilus) MWA 65.

Bat, Straw-colored Fruit (Eidolon helvum) ALE 11:87; MLS 100 (bw).

Bat, Vampire (Desmodus rotundus) ALE 11:128; BMA 229-231; DSM 60; HMA 29; HMW 185; JAW 38; MLS 103 (bw); SMW 44 (cp 26); TWA 144.

Bat, Whiskered (Myotis mystacinus) HMW 187; OBV 157.

Bateleur —See **Eagle, Bateleur.**

Batfish (Platax sp.) ALE 5:95; CAA 18 (bw); FEF 367, 368 (bw); HFI 32; HFW 175 (cp 85); WEA 63; WFW cps 313, 314.

Bats-in-the-belfry (Campanula trachelium) BFB cp 21; EFC cp 5; EPF pl 521; MBF cp 54; NHE 36; OWF 167; PFE cp 139 (1340); PWF 122c; WFB 227 (3).

Bayberry (Myrica cerifera) EGW 128; EPF pl 326; PFF 132; TGS 91; (M. pensylvanica) EEG 100; EFS 124; EPG 133; TGS 86.

Baza —See **Hawk, Crested.**

Beaconfish —See **Head-and-tail-light Fish.**

Bead Plant (Nertera granadensis) BIP 129 (cp 195); RDP 23, 276; (N. sp.) SNZ 105 (cp 320), 109 (cp 334).

Bead Tree —See **Chinaberry.**

Beak-sedge, Brown (Rhynchospora fusca) MBF cp 91; MFF pl 116 (1124); NHE 77.

Beak-sedge, White (Rhynchospora alba) MBF cp 91; MFF pl 116 (1123); NHE 77.

Bean, Hyacinth (Dolichos lablab) EGA 116; EGV 106; OFP 45; (D. lignosus) MWF 115; (D. gibbosus) LFT cp 84.

Bean Trefoil —See **Stinking Wood.**

Bear, Asiatic or Himalayan Black (Selenarctos thibetanus) ALE 12:123; DAA 74; HMW 149; MLS 268 (bw); SMW 189 (cp 120).

Bear, Brown (Ursus arctos) ALE 12:124; 13:163; BMA 34, 35; CAS 43 (bw); DAE 96-97; DSM 158; GAL 24 (bw); HMA 30, 31; HMW 149; LAW 533; LEA 557 (bw); MLS 269 (bw); SLP 78-79; SOS 214; TWA 81; WEA 65.

Bear, Koala —See **Koala.**

Bear, Malayan —See **Bear, Sun.**

Bear, Moon —See **Bear, Asiatic or Himalayan Black.**

Bear, Polar (Thalarctos maritimus or Ursus maritimus) ALE 12:123, 131; BMA 178 (bw), 179; CAS 14-15; CEU 290-291 (bw); DAE 141; HMA 30; HMW 150; JAW 107; LAW 533; LEA 556 (bw); LVA 61; LVS 106-107; MLS 271 (bw); PWC 96-97; SLS 243; SMW 190 (cp 121); VWA cp 15; WID 70.

Bear, Sloth (Melursus ursinus) ALE 12:123; CAS 185 (bw); DAA 75; DSM 156; HMW 151; JAW 108; MLS 273 (bw).

Bear, Spectacled (Tremarctos ornatus) ALE 12:123; DSM 159; LVA 27; LVS 98; MLS 267 (bw); PWC 148, 237 (bw); SLP 110; VWA cp 41.

Bear, Sun (Helarctos malayanus) ALE 12:123; DAA 125; DSM 157; HMW 151; MLS 272 (bw); SMW 201 (bw); TWA 173.

Bear, Syrian Brown (Ursus arctos syriacus) ALE 12:124; SMW 190 (cp 122).

Bearberry or Kinnikinnick (Arctostaphylos uva-ursi) AMP 137; EEG 140; EGC 122; EGE 112; EGW 90; ERW 97; GDS cp 10; HPW 126; MBF cp 55; MFF pl 77 (588); NHE 70; OWF 121; PFE cp 88 (925); PFF 275; PWF 41, 155j; TGF 156; WFB 173 (4).

Bearberry, Alpine or Black (Arctous alpinus) EWF cp 6d; MBF cp 55; MFF pl 77 (589); NHE 266.

Beard-grass (Bothriochloa ischaemum) NHE 243; PFE cp 182 (1808).

Beard-grass, Annual (Polypogon monspeliensis) MBF pl 95; MFF pl 123 (1233); PFE 534.

Bear's Breeches (Acanthus mollis) EGP 92; EPF pl 857; MWF 28; PFE cp 130 (1266); PFM pl 16 (397); TGF 236; (A. longifolius) HPW 250-1; TVG 77; (A. montanus) DEW 2:172 (cps 100, 101).

Bear's Breeches, Spiny (Acanthus spinosus) DFP 119 (cp 945); EWF cp 21f, OGF 157; PFE cp 130 (1267); PFW 12; SPL cp 185.

Bear's Foot —See **Hellebore.**

Bear's-ear (Primula auricula) DEW 1:213 (cp
118); DFP 19 (cp 148); EFC cp 21; EPF pl
260; ERW 130; NHE 278; OGF 15; PFE cp 90
(941); PFW 241; SPL 492; TVG 163; WFB
174.

Beauty Berry (Callicarpa sp.) DEW 2:186;
EFS 31, 98; EPG 109; EWF cp 104a; GDS cp
29; HSC 20, 22; PFW 304; TGS 340, 342.

Beauty-bush (Kolkwitzia amabilis) DFP 209
(cp 1668); EEG 98; EFS 120; ELG 117; EWF
cp 94d; GDS cp 269; HSC 72, 76; LFW 80 (cp
186); MWF 174; OGF 61; TGS 358; TVG 229.

Beaver (Castor fiber) ALE 11:257, 284; CAS
49 (bw); CEU 268 (bw); DAE 20; GAL 61
(bw); HMW 163; MLS 206 (bw); SMW 96 (cp
68).

Becard (Pachyrhamphus sp.) ALE 9:144;
CDB 146 (cp 592); SSA pl 45.

Bedbug (Cimex lectularius) ALE 2:182, 187;
BIN pl 48; CIB 113 (cp 10); CTI 20 (cp 12);
GAL 459 (bw); KIW 70 (bw); LAW 230; OBI
25; PEI 107 (bw); ZWI 126.

Bedstraw, Hedge or Smooth (Galium
mollugo) MBF cp 42; MFF pl 103 (813); NHE
188; OWF 93; PWF 78d; TEP 123; WFB 187
(3); WUS 355.

Bedstraw, Lady's or Yellow (Galium
verum) BWG 45; EGH 115; MBF cp 42; MFF
cp 50 (814); NHE 188; OWF 31; PWF 87g;
TEP 123; WFB 187 (7).

Bedstraw, Marsh (Galium palustre) MBF cp
42; MFF pl 103 (816); NHE 100; PFE cp 99
(1022); PWF 112a; WFB 187 (6).

Bedstraw, Northern (Galium boreale) MBF
cp 42; MFF pl 103 (812); NHE 188; PWF
121i; WFB 187 (4).

Bee (Osmia sp.) CIB 273 (cp 44); GAL 380 (cp
14); OBI 161; (Macropis labiata) CIB 272 (cp
43).

Bee, Bumble (Bombus sp.) ALE 2:512-514,
524; 11:321; CIB 276 (cp 45), 277 (cp 46); CTI
63 (cp 87); GAL 380 (cp 14); KIW 270 (cp
148); LAW 247; OBI 163; PEI 161 (cp 16a),
178, 179 (bw); PWI 156 (bw); ZWI 174.

Bee, Carpenter (Xylocopa violacea) CIB 276
(cp 45); CTI 62 (cp 86); GAL 380 (cp 14); PEI
180 (bw); ZWI 136.

Bee, Cuckoo or Parasitic (Psithyrus sp.)
CIB 277 (cp 46); LAW 247; OBI 163; PEI 180
(bw); (Nomada sp.) ALE 2:524; CIB 273 (cp
44); OBI 161; (Sphecodus sp.) CIB 273 (cp
44); OBI 159

Bee, Flower or Potter (Anthophora sp.) CIB
276 (cp 45); OBI 159.

Bee, Hairy-legged Mining (Dasypoda
plumipes) GAL 380 (cp 14); (D. hirtipes) CIB
272 (cp 43); OBI 159.

Bee, Honey (Apis sp.) ALE 2:26, 511, 524;
CAS 198; CIB 276 (cp 45); GAL 380 (cp 14);
KIW 273 (bw); LEA 114, 122, 150 (bw); OBI
163; PEI 174-177 (bw); PWI 95 (bw); WEA
68, 69; ZWI 200, 201.

Bee, Leaf-cutting (Megachile sp.) CIB 273
(cp 44); GAL 380 (cp 14); OBI 161; ZWI 134.

Bee, Long-horned or Solitary (Eucera sp.)
BIN pl 108; CIB 273 (cp 44); PEI 181 (bw).

Bee, Mason (Chalicodoma muraria) ALE
2:524; CIB 273 (cp 44); GAL 380 (cp 14).

Bee, Mining (Andrena sp.) CIB 273 (cp 44);
GAL 380 (cp 14); OBI 159; (Halictus sp.) CIB
273 (cp 44); GAL 380; OBI 159.

Beech, American (Fagus grandifolia) EET
147; MLP 66; OBT 156; PFF 139; SST cp 109;
TGS 105.

Beech, Antarctic (Nothofagus antarctica)
DEW 1:184; MLP 69; OBT 156.

Beech, Common or European (Fagus
sylvatica) DFP 202 (cp 1609); EEG 116; EGT
72; MTB 208 (cp 19); NHE 23; OBT 156,
front endpaper; TEP 278; TGS 87; WFB 32,
33 (3).

Beech (Hybrids) (Fagus sylvatica sp.) EET
146, 147; EGT 112; EGW 112; ELG 104; EPG
121; OBT 156.

Beech, New Zealand Black (Nothofagus
solandri) EET 148, 244; OBT 156; SNZ 86
(cp 251), 87 (cp 252).

Beech, Roble (Nothofagus obliqua) BKT cp
80; DEW 1:184; OBT 156; PWF 159j.

Beech, Southern or Raoul (Nothofagus
procera) DFP 214 (cp 1211); HPW 61; MTB
208 (cp 19); OBT 156; PWF 159i.

Beech Tuft or Porcelain Fungus
(Oudemansiella mucida) DLM 230; KMF 51;
LHM 107; ONP 125; PUM cp 77.

Bee-eater, Blue-bearded (Nyctyornis athertoni) KBA 176 (cp 23); WBV 123.

Bee-eater, Blue-cheeked (Merops superciliosus) BBE 185; FBE 189.

Bee-eater, Blue-tailed (Merops philippinus) DPB 209; KBA 176 (cp 23); WBV 123.

Bee-eater, Carmine (Merops nubicus) ABW 180; ALE 9:48; BBT 48, 50; WAB 157; WEA 68; (M. nubicoides) CAF 183; PSA 209 (cp 24).

Bee-eater, Chestnut-headed (Merops viridus) ALE 9:30; DPB 209; KBA 176 (cp 23); WBV 123; (M. leschenaulti) KBA 176 (cp 23).

Bee-eater, European (Merops apiaster) ABW 180; ALE 9:32, 48; BBB 272; BBE 185; CDB 128 (cp 502); CEU 61; DAE 138; FBE 189; FBW 77; LAW 474, 506; OBB 113; PBE 181 (cp 46); PSA 209 (cp 24); RBE 200.

Bee-eater, Green (Merops orientalis) CDB 129 (cp 503); FBE 189; KBA 176 (cp 23).

Bee-eater, Little (Merops pusillus or Melittophagus pusillus) CDB 128 (cp 501); PSA 209 (cp 24).

Bee-eater, Rainbow (Merops ornatus) BBT 16; CAU 97; CDB 129 (cp 504); HAB 135; MBA 28-29; MEA 40; MWA 65-67; RBU 176; SNB 118 (cp 58).

Bee-eater, Red-bearded or -breasted (Nyctyornis amictus) ALE 9:48; DAA 147; KBA 176 (cp 23); RBI 124; WAB 161.

Bee-eater, Red-throated (Merops bulocki or Melittophagus bulocki) ABW 180; ALE 9:48; BBT 49; CDB 128 (cp 499); GPB 212; WAB 157; WEA 68.

Bee-eater, White-fronted (Merops bullock-oides or Melittophagus bullockoides) CDB 128 (cp 498); GPB 213; PSA 209 (cp 24).

Beefsteak Plant (Acalypha wilkensiana) EFP 84; MEP 71; MWF 28; RDP 67.

Beefsteak Plant (Iresine) —See Bloodleaf.

Beefsteak Plant (Perilla frutescens) DFP 45 (cp 357); EGA 144; EGH 133.

Beefwood (Casuarina sp.) EET 243; EGE 117; EPF pl 325; HPW 62; MWF 75; PFF 127; PFW 73; SST cp 89.

Beet, Garden (Beta vulgaris) EVF 38, 83; PFF 155; TEP 163.

Beet, Sea (Beta vulgaris subsp. maritima) MBF cp 72; MFF pl 111 (212); OWF 57; NHE 135; PFE cp 10 (101); PWF 87f; WFB 47 (8).

Beetle —See also Weevil.

Beetle (Autocrates aeneus) ALE 2:236; RBT 69; (Dascillus cervinus) CIB 288 (cp 49); (Acmaeops collaris) GAL 276 (cp 6); (Akis italica) CTI 51 (cp 67); (Pimelia sp.) ZWI 215; (Tetraopes tetraophthalmus) KIW 91 (cp 53).

Beetle, Ant (Thanasimus formicarius) OBI 175; PEI 210 (bw).

Beetle, Asparagus (Crioceris asparagi) EPD 89; GAL 276 (cp 6); PEI 225 (cp 24a); (C. liliae) PWI 136 (bw).

Beetle, Atlas (Chalcosoma atlas) PEI 273 (bw); RBT 89.

Beetle, Bacon —See Beetle, Larder.

Beetle, Bark (Ips sp.) ALE 2:256; GAL 433 (bw); PEI 257-258 (bw); (Cucujus sp.) ALE 2:235; (Scolytus sp.) CIB 317 (cp 56); GAL 434 (bw); OBI 191; PEI 255 (bw).

Beetle, Bee Chafer or Banded Brush (Trichius fasciatus) ALE 2:246; CTI 45 (cp 59); OBI 183; PEI 272 (cp 29a), 278 (bw).

Beetle, Birch-leaf Roller (Deporaus betulae) ALE 2:265; LEA 126 (bw).

Beetle, Blister (Lytta vesicatoria) ALE 2:236; CIB 308 (cp 53); GAL 268 (cp 5); PEI 223 (bw).

Beetle, Bloody-nosed (Timarcha tenebricosa) ALE 2:255; CIB 316 (cp 55); OBI 187.

Beetle, Bombadier (Brachinus crepitans) ALE 2:216; CIB 284 (cp 47); GAL 268 (cp 5); OBI 169; ZWI 152.

Beetle, Burying —See Beetle, Sexton.

Beetle, Cardinal (Pyrochroa coccinea) CIB 305 (cp 52); GAL 268 (cp 5); LEA 160; (P. serraticornis) OBI 179.

Beetle, Carpenter's Longhorn (Ergates faber) PEI 224 (bw); RBT 125.

Beetle, Carpet or Museum (Anthrenus sp.) ALE 2:226; CIB 304 (cp 51); GAL 468 (bw); OBI 177; PEI 206, 207 (bw); ZWI 148.

Beetle, Cave-dwelling Ground (Leptodirus hohenwarti) ALE 2:225; CTI 41 (cp 53); LAW 201; PEI 200 (bw); RBT 67; (Sphodropsis ghilianii) LAW 200.

Beetle, Cellar or Churchyard (Blaps sp.) ALE 2:245; CIB 305 (cp 52); CTI 50 (cp 66); GAL 469 (bw); LAW 203; OBI 181.

Beetle, Cerulean Chafer (Hoplia coerulea) CTI 45 (cp 58); PWI 129; RBT 177.

Beetle, Checkered (Trichodes sp.) ALE 2:225; CTI 48 (cp 63); RBT 69.

Beetle, Click (Elater sp.) ALE 2:226; BIN cp 122; CIB 304 (cp 51); GAL 268 (cp 5); (Ampedus cinnabarinus) LAW 204; (Ctenicera virens) CTI 47 (cp 62).

Beetle, Cockchafer (Melolontha melolontha) CIB 304 (cp 51); GAL 290 (bw); LEA 152 (bw); PEI 267 (bw); ZWI 150; (Lepidiota bimaculata) RBT 83.

Beetle, Death-watch (Ptilinus pectinicornis) ALE 2:235; GAL 471 (bw); (Xestobium rufovillosum) CIB 305 (cp 52); GAL 471 (bw); OBI 179.

Beetle, Desert Ground (Anthia thoracica) ALE 2:215; RBT 65; (A. collaris) PWI 136 (bw).

Beetle, Dung (Phanaeus sp.) ALE 2:246; LAW 241; PEI 256 (cp 27b); RBT 81; (Aphodius sp.) CIB 289 (cp 50); OBI 181; (Heliocopris sp.) PEI 265 (bw); RBT 81; WEA 136.

Beetle, Dung or Sacred Scarab (Scarabaeus sacer or S. laticollis) ALE 2:250; BIN pl 118; CAS 90 (bw); LAW 234, 235; LEA 154 (bw); PEI 263 (bw); ZWI 146.

Beetle, Elephant (Megasoma elephas) PEI 271 (bw); RBT 95; (M. gyas) RBT 97; (Xylotrupes australicus) BIN pl 123.

Beetle, English Scarab (Copris lunaris) ALE 2:249; CIB 289 (cp 50); OBI 181.

Beetle, Eyed Ladybird (Anatis ocellata) CIB 308 (cp 53); PEI 192 (cp 19a).

Beetle, Fiddler or Violin (Mormolyce phyllodes) ALE 2:216; PEI 185 (bw); RBT 59.

Beetle, Fiery Searcher or Hunter (Calosoma sp.) ALE 2:215, 264; GAL 268 (cp 5); KIW 83 (cp 37); PEI 186; PWI 136 (bw); RBT 63.

Beetle, Flea (Phyllotreta sp.) ALE 2:277; CIB 316 (cp 55); GAL 276 (cp 6); OBI 187.

Beetle, Flower (Chelorrhina polyphemus) PEI 272 (cp 29b); RBT 111; (C. savagei) RBT 113; (Malachius sp.) GAL 268 (cp 5); OBI 175; (Oedemera sp.) ALE 2:236; CIB 305 (cp 52); OBI 179.

Beetle, Forest Dung (Geotrupes sp.) ALE 2:266; CIB 289 (cp 50); GAL 290 (bw), 291 (bw); OBI 181; PEI 263 (bw).

Beetle, Forest or Cylinder (Spondylis buprestoides) ALE 2:255; PEI 233 (bw).

Beetle, Four-spot Carrion (Xylodrepa quadripunctata) GAL 268 (cp 5); OBI 171; PEI 198 (bw).

Beetle, Fungus (Erotylus varians) ALE 2:235; (Cryptophagus saginatus) CIB 308 (cp 53); (Eumorphus marginatus) ALE 2:235.

Beetle, Furniture (Anobium punctatum) CIB 305 (cp 52); GAL 471 (bw); OBI 179.

Beetle, Garden Chafer (Chrysophora chrysochloa) PEI 269 (bw); RBT 87; (Phyllopertha horticola) CIB 305 (cp 51); GAL 268 (cp 5); OBI 183.

Beetle, Giraffe Stag (Cladognathus giraffa) PEI 260 (bw); RBT 117.

Beetle, Goliath (Goliathus sp.) ALE 2:246; CTI 43 (cp 56); PEI 281-283 (bw); RBT 101, 103, 105, 107.

Beetle, Grain (Zabrus sp.) ALE 2:216; ZWI 131.

Beetle, Great Diving or Water —See **Beetle, Predaceous Diving.**

Beetle, Great Oak Longhorn (Cerambyx cerdo) CTI 52 (cp 69); GAL 292 (bw); PEI 230 (bw); RBT 125; ZWI 138.

Beetle, Great Silver Water (Hydrophilus piceus or Hydrous piceus) ALE 2:216; CIB 285 (cp 48); GAL 298 (bw); OBI 171; PEI 195 (bw); ZWI 192.

Beetle, Ground (Carabus sp.) ALE 2:215; BIN pls 115, 116; CIB 284 (cp 47); CTI 38 (cps 46, 47), 39 (cp 48); GAL 268 (cp 5), 285 (bw); OB1 165; PEI 177 (cp 18a), 184 (bw); RBT 59, 63; WEA 176; ZWI 94.

Beetle, Ground (Various genera and species) ALE 2:51, 215, 216; CIB 284 (cp 47); CTI 39 (cp 49); LAW 197, 200, 240; OBI 165; PEI 177 (cp 18b); RBT 59, 61.

Beetle, Harlequin Longhorn (Acrocinus longimanus) CSA 115 (bw); PEI 239 (bw); RBT 133.

Beetle, Hercules—See Beetle, Rhinoceros (Dynastes sp.).

Beetle, Hide or Leather (Dermestes maculatus) CIB 304 (cp 51); OBI 177; (Necrobia sp.) CIB 305 (cp 52); OBI 175.

Beetle, Hister (Oysternus maximus) ALE 2:225; (Hister cadaverinus) CIB 288 (cp 49); (Saprinus maculatus) ALE 2:225.

Beetle, House Longhorn (Hylotrupes bajulus) CIB 309 (cp 54); GAL 469 (bw); OBI 185.

Beetle, Jewel (Chrysochroa sp.) CTI 46 (cp 60); RBT 75, 77; (Calodema sp.) RBT 77.

Beetle, Ladybird (Adalia sp.) CIB 308 (cp 53); LAW 205; OBI 177; PEI 203, 204 (bw); (Chilomenes lunata) ALE 2:275; WEA 217; (Hippodamia sp.) CTI 49 (cp 65); (Coccinella sp.) ALE 11:321; BIN pls 119, 120; CIB 308 (cp 53); EPD 54; GAL 268 (cp 5); OBI 177; PEI 203 (bw); PWI 183.

Beetle, Larder (Dermestes lardarius) ALE 2:226; CIB 304 (cp 51); GAL 467 (bw); OBI 177; PEI 207 (bw).

Beetle, Leaf (Chrysomela sp.) ALE 2:278; GAL 276 (cp 6); OBI 187; (Cryptocephalus sp.) CIB 316 (cp 55); PEI 224 (cp 23); (Mecistomela sp.) RBT 161; (Melasoma populi) ALE 2:278; CIB 316 (cp 55); PEI 241 (bw).

Beetle, Leaf or Reed (Donacia sp.) ALE 2:255; CIB 316 (cp 55); GAL 276 (cp 6); OBI 187.

Beetle, Lesser Stag (Dorcus parallelopipedus) CIB 289 (cp 50); GAL 289 (bw); OBI 183.

Beetle, Longhorn (Macrodontia sp.) CTI 52 (cp 70); PEI 208 (cp 21b), 225 (bw); RBT 131; (Rhagium sp.) GAL 276 (cp 6); OBI 185; PEI 229 (bw); (Sagra busqueti) ALE 2:255; LAW 240; PEI 240 (cp 25b) RBT 163; (Saperda sp.) CIB 309 (cp 54); GAL 276 (cp 6); OBI 185; PEI 236 (bw).

Beetle, Mealworm (Tenebrio molitor) ALE 2:245; CIB 305 (cp 52); GAL 469 (bw); LAW 201; OBI 181; PEI 217 (bw).

Beetle, Minotaur (Typhaeus typhoeus) ALE 2:249; CIB 289 (cp 50).

Beetle, Musk (Aromia moschata) CIB 309 (cp 54); GAL 276 (cp 6); OBI 185; RBT 125.

Beetle, Oil or Blister (Meloe sp.) ALE 2:236; BIN pl 121; CIB 308 (cp 53); GAL 268 (cp 5); KIW 84 (cp 40); LAW 204; LEA 155 (bw); OBI 179; PEI 222, 223 (bw); PWI 129; RBT 71.

Beetle, Pine Chafer (Polyphylla fullo) ALE 2:249; PEI 268 (bw); RBT 71.

Beetle, Powder-post (Lyctus sp.) CIB 305 (cp 52); GAL 467 (bw); OBI 179.

Beetle, Predaceous Diving (Acilius sulcatus) ALE 2:216; CIB 285 (cp 48); GAL 297 (bw); OBI 169; PEI 191, 192 (bw); (Dytiscus sp.) ALE 2:216; CIB 285 (cp 48); GAL 296 (bw); LAW 198; LEA 152 (bw); OBI 169; PEI 187-190 (bw); PWI 43 (bw); RBT 63; WEA 128; ZWI 192, 193.

Beetle, Raspberry (Byturus sp.) GAL 414 (bw); OBI 177.

Beetle, Red-brown Skipjack (Athous haemorrhoidalis) CIB 304 (cp 51); OBI 175.

Beetle, Rhinoceros (Dynastes sp.) ALE 2:250; CTI 43 (cp 55); KIW 137 (bw); LEA 160; PEI 257 (cp 28b), 270 (bw); RBT 91, 93; (Oryctes nasicornis) CTI 44 (cp 57); RBT 71.

Beetle, Rose Chafer (Cetonia aurata) CIB 304 (cp 51); GAL 268 (cp 5); OBI 183; PEI 276 (bw); PWI 133 (bw); (C. cupraea) LAW 203; (Dicranocephalus sp.) CIB 112 (cp 9); PEI 279 (bw); RBT 109; (Rhomborrhina japonica) RBT 115.

Beetle, Rove (Many genera and species) ALE 2:225; CIB 288 (cp 49); CTI 40 (cps 51, 52); GAL 268 (cp 5); OBI 173; PEI 196, 197.

Beetle, Sack (Clytra quadripunctata) CIB 316 (cp 55); GAL 276 (cp 6); (C. laeviscula) PEI 209 (cp 22a).

Beetle, Sailor —See Beetle, Soldier.

Beetle, Saw-toothed Grain (Oryzaephilus surinamensis) CIB 308 (cp 53); GAL 467 (bw).

Beetle, Sawyer —See Beetle, Tanner Longhorn.

Beetle, Screech or Squeak (Hygrobia hermanni) CIB 285 (cp 48); OBI 169; (H. tarda) ALE 2:216.

Beetle, Sexton (Necrophorus sp.) ALE 2:225; CIB 288 (cp 49); CTI 40 (cp 50); GAL 268 (cp 5), 286 (bw); KIW 84 (cp 39); LAW 237; OBI 171; PEI 198, 199 (bw); RBT 71; ZWI 146.

Beetle, Snout —See **Weevil.**

Beetle, Soldier (Cantharis sp.) CIB 304 (cp 51); GAL 268 (cp 5); OBI 175; PEI 208 (bw); (Themus generosus) ALE 2:225.

Beetle, Spider (Ptinus sp.) ALE 2:235; CIB 305 (cp 52); GAL 468; (bw); OBI 179; (Gibbium sp.) ALE 2:235; ZWI 225.

Beetle, Stag (Lucanus cervus) ALE 2:266; CIB 289 (cp 50); CTI 42 (cp 54); GAL 289 (bw); LAW 240; LEA 116 (bw); OBI 183; PEI 241 (cp 26), 259 (bw); PWI 134 (bw); RBT 119; WEA 348.

Beetle, Stag (Chiasognathus granti) ALE 2:246; PEI 261 (bw); RBT 117; (Neolamprima sp.) PEI 256 (cp 27a), 262 (bw); (Odontolabis sp.) RBT 117, 119, 121; (Systenocerus caraboides) GAL 268 (cp 5).

Beetle, Tanner Longhorn (Prionus coriarius) CIB 309 (cp 54); GAL 293 (bw); OBI 185; PEI 224 (bw); RBT 125.

Beetle, Tiger (Cicindela sp.) ALE 2:215, 263; CIB 284 (cp 47); GAL 268 (cp 5), 285 (bw); KIW 107 (bw); LAW 198; LEA 152 (bw), 160; OBI 165; PEI 176 (cp 17), 182, 183 (bw); PWI 135 (bw); RBT 63; ZWI 118; (Mantichora sp.) ALE 2:215; PEI 183 (bw).

Beetle, Timberman (Acanthocinus aedilis) GAL 276 (cp 6); OBI 185; RBT 125.

Beetle, Tortoise (Cassida sp.) CIB 316 (cp 55); GAL 276 (cp 6); KIW 90 (cp 50); LAW 206; OBI 189; PEI 244 (bw); (Aspidomorpha sp.) PEI 245 (bw); RBT 69; (Selenis spinifex) ALE 2:255.

Beetle, Wasp (Clytus arietis) CIB 309 (cp 54); LEA 121 (bw); OBI 185.

Beetle, Wasp Fan (Metoecus paradoxus) ALE 2:236; PEI 220.

Beetle, Water (Agabus sp.) OBI 169; (Dryops sp.) ALE 2:226.

Beetle, Weaver Longhorn (Lamia textor) CIB 309 (cp 54); PEI 234 (bw).

Beetle, Whirligig (Gyrinus sp.) CIB 285 (cp 48); OBI 169; PEI 193 (bw); (Orectogyrus bicostatus) ALE 2:216.

Beetle, Wireworm (Agriotes sp.) CIB 304 (cp 51); GAL 287 (bw); OBI 175.

Beetle, Wood-borer (Bupestris sp.) CIB 288 (cp 49); KIW 86 (cp 43); LAW 240; (Evides elegans) ALE 2:226; (Hylecoetus dermestoides) CIB 305 (cp 52); (Julodis sp.) ALE 2:226; LAW 240; (Lampra rutilans) ZWI 139.

Bee-wolf —See **Wasp, Bee-killer.**

Beggar-ticks (Bidens frondosa) PFF 323; WFB 238; WUS 379.

Begonia (Begonia sp.) BIP 31, 33; DEW 1:169, 170; DFP 52, 53; EDP 130; EFP 93; EGB 97; EGG 90; EMB 124; EPF pls 515, 516; ESG 96; EWF cp 174a; GFI 23; LFT cp 101; LFW 232; LHP 42, 43; MWF 54, 55; PFW 41, 42; RDP 92-95; TGF 158.

Begonia, Iron Cross (Begonia masoniana) DFP 52 (cp 414); EDP 139; EFP 93; LHP 41; RDP 95.

Begonia, Rex (Begonia rex) BIP 33 (cp 31); EFP 93; HPW 113; LHP 40; PFF 254; PFW 40; RDP 17.

Begonia Treebine (Cissus discolor) DEW 2:80; EDP 148; EFP 101; EGV 101; LHP 58; PFW 307; RDP 143.

Begonia, Tuberous (Begonia tuberhybrida) BIP 31 (cp 29); DFP 53, 54 (cps 419-430); EGB 58-59, 97; EGV 95; ESG 96; LFW 188, 189; MWF 55; OGF 125; SPL cp 329; TVG 183.

Begonia, Wax or Bedding (Begonia semperflorens) DFP 31 (cp 242); EGA 98; ESG 79, 81, 96; FHP 106; MWF 55; PFF 254; PFW 43; SPL 354; TVG 35.

Belgian Evergreen (Dracaena sanderiana) DFP 65 (cp 518); EGL 104; MWF 117; RDP 180.

Bellardia (Bellardia trixago) PFE cp 126 (1237); PFM cp 175.

Bellbird (Anthornis melanura) BNZ 45; CDB 193 (cp 837); FNZ 176 (cp 16); WAN 133; (Procnias alba) SSA pl 27; (P. averano) WEA 70.

Bellbird, Bare-throated (Procnias nudicollis) ALE 9:136, 144; CDB 147 (cp 594); GPB 243.

Bellbird, Crested (Oreoica gutturalis) HAB 249; SAB 46 (cp 22).

Bellbird, Three-wattled (Procnias tricarunculata) ABW 204; ALE 9:144.

Bellflower (Campanula portenschlagiana) DFP 4 (cp 26); PFW 62; TGF 188.

Bellflower, Alpine (Campanula alpina) EPF pl 523; NHE 284.

Bellflower, Bearded (Campanula barbata) EFC cp 5 (3); NHE 284; PFE cp 139 (1330); RWF 27; WFB 227 (6).

Bellflower, Chilean (Lapageria rosea) BIP 111 (cp 170); EGV 118; HPW 313; MEP 16; PFW 229; (Nolana sp.) DEW 2:147; DFP 45 (cp 354).

Bellflower, Chinese —See Balloon Flower (Platycodon).

Bellflower, Clustered (Campanula glomerata) BFB cp 2 (4); DFP 127 (cp 1011); MBF cp 54; MFF cp 4 (798); NHE 75; OWF 167; PFE cp 138 (1333); PFW 61 (5); PWF 122 b; SPL cp 474; TVG 81; WFB 227 (2).

Bellflower, Creeping (Campanula rapunculoides) EPF pl 519; HPW 255; MBF cp 54; MFF cp 4 (797); NHE 235; OWF 167; PFE cp 140 (1342); PWF 125 h.

Bellflower, Downy (Campanula bononiensis) NHE 235; PFE cp 140 (1343).

Bellflower, Dwarf (Campanula zoysii) DEW 2:192; DFP 4 (cp 27); NHE 284.

Bellflower, Italian (Campanula isophylla) BIP 47 (cp 55); DFP 55 (cp 440); EGG 95; EGV 98; FHP 110; LHP 50, 51; RDP 128; SPL cp 330.

Bellflower, Ivy-leaved (Wahlenbergia hederacea) MBF cp 54; MFF cp 7 (795); NHE 188; PWF 136a; WFB 227 (7).

Bellflower, Large or Giant (Campanula latifolia) BFB cp 1 (9); DFP 127 (cp 1013); MBF cp 54; MFF cp 4 (796); NHE 235; OWF 167; PFW 61 (2); PWF 136 b.

Bellflower, Milky (Campanula lactiflora) DFP 127 (cp 1012); OGF 157; PFW 61 (1); SPL cp 193.

Bellflower, Narrow-leaved or Peach-leaved (Campanula persicifolia) DFP 127 (cp 1015); EPF pl 520; LFW 21 (cps 40, 41); NHE 36; PFE cp 139 (1336); PFW 61 (4); TEP 257; TGF 188; TVG 83; WFB 227 (5).

Bellflower, Nettle-leaved —See Bats-in-the-belfry.

Bellflower, Rhomboidal (Campanula rhomboidalis) NHE 284; PFE cp 140 (1344).

Bellflower, Scheuchzer's (Campanula scheuchzeri) NHE 284; PFE cp 140 (1339).

Bellflower, Serbian (Campanula poscharskyana) EGC 26, 27, 125; MWF 71; TVG 145.

Bellflower, Spreading (Campanula patula) KVP cp 116; MBF cp 54; MFF cp 4 (800); NHE 188; OWF 167; PFE 411; TEP 147; WFB 227 (4).

Bellflower, Tussock (Campanula carpatica) DFP 3 (cp 23); EEG 132; ERW 105; MWF 71; PFW 62; SPL cp 192; TGF 188.

Bellflower, Yellow (Campanula thyrsoides) EWF cp 15 f; KVP cp 142; NHE 284; PFE cp 139 (1332).

Benne or Benniseed —See Sesame (Sesamum).

Bent, Silky (Apera spica-venti) DEW 2:263; MFF pl 133 (1231); NHE 244; (A. interrupta) MFF pl 121 (1232).

Bent-grass, Common (Agrostis tenuis) BWG 178 (bw); EGC 78-79, 114; MBF pl 95; NHE 195.

Bent-grass, Creeping or White (Agrostis stolonifera) MBF pl 95; MFF pl 131 (1230); NHE 195.

Bent-grass, Velvet (Agrostis canina) MBF pl 95; NHE 195; PFF 353.

Bergenia (Bergenia purpurascens) DFP 126 (cps 1002, 1003); EWF cp 99 d; PFW 272.

Bergenia, Heartleaf —See Saxifrage, Elephant-leaved.

Bergenia, Leather   See Pig Squeak.

Bergylt —See Haddock, Norway (Sebastes).

Bertolonia (Bertolonia sp.) EPF pl 713; RDP 99.

Be-still Tree —See Oleander, Yellow (Thevetia).

Bethlehem Sage —See Lungwort, Common (Pulmonaria).

Betony, Water —See Figwort, Water.

Betony, Wood (Betonica officinalis) MBF cp 69; MFF cp 29 (763); NHE 35; WFB 203 (9).

Betony, Wood (Stachys officinalis) AMP 159; BFB cp 17 (1); EGH 142; LFW 116 (cp 257); OWF 147; PWF 103 f; TVG 175.

Betony, Woolly —See Lamb's Ears

Bharal —See **Sheep, Blue.**

Biarum (Biarum tenuifolium) PFE cp 183 (1820); PFM cp 222.

Bib —See **Pouting** (Trisopterus).

Bichir (Polypterus sp.) ALE 4:140; HFW 35 (cp 5), 65 (bw); LAW 296; SFW 44 (pl 1); WFW cp 17.

Bichir, Ornate (Polypterus ornatipinnis) ALE 4:139; HFW 35 (cp 6); SFW 45 (pl 2).

Bidi-bidi, Red Mountain (Acaena microphylla) OGF 89; SNZ 128 (cp 398); TGF 108; (A. nova-zelandiae) SNZ 128 (cp 396); (A. viridior) SNZ 128 (cp 397).

Big Blood Stalk —See **Mycena, Bleeding.**

Big-eye (Priacanthus arenatus) ALE 5:94; HFI 24; HFW 89 (cp 46); (P. macracantha) ALE 5:79.

Bilberry or Blaeberry or Whortleberry (Vaccinium myrtillus) AMP 153; CTH 43 (cp 57); DEW 1:237; EPF pl 487, 488; MBF cp 55; MFF pl 77 (597); NHE 70; OFP 83; OWF 121; PFE 89 (936); PWF 67i, 155g; TEP 287; WFB 173 (1).

Billbergia (Billbergia sp.) EPF pl 1031; FHP 107; HPW 295; MEP 13; MWF 57; PFW 54; RDP 100.

Billy Buttons (Craspedia) —See **Woollyhead.**

Billy's Button (Geum) —See **Avens, Water.**

Bindweed, Black (Bilderdykia convolvulus) PWF 135f; WFB 43 (1).

Bindweed, Black (Polygonum convolvulus) MBF cp 73; MCW 25; MFF pl 102 (533); NHE 224; OWF 59; PFF 154; TEP 175; WUS 121.

Bindweed, Copse (Polygonum dumetorum) MBF cp 73; NHE 224.

Bindweed, Elegant (Convolvulus elegantissimus) PFE cp 100 (1039); PFM cp 135.

Bindweed, Field or Lesser (Convolvulus arvensis) BWG 118; CTH 46 (cp 65); DEW 2:110 (cp 71); EPF pl 784; MBF cp 61; MCW 80; MFF cp 27 (666); NHE 228; OWF 125; PFW 90; PWF 78; TEP 179; WFB 185 (3); WUS 291.

Bindweed, Great (Calystegia sepium) BFB cp 15; BWG 82; EPF pl 783; HPW 230; MBF cp 61; MFF pl 102; NHE 228; OWF 95; PWF 134b; SNZ 44 (cp 98); WFB 185 (1); (C. silvatica) EWF cp 18c; WFB 185 (1a).

Bindweed, Greater or Hedge (Convolvulus sepium) AMP 71; PFF 283; WUS 293.

Bindweed, Mallow-leaved (Convolvulus althaeoides) DFP 5 (37); EWF cp 36d; OGF 163; PFE cp 100 (1039); PFM cp 136.

Bindweed, Sea (Calystegia soldanella) BFB cp 1; KVP cp 104; MBF cp 61; MFF cp 27 (688); NHE 136; PFE cp 100 (1041); PWF 90c; SNZ 22 (cp 20); WFB 185 (2).

Binturong (Arctictis binturong) DAA 81; DSM 181; HMA 32; MLS 301 (bw); SMW 170 (bw).

Birch, Cut-leaved White (Betula pendula gracilis) EGT 100; ELG 102; EPG 107.

Birch, Downy or Hairy (Betula pubescens) EET 141; EPF pl 333; EWF cp 12d; MBF cp 76; NHE 69; OBT 13, 17; OWF 201; PWF 109i; SST cp 85.

Birch, Dwarf (Betula nana) MBF cp 76; NHE 69; PFE cp 4 (38); TEP 79; WFB 31 (7).

Birch, Manchurian (Betula costata) BKT cps 60, 66; DEW 1:186; EET 140.

Birch, Northern Chinese Red-barked (Betula albosinensis) DFP 183 (cp 1461); MTB 192 (cp 17).

Birch, Paper or Canoe (Betula papyrifera) EEG 113; EET 140, 141; EGT 47, 66, 100; EGW 93; OBT 153; PFF 138; SST cp 83; TGS 87.

Birch, River (Betula nigra) BKT cp 63; MLP 64; TGS 110.

Birch, Silver (Betula verrucosa) MFF pl 71 (558); OWF 201.

Birch, Silver or White (Betula pendula) BKT cps 59, 61, 65; DEW 1:108 (cps 61, 62); DFP 183 (cp 1462); EET 140, 141; EGW 94; EPF pl 332; HPW 59; MBF cp 76; MTB 192 (cp 17); NHE 23; OBT 13, 17; PFE cp 4 (36); PWF cp 20d; SST cp 84; TEP 277; TGS 110; WFB 31 (5).

Birch, Yellow (Betula lutea) EET 141; PFF 137.

Bird Flower (Crotalaria sp.) DEW 1:288 (cp 185); EGA 111; EWF cp 53d; LFT 170 (cp 84).

Bird of Paradise, Blue or Prince Rudolph's (Paradisaea rudolphi or Paradisornis rudolphi) ABW 233; ALE 9:470; WAW 11; WEA 71.

Bird of Paradise, Brown Sickle-billed
(Epimachus meyeri) ALE 9:479; CDB 221
(cp 991).

Bird of Paradise, Count Raggi's
(Paradisaea raggiana) ABW 233; BBT 4;
CDB 221 (cp 992); WAW 11; WEA 71.

Bird of Paradise, Greater (Paradisaea
apoda) ALE 9:491, 502; GPB 337; LEA 465
(bw); WAB 25, 206.

Bird of Paradise, King (Cicinnurus regius)
ABW 232; ALE 9:470; BBT 6; CDB 221 (cp
990); SLP 158.

Bird of Paradise, King of Saxony
(Pteridophora alberti) ABW 233; ALE
9:479; WAB 206.

Bird of Paradise, Lesser (Paradisea minor)
CAU 241; SLP 159.

Bird of Paradise, Magnificent (Diphyllodes
magnificus) ABW 233; ALE 9:491; BBT 5.

Bird of Paradise, Princess Stephanie's
(Astrapia stephaniae) CDB 221 (cp 989);
WAB 207.

Bird of Paradise, Superb (Lophorina
superba) ABW 232; ALE 9:470.

Bird of Paradise, Twelve-wired (Seleucides
melanoleuca) ALE 9:491; (S. ignotus) ABW
233.

Bird of Paradise, Waigeu or Wilson's
(Diphyllodes respublica) ABW 232; ALE
9:479.

Birdcatcher Tree (Pisonia umbellifera) CAU
132; RDP 319.

Bird-cherry (Prunus padus) DFP 220 (cp
1758); EET 177; MBF cp 26; MFF pl 75 (376);
NHE 21; OWF 195; PFE cp 49 (482); PWF
35i, 147e; TEP 272; TGS 206; TVG 237; WFB
115 (7).

Bird-of-Paradise Flower (Strelitzia
reginae) BIP 173 (cp 264); DEW 2:274 (cp
145), 275 (cp 149); EDP 126; EGG 143; EWF
cp 84a; FHP 146; GFI 27 (cp 3); HPW 296;
LFW 178 (cp 399); MEP 24; MLP 320; MWF
275; PFW 289; RDP 372; RWF 62; SPL cp
247.

Birdsfoot, Common or Least (Ornithopus
perpusillus) MBF cp 24; MFF cp 49 (318);
NHE 212; OWF 21; PWF 51f; WFB 129.

Birdsfoot, Orange (Ornithopus pinnatus)
MBF cp 24; MFF cp 49 (319).

Bird's-foot Trefoil —See **Trefoil,
Bird's-foot.**

Bird's-nest, Yellow (Monotropa hypopitys)
DEW 1:240, 241; EPF pl 473; HPW 130;
MBF cp 56; MFF cp 36 (604); NHE 34; PFE
cp 86 (917); PWF 89d; WFB 169 (8), 286.

Birthwort (Aristolochia clematitis) AMP 73;
DEW 1:132; KVP cp 62; MBF cp 76; MFF cp
63 (513); NHE 211; PFE cp 7 (76); PWF 114c;
SPL cp 382; WFB 39 (1).

Birthwort, Round-leaved (Aristolochia rotunda) EWF cp 31d; PFE cp 8 (77); PFM cp 9.

Bishop, Golden —See **Weaver, Napoleon.**

Bishop, Red —See **Weaver, Grenadier.**

Bishop's Cap (Astrophytum myriostigma)
ECS 14, 100; EPF cp VII; LCS cp 6; SCS cp 4;
SPL cp 128.

Bishop's Hat —See **Barrenwort.**

Bishop's-Weed —See **Goutweed.**

Bison, European (Bison bonasus or Bos
bonasus) ALE 13:163, 343; DAE 92-93; DSM
247; HMW 92; JAW 191; LVA 59; MLS 405
(bw); SLP 44, 45; SLS 85; VWA cp 21.

Bistort, Alpine (Polygonum viviparum) EPF
pl 320; EWF cp 5a; MBF cp 73; MFF pl 103
(528); NHE 276; PFE cp 9 (85); WFB 41 (4).

Bistort, Amphibious (Polygonum
amphibium) BFB cp 12 (9); KVP cp 2; MBF
cp 73; MFF cp 26 (530); NHE 97; OWF 127;
PFE cp 8 (83); PFF 153; PWF 170e; TEP 15;
WFB 41 (2); WUS 115.

Bistort, Snakeroot (Polygonum bistorta)
BFB cp 12; DFP 166 (cp 1325); EPF pl 319;
MBF cp 73; MFF cp 26 (529); NHE 183; OWF
127; PFE 64; PWF 56d; SPL cp 490; TEP 141;
WFB 41 (1).

Bitter Ash or **Bitter Damson** —See
Bitterwood (Quassia).

Bitter-cress —See also **Coral-wort.**

Bitter-cress, Hairy (Cardamine hirsuta)
BWG 64; MBF cp 7; MFF pl 91 (90); NHE
218; OWF 69; PWF 26c; WFB 95 (1).

Bitter-cress, Large (Cardamine amara) MBF
cp 7; MFF pl 91 (87); NHE 94; PFE cp 34
(310); PWF 40a; WFB 91 (2).

Bitter-cress, Narrow-leaved (Cardamine
impatiens) MBF cp 8; MFF pl 91 (88); NHE
30; PWF 83h; WFB 91 (4).

Bitter-cress, Wavy (Cardamine flexuosa)
MBF cp 7; MFF pl 91 (89); NHE 30; PWF cp
14d.

Bitterling (Rhodeus sericeus amarus) ALE
4:301, 327, 371; FEF 161 (bw); FWF 79; GAL
241 (bw); HFI 99; HFW 118 (bw); OBV 113;
SFW 245 (cp 58); WEA 72.

Bittern, Australian or Brown (Botaurus
poiciloptilus) BNZ 187; HAB 23; SNB 32 (cp
15).

Bittern, Black (Dupetor flavicollis) DPB 13;
HAB 23; KBA 112 (cp 11); SNB 32 (cp 15).

Bittern, Chinese Least —See **Bittern,
Yellow.**

Bittern, Cinnamon or Little (Ixobrychus
minutus) ABW 52; ALE 7:198; 11:257; BBB
269; BBE 35; DPB 13; FBE 39; HAB 23; NBB
41 (bw), 88 (bw); OBB 13; PEB 8 (cp 3); RBE
72; SBB 12 (pl 8); SNB 32 (cp 15); WAB
181; (I. cinnamomeus) KBA 112 (cp 11).

Bittern, Common or Eurasian (Botaurus
stellaris) ALE 7:198, 385; BBB 162; BBE 35;
DPB 13; FBE 39; FBW 61; LEA 348 (bw);
OBB 13; PBE 8 (cp 3); RBE 75; SBB 10 (pl 5);
WAB 124.

Bittern, Japanese —See **Bittern, Tiger.**

Bittern, Schrenck's Least (Ixobrychus
eurhythmus) DPB 13; KBA 112 (cp 12).

Bittern, Sun (Eurypyga helias) ABW 112;
ALE 8:100; CDB 76 (cp 240); GPB 133; LAW
503; SSA cp 3.

Bittern, Tiger (Gorsachius melanolophus)
ALE 7:198; DPB 13; KBA 112 (cp 11); (G.
goisagi) DPB 13.

Bittern, Yellow (Ixobrychus sinensis) DPB
13; KBA 112 (cp 11); SNB 32 (cp 15).

Bittersweet (Celastrus sp.) DFP 245 (cp
1959); EGV 99; GDS cp 62; HPW 179; HSC
25, 26; MLP 128; PFF 238; PFW 73; TGS 71.

Bittersweet Shell —See **Cockle, Dog.**

Bitterwood (Quassia amara) AMP 179; DEW
2:54 (cps 15, 16); HPW 199; MEP 66.

Bitterwort —See **Gentian, Great Yellow.**

Black Boys —See **Grass Tree** (Xanthorrhoea).

Black Bulgar or **Black-stud Fungus**
(Bulgaria inquinans) DEW 3:124; DLM 118;
LHM 45; ONP 147; RTM 235 (6).

Black Ruby (Barbus nigrofasciatus) CTF 26
(cp 19); FEF 193 (bw); HFI 101; SFW 256 (pl
61).

Black Widow (Gymnocorymbus ternetzi)
CTF 18 (cp 2); FEF 77 (bw); HFI 88; SFW 132
(pl 25).

Blackberry (Rubus fruticosus) BFB cp 9 (5);
CTH 31 (cp 32); EPF pl 402; EVF 114; LFW
128 (cp 290); MFF pl 76 (346); MLP 100;
NHE 20; OWF 79, 117; PFW 264; PWF 109o,
163n; TEP 271; WFB 109 (5).

Blackberry (Rubus ulmifolius) BWG 110,
111; EWF cp 9 b; HPW 142; HSC 110; MBF
cp 28; OFP 79.

Blackbird, European (Turdus merula) ALE
2:363; 9:290; BBB 54; BBE 263; CDB 167 (cp
706); FBE 257; GAL 126 (cp 2); HAB 174;
KBA 336 (cp 51); LAW 468; NBB 74, 79 (bw);
OBB 139; PBE 220 (cp 53); RBE 259; SAB 14
(cp 6); WAB 135 (bw); WBV 191.

Blackbird, Oriole (Gymnomystax mexicanus) CDB 206 (cp 909); DTB 77 (cp 38).

Blackbird, Red-breasted (Pezites militaris
or Leistes militaris) ABW 290; SSA cp 40.

Blackbuck (Antilope cervicapra) ALE
13:426; DAA 88; DSM 264; HMA 25; HMW
77; JAW 197; LAW 571; LEA 610 (bw); MLS
415 (bw); TWA 178 (bw); (Cervicapra
cervicapra) CAS 183.

Blackcap (Sylvia atricapilla) ABW 256; ALE
9:232; BBB 140; BBE 235; CDB 175 (cp 749);
DAE 133; FBE 233; GAL 128 (cp 3); OBB
151; PBE 229 (cp 56); RBE 268; WAB 108.

Blackcock —See **Grouse.**

Black-eyed Susan Vine (Thunbergia alata)
DEW 2:183; DFP 48 (cp 383); EDP 117; EGA
156; EGV 86, 140; FHP 148; MEP 147; MWF
283; PFW 295; RDP 40, 382; SPL cp 182.

Black-orchid —See **Orchid, Black Vanilla**

Blackthorn or **Sloe** (Prunus spinosa) BFB cp
9 (6); EET 172; EPF pls 422, 423; MFB cp 26;
MFF pl 75 (374); NHE 20; OFP 67; OWF 181;
PFE cp 49 (476); PWF 16a, 165f; TEP 128;
WFB 115 (5).

Bladder-nut (Staphylea pinnata) DEW 2:58
(cp 26), 76; EPF pl 665; GDS cp 465; HSC
115; NHE 22; WFB 141 (9); (S. colchica)
PFW 288; TGS 150; (S. holocarpa) HPW 190.

Bladder-seed (Physospermum cornubiense) MBF cp 36; MFF pl 99 (467); WFB 159 (8).

Bladder-senna (Colutea arborescens) EFS 102; EPF pl 456, 457; HSC 34, 36; MFF cp 64 (312); NHE 21; PFE cp 53 (525); SST cp 95; TGS 140; WFB 121 (6).

Bladderwort (Utricularia sp.) DEW 2:159; HPW 253; LFT cp 156; MBF cp 66; NHE 99.

Bladderwort, Greater (Utricularia vulgaris) DEW 2:158, 159; KVP cp 7; MBF cp 66; MFF cp 59 (740); NHE 99; OWF 31; PFE cp 132 (1281); PFF 298; PWF 115e; TEP 18; WFB 291 (6).

Bladderwort, Intermediate or Irish (Utricularia intermedia) MBF cp 66; NHE 99.

Bladderwort, Lesser or Small (Utricularia minor) MBF cp 66; MFF cp 59 (741); NHE 99; WFB 291 (6b).

Blanket-flower (Gaillardia aristata) DFP 143 (cps 1138, 1139); EEG 134; EWG 110; MWF 137; OGF 147; PFF 525; SPL cp 206; TGF 269; TVG 93.

Bleak (Alburnus alburnus) ALE 4:341; FWF 79; GAL 240 (bw); HFI 95; LAW 317; OBV 99; SFW 212 (pl 49); WFW cp 107.

Bleeding Heart —See also **Glory Bower** (Clerodendrum).

Bleeding-heart, Common (Dicentra spectabilis) DEW 1:140; DFP 139 (cp 1110); EEG 133; EPF pl 236; HPW 53; LFW 48 (cp 111); MLP 90; MWF 111; OGF 37; PFF 182; PFW 119; SPL cp 335; TGF 141; TVG 87.

Blenny (Petroscirtes sp.) WFW cp 425; (Tripterygion sp.) ALE 5:162; WFW cp 427.

Blenny, Butterfly (Blennius ocellaris) ALE 5:181; OBV 73; WFW cp 422.

Blenny, Common (Blennius pholis) CSF 151; CSS cp 31; OBV 73.

Blenny, Long-striped or Roux's (Blennius rouxi) ALE 5:162; HFI 43.

Blenny, Montagu's (Blennius montagui) CSS cp 31; HFI 44; OBV 73; (Coryphoblennius galerita) WFW cp 424.

Blenny, Peacock (Blennius pavo) ALE 5:162, 181; FEF 470 (bw); HFI 43.

Blenny, Snake (Lumpenus lampretaeformis) CSF 151; OBV 73.

Blenny, Tentacled (Blennius tentacularius) HFI 43; HFW 238 (bw).

Blenny, Tompot (Blennius gattorugine) CSS cp 31; OBV 73; WFW cp 419.

Blenny, Viviparous —See **Eelpout.**

Blenny, Yarrell's (Chirolophis ascanii or C. galerita) ALE 5:162; CSF 151; HFI 45; OBV 73.

Blesbok (Damaliscus dorcas) ALE 13:61; CAF 220; DDM 209 (cp 36); DSM 260; LVS 128; MLS 410 (bw); PWC 126; SLS 211; SMW 263 (bw); (D. albifrons) HMW 74; (D. pygargus) SLP 233.

Blewit, Narcissus (Tricholoma sulphureum) CTM 32; DEW 3:167; DLM 262; KMF 19; LHM 85; ONP 127; REM 115 (cp 83); RTM 104.

Blewit, Purple (Tricholomopsis rutilans) DLM 264; KMF 67; LHM 93; ONP 107; RTM 96.

Blewit, Wood (Tricholoma nudum) LHM 91; NFP 61; SMF 7.

Blewits (Lepista sp.) DLM 220; OFP 189; ONP 31, 134; PUM cps 73, 75, 76; RTM 119; SMG 160-162.

Blister Weed —See **Buttercup, Common Meadow** (Ranunculus).

Blood Flower (Asclepias currassavica) DEW 2:97 (cp 42); EGA 96; HPW 225; LFW 232 (cp 517); MEP 112.

Bloodfin (Aphyocharax rubripinnis) ALE 4:286; FEF 76 (bw); HFI 88; SFW 124 (cp 23).

Bloodleaf (Iresine herbstii) EGA 128; MWF 165; PFF 159; RDP 247.

Blotch-eye —See **Soldierfish** (Myripristis sp.).

Blowfish (Arothron sp.) ALE 5:257; FEF 528, 530; WFW cps 488-491.

Blue Boys (Pycnostachys urticifolia) EWF cp 80b; LFT cp 142.

Blue Bush or **Mottlecah** (Eucalyptus macrocarpa) EET 240; EWF cp 137e; WAW 119.

Blue Chalk Sticks (Kleinia repens) EGL 121; MWF 174.

Blue Curls —See **Self-heal, Common** (Prunella).

Blue Dawn Flower (Ipomoea learii) BIP 105 (cp 158); EWF cp 180d; MWF 165; PFW 91; SPL cp 168.

Blue Flame (Myrtillocactus geometrizans) ECS 132; LCS cp 113.

Blue Thimble Flower —See **Gilia, Blue.**

Bluebeard —See **Spirea, Blue.**

Bluebell Creeper (Sollya fusiformis) BIP 167 (cp 256); EWF cp 131f; (S. heterophylla) HPW 138.

Bluebell or Harebell (Wahlenbergia sp.) LFT cp 166; SNZ 129 (cps 399, 400), 161 (cps 510-512).

Bluebell or Hyacinth, Wild (Endymion non-scriptus) BFB cp 23 (4); BWG 133; DFP 91 (cp 727); EWF cp 24d; KVP cp 61; MBF cp 85; MWF 265; NHE 38; OWF 169; PWF 28b; TGF 13; WFB 267 (1).

Bluebell, Spanish (Endymion hispanicus) EGB 38; MWF 265; PFE 169 (1638); SPL cp 272; TVG 209; WFB 267 (1a).

Bluebell, Virginia (Mertensia virginica) DFP 159 (cp 1270); ESG 128; EWF cp 163f; EWG 125; LFW 212 (cp 475); OGF 37; PFF 286; PFW 49; TGF 173 (M. sp.) MLP 209.

Blueberry, Highbush (Vaccinium corymbosum) DFP 242 (cp 1933); EFS 142; EPF pl 489; EVF 115; HSC 119; OFP 83; PFF 277; (V. sp.) MLP 191; OGF 57.

Bluebird, Blue-backed or Palawan Fairy (Irena puella) ABW 245; ALE 9:184, 190; CAS 154; CDB 156 (cp 644); DPB 289; KBA 273 (cp 40); RBI 295; WAB 164.

Bluebird, Philippine Fairy (Irena cyanogaster) DPB 289; RBI 296.

Bluebottle —See **Cornflower.**

Blue-eyed Grass (Sisyrinchium bermudiana) DFP 27 (cp 209); EWG 25, 139; MFB cp 83; MFF cp 6 (1046); OGF 35; PWF 79l; WFB 273 (10).

Blue-eyed-Mary (Omphalodes verna) ERW 128; MFF cp 13 (648); PFE cp 101 (1047); PFW 51; PWF cp 19e; TGF 192; WFB 191 (2).

Bluefish (Pomatomus saltatrix) HFI 28; LMF 82.

Bluegrass, Canada —See **Meadow-grass, Flattened** (Poa).

Bluegrass, Kentucky —See **Meadow-grass, Common** (Poa).

Bluet, Mountain (Centaurea montana) DFP 128 (cp 1020); EEG 132; EFC cp 37; EPF pl 591; HPW 264; NHE 288; PWF 49i; SPL cp 476; TGF 236; WFB 249 (3).

Bluetail, Red-flanked (Tarsiger cyanurus) ALE 9:279; BBE 259; CDB 166 (cp 701); FBE 253; KBA 321 (cp 50); PBE 284 (cp 65).

Bluethroat (Luscinia svecica) ABW 254; ALE 9:279; BBB 268; BBE 257; CEU 247; FBE 253; FBW 83; (Erithacus svecicus) KBA 321 (cp 50); LAW 508; OBB 143; (Cyanosylvia svecica) GAL 126 (cp 2); PEB 221 (cp 54); RBE 263.

Bluets (Houstonia caerulea) ESG 63, 118; EWG 28, 118; MWF 156; PFF 303; TGF 188.

Blusher (Amanita rubescens) DLM 168; LHM 119; NFP 54; NHE 45; ONP 131; REM 61 (cp 29); RTM 21; SMF 35; TEP 223.

Boa Constrictor (Boa constrictor or Constrictor constrictor) ALE 6:376, 386; 8:359; ART 103; CSA 69; HRA 64; HWR 50, 124; LAW 427; LEA 319 (bw); SIR 153 (cp 73); SSW 46; VRA 211.

Boa, Cook's Tree (Boa cooki) HWR 21; SIR 150 (cp 69).

Boa, Emerald Tree (Boa canina) ERA 210; LAW 427; SIR 151 (cp 70); SSW 49.

Boa, Green Tree (Corallus caninus) ALE 6:375; HRA 66; MAR 152 (pl 62); VRA 217.

Boa, Madagascar (Acrantophis madagascariensis) ALE 6:375, 386; LVS 183.

Boa, Rainbow (Epicrates cenchris) ALE 6:384; LEA 319 (bw); SIR 152 (cp 71); SSW 47; VRA 215.

Boa, Rosy (Lichanura trivirgata) ALE 6:384; SIR 152 (cp 72).

Boa, Sand (Eryx sp.) ALE 6:376, 384; ERA 211; HRA 68; HWR 22; LAW 428; SIR 177 (bw); SSW 52.

Boa, Sanzinia (Sanzinia madagascariensis) HRA 65; SSW 53.

Boar, Wild (Sus scrofa) ALE 13:91, 94, 163; BMA 174 (bw); CAA 82-83 (bw); CAS 38, 278 (bw); CEU 55, 222; DAA 95; DAE 41; DDM 161 (cp 26); DSM 219; HMA 61; HMW 48; JAW 163; LAW 560; LEA 589 (bw); MLS 369 (bw); SMW 243 (bw); TWA 78; WEA 288, 393; (S. cristata) CAS 177.

Boar-fish (Capros aper) ALE 5:35; HFI 64.

Boat Lily (Rhoeo sp.) BIP 155 cp 237; EDP 144; EFP 141; HPW 280; MWF 254; RDP 16, 339.

Boatbill —See **Heron, Boat-billed.**

Bog Moss (Sphagnum sp.) DEW 3: cp 103; EPF pl 86; NFP 156; NHE 144; ONP 87, 89; PFF 92, 93; TEP 75.

Boga (Leporinus affinis) SFW 165 (cp 38); WFW cp 99.

Bogbean or Buckbean (Menyanthes trifoliata) AMP 39; BFB cp 15; DEW 2:123; EPF pl 677, 678; EWF cp 15a; HPW 231; KVP cp 24; MBF cp 59; MFF pl 69 (644); NHE 97; OWF 95; PFE cp 98 (1003); PFW 188; PWF 30d; TEP 31; TGF 157; WFB 179 (6).

Bog-rush, Black (Schoenus nigricans) MBF cp 91; MFF pl 116 (1122); NHE 77; PFE cp 185 (1847).

Bog-violet —See **Butterwort, Common** (Pinguicola).

Bois d'arc —See **Orange, Osage.**

Bokmakierie —See **Shrike, Bokmakierie.**

Boletinus (Boletinus sp.) DLM 171; SMG 80, 81.

Boletus (Suillus sp.) DEW 3: cp 70, 183; DLM 43, 255-258; KMF 71; PUM cp 7; SMG 78, 79, 82-90.

Boletus, Bitter (Boletus felleus) LHM 193; NHE 48; PFF 78; REM 166 (cp 134); RTM 179; TEP 217; (Tylopilus felleus) DLM 264; ONP 143; PMU cp 44; SMG 106.

Boletus, Brown-yellow (Boletus luteus) LHM 187; NHE 46; ONP 109; RTM 192; TEP 215.

Boletus, Cow (Boletus bovinus) LHM 189; ONP 109; REM 151 (cp 119); RTM 194; SMF 46.

Boletus, Edible (Boletus edulis) CTM 53; DEW 3: cp 72; DLM 94, 95, 104, 105, 107, 172; KMF 103; LHM 191; NFP 48; NHE 46; OFP 189; ONP 143; PFF 78; RTM 177, 304; SMG 96; TEP 217; (B. edulis var. reticulatus) ONP 143; PMU cp 42; REM 165 (cp 133); RTM 178; SMG 96.

Boletus, Elegant (Boletus elegans) CTM 59; LHM 187; NHE 46; ONP 109; REM 151 (cp 119); RTM 191; TEP 213.

Boletus, Granulated (Boletus granulatus) DLM 60; LHM 189; NHE 46; REM 152 (cp 120); RTM 193; TEP 217.

Boletus, Indigo (Gyroporus cyanescens) DLM 193; KMF 59; SMG 76.

Boletus, Lurid (Boletus luridus) CTM 62; DLM 174; LHM 193; NFP 48; NHE 48; PMU cp 39; REM 161 (129); RTM 184; SMF 48; SMG 93.

Boletus, Parasitic (Boletus parasiticus) DLM 174; KMP 95; LHM 191; ONP 154; RTM 227.

Boletus, Purple (Boletus purpureus) CTM 63; PMU cp 41; REM 164 (cp 132).

Boletus, Red-cracked (Boletus chrysenteron) CTM 56; DLM 172; LHM 189; NHE 46; ONP 143; REM 155 (cp 123); RTM 197; SMG 105; TEP 215.

Boletus, Rough-stemmed (Boletus scaber) LHM 195; NHE 46; ONP 115; PFF 78, REM 167 (cp 135); SMF 28; TEP 215.

Boletus, Satan's (Boletus satanas) CTM 64; DLM 73; NHE 48; REM 163 (cp 131); RTM 187; TEP 217.

Boletus, Sweet-chestnut (Boletus badius) LHM 191; NHE 46; ONP 109; REM 157 (cp 125); RTM 198; SMF 15; TEP 215.

Boletus, Velvet or Cracked (Xerocomus sp.) DEW 3: 182; KMF 103; PMU cps 28-30; PUM cp 8.

Boletus, Yellow-cracked (Boletus subtomentosus) CTM 56; DLM 175; LHM 189; NFP 48; NHE 46; ONP 143; REM 156 (cp 124); RTM 196.

Bomarea (Bomarea sp.) DFP 55 (cp 433); EWF cp 190b; MWF 58; PFW 21.

Bone Plant —See **Orchid, Vanda** (Vanda).

Bone-set —See **Comfrey, Common** (Symphytum).

Bongo (Boocercus euryceros) DDM 177 (cp 28); DSM 243; WEA 77; (Taurotragus euryceros) ALE 13:302.

Bonito, Common (Sarda sarda) CSF 147; HFI 51; LMF 52; OBV 27.

Bonito, Oceanic (Katsuwonus pelamis) ALE 5:191; CSF 146; HFW 229 (bw); LMF 52; OBV 27.

Bonnet Shell, (Phalium strigatum) CTS 34 (cp 46); GSS 64; (Casmaria vibex) GSS 65.

Bonobo —See **Chimpanzee, Dwarf or Pygmy.**

Bontebok —See **Blesbok.**

Booby, Abbott's (Sula abbotti) GPB 55; LVB 53.

Booby, Blue-faced —See **Booby, Masked.**

Booby, Blue-footed (Sula nebouxii) ABW 47; CDB 40 (cp 44); CSA 253; GPB 55, 57; HBG 65 (pl 2).

Booby, Brown (Sula leucogaster) ABW 47, 48; ALE 7:177;CAU 222 (bw); CDB 40 (cp 43); CSA 176 (bw); DPB 7; FBE 31; GPB 55; HAB 17; LEA 367 (bw); SAO 77.

Booby, Masked (Sula dactylatra) ABW 47; DPB 7; GPB 56; HAB 17; HBG 65 (pl 2); SAO 77; SSA pl 21; WEA 78.

Booby, Red-footed (Sula sula) ALE 7:177; BBT 122; CDB 40 (cp 45); DPB 7; GPB 55; HAB 17; HBG 65 (pl 2); SAO 77.

Boojum Tree (Idria columnaris) DEW 1: 209 (cp 108); LCS cp 324; MLP 156.

Boomslang —See **Snake, Boomslang.**

Boots —See **Pitcher Plant** (Sarracenia).

Borage (Borago officinalis) AMP 133; EGH 104; EPF pl 795; EVF 140; MFF cp 13 (65); MWF 59; OWF 171; PFE cp 103 (1054); PFM cp 139; PWF 125g; TGF 173; WFB 193 (3). (B. laxiflora) PWF 51.

Borage, Eastern (Trachystemon orientalis) MFF cp 13 (652); PFE cp 103 (1054); WFB 192.

Boronia (Boronia sp.) BIP 35 (cps 34, 35); EWF cp 138d; MWF 59; PFW 269; WAW 118.

Boswellia (Boswellia sp.) DEW 2:42, 54 (cp 17); HPW 197.

Botanical Wonder —See **Ivy, Japanese.**

Botfly, Horse (Gasterophilus intestinalis) ALE 2:403; CIB 224 (cp 33); GAL 380 (cp 14); OBI 139; PEI 524 (bw).

Botfly, Sheep (Oestrus ovis) CIB 225 (cp 34); OBI 139; PEI 529 (bw); ZWI 124.

Botia, Tiger —See **Loach, Clown.**

Bottlebrush (Callistemon coccineus) DEW 2:20; LFW 155 (cps 342, 343); (Callistemon sp.) EWF cp 137b; HPW 161; MWF 67.

Bottlebrush (Greyia sutherlandii) EPF pl 727; EWF cp 77f; MEP 75; PFW 133; (G. radlkoferi) LFT cp 99.

Bottlebrush, Lemon (Callistemon citrinus) BIP 47 (cp 52); DEW 2: 50 (cp 3); EGE 115; ELG 135; EPG 110; MWF 67; PFW 192; RDP 126; RWF 78; WAW 119.

Bottlebrush, One-sided (Calothamnus sp.) EPF pl 707; MWF 69.

Bottlebrush, Showy (Callistemon specious) EPF pl 706; MEP 97; MLP 171; SST cp 232.

Bottle-grass —See **Bristle-grass.**

Bottletree —See **Flame Tree of Australia.**

Boubou —See **Shrike, Boubou.**

Bougainvillea (Bougainvillea glabra) BIP 35 (cp 36); DEW 1: 109 (cp 64); EGV 85, 97; EWF cp 178b; LHP 46, 47; LFW 237 (cps 525, 527); MWF 60; PFF 160; PFW 195; RDP 102; SPL cp 156; TVG 217.

Bougainvillea (Bougainvillea sp.) ELG 148; EPG 108; FHP 107; GFI 31 (cp 4); LFW 226 (cp 508), 236 (cp 523); MEP 36.

Bougainvillea (Bougainvillea spectabilis) EPF cp IX B; HPW 69; MWF 60; PFM cp 11.

Bouncing Bett (Saponaria officinalis) AMP 91; BFB cp 5 (4); DEW 1:225; MBF cp 13; MFF cp 25 (162); NHE 225; OWF 107; PFE cp 16 (181); PFF 163; PFW 72; PWF 123f; SPL cp 406; TEP 41; TGF 71; WFB 60, 61 (8).

Bouvardia (Bouvardia sp.) BIP 37 (cp 37); MWF 60; PFW 267; SPL cp 67.

Bovista (Bovista sp.) DEW 3:187; DLM 130; LHM 219; NHE 47; ONP 37; RTM 231.

Bowerbird (Ailuroedus sp.) —See **Catbird.**

Bowerbird, Black-faced Golden (Sericulus aureus) ABW 231; ALE 9:492.

Bowerbird, Crestless (Amblyornis inornatus) ABW 230; WAB 199.

Bowerbird, Fawn-breasted (Chlamydera cerviniventris) HAB 159; SAB 78 (cp 38).

Bowerbird, Golden or Newton's (Prionodura newtoniana) ALE 9:492; CDB 220 (cp 987); HAB 161; MBA 10; SAB 78 (cp 38); WAB 199.

Bowerbird, Great (Chlamydera nuchalis) HAB 159; RBU 308; SAB 78 (cp 38); WAB 199.

Bowerbird, MacGregor's (Amblyornis macgregoriae) ABW 230; WAB 199.

Bowerbird, Orange-crested (Amblyornis subalaris) ABW 230; ALE 9:492.

Bowerbird, Regent (Sericulus chryso-
cephalus) ABW 230; ALE 9:492; GPB 336;
HAB 158; MEA 46; RBU 311; SAB 78 (cp 38);
WAB 183.

Bowerbird, Satin (Ptilonorhynchus
violaceus) ABW 230; BBT 7; CAU 26; CDB
221 (cp 988); GPB 335; HAB 159; SAB 78 (cp
38); WAB 181, 199; WEA 79; (P. maculatus)
MEA 46.

Bowerbird, Spotted (Chlamydera maculata)
ABW 230, 231; ALE 9:492; CDB 220 (cp
986); HAB 160; SAB 78 (cp 38); WAW 29
(bw).

Bowerbird, Tooth-billed (Scenopoeetes
dentirostris) ALE 9:492; HAB 163; LEA 465
(bw); SAB 78 (cp 38); WAB 198.

Bowerbird, Western (Chlamydera guttata)
HAB 158; SAB 78 (cp 38).

Bower-plant (Pandorea jasminoides) EGV
124; MEP 137; (P. brycei) LFW 160 (cp 357);
(P. pandorana) EWF cp 139c.

Box, Common (Buxus sempervirens) BFB cp
14 (8); DFP 184 (cp 1470); EEG 104; EET
186, 187; EFP 95; EGE 114; EPG 109; EWF
cp 31a; GDS cp 28; HPW 184; HSC 20, 22;
MBF cp 141; MFF pl 73 (265); NHE 21; OBT
37; PWF 18b, 159m; TGS 150; WFB 141 (4).

Box, Small or Littleleaf (Buxus microphylla)
EGW 95; EJG 104; EMB 111; ERW 103; TGS
152.

Box, Sweet or Christmas (Sarcococca
hookerana) EGW 33, 140; ESG 141; GDS cp
442; HSC 111, 112; TGS 247; (S. sp.) PFW 57;
TGS 261.

Box, Victorian or Mock Orange (Pittosporum undulatum) EWF cp 131g; MEP 44;
MWF 235.

Box-elder (Acer negundo) DEW 2:58 cp 24;
EET 189; EGT 91; EPF pl 732; MTB 333 (cp
32); MWF 28; NHE 22; OBT 188; PFF 242;
SST cp 72; TGS 169 bw.

Boxfish (Ostracion tuberculatus) ALE 5:257;
CTF 63 (cp 86); HFW 268 (cp 137); WFW cp
485; (Acantostracion tricornis) ALE 5:265;
(Lactophrys sp.) HFW 266 (cp 133); LEA
257; (Tetrosomus gibbosus) WFW cp 486.

Box-fish, Blue (Ostracion lentiginosum) FEF
523 (bw); HFI 14; HFW 267 (cps 134, 135);
LAW 353; LEA 263 (bw); WEA 80; WFW cp
484.

Brachystelma (Brachystelma barberiae)
EWF cp 78e; LFT 258 (cp 128).

Brachythecium (Brachythecium sp.) DEW 3:
Fig. 333; ONP 39, 59, 99; PFF 95.

Bracket Plant —See **Spider Plant.**

Bramble, Arctic (Rubus arcticus) PFE cp 43
(427); PWF 41i; WFB 109 (9).

Bramble, Stone or Rock (Rubus saxatilis)
MBF cp 29; MFF pl 76 (344); NHE 20; WFB
109 (7).

Brambling (Fringilla montifringilla) ABW
300; ALE 9:336, 392; BBB 148; BBE 279;
CDB 209 (cp 923); DPB 419; FBE 285; LEA
461 (bw); OBB 185; PBE 261 (cp 62); WAB
105.

Brandy-bottle —See **Water-lily, Yellow**

Brant (Branta bernicla) ALE 7:61, 284; BBB
219; BBE 47; FBE 49; LAW 487; OBB 33;
PBE 16 (cp 5), 20 (pl 7); RBE 108; WAB 101;
WEA 81.

Brasilocactus (Brasilocactus sp.) SCS cps
8,9; WWT 77.

Brass Buttons, New Zealand (Cotula
squalida) EGC 129; TGF 237; SNZ 113 (cp
345); (C. barbata) DFP 35 (cp 277).

Brazil-nut (Bertholletia excelsa) DEW 2:16;
EPF pl 703; OFP 31; PFF 262; SST cp 186.

Breadfruit, Mexican —See **Swiss Cheese
Plant** (Monstera).

Breadfruit Tree (Artocarpus communis or A.
altilis) DEW 1:104 (cp 53); HPW 97-5; MLP
59; PFF 148; SST cp 185; (A. heterophyllus) EET 221.

Bream (Abramis brama) ALE 4:341; FWF 85;
GAL 233 (bw); HFI 98; OBV 97; SFW 213 (pl
50); WFW cp 106; (Diplodes sp.) HFI 30;
LAW 337.

Bream, Black Sea (Spondyliosoma
cantharus) CSF 127; OBV 29; WFW cp 304.

Bream, Ray's (Brama brama) CSF 125; OBV
31.

Bream, Sea (Pagellus sp.) ALE 5:102; CSF
126; OBV 29.

Bream, Silver (Blicca bjoerkna) FEF 155
(bw); FWF 83; GAL 234 (bw); HFI 98; LAW
317; OBV 97; WEA 338; WFW cp 116.

Bream, Toothed (Dentex sp.) HFI 30; LAW
337; LMF 126.

Brick Tuft (Hypholoma sublateritium) CTM
43; DLM 201; LHM 147; NFP 83; ONP 141;
PFF 86; RTM 56.

Bridal Wreath (Spiraea prunifolia) EEG 102;
EFS 137; ELG 119; EPG 147; MWF 271; TGS
231; (S. vanhouttei) EFS 23, 127; EPF pl
393; LFW 132 (cp 300); TGS 231.

Bridewort —See **Spirea, Willow**.

Brill (Scophthalmus rhombus) CSF 175; OBV
55.

Bristle-bird, Eastern (Dasyornis brachypterus) HAB 196; SAB 4 (cp 1).

Bristle-bird, Rufous (Dasyornis broadbenti)
HAB 196; SAB 4 (cp 1); WAB 187.

Bristlemouth (Cyclothone sp.) ALE 4:266;
MOL 99; WFW cps 42, 43.

Bristlemouth —See **Pearl-side** (Maurolicus
sp.).

Bristle-grass, Green (Setaria viridis) EGC
71; MCW 21; MFF pl 123 (1254); NHE 243;
PFF 349; (S. sp.) NHE 243; WUS 83, 85.

Bristletail (Campodea sp.) ALE 2: 79, 97;
OBI 1; (Lepismachilis notata) PEI 28 (bw);
(Petrobius sp.) CSS pl XIII; WEA 82.

Brittle-star—See also **Basket-star** (Gorgonocephalus).

Brittle-star (Amphipholis sp.) ALE 3: 390;
CSS cp 26; OIB 181, 185 (Ophioderma long-
icauda) ALE 3:389; LAW 256; (Ophiura sp.)
ALE 3:390, 404; CSS cp 26; OIB 185.

Brittle-star, Common (Ophiothrix fragilis)
ALE 3:390, 405; CSS cp 26; MOL 88, 135
(bw); OIB 185; VWS 31 (bw).

Broadbill, Banded (Eurylaimus javanicus)
ALE 9:120; KBA 224 (cp 29); (E. steerii)
DPB 223.

Broadbill, Black-and-Red (Cymbirhynchus
macrorhynchus) KBA 224 (cp 29); RBI 140;
WBV 147.

Broadbill, Black-and-Yellow (Eurylaimus
ochromalus) KBA 224 (cp 29); RBI 143.

Broadbill, Cape (Smithornis capensis) ALE
9:120; CDB 139 (cp 557).

Broadbill, Dusky (Corydon sumatranus)
KBA 236 (bw); RBI 144.

Broadbill, Green (Calyptomena viridis)
ABW 196; ALE 9:120, 133; CDB 139 (cp
556); KBA 224 (cp 29); WEA 82.

Broadbill, Long-tailed (Psarisomus
dalhousiae) ABW 196; ALE 9:120; KBA 224
(cp 29); WBV 147.

Broadleaf (Griselinia littoralis) DFP 204 (cp
1626); SNZ 59 (cp 147).

Brolga —See **Crane, Australian or Brolga**.

Brome, Awnless or Smooth (Bromus inermis) NHE 196; PFF 360.

Brome, Barren (Bromus sterilis) BWG 176;
MWF pl 99; MFF pl 127 (1203); NHE 243;
TEP 205.

Brome, Drooping or Downy (Bromus tectorum) EWF cp 25b; MCW 16; NHE 243;
WUS 47.

Brome, Hairy (Bromus ramosus) MBF pl 99;
MFF pl 127 (1202); NHE 196; PFE cp 178
(1722).

Brome, Madrid (Bromus madritensis) MBF
pl 99; PFM pl 21 (418).

Brome, Meadow (Bromus commutatus)
HPW 289; MBF pl 99; NHE 243; WUS 41.

Brome, Rye (Bromus secalinus) NHE 243;
PFF 359; WUS 45.

Brome, Soft (Bromus mollis) MBF pl 99; MFF
pl 132 (1204); NHE 196.

Brome, Upright (Bromus erectus) MBF pl 99;
MFF pl 131 (1201); NHE 196; PFE cp 178
(1720).

Brooklime (Veronica beccabunga) MBF cp
64; MFF cp 11 (701); NHE 99; OWF 175; PFE
cp 124 (1225); PWF 74a; TEP 32; WFB 215 (9).

Brookweed (Samolus valerandi) MBF cp 58;
MFF pl 87 (627); NHE 186; OWF 93; PFE
304; PWF 126c; WFB 177 (4); (S. repens)
SNZ 22 (cp 19).

Broom (Cytisus sp.) BIP 67 (cp 93); DFP 6 (cps
45, 46); 194 (cp 1546); EGG 104; EHP 93;
GDS cps 131, 132; HSC 39, 41; NHE 72, 81;
OGF 59, 67; PFE cp 51 (504); PFM 5; RDP
172; TGS 39.

Broom (Sarothamnus scoparius) BFB cp 7 (6); EPF pl 451; MBF cp 21; MFF pl 47 (274); NHE 69; OWF 19; (Genista lydia) DFP 203 (cp 1623); GDS 58, 64; OGF 61.

Broom, Common or Scotch (Cytisus scoparius) AMP 81; DFP 194 (cp 1548); EGW 107; EWF cp 10f; GDS cps 137-142; HSC 39; LFW 39 (cp 88); MWF 103; OBT 36; OGF 59; PFE cp 51 (505); PFF 209; PWF 34c; TEP 131; TGS 135; TVG 223; WFB 121 (2).

Broom, Hedgehog (Erinacea anthyllis) HSC 50; PFE cp 52 (515).

Broom, Kew (Cytisus kewensis) EFS 107; EGC 129; ERW 110; GDS cp 133; TGS 132.

Broom, Mt. Etna (Genista aetnensis) GDS cp 210; HSC 58; OGF 61.

Broom, Pink (Notospartium sp.) EWF cp 130e; SNZ 110 (cp 336).

Broom, Purple (Cytisus purpureus) GDS cp 136; TGS 132.

Broom, Royal (Genista radiata) SPL cp 450; TGS 130.

Broom, Spanish or Weaver's (Spartium junceum) DEW 1:285 (cp 179); DFP 240 (cp 1914); EPF pl 449; GDS cp 459; HPW 150; HSC 114, 116; MWF 270; NHE 246; PFE cp 52 (515); PFM cp 56; SPL cp 117; TGS 135.

Broom, Spiny or Thorny (Calicotome sp. or Calycotome sp.) PFE cp 51 (498); PFM cp 55; RWF 33.

Broom, Warminster (Cytisus praecox) EEG 94; EFS 107; GDS cp 134, 135; HSC 39, 41; TGS 132.

Broom, Winged (Genista sagittalis or Chamaespartium sagittale) GDS cp 212; NHE 69; PFE cp 52 (513); TGF 146.

Broomrape (Orobanche sp.) DEW 2:158; EPF pl 868; HPW 248; KVP cp 94; MBF cp 66; MFF cp 36 (734); NHE 233; PFE cp 131 (1274); PFM cp 81.

Broomrape, Branched (Orobanche ramosa) EWF cp 38f; MBF cp 66; NHE 233; PFM cp 180.

Broomrape, Clove-scented (Orobanche caryophyllacea) MBF cp 66; MFF cp 36 (735); PFE cp 131 (1275), 399; WFB 219 (5).

Broomrape, Common or Lesser (Orobanche minor) MBF cp 66; MFF cp 36 (737); OWF 138; PFE cp 131 (1277); PWF 94a; WFB 219 (3), 286; WUS 345.

Broomrape, Greater (Orobanche rapum-genistae) MBF cp 66; MFF cp 36 (733); PFE 131 (1272).

Broomrape, Ivy (Orobanche hederae) DEW 2:173 (cp 103); MBF cp 66; NHE 36.

Broomrape, Purple (Orobanche purpurea) MBF cp 66; WFB 219 (4).

Broomrape, Tall (Orobanche elatior) MBF cp 66; MFF cp 36 (736); PWF 70f.

Broomsedge (Andropogon sp.) HPW 289; WUS 37.

Browallia (Browallia speciosa) BIP 37 (cp 39); DFP 55 (cp 435); EGA 100; EMB 124; ESG 98; FHP 108; RDP 109; TGF 205.

Browallia, Orange (Streptosolen jamesonii) BIP 173 (cp 266); DFP 82 (cp 653); EDP 116; EGV 138; FHP 147; MEP 132; MWF 276.

Brownweed —See Bartsia, Red (Odontites).

Bruang —See Bear, Sun.

Bruguiera (Bruguiera gymnorhiza) DEW 2:13, 14; HPW 158.

Brunnera (Brunnera macrophylla) DFP 126 (cp 1005); OGF 17; PFW 51; TGF 173.

Brush-finch —See Finch.

Brush-turkey —See Turkey, Brush.

Brussels Sprouts (Brassica oleracea gemnifera) EVF 12, 85; OFP 157; PFF 186.

Bryony (Bryonia sp.) AMP 73; BFB cp 10 (2); BWG 163; DEW 1:253; EPF pl 510; KVP cp 65, 65a; MBF cp 36;, MFF pl 76 (511); NHE 235; OWF 51; PFE cp 79 (815); PWF 82a, 173j; WFB 149 (5).

Bryony, Black (Tamus communis) AMP 51; BWG 172; MBF cp 84; MFF pl 76 (1056); NHE 241; OWF 51; PWF 83f, 173i; WFB 267 (8).

Bryozoan (Lophopus sp.) ALE 3:254; OIB 175.

Bryozoan, Freshwater (Cristatella sp.) ALE 3:254; OIB 175.

Bryozoan, Marine (Alcyonidium sp.) ALE 3:234; OIB 173.

Bryum (Bryum sp.) DEW 3: Fig. 324; NFP 107; ONP 59, 83; PFF 94.

Bubal —See **Hartebeest, Bubal or Red.**

Bubble Shell (Actaeon tornatalis) ALE 3:79; OIB 45; (Haminea sp.) ALE 3:79; OIB 45.

Bubble Shell, Paper (Akera bullata) ALE 3:90; OIB 45; (Hydatina physis) GSS 128; MOL 68; (H. albocincta) GSS 128.

Bucare Tree (Erythrina poeppigiana) EWF cp 170c; MEP 58; MLP 292.

Buckbean —See **Bogbean.**

Buckeye, Dwarf or Bottlebrush —See **Horse-chestnut, Dwarf.**

Buckeye, Sweet (Aesculus octandra) BKT cps 170, 171; DFP 181 (cp 1442); TGS 145.

Buckeye, Yellow (Aesculus flava) MTB 352 (cp 35); OBT 189.

Buckler-fern, Common (Dryopteris dilatata) BWG 158; MFF pl 138 (1295); NHE 80; ONP 187.

Buckler-fern, Crested (Dryopteris cristata) CGF 75; EGF 112; NHE 80; ONP 95; PFF 101.

Buckthorn (Rhamnus sp.) AMP 69; EEG 100; EFS 128; LFT cp 102; NHE 266; TGS 211.

Buckthorn, Alder (Frangula alnus) BFB cp 7 (8); MBF cp 20; MFF pl 71 (267); NHE 22; OBT 80; OWF 183; PWF 76a, 165m; TEP 285; WFB 141 (6).

Buckthorn, Common (Thamnus cathart-icus) AMP 69; MBF cp 20; MFF pl 71 (266); NHE 22; OWF 183; PFE cp 71 (723); PFF 245; PWF 60c, 165l; TGS 199; WFB 141 (5).

Buckthorn, Mediterranean (Rhamnus alaternus) PFE cp 71 (720); SST cp 159.

Buckthorn, Sea (Hippophae rhamnoides) AMP 49; BFB cp 14 (1); DEW 2:29; DFP 206 (cp 1647); EFS 116; EPF pl 700; HPW 169; MBF cp 75; MFF pl 73 (423); NHE 134; OWF 183; PFE cp 74 (761); PFW 104; PWF 35g, 165j; SST 127; TEP 67; TGS 247; WFB 141 (7).

Buckwheat (Fagopyrum esculentum) EPF pl 324; MFF pl 103 (537); PFE cp 9 (89); PFF 156; WFB 41 (7); (F. tataricum) NHE 224.

Buckwheat, Wild —See **Bindweed, Black** (Polygonum).

Buddleia (Buddleia sp.) DFP 184 (cps 1465, 1468); EWF cp 95 a, b; GDS cp 23; HPW 222; HSC 17; LFT 351; MWF 64.

Budgerigar (Melopsittacus undulatus) ABW 147; ALE 8:310; BBT 15; CAU 89; CDB 106 (cp 395); HAB 107; NBB 76; SBB 53 (pl 68), 54 (pl 68); SNB 116 (cp 57); WAB 192; WEA 84.

Buellia (Buellia sp.) DEW 3: Fig. 281; ONP 79, 169.

Buffalo, African or Cape (Syncerus caffer) ALE 13:344, 352-355; CAA 78 (bw); CAF 118 (bw); DSM 248; HMA 34; JAW 189; MLS 404 (bw); SMW 279 (cp 170); (Bubalus caffer caffer) TWA 20.

Buffalo, Dwarf —See **Anoa.**

Buffalo Grass —See **St. Augustine Grass**

Buffalo, Indian Water (Bubalus bubalis) CAS 269 (bw); DAA 96-97; DAE 25; DSM 249; HMA 34; HMW 90; JAW 184; LAW 565; MLS 399 (bw); PWC 135, 153 (bw); SMW 257 (bw); (B. arnee) ALE 13:285, 345; (B. bulaus) CAS 157; (Bos bubalus) TWA 164.

Buffledoorn (Burchella bubalina) EWF cp 79e; PFW 267.

Bug, Ambush (Phymata sp.) ALE 2:187; KIB 83 (pl 27); KIW 69 (bw); PWI 54 (bw), 115.

Bug, Bed —See **Bedbug.**

Bug, Fire —See **Firebug.**

Bug, Flat —See **Flatbug.**

Bug, Flower (Anthocoris nemorum) CIB 113 (cp 10); OBI 25.

Bug, Forest (Pentatoma rufipes) CIB 97 (cp 8); OBI 21.

Bug, Heath Assassin (Coranus subapterus) CIB 112 (cp 9); OBI 25.

Bug, Lace (Tingis sp.) ALE 2:187; CIB 112 (cp 9); OBI 25; (Galeatus maculatus) PEI 103 (bw).

Bug, Masked Hunter (Reduvius personatus) CIB 112 (cp 9); PEI 104 (bw).

Bug, May —See **Beetle, Cockchafer.**

Bug, Plant-eating (Phyllomorpha laciniata) ALE 2:187; CTI 21 (cp 14); PEI 102 (bw); PWI 54 (bw).

Bug, Red Assassin (Rhinocoris iracundus) ALE 2:182, 187; CIB 112 (cp 9); LAW 228; PEI 105 (bw); PWI 115.

Bug, Saucer (Ilyocoris cimicoides) CIB 128 (cp 11); OBI 31; PEI 110 (bw).

Bug, Shield —See Shield-bug.

Bug, Shore (Saldula saltatoria) CIB 113 (cp 10); OBI 29.

Bug, Squash (Coreus marginatus or Syromastes marginatus) CIB 97 (cp 8); PEI 100 (bw); (S. rhombeus) OBI 23.

Bug, Striped (Graphosoma italicum) CIB 97 (cp 8); PEI 98 (bw); PWI 115; (G. lineatum) ALE 2:187.

Bug, Water —See Waterbug.

Bugle, Blue or Geneva (Ajuga genevensis) EFC cp 53; EWF cp 20c; MBF cp 70; NHE 186; PFE cp 105 (1090).

Bugle, Common or Carpet (Ajuga reptans) BFB cp 12 (4); DFP 120 (cp 955); EEG 139; EGC 40-41, 120; EGH 93; ELG 146; EPF pl 805; ERW 92; ESG 90; LFW 3 (cp 6); MBF cp 70; MFF cp 10 (788); MWF 35; NHE 186; OWF 145; PFE cp 106 (1089); PWF 28 D; TEP 117; TGF 189; TVG 137; WFB 197 (1).

Bugle, Pyramidal (Ajuga pyramidalis) DFP 120 (cp 954); MBF cp 70; NHE 35; PFE cp 106 (1091); PWF 52 B.

Bugloss (Anchusa) —See Alkanet.

Bugloss, Purple Viper's (Echium lycopsis or E. plantagineum) DEW 2:162 (cp 79); DFP 37 (cp 292); EGA 116; EWF cp 37e; NHE 199; OGF 135; PFE cp 106 (1083); PFM cp 141; TVG 49; WFB 192.

Bugloss, Small or Lesser (Lycopsis arvensis) MBF cp 60; MFF cp 13 (654); NHE 229; OWF 171; PWF 101g; WFB 193 (5); (L. variegata) PFM cp 140.

Bugloss, Viper's (Echium vulgare) BFB cp 1 (8); EPF pl 802; EWF cp 20f; EWG 80; HPW 236; KVP cp 83; MBF cp 61; MCW 82; MFF cp 13 (665); MWF 121; NHE 229; OWF 71; PFE cp 104 (1082); PWF 85k; TEP 117; WFB 193 (1).

Bulbul (Hypsipetes sp.) CDB 155 (cp 639); DPB 283; KBA 253 (pl 36), 257 (cp 38); RBI 287; (Chloropsis jerdonii) LAW 508; (Phyllastrephus sp.) ABW 244; ALE 9:190; LVB 11; PSA 241 (cp 28).

Bulbul, Black-eyed or White-vented (Pycnonotus barbatus) ALE 9:190; CDB 155 (cp 640); FBE 217; GPB 256; NBA 55; PSA 241 (cp 28).

Bulbul, Black-headed or Blue-eyed (Pycnonotus atriceps) DPB 281; KBA 256 (cp 37); WBV 179.

Bulbul, Chinese or Light-vented (Pycnonotus sinensis) KBA 253 (pl 36); RBI 284.

Bulbul, Gray-cheeked (Criniger bres) DPB 281; KBA 257 (cp 38).

Bulbul, Olive-winged (Pycnonotus plumosus) DPB 281; KBA 256 (cp 37).

Bulbul, Red-eyed (Pycnonotus nigricans) CDB 155 (cp 642); PSA 241 (cp 28); (P. brunneus) KBA 256 (cp 37).

Bulbul, Red-vented (Pycnonotus cafer) CDB 155 (cp 641); KBA 253 (pl 36); SAB 14 (cp 6).

Bulbul, Red-whiskered (Pycnonotus jocosus) ABW 243; ALE 9:184, 190; KBA 253 (pl 36); LEA 445 (bw); SAB 14 (cp 6); WBV 179.

Bulbul, Somber (Andropadus importunus) CDB 155 (cp 638); PSA 241 (cp 28).

Bulbul, White-throated (Criniger flaveolus) ALE 9:184; KBA 257 (cp 38).

Bulbul, Yellow-vented (Pycnonotus goiavier) DPB 281; KBA 253 (pl 36).

Bullfinch (Pyrrhula pyrrhula) ABW 301; ALE 9:336; BBB 62; BBE 279; CDB 210 (cp 928); DAE 133; FBE 287; GAL 118 (cp 1); LAW 509; OBB 183; PBE 260 (cp 61); RBE 307; RBI 319; WAB 113; WEA 85; (P. erythaca) ALE 9:392; (P. leucogenys) DPB 419; (P. nipalensis) KBA 429 (bw).

Bullfinch, Trumpeter —See Finch, Trumpeter.

Bullfish —See Angelfish (Heniochus).

Bullfrog, African (Rana adspersa) ALE 5:455; HRA 99; (Pyxicephalus adspersus) ERA 81.

Bullfrog, Australian (Limnodynastes sp.) ERA 83; WAW 100.

Bullfrog, Indian (Rana tigrina) CAW 156 (bw); HRA 98.

Bullfrog, Malayan (Kaloula pulchra) ALE 5:410; CAW 175; HRA 101; HWR 103.

Bullhead (Cottus gobio) ALE 4:237; 5:65; FEF 508 (bw); FWF 127; GAL 228 (bw); HFI 58; OBV 101; SFW 676 (pl 185).

Bullhead, Brown (Ictalurus nebulosus) FEF 232 (bw); FWF 115; HFI 108; LEA 246 (bw).

Bullhead, Four-horned (Cottus quadricornis) CSF 167; HFI 58.

Bullhead, Norway (Taurulus lilljeborgi) CSF 165; OBV 67; (T. bulbalis) OBV 67.

Bulrush —See Reedmace (Typha).

Bulrush (Scirpus lacustris) MBF cp 91; MFF pl 115 (1117); NHE 105; SNZ 103 (cp 315); WFB 297 (9).

Bumblebee —See Bee, Bumble.

Bumblebee-fish —See Goby, Golden-banded or Wasp.

Bunting, Black-faced (Emberiza spodocephala) DPB 419; KBA 438 (bw); PBE 277 (cp 64).

Bunting, Black-headed (Emberiza melanocephala) ABW 300; ALE 9:341; BBE 295; CDB 197 (cp 859); FBE 277; PBE 277 (cp 64); RBI 308.

Bunting, Cinereous (Emberiza cineracea) BBE 293; FBE 279; PBE 277 (cp 64).

Bunting, Cirl (Emberiza cirlus) ALE 9:341; BBB 88; BBE 295; FBE 277; FNZ 208 (cp 18); OBB 187; PBE 277 (cp 64).

Bunting, Corn (Emberiza calandra) ALE 9:341; BBB 89; BBE 289; CDB 195 (cp 853); FBE 275; GAL 118 (cp 1); OBB 191; PBE 277 (cp 64); RBE 316.

Bunting, Cretzschmar's (Emberiza caesia) BBE 291; FBE 279; PBE 277 (cp 64).

Bunting, Golden-breasted (Emberiza flaviventris) CDB 196 (cp 856); PSA 317 (cp 40).

Bunting, House or Striped (Emberiza striolata) FBE 275; LAW 453.

Bunting, Lapland (Calcarius lapponicus) ABW 298; ALE 9:342; BBB 270; BBE 297; CDB 195 (cp 852); FBE 281; OBB 189; PBE 276 (cp 63).

Bunting, Little (Emberiza pusilla) BBB 270; BBE 289; DPB 419; FBE 281; KBA 438 (bw); OBB 189; PBE 277 (cp 64).

Bunting, Ortolan (Emberiza hortulana) ABW 300; ALE 9:341; BBB 268; BBE 291; CDB 197 (cp 857); FBE 279; OBB 187; PBE 277 (cp 64); RBE 319.

Bunting, Pine (Emberiza leucocephala) BBE 293; CDB 197 (cp 858); FBE 275; PBE 277 (cp 64).

Bunting, Red-headed (Emberiza bruniceps) BBE 293; FBE 277; PBE 277 (cp 64).

Bunting, Reed (Emberiza schoeniclus) ALE 9:341; BBB 182; BBE 293; CDB 197 (cp 860); FBE 281; GAL 118 (cp 1); OBB 189; PBE 276 (cp 63).

Bunting, Rock (Emberiza cia) ALE 9:341; BBE 291; CDB 196 (cp 854); FBE 275; LEA 457 (bw); PBE 277 (cp 64); (E. tahapisi) PSA 317 (cp 40).

Bunting, Rustic (Emberiza rustica) BBB 270; BBE 289; FBE 281; PBE 276 (cp 63).

Bunting, Siberian Meadow (Emberiza cioides) ABW 300; ALE 9:341; PBE 277 (cp 64).

Bunting, Snow (Plectrophenax nivalis) ABW 298; ALE 9:342; BBB 239; BBE 297; CDB 198 (cp 864); FBE 281; FBW 79; OBB 189; PBE 276 (cp 63); WAB 103 (bw).

Bunting, Yellow—See Yellowhammer.

Bunting, Yellow-breasted (Emberiza aureola) BBE 295; FBE 277; KBA 433 (cp 64); OBB 187; PBE 277 (cp 64).

Bunya-bunya (Araucaria bidwillii) DEW 1:41; SST cp 8.

Bur, New Zealand—See Bidi-bidi, Red Mountain.

Burbot (Lota lota) ALE 4:256, 456-457; FEF 264 (bw); FWF 111; GAL 229 (bw); HFI 76; OBV 93; WFW cp 187.

Bur-clover—See Medick.

Burdock, Great (Artium lappa); BWG 151; MBF cp 47; NHE 239; OWF 155; PFF 330; PWF 137d.

Burdock, Lesser or Common (Arctium minus) BFB cp 18 (6); EPF pl 580; MBF cp 47; MCW 99; MFF cp 31 (902); NHE 239; PWF 99m; TEP 203; WFB 245 (5); WUS 373.

Burdock, Woolly or Downy (Arctium tomentosum) NHE 239; PFE cp 152 (1472).

Bur-marigold, Nodding (Bidens cernua) MBF cp 45; MFF cp 58 (838); NHE 100; OWF 33; PFE cp 146 (1406); PWF 170a.

Bur-marigold, Trifid (Bidens tripartita) MBF cp 45; MFF cp 58 (839); NHE 100; PWF 170c; TEP 45; WFB 239 (6).

Burnet, Great (Sanguisorba officinalis); BFB cp 9; DEW 1:277 (cp 159); MBF cp 26; MFF cp 35 (367); NHE 179; PFE cp 45 (438); PWF 97g; TEP 143; WFB 107 (4).

Burnet, Salad (Sanguisorba minor or
Poterium sanguisorba) AMP 65; EGH 82,
138; EVF 140; HPW 142; MBF cp 26; MFF cp
35 (368); NHE 179; PWF 33f; WFB 107(5);
(S. obtusa) TGF 108.

Burnet-saxifrage (Pimpinella saxifraga)
MBF cp 38; MFF pl 99 (486); NHE 182; OWF
91; PWF 119l; TEP 115; WFB 159 (2).

Burnet-saxifrage, Greater (Pimpinella
major) MBF cp 38; MFF pl 99 (487); NHE 32;
PWF 127f.

Burning Bush (Dictamnus albus) DEW 2:39;
DFP 139 (cp 1111); EFC cp 22; EGH 112;
EPF pl 652; EWF cp 35d; LFW 49 (cp 115);
NHE 31; PFE cp 67 (688); PFF 228; TEP 111;
TGF 109; TVG 87; WFB 138 (9).

Bur-parsley (Caucalis sp.) EPF pl 759; MBF
cp 40; MFF pl 96 (464); NHE 221.

Bur-reed (Sparganium sp.) NHE 108; PFF
336.

Bur-reed, Branched (Sparganium erectum)
EPF pls 992, 993; KVP cp 23; MBF cp 88;
MFF pl 103 (1103); NHE 107; PFE cp 184
(1825); PWF 133g; TEP 38; WFB 297 (6).

Bur-reed, Floating (Sparganium
angustifolium) MBF cp 88; NHE 107.

Bur-reed, Small (Sparganium minimum)
MBF cp 88; NHE 107; (S. simplex) BFB cp
22 (5); MFF pl 103 (1104); WFB 291 (9).

Bur-reed, Unbranched (Sparganium
emersum) MBF cp 88; NHE 107; PWF 115f.

Bush Clockvine (Thunbergia erecta) BIP 175
(cp 270); MEP 147; MWF 283; (T. atriplici-
folia) LFT cp 157.

Bush Grass—See **Small-reed.**

Bushbaby—See **Galago.**

Bushbuck (Tragelaphus scriptus) ALE 13:
300-302; CAF 91; DDM 181 (cp 30); DSM
243; HMW 69; MLS 395 (bw); SMR 112 (cp
9); SMW 260 (bw); TWA 34; WEA 88.

Bushchat, Jerdon's (Saxicola jerdoni)
KBA 321 (cp 50); RBI 215.

Bushchat, Pied—See **Stonechat, Pied.**

Bushchat, Rufous—See **Warbler, Rufous.**

Bush-cricket, Bog (Metrioptera brachyptera)
LEA 130 (bw); OBI 13.

Bush-cricket, Conehead (Conocephalus sp.)
CIB 80 (cp 5); OBI 13.

Bush-cricket, Dark (Pholidiptera griseoaptera) CIB 80 (cp 5); OBI 13.

Bush-cricket, Green—See **Grasshopper,
Green.**

Bush-cricket, Oak (Meconema thalassinum)
CIB 80 (cp 5); OBI 13.

Bush-cricket, Speckled (Leptophyes punctatissima) CIB 80 (cp 5); OBI 13.

Bushdog (Speothos venaticus) ALE 12:283;
DSM 153; JAW 101; MLS 263 (bw); SMW
186 (cp 114).

Bush-hen (Amaurornis olivaceus) DPB 71;
SNB 64 (cp 31); (A. ruficrissus) HAB 56; (A.
akool) KBA 97 (cp 10).

Bushlark—See **Lark, Bush.**

Bushman's Poison—See **Wintersweet,
African.**

Bushmaster (Lachesis muta) ALE 6:468;
HRA 46; MAR 196 (pl 78); SIR 270 (bw);
SSW 135.

Bushpig (Potamochoerus porcus) ALE
13:93; CAF 130 (bw); DDM 161 (cp 26); DSM
220; HMA 60; HMW 50; LEA 590 (bw); MLS
368 (bw); SMR 85 (cp 6); SMW 244 (bw); (P.
koiropotamus) ALE 13:93; TWA 39.

Bush-shrike—See **Shrike.**

Bustard, Australian (Eupodotis australis)
CDB 76 (cp 243); HAB 59; SNB 58 (pl 28);
WAW 17 (bw); (Choriotis australis) RBU 20.

Bustard, Black-bellied (Lissotis melanogaster) ALE 8:443; CDB 77 (cp 245).

Bustard, Great (Otis tarda) ABW 114;
ALE 8:134; BBB 274; BBE 105; CDB 77 (cp
246); CEU 146 (bw); DAE 95; FBE 113; GPB
135; LEA 408 (bw); OBB 83; PBE 81 (pl 26);
RBE 127; SLP 36, 37.

Bustard, Great Indian (Choriotis nigriceps) LVB 37; LVS 219; PWC 200; WAB 31,
161.

Bustard, Houbara (Chlamydotis undulata)
ALE 8:134; BBE 105; FBE 113; PBE 81 (pl
26); WAB 121.

Bustard, Kori (Choriotis kori) ABW 114;
BBT 50, 52; CAF 86 (bw); CDB 76 (cp 242);
(Otis kori) PSA 161 (cp 18); (Ardeotis kori)
ALE 8:134; CAA 39 (bw); GPB 213; LAW
456; LEA 408 (bw).

Bustard, Little (Otis tetrax) BBB 269; BBE 105; CDB 77 (cp 247); FBE 113; OBB 83; PBE 81 (pl 26); (Tetrax tetrax) ALE 8:134.

Bustard, White-bellied (Eupodotis senegalensis) ALE 8:134; CAA 39 (bw); CDB 77 (cp 244).

Busy Lizzie (Impatiens walleriana) DEW 2:61 (cp 34); EDP 106, 107, 118; EGA 127; EGG 118; EGL 120; ESG 120; HPW 211; PFW 39; SPL cp 338; (I. petersiana) BIP 103 (cp 157); DFP 69 (552); RDP 246.

Butcher-bird, Black-backed (Cracticus mentalis) HAB 156; SAB 76 (cp 37).

Butcher-bird, Gray (Cracticus torquatus) ABW 229; ALE 9:469; CDB 220 (cp 983); GPB 334; HAB 156; LEA 464 (bw); RBU 287; SAB 76 (cp 37).

Butcher-bird, Pied (Cracticus nigrogularis) HAB 156; SAB 76 (cp 37).

Butcher's Broom (Ruscus aculeatus) AMP 131; DFP 237 (cp 1890); EWF cp 41f; HSC 109, 110; MBF cp 84; MFF pl 76 (994); OWF 191; PFM cp 234; PFW cp 18a, 174a; SPL cp 495; WFB 269 (8).

Butcher's Broom, Large (Ruscus hypoglossum) EPF pl 919; PFE 504.

Butter and Eggs—See also **Toadflax, Common** (Linaria sp.) and **Trefoil, Common Bird's-foot** (Lotus sp.).

Butterbur (Petasites hybridus) BFB cp 18 (9); DEW 2:201; EPF pls 562, 563; KVP cp 91; MBF cp 46; MFF cp 21 (855); NHE 101; OWF 123; PFE cp 149 (1440); PWF cp 15l; TEP 47; WFB 239 (1); (P. officinalis) AMP 109.

Butterbur, White (Petasites albus) NHE 285; PFE cp 149 (1441); TEP 47.

Buttercup (Ranunculus)—See also **Celadine, Crowfoot, Goldilocks, Spearwort.**

Buttercup, Bermuda (Oxalis pes caprae) MFF pl 63 (255); PFE cp 63 (639); PFM cp 79; SPL 367; WFB 133 (4).

Buttercup, Bulbous (Ranunculus bulbosus) BWG 17; MBF cp 3; MFF cp 39 (13); NHE 178; OWF 3; PWF 59h; WFB 69 (4).

Buttercup, Celery-leaved (Ranunculus scleratus) MBF 3; MFF cp 39 (20); NHE 91; OWF 5; PWF 75e; WFB 71 (2).

Buttercup, Common Meadow (Ranunculus acris) BFB cp 2 (2); BWG 16; DFP 168 (cp 1344); MBF cp 3; MCW 44; MFF cp 39 (11); NHE 178; OGF 39; OWF 3; PFF 169; PWF cp 21g; TEP 139; WFB 69 (3); WUS 187.

Buttercup, Corn (Ranunculus arvensis) MBF cp 3; MFF cp 39 (14); NHE 210; PWF 72d; WFB 71 (3).

Buttercup, Creeping (Ranunculus repens) BFB cp 2 (5); BWG 17; EGC 70; EPF pl 224; LWF 116 (cp 260), 117 (cp 261); MBF cp 3; MCW 45; MFF cp 39 (12); NHE 178; PWF 101i; TEP 38; TGF 86; WFB 69 (5); WUS 189.

Buttercup, Giant or Mountain (Ranunculus lyallii) CAU 280; EWF cp 128c; MLP 52; SNZ 153 (cps 480, 481).

Buttercup, Grassland (Ranunculus lappaceus) SNZ 153 (cp 482); WAW 118.

Buttercup, Hairy Alpine or Korikori (Ranunculus insignis) EWF cp 128b; SNZ 151 (cp 475).

Buttercup, Mountain (Ranunculus montanus) ERW 132; NHE 269; PFE cp 25 (241).

Buttercup, Pale Hairy (Ranunculus sardous) MBF cp 3; MFF cp 39 (15); NHE 210; PFE 105.

Buttercup, Persian or Turban (Ranunculus asiaticus) DFP 110 (cp 874); EGB 136; MWF 250; PFM cp 30; SPL cp 271.

Buttercup, Small-flowered (Ranunculus parviflorus) MBF cp 3; MFF cp 46 (16); OWF 3; PFE 105; WFB 71 (4).

Buttercup, Thora (Ranunculus thora) NHE 269; PFE cp 25 (234).

Buttercup Tree—See also **Cochlospermum.**

Buttercup Tree (Cassia corymbosa) BIP 49 (cp 59); MWF 74.

Buttercup, White (Ranunculus acontifolius) NHE 269; PFE 25 (246); TEP 235.

Butterfish (Pholis gunnellus) ALE 5:162; CSF 151; HFI 45; (Centronotus gunnellus) CSS cp 31.

Butterfly (Agrias sp.) ALE 2:360; KBM 163; PEI 459 (bw); SBM 106; (Charaxes sp.) CAF 177; KBM 139; PEI 452 (bw); SBM 108 (bw), 110; WDB cps 227, 237; ZWI 109.

Butterfly (Chlorippe sp.) CTB 61 (cp 128); KBM 157; (Ancyluris sp.) ALE 2:360; SBM 163; WDB cps 268 a,b; (Epiphile sp.) CSA 147; WDB cp 229f.

Butterfly (Cymothoe coccinata) KBM 139; LAW 241; WDB cp 216l; (C. herminia) WDB cp 225.

Butterfly (Delias sp.) ALE 2:360; KBM 123; PEI 497 (cp 48b); SBM 71; WDB cp 127k.

Butterfly (Heliconius sp.) KBM 157, 163, 174 (pl 60), 185; KIW 217 (cp 110); PEI 446 (bw); SBM 101; WDB cps 135, 136, 141, 142.

Butterfly (Perisama sp.) CTB 62 (cp 131); WDB cp 229; (Panacea prola) CTB 64 (cp 135); WDB cp 217; (Prepona sp.) SBM 102, 103; WDB cps 230c, 234, 236b.

Butterfly, Acraea (Acraea sp.) KBM 139; WDB cps 156, 157, 163.

Butterfly, Apollo (Parnassius sp.) ALE 13:463; CIB 149 (cp 16); CTB 29 (cps 32-35); CTI 29 (cp 31); GAL 320 (cp 7); KBM frontis., 40 (pl 10), 109; LAW 217; PEI 477 (bw), 480 (cp 45b); SBM 62, 63; WDB cps 94, 102.

Butterfly, Bath White (Pontia daplidice) CTB 30 (cp 38); FBW 41; OBI 55.

Butterfly, Berger's Clouded Yellow (Colias australis) PEI 497 (cp 48a); SBM 66.

Butterfly, Bhutan Glory (Armandia lidderdalii or Bhutanitis lidderdalii) ALE 2:360; CAS 127; PEI 480 (cp 45a); SBM 59.

Butterfly, Birdwing (Troides sp.) ALE 2:360; KBM 37, 45; LAW 241; PEI 481 (cp 46b), 481-483 (bw), 496 (cp 47b); SBM 52 (bw), 54; WDB cps 100c, 107f; (Ornithoptera priamus) CTB 63 (cp 133); WDB cp 91.

Butterfly, Birthwort (Zerynthia hypsipyle or Z. polyxena) ALE 2:327; PEI 449 (cp 44), 479.

Butterfly, Black Hairstreak (Strymonidia pruni) OBI 51; SBM 149 (bw).

Butterfly, Brimstone (Gonepteryx rhamni) ALE 2:354; CIB 157 (cp 18); CTB 32 (cp 49), 33 (cps 50, 51); GAL 320 (cp 7); OBI 57; WDB cp 118; (G. cleopatra) LAW 240.

Butterfly, Brown Argus (Aricia agestis) OBI 53; WDB cp 252.

Butterfly, Brown Hairstreak (Thecla betulae) OBI 51; PEI 433 (cp 42a); SBM 151-153; WDB cp 257.

Butterfly Bush (Buddleia davidii) BWG 141; DEW 2:101 (cp 52); DFP 184 (cp 1466); EFS 97; ELG 113; GDS cps 24, 25; HSC 19; LFW 16 (cp 30); MEP 106; MFF cp 12 (628); MWF 64; PFE cp 121 (1189); PFW 56; PFW 109; SPL cp 68; TGS 359.

Butterfly Bush, Fountain (Buddleia alternifolia) DFP 183 (cp 1464); EFS 97; EPG 108; GDS cp 22; HSC 19; TGS 367.

Butterfly Bush, Globe (Buddleia globosa) DFP 184 (cp 1467); GDS cp 26; PFW 56.

Butterfly, Cabbage White (Pieris brassicae) ALE 2:354; CTB 31 (cps 43, 44); LAW 212, 216; OBI 55; PEI 496 (bw); WDB cp 117.

Butterfly, Camberwell Beauty (Nymphalis antiopa) CIB 157 (cp 18); CTB 42 (cps 80, 81); CTI 32 (cp 36); GAL 324 (cp 8); KIW 196 (bw); OBI 49; SBM 78; WDB cp 197.

Butterfly, Chalk-hill Blue (Lysandra corydon) CIB 157 (cp 18); LAW 217; OBI 53; WDB cp 250.

Butterfly, Checkered Skipper (Carterocephalus palaemon) OBI 59; WDB cp 81.

Butterfly, Clouded Yellow (Colias croceus) ALE 2:311, 11:321; CIB 157 (cp 18); CTB 32 (cps 47, 48); GAL 320 (cp 7); OBI 57; WDB cp 121.

Butterfly, Comma (Polygonia c-album) CIB 156 (cp 17); CTB 41 (cps 78, 79); CTI 33 (cp 38); GAL 324 (cp 8); KBM 135 (pl 47); OBI 49; PEI 449 (bw); SBM 79; WDB cps 204, 205; (P. egea) CTB 41 (cps 76, 77).

Butterfly, Common Blue (Polyommatus icarus) ALE 2:348; CIB 157 (cp 18); CTB 26 (cps 22, 23), 27 (cps 26, 27); GAL 320 (cp 7); KIW 191 (bw); LAW 217; OBI 53; WDB cp 251.

Butterfly, Crow (Euploea sp.) KBM 123; WDB cp 130.

Butterfly, Dark Green Fritillary (Mesoacidalia aglaia) CTB 39 (cps 68, 69), 47 (cp 93); SBM 130.

Butterfly, Diadem (Hypolimnas dexithea) KBM 139; SBM 90; WDB cp 208a; (H. missippus); KBM 97; PEI 463 (bw); WDB cp 208j.

Butterfly, Dingy Skipper (Erynnis tages) CIB 157 (cp 18); OBI 57; WDB cp 85; (E. icelus) KBM 99.

Butterfly, Duke of Burgundy Fritillary (Hamearis lucina) CIB 157 (cp 18); OBI 47; SBM 162; WDB cp 267.

Butterfly, False Acraea (Pseudacraea boisduvalii) CAF 177; WDB cp 231h.

Butterfly Flower (Schizanthus sp.) BIP 161 (cp 247); DEW 2:154; EGA 153; EGG 141; LHP 193; MWF 264; PFW 283; TGF 205; TVG 71.

Butterfly, Fritillary (Argynnis sp.) ALE 2:353; GAL 320 (cp 7), 324 (cp 8); (Boloria sp.) KBM 40 (pl 10), 109; OBI 45; PEI 471 (bw); PWI 74 (bw); (Clossiana sp.) CIB 156 (cp 17); CTB 38 (cp 67); SBM 125 (bw); WDB cp 195.

Butterfly, Gatekeeper (Pyronia sp.) CTB 55 (cp 115); WDB cps 167, 168.

Butterfly, Glider (Neptis sp.) PEI 468 (bw); WDB cps 216b, 218.

Butterfly, Grayling or Wood-nymph (Hipparchia sp.) CTB 57 (cps 117, 118), 58 (cp 122); GAL 320 (cp7), 324 (cp 8); ZWI 162.

Butterfly, Great Banded Grayling (Brintesia circe) CTB 57 (cps 119, 120); WDB cp 165.

Butterfly, Green Hairstreak (Callophrys rubi) OBI 51; WDB cp 256.

Butterfly, Green-veined White (Pieris napi) CTB 31 (cps 40, 41); OBI 55; WDB cp 123.

Butterfly, Heath Fritillary (Melitaea sp.) ALE 2:323; CTB 37 (cps 63, 64), 45 (cp 90); GAL 324 (cp 8); OBI 47; PEI 472 (bw); WDB cps 190, 191.

Butterfly, Heath or Pearl-grass (Coenonympha sp.) CTB 53 (cps 107, 108); GAL 320 (cp 7), 324 (cp 8); OBI 43; WDB cps 169, 171, 172, 175.

Butterfly, High Brown Fritillary (Fabriciana adippe) CTB 47 (cp 94); WDB cp 193.

Butterfly, Holly Blue (Celastrina argiolus) OBI 51; WDB cp 249.

Butterfly, Indian Leaf (Kallima inachus) ALE 2:360; CTI 34 (cp 39); KBM 123; PEI 456 (bw); SBM 98, 99; WDB cp 160d.

Butterfly, Kaiser-I-Hind (Teinopalpus imperialis) PEI 480 (bw); SBM 60 (bw); WDB cp 96a.

Butterfly, Large Blue (Maculinea arion) CTB 25 (cps 20, 21); GAL 324 (cp 8); OBI 53.

Butterfly, Large Copper (Lycaena dispar) FBW 43; KIW 220 (cp 114); OBI 51; SBM 150; WDB cp 238.

Butterfly, Large or Poplar Admiral (Limenitis populi or Ladoga populi) CTB 36 (cp 58); GAL 324 (cp 8); PEI 470 (bw); SBM 112-114 (bw).

Butterfly, Large Skipper (Ochlodes venata) CIB 157 (cp 18); CTB 24 (cp 16); OBI 59; WDB cp 87.

Butterfly, Large Tortoise-shell (Nymphalis polychloros) CTB 42 (cps 82, 83); GAL 324 (cp 8); OBI 49; PEI 448 (bw); WDB cp 202.

Butterfly, Leaf-wing (Anaea sp.) SBM 94, 95; WDB cps 222, 230.

Butterfly, Long-tailed Blue (Lampides boeticus) FBW 69; WDB cp 248.

Butterfly, Map (Araschnia levana) ALE 2:353; GAL 324 (cp 8); PEI 433 (cp 42b); SBM 86, 87; WDB cp 163b.

Butterfly, Map-wing (Cyrestis rusca) KBM 123; (C. nivea) WDB cp 229e.

Butterfly, Marbled White (Melanargia galathea) CIB 149 (cp 16); CTB 59 (cp 123); OBI 41; PEI 438 (bw); SBM 146, 147; WDB cp 180.

Butterfly, Marsh Fritillary (Euphydryas aurinia) OBI 47; WDB cp 189; (E. cynthia) CTB 38 (cps 65, 66).

Butterfly, Meadow Brown (Maniola jurtina) CTB 55 (cps 113, 114); GAL 324 (cp 8); OBI 41; WDB cp 177.

Butterfly, Monarch (Danaus sp.) ALE 2:314, 328; BIN pls 78-81; CAA 15; CIB 149 (cp 16); CTB 56 (cp 116), 63 (cp 134); FBW 71; KBM 97, 193; KIW 201 (bw); LEA 122 (bw); PEI 473, 474 (bw); SBM 74; WDB cps 127, 129.

Butterfly, Morpho (Morpho sp.) ALE 2:327, 360; 8:359; KBM 151; KIW 218 (cp 112); LAW 166, 180-181; PEI 442-445 (bw); SBM 132-138; WDB cps 158, 161, 162, 164, 166.

Butterfly, Mourning Cloak —See **Butterfly, Camberwell Beauty.**

Butterfly, Nettle-tree (Libythea celtis) CIB 149 (cp 16); WDB cp 185a.

Butterfly, Orange-tip (Anthocharis cardamines) ALE 10:223; CIB 157 (cp 18); CTB 30 (cps 36, 37), 34 (cp 53); GAL 320 (cp 7); NHE 203; OBI 55; PEI 495 (bw); SBM 67; WDB cp 124; (A. genutia) KBM 147 (pl 51).

Butterfly, Owl (Caligo sp.) KBM 163; PEI 440, 441 (bw); SBM 140 (bw), 142; WDB cps 152, 153, 155.

Butterfly, Painted Lady (Vanessa cardui or Cynthia cardui) ALE 2:353; CTB 40 (cps 72, 73), 50 (cp 101); GAL 320 (cp 7); OBI 49; PEI 449 (bw); WDB cp 200.

Butterfly, Pale Clouded Yellow (Colias hyale) CTB 32 (cps 45, 46); GAL 320 (cp 7); OBI 57; WDB cp 119h.

Butterfly, Peacock (Nymphalis io or Inachis io) ALE 2:312; CIB 156 (cp 17); CTB 43 (cps 84, 85), 51 (cp 103), 52 (cp 104); CTI 31 (cps 34, 35) GAL 320 (cp 7); OBI 49; SBM 75; WDB cp 198; ZWI 163; (Anartia amathea) KIW 221 (cp 116).

Butterfly, Purple Emperor (Apatura iris) ALE 2:313, 353; CIB 156 (cp 17); CTB 35 (cps 54, 55); FBW 67; GAL 324 (cp 8); KBM 117; OBI 47; PEI 464, 465 (bw); SBM 122, 123; WDB cp 199; WEA 225; (A. ilia) CTB 35 (cps 56, 57); CTI 30 (cp 32).

Butterfly, Purple Hairstreak (Quercusia quercus or Thecla quercus) CIB 157 (cp 18); CTB 26 (cps 24, 25); OBI 51; WDB cp 258.

Butterfly, Red Admiral (Vanessa atalanta) ALE 2:327, 353; CIB 156 (cp 17); CTB 40 (cps 74, 75), 48 (cp 95); GAL 320 (cp 7); OBI 49; SBM 83; WDB cp 203.

Butterfly, Regent Skipper (Euschemon rafflesia) KBM 123; WDB cp 79a.

Butterfly, Ringlet (Aphantopus hyperanthus) CIB 149 (cp 16); OBI 41; WDB cp 170; WEA 225; (Erebia sp.) ALE 2:354; CTB 54 (cps 109, 110); GAL 324 (cp 8); KBM 109; OBI 43; PEI 452 (bw); WDB cps 173, 181.

Butterfly, Scarce Copper (Heodes virgaureae) GAL 324 (cp 8); SBM 159.

Butterfly, Scarce Swallow-tail (Iphiclides podalirius) ALE 2:25, 354; CTB 28 (cps 29, 31); PEI 492; SBM 26; WDB cp 93.

Butterfly, Silver-studded Blue (Plebejus argus) GAL 320 (cp 7); OBI 53; WDB cp 245.

Butterfly, Silver-washed Fritillary (Argynnis paphia) CIB 156 (cp 17); CTB 39 (cps 70, 71); GAL 324 (cp 8); OBI 45; SBM 128, 129 (bw); WDB cp 194.

Butterfly, Skipper (Pyrgus sp.) CIB 157 (cp 18); CTB 24 (cp 17), 25 (cps 18, 19); OBI 57; WDB cp 86; (Thymelicus sp.) OBI 59; WDB cps 83, 84.

Butterfly, Small Blue (Cupido minimus) OBI 53; WDB 247.

Butterfly, Small Copper (Lycaena phlaeas) CAS 198; CIB 157 (cp 18); GAL 320 (cp 7); KBM 117; OBI 51; SBM 155; WDB cp 239.

Butterfly, Small Tortoise-shell (Aglais urticae or Vanessa urticae) ALE 2:353; 10:223; CIB 156 (cp 17); CTB 43 (cps 86, 87); CTI 32 (cp 37); GAL 320 (cp 7); KIW 197; LEA 141 (bw); OBI 49; PEI 448 (cp 43b); SBM 82; WDB cp 196; WEA 225.

Butterfly, Small White (Pieris rapae) CIB 157 (cp 18); CTB 31 (cp 42); OBI 55; WDB cp 122.

Butterfly, Speckled Wood (Pararge aegeria) CTB 53 (cp 106); OBI 41; WDB cp 182.

Butterfly, Sulphur (Phoebis sp.) CTB 64 (cp 136); WDB cp 127l.

Butterfly, Swallow-tail (Graphium sp.) ALE 2:328; CTB 62 (cp 129); PEI 490 (bw); SBM 27, 38; WDB cps 97, 99.

Butterfly, Swallow-tail (Papilio sp.) ALE 2:327, 354; 13:285; BIN cps 86, 87; CIB 149 (cp 16); CTB 28 (cp 28), 61 (cps 126, 127); FBW 65; GAL 320 (cp 7); KBM 45, 59, 62 (pl 18), 97, 139, 173 (pl 59); KIW 186-188 (bw), 210 (cp 93); LAW 216; LEA 119 (bw); OBI 55; PEI 484-491 (bw), 496 (cp 47a); SBM 29 (bw), 30, 31, 43-47; WDB cps 92, 98, 99, 103, 105, 106.

Butterfly, Veined White (Aporia crataegi) CTB 30 (cp 39); KIW 189 (bw).

Butterfly, Wall Brown (Lasiommata megera or Pararge megera) CTB 53 (cp 105), 58 (cp 121); GAL 320 (cp 7); OBI 41; WDB cp 176.

Butterfly Weed (Asclepias tuberosa) EDP 116; EWG 89; LFW 186 (cp 420); SPL cp 190; TGF 157.

Butterfly, White Admiral (Limenitis camilla or Ladoga camilla) CIB 156 (cp 17); CTB 36 (cps 59, 60), 46 (cp 91); GAL 324 (cp 8); OBI 47; PEI 467 (bw); SBM 118; WDB cp 220.

Butterfly, White-letter Hairstreak (Strymonidia w-album) OBI 51; PEI 434 (bw); SBM 149 (bw); WDB 259.

Butterfly, Wood White (Leptidea sinapis) OBI 55; WDB cp 116.

Butterflyfish (Pantodon buchholtzi) ALE 4:285; FEF 70 (bw); HFI 129; HFW 101 (bw); SFW 49 (pl 4).

Butterflyfish (Chaetodon sp.) ALE 5:95, 96, 112; CTF 64 (cp 89); FEF 335, 373, 374, 376 (bw); HFI 33; HFW 90 (cps 48, 49), 96 (cps 59, 60); LEA 240; MOL 116, 124; WFW cps 315-331.

Butterflyfish, Emperor (Chaetodontoplus mesoleucus) ALE 5:112; FEF 375 (bw).

Butterflyfish, Long-nosed (Chelmon rostratus) ALE 5:112; CAA 18 (bw); CTF 63 (cp 87); FEF 382; HFI 33; HFW 91 (cps 50, 51); LAW 339; MOL 116; WFW cp 333; (Forcipiger longirostris) ALE 5:95; CAA 18 (bw); LAW 339.

Butterfly-flower or **Butterfly-lily** —See **Ginger-lily, White** (Hedychium).

Butternut (Juglans cinera) EDP 138; EET 137; EVF 138; LFO 106; OFP 29; PFF 133; TGF 81; (J. cordiformis) DEW 1:109 (cp 65).

Butterwort (Pinguicula macrophylla) EWF cp 164d; PFW 166; (P. moranensis) HPW 253.

Butterwort, Alpine (Pinguicula alpina) EPF pl 869; MBF cp 66; NHE 75; PFE 132 (1279).

Butterwort, Common (Pinguicula vulgaris) DEW 2:159; MBF cp 66; MFF cp 1 (739); NHE 75; OWF 139; PFE cp 132 (1280); PWF 51a; TEP 77; WFB 219 (7).

Butterwort, Large-flowered (Pinguicula grandiflora) EWF cp 21d; MBF cp 66; PFE cp 132 (1280); PFW 166; PWF 46b.

Butterwort, Pink or Pale (Pinguicula lusitanica) MBF cp 66; MFF pl 88 (738); OWF 139; WFB 219 (8).

Button-quail —See **Quail.**

Buttonweed (Cotula coronopifolia) MFF cp 52 (895); NHE 136; PFE cp 149 (1433); SNZ 25 (cp 28); WFB 239 (8).

Buzzard, African —See **Eagle, Bateleur.**

Buzzard, Black-breasted (Hamirostra melanosterna) GBP 234 (bw); HAB 35; RBU 83; SNB 44 (cp 21), 46 (cp 22).

Buzzard, Common (Buteo buteo) ALE 7:332, 356; BBB 131; BBE 79; DPB 45; FBE 77; GAL 152 (bw); KBA 64 (pl 5); OBB 39; PBE 53 (cp 18), 69 (pl 22); PSA 81 (pl 8), 128 (pl 13); RBE 8.

Buzzard, Gray Eagle (Geranoaetus melanoleucus) ALE 7:356; GBP 260 (bw).

Buzzard, Gray-faced (Butastur indicus) DPB 45; GBP 286 (bw); KBA 64 (pl 5); (B. teesa) SNB 248 (bw).

Buzzard, Honey (Pernis apivorus) ALE 7:346; BBB 267; BBE 79; DPB 37; FBE 77; GAL 152 (bw); GBP 219 (bw); GPB 101; KBA 64 (pl 5); OBB 43; PBE 59 (cp 18), 69 (pl 22); RBE 19.

Buzzard, Jackal (Buteo rufofuscus) GBP 266 (bw); PSA 81 (pl 8), 128 (pl 13).

Buzzard, Lizard (Kaupifalco monogrammicus) GBP cp after 112, 285 (bw); PSA 81 (pl 8), 128 (pl 13).

Buzzard, Long-legged (Buteo rufinus) BBE 79; DAE 116; FBE 77; LAW 477; PBE 53 (cp 18), 69 (pl 22).

Buzzard, Red-tailed (Buteo auguralis) ALE 7:356; GBP 267 (bw).

Buzzard, Rough-legged (Buteo lagopus) ALE 7:356; BBB 271; BBE 79; FBE 77; GPB 99; OBB 39; PBE 53 (cp 18), 69 (pl 22); WAB 102; WEA 88.

By-the-wind-sailor (Velella sp.) ALE 1:198, 271; CSS cp 2; LEA 43 (bw); MOL 59; OBV 11; WEA 89.

C

Cabbage, Bastard or Steppe (Rapistrum perenne) MFF cp 45 (58); NHE 216; PFE 135; (R. rugosum) NHE 216; PFE 135; WFB 89 (7).

Cabbage, Chinese —See **Pak-choi.**

Cabbage, Crested (Bunias erucago) NHE 217; PFE cp 32 (293); PFM pl 5 (334).

Cabbage, Dune or Isle of Man (Rhynchosinapis monensis) MBF cp 9; MFF cp 45 (50).

Cabbage, Field or Bargeman's (Brassica campestris) MCW 47; MFF cp 45 (47); OWF 11.

Cabbage, Hare's-ear —See **Mustard, Hare's-ear.**

Cabbage, Kerguelen (Pringlea antiscorbutica) EWF cp 1j; MPL 88; SAR 78.

Cabbage Palm (Cordyline indivisa) DFP 191 (cp 1523); SNZ 99 (cps 299, 300); SPL cp 78.

Cabbage, Purple or Violet (Moricandia arvensis) EWF cp 30e; HPW 121; PFE cp 37 (356); PFM cp 39.

Cabbage Tree (Cordyline australis) BKT cp 192; DFP 191 (cp 1522); MTB 385 (cp 40); MWF 93; OBT 200; RDP 157; SNZ 98 (cps 293-296), 99 (cps 297, 298, 301), 100 (cp 302).

Cabbage Tree, Spiked (Cussonia sp.) DEW 2:63 (cp 38); HPW 218; MWF 99.

Cabbage, Wallflower (Rhynchosinapis cheiranthos) MBF cp 9; NHE 215; WFB 89 (6).

Cabbage, Warty (Bunias orientalis) MFF cp 44 (77); WFB 87 (5).

Cabbage, Wild (Brassica oleracea) MBF cp 9; MFF cp 45 (46); MWF 62; NHE 215; OFP 157; SPL cp 416; WFB 89 (1).

Cacao or Cocoa Tree (Theobroma cacao) DEW 1:216 (cps 123-125); EET 223; EPF pls 606, 607; EWF cp 176a; HPW 92; MLP 142; OFP 113; PFF 252; SST cp 300; WWT 110.

Cachalot —See **Whale, Sperm.**

Cacique (Cacicus cela) ALE 9:381; CDB 206 (cp 908); GPB 313; (Xanthornis viridis) ABW 288.

Cactus (Gymnocalycium sp.) ECS 25, 120; EGL 115; LCS cps 54, 56, 58, 60-65; RDP 229; SCS cps 25-30; SPL cp 140, 141.

Cactus, Agave or Prism (Leuchtenbergia principis) ECS 126; SCS cp 34.

Cactus, Artichoke (Obregonia denegrii) ECS 134; SCS cp 50.

Cactus, Ball (Notocactus sp.) DEW 1:198; DFP 71 (cp 566); ECS 29; EGL 128; LCS cps 120, 125-127; RDP 279, 280; SCS cp 49.

Cactus, Barrel (Echinocactus sp.) DFP 65 (cp 520); ECS 112; EDP 130; EGL 105; LCS cp 42; LHP 189; PFW 58; RDP 186; SCS cp 15; SPL cps 133, 134.

Cactus, Bruch's Chin (Gymnocalycium bruchii) EMB 108; LCS cp 57.

Cactus, Button (Epithelantha micromeris) ECS 116; EHP 124; LCS cp 48; SCS cp 20.

Cactus, Candelabra —See **Euphorbia, Milk-striped.**

Cactus, Cereus (Cereus sp.) LCS cp 12; MLP 163, 166, 167; RDP 27.

Cactus, Christmas (Schlumbergera buckleyi) EDP 92; PFW 58; (S. bridesii) ECS 142; RDP 352.

Cactus, Cob (Lobivia hetrichiana) ECS 128; LCS cp 77; RDP 259; (L. famatimensis) EGL 125; LCS cp 76, 81; SCS cp 35.

Cactus, Cob (Lobivia sp.) EDP 138; EFP 122; LCS cps 72-82; MWF 188; SCS cps 35-37.

Cactus, Cottonball or Snowball —See **Cactus, Peruvian Old-man.**

Cactus, Crab or Christmas (Zygocactus truncatus) DFP 82 (cp 656); LCS cps 193-195; LFW 267 (cp 598); LHP 189; MWF 303; PFF 260; SCS cp 67.

Cactus, Easter (Rhipsalidopsis rosea or R. gaertneri) DFP 80 (cp 635); LCS cp 176; PFW 58; RDP 335; (R. sp.) ECS 15, 140; EWF cp 175a; HPW 64; PFW 59; RDP 336.

Cactus, Echinocereus (Echinocereus sp.) DFP 66 (cp 521); ECS 30; LCS cps 35-41; RDP 187; RWF 102; SCS cp 17.

Cactus, Empress (Nopalxochia phyllanthoides) LCS cp 121; MWF 214.

Cactus, Fishhook (Ferocactus sp.) ECS 118; LCS cps 52, 302; RDP 209; SPL cp 138.

Cactus, Golden Ball (Notocactus leninghausii) ECS 26; EFP 126; LCS cps 122, 124; RDP 280.

Cactus, Hedge (Browningia candelaris) LCS cp 301; RWF 88; (Espostoa melanostele) LCS cp 51; RWF 86.

Cactus, Hedgehog or Rainbow (Echinocereus pectinatus) ECS 113; LCS cp 39; RDP 187; SCS cp 16.

Cactus, Jungle —See **Euphorbia, Milk-striped.**

Cactus, Lamb's-tail (Wilcoxia schmollii) ECS 48; LCS cp 191; (W. poselgeri) ECS 30; LCS 190.

Cactus, Leafy —See **Barbados Gooseberry** (Pereskia).

Cactus, Old Man (Cephalocereus sp.) DFP 56 (cp 443); ECS 104; EDP 138; LCS cps 9-11; PFF 256; RDP 133; RWF 92; SCS cp 10.

Cactus, Orchid (Epiphyllum sp.) BIP 43 (cp 47); DFP 66 (cps 523, 524); ECS 115; EGG 107; FHP 118; LCS cps 46, 47; LFW 239 (cp 531); MEP 90; MWF 122; PFF 260; PFW 59; RDP 190, 191; SPL cp 135.

Cactus, Parodia (Parodia sp.) ECS 29, 136; LCS cps 146-156, 299; RDP 298; SCS cps 53-55.

Cactus, Peanut (Chamaecereus silvestrii) BIP cp 48; ECS 107; RDP 136; SCS cp 12.

Cactus, Peruvian Apple (Cereus peruvianus) ECS 46, 105; LCS cp 13; PFF 256; RDP 134; SPL cp 130; WWT 73, 74.

Cactus, Peruvian Old-man (Espostoa lanata) ECS 116; LCS cp 49, 50; RDP 194; SCS cp 24; WWT 82.

Cactus, Plaid (Gymnocalycium mihanovichii) EFP 116; HPW 64; LCS cp 59; RDP 229; (G. quehlianum) ECS 121; RDP 229; SCS cp 31.

Cactus, Prickly Pear —See **Fig, Barbary.**

Cactus, Rattail (Aporocactus flagelliformis) ECS 99; EDP 132; EFP 89; RDP 22, 40, 85; SCS cp 1; SPL cp 127.

Cactus, Rebutia or Crown (Rebutia sp.) BIP 41; ECS 28, 139; EMB 110; LCS cps 161-175; SCS cp 6; WWT 75.

Cactus, Rose (Pereskia grandiflora) LFW 262 (cp 585); MEP 90.

Cactus, Silver Bell (Notocactus scopa) EGS 134; LCS cp 123, 128; RDP 280.

Cactus, Star (Ariocarpus fissuratus) ECS 99; EHP 124; HPW 64; LCS cp 3; (A. retusus) EHP 125.

Cactus, Star (Astrophytum ornatum) LCS cp 7; SCS cp 5; SPL cp 129; (A. capricorne) LCS cps 5, 8; SCS cp 3.

Cactus, Strawberry (Hamatocactus setispinus) RDP 232; SCS cp 32.

Cactus, Strawberry Hedgehog (Echinocereus engelmanii) EGS 113; MEP 91.

Cactus, Thanksgiving or Claw (Schlumbergera truncata) EWF cp 175b; RDP 352; (S. zygocactus) EDP 132; FHP 143.

Cactus, Urchin or Hedgehog (Echinopsis multiplex) EFP 109; MWF 120; RDP 188; (E. sp.) ECS 114, 115; LCS cps 44, 45; SCS cp 19.

Caecilian, Mikan's (Siphonops annulatus) ALE 5:352; ART 18; CAW 14 (pl 1), 65 (cp 1); ERA 67; HRA 122.

Caecilian, Sao Thome (Schistometopum thomensis) ALE 5:352; CAW 65 (cp 2); LEA 276 (bw).

Caiman, Black (Melanosuchus niger) ALE 6:126; HRA 80; HWR 27; WEA 90.

Caiman, Smooth-fronted (Paleosuchus trigonatus) HRW 28; LEA 299 (bw); (P. palpebrosus) ALE 6: 126, 134.

Caiman, Spectacled (Caiman crocodylus) ALE 6:126; ART 144; CSA 18 (bw); ERA 146; LVS 167, 169; MAR 83 (pl 33); (C. sclerops) SIR 58 (cp 19).

Cajeput Tree (Melaleuca sp.) LFW 170 (cp 380); MEP 98; MLP 178; MWF 199.

Calabash Tree (Crescentia cujete) DEW 2:177; SST cp 279.

Caladium (Caladium sp.) BIP 45 (cp 50); EDP 124; EFP 95; EGB 56, 100; EGL 91; ESG 38, 98; LHP 48, 49; MEP 11; MWF 65; RDP 123; WWT 2.

Calamint, Common (Calamintha ascendens) MBF cp 68; MFF cp 29 (754); NHE 35, 281; OWF 145; PWF 121k.

Calamint, Lesser (Calamintha nepeta or C. nepetoides) DFP 126 (cp 1008); MBF cp 68; PFE cp 114 (1156); PWF 120c.

Calamint, Wood (Calamintha sylvatica) MBF cp 68; PWF 121k, 125j; WFB 205 (6).

Calathea (Calathea sp.) DEW 2:276 (cp 152); EWF cp 189b; HPW 300; PFW 186; RDP 124.

Calico Flower (Aristolochia elegans) BIP 27 (cp 20); DFP 52 (cp 410); EGV 93; HPW 41; LFW 228 (cp 509), 229 (cp 511); PFW 37.

Calico Hearts —See **Leopard's Spots.**

Calla Lily (Zantedeschia aethiopica) BIP 181 (cp 283); DEW 2:310; DFP 118 (cp 943); EGB 147; EPF pl 1005; EWF cp 88d; LFW 150 (cp 334); MWF 300; PFF 365; PFW 34; RWF 61; SPL cp 514.

Calla Lily, Golden (Zantedeschia elliottiana) FHP 151; LFW 151 (cp 336); PFW 34; RDP 397; (Z. sp.) LFT cps 6, 7.

Calla Lily, Pink (Zantedeschia rehmannii) EGB 147; EWF cp 88e; LFT cp 3; MWF 300; RDP 397; (Z. sp.) CAF 224; EDP 126; FHP 151; LFT cps 4-7; LFW 151 (cp 336); PFW 34; RDP 396, 397.

Caltrops, Small or Land —See **Maltese Cross** (Tribulus).

Calycella (Calycella citrina) DLM 119; ONP 145.

Camel, Arabian (Camelus dromedarius) ALE 3:131, 132; BMA 40-43; DSM 224; HMA 35; HMW 54; JAW 167; LAW 562; LEA 593 (bw); MLS 381 (bw); SMW 247 (bw); TWA 166.

Camel, Bactrian (Camelus bactrianus ferus) ALE 13:132; CAS 71, 77, 109; DSM 224; HMA 35; HMW 55; JAW 168; LEA 592 (bw); LVA 45; MLS 380 (bw); PWC 197, 244 (bw); SMW 236 (cp 149); TWA 167; VWA cp 20.

Camellia (Camellia hybrids) DFP 187 (cps 1489, 1490); EGE 71; EWF cp 101f; GDS cps 39, 40; HPW 83; HSC 21; MWF 70; OGF 27; PFW 292, 293; TGS 230.

Camellia (Camellia japonica) BIP 47 (cp 53); DFP 185 (cps 1479, 1480), 186 (cps 1481-1486); EDP 142; EEG 105; EGE 116; EGW 97; EJG 104; ELG 135; EPF pl 471; EPG 110; ESG 99; FHP 109; GDS cps 41-48; HSC 23; MEP 86; MWF 70; OGF 27; PFW 292-294; RDP 127; SPL 71; SST 233; TGS 230.

Camellia (Camellia williamsii) BIP 47 (cp 54); DFP 187 (cps 1491-1494); GDS cps 49-51; HSC 21, 23; MWF 71.

Camellia, Netvein (Camellia reticulata) DFP 186 (cps 1487, 1488); EGG 94; EWF cp 101c.

Camphor Tree (Cinnamomum camphora) EGE 118; EJG 109; ELG 126; SST cp 273; WWT 104.

Campion —See also **Catchfly.**

Campion, Bladder (Silene vulgaris) DEW 1:207; MBF cp 13; MFF cp 85 (146); NHE 185; PFE cp 14 (169); PWF 44d, 66b; TEP 107; WFB 59 (1); (S. cucubalus) MCW 40; OWF 73.

Campion, Moss (Silene acaulis) EPF pl 305; EWF cp 3a; KVP cp 126; MBF cp 13; MFF cp 23 (150); NHE 277; PFE cp 16 (170); PWF 149j; TEP 90; WFB 59 (5).

Campion, Red (Melandrium dioicum) EFC cp 6; MFF cp 14 (155); (M. rubrum) BFB cp 5; OWF 73, 107.

Campion, Red (Silene dioica) HPW 68; MBF cp 14; NHE 225; PFE cp 15 (165); PWF 31h; WFB 61 (1).

Campion, Rose (Lychnis coronaria) DFP 157 (cp 1250); EWF cp 32e; LFW 92 (cp 211); PFW 72; SPL cp 227; TGF 71.

Campion, Sea (Silene maritima) BFB cp 5 (9); KVP cp 102; MFF pl 85 (147); OWF 73.

Campion, White (Lychnis alba) MCW 38; PFF 162; TGF 71; WUS 161.

Campion, White (Melandrium album) MLP pl 85 (156); OWF 73.

Campion, White (Silene alba) EPF pl 303; MBF cp 14; NHE 225; PWF 49f; TEP 141; WFB 61 (2).

Canarina (Canarina sp.) EWF cps 43b, 55a; HPW 255; PFW 64.

Canary, Bush —See **Yellowhead.**

Canary, Cape (Serinus canicollis) NBA 97; PSA 317 (cp 40).

Canary Creeper (Senecio tamoides) EWF cp 83a; MEP 152.

Canary Creeper or **Canary Bird Flower** (Tropaeolum peregrinum) EGA 159; EWF cp 171a; OGF 163; PFW 299.

Canary, Yellow (Serinus canaria) FBE 293; GAL 118 (cp 1); OBB 177; RBE 304; (S. gularis) PSA 317 (cp 40); (S. flaviventris) CDB 210 (cp 930); PSA 317 (cp 40).

Canary, Yellow-fronted (Serinus mozambicus) ALE 9:382; NBA 97; PSA 317 (cp 40).

Canary-grass (Phalaris canariensis) MFF pl 124 (1247); NHE 196; PFE cp 181 (1795); PFF 351.

Canary-grass, Reed (Phalaris arundinacea) DFP 164 (cp 1310); MBF pl 95; MFF pl 134 (1246); NHE 107; PFF 351; TEP 37; TVG 129; WUS 79.

Cancerbush or **Kankerbos** (Sutherlandia frutescens) EWF cp 72b; MEP 62.

Candelabra Flower (Ceropegia stapeliiformis) HPW 225; LCS cp 218; RWF 58; SCS cp 73.

Candle Plant (Plectranthus oertendahlii) MWF 237; RDP 322, 323.

Candlewick —See **Mullein, Common**

Candytuft, Annual or Wild (Iberis amara) MBF cp 10; MFF pl 92 (67); MWF 161; NHE 216; PFE cp 36 (348); PWF 100c; TVG 53; WFB 97 (1).

Candytuft, Evergreen or Perennial (Iberis sempervirens) DFP 148 (cp 1182); EEG 136; EGC 136; EGW 116; EPF pl 244; ERW 122; LFW 73 (cp 168); SPL cp 362.

Candytuft, Globe (Iberis umbellata) DFP 39 (cp 310); EGA 126; TGF 77; TVG 53; WFB 96 (1b).

Candytuft, Rock (Iberis saxatilis) DFP 12 (cp 52); TVG 155.

Cane-rat (Thryonomys swinderianus) ALE 11:426; DDM 49 (cp 4); MLS 235 (bw); SMR 129 (cp 10); SMW 142 (cp 92).

Canker —See **Poppy, Corn** (Papaver).

Canna (Canna sp.) DEW 2:278 (cp 155); DFP 56 (cp 441); EGB 101; EPF pl 1038; ERW 140; HPW 299; LFW 242 (cps 537-539); 243 (cps 540-541); MEP 28; MWF 72; PFF 390; PFW 65; SPL cp 356; TGF 44; TVG 183.

Cannonball Tree (Couroupita guianensis) EWF cp 179b; LFW 244 (cp 542); MEP 94; MLP 175; SST cp 97.

Cantaloupe or **Muskmelon** (Cucumis melo) EMB 146; EVF 97; OFP 119; PFF 309.

Canterbury Bells (Campanula medium) DEW 2:193; DFP 32 (cp 255); EGA 102; LFW 20 (cps 38, 39); MFF cp 4 (801); MWF 71; PFF 310; PFW 61 (3); SPL cp 331; TGF 188; TVG 39.

Cape Gooseberry—See **Cherry, Bladder or Chinese Lantern** (Physalis).

Cape Poison (Boophone disticha) DEW 2:247; EWF cp 86b; LFT 62 (cp 30).

Caper Bush (Capparis spinosa) CTH 30 (cp 30); DEW 1:146; EWF cp 148d; HPW 118; NHE 139; OFP 133; PFE cp 31 (283); PFW 66; SPL cp 444.

Capercaillie (Tetrao urogallus) ABW 88; BBB 152; BBE 97; CDB 67 (cp 188); DAE 125; FBE 99; FBW 21; GPB 114; OBB 53; PBE 164 (cp 43); RBE 44; WAB 105; WEA 93.

Capuchin, Black-capped (Cebus apella) ALE 10:337; DSM 76; MLS 132 (bw); TWA 129.

Capuchin, Black-capped or Weeper (Cebus nigrivittatus) ALE 10:337; DSM 76; SMW 84 (cp 42).

Capuchin, White-fronted (Cebus albifrons) ALE 10:337; DSM 76.

Capuchin, White-throated (Cebus capucinus) ALE 10:337; CSA 19; DSM 76; HMA 71; HMW 26; JAW 43; LAW 614; WEA 93.

Capybara (Hydrochoerus hydrochoeris) ALE 8:359; 11:424, 438; DSM 135; LAW 595; MLS 227 (bw); SMW 140 (cp 89); (H. capybara) HMA 35; HMW 178.

Caracal (Felis caracal) DAA 66; DDM 132 (cp 21); DSM 189; LAW 543; SMR 65 (cp 4); SMW (cp 108); (Caracal caracal) ALE 12:284, 297; LEA 569 (bw).

Caracara (Phalcoboenus sp.) ALE 7:403; GPB 378 (bw); SSA pl 22; WFI 37, 135 (bw), 153.

Caracara, Yellow-headed (Milvago chimachima) ALE 7:403; GBP cp after 160, 377 (bw).

Caralluma (Caralluma sp.) ECS 103; EPF pl 695; EWF cp 78f; LCS cp 231; LFT 259 (cp 129), 262 (cp 130); RWF 60; SCS cp 71.

Carambola Tree (Averrhoa carambola) MWF 49; OFP 103.

Caraway, Common (Carum carvi) EGH 106; EPF pl 762; EVF 140; MBF cp 37; MFF pl 96 (483); NHE 221; OFP 139; PFF 265; PWF 43h; WFB 157 (8).

Caraway, Tuberous —See **Earth-nut, Great.**

Caraway, Whorled (Carum verticillatum) MBF cp 37; MFF pl 98 (482); WFB 157 (9).

Cardamon (Elettaria cardamonum) AMP 43; BIP 73 (cp 107); OFP 131; RDP 189.

Cardinal Flower (Lobelia cardinalis) DEW 2:176 (cp 111); ERW 143; ESG 85, 125; EWF cp 156b; EWG 30, 123; HPW 255; MWF 188; PFF 310; SPL cp 224; TGF 236; (L. fulgens) DFP 156 (cp 1243); OGF 121; PFW 63.

Cardinal Flower (Rechsteineria cardinalis) BIP 151 (cp 230); EGB 137; FHP 140; PFW 129.

Cardinal, Red-capped (Paroaria gularis) SSA cp 42; WAB 91.

Cardinal, Red-crested (Paroaria coronata or P. cucullata) ALE 9:342; CDB 199 (cp 869).

Cardinal, Yellow (Gubernatrix cristata) ALE 9:334, 342; CDB 198 (cp 868).

Cardinal's Guard (Pachystachys coccinea) BIP
107 (cp 163) EWF cp 184c; PFW 13.

Cardoon —See Artichoke, Wild.

Caribou —See Reindeer.

Carissa —See Wintersweet, African.

Carludovica (Carludovica sp.) EPF pl 108;
HPW 305.

Carnation or Pink, Clove (Dianthus caryophyllus) DFP 36, 63, 64, 137, 138; EDP 118;
EGA 114; EGG 105; LFW 47 (cp 108); MBF cp
13; MWF 110; PFF 164; PFW 70; SPL cp 334;
TGF 60; TVG 47; WFB 65 (2).

Carnival Bush (Ochna atropurpurea) BIP 131
(cp 198); EWF cp 77b; HPW 84 (O. serrulata)
MWF 216; PFW 198; (O. sp.) EWF cp 77h; LFT
cp 109; MEP 85.

Carob (Ceratonia siliqua) AMP 59; DEW 1:305;
NHE 294; PFE cp 49 (486); SST cp 190.

Carolina Allspice (Calycanthus floridus) DEW
1:57 (cp 21); EFS 99; EPF pl 367; ESG 99; HSC
21, 22; LFW 198 (cp 448); SPL cp 69; TGS 151;
(C. occidentalis) EWF cp 150a; HPW 35; PFW
60; SPL cp 70; TVG 219.

Carp, Common (Cyprinus carpio) ALE 4:333;
FBW 35; FEF 167, 170, 173 (bw); FWF 95, 97,
99, 101; GAL 231 (bw); HFI 93; HFW 119 (bw);
LAW 316; LEA 243, 244 (bw); OBV 113; SFW
213 (pl 50); WEA 95.

Carp, Crucian (Carassius carassius) ALE
4:327; FEF 162 (bw); FBW 35; FWF 93; GAL
239 (bw); HFI 93; OBV 113; SFW 221 (pl 52).

Carp, Toothed —See Toothcarp.

Carpet Shell (Venerupis sp.) CSS pl XIX; OIB
75.

Carpetweed or Chickweed, Indian (Mollugo
verticillata) EGC 70; PFF 166; WUS 151.

Carpincho —See Capybara.

Carrion Flower (Stapelia gigantea) MEP 114;
MWF 272; RDP 368; SPL cp 154; (S. nobilis)
ECS 146; EWF cp 78h; LCS cp 296; LFT 266 (cp
132); MLP 205.

Carrion Flower (Stapelia variegata) BIP 171
(cp 261); DEW 2:104 (cps 58, 59); EPF pl 692;
LCS cp 293; PFW 38; RDP 369; SCS cp 96;
WWT 47; (S. sp.) CAF 201; EPF pl 694; LCS cp
297.

Carrot (Daucus carota var. sativa) EMB 148;
EVF 86; OFP 175; PFF 268.

Carrot, Moon —See Moon Carrot (Seseli).

Carrot, Wild —See Queen Anne's Lace.

Carrotwood or Tuckeroo (Cupaniopsis anacardioides) EGE 120; HPW 193.

Cartwheel Flower —See Hogweed, Giant.

Cascadura (Hoplosternum thoracatum) FEF
249 (bw); HFI 113.

Cashew (Anacardium occidentale) DEW 2:67;
HPW 198; MLP 127, 130; OFP 31; PFF 236;
SST cp 268.

Cassava (Manihot esculenta) DEW 1:268; SST
cp 287; (M. palmata) WWT 108; (M. ultissima)
EPF pl 626; MWF 197; OFP 180.

Cassia, Feathery or Silvery (Cassia artemisioides) EWF cp 130g; MWF 74; (C. mimosoides)
LFT cp 79.

Cassowary, Bennett's (Casuarius bennetti)
ABW 19; ALE 7:88; RBU 4.

Cassowary, Double-wattled (Casuarius
casuarius) ABW 19; ALE 7:88, 97; BBT 20, 21;
CDB 33 (cp 3); HAB 3; LAW 500; LEA 343 (bw),
368; SNB 135 (bw); WAB 182; WAW 8 (bw), 9;
WEA 95.

Cassowary, One-wattled (Casuarius unappendiculatus) ALE 7:88; RBU 3.

Cast-iron Plant (Aspidistra elatior or A. lurida)
BIP 29 (cp 23); EFP 90; EPF cp XXIII B; LHP
34-35; PFF 378; SPL cp 4; RDP 88.

Castor-oil Plant or Castor Bean (Ricinus
communis) AMP 71; CTH 26 (cp 20); DEW
1:267; DFP 46 (cps 366, 367); EGA 150; EPF pls
620, 621, 623; MEP 73; MLP 121; PFE cp 65
(668); PFF 233; SPL cp 400; TGF 109.

Cat, Bay (Felis badia or Profelis badia) ALE
12:314; DAA 118.

Cat, Black-footed (Felis nigripes) ALE 12:314;
DDM 132 (cp 21); LVS 76.

Cat, Caffer Wild (Felis libyca) ALE 12:284, 297;
BMA 47; DAA 66; DDM 132 (cp 21); HMA 37;
SMR 65 (cp 4); (F. cattus) DAE 45.

Cat, Chinese Desert (Felis bieti) ALE 12:284;
DAA 64.

Cat, Fishing (Felis viverrina or Prionailurus
viverrinus) ALE 12:323; DAA 64; JAW 129.

Cat, Flat-headed (Felis planiceps or Ictailurus
planiceps) ALE 12:299, 314; DAA 118.

Cat, Golden (Felis aurata, F. temmincki or
Profelis temmincki) ALE 12:323; DAA 118;
DDM 133 (cp 22); DSM 187; MLS 13 (bw).

Cat, Jaguarundi or Otter (Felis yagouarondi)
DSM 185; HMA 36, 37; JAW 131; MLS 315
(bw); SMW 167 (bw); (Herpailurus yagouaroundi) ALE 12:300, 324.

Cat, Jungle or Swamp (Felis chaus) ALE
12:284; DAA 63; DDM 132 (cp 21).

Cat, Leopard (Felis bengalensis or Prionailurus
bengalensis) ALE 12:299; DAA 118; DSM 186.

Cat, Marbled (Felis marmorata or Pardofelis
marmorata) ALE 12:313, 323; DAA 115.

Cat, Margay (Felis tigrina or F. wiedi) ALE
12:300, 323; DSM 184; SMW 164 (bw).

Cat, Native —See **Dasyure** or **Tiger-cat.**

Cat, Pallas' (Felis manul) DAA 64; DSM 186;
(Otocolobus manul) ALE 12:298, 323.

Cat, Pampas (Felis colocolo) DSM 185;
(Lynchailurus pajeros braccatus) ALE 12:324.

Cat, Rusty-spotted (Felis rubiginosa or
Prionailurus rubiginosus) ALE 12:314; DAA
66.

Cat, Sand (Felis margarita) ALE 12:284; DAA
13; DDM 132 (cp 21); JAW 127; SMW 180 (cp
104.

Cat, Serval (Felis serval or Leptailurus serval)
ALE 12:297, 323; CAA 95 (bw); CAF 101 (bw);
DDM 133 (cp 22); DSM 187; HMW 126; JAW
128; LEA 569 (bw); MLS 312 (bw); SMR 65 (cp
4); SMW 167 (bw); TWA 46; WEA 331.

Cat, Tiger —See **Tiger-cat.**

Cat, Wild (Felis sylvestris) ALE 12:297; 13:163;
DSM 186; FBW 27; GAL 23 (bw); HMW 125;
LEA 568 (bw); MLS 309 (bw); OBV 173; TWA
80; (F. ocreata) SMW 165 (bw).

Catalufa —See **Big-eye.**

Catathelasma (Catathelasma imperialis) DLM
176; SMG 143.

Catbird, Green (Ailuroedus crassirostris) ABW
231; ALE 9:492; HAB 163; LAW 507; MEA 47;
RBU 312; SAB 78 (cp 38); WAB 198.

Catbird, Spotted (Ailuroedus melanotis) HAB
163; SAB 78 (cp 38).

Catbird, Tooth-billed —See **Bowerbird,
Tooth-billed.**

Catchfly —See also **Campion.**

Catchfly (Silene colorata) EWF cp 32b, c; PFE cp
15 (174); PFM cp 16.

Catchfly, Alpine (Lychnis alpina) KVP cp 143;
MBF cp 14; NHE 276; PFE cp 14 (162); WFB 61
(6).

Catchfly, Berry (Cucubalus baccifer) NHE 33;
PFE cp 15 (177); PWF 123i; WFB 62, 63 (3).

Catchfly, Drooping (Silene pendula) DFP 47
(cp 375); EGA 154; OGF 135; SPL cp 467.

Catchfly, Forked (Silene dichotoma) NHE 226;
WFB 59 (10).

Catchfly, Italian (Silene italica) MBF cp 13;
PFE 90.

Catchfly, Night-flowering or Night-scented
(Silene noctiflora) MBF cp 14; MCW 41; MFF
pl 85 (154); NHE 225; PFF 163; PWF 137b;
WFB 61 (3); WUS 165.

Catchfly, Nodding or Nottingham (Silene
nutans) EWF cp 13c; KVP cp 86; MBF cp 13;
MFF pl 85 (152); NHE 184; OWF 73; PFE cp 15
(167); PWF 79f; TEP 109; WFB 59 (2).

Catchfly, Red German or Sticky (Lychnis
viscaria or Viscaria vulgaris) DFP 157 (cp
1253); EWF cp 13e; MBF cp 14; MFF cp 14
(153); NHE 73; OWF 107; PFE 14 (161); PWF
48d; TVG 110; WFB 61 (6).

Catchfly, Rock (Silene rupestris) NHE 277;
PFE 90; WFB 59 (8).

Catchfly, Sand or Striated (Silene conica) MFF
cp 22 (148); NHE 73; PWF 72c; WFB 63 (2).

Catchfly, Small-flowered (Silene gallica) KVP
cp 182; MBF cp 13; NHE 226; PFE cp 16 (174);
PWF 111d; WFB 63 (1); (S. anglica) MFF pl 85
(149).

Catchfly, Spanish or Breckland (Silene otites)
MBF cp 13; MFF pl 78 (151); NHE 184; PFE cp
15 (168); PWF 162b; WFB 59 (4).

Catchfly, Sweet-William (Silene armeria) EGA
154; LFW 221 (cp 497); NHE 226; PFE 90; PWF
97i; TGF 60; WFB 59 (6).

Caterpillar (Various genera and species) ALE
2:324, 325; CIB 324-333 (cps 57-60); CSA 141;
GAL 340 (cp 10), 349 (pl 11); KBM 9, 50 (pl 14);
KIW 160 (bw), 161 (cps 62, 63), 172-173 (cps
82-84); LAW 214, 215; PEI 289 (cp 32a), 338,
340, 342 (bw); SBM 167, 178, 179, 249-254,
318-320; ZWI 128, 160, 161, 227.

Catfish (Auchenoglanis occidentalis) FEF 229
(bw); LAW 320; SFW 344 (cp 87); (Brochis
coeruleus) ALE 4:411; FEF 248 (bw).

Catfish (Leiocassis siamensis) ALE 4:393; FEF
230 (bw); (L. poecilopterus) SFW 344 (cp 87).

Catfish (Otocinclus sp.) ALE 4:412; FEF 250
(bw); SFW 380 (pl 95), 385 (pl 100).

Catfish (Pimelodus sp.) HFI 112; WFW cp 166; (Pimelodella sp.) FEF 233, 234 (bw); SFW 380 (pl 95), 423 (bw).

Catfish, American —See **Bullhead, Brown.**

Catfish, Armored (Corydoras sp.) CTF 33 (cps 32, 33); HFW 123 (bw).

Catfish, Armored (Callichthys callichthys) HFI 113; WFW cp 171; (C. lichthys) ALE 4:411; (Loricaria parva) ALE 4:412; HFI 111; WFW cp 169; (L. filamentosa) FEF 250 (bw); WFW cp 168.

Catfish, Bowline —See **Corydoras, Arched.**

Catfish, Eel (Channallabes apus) HFI 109; HFW 125 (bw).

Catfish, Electric (Malapterurus electricus) CAA 17 (bw); FEF 238 (bw); HFI 112; HFW 43 (cp 23), 131 (bw); SFW 344 (cp 87); WFW cp 158.

Catfish, European —See **Wels, European.**

Catfish, Glass or Ompok (Ompok bimaculatus) HFI 107; HFW 127 (bw); LAW 320; WFW cp 154.

Catfish, Indian Glass (Kryptopterus bicirrhis) ALE 4:388, 393; CTF 34 (cp 34); FEF 227 (bw); HFI 107; SFW 337 (cp 86); WFW cp 153.

Catfish, Marine (Plotosus anguillaris or P. lineatus) ALE 4:385; FEF 229 (bw); HFI 110; LAW 319; WFW cp 165.

Catfish, Pimelodid or Tiger (Pseudo-platystoma fasciatum) ALE 4:387; HFW 128 (bw

Catfish, Sea (Anarhichas lupus) ALE 5:162, 181; CSF 153; FEF 472 (bw); HFI 44; HFW 216 (cp 103), 241; MOL 106 (A. minor) CSF 153.

Catfish, Shovel-nosed (Sorubim lima) ALE 4:400; HFI 112; WFW cp 167.

Catfish, Stinging (Heteropneustes fossilis) ALE 4:393; FEF 232 (bw); HFI 109.

Catfish, Striped Dwarf (Mystus vittatus) ALE 4:393; HFI 107.

Catfish, Upside-down (Synodontis nigri-ventris) CTF 34 (cp 35); FEF 236 (bw); HFI 111; SFW 380 (pl 95); WEA 96; WFW cp 161; (S. angelicus) HFW 44 (cp 24).

Catfish, Upside-down (Synodontis sp.) ALE 4:399; FEF 237 (bw); HFI 111; HFW 130 (bw); LAW 322; SFW 345 (cp 88), 372 (pl 93); 373 (pl 94); WFW cps 159, 160.

Cathedral Bells —See **Cup and Saucer Vine.**

Cathedral Windows —See **Peacock Plant.**

Catmint (Nepeta cataria) EGH 127; EHP 20; MBF cp 68; MFF pl 112 (778); NHE 230; PFE cp 109 (1116); PFF 288; PWF 129h; WFB 199 (5).

Catmint (Nepeta mussinii) EGC 141; LFW 103 (cp 230); MWF 211; TGF 189; (N. longibracteata) EWF cp 105a.

Catmint, Garden (Nepeta x faassenii) DFP 160 (cp 1278); LFW 93 (cp 214); OGF 153; PWF 66d.

Catmint, Hairless (Nepeta nuda) PFE 352; WFB 199 (6).

Cat's Claw (Doxanthus unguis-cati) MEP 135; MWF 117.

Cat's Tail —See **Chenille Plant.**

Cat's-ear, Common (Hypochoeris radicata) BFB cp 21 (9); BWG 52; MBF cp 50; MCW 116; MFF cp 54 (727); NHE 190; OWF 37; PWF 75g; WFB 255 (4).

Cat's-ear, Giant (Hypochoeris uniflora) EPF pl 599; NHE 289; PFE cp 158 (1521).

Cat's-ear, Smooth (Hypochoeris glabra) MBF cp 50; MFF cp 55 (928); NHE 239.

Cat's-ear, Spotted (Hypochoeris maculata) MBF cp 50; MFF cp 25 (929); NHE 37; PWF 69j; WFB 255 (7).

Cat's-foot (Antennaria dioica) EPF pl 538; MBF cp 44; MFF pl 94 (873); NHE 76, 285; OWF 127; PFE cp 143 (1378); PWF 68d; TEP 259; WFB 235 (1).

Cat's-tail, Alpine (Phleum alpinum) MBF pl 95; NHE 292.

Cat's-tail, Purple-stem or Pointed (Phleum phleoides) MBF pl 95; MFF pl 122 (1237; (P. paniculatum) NHE 245.

Cat's-tail, Sand (Phleum arenarium) MBF pl 95; MFF pl 122 (1238); NHE 138; TEP 66.

Cattail —See **Reedmace** (Typha).

Cattle (Bos taurus) ALE 13:346, 351; DAE 68-70; HMW 95; LEA 604 (bw); OBV 151; PFF 727-729 (bw); WEA 98.

Cattle, Brahman (Bos indicus) ALE 13:345; HMA 102; JAW 185; LEA 605 (bw); SMW 278 (cp 166).

Cattle, Wild —See **Banteng** and **Gaur.**

Caudo (Phalloceros caudomaculatus) FEF 311 (bw); HFI 69; SFW 285 (cp 72), 477 (pl 128).

Cauliflower (Brassica oleracea botrytis) EVP 84, 87; PFF 186.

Cavally, Golden —See Jack, Yellow.

Cave-fish, Blind —See Characin, Blind.

Cavy —See Guinea-pig and Mara.

Cedar, Atlantic or Atlas (Cedrus atlantica) BKT cp 40; DEW 1:36, 54 (cp 12); DEW 2:213 (cp 123); DFP 251 (cp 2003); EEG 124; EET 104; EGE 91; EGW 99; ELG 121; EPF pls 134, 135, 137; MTB 93 (cp 8); MWF 77; OBT 120; SST cp 10; TGS 24; TVG 251.

Cedar, Deodar (Cedrus deodara) EEG 124; EET 105; EGE 44, 45, 91; EGW 99; MTB 93 (cp 8); MWF 77; OBT 120; SST cp 11; TGS 23.

Cedar, Japanese (Cryptomeria japonica) BKT cp 16; DEW 1:55 (cp 16); DFP 252 (cp 2011); EET 96, 97; EGE 94; EGW 106; EJG 111; EMB 112; ERW 109; GDS cps 129, 130; MTB 84 (cp 5); OBT 101, 124; SST cp 16; TGS 38.

Cedar of Lebanon (Cedrus libani) BKT cp 39, 41; EET 104, 105; EGE 92; MTB 93 (cp 8); OBT 100, 120; PWF 171m; SST cp 12; TGS 24; (C. libanotica) MLP 37.

Cedar, Port Orford —See Cypress, Lawson.

Cedar, White (Chamaecyparis thyoides) GDS cp 74; NFP 151; PFF 125.

Celandine (Chelidonium majus) AMP 177; BFB cp 4 (9); BWG 20; CTH 28 (cp 25); DEW 1:138; EFC cp 19; EPF pl 230; MBF cp 5; MFF cp 38 (42); NHE 214; OWF 7; PFF 180; PWF 43f; TEP 189; WFB 81 (9).

Celandine, Lesser or Pilewort (Ranunculus ficaria) AMP 173; BFB cp 2 (8); BWG 18, 20; DFP 23 (cp 177); EFC cp 48; EPF pl 228; HPW 48; KVP cp 56; MBF cp 3; MFF cp 38 (22); NHE 178; OGF 5; OWF 5; PFE cp 25 (233); PFM cp 29; PWF cp 17e; TEP 235; WFB 69 (8).

Celeriac (Apium graveolens var. rapaceum) EVP 88; OFP 175.

Celery (cultivated) (Apium graveolens var. dulce) EVP 88; OFP 149; PFF 267.

Celery, Wild (Apium graveolens) MBF 37; MFF pl 98 (474); NHE 135; OWF 147; WFB 163 (3); (A. australe) SNZ 20 (cp 12).

Centaury, Common (Centaurium erythraea) AMP 19; CTH 42 (cp 56); EPF pl 687; MBF cp 58; MFF cp 22 (635); NHE 185; PFE cp 95 (986); PWF 116a; TEP 251; WFB 181 (1); (C. minus) BFB cp 13; OWF 125.

Centaury, Guernsey (Exaculum pusillum) MBF 58; MFF cp 22 (633); WFB 181 (6).

Centaury, Lesser (Centaurium pulchellum) MBF 58; MFF cp 22 (634); NHE 185; WFB 181 (2).

Centaury, Seaside (Centaurium littorale) MBF cp 58; PWF 128b; WFB 180 (1a); (C. maritimum) KVP cp 172; (C. vulgare) NHE 136.

Centipede, Brown (Lithobius sp.) ALE 1:510; 2:51; GAL 477 (bw); GSP 144; OIB 153, 155; PEI 21 (bw); PWI 11 (bw).

Centipede, Garden (Scutigerella sp.) EPD 111; GSP 141; OIB 149.

Centipede, Giant (Scolopendra sp.) ALE 1:510; CSA 144 (bw); GSP 145; LAW 160, 161; PEI 48 (cp 5a).

Centipede, House (Scutigera coleoptrata) ALE 1:496; GSP 143; OIB 149.

Centipede, Luminous (Geophilus sp.) ALE 1:510; 2:51; GAL 477 (bw); PEI 22 (bw).

Centradenia (Centradenia sp.) BIP 51 (cp 60); DEW 2:51 cp 6; PFW 188.

Century Plant, American (Agave americana) DEW 2:217 (cp 129); ECS 94; LCS cps 200-201; MEP 20; MWF 33; PFE cp 171 (1660); PFM cp 255; PFW 15; RDP 75; SPL cp 469.

Century Plant, Queen Victoria (Agave victoria-reginae) EDP 137; EFP 86; EGL 83; LCS cp 205; RDP 14, 75; SCS cp 70.

Century Plant, Thread-bearing (Agave filifera) EPF pl 935; LCS cp 199.

Cep —See Boletus, Edible.

Ceratium (Ceratium sp.) NFP 21; PFF 60.

Ceratozamia (Ceratozamia mexicana) DEW 1:50 (cp 2); EPF pl 174.

Ceriman —See Swiss Cheese Plant (Monstera).

Cerith —See Hydroid (Clava sp.)

Ceropegia (Ceropegia sp.) DEW 2:106 (cp 61), 127; ECS 13; EPF pl 693; EWF cp 60b; LCS cp 217; LFT 255 (cp 127), 261; RWF 58, 59; SCS 72.

Cestrum (Cestrum sp.) DFP 56 (cp 445); MWF 79; PFW 283.

Cestrum, Orange (Cestrum aurantiacum) BIP 51 (cp 62); DFP 56 (cp 444); MWF 79.

Chaca (Chaca chaca) ALE 4:393; WFW cp 157.

Chaffinch (Fringilla coelebs) ALE 9:392; 13:163; BBB 61; BBE 279; CDB 209 (cp 922); FBE 285; FNZ 208 (cp 18); GAL 118 (cp 1); GPB 315; OBB 185; PBE 261 (cp 62); RBE 315.

Chaffweed (Anagallis minima) MBF cp 58; NHE 228; WFB 177 (7); (Centunculus minimus) MFF pl 105 (625); OWF 125.

Chain of Love —See **Coral Vine.**

Chalice Vine (Solandra guttata) MLP 214; MWF 267; (S. sp.) EGV 136; MEP 130; PFW 282.

Chalk Plant or **Chalkwort** —See **Baby's-breath.**

Chameleon (Chamaeleo sp.) ALE 6:212, 226, 238, 243; CAF 186; HRA 40; HWR 31; LEA 289; WEA 99.

Chameleon, Common (Chamaeleo chamaeleo) ALE 6:243; ART 161, 199; CAA 116 (bw); HRA 39; LAW 416; LEA 312 (bw); SIR 90 (bw).

Chameleon, Dwarf (Chamaeleo pumila) ALE 6:243; CAA 106 (bw); SIR 92 (bw).

Chameleon, Elliot's Dwarf (Chamaeleo bitaeniatus ellioti) ALE 6:225; ERA 162, 163; LBR 48, 49.

Chameleon, Fischer's (Chamaeleo fischeri) ALE 6:238; CAA 116 (bw); LEA 313 (bw); SIR 94 (bw); VRA 175.

Chameleon, Flap-necked (Chamaeleo dilepis) ALE 6:243; ART 100; CAA 108; HWR 32; LAW 417; LEA 311 (bw); MAR 114 (pl 46); SIR 147 (cp 63); VRA 173.

Chameleon, Giant One-horned or Meller's (Chamaeleo melleri) ALE 6:238; SIR 146 (cp 62); VRA 179.

Chameleon, Jackson's or Three-horned (Chamaeleo jacksoni) ALE 6:238; ART 116; CAA 117 (bw); HWR 49; LAW 395; LBR 55 (cps 84, 85); MAR 119 (cp VIII); SIR 146 (cp 61); VRA 177.

Chameleon, Leaf (Brookesia spectrum) ALE 6:226, 243; MAR 119 (cp VIII).

Chamois (Rupicapra rupicapra) ALE 13:463; CEU 56 (bw); DAE 52; DSM 269; HMA 38; HMW 80; JAW 202; LAW 517; MLS 424 (bw); TWA 77.

Chamomile (Anthemis sp.) DFP 2 (cp 12), 121 (cp 967); EWF cp 40c; NHE 236; PFM cp 203.

Chamomile, Corn or Field (Anthemis arvensis) KVP cp 36; MBF cp 45; MFF pl 100 (884); NHE 236; OWF 99; PWF 101e; TEP 181.

Chamomile, German (Matricaria chamomilla) AMP 147; EGH 123; PFF 325; SPL cp 396; TGF 252.

Chamomile, Roman or English (Anthemis nobilis) AMP 157; EGC 121; EGH 99; TVG 139.

Chamomile, Scentless — See **Mayweed, Scentless.**

Chamomile, Stinking or Mayweed (Anthemis cotula) MBF cp 45; MCW 98; NHE 236; PFF 326; WUS 371.

Chamomile, Sweet (Chamaemelum nobile) MBF cp 45; MFF pl 100 (885); NHE 236; PFE cp 147 (1413); PWF 120b.

Chamomile, Wild (Matricaria recutita) CTH 58 (cp 93); EPF pl 554; MBF cp 46; NHE 236; PFF 325; PWF 37g; TEP 181.

Chamomile, Yellow (Anthemis tinctoria) AMP 157; DFP 121 (cp 968); 122 (cps 969, 970); EFC cp 36; EGP 96; LFW 13 (cp 27); MBF cp 45; MFF cp 56 (883); MLP 237; MWF 42; NHE 236; PFE cp 147 (1409); PWF 113i; WFB 237 (8).

Chanchita —See **Cichlid, Chameleon.**

Chandelier Plant —See **Friendly Neighbor.**

Chanterelle (Chantarellus cinereus) DLM 144; LHM 59; PUM cp 150; RTM 223; (C. infundibuliformis) DLM 144; LHM 59; ONP 133; SMG 132.

Chanterelle, False (Hygrophoropsis aurantiaca) DLM 197; LHM 185; ONP 103; REM 125 (cp 93); RTM 118.

Chanterelle, Scaly (Cantharellus floccosus) RTM 224; SMG 126.

Chanterelle, Yellow (Cantharellus cibarius) CTM 51; DEW 3:160; DLM 144; EPF pl 66; KMF 95; LHM 59; NFP 76; NHE 45; OFP 189; ONP 123; PFF 84; PMU cp 78; REM 123 (cp 91); RTM 224; SMF 9; SMG 130, 131; TEP 219.

Char, Alpine (Salvelinus alpinus) FBW 37; FWF 41; HFI 120; OBV 111.

Char, Brook —See **Trout, Brook.**

Characin (Hemiodus sp.) ALE 4:291; FEF 122 (bw).

Characin, Ansorge's (Neolebias ansorgei) ALE 4:308; FEF 136 (bw); HFI 90; SFW 176 (cp 39).

Characin, Blood —See **Tetra, Jewel.**

Characin, Darter (Characidium fasciatum) FEF 122 (bw); SFW 161 (pl 36).

Characin, Long-finned (Alestes longipinnis) ALE 4:308; FEF 112 (bw); SFW 156 (pl 31); WFW cp 49.

Characin, One-striped African (Nannaethlops unitaeniatus) ALE 4:308; FEF 135 (bw); SFW 201 (pl 46).

Characin, Pike (Boulengerella sp.) ALE 4:307; HFI 86; HFW 37 (cp 10); (Hepsetus odoe) FEF 113 (bw); SFW 160 (pl 34).

Characin, Red-eyed (Arnoldichthys spilopterus) FEF 153; SFW 112 (pl 19); WFW cp 51.

Characin, Spraying (Copeina arnoldi or Copella arnoldi) ALE 4:291; FEF 108 (bw); HFI 87; SFW 160 (pl 33), 164 (cp 37).

Characin, Sword-tail (Corynopoma riisei) ALE 4:286; FEF 102 (bw); SFW 156 (pl 31).

Characin, Three-striped African (Nannaethiops tritaeniatus) ALE 4:308; HFI 90; SFW 201 (pl 46).

Characin, Yellow Congo (Alestopetersius caudalis) ALE 4:308; FEF 113 (bw).

Chard, Swiss (Beta vulgaris cicla) EVF 89; OFP 161.

Charlock (Sinapis arvensis) BFB cp 4 (2); BWG 21; MCW 56; MFF cp 45 (51); NHE 215; OWF 9; PWF cp 23g; TEP 173; WFB 89 (4).

Charlock, White —See **Radish, Wild.**

Charsa —See **Marten, Yellow-throated.**

Chaste Tree (Vitex agnus-castus) AMP 119; EFS 27, 145; ELG 120; EPG 152; HPW 237; LFW 104 (cps 232, 234); NHE 109; PFE cp 106 (1087); TGS 327; (V. sp.) DEW 2:186; SNZ 77 (cps 216, 217).

Chat, Bush —See **Bushchat.**

Chat, Crimson (Ephthianura tricolor) HAB 251; MBA 30, 31; RBU 224; SAB 48 (cp 23).

Chat, Familiar (Cercomela familiaris) CDB 162 (cp 676); PSA 256 (cp 29).

Chat, Gibber (Ashbyia lovensis) SAB 48 (cp 23); WAB 197.

Chat, Mocking (Thamnolaea cinnamomeiventris) CDB 167 (cp 702); PSA 256 (cp 29).

Chat, Mountain (Oenanthe monticola) CDB 165 (cp 696); PSA 256 (cp 29).

Chat, Orange (Ephthianura aurifrons) CDB 177 (cp 760); HAB 251; MBA 27, 30; MEA 48; SAB 48 (cp 23).

Chat, Pied (Oenanthe pileata) ALE 9:280; PSA 256 (cp 29); (O. picata) RBI 211.

Chat, Robin —See **Robin, Cape.**

Chat, Rose-breasted (Granatellus pelzelni) ABW 285; ALE 9:371; SSA cp 14.

Chat, Stone —See **Stonechat.**

Chat, White-faced (Ephthianura albifrons) HAB 251; MWA 110; SAB 48 (cp 23).

Chat, Yellow (Ephthianura crocea) HAB 251; SAB 48 (cp 23).

Chayote (Sechium edule) HPW 116; OFP 123.

Cheat —See **Brome, Rye or Downy.**

Cheetah (Acinonyx jubatus) ALE 12:340, 355; BMA 51, 52; CAA 96; CAF 88, 143 (bw); DAA 69; DDM 140 (cp 23); DSM 195; HMA 36, 38; HMW 128; JAW 141; LAW 545; LEA 574-575 (bw); LVA 23; LVS 91; MLS 323 (bw); PWC 240 (bw); SLP 220; SMR 80 (cp 5); SMW 183 (cp 109); TWA 44; VWA cp 4.

Chenille Plant (Acalypha hispida) BIP 17 (cp 3); DFP 51 (cp 401); EPF pl 619; FHP 102; GFI 39 (cp 6); MEP 71; MWF 28; PFW 117; RDP 67; SPL cp 35; (A. sp.) HPW 186-3.

Chenille Weed (Dasya pedicellata) NFP 17; PFF 64.

Chenopodium (Chenopodium sp.) AMP 55; DEW 1:161 (cp 73); DFP 33 (cp 263); MBF cp 72; NHE 227.

Cherimoya —See **Custard Apple.**

Cherry, Bladder or Winter or Ground (Physalis alkekengi) AMP 133; DEW 2:150; EDP 117; HPW 228; LFW 122 (cp 247); NHE 232; PFE cp 119 (1178); SPL cp 241; TGF 205; TVG 117.

Cherry, Brush or Scrub (Eugenia sp.) EPF pl 705; EWF cp 147d; MWF 129; SNZ 63 (cps 163, 164.

Cherry, Christmas or Winter (Solanum capsicastrum) BIP 167 (cp 254); LHP 170, 171; RDP 23, 365.

Cherry, Conradine's (Prunus conradine) DFP 220 (cp 1755); EWF cp 92d; OBT 176.

Cherry, Flowering (Prunus serrulata) BKT cp 116; DEW 1:277; DFP 221 (cp 1763, 1764); EET 178, 179; EGT 136, 137; EJG 139; ELG 109; EPG 139; LFW 8 (cp 187), 112 (cp 249); MWF 245; PFF 206; PFW 263; TGS 198.

Cherry, Manchurian (Prunus maackii) EET 177; MTB 304 (cp 27).

Cherry Pie —See **Heliotrope.**

Cherry, St. Lucie's or Mahaleb (Prunus mahaleb) BKT cp 118; EFC cp 58; NHE 21; PFE cp 49 (484).

Cherry, Sour (Prunus cerasus) AMP 41; EVF 117; MBF cp 26; NHE 21; SST cp 217.

Cherry, Sweet or Wild (Prunus avium) BFB cp 9 (2); BKT cp 115; DFP 220 (cp 1753); EET 176; EPF pls 419, 420; EVF 117; MBF cp 26; MFF pl 75 (376); NHE 21; OBT 20, 24, 176; OWF 195; PFE cp 49 (480); PFF 207; PWF cp 25k; SPL cp 317; SST cp 216; TEP 128; WFB 115 (6).

Cherry, Yoshino or Tokyo (Prunus yedoensis) EGT 138; EJG 140; EWF cp 92b; OBT 177.

Cherry-plum (Prunus cerasifera) DFP 220 (cp 1754); EET 173; MFF pl 75 (375); MTB 304 (cp 27); OBT 180; OFP 69; OWF 181; PFE cp 48 (477); PWF cp 16d; TGS 198.

Chervil (Chaerophyllum temulentum) MBF cp 38; MFF pl 99 (456); NHE 32, 222; OWF 91; PWF 65i; WFB 157 (2); (Anthricus cerefolium) EGH 99; EVF 141; OFP 147.

Chervil, Bur (Anthricus caucalis); MBF cp 38; MFF pl 99 (457); NHE 222; WFB 157 (3).

Chervil, Giant—See **Sweet Cicely** (Myrrhis).

Chess—See **Brome, Rye or Downy.**

Chestnut, Cape (Calodendron capense) EWF cp 77e; LFT cp 90; MWF 68.

Chestnut, Chinese (Castanea mollissima) EDP 138; EGT 42, 103; EPG 112; EVF 135; TGS 105.

Chestnut, Lowveld (Sterculia murex) EWF cp 57d; LFT cp 108; (S. sp.) DEW 1: 219 (cp 130); HPW 92; MEP 82.

Chestnut, Moreton Bay (Castanospermum australe) MEP 55; MWF 74; OFP 31.

Chestnut, Spanish or Sweet (Castanea sativa) AMP 123; DEW 1:107 (cp 60), 185; EET 148, 149; EFP pl 335, 336; HPW 61; MBF cp 77; MTB 208 (cp 19); NHE 294; OBT 21, 25; OFP 27; PFE cp 5 (46); PWF 109e, 175g; SST cp 189; TGS 87; WFB 33 (4).

Chestnut Vine (Tetrastigma voinierianum) BIP 175 (cp 269); RDP 381; (T. obtectum) HPW 189.

Chevelure—See **Fern, Southern Maidenhair.**

Chevrotain, Indian (Tragulus meminna) ALE 13:134; CAS 181 (bw); DAA 89; DSM 226; LEA 596 (bw).

Chevrotain, Malayan (Tragulus javanicus) ALE 13:134; DAA 108; HMW 58; MLS 382 (bw).

Chevrotain, Water (Hyemoschus aquaticus) ALE 13:134; DDM 257 (cp 42); DSM 226; HMA 43.

Chickweed—See also **Mouse-ear** (Cerastium).

Chickweed, Alpine Mouse-ear (Cerastium alpinum) MBF cp 14; MFF pl 85 (166); NHE 277; PFE cp 12 (142).

Chickweed, Arctic Mouse-ear (Cerastium arcticum) EPF pl 301; MBF cp 14; PWF 154c.

Chickweed, Common (Stellaria media) BWG 67; EGC 70; MBF cp 15; MCW 43; MFF pl 86 (169); NHE 225; OWF 75; PFF 161; PWF cp 14a; WFB 55 (2).

Chickweed, Common Mouse-ear (Cerastium fontanum) MBF cp 14; PFE cp 12 (144); PWF 91g; WFB 55 (6).

Chickweed, Field Mouse-ear (Cerastium arvense) EPF pl 300; KVP cp 96; MBF cp 14; MFF pl 85 (165); NHE 184; PWF 52a; TEP 107; WFB 55 (5); WUS 157.

Chickweed, Indian—See **Carpetweed.**

Chickweed, Jagged or Umbellate (Holosteum umbellatum) MBF cp 14; NHE 184; PFE 81; WFB 55 (4).

Chickweed, Mouse-ear (Cerastium vulgatum) EGC 70; EWG 77; MCW 37; OWF 75; PFF 162; WUS 159.

Chickweed, Sticky (Cerastium glomeratum) MBF cp 14; NHE 226; OWF 75; WFB 55 (7).

Chickweed, Upright (Moenchia erecta) MBF cp 14; MFF pl 86 (174); NHE 225; OWF 75; WFB 55 (8).

Chickweed, Water or Great (Myosoton aquaticum) MBF cp 15; MFF pl 86 (168); NHE 97; OWF 73; PFE cp 12 (145); PWF 78a; WFB 55

Chicory (Cichorium intybus) AMP 37; BFB cp 21; CTH 64 (cp 104); EFC cp 38-2; EGH 108; EPF pl 597; EWF cp 22d; HPW 264; KVP cp 88; MBF cp 50; MCW 106; MFF cp 9 (924); NHE 239; OFP 111, 151; OWF 179; PFE cp 158 (1512); PFF 332; PWF 114a; TEP 201; WFB 251 (5); WUS 395.

Chiffchaff (Phylloscopus collybita) ALE 9:259; BBB 142; BBE 243; CDB 174 (cp 743); FBE 237; GAL 128 (cp 3); OBB 155; PBE 244 (cp 57).

Chimney Sweep—See **Ibis, Sacred or White.**

Chimpanzee (Pan troglodytes or P. satyrus) ALE 11:25-35; BMA 53-55; CAA 90 (bw), 100; CAF 135; DDM 80 (cp 11); DSM 97; HMA 39; HMW 13; JAW 62; LAW 623-625; LEA 513; LVS 44-45; MLS 161 (bw); SMW 88 (cp 51); TWA 14 (bw); WEA 100, 101.

Chimpanzee, Pygmy (Pan paniscus) ALE 11:30 (cp 6), 34 (cp 10); LEA 513; PWC 234 (bw).

Chinaberry (Melia azedarach); DEW 2:46; EGT 44, 127; HPW 201; MEP 66; MWF 199; PFM cps 86, 87; SST cp 256; TGS 294.

Chincherinchee (Ornithogalum thyrosoides) DFP 109 (cp 870); LFW 24 (cp 49); MWF 218; SPL cp 233.

Chinchilla (Chinchilla laniger) ALE 11:438; DSM 132; HMA 39; JAW 85; LEA 525 (bw); LVA 31; TWA 139; VWA cp 40; (C. chinchilla) HMW 176; SMW 153 (bw); (C. brevicaudata) MLS 232 (bw).

Chinchilla, Mountain (Lagidium peruanum) HMW 176; SMW 153 (bw); (L. viscacia) ALE 11:438.

Chinese Date—See **Jujube** (Zizyphus).

Chinese Evergreen (Aglaonema pseudobracteata) LHP 20; MWF 34.

Chinese Evergreen, Spotted (Aglaonema commutatum) DFP 51 (cp 403); EFP 87; EGL 83.

Chinese Hat Plant (Holmskioldia sanguinea) GFI 43 (cp 7); LFW 168 (cp 378); MEP 122; MWF 156.

Chinese Lantern (Physalis franchetti) DFP 165 (cp 1318); EPF pls 822, 823; MWF 232; PFW 284; (P. pruinosa) OFP 127; (Abutilon milleri) MWF 26; TVG 217.

Chinese Paper Plant—See **Rice Paper Plant** (Tetrapanax).

Chinese Parasol Tree (Firmiana simplex) EGT 114; EJG 116; (F. colorata) EWF cp 116b.

Chinese Scholar Tree—See **Pagoda Tree**.

Chirita (Chirita sp.) EGL 94; EWF cp 122f; FHP 111.

Chironia (Chironia sp.) HPW 223; LFT 246 (cp 122).

Chiru—See **Antelope, Tibetan**.

Chital—See **Deer, Axis**.

Chiton (Acanthochitona sp.) ALE 3:36; CSS pl XIV; LAW 116; OIB 33; (Lepidochitona sp.) ALE 3:36; CSS pl XIV; OIB 33.

Chives (Allium schoenoprasum) DEW 2:234; DFP 84 (cp 668); EGH 95; EPF pl 890; EVF 141; LFW 8 (cp 14); MBF cp 85; MFF cp 32 (1014); NHE 191; OFP 167; PFE cp 165 (1603); PWF 80e; SPL cp 252; WFB 265 (4).

Chlorophonia (Chlorophonia sp.) ABW 293; ALE 9:361; SSA cp 40.

Chlorosplenium (Chlorosplenium sp.) DLM 82, 119; LHM 43; SMF 56.

Chocolate-creams (Trichodesma physaloides) EWF cp 63c; LFT cp 139.

Chokeberry, Red (Aronia arbutifolia) EFS 33, 96; EGW 90; EPG 105; ESG 93; TGS 246.

Choliba—See **Owl, Tropical Screech**.

Chough (Pyrrhocorax pyrrhocorax) ABW 223; ALE 9:469; BBB 263; BBE 217; FBE 307; FBW 31; OBB 127; PBE 197 (cp 50); RBE 236; WAB 135.

Chough, Alpine (Pyrrhocorax graculus) ALE 9:469; 13:463; BBE 217; CDB 224 (cp 1008); FBE 307; PBE 197 (cp 50).

Chough, Ground—See **Jay, Ground**.

Chough, White-winged (Corcorax melanorhamphus) ABW 228; ALE 9:469; HAB 153; RBU 292; SAB 18 (cp 8).

Chowchilla—See **Log-runner**.

Chukar (Alectoris graeca) BBE 99; FBE 103; PBE 157 (cp 42); (A. chukar) BBE 99; FBE 103.

Christmas Bell, New Zealand (Alstroemeria pulchella) MWF 38; PFW 21.

Christmas Bells (Blandfordia sp.) EWF cp 143a; MWF 57; WAW 119.

Christmas Bush (Ceratopetalum gummiferum) EWF cp 133b; MEP 45; MWF 78.

Christmas Cheer (Sedum rubrotinctum) ECS 143; EGC 38; MWF 266; RDP 356.

Christmas Rose (Helleborus niger) AMP 53; DEW 1:117; DFP 146 (cps 1164, 1165); EGW 115; EPF pl 201; ESG 116; HPW 47; MLP 54; NHE 268; OGF 5; PFE cp 19 (201); PFW 250; RWF 19; TGF 61; TVG 97.

Christmas Tree (Nuytsia floribunda) EWF cp 135c; HPW 174; MLP 73.

Christmas Tree, New Zealand—See **Rata** (Metrosideros).

Christ's-thorn (Paliurus spina-christi) DEW 2:52 (cp 9), 77; PFM 35 (pl 9); (P. virgatus) HPW 188.

Chromide, Green (Etroplus suratensis) FEF 437 (bw); SFW 581 (pl 158).

Chromide, Orange (Etroplus maculatus) FEF 437 (bw); HFI 36; SFW 581 (pl 158).

Chrysanthemum (Chrysanthemum sp.) AMP
179; BIP 53 (cps 65-67); DFP 56-59, 129-131;
EGH 107; EGP 72-77; EWF cp 39d; MWF 84,
85; NHE 76, 287; OFP 145; OGF 181; PFE cp
148 (1424); PFW 78, 79; SPL cps 422-424; TEP
257; TGF 252; TVG 83, 147.

Chrysanthemum, Alpine (Chrysanthemum
alpinum) EPF pl 557; NHE 287.

Chrysanthemum, Annual (Chrysanthemum
carinatum) DFP 33 (cp 264); EGA 106; OGF 87;
TVG 41.

Chrysanthemum, Florist's (Chrysanthemum
morifolium) EDP 131; EGG 96; EJG 108, 109;
LFW 28, 30-33; PFW 327; RDP 141.

Chub (Leuciscus cephalus or Squalius cephalus)
ALE 4:341; FWF 61, 63; GAL 243 (bw); HFI 95;
LAW 317; OBV 99; SFW 220 (pl 51).

Chufa —See **Nutsedge.**

Chupachupa —See **Glory Flower** (Eccremocarpus).

Church Steeples —See **Agrimony, Common.**

Cicada (Cicada sp.) PEI 116, 124 (bw);
(Cicadatra atra) PWI 58 (bw); (Melampsalta
sp.) BIN pls 65, 66, cp 67; (Munsa clypealis)
LAW 242-243.

Cicada, Forest or Mountain (Cicadetta
montana) ALE 2:180-181; CIB 129 (cp 12); OBI
33; PEI 116 (bw); ZWI 184.

Cicada-bird (Edoliisoma tenuirostre) HAB 147;
(Coracina tenuirostris) SAB 12 (cp 5).

Cichlid (Aequidens sp.) ALE 5:128; CTF 52 (cp
67); FEF 401-407 (bw); HFI 35; SFW 548 (pl
147), 553 (cp 150), 589 (pl 160); WFW cp 354.

Cichlid (Crenicara sp.) ALE 5:128; SFW 580 (pl
157); (Geophagus sp.) ALE 5:127, 132; SFW
581 (pl 158); WFW cps 360, 361; (Lamprologus
sp.) FEF 457; WFW cp 367; (Labeotropheus sp.)
CTF 52 (cp 66); FEF 445, 456, 463.

Cichlid (Pelmatochromis sp.) ALE 5:128; CTF
57 (cp 75); FEF 428, 447-449, 451, 452; HFI 36;
SFW 553 (cp 150); WFW cps 370-372.

Cichlid, African (Nannochromis sp.) FEF 427,
444 (bw); SFW 552 (cp 149).

Cichlid, Banded (Cichlasoma severum) FEF
424 (bw); SFW 601 (cp 166); WFW cp 358.

Cichlid, Barred or Flag (Cichlasoma festivum)
ALE 5:128; CTF 55 (cp 72); FEF 418, 419 (bw);
HFI 35; SFW 573 (pl 156); WFW cp 357.

Chichlid, Chameleon (Cichlasoma facetum)
ALE 5:127; SFW 564 (pl 153).

Chichlid, Dwarf (Apistogramma sp.) ALE
5:128; FEF 408 (bw), 411-415 (bw); HFI 36;
SFW 549 (pl 148), 552 (cp 149); WFW cp 355.

Cichlid, Firemouth (Cichlasoma meeki) ALE
5:127, 132; CTF 55 (cp 71); FEF 420, 421 (bw);
HFI 35; SFW 501 (cp 136).

Cichlid, Golden-eyed Dwarf (Nannacara
anomala) ALE 5:128; FEF 426, 429 (bw); SFW
501 (cp 136).

Cichlid, Hellabrunn (Cichlasoma hellabrunni)
ALE 5:127; SFW 560 (cp 151).

Cichlid, Jewel —See **Jewelfish.**

Cichlid, Oscar's or Velvet (Astronotus ocellatus) CTF 56 (cp 73); FEF 416 (bw); SFW 560 (cp
151).

Cichlid, Perch (Cichla ocellaris) ALE 5:128;
HFI 36.

Cichlid, Pike (Crenicichla sp.) ALE 5:127, 132;
SFW 580 (pl 157).

Cichlid, Turquoise-gold (Pseudotropheus
auratus) ALE 5:128; CTF 51 (cp 65); FEF 445;
WFW cps 375, 376.

Cigar Plant or **Cigar Flower** (Cuphea sp.) DFP
35 (cp 279); EGA 112; EGL 100; EWF cp 156a;
FHP 116; HPW 157; MEP 93; MWF 98; PFF
261; RDP 164.

Cineraria (Senecio cruentus) EDP 123; EGA
153; FHP 144; LFW 133 (cps 301, 302); MWF
266; PFW 88; RDP 142; SPL cp 56; TGF 268.

Cinnamon (Cinnamomum zeylandicum) AMP
21; CTH 26 (cp 21); EET 83; EPH pl 365; OFP
131; SST cp 274; (C. burmani) DEW 1:89; (C.
litseifolium) HPW 37.

Cinnamon Vine —See **Yam.**

Cinquefoil (Potentilla sp.) DFP 19 (cp 146), 166
(cp 1328), 167 (cp 1329); EPF pl 406; ERW 130;
EWF cp 2a, b; HPW 142; MWF 240; NHE 71,
211, 272; OGF 107; PFE cp 46 (445); SPL cp
315; TGF 108.

Cinquefoil, Alpine (Potentilla crantzii) MBF cp
27; NHE 272; PFE 161.

Cinquefoil, Bush (Potentilla arbuscula) EFS
24, 78, 127; EWF cp 92; GDS cps 326, 327.

Cinquefoil, Creeping (Potentilla reptans) AMP
63; BWG 33; MBF cp 27; MFF cp 42 (357); NHE
211; OWF 17; PWF 104d.

Cinquefoil, Hoary or Silvery (Potentilla argentea) MBF cp 27; MCW 59; MFF cp 42 (353); NHE 211; PWF 143f; WFB 113 (4).

Cinquefoil, Least (Sibbaldia procumbens) MBF cp 27; MFF cp 42 (358); NHE 271; PFE 161; WFB 113 (8).

Cinquefoil, Marsh (Potentilla palustris) EWF cp 9c; KVP cp 12; MBF cp 27; MFF cp 25 (349); NHE 92; OWF 117; PFE cp 46 (446); PWF 107h; TEP 77; WFB 111 (4).

Cinquefoil, Norwegian (Potentilla norvegica) EFC cp 56; MBF cp 27; MCW 60; NHE 92; PFE 61.

Cinquefoil, Procumbent (Potentilla anglica) MBF cp 27; NHE 92.

Cinquefoil, Rock or White (Potentilla rupestris) MBF cp 27; MFF pl 80 (351); NHE 27; PFE cp 46 (448); PWF 73e; WFB 111 (3).

Cinquefoil, Shrubby (Potentilla fruticosa) BFB cp 9 (10); DFP 219 (cps 1745-1750); EPF pl 404; GDS cps 328-331; HSC 94, 97; MBF cp 27; MFF cp 42 (348); OGF 161; OWF 17; PFE cp 46 (457); PFW 262; PWF 61k; SPL cp 316; TGS 294; TVG 235; WFB 113 (1).

Cinquefoil, Spring (Potentilla tabernae-montani) MBF cp 27; MFF cp 42 (355); NHE 211; PFE cp 46 (451); PWF 42c; TEP 109; WFB 113 (3).

Cinquefoil, Sulphur or Upright (Potentilla recta) DFP 167 (cp 1330); MCW 61; MFF cp 42 (354); NHE 71; PWF 143g; WFB 113 (4); WUS 221.

Cinquefoil, White (Potentilla alba) NHE 71; PFE 161; TEP 241.

Cirsium (Cirsium rivulare atropurpureum) DFP 132 (cp 1052); OGF 115.

Cissus (Cissus sp.) EGV 82; EWF cp 58d; HPW 189; LFT 207 (cp 103); RDP 143.

Cisticola, Bright-capped or Golden-headed (Cisticola exilis) CDB 173 (cp 735); DPB 315; HAB 182; KBA 353 (cp 54); LAW 461; RBU 223; SAB 26 (cp 12).

Cisticola, Fan-tailed —See **Warbler, Fan-tailed.**

Cistus —See also **Rockrose.**

Cistus (Cistus sp.) DEW 1:166 (cp 83); EPF pl 495; GDS cp 7; HPW 108; HSC 27, 29; MWF 85; OGF 65; PFM cps 100, 101, 106, 107; PFW 75; TGS 150.

Cistus, Gray-leaved (Cistus albidus) EWF cp 34c; PFE cp 77 (788); PFM cp 99.

Cistus, Gum (Cistus ladaniferus) DFP 190 (cp 1513); EWF cp 34b; HPW 108; PFE cp 77 (793); PFM cps 104, 105.

Cistus, Laurel-leaved (Cistus laurifolius) DFP 190 (cp 1514); PFE cp 77 (794).

Cistus, Narrow-leaved (Cistus monspeliensis) NHE 81; PFE cp 77 (791); PFM cp 103.

Cistus, Sage-leaved (Cistus salvifolius) DEW 1:242; PFE cp 77 (790); PFM cp 102; SPL cp 478.

Citron (Citrus medica) DEW 2:41; MLP 122; OFP 89.

Civet, African (Civettictis civetta) DSM 179; HMA 40; JAW 122; TWA 50; (Viverra civetta) ALE 12:138; DDM 85 (cp 14); MLS 299 (bw); SMR 48 (cp 3).

Civet, Aquatic or Water (Osbornictis piscivora) ALE 12:138; DDM 92 (cp 15).

Civet, Large Indian (Viverra zibetha) ALE 12:138; DAA 76; HMW 131.

Civet, Malagasy (Fossa fossa) ALE 12:158; LVS 80.

Civet, Otter (Cynogale bennetti) ALE 12:158; CAS 262 (bw); DAA 122; LEA 565 (bw); LVS 81; MLS 302 (bw).

Civet, Palm or Tree —See **Palm-civet.**

Civet, Small Indian (Viverricula indica) ALE 12:147; DAA 76; DSM 179.

Cladanthus (Cladanthus arabicus) EGA 107; PFM cp 201.

Cladonia (Cladonia sp.) DEW 3:221, cp 80; EPF pl 80; NFP 95, 99; NHE 44; ONP 29, 45, 51, 69, 75, 85, 175; PFF 74.

Clam, File (Lima sp.) ALE 3:141; LAW 118; OIB 85.

Clam, Giant (Tridaena sp.) ALE 3:161, 167; CAU 135 (bw); CTS 60 (cp 115), 61 (cp 116); GSS 140; LEA 92 (bw).

Clam, Giant or Horse's Hoof (Hippopus sp.) ALE 3:167; CTS 62 (cp 119); VWS 93.

Clam, Nut (Nucula sp.) ALE 3:139; OIB 83.

Clam, Otter Shell (Lutraria sp.) CSS pl XVIII; GSS 151; OIB 81.

Clam, Razor (Solen sp.) ALE 3:177; CSS cp 23; GSS 154; LAW 119; OIB 81; (Ensis sp.) ALE 3:177; CSS cp 23; GSS 154; OIB 81; WEA 307.

Clam, Red-nose (Hiatella arctica) ALE 3:177; CSS pl XX; OIB 87.

Clam, Surf —See **Trough Shell.**

Clam, Venus (Venus sp.) ALE 3:168; CSS pl XIX; GSS 146; MOL 71; OIB 75.

Clary—See also **Sage.**

Clary (Salvia sclarea) AMP 33; DFP 170 (cp 1359); EFC cp 51; EPF pl 812; PFE cp 113 (1144).

Clary, Meadow (Salvia pratensis) AMP 35; BFB cp 16 (7); DEW 2:189; EFC cp 50; KVP cp 76; LFW 220 (494); MBF cp 68; MFF cp 10 (758); NHE 186; OWF 177; PFE cp 114 (1147); PWF 69h; TEP 119; WFB 203 (1).

Clary, Wild (Salvia horminoides) MBF cp 68; MFF cp 10 (759); OWF 177; WFB 203 (3); (S. verbenaca) PFM cp 156; PWF 69e.

Clavaria (Clavaria sp.) CTM 70, 71; DEW 3: cp 52; DLM 61, 145; EPF pl 48; LHM 55; NFP 46; NHE 47; PFF 77; PMU cps 8, 9; PUM 155, 156; REM 177 (cp 145); RTM 218-222, 245; SMF 61; SMG 120, 121.

Clavaria, Beautiful (Clavaria formosa) CTM 70; PMU cp 9; REM 176 (cp 144); RTM 218.

Clavaria, Golden (Clavaria aurea) PMU cp 10; RTM 218; SMG 118.

Clavulina (Clavulina sp.) DEW 3:159; DLM 146; KMF 107; ONP 37, 153.

Clavulinopsis (Clavulinopsis sp.) DLM 146; ONP 37, 153.

Claytonia, Perfoliate —See **Spring Beauty.**

Claytonia, Pink (Montia sibirica) MBF cp 16; PFE cp 11 (124); PWF 32a; WFB 438.

Clayweed —See **Coltsfoot** (Tussilago).

Cleaner-fish —See **Wrasse, Cleaner.**

Clearwing (Aegeria sp.) ALE 2:327, 347; OBI 115; PEI 273 (cp 30b); (Sciapteron tabaniformis) PWI 156 (bw); (Synanthedon tipuliformis) CIB 160 (cp 19).

Clearwing, Hornet (Sesia apiformis) CIB 160 (cp 19); GAL 332 (cp 9); KBM 87; KIW 162 (cp 65); OBI 115; PEI 273 (cp 30a); SMB 342; WDB cp 28a.

Cleavers (Galium aparine) BWG 90; MBF cp 42; MFF pl 102 (817); NHE 100; OWF 93; PFF 301; PWF 64a, 134a; TEP 199; WFB 187 (8); WUS 353.

Clematis (Clematis hybrids) DFP 246 (cps 1967-1973); EGV 15-17, 102; EPG 115; ESG 102; GDS cp 93-99; HSC 30-32; LFW 34 (cps 81, 82, 84); PFE cp 24 (229); PFW 254, 255; SPL cps 75, 160, 161; TGS 70; TVG 221.

Clematis (Clematis sp.) DFP 132 (cp 1053), 245 (cp 1960), 246 (cps 1962, 1965); EWF cp 128d; GDS cp 88; LFT cps 69, 70; OGF 171; PFF 172; SNZ 29 (cp 43), 69 (cps 184, 187), 120 (cps 369, 370); TGS 57.

Clematis, Alpine (Clematis alpina) EFC cp 43; HPW 48; NHE 268; PFE cp 23 (227); WFB 73 (7).

Clematis, Fragrant (Clematis flammula) PFE cp 24 (225); SPL cp 159.

Clematis, Golden (Clematis tangutica) DFP 246 (cp 1966); EGV 18; GDS cp 89; HSC 30; TGS 70.

Clematis, Mountain (Clematis montana) DEW 1:125; DFP 246 (cp 1963); MWF 87; OGF 63; SPL cp 163; TGS 59.

Clematis, Orange or Lemon Peel (Clematis orientalis) DFP 246 (cp 1964); EGV 102; EJG 110; GDS cp 87; OGF 165; PFW 255.

Clematis, Purple (Clematis viticella) GDS cps 91, 92; HSC 30; NHE 178; TGS 57.

Clematis, Scarlet (Clematis texensis) GDS cp 90; HSC 30; TGS 59.

Clematis, Upright (Clematis recta) DFP 132 (cp 1054); NHE 26.

Clematopsis (Clematopsis sp.) EWF cp 52d; LFT 139 (cp 69).

Clerodendrum (Clerodendrum bungei) DFP 190 (cp 1517); EWF cp 104f; GDS cp 100; HSC 31, 32; SPL cp 76; (C. sp.) DEW 2:175 (cp 107); LFT 282 (cp 140); MEP 120, 121; MWF 88.

Climbing-perch —See **Perch.**

Clitocybe (Clitocybe sp.) CTM cps 24, 26, 27; DEW 3: cp 61, 166; DLM 74, 176-179; KMF 19, 91, 123; LHM 93-99; NFP 25, 62, 63; ONP 35, 107, 130; PFF 82; PMU cps 76-87; PUM cps 46-48; REM cps 94-100; RTM 111-118; SMF 40; SMG pls 99-103.

Cloth of Gold —See **Lantana, Common.**

Cloudberry (Rubus chamaemorus) EPF pl 403; EWF cp 2d; MBF cp 29; MFF pl 84 (343); NHE 71; OFP 79; OWF 81; PFE cp 44 (426); PWF 88c, 157f; WFB 109 (8).

Clove (Eugenia carophyllus) AMP 117; EET
224; OFP 133.

Clove Tree (Syzygium aromaticum) DEW 2:18;
SST cp 296; (S. sp.) LFT cp 116; MEP 99.

Clover—See also **Trefoil** and **Hop-trefoil**
(Trifolium sp.).

Clover, Alsike (Trifolium hybridum) BFB cp 8
(3); BWG 107; MBF cp 22; MFF cp 19 (300);
OWF 115; PFE cp 59 (601); PFF 211.

Clover, Crimson (Trifolium incarnatum) MBF
cp 22; MFF cp 19 (291); PFE cp 59 (595); PFF
212; WFB 131 (7).

Clover, Flat-headed or Clustered (Trifolium
glomeratum) MBF cp 22; MFF cp 19 (298).

Clover, Hare's-foot (Trifolium arvense) BWG
71; EWF cp 10c; MBF cp 22; MFF cp 19 (292);
NHE 212; OWF 115; PFE cp 60 (594); PWF
79k; WFB 131 (6).

Clover, Lesser Yellow (Trifolium dubium)
BWG 32; MBF cp 23; MFF cp 48 (304); NHE
180.

Clover, Oval-headed (Trifolium alpestre) EPF
pl 462; NHE 28.

Clover, Mountain (Trifolium montanum) DEW
1:313; EPF pl 464; NHE 180; PFE cp 59 (601);
WFB 131 (2).

Clover, Red or Purple (Trifolium pratense)
BFB cp 8 (7); BWG 108; CTH 32; MBF cp 22;
MFF cp 19 (286); NHE 180; OWF 115; PFF 211;
PWF 53i; TEP 165; WFB 131 (1); (T. purpure-
um) PFM cp 66.

Clover, Rough (Trifolium scabrum) MBF cp 22;
MFF cp 19 (294); NHE 180; PFE 206.

Clover, Sea or Teazel-headed (Trifolium
squamosum) MBF cp 22; MFF cp 19 (289);
OWF 115.

Clover, Soft or Knotted (Trifolium striatum)
MBF cp 22; MFF cp 19 (293); NHE 180; PWF
50f; WFB 131 (5).

Clover, Star (Trifolium stellatum) MFF cp 19
(290); PFE cp 60 (602); PFM cp 65.

Clover, Strawberry (Trifolium fragiferum)
MBF cp 23; MFF cp 19 (302); NHE 180; OWF
115; PWF 104c; WFB 131 (4).

Clover, Subterranean (Trifolium subterraneum) MBF cp 22; MFF pl 79 (296).

Clover, Suffocated (Trifolium suffocatum)
MBF cp 22; MFF pl 79 (299); WFB 131 (9).

Clover, Sulphur (Trifolium ochroleucon) MBF
cp 22; MFF pl 79 (287); NHE 179.

Clover, Twin-flowered (Trifolium bocconei)
MBF cp 22; MFF pl 79 (295).

Clover, Upright (Trifolium strictum) MBF cp
22; MFF cp 19 (297).

Clover, White or Dutch (Trifolium repens)
BWG 72; EGC 148; MBF cp 22; MFF pl 79
(301); NHE 180; OWF 83; PFF 211; PWF 55f;
TEP 143; WFB 131 (3).

Clover, Zigzag (Trifolium medium) DEW 1:312;
MBF cp 22; MFF cp 19 (288); NHE 179; OWF
115; PFE cp 59 (607); PWF 80d.

Clown-fish —See **Anemone-fish.**

Clubmoss (Lycopodium sp.) CGF 219-233; DEW
3: cps 118-120; 252; EPF pl 94; MFF pl 135
(1261-1264); NFP 122, 123; NHE 80; ONP 57,
93; PFF 108; SNZ 93 (cp 277), 100 (cp 304), 149
(cp 469); TEP 229.

Clubmoss (Selaginella sp.) BIP 163 (cp 249);
CGF 237, 239; DEW 3: cp 121; 253; DFP 81 (cp
647); EGF 141; EGL 143; EPF pls 95, 96; MFF
pl 135 (1265); NFP 125, 126; NHE 108; ONP
57; PFF 109; RDP 358; SPL 34.

Clubmoss, Stagshorn (Lycopodium clavatum)
AMP 139; CGF 221; DEW 3: cp 117; 252; EPF
pl 94; MFF pl 135 (1263); NFP 124; NHE 80;
ONP 57; PFF 109; TEP 229.

Club-rush (Blysmus sp.) MBF cp 91; MFF pl 115
(1116); NHE 138.

Club-rush, Bristle (Scirpus setaceus) MBF cp
91; MFF pl 116 (1118); NHE 105; PFE 564.

Club-rush, Floating (Scirpus fluitans) MBF cp
91; MFF pl 116 (1119); NHE 105; PFE 564.

Club-rush, Grayish or Glaucous (Scirpus
tabernaemontani) DFP 171 (cp 1367); MBF cp
91; PFE cp 185 (1843).

Club-rush, Round-headed (Scirpus holoschoenus) MFF pl 115 (1115); NHE 105; PFE
cp 185 (1841).

Club-rush, Sea (Scirpus maritimus) EPF pl 958;
MBF cp 91; MFF pl 115 (1113); NHE 138; PFE
cp 185 (1839); TEP 63.

Club-rush, Sharp (Scirpus americanus) NHE
105; SNZ 24 (cp 25); (S. nodosus) SNZ 33 (cp 59).

Club-rush, Wood (Scirpus sylvaticus) MBF cp
91; MFF pl 115 (1114); NHE 105; PFE cp 185
(1840); TEP 35.

Clusia (Clusia sp.) EWF cp 177d; MEP 86.

Clusterberry (Cotoneaster frigida) MWF 95; OBT 136.

Cnestis (Cnestis sp.) DEW 2:60 (cp 30); HPW 192.

Cnidium (Cnidium dubium) NHE 182; WFB 165 (7).

Coach-horse, Devil's (Ocypus olens) CIB 288 (cp 49); OBI 173; (Staphylinus olens) GAL 285 (bw); LAW 202; RBT 63.

Coachman —See Moorish Idol.

Coalfish —See Pollack.

Coati, Ring-tailed or Red (Nasua nasua) ALE 8:359; 12:94, 113; BMA 59, 60; DSM 162; HMW 146; MLS 276 (bw).

Coati, White-nosed (Nasua narica) ALE 12:94; HMA 41; HMW 146; LEA 558 (bw); SMW 193 (bw); TWA 138.

Cob or Cobnut —See Filbert or Hazel.

Cobia (Rachycentron canadus) HFI 28; HFW 184 (bw); LMF 82.

Cobra, Black-necked Spitting (Naja nigricollis) ART 213; FSA 144 (pl 21), 164 (cp 25); SSW 102.

Cobra, Cape (Naja nivea) ERA 213; FSA 129 (cp 16), 145 (pl 22); MAR 155 (cp XII).

Cobra, Egyptian (Naja haje) ALE 6:430; ERA 214; FSA 129 (cp 16), 145 (pl 22); HRA 49; SSW 101.

Cobra, Hooded or Indian (Naja naja) ALE 6:443; ART 216; DAA 153; ERA 213; HRA 49; LAW 397, 429; LBR 16; LEA 326 (bw); MAR 173 (pl 69); SIR 234, 235 (bw); VRA 237; WEA 105.

Cobra, King (Ophiophagus hannah) ALE 6:430; SSW 100; VRA 239.

Cobra Plant —See Pitcher-plant, California.

Cobra, Ring-necked Spitting (Hemachatus haemachatus) ALE 6:446; FSA 145 (pl 22), 164 (cp 25); SSW 103.

Coca (Erythroxylon coca) AMP 117; DEW 2:60 (cp 32); EPF pl 637; SST cp 281.

Coccinia (Coccinia sp.) EWF cp 79f; HPW 116.

Cochliostema (Cochliostema jacobianum) EWF cp 187a; PFW 77.

Cochlospermum (Cochlospermum sp.) DEW 1:165 (cp 81), 166 (cp 85); GFI 127 (cp 28); HPW 107; MEP 87.

Cochoa, Green (Cochoa viridis) KBA 273 (cp 40); RBI 196; WBV 191.

Cochoa, Purple (Cochoa purpurea) ALE 9:280; KBA 273 (cp 40).

Cock, Water —See Watercock.

Cockatiel (Nymphicus hollandicus or Leptolophus hollandicus) ALE 8:309; CDB 107 (cp 398); GPB 178; HAB 108; MBA 40; SNB 110 (cp 54); WAB 192; WAW 12.

Cockatoo, Gang-gang (Callocephalon fimbriatum) ALE 8:293, 300; CDB 105 (cp 389); HAB 92; RBU 135; SNB 108 (cp 53); WAW 12.

Cockatoo, Glossy Black (Calyptorhynchus lathami) HAB 94; SNB 108 (cp 53).

Cockatoo, Leadbeater's or Pink (Cacatua leadbeateri) ABW 198; ALE 8:300; BBT 14; RBU 136; SNB 108 (cp 53).

Cockatoo, Little (Cacatua sanguinea) SNB 108 (cp 53); WAW 12.

Cockatoo, Palm (Probosciger aterrimus) ABW 148; ALE 8:300; HAB 93; RBU 128; SNB 108 (cp 53); WAW 13 (bw).

Cockatoo, Red-tailed Black (Calyptorhynchus magnificus) RBU 131; SNB 108 (cp 53); (C. banksii) HAB 93.

Cockatoo, Roseate (Cacatua roseicapilla) ALE 8:300; CDB 104 (cp 388); GPB 180; HAB 92; MWA 50, 59; RBU 139; SBB 51 (pl 62); WAB 181; (Eolophus roseicapillus) MBA 38; SNB 108 (cp 53).

Cockatoo, Slender-billed (Cacatua tenuirostris) ALE 8:300; HAB 96; RBU 140; SNB 108 (cp 53); WEA 105.

Cockatoo, Sulphur-crested (Cacatua galerita) ABW 148; ALE 8:300; BBT 11; HAB 96; LEA 423 (bw); MBA 40; NBB 24 (bw); SBB 51 (pl 61); SNB 108 (cp 53); WAB 184; WEA 106.

Cockatoo, White-tailed Black (Calyptorhynchus baudinii) ALE 10:89; HAB 94; RBU 132; SNB 108 (cp 53).

Cockatoo, Yellow-crested —See Cockatoo, Sulphur-crested.

Cockatoo, Yellow-tailed (Calyptorhynchus funereus) ALE 8:300; HAB 95; SNB 108 (cp 53).

Cockatoo-parrot (Chalcopsitta sp.) ALE 8:299; SBB 54 (pl 69).

Cockchafer —See Beetle, Cockchafer.

Cockle (mollusk) (Cardium sp.) ALE 3:168; CSS pl XVIII; GSS 144; OIB 77.

Cockle, Dog (Glycymeris sp.) ALE 3:139; CSS pl XVII; CTS 56 (cp 104); OIB 83.

Cockle, Edible (Cardium edule or Cerastoderma edule) ALE 3:168; CSS pl XVIII; GSS 144; MOL 71; OIB 77; WEA 107.

Cockle, Little (Cardium exiguum or Parvicardium exiguum) CSS pl XVIII; OIB 77.

Cockle, White (plant) —See **Campion, White** (Lychnis alba).

Cocklebur, Common (Xanthium strumarium) EPF pl 594; NHE 238; PFE cp 146 (1401); PFF 319.

Cocklebur, Spiny (Xanthium spinosum) MFF pl 107 (841); NHE 238; PFE cp 146 (1402); PWF 125e; WFB 231 (9).

Cockleburr —See **Agrimony, Common.**

Cock-of-the-Rock, Andean or Peruvian (Rupicola peruviana) ABW 205; ALE 9:133; CDB 147 (cp 595); DTB 61 (cp 30; GPB 242; SSA cp 37.

Cock-of-the-Rock, Golden or Guianan (Rupicola rupicola) ABW 205; ALE 9:144; BBT 96; LVS 237; NBB 28; WAB 91.

Cockroach, American (Periplaneta americana) ALE 2:126; CIB 96 (cp 7); GAL 307 (bw); KIW 34 (cp 3); LEA 132 (bw); OBI 11; PEI 76 (bw); PWI 16 (bw).

Cockroach, Common or Oriental (Blatta orientalis) ALE 2:126; CIB 96 (cp 7); GAL 306 (bw), 307 (bw); OBI 11; PEI 77 (bw); ZWI 224.

Cockroach, Dusky (Ectobius lapponicus) CIB 96 (cp 7); GAL 307 (bw); OBI 11.

Cockroach, German (Blatella germanica or Phyllodromia germanica) ALE 2:126; CIB 96 (cp 7); GAL 306 (bw); PEI 76 (bw); ZWI 224.

Cockscomb (Celosia sp.) DEW 1:205; DFP 33 (cps 257, 258); EDP 138; EGA 103; EPF pl 315; LFW 156 (cp 345), 157 (cps 346, 347), 158 (cps 348, 349), 159 (cps 351, 352); MWF 77; SPL cps 419, 420; TGF 58; TVG 39.

Cocksfoot (Dactylis glomerata) BWG 175; EPF pl 1066; MBF pl 97; MFF pl 133 (1193); NHE 195; PFF 360; TEP 151.

Cocoa Tree —See **Cacao** (Theobroma).

Coconut, Double or Coco de Mer (Lodoicea maldivica) EET 235; EPF pl 1012; EWF cp 191g.

Cod, Antarctic (Trematomus bernacchi) HFI 49; HFW 176 (cp 89); MOL 104.

Cod, Atlantic (Gadus callarias or G. morhua) ALE 4:431; CSF 98, 99; FEF 262 (bw); HFI 77; HFW 154 (bw); LAW 327; LEA 250 (bw); LMF 155 (bw); MOL 106; OBV 35; VWS 14 (bw).

Codonopsis (Codonopsis ovata) DFP 5 (cp 36); EWF cp 97f; PFW 64.

Coelacanth (Latimeria chalumnae) ALE 5:271; HFI 133; LAW 295; LEA 265 (bw); WFW cp 501.

Coendou —See **Porcupine.**

Coffee Tree (Coffea arabica) AMP 127; BIP 59 (cp 77); DEW 2:107 (cps 64, 65); EPF pl XX a; EWF cp 60d; FHP 113; HPW 258; MLP 226; OFP 111; PFF 302; RDP 152; SST cp 277; WWT 106; 109.

Cohosh, Blue (Caulophyllum thalictroides) EWG 95; TGF 86.

Cola Nut or **Kolanut** (Cola sp.) AMP 127; DEW 214 (cps 119-121), 215 (cp 122); HPW 92; SST cp 278.

Coleto —See **Mynah, Coleto.**

Coleus (Coleus sp.) BIP 59 (cp 78); DFP 60 (cps 478, 479); EDP 148; EFP 102; EGA 108; EGL 96; EHP 139; ESG 39, 103; EWF cp 63a; HPW 238; LFT 287 (cp 143); MWF 91; OGF 125; PFF 288; PFW 153; RDP 26, 153; SPL cp 10, front; TGF 205; TVG 43.

Coleus, Summer —See **Beefsteak Plant** (Perilla).

Colletia (Colletia sp.) DEW 2:79; GDS cp 101; HPW 188; HSC 33.

Collybia (Collybia sp.) DLM 180, 181; LHM 101, 103; NFP 65, 66; ONP 123, 133; PFF 82; PUM cps 80, 81a; REM 135-137; RTM 88-91; SMG 163-167.

Colobus, Black-and-White (Colobus angolensis) DDM 65 (cp 10); SMR 16 (cp 1).

Colobus, Northern Black-and-White (Colobus abyssinicus) ALE 10:440, 456; DDM 65 (cp 10); DSM 91; HMA 8; HMW 17; SMW 77 (bw).

Colobus, Olive (Colobus verus) DDM 65 (cp 10); (Procolobus verus) DSM 91.

Colobus, Red (Colobus badius) ALE 10:445, 456; DDM 65 (cp 10); DSM 91; SLS 195; (C. kirkii) LVS 47; (C. pennanti) DDM 65 (cp 10).

Colobus, Southern Black-and-White (Colobus polykomos) ALE 10:456; CAA 91; CAF 278; DDM 65 (cp 10); JAW 58; LAW 620; MLS 157 (bw).

Coltrichia (Coltrichia perennis) DLM 146; LHM 63; ONP 47; SMF 65.

Coltsfoot (Tussilago farfara) AMP 99; BFB cp 19 (9); BWG 48; CTH 54 (cp 83); EGH 147; EPF pl 561; MBF cp 46; MFF cp 55 (854); NHE 101; OWF 41; PFE cp 149 (1439); PFF 329; PWF 15j, 135j; TEP 182; WFB 240, 241 (5).

Coltsfoot, Alpine (Homogyne alpina) EWF cp 22g; NHE 287; PFE cp 150 (1443); TEP 259; WFB 241 (6).

Colugo (Cynocephalus volans or Galeopithecus volans) ALE 11:83, 84; DSM 55; HMA 52; HMW 194; MLS 89 (bw); (C. variegatus) CAS 245; DAA 142; JAW 36.

Columbine (Aquilega hybrid) DFP 122 (cps 973, 974); EGP 96; LFW 184 (cps 410-412), 185 (cps 413-415; OGF 39; PFW 253; SPL cp 189; TVG 79.

Columbine, American (Aquilega canadensis) EEG 130; ESG 91; EWG 27, 86.

Columbine, Common (Aquilega vulgaris) AMP 71; BWG 136; EWF cp 8f; KVP cp 64; MBF cp 4; MFF cp 12 (25); MWF 44; NHE 26; PFE cp 27 (253); PWF 72a; TEP 235; WFB 75 (1).

Columbine, Fan (Aquilega flabellata) EMB 135; ERW 96.

Columbine, Golden (Aquilega chrysantha) DEW 1:119; MLP 49; TGF 141.

Columnea (Columnea sp.) BIP 59 (cp 79); DFP 60 (cp 480); EGG 100; EGL 97; EGV 104; EWF cp 185f, h; FHP 114; HPW 247; LHP 68, 69; MEP 140; MWF 91; RDP 154, 155; SPL 42.

Colza —See Rape.

Comb-jelly (Beroe sp.) ALE 1:271; CSS cp 2; LAW 69; (Pleurobrachia sp.) ALE 1:271; CSS cp 2; LEA 53 (bw); MOL 136 (bw); OIB 17.

Combretum (Combretum sp.) EWF cp 61b; HPW 165; LFT 231 (cp 115); LFW 233 (cps 520, 521); MEP 95; MFW 92.

Comfrey, Common (Symphytum officinale) AMP 157; BFB cp 15 (8); BWG 142; DEW 1:145; DFP 175 (cp 1393); EGH 143; MBF cp 60; MFF pl 94 (650); NHE 186; OWF 131; PWF 40c, 59i; TEP 43; WFB 189 (1).

Comfrey, Eastern (Symphytum orientale) PFE cp 101 (1053); PWF cp 23n.

Comfrey, Tuberous (Symphytum tuberosum) EPF pl 793; MBF cp 60; PFE cp 102 (1053); TEP 251; PWF 42l; (S. peregrinum) OGF 17.

Conch, Spider or Scorpion (Lambis sp.) ALE 3:71; CTS 22-23 (cps 16-19), 24 (cps 20, 22); GSS 46, 47.

Conch, True (Strombus sp.) CTS 24 (cp 21); GSS 16, 44, 45.

Condor, Andean (Vultur gryphus) ABW 72, ALE 7:334; GBP 39 (bw), 197 (bw); LVS 243; SLP 111; WAB 79.

Cone Shell (Conus sp.) ALE 3:56, 73; CTS 52-54 (cps 91-98); GSS 21, 109-119; MOL 67; VWS 91, 94 (bw).

Conebill, Blue-backed (Conirostrum sitticolor) DTB 79 (cp 39); SSA cp 39.

Coneflower (Rudbeckia laciniata) DFP 170 (cp 1353); EPF pl 542; OGF 147; PFE cp 146 (1403); PFF 320; PFW 83; PWF 145e; SPL cp 244; TGF 285; TVG 121; WFB 237 (5).

Cone-lizard—See Skink, Blue-tongued.

Conger—See Eel, Conger.

Constrictor, Boa—See Boa Constrictor.

Conure, Golden—See Parakeet, Golden.

Convolvulus (Convolvulus sp.) — LFT 274 (cp 136); PFM cp 138.

Convolvulus, Dwarf (Convolvulus tricolor) DFP 34 (cp 272) 35 (cp 273); EGA 110; EWF cp 36e; OGF 133; PFE cp 100 (1035); PFM cp 133; PFW 90; TGF 185; TVG 43.

Convolvulus, Pink (Convolvulus cantabricus) PFE 326; PFM cp 137.

Convolvulus, Shrubby (Convolvulus cneorum) BIP 61 (cp 82); DFP 191 (cp 1521); HSC 34; PFW 89.

Coot, Bald—See Swamphen.

Coot, Black or Common (Fulica atra) ALE 8:98-99; 11:257; BBB 213; BBE 109; CDB 74 (cp 225); DPB 73; FBE 117; FNZ 107 (bw); HAB 56; KBA 80 (pl 7); LAW 460; OBB 59; PBE 52 (cp 17); SNB 64 (cp 31); WBV 75; WEA 110, 111.

Coot, Crested or Red-Knobbed (Fulica cristata) BBE 109; CAF 256; CDB 74 (cp 226); FBE 117; PBE 52 (cp 17); PSA 145 (pl 16).

Copeina, Red-spotted (Copeina guttata) FEF 110 (bw); HFI 87; SFW 157 (pl 32).

Copepod (Various genera and species) ALE 1:443, 451; LAW 155; MOL 60, 77; OBV 21; OIB 127, 135, 141; WEA 111.

Copiapoa (Copiapoa sp.) LCS 19, 20; SCS cp 13

Copper Leaf—See Beefsteak Plant.

Copper Tips (Crocosmia aurea) LFT 75 (cp 37); TGF 44.

Coprosma (Coprosma sp.) EWF cp 2j; MLP 231; MWF 93; SNZ 20, 21, 39, 40, 56, 78, 79, 100, 103, 120, 139, 145, 148, 149, 162.

Coquette—See Hummingbird.

Coral (Dendrophyllia sp.) LAW 66; VWS 76 (bw).

Coral Berry (Ardisia crenata) DEW 1:232; MWF 45; RDP 86, 87.

Coral, Cup (Caryophyllia sp.) CSS cp 7; OIB 19; VWS 85.

Coral Gem (Lotus bertholetii) BIP 115 (cp 174); EGC 138; MWF 190.

Coral Necklace—See Illecebrum.

Coral Plant (Jatropha sp.) —See Jatropha.

Coral Plant (Russelia sp.)—See Fountain Plant.

Coral, Precious (Corallium rubrum) ALE 1:242; (C. nobile) MOL 54.

Coral, Stag's Horn (Acropora sp. or Madrepora sp.) LEA 51 (bw); VWS 72-74, 79, 80 (bw).

Coral Tree or Coral Bean (Erythrina sp.) EFS 110; EHP 96; EWF cp 72 c; HPW 150; LFT cp 84; MEP 58; MWF 126; SST cp 244.

Coral Vine (Antigonon leptopus) EGV 42, 93; GPI 47 (cp 8); LFW 231 (cp 516); MEP 35; MWF 43; PFW 236.

Coralberry (Symphoricarpos orbiculatus) OGF 189; PFF 303; TGS 382.

Coral-fish, Beaked—See Butterflyfish, Long-nosed (Chelmon sp.).

Coralita—See Coral Vine.

Coral-root (Corallorhiza)—See Orchid, Coral-root.

Coral-wort (Cardamine)—See also Bitter-cress.

Coral-wort or Coralroot (Dentaria bulbifera) BFB cp 4 (3); MFF cp 25 (91).

Coral-wort, Common (Cardamine bulbifera) EFC cp 8; MBF cp 8; NHE 30; PFE cp 34 (308); WFB 91 (3).

Coral-wort, Five-leaved (Cardamine penta-phyllos) DEW 1:170 (cp 96); NHE 274; PFE cp 34 (309).

Coral-wort, Pale (Cardamine enneaphyllos) NHE 30; PFE cp 33 (309); WFB 90.

Corb (Corvina nigra) FEF 359 (bw); HFI 29.

Cord-grass (Spartina townsendii) MBF pl 95; MFF pl 125 (1250); PFE cp 181 (1796); TEP 66; (S. pectinata) TVG 129.

Cordon Bleu, Red-cheeked (Uraeginthus bengalus) ABW 303; ALE 9:440; BBT 67; CDB 213 (cp 946); GPB 321.

Corella—See Cockatoo.

Coriander (Coriandrum sativum) EGH 109; EVF 142; MFF pl 96 (465); OFP 139; PFE 280; PFF 266; WFB 157 (6).

Coriaria (Coriaria sp.) EHP 107; HPW 200; PFE cp 68 (702); SNZ 41 (cps 83, 84).

Cork Tree, Amur (Phellodendron amurense) EEG 120; EGT 132; EGW 131; SST cp 140; TGS 326.

Cormorant, Black or Great—See Cormorant, Common or European.

Cormorant, Black-faced—See Cormorant, White-breasted.

Cormorant, Blue-eyed or Antarctic (Phalacrocorax atriceps) CDB 40 (cp 48); SAO 83; SAR 82-83, 86, 117; WAB 222.

Cormorant, Cape (Phalacrocorax capensis) PSA 32 (cp 1); SAO 83.

Cormorant, Common or European (Phalacrocorax carbo) ABW 40; ALE 7:61, 167, 255; BBB 250; BBE 33; CDB 41 (cp 50); DPB 9; FBE 33; FNZ 97 (pl 8); HBA 14; KBA 32 (pl 1); LAW 463; OBB 11; PSA 32 (cp 1); RBE 64; SAO 83; SBB 24 (pl 23); SNB 26 (cp 12); WAW 21 (bw); WBV 41; WEA 113.

Cormorant, Flightless (Nannopterum harrisi) CDB 40 (cp 46); GPB 59; HBG 74 (bw); LVB 27; LVS 200; PWC 169; WAB 217; (Phalacrocorax harrisi) BBT 119.

Cormorant, Guanay or Peruvian (Phalacrocorax bougainvillii) ABW 40; CDB 40 (cp 49); CSA 228, 234 (bw); WAB 79.

Cormorant, Kerguelen or King (Phalacrocorax albiventer) CSA 288; SAR 118; WAN 116, 117; WFI 100 (bw).

Cormorant, Little Pied (Phalacrocorax melanoleucus) ALE 7:167; FNZ 97 (pl 8); HAB 14, 15; SNB 26 (cp 12).

Cormorant, Pygmy (Phalacrocorax pygmaeus) ALE 7:167; BBE 33; FBE 33.

Cormorant, Red-faced (Phalacrocorax urile) ABW 44; ALE 8:229.

Cormorant, Reed (Phalacrocorax africanus)
CDB 40 (cp 47); FBE 33; PSA 32 (cp 1); SAO 83.

Cormorant, Rock (Phalacrocorax magellanicus) SAO 83; WFI 101.

Cormorant, White-breasted (Phalacrocorax
fuscescens) CAU 48-49 (bw); RBU 99; SNB 26
(cp 12).

Cormorant, Yellow-faced—See **Shag, Pied.**

Corn (Zea mays) AMP 139; DEW 2:268; EGA
163; EMB 150; EPF pl 1069-1071; EVF 17, 91;
MWF 301; OFP 7; PFF 341-343; PFW 133; TEP
167.

Corn Cockle (Agrostemma githago) BFB cp 5
(3); DFP 29 (cp 226); EPF pl 302; KVP cp 43;
MBF cp 14; MFF cp 14 (158); NHE 225; OGF
87; OWF 107; PFE cp 16 (163); PFF 162; PWF
139f; TEP 175; WFB 61 (7); WUS 155.

Corn Salad, Smooth-fruited (Valerianella
dentata) MBF 43; NHE 235; TEP 179.

Corncrake—See **Crake, Corn.**

Cornel—See **Dogwood** (Thelycrania sp.).

Cornel, Bentham's—See **Dogwood, Evergreen.**

Cornel, Dwarf (Cornus suecica); DEW 2:82;
EWF cp 5e; KVP cp 155, 155a; PFE cp 82 (847);
WFB 155 (3); (Chamaepericlymenum sueci-
cum) MBF cp 41; MFF pl 84 (449); NHE 74;
OWF 192.

Cornelian Cherry or **Cornel** (Cornus mas)
AMP 63; BKT cp 174; CTH 39 (cp 51); DEW
2:81; DFP 191 (cp 1528); EEG 93; EFS 21, 103;
EPF pls 743, 744; GDS cp 109; NHE 23; OBT
149; PFE cp 82 (845); PFW 93; PWF cp 16 e;
SPL cp 257; SST cp 197; TEP 132; TVG 220;
WFB 155 (2).

Cornflower (Centaurea cyanus) BFB cp 20
(8); CTH 62 (cp 100); DFP 33 (cp 259); EDP 122;
EFC cp 37; EGA 104; EGG 96; EPF pl 589;
KVP cp 39; LFW 22 (cp 43); MBF cp 49; MFF cp
9 (918); MWF 78; NHE 240; OGF 87; OWF 179;
PFE cp 156 (1501); PFF 330; PWF 110d; SPL cp
194; TEP 181; TGF 236; WFB 249 (4).

Corokia (Corokia sp.) HPW 168; HSC 36, 37;
SNZ 42 (cps 87-91), 166 (cp 529).

Corpse-flower—See **Toothwort.**

Cortinarius (Cortinarius sp.) CTM 44-47; DLM
66, 185-189; KMF 19; LHM 163-173; ONP 115,
119; PUM cps 86-125; REM cps 60-69; RTM
00-78, 3MF 40, 53; 3MG 203.

Cortinarius, Violet (Cortinarius violaceus)
DEW 3:177; DLM 189; LFO 26; NFP 79; PFF
86; REM 97 (cp 65); RTM 76; SMG 202; (C.
alboviolaceus) CTM 46; DLM 185; KMF 19;
LHM 165; REM 96 (cp 64) RTM 74; SMF 23.

Corydalis (Corydalis sp.) DFP 5 (cp 40); EWF cp
91b; OGF 45; PFW 119; TGF 141.

Corydalis, Climbing (Corydalis claviculata)
MBF cp 6; MFF pl 102 (43); NHE 29; OWF 83;
PFE cp 30 (277); PWF 111a; WFB 79 (4).

Corydalis, Hollow (Corydalis cava) AMP 125;
EFC cp 11; EPF pl 235.

Corydalis, Purple (Corydalis solida); DEW 1:
139; DFP 87 (cp 691); EWF cp 11c; NHE 29;
PFE cp 30 (279); SPL cp 258; WFB 79 (3).

Corydalis, Yellow (Corydalis lutea) BWG 21;
ESG 105; HPW 53; MBF cp 6; MFF cp 43 (44);
OWF 31; PFE cp 30 (278); PWF 48a; SPL cp
447; WFB 79 (2).

Corydoras, Arched (Corydoras arcuatus) ALE
4:411; FEF 239 (bw); SFW 385 (pl 99), 447 (bw);
WFW cp 172.

Corydoras, Black-spotted (Corydoras melanistius) ALE 4:411; SFW 381 (pl 96), 443 (bw);
WFW cp 174.

Corydoras, Leopard (Corydoras julii) FEF 240
(bw); SFW 384 (pl 97), 447 (bw); WFW cp 173.

Corydoras, Myers' (Corydoras myersi) FEF
242 (bw); HFI 113; SFW 385 (pl 99), 447 (bw).

Corydoras, Peppered (Corydoras paleatus)
FEF 245 (bw); HFI 113; SFW 318 (pl 96), 447
(bw).

Coryphantha (Coryphantha sp.) ECS 109;
LCS cps 21-27.

Cosmos (Cosmos bipinnatus) DFP 35 (cps 275,
276); EGA 111; LFW 245 (cp 245); MWF 94;
OGF 137; PFF 323; SPL cp 199; TVG 45.

Cosmos, Yellow (Cosmos sulphureus) EGA 111;
MWF 94; OGF 137; TVG 45.

Costus (Costus sp.) DEW 2:276 (cp 150); EWF cp
68b; HPW 298; LFW 179 (cp 402); MEP 28;
MLP 320; PFW 94.

Cotinga (Cotinga sp.) ALE 9:133, 144; CDB
146 (cp 591) SSA cp 37; (Carpodectes hopkei)
SSA pl 45.

Cotinga, Black-necked Red (Phoenicircus
nigricollis) ABW 204; ALE 9:144; SSA cp 37.

Cotinga, Pompadour (Xipholena punicea)
ABW 205; ALE 9:144; SSA cp 37.

Cotinga, Swallow-tailed (Phibalura flavirostris) ALE 9:144; SSA cp 11.
Cotoneaster (Cotoneaster sp.) GDS cp 127;
MEP 46; NHE 20, 267.
Cotoneaster, Bearberry (Cotoneaster
dammeri) EEG 141; EGC 128; EGE 119; EGW
104; EPG 116.
Cotoneaster, Common or Wild (Cotoneaster
integerrimus) MBF cp 31; NHE 267; PFE cp
48 (473); PWF 35h; TEP 95; WFB 115 (8).
Cotoneaster, Cranberry (Cotoneaster
apiculatus) EEG 94; EFS 105; ELG 114; TGS
287.
Cotoneaster, Fishbone (Cotoneaster horizontalis) DFP 193 (cp 1541); EPF cp XII b; GDS cps
123-126; HSC 38, 40; LFW 36 (cp 86); MWF 95;
TGS 279; TVG 221.
Cotoneaster, Many-flowered (Cotoneaster
multiflora) EFS 47, 105; EPF pl 428; TGS 289.
Cotoneaster, Willowleaf (Cotoneaster
salicifolius) HPW 143; SPL cp 310; TGS 288.
Cotoneaster, Wintergreen (Cotoneaster conspicuus) DFP 193 (cp 1537); GDS cps 119-121;
HSC 38, 40.
Cotton (Gossypium sp.) DEW 1:259; EPF pls
610, 611; MLP 144.
Cotton, Levant (Gossypium herbaceum) LFT cp
106; PFE cp 73 (750).
Cotton, Upland (Gossypium hirsutum) DEW
1:219 (cp 131); PFF 249.
Cotton-grass, Common (Eriophorum angust-
ifolium) EPF pl 954; MBF cp 91; MFF pl 115
(1107); NHE 77; TEP 77; (E. sp.) MLP 299,
315.
Cotton-grass, Hare's-tail (Eriophorum vagin-
atum) EPF pl 957; MBF cp 91; MFF pl 115
(1108); NHE 77; PFE cp 184 (1835); PWF 67h;
TEP 77.
Cotton-grass, Scheuchzer's (Eriophorum
scheuchzeri) DEW 2:290; NHE 291; RWF 31.
Cottonweed (Otanthus maritimus) MBF cp 45;
MFF cp 52 (890); WFB 239 (7).
Cottus—See Bullhead.
Cotyledon (Cotyledon sp.) DFP 6 (cp 41); EPF pl
379; LCS cps 228, 229; LFT 154 (cp 76); RDP
157.
Coucal (Coua sp.) ALE 10:283; WAB 30.

Coucal, Black (Centropus melanops) ALE
8:370; DPB 165; (C. grilli) ABW 152.
Coucal, Greater (Centropus sinensis) CDB 111
(cp 419); DAA 149; DPB 165; KBA 192 (pl 25);
WAB 168; WBV 117.
Coucal, Lesser (Centropus bengalensis) DPB
165; KBA 192 (pl 25).
Coucal, Pheasant (Centropus phasianinus)
GPB 185; HAB 117; SNB 122 (cp 60); WAB
181; WAW 30 (bw).
Coucal, Senegal (Centropus senegalensis) CDB
111 (cp 418); FBE 175.
Coucal, White-browed or Burchell's (Centropus superciliosus) ALE 8:443; BBT 41; CDB
111 (cp 420); PSA 192 (cp 21).
Couch—See also Quackgrass.
Couch, Bearded (Agropyron caninum) MBF pl
99; MFF pl 124 (1207); NHE 40, 242.
Couch, Sand (Agropyron junceiforme) MBF pl
99; MFF pl 125 (1210); NHE 138; PFE cp 178
(1727); TEP 65.
Couch, Sea (Agropyron pungens) MBF pl
99; MFF pl 125 (1209).
Cougar (Felis concolor or F. cougar) ALE 12:299,
300, 324; BMA 188; DSM 191; HMA 65; HMW
122; JAW 132; LEA 572 (bw); LVA 35; MLS
316 (bw); SMW 181 (cp 105); TWA 120; WEA
300.
Coughwort—See Coltsfoot (Tussilago).
Courser (Rhinoptilus sp.) ALE 8:443; PSA 145
(pl 16); (Hemerodromus africanus) CDB 90 (cp
312).
Courser, Australian—See Pratincole,
Australian.
Courser, Cream-colored (Cursorius cursor)
ABW 126; ALE 8:158, 159, 170; BBB 273; BBE
137; FBE 143; OBB 83; PBE 84 (pl 27), 100 (cp
31).
Courser, Indian (Cursorius coromandelicus)
ABW 126; BBT 43; RBI 63; WEA 114.
Courser, Temminck's (Cursorius temminckii)
CDB 90 (cp 309); GPB 156.
Cow—See Cattle.
Cow Basil (Vaccaria pyramidata) MBF cp 13;
NHE 225; WFB 63 (6).
Cowbane (Cicuta virosa) MBF cp 37; MFF pl 98
(480); NHE 95; PWF 136f; TEP 31; WFB 161
(7).

Cowberry (Vaccinium vitis-idaea) AMP 137; CTH 43 (cp 58); DEW 1:211 (cp 112); DFP 243 (cp 1937); EMB 64; EPF pl 490; EWG 144; MBF cp 55; MFF pl 77 (596); MLP 191; NHE 70; OFP 83; OWF 120; PFE cp 89 (935); PWF 41g, 155k); TEP 288; TGS 39; WFB 173 (2).

Cowbird, Glossy or Shining (Molothrus bonariensis) ABW 290; ALE 9:381; SSA pl 29.

Cowfish—See **Boxfish.**

Cow-lily—See **Water-lily** (Nuphar sp.).

Cow-parsley (Anthricus sylvestris) BWG 75; MBF cp 38; MFF pl 99 (458); NHE 222; OWF 87; PWF cp 23j; WFB 157 (1).

Cow-parsnip Heracleum maxima) EDP 128; EWG 56, 116.

Cowpea (Vigna sp.) LFT cp 81; OFP 45; PFF 220.

Cowry (Cypraea sp.) ALE 3:55, 72; CTS 26-32 (cps 26-42); GSS 17, 20, 21, 53-62; MOL 68; VWS 92 (bw); (Trivia sp.) CSS pl XV; CTS 26 (cp 25); OIB 37.

Cow's Horn (Euphorbia grandicornis) ECS DEW 1:221 (cp 138); EPF pl 630; SCS cp 79.

Cowslip (Primula veris) AMP 89; DEW 1:229; EGH 134; ERW 131; MBF cp 57; MFF cp 40 (612); NHE 33; OWF 27; PWF cp 21h; TEP 249; TGF 171; WFB 175 (3).

Cowslip, American (Dodecatheon sp.) —See **Shooting Star.**

Cowslip, Cape (Lachenalia aloides) BIP 111 (cp 168); DFP 70 (cp 554); EGB 124; EGG 122; MWF 176; PFW 174; (L. pendula) FHP 130; (L. tricolor) MWF 176.

Cowslip, Virginia—See **Bluebell, Virginia** (Mertensia).

Cow-wheat, Blue-topped (Melampyrum nemorosum) NHE 35; PFE cp 128 (1259).

Cow-wheat, Common (Melampyrum pratense) MBF cp 65; MFF cp 60 (726); NHE 187; OWF 25; PWF 84a; WFB 217 (7).

Cow-wheat, Crested (Melampyrum cristatum) DEW 2:156; MBF cp 65; MFF cp 60 (724); NHE 36; PFE 384; PWF 144a; WFB 217 (8).

Cow-wheat, Field (Melampyrum arvense) KVP cp 42; MBF cp 65; MFF cp 60 (725); NHE 200, PFE cp 128 (1258).

Cow-wheat, Wood (Melampyrum sylvaticum) MBF cp 65; MFF cp 60 (727); NHE 35.

Coypu—See **Nutria.**

Crab, Common Shore (Carcinus maenas) CSS cp 15; LAW 153; LEA 166 (bw), 176; OIB 125.

Crab, Edible or Rock (Cancer pagurus) ALE 1:473; CSF 203; CSS pl XII; MOL 83; OIB 123.

Crab, Fiddler (Uca sp.) ALE 1:476; LEA 176 (bw); WEA 149.

Crab, Fresh-water (Astacus astacus) ALE 11:105; GAL 493 (bw); LAW 146; (A. fluviatilis) OIB 133.

Crab, Ghost (Ocypode sp.) ALE 1:476; WEA 164.

Crab, Hermit (Paguristes sp.) ALE 1:494; LAW 142; MOL 84; VWS 52.

Crab, Hermit or Soldier (Pagurus bernhardus or Eupagurus bernhardus) CSS cp 15; LAW 157; LEA 175 (bw); OIB 123; WEA 189; (Pagurus japonicus) MOL 84.

Crab, Mitten (Eriocheir sinensis) ALE 1:476; GAL 494 (bw); (E. japonicus) MOL 83.

Crab, Oyster or Pea (Pinnotheres pisum) ALE 1:493; OIB 125.

Crab, Porcelain (Porcellana sp.) ALE 1:476; CSS cp 14; OIB 125; (Petrolisthes sp.) VWS 48.

Crab, Robber (Birgus latro) ALE 1:493; LEA 175 (bw).

Crab, Rock (Galathea sp.) ALE 1:493; CSS cp 14; MOL 84; OIB 123.

Crab, Spider (Macropodia sp.) CSS cp 16; LEA 77 (bw); OIB 125; (Stenorhynchus seticornis) WEA 117, 345.

Crab, Spiny Spider (Maia squinado) CSF 202; CSS cp 16.

Crab, Swimming (Portunus sp.) ALE 1:493; CSS cp 15; MOL 83, 92; (Neptunus pelagicus) VWS 48.

Crabapple (Malus sylvestris) BFB cp 10 (9); EET 170, 171; EFC cp 58; EPF pl 433; MBF cp 31; MFF cp 17 (387); MTB 289 (cp 26); NHE 20; OBT 28; OWF 181; PWF 25m, 172a, b; SST cp 208; WFB 115 (1).

Crabapple, Japanese Flowering (Malus floribunda) DFP 213 (cp 1699); EET 170; EGE 34; EJG 128; MWF 195; OBT 181; SPL cp 313; TGS 183.

Crabapple, Purple Flowering (Malus purpurea lemoinei) EET 171; EGT 127; EPG 132.

Crabapple, Red Flowering (Malus purpurea) MWF 195; PFW 261; SPL cp 314; TGS 197.

Crabapple, Siberian (Malus baccata) OBT 181; OFP 47; TGS 201; (M. robusta) DFP 213 (cp 1703).

Crab-grass (Digitaria sanguinalis) MFF pl 126 (1253); NHE 243; PFF 347; WUS 59.

Crabgrass, Silver (Eleusine) —See **Goose-grass.**

Crab-grass, Smooth (Digitaria ischae-mum) EGC 70; MCW 17; NHE 243; WUS 57.

Crab-spider —See **Spider, Crab.**

Crake, Baillon's or Lesser Spotted (Porzana pusilla) ALE 8:83; BBB 269; BBE 107; DPB 73; FBE 115; FNZ 107 (bw); HAB 55; KBA 97 (cp 10); OBB 59; PBE 52 (cp 17); SNB 62 (cp 30); WBV 75.

Crake, Band-bellied (Porzana paykullii) DPB 73; KBA 97 (cp 10).

Crake, Black (Limnocorax flavirostra) ALE 8:83; CDB 74 (cp 229); PSA 160 (cp 17); WAB 139.

Crake, Buff-spotted (Sarothrura elegans) ALE 8:443; PSA 160 (cp 17).

Crake, Corn (Crex crex) ALE 8:83, 10:223; BBB 73; BBE 107; CDB 74 (cp 224); FBE 115; OBB 57; PBE 52 (cp 17); RBE 123; SNB 62 (cp 30).

Crake, Little or Lesser (Porzana parva) ALE 7:255; 8:83; BBB 275; BBE 107; FBE 115; OBB 59; PBE 52 (cp 17).

Crake, Malay Banded or Red-legged (Rallina fasciata) DPB 71; KBA 97 (cp 10); SNB 62 (cp 30).

Crake, Marsh —See **Crake, Baillon's or Lesser Spotted.**

Crake, Paint-billed (Neocrex erythrops) HBG 80 (pl 3); SSA cp 2.

Crake, Philippine Banded (Rallina eurizonoides) DPB 71; KBA 97 (cp 10).

Crake, Ruddy-breasted (Porzana fusca) ABW 107; DPB 73; KBA 97 (cp 10); SNB 273 (bw).

Crake, Spotless (Porzana tabuensis) BNZ 169; DPB 73; FNZ 106 (bw); HAB 55; SNB 62 (cp 30); (P. plumbea) CDB 75 (cp 234).

Crake, Spotted (Porzana porzana) ABW 107; ALE 8:97; 11:257; BBB 169; BBE 107; FBE 115; OBB 59; PBE 52 (cp 17); (P. fluminea) HAB 55; SNB 62 (cp 30); WAB 200.

Crake, White-browed (Porzana cinerea) KBA 97 (cp 10); SNB 62 (cp 30); (Polio-limnus cinereus) ALE 8:84; DPB 71; HAB 55.

Cramp Balls (Daldinia concentrica) DLM 121; LHM 47; NFP 39; ONP 147; RTM 245; SMF 60.

Cranberry (Vaccinium oxycoccus) EPF pls 491, 492; MBF cp 55; MFF 23 (599); NHE 70; OFP 83; PFE cp 88 (937); PFF 277; PWF 80f, 155f; TEP 80; WFB 172.

Cranberry, Mountain —See **Cowberry.**

Crane, Australian or Brolga (Grus rubi-cunda) ALE 8:120; CAU 155; CDB 73 (cp 222); HAB 54; MBA 52; RBU 23; SNB 30 (cp 14); WAB 201; WAW 16 (bw).

Crane, Blue (Anthropoides paradisea) ALE 8:120; CDB 72 (cp 214); PSA 161 (cp 18).

Crane, Common (Grus grus or Megalornis grus) ALE 7:385; BBB 271; BBE 103; CDB 73 (cp 219); FBE 111; GAL 170 (bw); GPB 125; KBA 114 (bw); LEA 403 (bw); OBB 15; PBE 8 (cp 3); RBE 116; SBB 46 (pl 55).

Crane, Crowned (Balearica pavonina) ABW 103; ALE 8:118; CAA 28 (bw), 37; CAF 113; CDB 72 (cp 216); GPB 127; LEA 344 (bw); PSA 161 (cp 18); SBB 48 (cp 58); WAB 139.

Crane, Demoiselle (Anthropoides virgo) ABW 103; ALE 8:120; BBE 103; CDB 72 (cp 215); FBE 111; PBE 8 (cp 3); RBE 119; SBB 47 (pl 57); WAB 111; WEA 116.

Crane Flower —See **Bird-of-paradise Flower** (Strelitzia).

Crane, Hooded (Grus monacha) ALE 8:117; LVB 31.

Crane, Japanese or Manchurian (Grus japonensis) ALE 8:117-119; CDB 73 (cp 220); LVB 31; LVS 214-215; PWC 221 (bw); SLP 178, 179.

Crane, Sarus (Grus antigone or Megalornis antigone) ABW 104; ALE 8:119, 120; CAS 163; CDB 72 (cp 217); DPB 25; KBA 114 (bw); LAW 479; LEA 404 (bw); SBB 46 (pl 56); SNB 30 (cp 14); WAB 176; WEA 118.

Crane, Siberian White (Grus leucogeranus or Megalornis leucogeranus) ALE 8:120; BBE 103; CAS 161; CDB 73 (cp 221); FBE 111; LVB 31; WAB 29.

Crane, Wattled (Bugeranus carunculatus) ALE 8:120; CAF 60 (bw); CDB 72 (cp 218); PSA 161 (cp 18).

Crane, White-necked (Grus vipio) ABW 104; ALE 8:117.

Cranesbill (Geranium sp.) DFP 10 (cp 78), 143 (cps 1143-1145), 144 (cp 1150); LFT cp 88; MWF 141; NHE 31; OGF 37, 43; PFW 123; SNZ 129 (cp 401), 164 (cp 522); TGF 109; TVG 95.

Cranesbill, Bloody or Blood-red (Geranium sanguineum) BFB cp 6 (7); DFP 144 (cp 1148); EEG 134; EMB 138; EPF pl 646; ERW 118; HPW 208; KVP cp 118; LFW 54 cp 125, 55 cp 128; MBF cp 19; MFF cp 16 (239); NHE 31; OWF 129; PFE cp 63 (641); PWF 45h; TEP 111; TGF 109; WFB 135 (2).

Cranesbill, Broad-leaved or Knotted (Geranium nodosum) MFF cp 16 (237); PFE cp 63 (642); PWF 123j.

Cranesbill, Cut-leaved (Geranium dissectum) BWG 104; EGC 71; MBF cp 19; MFF cp 16 (2431); NHE 221; OWF 129; PWF 49g; WFB 135 (9).

Cranesbill, Dove's-foot (Geranium molle) BWG 105; MBF cp 19; MFF cp 16 (244); NHE 221; OWF 129; PFF 225; PFM cp 82; PWF 53h; WFB 135 (8).

Cranesbill, Dusky (Geranium phaeum) EWF cp 14d; MBF cp 19; MFF cp 35 (238); PFE cp 64 (646); PWF 45g; WFB 135 (3c).

Cranesbill, French (Geranium endressi) MFF cp 16 (235); PWF 104a; WFB 135 (3).

Cranesbill, Italian or Rock (Geranium macrorrhizum) DFP 144 (cp 1146); MFF cp 16 (240); OGF 37; PFE cp 64 (647); PWF 45l.

Cranesbill, Long-stalked (Geranium columbianum) MFF cp 16 (242); NHE 73; PWF 69i.

Cranesbill, Marsh (Geranium palustre) NHE 95; WFB 135 (4).

Cranesbill, Meadow (Geranium pratense) BFB cp 6 (10); DFP 144 (cp 1147); EPF pl 642, 644; EWF cp 14b; KVP cp 14; MBF cp 19, MFF cp 12 (233); NHE 181; OWF 169; PWF 66a; WFB 135 (1).

Cranesbill, Mountain or Hedgerow (Geranium pyrenaicum) BFB cp 6 (9); MBF cp 19; MFF 16 (241); PFE cp 63 (642); PWF 37i; WFB 135 (5).

Cranesbill, Pencilled or Streaked (Geranium versicolor) MBF cp 19; MFF cp 16 (236); WFB 135 (3a).

Cranesbill, Round-leaved (Geranium rotundifolium) MBF cp 19; NHE 220; PWF 45k; WFB 134, 135 (8a).

Cranesbill, Shining (Geranium lucidum) BFB cp 6 (5); MBF cp 19; MFF cp 16 (246); OWF 129; PFE cp 63 (651); PWF 38c; WFB 135 (7).

Cranesbill, Small-flowered (Geranium pusillum) MBF cp 19; MFF cp 16 (245); PWF 45j; WFB 134, 135 (8b).

Cranesbill, Tuberous (Geranium tuberosum) EWF cp 32f; PFM cp 83.

Cranesbill, Wood (Geranium sylvaticum) BFB cp 6 (1); DEW 2:62 (cp 36); DFP 10 (cp 79); EFC cp 14; MBF cp 19; MFF cp 12 (234); NHE 31; PFE cp 63 (645); PWF 45e; WFB 135 (1a).

Crape Myrtle (Lagerstroemia indica) BKT cp 153; DFP 209 (cp 1670); EEG 98; EFS 121; EGW 122; EJG 125; ELG 105, 123; EPG 128; LFW 172 (cp 387); MWF 177; PFW 178; SPL cp 98; SST cp 249; TGS 151; (L. speciosa) DEW 2:10; EWF cp 114b; MEP 93.

Crassula (Crassula sp.) DEW 1:280 (cp 165); DFP 61 (cp 484); ECS 12; EPF pls 377, 380; LCS cps 235-239; LFT 150 (cp 74); MBF cp 33; RDP 158.

Crassula, Club Moss (Crassula lycopodioides) RDP 158; SPL cp 131.

Crawfish, Marine —See **Lobster, Spiny.**

Crayfish (Astacus sp.) —See **Crab, Freshwater.**

Cream of Tartar —See **Baobob.**

Creeper, Brown (Finschia novaeseelandiae) BNZ 31; FNZ 193 (cp 17); (Certhia familiaris) —See **Tree-creeper, Brown or Common.**

Creeper, Gray Spotted (Salpornis spilonotus) ABW 239; ALE 9:319; RBI 255.

Creeper, Plain-headed (Rhabdornis inornatus) DPB 261; WAB 171.

Creeper, Rangoon —See **Rangoon Creeper** (Quisqualis).

Creeper, Silver-vein (Parthenocissus henryana) HSC 86, 89; MWF 222; PFW 307.

Creeper, Stripe-headed (Rhabdornis mystacalis) ABW 239; ALE 9:319; DPB 261.

Creeper, Tree —See **Tree-creeper.**

Creeper, Virginia —See **Virginia Creeper.**

Creeper, Wall —See **Wall-creeper.**

Creeping Jenny or **Creeping Charlie** —See **Moneywort.**

Crepidotus (Crepidotus sp.) DLM 189, 190; LHM 177; ONP 145; RTM 127.

Cress, Garden or Upland (Lepidium sativum) EGH 121; NHE 216; OFP 153; WFB 99 (3).

Cress, Hoary (Cardaria draba) EPF pl 242; MBF cp 10; MFF pl 94 (65); NHE 216; PFE cp 36 (353); PWF 40d; TEP 190; WFB 99 (1); WUS 201.

Cress, Shepherd's (Teesdalia nudicaulis) MBF cp 10; MFF pl 91 (71); NHE 217; WFB 97 (3).

Cress, Thale (Arabidopsis thaliana) MBF cp 8; MFF pl 91 (110); NHE 218; OWF 71; PWF 14c, 79h; WFB 95 (4).

Cress, Winter (Barbarea vulgaris) BFB cp 4 (5); BWF 22; EGH 103; MBF cp 7; MFF cp 43 (92); NHE 218; OWF 11; PFF 189; PWF 36c; WFB 87 (7);WUS 191.

Cricket, Bush —See **Bush-cricket.**

Cricket, Field (Gryllus sp.) ALE 11:321; BIN pl 18; CIB 80 (cp 5); GAL 308 (bw); KIW 22; (bw); LAW 194; OBI 11; PEI 57, 58 (bw); PWI 16 (bw).

Cricket, Greenhouse (Tachysines asynamorus) CIB 81 (cp 6); PEI 49 (bw).

Cricket, House (Acheta domesticus) ALE 2:126; CIB 80 (cp 5); GAL 308 (bw); OBI 11.

Cricket, Mole (Gryllotalpa sp.) ALE 2:125; 11:321; BIN pls 19, 20; CIB 80 (cp 5); GAL 309 (bw); LAW 195; LEA 130 (bw); OBI 11; PEI 59 (bw); PWI 101.

Cricket, Water —See **Water-cricket.**

Cricket, Wood (Nemobius sylvestris) CIB 80 (cp 5); OBI 11.

Crinodonna (Crinodonna corsii) DFP 87 (cp 692); EGB 105; FHP 114.

Crocias, Gray-crowned —See **Sibia, Langbian.**

Crocodile, African Long-nosed (Crocodylus cataphractus) HRA 75; MAR 91 (pl 35).

Crocodile, American (Crocodylus acutus) ALE 6:139; HRA 77.

Crocodile, Broad-fronted (Osteolaemus tetraspis) ALE 6:134, 139; HRA 77; SIR 48 (bw).

Crocodile, Estuarine or Saltwater (Crocodylus porosus) ALE 6:134, 139; COG cp 1; ERA 147-149; HRA 76; LVS 165; MEA 15; WAW 95 (bw), 97.

Crocodile, Gavial —See **Gavial.**

Crocodile Jaws (Aloe humilis) ECS 97; LCS cps 209-210.

Crocodile, Johnstone's (Crocodylus johnstoni) LVS 163; WAW 97.

Crocodile, Mugger (Crocodylus palustris) SIR 65; VRA 161.

Crocodile, New Guinea (Crocodylus novaeguineae) LBR 118; WAW 94 (bw).

Crocodile, Nile (Crocodylus niloticus) ALE 6:131-134, 139; ART 156; CAA 120-121 (bw); ERA 151; HRA 75; LAW 435; LEA 298 (bw); LBR 12-13, 116-117; LVS 164-165; MAR pl 34; SIR cp 20.

Crocodile, Siamese (Crocodylus siamensis) LEA 298 (bw); SIR 61 (cp 21).

Crocus (Crocus sp.) DFP 88-90 (cps 697-707, 716); EDP 120; EGB 53, 107, 108; EGG 101; EGW 105, 106; EMB 99; EPF pl 937, cp XXV; EWF cps 42h, 51f; FHP 115; HPW 311; MLP 262; OGF 11; PFF 387; PFM cp 270; PFW 148, 149; RDP 160; TVG 185, 187.

Crocus, Autumn —See also **Saffron, Meadow.**

Crocus, Autumn (Colchium sp.) DFP 86 (cps 687-690); HPW 313; NHE 290; OGF 183; TVG 185.

Crocus, Autumn (Crocus nudiflorus) DFP 90 (cp 715); MFF cp 34 (1051); OWF 163; PFE cp 174 (1676).

Crocus, Purple (Crocus albiflorus) PFE cp 174 (1678); RWF 18; WFB 273 (9); (C. purpureus) MBT cp 83; MFF cp 6 (1052); NHE 192; PWF cp 19k.

Crocus, Spring (Crocus vernus) DEW 2:224 (cp 142); DFP 90 (cp 714); LFW 39 (cp 90); PFW 149; SPL cp 259; TGF 12.

Crocus, Warren or Sand (Romulea columnae) MBF cp 83; MFF cp 81 (1053); WFB 273 (8); (R. bulbocodium) DFP 110 (cp 875); PFM cp 268.

Crocus, Yellow (Crocus flavus) HPW 311; LFW 39 (cp 89); PFM cp 271.

Crombec, Long-billed (Sylvietta rufescens) CDB 176 (cp 755); PSA 272 (cp 31).

Crossandra (Crossandra sp.) BIP 63 (cp 88); DEW 2:171 (cp 97); EWF cp 64a.

Crossberry —See Starflower.

Crossbill, Common or Red (Loxia curvirostra) ALE 9:382; BBB 155; BBE 287; CDB 209 (cp 926); DPB 419; FBE 297; FBW 29; GAL 118 (cp 1); OBB 181; PBE 260 (cp 61); RBE 312; RBI 320; WBV 221; WEA 120.

Crossbill, Parrot or Scottish (Loxia pytyopsittacus) BBE 287; FBE 297; FBW 29; GPB 318; OBB 181; PBE 260 (cp 61).

Crossbill, Two-barred or White-winged (Loxia leucoptera) ABW 301; BBB 271; BBE 287; FBE 297; OBB 181; PBE 260 (cp 61); WAB 38.

Cross-vine (Bignonia capreolata) EGV 96; HPW 249; TGS 39.

Crosswort (Cruciata laevipes) MBF cp 42; PFE cp 99 (1029); PWF 59j; WFB 187 (2); (C. chersonensis) NHE 234; (Galium cruciata) MFF cp 50 (811); OWF 31.

Croton (Codiaeum variegatum) BIP 57 (cp 76); DFP 60 (cp 476); EDP 148; EFP 23, 101; EGE 119; EGL 96; ELG 136; LHP 66, 67; MEP 69; MWF 89; PFF 233; RDP 16, 150, 151; SPL cp 9.

Croton (Croton sp) HPW 186; LFT 190 (cp 94); WUS 245.

Crow, Bald —See Rockfowl, Gray-necked.

Crow, Carrion (Corvus corone corone) ALE 9:498, 501; BBB 79; BBE 219; FBE 311; OBB 125; PBE 309 (bw); RBE 227.

Crow, Hooded (Corvus corone cornix) ABW 224; ALE 9:498; BBB 122; BBE 219; CDB 222 (cp 998); FBE 311; OBB 125; PBE 309 (bw); RBE 228.

Crow, Indian House (Corvus splendens) ALE 9:498; GPB 341; KBA 277 (pl 42); SAB 80 (cp 39).

Crow, King —See Drongo, Black.

Crow, Large-billed (Corvus macrorhynchos) DPB 253; KBA 277 (pl 42).

Crow, Little (Corvus enca) DPB 253; (C. bennetti) HAB 150; SAB 80 (cp 39).

Crow, Pied (Corvus albus) ALE 9:498; PSA 161 (cp 18).

Crowberry (Empetrum nigrum) MBF cp 55; MFF pl 77 (605); NHE 70; OWF 120; PFE 304; PWF 155i; WFB 173 (10); (E. rubrum) HPW 129.

Crowea (Crowea saligna) EWF cp 138b; HPW 203.

Crowfoot —See also Buttercup (Ranunculus).

Crowfoot, Alpine (Ranunculus alpestris) EPF pl 226; NHE 269.

Crowfoot, Common Water (Ranunculus aquatilis) EPF pl 227; MBF cp 2; MFF pl 80 (21); NHE 91; OWF 67; PFF 169; PWF 46e; TEP 15; WFB 71 (7).

Crowfoot, Glacier (Ranunculus glacialis) DFP 23 (cp 178); EWF cp 1a; KVP cp 131; NHE 269; PFE cp 26 (247); WFB 71 (6)..

Crowfoot, Ivy-leaved (Ranunculus hederaceus) MBF cp 2; NHE 91; WFB 71 (8).

Crowfoot, Parnassus-grass (Ranunculus parnassifolius) EPF pl 225; NHE 269; PFE cp 25 (248); TVG 169.

Crowfoot, Pyrenean (Ranunculus pyrenaeus) KVP cp 132; NHE 269.

Crowfoot, Rigid-leaved (Ranunculous circinatus) EWF cp 8a; KVP cp 6; MBF cp 2; NHE 91.

Crown Imperial (Fritillaria imperialis) DEW 2:233; DFP 93 (cps 737, 738); EGB 55, 115; EPF 896, 899, cp XXIIIa; PFW 169; TGF 38b; TVG 195.

Crown of Thorns (echinoderm) —See Starfish (Acanthaster sp.).

Crown of Thorns (plant) (Euphorbia milli) EWF cp 58a; LCS cp 250; LFW 162 (cp 361); LHP 189; MEP 70; MWF 130; PFW 115; RDP 197; SPL cp 136; (E. splendens) EGL 108; EPF pl 628; MLP 114.

Crown-vetch (Coronilla varia) EEG 141; EGC 128; EPF pl 463; KVP cp 110; MFF cp 25 (320); NHE 212; PFE cp 62 (627); PWF 103j; TEP 111; TGF 146; WFB 119 (7).

Crows-toes —See Trefoil, Common Bird's-foot.

Crumble-caps, Trooping (Coprinus disseminatus) DLM 36, 183; KMF 59; LHM 141; ONP 139; RTM 52; SMF 45.

Cuckoo (Coccyzus sp.) HBG 97; SSA pl 24.

Cuckoo, Banded Bay (Cacomantis sonneratii) DPB 157; KBA 161 (cp 22).

Cuckoo, Black-eared (Chrysococcyx
osculans) HAB 119; SNB 118 (cp 58), 120 (cp
59).

Cuckoo, Bronze (Chrysococcyx basalis) KBA
161 (cp 22); SNB 120 (cp 59); (Chalcites
basalis) HAB 121.

Cuckoo, Brush (Cacomantis variolosus) DPB
157; HAB 121; SNB 118 (cp 58).

Cuckoo, Channel-billed (Scythrops novaehollandiae) HAB 121; RBU 108; SNB 122 (cp
60).

Cuckoo, Chestnut-breasted (Cacomantis
castaneiventris) HAB 121; SNB 118 (cp 58).

Cuckoo, Common (Cuculus canorus) ABW
152; ALE 8:346, 369; 13:163; BBB 96; BBE
171; CDB 112 (cp 425); DAE 136, 137; DPB
153; FBE 175; GAL 148 (cp 4); GPB 183;
KBA 192 (pl 25); LEA 432; OBB 117; PBE
181 (cp 46); RBE 191; WEA 121.

Cuckoo, Didric (Chrysococcyx caprius) ALE
8:369; CDB 112 (cp 421); PSA 192 (cp 21).

Cuckoo, Drongo (Surniculus lugubris) DPB
157; KBA 192 (pl 25).

Cuckoo, Emerald (Chrysococcyx cupreus)
ABW 152; ALE 8:369; PSA 192 (cp 21); (C.
maculatus) KBA 161 (cp 22).

Cuckoo, Fan-tailed (Cacomantis pyrrhophanus) HAB 121; SNB 118 (cp 58).

Cuckoo, Golden Bronze (Chalcites
plagosus) HAB 121; (Chrysococcyx
plagosus) SNB 120 (cp 59).

Cuckoo, Great Spotted (Clamator glandar-
ius) ALE 8:369; BBE 171; CDB 112 (cp 422);
FBE 175; PBE 181 (cp 46); PSA 192 (cp 21);
RBE 192.

Cuckoo, Ground (Carpococcyx renauldi)
CDB 111 (cp 417); KBA 192 (pl 25); (C.
radiceus) ALE 8:370; (Neomorphus geoffroyi) ALE 8:370.

Cuckoo, Guira (Guira guira) ALE 8:370; SSA
pl 24.

Cuckoo, Hawk (Cuculus fugax) DPB 153;
KBA 192 (pl 25); (C. vagans) KBA 192 (pl
25); (C. sparverioides) DPB 153; KBA 192
(pl 25); (Hierococcyx fugax) ALE 8:369.

Cuckoo, Jacobin or Pied (Clamator
jacobinus) KBA 192 (pl 25); PSA 192 (cp 21).

Cuckoo, Little Bronze (Chalcites minutillus)
HAB 121; (Chrysococcyx minutillus) SNB 120
(cp 59).

Cuckoo, Long-tailed (Eudynamis taitensis)
BNZ 71; FNZ 193 (cp 17); SNB 122 (cp 60).

Cuckoo, Malayan Bronze (Chrysococcyx
malayanus) DPB 153; KBA 161 (cp 22).

Cuckoo, Oriental (Cuculus saturatus) BBE
171; DPB 153; FBE 175; HAB 118, 119; RBU
104; SNB 118 (cp 58).

Cuckoo, Pallid (Cuculus pallidus) CDB 112
(cp 426); HAB 118; SNB 118 (cp 58).

Cuckoo, Plaintive (Cacomantis merulinus)
DPB 157; KBA 161 (cp 22); WBV 117.

Cuckoo, Red-chested (Cuculus solitarius)
ALE 8:369; NBA 43; PSA 192 (cp 21).

Cuckoo, Red-winged Indian or Crested
(Clamator coromandus) ABW 152; ALE
8:369; DAA 145; DPB 153; KBA 161 (cp 22).

Cuckoo, Rough-crested (Phoenicophaeus
superciliosus) DPB 157; RBI 79.

Cuckoo, Scale-feathered (Phoenicophaeus
cummingi) DPB 157; RBI 80.

Cuckoo, Shining Bronze (Chalcites lucidus)
BNZ 79; FNZ 193 (cp 17); HAB 121; RBU
107; (Chrysococcyx lucidus) SNB 120 (cp
59).

Cuckoo, Squirrel (Piaya cayana) CDB 113
(cp 428); SSA pl 24; WAB 91.

Cuckoo, Violet (Chrysococcyx xanthorhynchus) DPB 153; KBA 161 (cp 22).

Cuckoo-dove (Macropygia unchall) KBA 113
(cp 12); WBV 109; (M. phasianella) DPB
135; (M. ruficeps) WBV 109; (M. amboinensis) SNB 104 (cp 51).

Cuckoo-falcon —See **Hawk, Crested.**

Cuckoo-flower (Cardamine pratensis) BFB
cp 4; EFC cp 8; EPF pl 243; KVP cp 33; MBF
cp 7; MFF cp 25 (86); NHE 181; OWF 113;
PFE cp 34 . (310); PWF cp 19f; TEP 139;
WFB 91 (1).

Cuckoo-pint (Arum maculatum) BFB cp 22
(6); BWG 173; CTH 21 (cp 10); DEW 2:312;
EPE pl 1009; HPW 308; KVP cp 47, 47a;
MBF cp 88; MFF cp 35 (1097); NHE 41; OWF
63; PWF 40b, 134d; TEP 264; WFB 270, 271
(7).

Cuckoo-roller (Leptosomus discolor) ALE 9:57; 10:283; WAB 159.

Cuckoo-shrike, Bar-bellied (Coracina striata) DPB 237; KBA 252 (pl 35).

Cuckoo-shrike, Barred (Coracina lineata) ABW 219; ALE 9:450; HAB 145.

Cuckoo-shrike, Black (Campephaga phoenicia) ABW 219; ALE 9:450; PSA 240 (cp 27); (C. sulphurata) CDB 154 (cp 636); (Coracina coerulescens) DPB 237.

Cuckoo-shrike, Black-faced or Large (Coracina novaehollandiae) CDB 154 (cp 637); HAB 145; KBA 252 (pl 35); SAB 12 (cp 5).

Cuckoo-shrike, Ground (Pteropodocys maxima) HAB 145; MBA 26; RBU 315; SAB 12 (cp 5).

Cuckoo-shrike, Papuan (Coracina papuensis) HAB 145; SAB 12 (cp 5).

Cucumber (Cucumis sativa) EPF pl 513; EVF 38, 92; OFP 117; PFF 308; (C. anguria) OFP 117; (C. sativus) EMB 146.

Cucumber, Bitter (Citrullus sp.) MLP 232; PFE cp 78 (812).

Cucumber Slice (Macrocystidia cucumis) DLM 222; LHM 109.

Cucumber, Squirting (Ecballium elaterium) AMP 69; CTH 53 (cp 80); EPF pl 514; EWF cp 34e; NHE 246; PFE cp 79 (811); PFM cp 188.

Cudweed (Filago sp.) MBF cp 44; MFF pl 101 (864-866); NHE 189, 237; OWF 33; PFE cp 143 (1375); PWF 137a, f; WFB 235 (2,3).

Cudweed (Gnaphalium sp.) MBF cp 44; MFF cp 52 (871); NHE 285; SNZ 108 (cp 329), 109 (cp 335).

Cudweed, Dwarf or Creeping (Gnaphalium supinum) MBF cp 44; MFF cp 37 (868); NHE 285.

Cudweed, Jersey (Gnaphalium luteoalbum) MFF cp 52 (870); NHE 189; WFB 235 (5).

Cudweed, Upright False (Micropus erectus) NHE 237; WFB 235 (8).

Cudweed, Wayside or Marsh (Gnaphalium uliginosum) MBF cp 44; MFF cp 37 (869); NHE 189; OWF 33; PFE cp 144 (1380); PFF 317; PWF 108; WFB 235 (4).

Cudweed, Wood or Heath (Gnaphalium sylvaticum) MBF cp 44; MFF cp 37 (867); NHE 37; TEP 259; WFB 235 (6).

Cumin (Cuminum cyminum) EGH 110; OFP 139.

Cunonia (Cunonia capensis) HPW 137; SST cp 99.

Cup and Saucer Vine (Cobaea scandens) DEW 2:143; DFP 247 (cp 1975); EGA 108; EGV 104; EPF pl 790; EWF cp 181a; MEP 118; MWF 89; PFW 76; SPL cp 425; (C. hookerana) EWF cp 181d.

Cup of Gold —See **Chalice Vine** (Solandra) and **Golden Trumpet.**

Cup-fungus, Early (Galactinia vesiculosa) KMF 135; REM 185 (cp 153).

Cupidone —See **Succory, Blue** (Catananche).

Cupid's Bower —See **Magic Flower.**

Cupid's Dart —See **Succory, Blue.**

Cupid's Paintbrush (Emilia sp.) EGA 117; EWF cp 65c.

Cups-and-saucers (Holmskioldia tettensis) EWF cp 80; LFT cp 140.

Curassow (Crax sp.) ABW 86; ALE 7:436; CDB 65 (cp 178); SSA cp 1; WAB 27; WEA 122; (Pauxi pauxi) ALE 7:436.

Curassow, Razor-billed (Mitu mitu) ABW 86; ALE 7:436.

Curlew, Bush or Scrub (Burhinus magnirostris) HAB 71; RBU 35; SNB 58 (pl 58); WAW 30 (bw).

Curlew, Common or Eurasian (Numenius arquata) ALE 7:255; 8:154-155; BBB 118; BBE 121; CDB 87 (cp 289); DPB 87; FBE 139; KBA 128 (pl 13), 149 (pl 18); LEA 333 (bw); NBB 89 (bw); OBB 69; PBE 92 (pl 29), 101 (cp 32); PSA 145 (pl 16); RBE 139; SNB 84 (cp 41), 86 (pl 42); WBV 75.

Curlew, Long-billed (Numenius madagascariensis) DPB 87; HAB 67; SNB 84 (cp 41), 86 (pl 42).

Curlew, Slender-billed (Numenius tenuirostris) BBE 121; FBE 139.

Curlew, Stone (Burhinus oedicnemus) ABW 125; ALE 8:158, 159; BBB 94; BBE 137; CAS 89; CDB 89 (cp 308); CEU 196 (bw); FBE 143; GPB 155; KBA 149 (pl 18); OBB 83; PBE 81 (pl 26), 85 (pl 28); RBE 156; WAB 169; WBV 91.

Currant, Black (Ribes nigrum) AMP 49; MBF cp 33; NHE 19; OFP 81; WFB 141 (1a).

Currant, Mountain or Alpine (Ribes alpinum) MBF cp 33; MFF pl 78 (413); MWF 254; NHE 19; PFE 43 (414); TGS 234.

Currant, Red (Ribes rubrum or R. sylvestre) CTH 31 (cp 33); EPF pls 373, 375; MBF cp 33; MFF pl 78 (412); PFE cp 43 (414); WFB 141 (1).

Currant, Upright Red (Ribes spicatum) MBF cp 33; NHE 19.

Currawong, Black (Strepera fuliginosa) HAB 157; RBU 283; SAB 80 (cp 39).

Currawong, Clinking (Strepera arguta) HAB 157; SAB 80 (cp 39).

Currawong, Gray (Strepera versicolor) HAB 157; SAB 80 (cp 39).

Currawong, Pied (Strepera graculina) ALE 9:469; CDB 220 (cp 985); HAB 157; SAB 80 (cp 39); WAB 203.

Cuscus —See also **Phalanger.**

Cuscus, Gray (Phalanger orientalis) ALE 10:99 (cp 1), 106; LVA 35.

Cuscus, Spotted (Phalanger maculatus or P. nudicaudatus) ALE 10:99 (cp 1), 102; BMA 68, 69; CAS 287; CAU 114, 242 (bw); DSM 31; HMW 203; JAW 25; MLS 57 (bw); SMW 36 (cp 6); TWA 203 (bw).

Cushionflower or **Corkwood** (Hakea sp.) DFP 204 (cp 1627); MEP 32; MLP 73.

Cushion-star (Asterina sp.) —See **Starlet.**

Cushion-star (Porania sp.) —See **Starfish, Cushion.**

Cusimanse —See **Mongoose.**

Cusk (Brosme brosme) ALE 4:431; CSF 117; HFI 78.

Custard Apple (Annona cherimola) DEW 1:82; MWF 42; OFP 97; SST cp 183.

Cutia (Cutia nipalensis) KBA 288 (cp 45); WBV 205.

Cutting-grass —See **Cane-rat.**

Cuttlefish, Common (Sepia officinalis) ALE 3:187, 224; CSS pl XXI; LEA 94 (bw); MOL 75; OIB 93; WEA 123.

Cuttlefish, Little (Sepiola atlantica) CSS pl XXI; OIB 93; (S. rondeleti) ALE 3:187.

Cyclamen (Cyclamen hederifolium) HPW 135; MBF cp 57; MFF cp 32 (617); PFE cp 92 (958); PFW 240; PWF 168d; (C. coum) DFP 90 (cp 719); EGW 107; PFW 240.

Cyclamen, Common (Cyclamen purpurascens) EPF pl 263; NHE 278; PFE cp 92 (959); TVB 187; WFB 177 (2).

Cyclamen, European (Cyclamen europaeum) DEW 1:230; KVP cp 113; PFW 240; SPL cp 479; TEP 249.

Cyclamen, Florist's (Cyclamen persicum) BIP 65 (cp 91); DFP 91 (cp 722); EDP 118; EGG 102; EWF cp 48g; FHP 18-21, 116; LFW 39 (cps 91, 92); LHP 76, 77; MWF 100; PFM cp 126; PFW 238; RDP 167; SPL cp 44.

Cyclamen, Neapolitan (Cyclamen neapolitanum) DFP 91 (cp 721); EGB 60, 61, 109; MWF 100; OGF 177; PFM cp 124; PFW 240.

Cyclamen, Repand (Cyclamen repandum) DFP 91 (cp 724); EWF cp 33d; PFE cp 92 (960); PFM cp 123.

Cyperus, Brown or Black (Cyperus fuscus) MBF cp 91; MFF pl 116 (1121); NHE 104; PFE 564; (C. compressus) HPW 292.

Cyphal (Cherleria sedoides) MBF cp 16; MFF pl 89 (185).

Cypress, Arizona (Cupressus arizonica or C. glabra) DFP 252 (cp 2013); EET 91; EGE 95; MTB 64 (cp 3); NFP 151; OBT 129; SST cp 17.

Cypress, Bald or Swamp (Taxodium distichum) BKT cps 9, 10; DEW 1:39; DFP 256 (cp 2041); EET 22, 98; EGE 107; EFP pls 147, 148; MLP 36, 42; MTB 85 (cp 6); MWF 280; OBT 101, 125; PFF 123; SST cp 40; TGS 22.

Cypress, Hinoki (Chamaecyparis obtusa) DFP 252 (cps 2009, 2010); EEG 105; EGE 93; ELG 131; EMB 111; EPF pl 160; ERW 106; MTB 64 (cp 3); OBT 129; SST cp 14; TGS 22; TVG 251; (C. formosensis) EET 89.

Cypress, Italian or Mediterranean (Cupressus sempervirens) AMP 175; BKT cps 51, 54; DEW 1:42, 43; EET 90; EGE 96; EPF pls 158, 159; MWF 99; OBT 129; PFE cp 2 (11); PFF 125; SST cp 19; TGS 22.

Cypress, Lawson (Chamaecyparis lawsoniana) BKT cp 36; DFP 251 (cps 2004-2008); EET 88, 89; GDS cps 68-70; MTB 64 (cp 3); MWF 81; OBT 104, 113; PWF 171k; SST cp 13; TVG 251; WFB 25 (5).

Cypress, Leyland (Cupressocyparis leylandii) BKT cp 52; DFP 252 (cp 2012); EET 91; EGE 95; OBT 129; PWF 171l.

Cypress, Monterey (Cupressus macrocarpa) EET 90; MBT 64 (cp 3); OBT 104, 129; SST cp 18.

Cypress, Nootka (Chamaecyparis nootkatensis) BKT cps 53, 55; EET 88, 89; EPF pl 161; OBT 129.

Cypress, Sawara (Chamaecyparis pisifera) EET 88, 89; EGE 93, 94; EGW 100; EJG 107; GDS cps 72, 73; MTB 64 (cp 3); OBT 129; SST cp 15; TGS 17.

Cypress, Summer (Kochia scoparia) DFP 40 (cp 314); EGA 128; HPW 73; MCW 33; PFF 157; TGF 58; TVG 55; WUS 137; (K. laniflora) NHE 226.

Cystoderma (Cystoderma sp.) DLM 190; LHM 129; ONP 103; RTM 40.

Cytinus (Cytinus hypocistis) DEW 1:161 (cp 71); EWF cp 31g; KVP cp 163; PFE cp 8 (78); PFM cp 10; RWF 34; (C. sanguineus) HPW 177.

D

Dab (Limanda limanda) CSF 183; HFI 18; OBV 49; WEA 124; (Pleuronectes limanda) CSS pl XXIV.

Dab or Dabb-lizard —See **Lizard, Spiny-tailed or Palm.**

Dabchick —See **Grebe, Little.**

Dace (Leuciscus leuciscus) FEF 143 (bw); FWF 75; GAL 243 (bw); HFI 95; OBV 103; WFW cp 131.

Dacnis (Dacnis lineata) CDB 203 (cp 88); (D. berlepschi) SSA cp 39; (Xenodacnis parina) SSA cp 16.

Dacrymyces (Dacrymyces sp.) DLM 141; LHM 225; ONP 149.

Daddy-long-legs —See **Fly, Crane,** and **Harvestman.**

Daffodil, Cyclamen-flowered (Narcissus cyclamineus) DFP 105 (cp 833); EGB 132, 133; EMB 103; OGF 11; PFW 23 (3).

Daffodil, Hoop Petticoat (Narcissus bulbocodium) DFP 104 (cps 830-832); EGB 133; EMB 103; EWF cp 42d; HPW 315; MWF 208; PFE cp 173 (1666); PFW 23.

Daffodil, Sea —See **Lily, Sea.**

Daffodil, Wild (Narcissus pseudonarcissus) BFB cp 23 (2); DFP 105 (cp 839); EPF pl 933; LFW 98, 99 (cps 222-226); MBF cp 83; MFF cp 63 (1045); NHE 192; OWF 29; PFF 383; PWF cp 19h; TGF 12; WFB 271 (1).

Daffodil, Winter or Autumn (Sternbergia lutea) DFP 111 (cp 884); EGB 140; MWF 274; OGF 183; PFE cp 172 (1664); PFM cp 253; SPL cp 437; TGF 12; (S. sp.) PFE cp 172 (1664); PFM cp 254.

Dais (Dais cotinifolia) LFT 230 (cp 114); MEP 92.

Daisy, African (Arctotis sp.) DFP 30 (cps 238-239); EGA 95; EWF cp 83b; MWF 45; OGF 139; TGF 269.

Daisy, African (Dimorphotheca) —See **Marigold, Cape.**

Daisy, Atlas (Anacyclus depressus) DFP 1 (cp 4); EMB 135; ERW 94; TVG 139; (A. atlanticus vestitus) OGF 91.

Daisy, Barberton or Transvaal (Gerbera jamesonii) BIP 91 (cp 137); EGG 112; EPF pls 593, 595; LFT cps 170, 174; LFW 19 (cps 36, 37); MEP 153; MWF 141; SPL cp 208; TGF 269.

Daisy, Blue (Felicia amelloides) EDP 122; EGA 118; FHP 121; MWF 133; TGF 252; (F. angustifolia) MWF 133; (F. muricata) LFT cp 171.

Daisy, Boston —See **Marguerite.**

Daisy, Crown (Chrysanthemum coronarium) DFP 34 (cp 265); EGA 106; KVP cp 168; OFP 155; PFE cp 148 (1425); PFM cps 199, 200.

Daisy, English or Lawn (Bellis perennis) BWG 96; DFP 31 (cp 243); EGA 98; EPF pl 532; ERW 102; EWF cp 22c; HPW 266; LFW 18 (cp 34); MBF cp 44; MFF pl 100 (881); MWF 55; NHE 190; OGF 33; OWF 99; PFF 314; PWF cp 15 I; SPL cp 355; TGF 253; TVG 35; WFB 233 (1).

Daisy, European Michaelmas (Aster amellus) DFP 123 (cp 987); MWF 48; NHE 76; OGF 123; PFE cp 142 (1364); TVG 79.

Daisy, Everlasting or Paper (Acroclinium roseum or Helipterum roseum) MWF 30; TVG 51.

Daisy, False or Alpine (Bellidastrum michelii) PFE cp 142 (1362); KVP cp 136.

Daisy, Mat (Raoulia sp.) DFP 23 (cp 180); EMB 64; SNZ 117 (cps 359, 360), 118 (cps 361, 363), 119 (cps 366-368), 124 (cp 383), 126 (cps 390, 391).

Daisy, Michaelmas (Aster novi-belgii) DFP 124 (cps 986-991); EDP 121; MLP 237; OGF 179; PWF 113f; SPL cp 415; TGF 253; TVG 81; WFB 233 (5).

Daisy, Moon or Ox-eye —See **Marguerite.**

Daisy, Namaqualand —See **Marigold, Cape** (Dimorphotheca sp.) and **Monarch-of-the-Veldt** (Venidium).

Daisy, New Zealand Mountain (Celmisia sp.) CAU 282; DFP 4 (cps 30-32); EWF cp 140d; MLP 243; SNZ 106 (cp 325); 112 (cp 344), 151 (cp 474), 155 (cp 489), 156 (cps 491-494), 157 (cps 495-498), 158 (cps 499-500).

Daisy, Paris (Chrysanthemum sp.) —See **Marguerite.**

Daisy, Paris or Sunshine (Gamolepis sp.) EGA 120; MWF 138.

Daisy, Shasta (Chrysanthemum maximum) DEW 2:201; DFP 132 (cp 1049); EDP 124; LFW 26 (cps 53, 54); MWF 84; PFF 327; SPL cp 358; TGF 252; TVG 83.

Daisy, Swan River (Brachycome iberidifolia) DFP 31 (cp 244); EGA 98; EWF cp 140b; OGF 141; TGF 269; TVG 35.

Daisy, Trailing African (Osteospermum fruticosum) EEG 145; EGC 142.

Daisy-Bush or **Tree Daisy** (Olearia sp.) DFP 215 (cp 715); GDS cp 300; HSC 82, 85; SNZ 28, 46-48, 53, 60, 96, 107, 145, 147, 150, 163, 164.

Dalechampia (Dalechampia roezliana) EPF pls 622, 624; EWF cp 178d; (D. spathulata) DEW 1:275 (cps 151, 152).

Dame's-rocket or **Dame's voilet** (Hesperis matronalis) BWG 137; EWG 81; MBF cp 8; MFF pl 93 (103); PFE cp 31 (296); PFF 187; PWF 79e; TGF 77; WFB 91 (6); (H. tristis) NHE 218.

Damselfish (Dascyllus sp.) ALE 5:133, 137; HFI 37; HFW 202 (bw); WFW cp 389; (Pomacentrus sp.) ALE 5:133; HFI 37; MOL 19; WFW cp 390.

Damselfish, Blue-green (Chromis chromis) LAW 340; WFW cp 388.

Damselfly —See also **Dragonfly.**

Damselfly (Lestes sp.) ALE 2:90; CIB 69 (cp 2); CTI 17 (cp 1); GAL 364 (cp 13); OBI 3; PEI 38, 41 (bw); (Ischura elegans) CIB 69 (cp 9); OBI 3.

Damselfly, Agrion (Agrion sp.) CIB 69 (cp 2); GAL 364 (cp 13); LEA 129 (bw); OBI 3; PEI 38, 39 (bw).

Damselfly, Coenagrion (Coenagrion puella) CIB 69 (cp 2); GAL 364 (cp 13); PEI 40 (bw).

Damselfly, White-legged (Platycnemis pennipes) ALE 2:91; CIB 69 (cp 2); GAL 364 (cp 13).

Dancing Bones (Hatiora salicorniodes) ECS 122; EDP 92; LCS cp 70.

Dancing-bird —See **Manakin.**

Dandelion (Taraxacum officinale) AMP 37; BFB cp 21 (6); BWG 57; CTH 60; DEW 2:307; EFC cp 40; EGC 71; EGH 144; EWG 76; MBF cp 53; MCW 129; MFF cp 55 (956); NHE 100, 190, 288; OFP 111; OWF 41; PFE cp 159 (1535); PFF 333; PWF cp 21i; SPL cp 408; TEP 147; WFB 255; WUS 439.

Danesblood —See **Bellflower, Clustered.**

Danewort (Sambucus ebulus) EPF pl 769; MBF cp 41; MFF pl 95 (819); NHE 25; OWF 193; PFE cp 133 (1295); PWF 119k, 163m; TEP 257; WFB 223 (5).

Danio (Brachydanio kerri) ALE 4:318; FEF 178 (bw).

Danio, Giant (Danio malabaricus) ALE 4:318; FEF 183 (bw); HFI 103; SFW 269 (pl 67); (D. sp.) WFW cps 124, 125.

Danio, Pearl (Brachydanio albolineatus) ALE 4:318; HFI 103; SFW 269 (pl 68); WFW cp 117.

Danio, Spotted (Brachydanio nigrofasciatus) ALE 4:318; FEF 178 (bw); SFW 268 (pl 65); WFW cp 118.

Danio, Zebra (Brachydanio rerio) ALE 4:318; CTF 25 (cp 16); FEF 179 (bw); HFI 103; SFW 269 (pl 68); WFW cp 119.

Daphne, Rose —See **Garland Flower.**

Daphne, Sweet or Winter (Daphne odora)
DFP 195 (cp 1555); EGW 108; EJG 113; FHP
117; GDS cp 148; HSC 45; MWF 106; OGF
27; PFW 296; TGF 247.

Darkie Charlie (Dalatias licha) MOL 101;
OBV 9.

Darter, Australian (Anhinga novaehollandiae) HAB 15; RBU 96.

Darter, Indian or Oriental (Anhinga
melanogaster) CAS 160 (bw); KBA 32 (pl 1);
LEA 366 (bw).

Darwinia (Darwinia sp.) EWF cp 137f; MWF
106.

Dassie —See **Hyrax.**

Dasyure, Common or Spotted (Dasyurus
viverrinus) HMW 199; LVS 31; SMW 20
(bw); TWA 204 (bw); VWA cp 32; (D.
geoffroii) MEA 92.

Dasyure, Eastern (Dasyurus quoll) ALE
10:71; JAW 21; LVA 57; WAW 52 (bw).

Dasyure, Little Northern (Satanellus
hallucatus) DSM 26; MLS 52 (bw).

Date-plum (Diospyros lotus) BKT cp 179;
DEW 1:255.

Datura, Scarlet (Datura sanguinea) EHP
146; MEP 128; (D. rosei) EWF cp 183d.

Day Lily (Hemerocallis sp.) DFP 146 (cps
1167-1171); EDP 114; EEG 135; EGC 135;
EGP 64-67, 122; EJG 119; EPF pl 885; ESG
116; LFW 60, 61 (cps 141-145), 84 (cp 195);
MWF 153; NHE 191; OGF 73; PFE cp 164
(1597); PFF 374; SPL cps 287, 288; TGF 29;
TVG 99.

Day-flower (Commelina sp.) BIP 59 (cp 80);
DEW 2:280 (cp 158); HPW 280; LFT 18 (cp
8); PFF 369.

Dead Man's Fingers (fungus) (Xylaria sp.)
DLM 129; KMF 63; NFP 39; PFF 73.

Dead Men's Bells —See **Foxglove.**

Dead Men's Fingers (coral) (Alcyonium sp.)
ALE 1:45, 240, 271; CSS cp 5; LAW 62; OIB
19.

Dead-nettle, Cut-leaved (Lamium
hybridum) ESG 63; MBF cp 70; NHE 231.

Dead-nettle, Henbit (Lamium amplexicaule)
EGC 72; MBF cp 70; MFF cp 28 (770); NHE
231; OWF 149; PFE cp 111 (1127); PWF 50a;
WFB 201 (3); WUS 315.

Dead-nettle, Pyrenean —See **Dragonmouth** (Harminum).

Dead-nettle, Red or Purple (Lamium
purpureum) BFB cp 12; BWG 124; MBF cp
70; MFF cp 28 (771); NHE 231; OWF 149;
PWF 15f, 138c; WFB 201 (2).

Dead-nettle, Spotted (Lamium maculatum)
DEW 2:188; DFP 154 (cp 1231); EFC cps 53,
54; EGH 119; NHE 231; PFE cp 111 (1130);
PFW 151; PWF cp 19g; WFB 201 (1a).

Dead-nettle, White (Lamium album) AMP
67; BWG 85; EPF pls 810, 811; EWF cp 20d;
MBF cp 70; MFF pl 112; NHE 231; OWF 97;
PWF cp 15g; TEP 195; WFB 201 (1).

Death Cap (Amanita phalloides) CTH 18;
DEW 3: cp 63; DLM 74, 168; KMF 11; LHM
117; NFP 51; NHE 47; ONP 119; PMU cp
119; REM 64 (cp 32); RTM 26, 27, 269; SMF
36; TEP 223.

Death Cap, False (Amanita citrina) CTM 23;
DLM 63, 166; LHM 119; NHE 47; ONP 119;
PMU cp 118; REM 63 (cp 31); RTM 26, 244;
SMF 37; SMG 181; TEP 223.

Decaisnea (Decaisnea fargesii) DEW 1:95;
DFP 195 (cp 1559); GDS cp 151; HPW 50;
HSC 43; PFW 156.

Deer, Altai —See **Deer, Red.**

Deer, Axis (Axis axis) ALE 13:211, 285;
DAA 71; DSM 229; HMA 41; HMW 61; LEA
600 (bw); SMW 240 (cp 154); (Cervus axis)
CAS 183.

Deer, Barasingha —See **Barasingha.**

Deer, Barking —See **Muntjac.**

Deer, Brocket (Mazama americana or M.
gouazoubira) ALE 13:223; DSM 237; HMA
33.

Deer, Brow-antlered or Eld's (Cervus eldi)
ALE 13:213; DAA 92; (C. e. siamensis) LVA
51.

Deer, Caucasian Red —See **Maral.**

Deer, Fallow (Dama dama) ALE 13:210; DAE
56-57; DSM 229; FBW 17; HMA 43; HMW
61; LAW 564; LEA 476, 477, 597 (bw); MLS
385 (bw); OBV 147; SMW 240 (cp 156); TWA
69 (bw).

Deer, Hog (Axis porcinus) ALE 13:211; DAA
85.

Deer, Japanese —See **Sika, Formosan.**

Deer, Marsh (Blastocerus dichotomus) DSM
236; HMW 64; LAW 563; (Odocoileus
dichotomus) ALE 13:222.

Deer, Muntjac —See **Muntjac.**

Deer, Musk (Moschus moschiferus) ALE
13:209; DAA 44; DSM 227; HMA 41; HMW
58; MLS 383 (bw).

Deer, Pampas (Ozotoceros bezoarcticus) ALE
13:223; DSM 236; LVA 31; SMW 252 (bw);
(Odocoileus bezoarcticus) ALE 13:223.

Deer, Pere David's (Elaphurus davidianus)
ALE 13:218; CAA 144 (bw); DAA 138; LVA
51; MLS 387 (bw); PWC 245 (bw); SLP 47;
SMW 250 (bw); VWA cp 22.

Deer, Persian Fallow (Dama mesopotamica)
ALE 13:210; LVA 11; PWC 244 (bw); SLS
153.

Deer, Pudu (Pudu pudu) ALE 13:223; DSM
237; HMA 41; HMW 65; SMW 252 (bw).

Deer, Red (Cervus elaphus var.) ALE 13:163,
185-187, 215-217; CAS 44-45, 53 (bw); DAA
42, 90; DAE 36; DSM 230; HMA 43; HMW
60; JAW 173; LAW 563; LEA 594 (bw); LVS
136; MLS 386 (bw); OBV 143; PWC 245 (bw);
SMW 240 (cp 155); TWA 74, 75; WEA 308.

Deer, Roe (Capreolus capreolus) ALE 13:163,
191, 219; DAE 38; DSM 230; HMW 64; LAW
517; LEA 598 (bw); MLS 391 (bw); OBV 145;
SMW 274 (cp 159); TWA 76; (C. pygargus)
DAA 19.

Deer, Sambar —See **Sambar.**

Deer, Sika —See **Sika.**

Deer, Swamp (Cervus duvauceli) ALE
13:213; DAA 92; LVS 127; PWC 158-159;
(Capreolus duvauceli) DSM 231.

Deer, Tufted (Elaphodus cephalophus
cephalophus) ALE 12:103; 13:209; (E.
michianus) DAA 44.

Deer, Water (Hydropotes inermis) ALE
13:219; CAA 153 (bw); DAA 44; DAE 58;
FBW 25; HMW 64; MLS 390 (bw); OBV 149;
SMW 254 (bw).

Deer-grass (Scirpus cespitosus) DEW 2:272;
MBF cp 91; MFF pl 116 (1109).

Defassa —See **Waterbuck, Defassa or
Sing-sing.**

Desert Candle —See **Lily, Foxtail.**

Desert Pea, Sturt's (Clianthus formosus)
BIP 57 (cp 73); DFP 60 (cp 473); EWF cp
129d; MLP 107; PFW 164; WAW 120.

Desert Rose —See **Impala Lily** (Adenium).

Desfontainea (Desfontainea spinosa) BIP 69
(cp 97); DEW 2:100 (cp 48); DFP 196 (cp
1561); HSC 43, 45; (D. hookeri) EHP 151.

Desman, Pyrenean (Galemys pyrenaicus)
ALE 10:190; DSM 52; HMA 44; (Desmana
pyrenaica) DAE 89.

Desman, Russian (Desmana moschata) ALE
10:190; MLS 85 (bw).

Destroying Angel (Amanita verna) CTM 19;
DLM 169; NFP 52; NHE 47; PFF 81; REM 65
(cp 33); RTM 28, 269; SMG 175, 176.

Destroying Angel (Amanita virosa) CTM 19;
DLM 169; LHM 117; NFP 52; NHE 47; ONP
131; REM 65 (cp 33); RTM 29, 269.

Deutzia (Deutzia sp.) DFP 196 (cps 1562,
1565); EWF cp 99e; GDS 153; HSC 43; PFW
229; (D. rosea) DFP 196 (cp 1563); GDS cp
154; TVG 223.

Deutzia, Fuzzy (Deutzia scabra) DEW 1:292;
DFP 196 (cp 1564); EFS 109; EPF pl 370;
EPG 117; HSC 43, 45; MWF 109; SPL cp 82;
TGS 182.

Deutzia, Lemoine (Deutzia lemoinei) EFS
108; TGS 179.

Deutzia, Slender (Deutzia gracilis) EEG 95;
ELG 115; LFW 44 (cps 102, 103); MWF 109;
OGF 59; TGS 182.

Devil, Mountain or Thorny (Moloch
horridus) ALE 6:211, 237; CAU 183; COG cp
16; ERA 186, 187; HRA 34; HWR 103, 105;
LAW 412; MAR cp VII, pl 45; MEA 22; MWA
58; SIR 89; VRA 171; WAW 77 (bw).

Devil Ray or Devilfish —See **Manta,
Atlantic.**

Devil, Tasmanian—See **Tasmanian Devil.**

Devil's Backbone (Pedilanthus) —See
Redbird Cactus and **Lifeplant** (Bryophyllum sp.).

Devil's Breeches (Kohleria bogotensis) BIP
109 (cp 167); EGL 122; EWF cp 185e; MEP
141.

Devil's Claw (Phyteuma comosum) DEW
2:194; DFP 18 (cp 137); EWF cp 15b; NHE
284; PFE cp 140 (1352); SPL cp 460; TVG
161.

Devil's Coach-horse —See **Coach-horse, Devil's.**

Devil's Shoe Lace —See **Sea Lace.**

Devil's Tongue (Ferocactus latispinus) LCS cp 53; RDP 209.

Dewberry (Rubus caesius) MBF cp 29; MFF pl 76 (347); NHE 21; OFP 79; OWF 77; PFE cp 44 (429); PWF 62a, 147d.

Dewflower —See **Redondo Creeper.**

Dhak Tree —See **Flame-of-the-Forest.**

Dhole, Indian (Cuon alpinus) ALE 12:283; DAA 70; DSM 147; HMA 44, 45; JAW 102; MLS 264 (bw); SMW 199 (bw).

Diamond-bird —See **Pardalote.**

Diapensia (Diapensia lapponica) EWF cp 6a; HPW 131; MBF cp 57; MFF pl 77 (603); MLP 186; NHE 280; PFE cp 87 (911); WFB 169 (7).

Dibatag (Ammodorcas clarkei) ALE 13:426; DDM 225 (cp 38); VWA cp 9.

Dibbler —See **Mouse, Marsupial.**

Dichondra (Dichondra repens) EGC 24, 25, 33, 38, 130; ESG 108; HPW 230.

Dicranum (Dicranum sp.) EPF pl 90; ONP 83.

Didierea (Didierea madagascariensis) DEW 1:110 (cp 67), 201; EWF cp 70b; HPW 74.

Dieffenbachia —See **Dumb Cane.**

Dik-dik, Long-snouted (Rhynchotragus kirki) ALE 13:426; DDM 257 (cp 42); (R. guentheri) DDM 257 (cp 42); (Madoqua kirki) DSM 263; HMA 44.

Dik-dik, Phillips' (Madoqua saltiana) DDM 257 (cp 42); DSM 263.

Dikkop, Cape —See **Thick-knee, Spotted.**

Dill (Anethum graveolens) DEW 2:85; EGH 82-83, 98; EVF 142; OFP 139, 147.

Dingo (Canis dingo) ALE 12:257; BMA 72, 73; HMA 45; HMW 141; JAW 141; LEA 549 (bw); MWA 53 (bw); SMW 186 (cp 115); TWA 185; WAW 64-65 (bw); WEA 129.

Dioch, Red-billed —See **Quelea, Red-billed.**

Dionysia (Dionysia sp.) DFP 7 (cp 55); EWF cp 48 e, f.

Dipelta (Dipelta floribunda) EWF cp 94e; HSC 43.

Dipladenia (Dipladenia splendens or Mandevilla splendens) BIP 117 (cp 179); EGG 106; EGV 84, 122; EWF cp 180e; PFW 28; (D. sp.) BIP 71 (cp 100); FHP 117; LHP 82, 83; RDP 178.

Dipper (Cinclus leucocephalus) SSA pl 29; (C. pallasii) KBA 385 (pl 58).

Dipper, Eurasian (Cinclus cinclus) ABW 246; ALE 9:190, 11:105; 13:463; BBB 183; BBE 221; CDB 159 (cp 662); FBE 273; GAL 126 (cp 2); LEA 447 (bw); OBB 135; PBE 212 (cp 51); RBE 251; RBI 228; WAB 123.

Disanthus (Disanthus cercidifolius) DFP 196 (cp 1566); GDS cp 155; HSC 46, 48; SST cp 104.

Discus, Brown or Green (Symphysodon sp.) ALE 5:121, 131, 132; FEF 435, 436; HFI 35; HFW 199; LAW 345; SFW 600 (cp 165); WFW cps 378, 379.

Dissotis (Dissotis princeps) EWF cp 59b; LFT cp 117.

Distichodus (Distichodus rostratus) FEF 134 (bw); SFW 204 (pl 47).

Distichodus, Six-banded (Distichodus sexfasciatus) ALE 4:308; HFI 90; HFW 37 (cp 9); SFW 200 (pl 45).

Dittander (Lepidium latifolia) MBF cp 10; MFF pl 93 (62); NHE 134; PWF 117e; WFB 93 (6).

Dittany —See **Burning Bush** (Dictamnus).

Dittany of Crete (Origanum dictamnus) EGH 129; EPF pl 815.

Diucon, Fire-eyed (Pyrope pyrope) SSA pl 28; WFI 228 (bw).

Diver —See **Loon.**

Dock, Broad-leaved (Rumex obtusifolius) BWG 169; MBF cp 74; MFF pl 108 (545); NHE 224; OWF 59; PFF 152; PWF 65j; WFB 45 (3).

Dock, Curled (Rumex crispus) EGC 71; MBF cp 74; MCW 30; MFF pl 108 (544); NHE 183; OWF 59; PFE cp 9 (97); PFF 15; PWF 127k; WUS 131.

Dock, Fiddle (Rumex pulcher) MBF cp 74; MFF pl 108 (546); NHE 224; PFE 67; WFB 45 (5).

Dock, Golden (Rumex maritimus) MBF cp 74; MFF pl 108 (548); NHE 96; PFE 67.

Dock, Great Water (Rumex hydrolapathum) MBF cp 74; NHE 96; PFE cp 9 (96); PWF 127j; WFB 45 (6).

Dock, Marsh (Rumex palustris) MBF cp 74; PFE 67; PWF 107h; WFB 45 (7).

Dock, Red-veined or Wood (Rumex sanguineus) MBF cp 74; NHE 96; PFE cp 9 (98); PWF 127i.

Dock, Sharp or Clustered (Rumex conglomeratus) MBF cp 74; NHE 96; OWF 59; PWF 127e; WFB 45 (4).

Dock, Shield (Rumex scutatus) EGH 136; NHE 224; PFE 67.

Dock, Water (Rumex aquaticus) NHE 96; PFE 67.

Dodder (Cuscuta sp.) NHE 97, 228; WUS 295, 297, 299.

Dodder, Common or Lesser (Cuscuta epithymum) MBF cp 61; MFF pl 102 (669); NHE 228; OWF 125; PFE 101 (1044); PWF 156d; WFB 185 (4).

Dodder, Greater (Cuscuta europaea) EPF pl 786; MBF cp 61; NHE 228; PWF 117f.

Dog, Bush —See **Bushdog.**

Dog, Cape Hunting (Lycaon pictus) ALE 12:274, 283, 13:391; CAA 84 (bw) CAF 157 (bw); DDM 113 (cp 20); DSM 146; HMA 45; HMW 145; JAW 103; LAW 532; MLS 265 (bw); SLP 222; SMR 33 (cp 2); SMW 199 (bw); TWA 49; WEA 128.

Dog, Indian Wild —See **Dhole, Indian.**

Dog, Raccoon-like (Nyctereutes procyonoides) ALE 12:257; CAS 55 (bw); DAA 140; DAE 81; DSM 149; HMA 45; HMW 145; MLS 261 (bw).

Dog, Wild —See **Dog, Cape Hunting.**

Dogfish, Black-mouthed (Galeus melastomus) CSF 47; OBV 7.

Dogfish, Greater Spotted (Scyliorhinus stellaris) ALE 4:104; HFI 149; LAW 286; OBV 9.

Dogfish, Lesser Spotted (Scyliorhinus caniculus) ALE 4:104, 114; CSF 46; FEF 17; HFI 149; OBV 9; WEA 130.

Dogfish, Piked or Spiny (Squalus acanthias) ALE 4:104; CSF 45; HFI 141; OBV 7.

Dog's Tooth Violet (Erythronium denscanis) DFP 92 (cp 731); EPF pl 895; ERW

116; MWF 126; PFE cp 168 (1628); PFW 175; SPL cp 483.

Dog's-tail, Crested (Cynosurus cristatus) MBF pl 97; MFF 121 (1194); NHE 194; TEP 151.

Dog's-tail, Rough (Cynosurus echinatus) MFF pl 124 (1195); PFE cp 179 (1748).

Dogwinkle (Nucella sp.) ALE 3:56; CSS pl XVI; GSS 79; OIB 43.

Dogwood (Thelycrania sanguinea) MBF cp 41; MFF pl 74 (448).

Dogwood, Common (Cornus sanguinea) EET 198, 199; EPF pl 745, 746; NHE 22; OBT 32; OWF 193; PFE cp 82 (846); PWF 58c, 164i; TEP 287; WFB 155 (1).

Dogwood, Evergreen (Cornus capitata) MWF 94; SST 238.

Dogwood, Japanese or Oriental (Cornus kousa) BKT cp 173; DFP 191 (cp 1527); EET 199; EGT 43, 107; EJG 110; ELG 103; GDS cp 106; LFW 191 (cp 431); PFW 92; SPL cp 81; SST cp 240.

Dogwood, Tartarian or Siberian (Cornus alba) DFP 191 (cps 1524, 1525); EEG 93; EFS 23, 33, 103; EGW 102; EPF pl 747; GDS cp 102; HSC 34, 36; OGF 189; TGS 151.

Dolichothele (Dolichothele sp.) LCS cp 29; RDP 179; SCS cp 14.

Dollarbird (Eurystomus orientalis) ALE 9:57; DPB 209; GPB 215; HAB 135; KBA 177 (cp 24); LAW 511; RBU 179; SNB 118 (cp 58).

Dolphin (fish) (Coryphaena hippurus) ALE 5:102; HFI 26; HFW 187 (bw); LMF 86; (C. equiselis) MOL 94.

Dolphin (mammal) (Stenella sp.) ALE 11:480; OBV 183; (Stenodelphis blainvillei) ALE 11:479; (Cephalorhynchus commersonii) ALE 11:480.

Dolphin, Amazon (Inia geoffroyensis) DSM 141; HMA 47.

Dolphin, Bottle-nosed (Tursiops truncatus) ALE 11:480, 483, 499, 500; BMA 30-33; JAW 88; LEA 547 (bw); MLS 243 (bw); OBV 185; (T. aduncus) ALE 11:480.

Dolphin, Common (Delphinus delphis) ALE 11:480; DSM 140; HMA 47; HMW 107; LEA 546 (bw); MLS 242 (bw); OBV 183; SMW 229 (cps 134, 135); TWA 151 (bw).

Dolphin, Ganges River (Platanista gangetica) ALE 11:479; DSM 141; HMA 47.

Donkey's Tail (Scdum morganianum) EDP 92, 132; EFP 142; EGP 134; ELG 50; EWF cp 157c; RDP 27, 357; SPL cp 153.

Doorweed —See **Knotweed, Prostrate** (Polygonum).

Dormouse, Common or Hazel (Muscardinus avellanarius) ALE 2:363; 11:382; DAE 28; DSM 128; FBW 73; GAL 60 (bw); HMW 165; LAW 592; MLS 220 (bw); OBV 133; SMW 138 (cp 86); TWA 99 (bw); WEA 132.

Dormouse, Edible or Fat (Glis glis) ALE 11:382, 13:163; DAE 28; DSM 128; FBW 95; GAL 60 (bw); HMA 48; HMW 165; LAW 592; LEA 533 (bw); MLS 219 (bw); OBV 133; SMW 138 (cp 85).

Dormouse, Forest or Russian (Dryomys nitedula) ALE 11:382; DAE 82; HMW 165.

Dormouse, Garden (Eliomys quercinus) ALE 11:381, 382; DAE 28; GAL 60 (bw); HMW 165.

Dorstenia (Dorstenia sp.) DEW 1:102 (cp 50); EPF pl 355; EWF cp 53a.

Dory, John (Zeus faber) ALE 5:36; CSF 122; HFI 64; HFW 159 (bw); LEA 252 (bw); OBV 41; VWS 14 (bw); WFW cp 224.

Dorycnium (Dorycnium sp.) NHE 28, 199; PFE cp 60 (609, 610).

Dotterel (Eudromias morinellus) ALE 8:160, 169; BBB 121; BBE 117; CDB 81 (cp 263); FBE 123; OBB 63; PBE 93 (pl 30), 100 (cp 31); (Charadrius morinellus) RBE 132; (C. obscurus) BNZ 125.

Dotterel, Australian (Peltohyas australis) SNB 80 (cp 39), 82 (pl 40); WAB 197.

Dotterel, Black-fronted (Charadrius melanops) CDB 81 (cp 259); HAB 67; MBA 52; SNB 80 (cp 39), 82 (pl 40).

Dotterel, Double-banded (Charadrius bicinctus) BNZ 125; HAB 67; SNB 76 (cp 37), 78 (pl 38).

Dotterel, Hooded (Charadrius cucullatus) CDB 80 (cp 257); HAB 67; SNB 80 (cp 39), 82 (pl 40).

Dotterel, Large —See **Sandplover, Greater.**

Dotterel, Oriental —See **Plover, Oriental.**

Dotterel, Red-capped —See **Plover, Kentish.**

Dotterel, Red-kneed (Charadrius cinctus) HAB 67; SNB 80 (cp 39), 82 (pl 40).

Dotterel, Rufous-chested (Zonibyx modestus) SSA pl 21; WFI 154.

Double-tail (Japyx sp.) ALE 2:73; LAW 186; (Catajapyx confusus) PEI 27 (bw).

Douglasia (Douglasia vitaliana) DFP 8 (cp 57); TGF 174.

Douroucouli —See **Ape, Night.**

Dove, Barbary (Streptopelia risoria) BBB 280; FBE 173; OBB 105.

Dove, Barred Ground (Geopelia striata) ALE 8:259; CDB 101 (cp 363); DPB 137; KBA 160 (cp 21); SNB 102 (cp 50); (G. placida) HAB 84.

Dove, Bar-shouldered (Geopelia humeralis) CDB 100 (cp 362); HAB 84; SNB 102 (cp 50.

Dove, Cape or Namaqua (Oena capensis) ALE 8:259; CDB 101 (cp 369); FBE 173; PSA 177 (cp 20); WAB 149.

Dove, Cape Turtle (Streptopelia capicola) NBA 39; PSA 177 (cp 20).

Dove, Collared (Streptopelia decaocto) ALE 8:246; BBB 51; BBE 169; CDB 102 (cp 375); DAE 135; FBE 173; FBW 81; OBB 105; PBE 80 (pl 25); WAB 137.

Dove, Cuckoo —See **Cuckoo-dove.**

Dove, Diamond (Geopelia cuneata) ALE 8:259; CDB 99 (cp 361); HAB 84; SNB 102 (cp 50).

Dove, Emerald or Green-winged (Chalcophaps indica) ALE 8:259; BBT 9; CDB 97 (cp 351); DPB 137; KBA 160 (cp 21); SNB 104 (cp 51); WAB 172; WBV 109; (C. chrysochlora) HAB 83.

Dove, Fruit (Ptilinopus superbus) ALE 8:243; DPB 127; SNB 104 (cp 51); (P. melanospila) CDB 102 (cp 373); DPB 127; (P. jambu) KBA 113 (cp 12); NBB 34; (P. luteovirens) WAB 173.

Dove, Galapagos (Zenaida galapagoensis) CDB 103 (cp 382); CSA 253; HBG 97 (cp 6).

Dove, Java or Philippine Turtle (Streptopelia bitorquata) ABW 143; DPB 135.

Dove, Laughing or Palm (Streptopelia senegalensis) ALE 8:246; BBE 169; CDB 102 (cp 377); FBE 173; NBA 37; PBE 284 (cp 65); PSA 177 (cp 20); SNB 102 (cp 50).

Dove, Nicobar —See **Pigeon, Nicobar.**

Dove, Oriental or Rufous Turtle (Streptopelia orientalis) BBE 169; FBE 173; KBA 160 (cp 21); WBV 109.

Dove, Peaceful —See **Dove, Barred Ground.**

Dove, Quail (Geotrygon montana) CDB 101 (cp 364); WAB 83; (G. saphirina) SSA cp 31; (G. veraguensis) SSA cp 31.

Dove, Red Turtle (Streptopelia tranquebarica) DPB 135; KBA 160 (cp 21); WBV 109.

Dove, Red-eyed Turtle (Streptopelia semitorquata) ALE 8:443; CDB 102 (cp 376).

Dove, Ring —See **Pigeon, Wood.**

Dove, Rock —See **Pigeon, Common.**

Dove, Spinifex —See **Pigeon, Plumed.**

Dove, Spotted or Lace-necked (Streptopelia chinensis) CDB 102 (cp 374); DPB 135; KBA 160 (cp 21); SNB 102 (cp 50); WBV 109.

Dove, Stock (Columba oenas) BBB 75; BBE 167; CDB 99 (cp 357); FBE 171; OBB 103; PBE 80 (pl 25).

Dove Tree (Davidia involucrata) BKT cp 148; DFP 195 (cp 1558); EEG 115; EET 196; EGT 110; EPF pl 742; EPG 117; MTB 368 (cp 37); OBT 149, end paper; PFW 100; SST cp 242; TGS 198.

Dove, Turtle (Streptopelia turtur) ALE 8:246; BBB 136; BBE 169; CDB 103 (cp 378); DAE 135; FBE 173; GPB 175; LEA 420 (bw); OBB 105; PBE 80 (pl 25); RBE 188; (S. reichenowi) LVB 11; (S. picturata rostrata) LVS 226.

Dove, Wood (Turtur chalcospilos) CDB 103 (cp 380); PSA 177 (cp 20); (T. abyssinicus) CDB 103 (cp 379).

Dove, Zebra —See **Dove, Barred Ground.**

Dovekie (Plautus alle) ABW 138; ALE 8:211; BBB 277; BBE 163; DAE 31; FBE 165; GPB 170; PBE 125 (pl 36); SAO 143.

Dowitcher (Limnodromus semipalmatus) DPB 97; KBA 128 (pl 13), 149 (pl 18); (L. griseus) ALE 8:170.

Dracaena, Florida Beauty (Dracaena godseffiana) EDP 146; EFP 108; MWF 117.

Dracaena, Fragrant (Dracaena fragrans) DFP 65 (cp 516); EPF pl 920; LHP 84; PFF 383; SPL cp 18; (D. fragrans massangeana) EDP 145; EFP 107; RDP 180.

Dracaena, Gold Dust (Dracaena surculosa) DEW 2:214 (cp 124); PFW 17; RDP 180.

Dracaena, Palm (Dracaena deremensis) BIP 73 (cp 102); DFP 65 cp 515; PFW 17; RDP 181; SPL cp 17.

Dracaena, Red-margined (Dracaena marginata) BIP 73 (cp 103); DFP 65 (cp 517); EDP 29; EFP 18, 54, 108; LHP 84; RDP 181.

Dragon, Angle-headed —See **Lizard, Angle-headed.**

Dragon, Bearded —See **Lizard, Bearded.**

Dragon, Komodo —See **Monitor, Komodo Dragon.**

Dragon Plant or **Dragon, Green** (Dracunculus) —See **Arum, Dragon.**

Dragon Tree (Aloe sp.) —See **Quiver.**

Dragon Tree (Dracaena draco) EDP 27; EFP 14; EPF pl 917; MLP 261; PFW 15; RDP 181; SST cp 280.

Dragon, Two-lined (Diporiphora bilineata) ALE 6:210; COG cp 21.

Dragonet, Common (Callionymus lyra) CSF 137; HFI 48; HFW 237 (bw); OBV 71; WEA 133.

Dragonet, Spotted (Callionymus maculatus) ALE 5:161; CSF 137; HFI 48; OBV 71.

Dragonfish (Pegasus sp.) ALE 5:40; HFI 9.

Dragonfly —See also **Damselfly.**

Dragonfly (Calopteryx sp.) ALE 2:83, 89; 11:257; KIW 55 (bw); LAW 189; PWI 19; (Crocothemis nigrifrons) BIN cp 8; (Orthetrum sp.) ALE 2:84; CIB 76 (cp 3); (Pyrrhosoma sp.) ALE 2:83; CIB 69 (cp 2); LAW 187; OBI 3; (Cordulegaster sp). CIB 76 (cp 3); GAL 364 (cp 13); OBI 5; (Neurobasis chinensis) CAS 265.

Dragonfly, Aeschna (Aeschnna sp.) ALE 2:84, 89, 91; CIB 76 (cp 3); GAL 364 (cp 13); LAW 188, 189; OBI 5; PEI 42-44 (bw), 49 (cp 6), 64 (cp 7); WEA 134.

Dragonfly, Broad-bodied Libellula (Libellula depressa) ALE 2:84; CIB 76 (cp 3); GAL 364 (cp 13); OBI 7; PEI 46, 47 (bw).

Dragonfly, Club-tailed (Gomphus vulgatissimus) CIB 76 (cp 3); GAL 364 (cp 13).

Dragonfly, Downy Emerald (Cordulia sp.) CIB 76 (cp 3); GAL 364 (cp 13); OBI 7.

Dragonfly, Emperor (Anax imperator) CIB 76 (cp 3); GAL 364 (cp 13); OBI 5; PEI 45; ZWI 108.

Dragonfly, Sympetrum (Sympetrum sp.) CIB 76 (cp 3); KIW 39 (cp 12); OBI 7.

Dragonhead (Dracocephalum sp.) MCW 84; OGF 153; TGF 204; WFB 198.

Dragonmouth (Horminum pyrenaicum) EWF cp 37a; NHE 281; PFE cp 114 (1151).

Dragonroot or **Green Dragon** (Arisaema dracontium) ERW 98; EWG 87.

Dragon's Teeth (Tetragonolobus martimus) MFF cp 49 (311); NHE 180; PFE 206; PWF 139e; WFB 129 (3).

Drill (mammal) —See **Mandrill.**

Drill, Oyster (Urosalpinx cinerea) CSS pl XVI; OIB 43.

Dromedary —See **Camel, Arabian.**

Drongo, Ashy or **Gray** (Dicrurus leucophaeus) DPB 247; KBA 276 (pl 41).

Drongo, Black (Dicrurus macrocercus) ABW 220; GPB 332; KBA 276 (pl 41); WEA 135.

Drongo, Crow-billed (Dicrurus annectans) DPB 247; KBA 276 (pl 41).

Drongo, Fork-tailed (Dicrurus adsimilis) CDB 218 (cp 979); PSA 240 (cp 27); WBV 155.

Drongo, Racket-tailed (Dicrurus paradiseus) ALE 9:450; KBA 276 (pl 41); WAB 166; (D. remifer) KBA 276 (pl 41).

Drongo, Spangled (Dicrurus hottentottus) ALE 9:450; DPB 247; KBA 276 (pl 41); RBU 316; (D. bracteatus) SAB 72 (cp 35); (Chibea bracteata) HAB 148.

Dropwort (Filipendula vulgaris or F. hexapetala) DFP 142 (cp 1135); EGH 113; EWF cp 9d; MBF cp 26; MFF pl 94 (341); NHE 179; PFE cp 44 (422); PWF 101k; TGF 108.

Dropwort, Common Water (Oenanthe fistulosa) MBF cp 39; MFF pl 98 (493); NHE 95; OWF 89; PFE cp 84 (873); PWF 156f; WFB 163 (4).

Dropwort, Fine-leaved Water (Oenanthe aquatica) MBF cp 39; MFF pl 98 (496); NHE 95; OWF 89; PWF 110f; TEP 29; WFB 163 (6).

Dropwort, Hemlock Water (Oenanthe crocata) MBF cp 39; MFF pl 98 (495); OWF 89; PFE cp 84 (872); PWF 54e; WFB 161 (4).

Dropwort, Parsley Water (Oenanthe lachenalii) MBF cp 39; MFF pl 98 (494); OWF 89; PWF 126b; WFB 163 (5).

Drunken Sailor —See **Rangoon Creeper** (Quisqualis).

Drupe, Purple (Drupa morum) CTS 42 (cp 69); GSS 79.

Dryad's Saddle (Polyporus squamosus) DLM 156; KMF 15; LHM 75; ONP 129; REM 172 (cp 140); RTM 202; SMF 70; (P. sp.) CTM 66; DLM 156; LHM 75; PFF 79; REM 173, 174; RTM 203-209, 212, 263.

Dryandra (Dryandra sp.) EWC 134c; MWF 118.

Duck —See also specific names (i.e., **(Gadwall, Mallard, Old-squaw, Widgeon, etc.).**

Duck, African Black (Anas sparsa) ALE 3:307; PSA 64 (pl 5).

Duck, Auckland Island (Anas aucklandica aucklandica) BNZ 219; WAN 119; (A. a. chlorotis) LVB 41.

Duck, Australian Black (Anas superciliosa) ALE 7:307; CAU 155; FNZ 161 (cp 15); HAB 28; SNB 39 (cp 18); WAN 119.

Duck, Australian White-eyed (Aythya australis) ALE 7:307; HAB 32; SNB 38 (cp 18).

Duck, Bahama (Anas bahamensis) ALE 7:303; GPB 80; HBG 80 (pl 3).

Duck, Black-headed (Heteronetta atricapilla) ALE 7:307; WAB 95 (bw).

Duck, Blue or **Mountain** (Hymenolaimus malacorhynchos) ALE 7:307; BNZ 233; FNZ 161 (cp 15); SLP 151.

Duck, Blue-billed Stiff-tail (Oxyura australis) CDB 54 (cp 124); HAB 31; SNB 40 (cp 19).

Duck, Brazilian —See **Teal, Brazilian.**

Duck, Comb (Sarkidiornis melanotus) ALE 7:318; CDB 54 (cp 127); PSA 64 (pl 5).

Duck, Eider —See **Eider.**

Duck, Ferruginous (Aythya nyroca) ALE 7:307; BBB 269; BBE 59; FBE 59; OBB 23; PBE 29 (cp 10), 41 (pl 14), 49 (pl 16); RBE 95.

Duck, Freckled (Stictonetta naevosa) ALE 7:294; HAB 32; SNB 36 (cp 17).

Duck, Harlequin (Histrionicus histrionicus) ABW 69; ALE 7:309; BBE 61; CDB 52 (cp 116); FBE 63; GPB 84; PBE 36 (cp 11), 41 (pl 14), 49 (pl 16); RBE 96.

Duck, Knob-billed —See **Duck, Comb.**

Duck, Long-tailed —See **Old-squaw.**

Duck, Mallard —See **Mallard.**

Duck, Mandarin (Aix galericulata) ABW 70; ALE 7:317; BBB 280; BBE 57; BBT 44; CAS 154; CDB 49 (cp 93); FBE 51; FBW 23; OBB 21; PEB 28 (cp 9); RBI 71; SBB 36 (pl 39); WAB 161, 175; WEA 135.

Duck, Maned —See **Goose, Maned.**

Duck, Muscovy (Cairina moschata) ABW 68; ALE 7:318; CDB 51 (cp 107); OBB 21; SBB 35 (pl 38).

Duck, Musk (Biziura lobata) ALE 7:320; 10:89; CDB 50 (cp 102); HAB 31; RBU 68; SNB 40 (cp 19).

Duck, Paradise —See **Shelduck, Paradise.**

Duck, Patagonian Crested (Lophonetta specularioides) ALE 7:307; CDB 52 (cp 117); SSA cp 2; WFI 119.

Duck, Pink-eared (Malacorhynchos membranaceus) ALE 7:294; HAB 32; RBU 67; SNB 40 (cp 19).

Duck, Pintail —See **Pintail.**

Duck, Spotbill (Anas poecilorhyncha) ALE 7:306; DPB 29.

Duck, Steamer (Tachyeres sp.) ABW 68; ALE 7:294; CDB 54 (cp 128); SSA pl 21; WFI 19 (bw), 119, 122 (bw).

Duck, Teal —See **Teal.**

Duck, Torrent (Merganetta armata) ABW 68; ALE 7:294; GPB 81; SSA cp 2; WAB 93.

Duck, Tree —See **Tree-duck.**

Duck, Tufted (Aythya fuligula) ALE 7:61, 255, 308; BBB 202; BBE 59; CDB 50 (cp 101); DPB 33; FBE 59; GPB 82; OBB 25; PBE 29 (cp 10), 41 (pl 14), 49 (pl 16).

Duck, Whistling —See **Tree-duck.**

Duck, White-headed Stiff-tail (Oxyura leucocephala) ALE 7:320; BBE 67; FBE 67; GPB 85; PBE 37 (cp 12).

Duck, Yellow-billed (Anas undulata) ALE 7:307; CAF 244; PSA 49 (cp 4), 64 (pl 5).

Duckbill —See **Platypus.**

Duckweed, Common or Lesser (Lemna minor) DEW 2:315; HPW 307; MBF cp 88; MFF pl 104 (1100); NHE 107; PFE cp 184 (1823); PFF 367; WFB 293 (7).

Duckweed, Gibbous or Fat (Lemna gibba) HPW 307; MBF cp 88; MFF pl 104; NHE 107.

Duckweed, Great (Lemna polyrhiza) DEW 2:315; HPW 307; MBF cp 88; MFF pl 104 (1098); NHE 107; PFE cp 184 (1822).

Duckweed, Ivy (Lemna triscula) HPW 307; MBF cp 88; MFF pl 104; NHE 107; PFF 366; WFB 293 (8).

Duckweed, Least or Rootless (Wolffia arrhiza) HPW 307; MFF pl 104 (1102); NHE 107; WFB 293 (9).

Dugong (Dugong dugong) ALE 12:510; DAA 127; DDM 148 (bw); DSM 208; JAW 152; SLS 251; (Halicore australe) HMW 102.

Duiker, Banded (Cephalophus zebra) ALE 13:401; DDM 241 (cp 40); LAW 568; WEA 136.

Duiker, Bay (Cephalophus dorsalis) ALE 13:401; DDM 256 (cp 41); DSM 253.

Duiker, Black (Cephalophus niger) ALE 13:401; DDM 241 (cp 40).

Duiker, Black-fronted (Cephalophus nigrifrons) ALE 13:401; DDM 256 (cp 41).

Duiker, Blue (Cephalophus monticola) ALE 13:401; DDM 241 (cp 40); DSM 253; SMR 85 (cp 6); (Philantombia monticola) LAW 568.

Duiker, Gray or Grimm's (Sylvicapra grimmia) ALE 13:401; DDM 256 (cp 41); DSM 252; MLS 407 (bw); SMR 92 (cp 7).

Duiker, Jentink's (Cephalophus jentinki) DDM 241 (cp 40); LVS 126.

Duiker, Red (Cephalophus natalensis) CAF 129; DDM 256 (cp 41); HMA 48; TWA 27.

Duiker, Red-flanked (Cephalophus rufilatus) DDM 241 (cp 40); HMW 76.

Duiker, Yellow-backed or Giant (Cephalophus sylvicultor) ALE 13:401; DDM 241 (cp 40); DSM 252; HMW 76; SMR 85 (cp 6).

Duke of Argyll's Tea-Plant or Tea-Tree —See **Matrimony Vine.**

Dumb Cane (Pedilanthus sp.) —See **Redbird Cactus.**

Dumb Cane (Dieffenbachia amoena) BIP cp 99; DFP 64 (cp 511); LHP 80, 81; SPL cp 14; (D. leoniae) MWF 113.

Dumb Cane, Spotted or Variable (Dieffenbachia picta) DFP 65 (cp 513); EDP 112; EFP 52, 53, 106; EPF pl 1004; RDP 114; SPL cp 15; (D. maculata) MEP 12; RDP 176; (D. exotica) DFP 64 (cp 512); EGL 103; RDP 27, 176.

Dung Roundhead (Stropharia semiglobata)
DLM 51, 255; LHM 145; ONP 31; RTM 54.

Dunlin (Calidris alpina) ALE 7:61, 255; BBB
224; BBE 129; CDB 85 (cp 277); FBE 127;
GPB 145; KBA 148 (pl 17); NBB 19; OBB 77;
PBE 85 (pl 28), 100 (cp 31), 117 (cp 34); SNB
66 (cp 32); WAB 103; WBV 87; (Erolia
alpina) ABW 120.

Dunnart —See **Mouse, Fat-tailed Pouched.**

Dunnock —See **Sparrow, Hedge.**

Durian (Durio zibethinus) EWF cp 115a;
HPW 93; OFP 103; WWT 117.

Dusty Miller —See **Ragwort, Silver**
(Senecio sp.) and **Snow-in-Summer**
(Cerastium sp.).

Dutchman's Breeches (Dicentra cucullaria)
ESG 107; EWG 60, 82, 103; LFW 193 (cp
435); PFF 181; PFW 119.

Dutchman's Pipe (Aristolochia durior) ESG
92; HSC 13, 14; TGS 70; (A. brasiliensis)
SPL cp 155; (A. macrophyllum) TVG 217; (A.
saccata) EWF cp 110d; PFW 36; (A. salpinx)
PFW 37.

Duvalia (Duvalia sp.) LCS cp 240; LFT cp 133.

Dyckia (Dyckia sp.) EDP 132; FHP 119; MEP
14; MWF 119; RDP 183.

Dyer's- weed —See **Woad.**

Dzeggetai —See **Ass, Asian or Indian Wild.**

E

Eagle, African Fish or Sea (Haliaeetus
vocifer) ALE 7:376; CDB 58 (cp 152); GBP
327 (bw); PSA 80 (pl 7), 96 (cp 9); WAB 156.

Eagle, Australian Little (Haliaeetus
morphnoides or Hieraetus morphnoides)
HAB 44; LEA 386 (bw); MBA 42; MWA 59;
SNB 44 (cp 21), 52 (pl 25).

Eagle, Bateleur (Terathopius ecaudatus)
ABW 79; ALE 7:398; 8:443; CAA 45; CDB 61
(cp 165); GBP cp after 160, 358-360 (bw);
PSA 80 (pl 7), 96(cp 9); WAB 146; WEA 137.

Eagle, Black or Verreaux's (Aquila
verreauxi) ALE 7:375; CDB 56 (cp 137); FBE
81; GBP cp after 112; PSA 80 (pl 7), 96 (cp 9);
WAB 139.

Eagle, Bonelli's (Hieraetus fasciatus) ALE
7:366; BBE 77; DAE 117; FBE 79; GBP 311
(bw); KBA 91 (bw); PBE 69 (pl 22).

Eagle, Booted (Hieraetus pennatus) ALE
7:366; BBE 77; CDB 59 (cp 155); FBE 79;
KBA 91 (bw); PBE 69 (pl 22).

Eagle, Changeable Hawk (Spizaetus
cirrhatus) DPB 51; KBA 48 (pl 3), 64 (pl 5);
WBV 57.

Eagle, Crowned (Stephanoaetus coronatus)
ALE 7:366; CAF 230; CDB 61 (cp 164); GBP
307 (bw); (Polemaetus coronatus) PSA 80 (pl
7), 96 (cp 9).

Eagle, Golden (Aquila chrysaetos) ABW 78;
ALE 7:327; 13:463; BBB 112; BBE 73; CDB
55 (cp 135); FBE 79; FBW 39; GAL 159 (bw);
GBP cp before 65, cp after 112, 314 (bw);
GPB 94; LAW 475; OBB 37; PBE 68 (pl 21),
83 (bw); RBE 3; WAB 122; WEA 170.

Eagle, Gray-headed Fishing (Icthyophaga
ichthyaetus) ALE 7:376; DPB 41; GBP 330
(bw); KBA 49 (pl 4).

Eagle, Greater Spotted (Aquila clanga) ALE
7:375; BBE 75; DAE 88; FBE 81; KBA 44 (pl
4); RBE 7; WBV 57.

Eagle, Harpy (Harpia harpyja) ABW 79; ALE
7:366; GBP 161 (bw), 294 (bw); WAB 85.

Eagle, Imperial —See also **Vulture,
Bearded.**

Eagle, Imperial (Aquila heliaca) ALE 7:375;
BBE 75; CEU 36 (bw); FBE 79; GAL 160
(bw); GBP 316 (bw); GPB 95; RBE 4; SLP 41;
WAB 29, 117.

Eagle, Indian Black (Ictinaetus malayensis)
ALE 7:375; GBP 322 (bw); KBA 49 (pl 4).

Eagle, Lesser Spotted (Aquila pomarina)
ALE 7:375; BBE 75; FBE 81.

Eagle, Long-crested (Lophoaetus occipitalis) ALE 7:366; CDB 59 (cp 156); GBP 306
(bw).

Eagle, Martial (Polemaetus bellicosus) ALE 7:366; CAF 95; CDB 60 (cp 161); GBP 111 (bw), cp after 112, 309 (bw); GPB 96; PSA 80 (pl 7), 96 (cp 9).

Eagle, Monkey-eating (Pithecophaga jefferyi) ALE 7:327, 365, 375; DPB 49; GBP 297 (bw); GPB 96-97; LVS 247; PWC 98, 219 (bw); SLS 175; WAB 31, 170.

Eagle, New Guinea Harpy (Harpyopsis novaeguineae) ALE 7:366; GBP 298 (bw).

Eagle, Ornate Hawk (Spizaetus ornatus) ABW 80; ALE 7:366; GBP 34 (bw), cp before 161, 301 (bw); SSA pl 43; WAB 97.

Eagle, Pallas' Sea (Haliaeetus leucoryphus) ALE 7:376; BBE 73; DAA 148; FBE 71; KBA 49 (pl 4).

Eagle, Rufous-bellied (Hieraetus kienerii) DPB 51; KBA 91 (bw).

Eagle, Serpent (Spilornis sp.) DPB 41; KBA 48 (pl 3), 49 (pl 4), 64 (pl 5); WBV 57; (Dryotriorchis sp.) GBP 366 (bw).

Eagle, Short-toed (Circaetus gallicus) ALE 7:398; BBE 77; CAS 75; CEU 48 (bw); FBE 73; GBP 361 (bw); KBA 64 (pl 5); PBE 69 (pl 22); WAB 169.

Eagle, Snake (Circaetus cinereus) CDB 57 (cp 142); GPB cp after 112, 170-175 (bw); PSA 80 (pl 7).

Eagle, Spanish Imperial (Aquila heliaca adelberti) LVB 47; LVS 241; PWC 20; SLS 89.

Eagle, Steller's Sea (Haliaeetus pelagicus) ALE 7:376; 8:229; GBP 325 (bw).

Eagle, Steppe (Aquila nipalensis) ALE 7:375; KBA 49 (pl 4).

Eagle, Tawny (Aquila rapax) BBE 75; CAS 106 (bw); CDB 55 (cp 136); FBE 81; PSA 80 (pl 7), 96 (cp 9); WAB 110.

Eagle, Verreaux's —See **Eagle, Black or Verreaux's.**

Eagle, Wahlberg's (Aquila wahlbergi) CDB 56 (cp 138); PSA 80 (pl 7).

Eagle, Wedge-tailed (Aquila audax or Uroaetus audax) ALE 10:89; CAU 163 (bw); CDB 55 (cp 134); GBP 319 (bw); HAB 45; SNB 46 (cp 22), 54 (pl 26); WAB 193; WAW 33 (bw).

Eagle, Whistling —See **Kite, Whistling.**

Eagle, White-bellied Sea (Haliaeetus leucogaster) ALE 7:376; CDB 58 (cp 151); DPB 41; GBP 326 (bw); HAB 46; KBA 49 (pl 4); SNB 46 (cp 22), 54 (pl 26); WBV 57.

Eagle, White-tailed or Gray Sea (Haliaeetus albicilla) ALE 7:376; BBB 271; BBE 73; CAS 46 (bw); CDB 58 (cp 150); DAE 118; FBE 71; GAL 158 (bw); GBP 160 (bw), 325 (bw); OBB 37; PBE 68 (pl 21); RBE 16.

Earth Nut or **Earth Almond** (Cyperus) —See **Nutsedge.**

Earth Star (Cryptanthus bivittatus) EDP 144; EGG 101; EGL 99; LHP 74; RDP 162.

Earth Star (Cryptanthus fosterianus) BIP 65 (cp 89); DFP 61 (cp 485); LHP 75; PFW 54.

Earth Star (Cryptanthus zonatus) DEW 2:255; EPF pl 1035; FHP 115; RDP 163.

Earth Star (Geastrum sp.) DEW 3: cp 76, 188; DLM 32, 132, 133; EPF pls 70, 71; KMF 127; LHM 221; NFP 88; ONP 155; PFF 88; RTM 232; SMF 59.

Earthball (Scleroderma sp.) CTM 74; DEW 3:186; DLM 138; KMF 95; LHM 223; NHE 48; ONP 155; PFF 89; PMU cp 62; REM 179 (cp 147); RTM 227; SMG 49, 50; TEP 225.

Earthcreeper, Striated (Upucerthia serrana) ABW 199; ALE 9:125.

Earthnut —See **Pignut.**

Earth-nut, Great (Bunium bulbocastanum) MBF cp 37; MFF pl 96 (484); NHE 223.

Earth-smoke —See **Fumitory, Common.**

Earth-tongue (Geoglossum sp.) DEW 3:124; LHM 45; NFP 37; RTM 239.

Earthworm (Lumbricus sp.) ALE 2:51; GAL 519 (bw); OIB 113.

Earthworm, Brandling (Eisenia foetida) LAW 103; (E. rosea) OIB 113.

Earwig (Labia minor) CIB 96 (cp 7); OBI 9; (Labidura riparia) CTI 22 (cp 17).

Earwig, Common (Forficula auricularia) ALE 2:115; CIB 96 (cp 7); GAL 310 (bw); LAW 197; LEA 115 (bw); OBI 9; PEI 71 (bw); PWI 53 (bw); WEA 139.

East Indian Arrowroot —See **Bat Flower.**

Easter-ledges —See **Bistort, Snakeroot** (Polygonum).

Echeveria (Echeveria sp.) BIP 73 (cps 105, 106); DFP 65 (cp 519); ECS 112; EGL 104; EMB 106; EWF cp 157a; HPW 145; LCS cps 244, 246; LFP 188; MEP 142; MWF 119; PFW 95; RDP 184, 185; SCS cp 78; SPL cp 132.

Echidna (Tachyglossus aculeatus or T. setosus) ALE 10:41, 42, 89; CAU 63 (bw); HMA 72; JAW 18; LAW 524; LEA 481 (bw); SLP 139; WAW 69 (bw); WEA 141; (Echidna aculeata) TWA 200.

Echidna, Long-beacked (Zaglossus bruijni) ALE 10:41; CAU 248 (bw); DSM 21; HMW 208; MLS 38 (bw).

Echium, Canary Islands (Echium bourgaeanum) RWF 50; (E. fastuosum) MWF 121; (E. wildpretii) EWF cp 44a.

Edelweiss (Leontopodium alpinum) DFP 13 (cp 98); EPF pl 539; ERW 125; MLP 242; PFE cp 144 (1379); PFW 84; RWF 22; SPL cp 455; TEP 95; TGF 45; (L. haplophylloides) HPW 266.

Edelweiss, Brazilian (Rechsteineria leucotricha) EGG 137; EGL 139.

Edelweiss, North Island (Leucogenes leontopodium) DFP 13 (cp 100); SNZ 159 (cp 503).

Eel, Common or European (Anguilla anguilla) ALE 4:167; CSF 83, 85; CSS pl XXII; FWF 109; GAL 225 (bw); HFI 81; HFW 47 (cp 30); OBV 87; SFW 93 (pl 14).

Eel, Conger (Conger conger) ALE 4:167; CSF 87; CSS pl XXII; FEF 259; HFI 83; LAW 304; OBV 85; WEA 110.

Eel, Cusk —See **Cusk-eel.**

Eel, Electric (Electrophorus electricus) ALE 4:317; FEF 139 (bw); HFI 92; HFW 114 (bw); LAW 319; LEA 217 (bw); SFW 337 (cp 86); WFW cp 104.

Eel, Gulper —See **Gulper-eel.**

Eel, Moray (Muraena sp.) ALE 4:168, 174; FEF 258 (bw); HFI 82; HFW 44 (cp 25); LAW 303; OBV 85; WFW cps 19, 21, 24; (Lycodontis sp.) ALE 4:175; CAU 140; WFW cp 22.

Eel, Sand —See **Sand-eel.**

Eel, Snipe —See **Snipe-eel** (Nemichythys).

Eel, Spiny (Mastacembelus sp.) ALE 5:230; HFI 17; HFW 272 (cp 145); SFW 833 (bw); WFW cps 463, 464; (Macrognathus aculeatus) FEF 516 (bw); WFW cp 462.

Eel, Swamp (Synbranchus marmoratus) ALE 5:45; (S. afer) HFI 61.

Eel-grass (Zostera marina) MBF cp 90; MFF pl 68 (971); MLP 268; NHE 137; PFF 339; WFB 295 (8).

Eelpout (Zoarces viviparous) ALE 4:432; CSF 155; CSS cp 31; HFI 46.

Egg Plant (Solanum melongena) AMP 39; EMB 150; EVF 38, 40, 93; OFP 127; PFF 292; PFW 282.

Eggs and Bacon —See **Toadflax.**

Egret, Cattle (Ardeola ibis) ALE 7:197; BBE 37; CAA 29 (bw); CEU 41 (bw); FNZ 113 (pl 10); GPB 67; HAB 22; LEA 370 (bw); PSA 33 (cp 2); SNB 28 (cp 13).

Egret, Cattle (Bubulcus ibis) ABW 51; ALE 13:40, 355; CDB 43 (cp 56); DPB 15; FBE 37; HBG 96 (cp 5); KBA 32 (pl 1), 112 (cp 11); PBE 9 (cp 4); WAB 218; (B. egret) CAF 16.

Egret, Chinese (Egretta eulophotes) DPB 15; KBA 32 (pl 1); SNB 28 (cp 13).

Egret, Common or Great White (Egretta alba) BBE 37; BNZ 183; CAU 278 (bw); CDB 43 (cp 58); CSA 157; DPB 15; FBE 37; FNZ 113 (pl 10); HAB 22; KBA 32 (pl 1); MBA 52; PBE 9 (cp 4); PSA 33 (cp 2); RBE 71; SLP 38; SNB 28 (cp 13); WBV 41; (Casmerodius albus) ABW 51; ALE 7:186, 195; CAA 32 (bw); HBG 96 (cp 5).

Egret, Little (Egretta garzetta) ALE 7:197; BBB 273; BBE 37; CEU 59 (bw); DPB 15; FBE 37; FNZ 113 (pl 10); HAB 23; KBA 32 (pl 1); LEA 369 (bw); NBB 40 (bw), 86 (bw); PBE 9 (cp 4); PSA 33 (cp 2); RBE 68; SBB 8 (pl 3); SNB 28 (cp 13); WBV 41; (E. egretta) OBB 15.

Egret, Pacific Reef (Egretta sacra) BNZ 183; CDB 44 (cp 59); DPB 15; KBA 32 (pl 1); SNB 28 (cp 13), 30 (cp 14).

Egret, Plumed or Lesser (Egretta intermedia) DPB 15; HAB 23; KBA 32 (pl 1); MEA 45; SNB 28 (cp 13).

Egyptian Paper Plant (Cyperus papyrus) CAF 105; DFP 62 (cp 492); EPF pl 956; MLP 301; MWF 102; PFF 361; RDP 170; SPL cp 500.

Egyptian Star Cluster (Pentas lanceolata) DEW 2:134; FHP 138; LFW 163 (cp 363); MWF 225; PFW 267; RDP 306.

Ehretia (Ehretia sp.) HPW 234; LFT cp 139; MEP 119.

Eider, Common (Somateria mollissima) ABW 71; ALE 7:309; 8:187, 229; BBB 230; BBE 43, 65; FBE 61; LEA 380 (bw); OBB 27; PBE 36 (cp 11), 41 (pl 14), 49 (pl 16); RBE 99; SLP 35; WAB 128.

Eider, King (Somateria spectabilis) ALE 7:309; BBB 277; BBE 65; FBE 61; PBE 36 (cp 11), 41 (pl 14), 49 (pl 16).

Eider, Spectacled (Somateria fischeri) ALE 7:307; 8:229; FBE 61; (Arctonetta fischeri) ABW 71.

Eider, Steller's (Polysticta stelleri) ABW 69; ALE 7:309; 8:229; BBE 65; CDB 54 (cp 126); FBE 61; PBE 36 (cp 11).

Elaeocarpus (Elaeocarpus sp.) EWF cp 116d, cp 146c; HPW 90.

Eland, Cape or Common (Taurotragus oryx) ALE 13:61, 302; CAA 72; DDM 177 (cp 28); DSM 245; HMA 48; HMW 70; JAW 182; MLS 397 (bw); SLP 234; SMR 112 (cp 9); SMW 281 (cp 173); TWA 21; WEA 143.

Eland, Giant (Taurotragus derbianus) ALE 13:302; DDM 177 (cp 28); LVA 13; PWC 246 (bw).

Elder, Alpine or Red (Sambucus racemosa) AMP 135; DFP 238 (cp 1898); GDS cp 438; NHE 25; PFE cp 134 (1297); TEP 290; TGS 327.

Elder, Dwarf —See **Danewort** (Sambucus).

Elder, European (Sambucus nigra) BFB cp 11 (5); BWG 92; DEW 2:139; EET 199; EPF pls 767, 768; HSC 110, 112; MBF cp 41; MFF pl 74 (820); NHE 25; OBT 33; OFP 191; OWF 193; PWF 39e, 163e; SPL cp 405; SST cp 166; TEP 290; TGS 331.

Elecampane (Inula helenium) AMP 93; CTH 60 (cp 96); EFC cp 34; EGH 117; MBF cp 45; MFF cp 56 (858); PFE cp 144 (1388); PFF 318; WFB 237 (3).

Elegia (Elegia juncea) EWF cp 88a; HPW 284.

Elephant, African (Loxodonta africana) ALE 12:464, 486, 487; BMA 80-84; CAA 60-63 (bw), 68; CAF 114 (bw), 115 (bw), 116-117, 138 (bw), 167 (bw); DSM 205; HMA 8, 49, 50; HMW 99; JAW 150; LAW 575; LEA 580 (bw); MLS 337 (bw); PWC 166 (bw); SLP 246-248; SMW 287 (cp 189); WEA 144-146; (Elaphus africanus) TWA 54.

Elephant, Asian or Indian (Elaphus indicus or E. maximus) ALE 12:464, 485; 13:285; BMA 82-83; CAS 170-171 (bw), 202-203; DAA 99; DSM 204; HMA 50, 51; HMW 98; JAW 149; LAW 574; LEA 579 (bw); LVA 43; MLS 338 (bw); SLS 136-137; SMW 288 (cp 190); TWA 170; WEA 145.

Elephant Bush (Portulacaria afra) ECS 138; MWF 240.

Elephant Ear —See **Velvet Leaf**.

Elephant-foot Tree (Beaucarnea recurvata) ECS 101; EDP 137; EFP 92.

Elephant's Ears —See also **Lily, Cunjevoi** (Alocasia sp.) and **Caladium**.

Elephant's Ears (Colocasia esculenta var. antiquorum) EGB 104; MLP 289; OFP 181; SPL cp 498.

Elephant-shrew, Checkered-back (Rhynchocyon cirnei) ALE 10:180; DDM 32 (cp 1); DSM 49.

Elephant-shrew, Forest (Petrodomus sultani) ALE 10:180; HMW 193.

Elephant-shrew, Four-toed (Petrodromus tetradactylus) DDM 32 (cp 1); DSM 48.

Elephant-shrew, North African (Elephantulus rozeti) ALE 10:180; CAA 87 (bw); HMW 193; LAW 598; MLS 82 (bw).

Elephant-shrew, Short-eared (Macroscelides proboscideus) ALE 10:180; LEA 492 (bw).

Elephant-snout—See **Mormyrid, Elephant-snout**.

Elk, European (Alces alces) ALE 11:257; 13:190, 224; BMA 154-157; DAE 106; DSM 232; HMA 72; HMW 63; JAW 176; MLS 388 (bw); SLP 90.

Elm, Chinese (Ulmus parvifolia) EET 155; EGT 147; EGW 145; EJG 148; EPG 151; TGS 134; (U. parvifolia pendens) EGE 147; ELG 130.

Elm, Dutch (Ulmus hollandica) MTB 241 (cp 22); OBT 160, 161; TGS 134.

Elm, English (Ulmus procera) BKT cps 98, 101; EET 154, 156, 157; MBF cp 76; MFF cp 17 (556); MTB 241 (cp 22); OBT 12, 16; OWF 199, 203, 204; PWF 13l, 25i, 159g; TGS 134; (U. campestris) HPW 96; SST 179.

Elm, Fluttering or European White (Ulmus laevis) EPF pls 356, 357; NHE 24; PFE 51; SST cp 180; TEP 269.

Elm, Smooth-leaved or Wheatley (Ulmus carpinifolia) BKT cp 100; DFP 242 (cp 1932); EET 155, 156; EGT 146; MTB 241 (cp 22); (U. minor) GDS cp 483; NHE 24; OBT 160, 161.

Elm, Wych (Ulmus glabra) BKT cp 99; DEW 1:157; EET 154, 155; MBF cp 76; MTB 241 (cp 22); NHE 24; OBT 12, 16, 160, 161; OWF 199; PFE cp 5 (56); PWF 13k, 25h, 159h; TEP 270; WFB 34, 35 (1).

Elsholtzia (Elsholtzia stauntonii) HSC 46; TGS 342.

Emerald (Amazilia franciae) SHU 44; SSA cp 33; (A. iodura) SHU 74, 121; (A. leucogaster) ALE 8:453; (A. tzacatl) SHU 73, 106; (Chlorostilbon sp.) SHU 41, 124; SSA cp 32.

Emperor Fish —See **Angelfish** (Pomacanthus sp.).

Empress Tree (Paulownia imperialis or P. tomentosa) DFP 216 (cp 1723); EGT 131; EWF cp 103b; MTB 384 (cp 39); OBT 197; PFW 280; SST cp 257; TGS 358.

Emu (Dromaius novaehollandiae or Dromiceius novaehollandiae) ABW 19; ALE 7:88; CAU 92-93 (bw); CDB 34 (cp 4); GPB 33; HAB 2; LEA 358 (bw); MEA 53; MWA 52 (bw); RBU 11; SNB 135 (bw); WAB 193; WAW 6 (bw).

Emu Bush (Eremophila sp.) EWF cp 132a; HPW 242.

Emu-wren, Mallee (Stipiturus mallee) HAB 191; SAB 20 (cp 9).

Emu-wren, Rufous-crowned (Stipiturus ruficeps) HAB 191; MWA 18, 55; SAB 20 (cp 9); WAB 196.

Emu-wren, Southern (Stipiturus malachurus) ABW 258; ALE 9:259; HAB 191; RBU 211; SAB 20 (cp 9).

Enkianthus, Redvein (Enkianthus campanulatus) DFP 197 (cp 1574); EFS 110; ELG 115; EPF pl 479; ESG 109; EWF cp 101b; GDS cp 162; HSC 46, 49; MWF 121; PFW 108; SPL cp 83; TGS 407.

Enkianthus, White (Enkianthus perulatus) EJG 114; GDS cp 163; (E. cernuus) DFP 197 (cps 1575, 1576).

Entoloma (Entoloma sp.) DLM 190, 191; NFP 78; PMU cps 91, 92; PUM cps 83, 84; RTM 82, 83, 287.

Epaulette Tree (Pterostyrax hispida) EGT 138; TGS 407.

Epiplatys (Epiplatys sp.) ALE 4:450; CTF 36 (cp 38); FEF 243, 244, 284, 286 (bw); HFI 68; SFW 441 (pl 116); WFW cp 200.

Eranthemum, False (Pseuderanthemum sp.) BIP 149 (cp 228); RDP 330.

Erinus, Alpine (Erinus alpinus) DEW 2:161 (cp 78); HPW 244; MFF cp 28 (699); NHE 281; OGF 43; PFE cp 127 (1235); PWF 42a; TGF 221; WFB 213 (5).

Ermine —See **Weasel, Short-tailed.**

Eryngo —See also **Sea Holly.**

Eryngo (Eryngium sp.) DEW 2:89; DFP 141 (cps 1126, 1127); EDP 139; HPW 220; NHE 95; OGF 155; TVG 89.

Eryngo, Alpine (Eryngium alpinum) DFP 141 (cp 1124); EWF cp 17d; NHE 275; OGF 155; SPL cp 204.

Eryngo, Blue (Eryngium amethystinum) NHE 199; PFE cp 83 (856); TGF 125.

Eryngo, Field (Eryngium campestre) EPF pl 757; KVP cp 93; MBF cp 36; MFF pl 106 (455); NHE 221; PFE cp 83 (855); PWF 143h; TEP 115.

Escallonia (Escallonia sp.) DFP 200 (cps 1595, 1596); EGE 123; GDS cp 182; HSC 50, 52; MWF 127; OGF 159; PFW 113.

Escargot—See **Snail, Edible.**

Escuerzo, Painted (Ceratophrys ornata) ALE 5:449, 455; CAW 77 (cp 27), 91 (bw); ERA 96; HRA 105; MAR pl 6.

Espeletia (Espeletia sp.) MLP 249; RWF 84.

Eucalyptus —See **Gum Tree.**

Eucryphia (Eucryphia sp.) DFP 201 (cps 1601, 1602); GDS cps 184, 185; OBT 137; OGF 171; PFW 113.

Euglenoid (Euglena sp.) ALE 1:88; LEA 10, 15, 16 (bw).

Eulalia (Miscanthus sinensis) DFP 160 (cp 1275); EGW 127; EJG 129; MWF 201; TVG 129.

Eulophia (Eulophia quartiniana) EWF cp 69c; PFW 212.

Euonymus (Euonymus sp.) DEW 2:64 (cp 41); DFP 201 (cp 1606); EFS 110; EWF cp 101a; HSC 53; NHE 22, 266; TGS 176.

Euphonia (Euphonia sp. or Tanagra sp.) ABW 291; CDB 201 (cp 880); SSA cp 40.

Euphorbia (Euphorbia sp.) DEW 1:264, 266; DFP 142 (cp 1131); EPF pls 631-635; EWF cp 58; HPW 186; LFT cps 95-97; LCS cps 247-253, 321; MLP 117; MWF 130; NHE 33, 96, 224; PFM cps 91, 92; PFW 115; SPL cps 86, 137; TGF 122; WWT 75.

Euphorbia, Globe or Basketball (Euphorbia obesa) ECS 13, 117; EPF pl 632; LCS cp 256; RWF 54; SCS cp 81.

Euphorbia, Milk-striped (Euphorbia lactea) EDP 27, 137; EFP 111; GFI 87 (cp 18).

Evening-primrose (Oenothera sp.) DFP 161 (cp 1287); MLP 169; MWF 216; NHE 93; OGF 127; PFE cp 81 (834).

Evening-primrose, Common (Oenothera biennis) BFB cp 10 (8); DEW 2:23; EFC cp 18 (2); EPF pls 717, 718; HPW 163; KVP cp 87; MCW 75; NHE 220; PFE cp 81 (831); PFF 263; SPL cp 230.

Evening-primrose, Fragrant (Oenothera stricta) MFF cp 62 (437); PWF 111c; WFB 149 (8).

Evening-primrose, Large-flowered (Oenothera erythrosepala) BWG 37; MFF cp 62 (436); PWF 156b; WFB 149 (7).

Everlasting (Helichrysum milfordiae) DFP 11 (cp 87); OGF 91; TVG 153.

Everlasting (Helichrysum bracteatum) DFP 39 (cp 305); EDP 131; EGA 123; EPF pl 541; EWF cp 140e; LFW 164 (cps 365-367); MWF 150; OGF 191; PFF 331; PFW 86; SPL cp 211; TGF 45.

Everlasting (Helichrysum sp.) DFP 145 (cp 1159); EWF cp 65a; GDS cp 230; LFT cp 174; OGF 43, 65; PFE cp 144 (1385); PFM cps 194, 195; SNZ 49 (cp 112), 123 (cp 380), 124 (cp 384), 154 (cp 486).

Everlasting (Helipterum sp.) EWF cp 83d; MWF 152.

Everlasting, Italian (Helichrysum italicum) AMP 141; NHE 246.

Everlasting, Swan River (Helipterum manglesii) DFP 39 (cp 307); EGA 124; TGF 45.

Everlasting, Winged (Ammobium alatum) DFP 30 (cp 234); EGA 93; EWF cp 140c; OGF 191; TGF 45.

Everlasting, Yellow (Helichrysum arenarium) NHE 76; PFE 432; WFB 235 (7).

Eyebright (Euphrasia sp.) DEW 2:157; EPF pl 847; MBF cp 64; MFF pl 90 (728); NHE 74, 187, 282; OWF 93; PFE cp 127 (1242, 1244); PWF 93g; SNZ 152 (cp 477), 159 (cps 502, 505); WFB 217 (1).

F

Fabiana (Fabiana imbricata) DFP 201 (cp 1608); GDS cp 193; HSC 51, 56; OGF 65.

Fair Maid of February —See **Snowdrop.**

Fair Maids of France —See **Buttercup, White** (Ranunculus).

Fairy —See **Hummingbird.**

Fairy Stool (Coriolus versicolor) CTM 65; DEW 3: cp 56; DLM 147; KMF 39; RTM 209.

Fairy Thimbles —See **Foxglove.**

Fairy-borage, Dwarf —See **King of the Alps.**

Fairy's Thimble (Campanula cochlearifolia) DFP 3 (cp 24); EPF pl 522; NHE 284; PFE cp 140 (1338); PFW 62; SPL cp 473.

Falcon, Amur (Falco amurensis) KBA 65 (pl 6); PSA 97 (cp 10).

Falcon, Barred Forest (Micrastur ruficollis) CDB 63 (cp 171); SSA pl 43.

Falcon, Black (Falco subniger) HAB 37; SNB 48 (cp 23), 56 (pl 27).

Falcon, Brown (Falco berigora) CDB 62 (cp 167); GBP 136 (bw), cp after 160, 397 (bw); HAB 37; SNB 48 (cp 23), 56 (pl 27).

Falcon, Collared Forest (Micrastur semitorquatus) ALE 7:403; GBP 373 (bw); WAB 91.

Falcon, Desert (Falco pelegrinoides) ALE 7:417; FBE 93.

Falcon, Eleonora's (Falco eleonorae) ALE 7:417; BBE 89; FBE 93; LAW 481; PBE 61 (cp 20); WAB 150.

Falcon, Gray (Falco hypoleucus) HAB 37; SNB 48 (cp 23), 56 (pl 127).

Falcon, Hobby —See Hobby.

Falcon, Lanner (Falco biarmicus) ALE
7:404; BBE 87; CDB 62 (cp 168); FBE 91;
GBP cp before 65, 391 (bw); PBE 61 (cp 20);
PSA 97 (cp 10); WAB 152.

Falcon, Laughing (Herpetotheres cachinnans) ALE 7:403; GBP cp after 144, 371
(bw); SSA pl 43.

Falcon, Little (Falco longipennis) HAB 36;
SNB 48 (cp 23), 56 (pl 27).

Falcon, New Zealand (Falco novaeseelandiae) BNZ 113; FNZ 99 (bw); WAN 120;
(Nesierax novaeseelandiae) ALE 7:404.

Falcon, Peregrine (Falco peregrinus) ABW
84; ALE 7:417; BBE 87; CDB 62 (cp 169);
DPB 51; FBE 93; GAL 154 (bw); GBP cp after
96, cp before 113, 143 (bw), cp after 160, 187
(bw), 393 (bw); HAB 36; HBG 81 (pl 4); KBA
65 (pl 6); LVB 51; OBB 47; PBE 61 (cp 20), 73
(pl 24); RBU 79; SNB 48 (cp 23), 56 (pl 27);
WAB 35; WAW 32 (bw); WBV 61.

Falcon, Pygmy (Polihierax semitorquatus)
ALE 7:403; CDB 63 (cp 172); GBP 384 (bw);
PSA 97 (cp 10); (P. insignis) KBA 48 (pl 3),
65 (pl 6).

Falcon, Red-footed (Falco vespertinus) ALE
7:423; BBB 275; BBE 91; FBE 95; PBE 61 (cp
20), 73 (pl 24); RBE 31; WAB 100.

Falcon, Red-headed or -necked (Falco
chiquera) ALE 7:418; PSA 97 (cp 10).

Falcon, Saker (Falco cherrug) ALE 7:404;
BBE 87; FBE 91; PBE 61 (cp 20); RBI 76.

Falcon, Sooty (Falco concolor) ALE 7:418;
FBE 93.

Falconet, Red-legged (Microhierax caerulescens) ABW 84; ALE 7:403; GBP 386 (bw);
KBA 48 (pl 3), 65 (pl 6); WAB 161; (M.
erythrogonys) DPB 51.

Falconet, Spot-winted (Spiziapteryx
circumcinctus) ALE 7:403; GBP 383 (bw);
SSA pl 22.

False Jerusalem Cherry —See Cherry,
Christmas (Solanum).

False-asphodel, Common —See Asphodel,
German.

False-brome, Chalk (Brachypodium
pinnatum) DEW 2:258; MBF pl 99; MFF pl
132 (1206); NHE 195.

False-brome, Wood (Brachypodium
sylvaticum) MBF pl 99; MFF pl 132 (1205);
NHE 40.

Falseflax, Smallseed (Camelina microcarpa) NHE 219; WUS 197.

False-helleborine, Black (Veratrum
nigrum) NHE 290; PFE 162 (1586).

False-helleborine, White (Veratrum album)
AMP 163; EFC cp 27; NHE 290; PFE cp 162
(1586); WFB 261 (6); (V. lobelianum) EPF pl
878.

False-sedge (Kobresia simpliciuscula) MBF
cp 91; MFF pl 115 (1125); NHE 290, 291.

Fanaloka —See Civet, Malagasy.

Fantail, Black or Gray (Rhipidura fuliginosus) BNZ 41; CDB 183 (cp 790); FNZ 199
(bw); RBU 239; SAB 40 (cp 19).

Fantail, Northern (Rhipidura setosa) HAB
210; (R. rufiventris) SAB 40 (cp 19).

Fantail, Pied or Malaysian (Rhipidura
javanica) BNZ 41; DPB 333; KBA 384 (pl
57); WBV 179.

Fantail, Rufous (Rhipidura rufifrons) CAU
19; CDB 183 (cp 791); HAB 210; MBA 13;
RBU 240; SAB 40 (cp 19); (R. phoenicura)
GPB 286.

Fanworm, Twin (Bispira volutacornis) CSS
cp 13; VWS 59; WEA 148.

Fat Hen (Chenopodium album) BWG 160;
DEW 1:203; EGC 72; MBF cp 71; MCW 32;
MFF pl 110 (207); NHE 228; OWF 55; PFE cp
10 (105); PFF 156; PWF 166a; TEP 191;
WFB 49 (2); WUS 133.

Father Lasher (Cottus scorpius) HFI 58;
OBV 67.

Fatsia, Japanese (Fatsia japonica or Aralia
japonica) BIP 81 (cp 122); DFP 202 (cp 1610);
EDP 126; EFP 19, 55, 112; EGE 125; EJG
115; ELG 138; EPF pl 753; GDS cps 194, 195;
HSC 54, 56; MWF 132; RDP 201, 202; SST cp
111; TGS 246.

Featherback (Notopterus chitala) CTF cp 1;
HFW 101 (bw); LAW 305.

Featherback, False —See Knifefish, Black
African.

Feather-grass, Common (Stipa pennata)
DEW 2:264; DFP 174 (cp 1390); NHE 245;
PFE cp 182 (1786); TEP 126.

Feather-grass, Hair-like (Stipa capillata) HPW 287; NHE 245; TEP 126; (S. sp.) DFP 174 (cp 1389); EPF pl 1057.

Feather-star (Antedon sp.) ALE 3:282, 291; CSS cp 26; LAW 257; LEA 197 (bw); MOL 90; OIB 181.

Felwort, Autumn —See **Gentian, Autumn.**

Felwort, Marsh (Swertia perennis) EPF pl 686; NHE 185; PFE cp 95 (1002); WFB 183 (9).

Fenestraria (Fenestraria rhodalophylla) DEW 1:197; ECS 13; SCS cp 83.

Fennec —See **Fox, Fennec.**

Fennel, Common (Foeniculum vulgare) AMP 155; BFB cp 11; BWG 41; DFP 143 (cp 1137); EGH 114; EPF pl 763; EVF 143; EWF cp 17c; MBF cp 39; MFF cp 51 (498); OFP 139; OWF 47; PFF 266; PWF 119i; WFB 167 (4).

Fennel, Giant (Ferula communis) DEW 2:62 (cp 35); PFE cp 85 (897); PFM cp 115.

Fennel-flower (Nigella sativa) EGH 128; PFE 18 (203); (N. arvensis) NHE 210; (N. ciliata) EWF cp 27c.

Fenugreek (Trifolium ornithopodioides) MBF cp 22; WFB 131 (8).

Fenugreek (Trigonella foenum-graecum) EGH 146; PFE cp 58 (580); PFF 212; WFB 127 (8); (T. sp.) MFF pl 79 (276); PFE cp 57 (577).

Fer-de-Lance (Bothrops atrox) ERA 218, 219; HRA 47; SSW 134.

Fern, Adder's-tongue (Ophioglossum sp.) CGF 183; NFP 127; SAR 81; WWT 36.

Fern, Adder's-tongue (Ophioglossum vulgatum) CGF 183; DEW 3:262 (cp 129); EGF 126; MFF pl 136 (1306); NFP 127; NHE 198; ONP 191; PWF 46f.

Fern, Angola Stag's-horn (Platycerium angolense) DEW 3:268 (cp 148); EGF 131; (P. grande) RDP 320; LHP 152.

Fern, Bead —See **Fern, Sensitive.**

Fern, Beech (Thelypteris phegopteris) CGF 83; MFF pl 139 (1300); NFP 140; NHE 43; TEP 231.

Fern, Bird's-nest (Asplenium nidus) DEW 3: (cps 147, 150), Fig.374; EDP 133; EGP 96; EPF pl 102; RDP 10, 89; (A. nidus-avis) DFP 52 (cp 412); EFP 91; LHP 36, 37.

Fern, Boston (Nephrolepis exaltata) EDP 93, 101, 133, 140; EFP 16, 49, 125; EGF 80, 82-83, 124; LHP 132-133; MWF 211; NFP 142; RDP 13; SPL cp 24.

Fern, Brake or Bracken (Pteridium aquilinum) BWG 157; CGF 135; DEW 3:281; EGF 138; EPF pls 103, 104; MFF pl 138 (1277); NFP 134; NHE 43; ONP 185; PFF 98; SNZ 4 (cp 86); TEP 231; WUS 9.

Fern, Bulblet Bladder (Cystopteris bulbifera) CGF 157; EGF 107; NFP 141; PFF 103.

Fern, Button (Pellaea rotundifolia) EDP 113; EGF 128; EGL 135; RDP 12, 304.

Fern, Cliff-brake (Pellaea atropurpurea) CGF 137; EGF 14, 128; NFP 135.

Fern, Cretan Brake (Pteris cretica) DEW 3:281; EFP 139; EGF 68, 139; LHP 154; NHE 245; RDP 331; SPL 31.

Fern, Curly-Grass (Schizaea pusilla) CGF 165; EGF 14; NFP 131.

Fern, Delta Maidenhair (Adiantum raddianum) EGF 92; EGL 80; RDP 70.

Fern, Elephant's-ear —See **Fern, Angola Stag's-horn.**

Fern, Fan Maidenhair (Adiantum tenerum wrightii) EDP 128; EFP 85; EGF 92.

Fern, Fiji or Rabbit's-foot (Davallia fejeensis) EDP 128; EFP 106; EGF 108; RDP 173.

Fern, Fragile or Brittle Bladder (Cystopteris fragilis) CGF 159; EGF 107; MFF pl 139 (1291); NFP 141; NHE 245; ONP 73; TEP 89.

Fern, Golden Polypody (Polypodium aureum or Phlebodium aureum) EFP 137; EGF 133; NFP 133; RDP 326.

Fern, Grape (Botrychium sp.) CGF 185-193; EGF 13, 100; NFP 128; NHE 80, 198; PFF 106.

Fern, Green Cliff-brake (Pellaea viridis) EGF 129; RDP 304.

Fern, Hard or Deer (Blechnum spicant) DEW 3: cp 146; EGF 99; EPF pl 10; MFF pl 140 (1281); NHE 43; ONP 189; TEP 232; TVG 135.

Fern, Hartford (Lygodium palmatum) CGF 163; EGF 14, 120; NFP 130.

Fern, Hart's-tongue (Phyllitis scolopendrium or Asplenium scolopendrium) CGF 61; DEW 3: cps 144, 145; EGF 129; EPF pl 113; MFF pl 140 (1282); NFP 143; NHE 43; ONP 190; RDP 316; SPL cp 459; TVG 135.

Fern, Hay-scented (Dennstaedtia punctilobula) CGF 117; EGF 108; ESG 107; NFP 141; PFF 103.

Fern, Holly (Cyrtomium falcatum) EFP 105; EGF 106; EGL 103; EJG 113; RDP 171.

Fern, Japanese Climbing (Lygodium japonicum) DEW 3: Fig. 350; EGF 120; EGV 81, 121; EJG 127.

Fern, Japanese Felt (Cyclophorus lingua) EGF 106; EJG 96, 112.

Fern, Japanese Painted (Athyrium georingianum pictum) EGF 97; EJG 96, 101; ESG 95.

Fern, Jersey (Anogramma leptophylla) MFF pl 139 (1279); (A. chaerophylla) EGF 95.

Fern, Killarney or Filmy or Bristle (Trichomanes sp.) CGF 167; DEW 3:355; EGF 15, 143; MFF pl 139 (1275); NFP 130.

Fern, Lady (Athyrium filix-femina) CGF 111; EGF 96; MFF pl 138 (1290); NFP 137; NHE 43; ONP 189; PFF 99; TEP 231.

Fern, Leather (Acrostichum aureum) DEW 3: Fig. 360; EGF 90; NFP 133.

Fern, Maidenhair (Adiantum reniforme) DEW 3: cp 136; EMB 119.

Fern, Male (Dryopteris felix-mas) AMP 57; CGF 67; CTH 20 (cp 6); DEW 3: cp 141, Fig. 368, 369; EFC cp 20; EGF 113; EPF pl 114; MFF pl 138 (1294); NFP 138; NHE 43; ONP 185; TEP 231; TVG 135.

Fern, Marsh (Thelypteris palustris) CGF 85; MFF pl 139 (1299); NFP 139; NHE 109; PFF 102.

Fern, Mexican Tree (Cibotium schiedei) EFP 100; EGF 104; (C. glaucum) EGF 103.

Fern, Mosquito (Azolla caroliniana) CGF 179; EPF pl 117; NFP 146; NHE 108.

Fern, Mother or Parsley (Asplenium bulbiferum) EGF 95; ESG 94; LHP 36; RDP 89.

Fern, Mountain Bladder (Cystopteris montana) MFF pl 139 (1292); NHE 293.

Fern, Northern Maidenhair (Adiantum pedatum) CGF 141; EGF 14, 92; ESG 89; NFP 135; PFF 98.

Fern, Ostrich (Matteuccia struthiopteris) CGF 119; EGF 121; EPF pl 109; NHE 43; PFF 104; TEP 232; TVG 135.

Fern, Parsley (Cryptogramma crispa) EGF 104; MFF pl 139 (1278); NHE 293; ONP 55.

Fern, Polystichum (Polystichum sp.) CGF 127, 129; DEW 3: cp 138; EFP 138; EGF 46, 47, 136, 137; EGW 134; EPF pl 115; MFF pl 138 (1296, 1297); NFP 140; NHE 293; ONP 55, 187; PFF 100; RDP 328; SAR 80.

Fern, Ribbon or Stove —See **Fern, Cretan Brake.**

Fern, Rosy Maidenhair (Adiantum hispidulum) EGF 92; RDP 70.

Fern, Royal (Osmunda regalis) CGF 169; DEW 3: cp 135, Fig. 349; EGF 13, 127; EPF pl 106; ESG 132; MFF pl 138 (1274); NFP 129; NHE 80; ONP 95; PFF 105.

Fern, Rusty-back (Ceterach officinarum) AMP 135; EGF 102; EPF pl 108; MFF pl 140 (1289); NHE 245; ONP 73; SPL cp 446.

Fern, Sensitive (Onoclea sensibilis) CGF 121; EGF 124; ESG 131; NFP 145; PFF 104.

Fern, Shoestring (Vittaria lineata) EGF 14; NFP 134.

Fern, Southern Maidenhair (Adiantum capillis-veneris) AMP 143; CGF 143; EGF 91; LHP 17; MFF 139 (1280); MWF 32; NFP 135; NHE 293; ONP 1; RDP 26, 70; SPL cp 1.

Fern, Spear-leaved or Hand (Doryopteris pedata palmata) EDP 112; EGF 111.

Fern, Spider Brake (Pteris multifida) EFP 139; EGE 139; LHP 154.

Fern, Stag's-horn (Platycerium bifurcatum or P. alcicorne) DEW 3:269 (cp 149); EDP 136; EFP 136; EGF 132; EPF pl 112; LHP 153; MWF 236; RDP 321; SPL cp 30; WWT 36.

Fern, Tasmanian Tree (Dicksonia antarctica) BKT cp 1; EGF 109; MWF 112; WWT 39; (D. squarrosa) EGF 15.

Fern, Tongue —See **Fern, Japanese Felt.**

Fern, Venus-hair —See **Fern, Southern Maidenhair.**

Fern, Water (Azolla filiculoides) DEW 3: cp
157; EGP 67, 98; EMB 120; MFF pl 104
(1304); ONP 97.

Fern, Whiskbroom (Psilotum nudum) EPF
pl 92; NFP 119; PFF 110; (P. triquetrum)
DEW 3: Fig. 338, cp 123.

Fernbird (Bowdleria punctata) BNZ 37; WAB
209.

Fernwren —See **Wren, Fern.**

Ferret or **Fitchew** —See **Polecat,
European.**

Ferret-badger —See **Badger.**

Fescue (Festuca sp.) EWF cp 7b; NHE 291,
292; PFF 358.

Fescue, Blue (Festuca ovina glauca) ECC 30,
32, 132; EGW 112; ESG 112; TVG 129.

Fescue, Creeping or Red (Festuca rubra)
EGC 79, 117; MBF pl 98; MFF pl 130 (1175);
NHE 195.

Fescue, Darnel (Catapodium marinum) MBF
pl 98; MFF pl 128 (1186).

Fescue, Giant (Festuca gigantea) MBF pl 98;
pl 132 (1174); NHE 40.

Fescue, Meadow (Festuca pratensis) MBF pl
98; MFF pl 132 (1173); NHE 195; TEP 151.

Fescue, Rat's-tail (Vulpia myuros) MBF pl
98; MFF pl 132 (1181); NHE 244.

Fescue, Sheep's (Festuca ovina) DFP 142 (cp
1134); MBF pl 98; MFF pl 130 (1176); NHE
244.

Fescue, Squirrel-tail (Vulpia bromoides)
MBF pl 98; MFF pl 129 (1180); NHE 244.

Fescue, Tall (Festuca arundinacea) EGC 117;
MBF pl 98; NHE 195; PFE cp 180 (1757).

Fescue, Viviparous (Festuca vivipara) EWF
cp 25c; MFF pl 129 (1177).

Fescue, Wood (Festuca altissima) MBF pl 98;
NHE 40.

Feverfew (Chrysanthemum parthenium)
AMP 57; BWG 97; EGA 106; EGG 97; LFW
36 (cp 85); MBF cp 46; MFF pl 100 (893);
MWF 84; OWF 99; PFE cp 148 (1429); SPL cp
389.

Feverfew (Tanacetum parthenium) PWF
101j; WFB 241 (2).

Fieldfare (Turdus pilaris) ALE 9:290; BBB
82; BBE 265; DAE 134; FBE 257; GPB 270;
OBB 137; PBE 220 (cp 53).

Fig, Banjo or Fiddle-leaved (Ficus lyrata)
EFP 16, 113; LHP 94; MWF 133; RDP 210;
SPL cp 20; SST cp 114.

Fig, Barbary (Opuntia ficus-indica) PFE cp
79 (818); PFF 255; PFM cps 110, 111; SPL cp
148.

Fig, Common (Ficus carica) BKT cp 97; DEW
1:160; EET 221; EGT 113; EPF pls 353, 354;
EVP 122; HPW 97; NHE 294; OBT 200; OFP
95; PFF 149; PFM cp 7; SST cp 203.

Fig, Creeping or Climbing (Ficus pumila)
BIP 85 (cp 126); DFP 67 (cp 531); EDP 29;
EGL 110; EGV 6, 39, 78, 108; EMB 125; LHP
96, 97; MWF 133; RDP 210; SPL 165.

Fig, Mistletoe (Ficus deltoides) DEW 1: cp 51;
LHP 92; RDP 210.

Fig, Red Hottentot (Carpobrotus acinaciformis) KVP cp 165; MEP 37; PFE cp 11
(120); PFM cp 12; SPL 147; (C. chiliensis)
MWF 73.

Fig, Sycamore (Ficus sycomorus) DEW 1:103
(cp 52); SST cp 116.

Fig, Weeping (Ficus benjamina) BIP 85 (cp
128); DFP 66 (cp 528); EGE 126; EFP 113;
LHP 93; MWF 133; RDP 27, 211; SST cp 113.

Fig, Yellow Hottentot (Carpobrotus edulis)
ECS 131; EGC 126; KVP cp 166; MFF cp 15
(201); PFM cp 13; WFB 43 (5).

Figbird, Southern (Sphecotheres vieilloti)
HAB 162; SAB 72 (cp 35).

Figbird, Yellow (Sphecotheres flaviventris)
ABW 222; ALE 9:450; HAB 162; SAB 72 (cp
35); WAB 183.

Fighting-fish, Siamese (Betta splendens)
ALE 5:219, 229; CTF 44 (cps 53, 54); FEF
525; HFI 53; HFW 217 (cp 104); LAW 280;
SFW 660 (cp 181); WFW cp 454.

Fig-parrot —See **Parrot, Fig.**

Figwort —See also **Celandine, Lesser**
(Ranunculus).

Figwort (Scrophularia sp.) MBF cp 63; NHE
99; PFE cp 123 (1216), 384; PFM cp 178;
WFB 209 (7).

Figwort, Balm-leaved (Scrophularia scorodonia) MBF cp 63; PFE cp 123 (1215); PWF
162d; WFB 209 (6).

Figwort, Common (Scrophularia nodosa)
DFP 171 (cp 1368); KVP cp 21; MBF cp 63;
MFF cp 37 (692); NIIE 99; OWF 141; PWF
99h; WFB 209 (5).

Figwort, Water (Scrophularia aquatica)
MBF cp 63; MFF cp 37 (693); NHE 99; PWF
75k.

Figwort, Yellow (Scrophularia vernalis)
EPF pl 841; MFF cp 61 (694); NHE 35; WFB
209 (8).

Filbert —See also **Hazel.**

Filbert (Corylus maxima) EET 144; EWF cp
12c; GDS cp 115; OBT 152; OFP 27.

Filefish, Fringed (Monacanthus sp.) ALE
5:256; LEA 218 (bw), 257.

Filefish, Red-toothed (Odonus niger) ALE
5:255; FEF 520 (bw); WFW cp 474.

Filmy Fern, Tunbridge (Hymenophyllum
sp.) DEW 2: Fig. 354; EGF 20, 118; MFF pl
140 (1276); NHE 293; ONP 61.

Finch, Black-throated (Melanodera
melanodera) SSA pl 30; WFI 99 (bw);
(Poephila cincta) HAB 259; SAB 70 (cp 34).

Finch, Cactus Ground (Geospiza scandens)
ALE 9:351; HBG 112 (cp 7); (G. conirostris)
HBG 12 (cp 7).

Finch, Chestnut-breasted (Lonchura
castaneothorax) ALE 9:439; CDB 212 (cp
939); HAB 261; SAB 70 (cp 34).

Finch, Citril (Serinus citrinella) BBE 281;
FBE 293; PBE 261 (cp 62).

Finch, Crimson (Neochmia phaeton) ALE
9:439; CDB 212 (cp 940); HAB 257; SAB 68
(cp 33); (Rhodospingus cruentus) SSA cp 18.

Finch, Cut-throat (Amadina fasciata) ALE
9:439; GPB 323; LAW 509; PSA 309 (cp 38).

Finch, Double-barred (Stizoptera bichenovii) HAB 261; SAB 68 (cp 33).

Finch, Firetail —See **Firetail.**

Finch, Gouldian (Chloebia gouldiae) ALE
9:439; HAB 259; SAB 70 (cp 34); (Poephila
gouldiae) ABW 303; CDB 212 (cp 942); RBU
319.

Finch, Large Ground (Geospiza magnirostris) ABW 297; ALE 9:351; BBT 117;
HBG 112 (cp 7).

Finch, Long-tailed (Poephila acuticauda)
ALE 9:439; HAB 259; SAB 70 (cp 34).

Finch, Masked (Poephila personata) HAB
260; SAB 70 (cp 34).

Finch, Medium Ground (Geospiza fortis)
BBT 116; CDB 209 (cp 924); GPB 320; HBG
112 (cp 7).

Finch, Melba (Pytilia melba) ABW 303; ALE
9:429; PSA 316 (cp 39).

Finch, Ochre-breasted Brush (Atlapetes
semirufous) CDB 208 (cp 917); DTB 141 (cp
70); SSA cp 42.

Finch, Painted (Emblema picta) ALE 9:439;
HAB 256; SAB 68 (cp 33).

Finch, Parrot —See **Parrot-finch.**

Finch, Pictorella (Heteromunia pectoralis)
HAB 261; SAB 70 (cp 34).

Finch, Plum-headed (Aidemosyne modesta)
CDB 211 (cp 935); HAB 257; SAB 68 (cp 33).

Finch, Plush-capped (Catamblyrhynchus
diadema) ABW 292; ALE 9:362; DTB 137 (cp
68); GPB 305; SSA cp 16.

Finch, Red-browed (Aegintha temporalis)
ALE 9:439; HAB 259; SAB 68 (cp 33).

Finch, Red-headed (Amadina erythrocephala) CDB 211 (cp 936); PSA 309 (cp 38).

Finch, Rose —See **Rosefinch.**

Finch, Saffron (Sicalis flaveola) ABW 299;
ALE 9:342; CSA 41; SSA cp 42.

Finch, Scarlet (Haematospiza sipahi) KBA
433 (cp 64); RBI 315.

Finch, Scarlet Rose —See **Grosbeak,
Scarlet.**

Finch, Seed —See **Seedfinch.**

Finch, Small Insectivorus Tree (Camarhynchus parvulus) ALE 9:351; HBG 113 (cp
8).

Finch, Snow (Montifringilla nivalis) ALE
9:416; 13:463; BBE 297; CDB 214 (cp 953);
CEU 162; FBE 301; PBE 276 (cp 63); WAB
122; (M. adamsi) RBI 304.

Finch, Spice (Lonchura punctulata) ALE
9:439; DPB 413; HAB 261; KBA 401 (pl 60);
SAB 70 (cp 34); WBV 221.

Finch, Star (Bathilda ruficauda) HAB 257;
SAB 68 (cp 33).

Finch, Tanager (Oreothraupis arremonops)
DTB 143 (cp 71); SSA cp 42.

Finch, Trumpeter (Rhodopechys githaginea) BBE 285; FBE 289; PBE 284 (cp 65).

Finch, Vegetarian Tree (Platyspiza
crassirostris) ALE 9:351; HBG 112 (cp 7).

Finch, Warbler (Certhidea olivacea) ALE
9:351; HBG 113 (cp 8).

Finch, Warbling (Poospiza sp.) ALE 9:342; SSA cp 16.

Finch, Widow —See Widow-finch.

Finch, Woodpecker (Camarhynchus pallida) ABW 296; CSA 257 (bw); HBG 113 (cp 8); (Cactospiza pallida) ALE 9:333, 351.

Finch, Yellow-rumped (Lonchura flavi-prymna) HAB 261; SAB 70 (cp 34).

Finch, Zebra (Taeniopygia castanotis or T. guttata) ALE 9:439, 10:89; CDB 213 (cp 945); HAB 261; LAW 509; SAB 68 (cp 33); WEA 2.

Finch-lark (Eremopterix sp.) FBE 199; PSA 240 (cp 27); WAB 153.

Finfoot, African or Peter's (Podica sene-galensis) ABW 112; ALE 8:100; GPB 132.

Finfoot, Masked (Heliopais personata) KBA 119 (bw); WAB 174.

Finger Tree —See Milkbush.

Fingerfish, Common (Monodactylus argenteus) ALE 5:111; CTF 49 (cp 62); FEF 364 (bw); HFI 30; HFW 175 (cp 86); SFW 540 (pl 145).

Finger-grass —See Crab-grass (Digitaria).

Fir, Balsam (Abies balsamea) PFF 121; SST cp 2.

Fir, Caucasian or Nordmann (Abies nordmanniana) BKT cp 25; EET 102; NHE 294; OBT 116; SST cp 4; TGS 35 (bw).

Fir, Chinese (Cunninghamia lanceolata) BKT cp 14; DEW 1:40; EGE 94; EJG 112; EPF pl 152; MTB 84 (cp 5); TGS 38.

Fir, Colorado White or Silver (Abies concolor) BKT cp 18; EEG 123; EET 101; EGE 90; EGW 88; ELG 121; EPF pl 123; EPG 102; MTB 92 (cp 7); PFF 121; TGS 23.

Fir, Common or European Silver (Abies alba) BKT cp 19; DEW 1:33; DFP 251 (cp 2001); EET 102; EPF pls 121, 122; MTB 92 (cp 7); NHE 18; OBT 116; SST cp 1; TEP 267; WFB cp 25(1).

Fir, Douglas (Pseudotsuga menziesii) BKT cps 13, 26, 28; DEW 1:34; EEG 128; EET 54, 55, 114, 115; EGW 136; EPF pl 125; MTB 145 (cp 12); NHE 18; OBT 100, 109; PWF 169e; SST cp 37; WFB cp 25 (1a).

Fir, Grand or Giant (Abies grandis) BKT cp 24; EET 100; MTB 92 (cp 7); OBT 108; TGS 35 (bw).

Fir, Greek (Abies cephalonica) EET 103; PFE cp 1-1; PFM cp 5; SST cp 3.

Fir, Korean Silver (Abies koreana) BKT cps 20, 21; DEW 1:54 (cp 13); EET 102, 103; OBT 116.

Fir, Noble (Abies procera) BKT cp 22; EET 100; EPF pl 124; MPL 34; MTB 92 (cp 7); OBT 100, 108; PWF 171h.

Fir, Spanish (Abies pinsapo) EET 103; OBT 117; SST cp 5.

Fir, Veitch's Silver (Abies veitchii) BKT cp 23; OBT 117; TGS 35 (bw).

Fire Bush, Chilean (Embothrium cocci-neum) DFP 197 (cps 1572, 1573); GDS cps 160, 161; HSC 47; MEP 32; OBT 141.

Fire Dragon Plant —See Beefsteak Plant.

Fireback —See Pheasant.

Firebrat (Thermobia domestica) OBI 1; PEI 31 (bw).

Firebug, Common (Pyrrhocoris apterus) ALE 2:187; CIB 112 (cp 9); PEI 104 (bw).

Firecracker Flower (Crossandra infundi-buliformis) EGL 99; FHP 115; LHP 72, 73; MEP 144; MWF 97; PFW 14; RDP 161; SPL cp 43.

Firecracker Plant —See also Fountain Plant.

Firecracker Plant (Echeveria setosa) EPF pl 383; EWF cp 157d; LCS cp 245; RDP 184.

Firecracker Vine (Manettia bicolor) MWF 196; (M. inflata) BIP 117 (cp 180); EGV 122; FHP 133; RDP 263.

Firecrest (Regulus ignicapillus) ALE 9:259; BBB 267; BBE 247; FBE 239; FBW 85; OBB 159; PBE 212 (cp 51).

Firefinch, Blue-billed (Lagonosticta rubricata) ALE 9:440; CDB 211 (cp 938); PSA 316 (cp 39).

Firefish (Pterois sp.) —See Lionfish.

Firefly (Luciola italica) ALE 2:225; (Lamprocera latreillei) RBT 69; (Photinus pyralis) KIW 84 (cp 38).

Firemouth —See Cichlid, Firemouth.

Firetail (Zoneaeginthus bellus) HAB 256; (Z. oculatus) HAB 256; LEA 462 (bw); MWA 86; (Emblema oculata) SAB 68 (cp 33).

Firethorn, Gibb's (Pyracantha atlantiodes) GDS cp 337; HSC 94; OGF 187; TGS 282.

Firethorn, Orange (Pyracantha angustifolia) MEP 47; MWF 247.

Firethorn, Scarlet (Pyracantha coccinea) DFP 222 (cp 1769); EDP 116; GDS cp 338; HSC 94; MWF 247; PFF 199; SPL cp 321; TGS 279; TVG 239; (P. coccinea lalandei) EGE 141; ELG 143; EPG 140.

Firethorn, Yellow (Pyracantha rogersiana) DFP 222 (cp 1770); GDS cp 340; HSC 94; MWF 247; OGF 187.

Fireweed (Chamaenerion) —See **Willow-herb, Rosebay.**

Fireweed (Epilobium) —See also **Willow-herb.**

Fireweed (Epilobium angustifolium) BWG 113; DEW 2:50 (cp 5); EWF cp 4f; EWG 106; KVP cp 121; MLP 169; PFE cp 80 (835); PFF 262; PWF 95h; SPL cp 482; TEP 245; TGF 125; WFB 151 (1).

Firewheel Tree (Stenocarpus sinuatus) BIP 171 (cp 262); MEP 34; MWF 273; RDP 370; SST cp 265; (S. salignus) DEW 2:94.

Firewheels —See **Indian Blanket** (Gaillardia).

Firewood-gatherer (Anumbius annumbi) ALE 9:125; SSA pl 25.

Fish-eagle —See **Eagle.**

Fishing-frog —See **Frogfish.**

Five-cornered Fruit —See **Carambola Tree.**

Five-faced Bishop —See **Moschatel.**

Flag, Blue or Purple (Iris versicolor) DFP 151 (cp 1205); EWF 120; PFF 386; WFB 273 (2).

Flag, Sweet (Acorus calamus) AMP 45; BFB cp 22; EGH 92; EHP 73; EPF pls 996-997; MBF cp 88; MFF pl 106 (1096); NHE 107; PFE cp 183 (1816); PFF 366; TEP 37; WFB 297 (5).

Flag, White-striped Sweet (Acorus gramineus) EFP 85; RDP 26, 69.

Flag, Yellow (Iris pseudacorus) BFB cp 1 (2); DEW 2:251; DFP 150 (cp 1198); EFC cp 15; KVP cp 16; MBF cp 83; MFF cp 59 (1050); NHE 104; OWF 29; PWF 77i, 141i; SPL 503; TEP 35; TVG 131; WFB 273 (3).

Flamboyante —See **Poinciana, Royal.**

Flame Flower, Scottish (Tropaeolum speciosum) DFP 250 (cp 1994); PFW 299.

Flame Gold Tips (Leucadendron discolor) DEW 2:96, 114; PFW 246.

Flame of the Forest —See **Tulip Tree, African** (Spathodea).

Flame of the Woods (Ixora coccinea) EGE 131; FHP 128; MEP 149; RDP 247.

Flame Pea (Chorizema cordatum) EGV 100; MWF 83; PFW 165; (C. ilicifolium) BIP 53 (cp 64); EWF cp 130b.

Flame Tree (Poinciana) —See **Poinciana, Royal.**

Flame Tree of Australia (Brachychiton sp.) EET 244; MEP 82; MWF 61; PFW 288; SST cp 86; WAW 106.

Flame Vine (Pyrostegia ignea or P. venusta) EGV 132; EWF cp 180b; LFW 251 (cp 558); MEP 137; MWF 248.

Flame Vine, Mexican (Senecio confusus) FHP 144; MEP 152.

Flame-of-the-forest (Butea frondosa) LFW 154 (cp 341); MEP 50; MWF 64.

Flame-violet (Episcia cupreata) BIP 75 (cp 109); EGG 108; EGL 107; EGV 107; FHP 118; MEP 140; MLP cp 210; PFW 130; RDP 192.

Flaming Sword (Vriesia splendens) BIP 181 (cp 281); DEW 2:223 (cp 141); LHP 185; MEP 14; RDP 393; SPL cp 60; (V. sp.) BIP 147 (cp 223); EGL 148; FHP 151; LHP 184; MLP 292; MWF 295; RDP 392.

Flamingo, Andean (Phoenicoparrus andinus) ALE 7:238; CDB 49 (cp 90); SSA cp 3.

Flamingo, Chilean (Phoenicopterus ruber chilensis) ALE 7:238; SBB 21 (pl 19).

Flamingo Flower (Anthurium andreanum) BIP 25 (cp 16); DFP 51 (cp 408); EDP 127; EGG 88; EPF pl 1000; HPW 308; LFW 230 (cp 514); LHP 24; MEP 12; MWF 42; PFW 35; RWF 83; SPL cp 38.

Flamingo Flower (Anthurium scherzeranum) EGL 86; FHP 105; LHP 24, 25; MLP 291; MWF 42; PFW 34; RDP 83.

Flamingo, Greater (Phoenicopterus ruber) ABW 61; ALE 7:238; BBB 273; BBE 41; BBT 56-57; CAA 34-35 (bw), 48; CDB 49 (cp 91); FBE 43; GAL 170 (bw); GPB 76, 77; HBG 80 (pl 3); LAW 460; PBE 9 (cp 4); WAB 139; (P. antiquorum) NBB 46.

Flamingo, Lesser (Phoeniconaias minor) ALE 7:238, 243; CAF 255 (bw), 256-257; CDB 49 (cp 89); LEA 369; WAB 155; WEA 150.

Flannel Bush (Fremontia californica) DFP 248 (cp 1977); GDS cp 202; HSC 55, 57; PFW 288.

Flannel Flower (Actinotus helianthi) EWF cp 133c; MWF 31.

Flannel-leaf —See **Mullein, Common.**

Flatbug (Aradus sp.) CIB 97 (cp 8); OBI 23.

Flathead (Platycephalus indicus) ALE 5:46; WFW cp 254.

Flax (Linum sp.) DFP 156 (cp 1242); MBF cp 19; NHE 73, 275; PFE cp 65 (664); PFM cps 84, 85; SNZ 26 (cp 32); TGF 109.

Flax, Austrian (Linum austriacum) EPF pl 643; EWF cp 18a.

Flax, Cultivated (Linum usitatissimum) AMP 61; LFW 90 (cp 210); MFF cp 6 (230); PFF 223; TEP 165.

Flax, Dwarf Flowering (Linum elegans) DFP 14 (cps 105, 106); EMB 139.

Flax, Flowering or Scarlet (Linum grandiflorum) DFP 42 (cp 329); EGA 132; HPW 207; OGF 133.

Flax, Mountain (Phormium colensoi) EWF cp 144b; SNZ 25 (cp 29), 142 (cp 445).

Flax, Narrow-leaved (Linum tenuifolium) DEW 2:59 (cp 28); NHE 181.

Flax, New Zealand (Phormium tenex) DFP 217 (cp 1732); EPF pl 886; MWF 231; SNZ 102 (cp 310).

Flax, Pale (Linum bienne) MBF cp 19; MFF cp 6 (229); OWF 169.

Flax, Perennial (Linum perenne) DEW 2:30; PFE cp 65 (659); PWF 69j; SPL cp 223; TVG 109; WFB 133 (7).

Flax, Purging or Fairy (Linum catharticum) MBF cp 19; MFF pl 87 (231); NHE 181; OWF 83; PWF 69f; WFB 133 (8).

Flax, Yellow (Linum flavum) NHE 221; PFE 220; WFB 132, 133 (9).

Flax, Yellow (Reinwardtia sp.) BIP 153 (cp 232); EWF cp 119b; HPW 207; MWF 251; PFW 177.

Flax-seed —See **Allseed.**

Flea, Chigoe (Tunga penetrans) ALE 2:404; ZWI 126.

Flea, Dog (Ctenocephalides canis) GAL 458 (bw); PEI 531 (bw); ZWI 126.

Flea, Human (Pulex irritans) ALE 2:404; GAL 458 (bw); OBI 19; PEI 530 (bw); ZWI 126.

Flea, Sand (Orchestria gammarella) CSS pl IX; WEA 45.

Flea, Snow (Boreus hyemalis) ALE 2:302; GAL 475 (bw); OBI 37.

Flea, Water —See **Water-flea.**

Fleabane (Erigeron sp.) DFP 8 (cps 63, 64), 140 (cps 1118-1120), 141 (cps 1121-1123); ERW 115; MCW 110-112; MFF pl 100 (879); MWF 124; NHE 284, 285; OGF 123; TEP 203; TGF 268; TVG 89.

Fleabane, Alpine (Erigeron alpinus) NHE 285; WFB 233 (7); (E. borealis) MBF cp 44; MFF cp 23 (878).

Fleabane, Annual (Erigeron annus) NHE 189; PFF 316; WUS 403.

Fleabane, Blue (Erigeron acer) MBF cp 44; MFF cp 8 (877); NHE 237; PFE cp 143 (1369); PWF 150a; WFB 233 (6); (E. acris) OWF 155.

Fleabane, British (Inula britannica) NHE 189; PFE cp 145 (1391).

Fleabane, Canadian (Conyza canadensis) BWG 93; MBF cp 44; MFF pl 100 (880); NHE 237; PWF 130c; WFB 231 (5); WUS 401.

Fleabane, Common (Pulicaria dysenterica) BFB cp 19 (1); KVP cp 28; MBF cp 45; MFF cp 56 (862); NHE 100; OWF 39; PFE cp 145 (1393); PWF 144c; WFB 237 (1).

Fleabane, Irish (Inula salicina) MFF cp 56 (859); NHE 189; WFB 237 (2).

Fleabane, Lesser or Small (Pulicaria vulgaris) MBF cp 45; MFF cp 52 (863); NHE 100; WFB 235 (9).

Fleawort —See also **Plantain, Branched.**

Fleawort, Field (Senecio integrifolius) MBF cp 47; MFF cp 56 (850); PWF 73j; WFB 243 (8).

Fleawort, Marsh (Senecio palustris) NHE 189; WFB 243 (3).

Fleece Flower, Himalayan (Polygonum affine) DFP 18 (cp 144); EWF cp 96c; PFW 236; TVG 161.

Fleur-de-lis (Iris florentia) AMP 45; PFM cp 259.

Fleur-de-lis of France —See **Flag, Yellow.**

Flicker, South American (Colaptes rupicola) CSA 199 (bw); WAB 93; (C. campestris) SSA cp 9.

Flixweed (Descurainia sophia) MBF cp 8; MCW 50; MFF cp 43 (112); NHE 215; OWF 11; PWF 146b; WFB 87 (3a); (D. pinnata) WUS 205.

Floating Hearts —See **Water-lily, Fringed.**

Florican (Sypheotides indica) ALE 8:134; RBI 59; (Eupodotis bengalensis) KBA 120 (bw).

Floss Flower (Ageratum houstonianum) DFP 29 (cp 225); EDP 84; EGA 30, 32, 33, 91; EGP 140; LFW 265 (cp 593); MWF 33; OGF 141; PFF 314; SPL cp 351; TGF 237; TVG 33.

Floss-silk Tree (Chorisia speciosa) MEP 81; MLP 141; (C. ventricosa) MLP 148.

Flounder (Platichthys sp.) CSF 189; CSS pl XXIV; FEF 512 (bw); GAL 230 (bw); HFI 18; LMF 160; OBV 47; WEA 151; WFW cp 468.

Flounder, Peacock (Bothus lunatus) HFI 19; HFW 269 (cp 138); MOL 107; WFW cp 466.

Flower of Tigris —See **Tiger Flower.**

Flower-of-an-Hour —See **Ketmia, Bladder.**

Flower-of-Jove (Lychnis flos-jovis) DFP 157 (cp 1251); PFE cp 14 (159); SPL cp 457.

Flowerpecker (Prionochilus sp.) DPB 393; KBA 432 (cp 63).

Flowerpecker, Fire-breasted (Dicaeum ignipectus) DPB 401; KBA 432 (cp 63); RBI 260.

Flowerpecker, Orange-bellied (Dicaeum trigonostigma) DPB 401; KBA 432 (cp 63); WBV 215.

Flowerpecker, Scarlet-backed (Dicaeum cruentatum) ABW 279; ALE 9:372; KBA 432 (cp 63); RBI 259; WBV 215.

Flowerpiercer (Diglossa sp.) ABW 292; ALE 9:362; CDB 203 (cp 889); DTB 81 (cp 40); SSA cp 39; WAB 93 (bw).

Fluellen, Round-leaved (Kickxia spuria) MBF cp 62; MFF cp 61; NHE 232; OWF 25.

Fluellen, Sharp-leaved (Kickxia elatine) MBF cp 62; MFF cp 61 (690); NHE 232; PWF 118c; WFB 211 (8).

Flufftail, Buff-spotted —See **Crake, Buff-spotted.**

Fluke, Liver (Fasciola hepatica) ALE 1:295; LEA 48; OIB 25.

Fly, Alder (Sialis lutaria) ALE 2:301; CIB 148 (cp 15); LAW 210; OBI 37; PEI 288 (bw); WEA 42; (S. flavilatera) PWI 33.

Fly, Bee (Bombylius sp.) ALE 2:394; BIN pl 105; CIB 204 (cp 31); GAL 380 (cp 14); OBI 127; PEI 513 (bw).

Fly, Black (Simulium sp.) ALE 2:394; CIB 197 (cp 30); OBI 125.

Fly, Blow or Bluebottle (Calliphora sp.) ALE 2:403; CIB 240 (cp 35); OBI 139; WEA 76; ZWI 144.

Fly, Bot —See **Botfly.**

Fly, Caddis (Many genera and species) CIB 193 (cp 28), 196 (cp 29); KIW 94 (cp 59); LAW 212; OBI 39; PEI 306-308 (bw); WEA 91.

Fly, Carrot (Psila rosae) CIB 224 (cp 33); EPD 116; GAL 413 (bw); OBI 135.

Fly, Chalcid (Pteromalus puparum) CIB 257 (cp 38); OBI 147.

Fly, Cluster (Pollenia rudis) CIB 240 (cp 35); OBI 139.

Fly, Crane (Tipula sp.) ALE 2:394; CIB 197 (cp 30); GAL 397 (bw); LEA 143 (bw); OBI 123; PEI 498 (bw); PWI 82 (bw); (Ptilogyna ramicornis) BIN pl 97.

Fly, Damsel —See **Damselfly.**

Fly, Dance or Empid (Empis tessellata) CIB 204 (cp 31); GAL 400 (bw); OBI 129; (Hilara maura) CIB 204 (cp 31).

Fly, Deer or Louse (Lipoptena cervi) ALE 2:403; GAL 459 (bw).

Fly, Dragon —See **Dragonfly.**

Fly, Drone (Eristalis sp.) ALE 2:394; CIB 205 (cp 32); KIW 259 (cp 125); OBI 131; PEI 517 (bw); ZWI 164.

Fly, Dung (Scatophaga stercoraria) ALE 2:403; CIB 240 (cp 35); OBI 137.

Fly, Fever (Dilophus febrilis) CIB 197 (cp 30); OBI 125.

Fly, Flesh (Sarcophaga sp.) ALE 2:403; CIB 240 (cp 35); GAL 396 (bw); OBI 139; PWI 82 (bw); ZWI 144.

Fly, Flower or Hover (Syrphus sp.) BIN pl
   100; CIB 205 (cp 32); GAL 380 (cp 14); LAW
   221; OBI 131; PEI 516 (bw); PWI 183;
   (Volucella sp.) ALE 2:394; CIB 205 (cp 32);
   CTI 37 (cp 43); LAW 247; LEA 129; OBI 131;
   PEI 517 (bw); PWI 76 (bw).
Fly, Forest (Hippobosca equina) CIB 240 (cp
   35); GAL 459 (bw); LAW 222; OBI 141; PEI
   521 (bw); ZWI 124.
Fly, Frit (Oscinella frit) ALE 2:403; CIB 225
   (cp 34); GAL 424 (bw).
Fly, Fruit (Rhagoletis cerasi) ALE 2:403;
   GAL 418 (bw); PEI 519 (bw).
Fly, Fruit (Vinegar) (Drosophila sp.) ALE
   2:403; CIB 225 (cp 34); CTI 36 (cp 42); GAL
   396 (bw); OBI 135.
Fly, Gad —See **Gadfly.**
Fly, Greenbottle (Lucilia sp.) ALE 2:403;
   CIB 240 (cp 35); GAL 380 (cp 14); KIW 260
   (cp 128); LAW 221; LEA 129; OBI 139; PEI
   527 (bw); PWI 159; ZWI 145.
Fly, Hair —See **Hairfly.**
Fly, Hedgehog or Parasitic (Echinomyia
   sp.) OBI 137; PEI 527 (bw).
Fly, Holly Leaf-miner (Phytomyza illicis)
   CIB 225 (cp 34); OBI 135.
Fly, Horse (Tabanus sp.) ALE 2:387, 394;
   CIB 204 (cp 31); CTB 35 (cps 40, 41); KIW
   223 (cps 120, 121), 224 (cp 122); LAW 220;
   OBI 127; PEI 510 (bw); PWI 81 (bw);
   (Haematopota pluvialis) ALE 2:394; CIB
   204 (cp 31); OBI 127; PEI 510 (bw);
   (Chrysops sp.) ALE 2:394; CIB 204 (cp 31);
   GAL 380 (cp 14); OBI 127; PEI 509 (bw).
Fly, House (Musca domestica) ALE 2:403;
   CIB 240 (cp 35); CTI 37 (cp 44); GAL 395
   (bw); KIW 261 (cp 130); OBI 141; PEI 526
   (bw).
Fly, Ichneumon —See **Wasp, Ichneumon.**
Fly, Kelp or Seaweed (Coelopa frigida) CIB
   224 (cp 33); OBI 135.
Fly, Large Narcissus (Merodon equestris)
   KIW 259 (cp 127); OBI 133.
Fly, Little House (Fannia canicularis) ALE
   2:403; GAL 395 (bw); OBI 141; PEI 525
   (bw).
Fly, Long-headed or -legged (Dolichopus
   sp.) CIB 205 (cp 32); GAL 391 (bw); OBI 129.
Fly, Mantis (Mantispa styriacus) ALE
   2:302; CIB 145 (cp 14).

Fly, May (Ephemera sp.) ALE 2:74; 11:105;
   CIB 68 (cp 1); GAL 388 (bw); LEA 128 (bw);
   OBI 9; PEI 32 (bw); WEA 247; ZWI 190;
   (Cloeon dipterum) CIB 68 (cp 1); OBI 9; PEI
   32 (bw); (Palingenia longicauda) CIB 69 (cp
   2).
Fly, Moth —See **Midge, Owl.**
Fly, Mushroom —See **Gnat, Fungus.**
Fly, Reed Gall (Lipara lucens) CIB 225 (cp
   34); OBI 135.
Fly, Robber (Asilus sp.) ALE 2:386, 394; CIB
   204 (cp 31); GAL 380 (cp 14); KBM 83 (pl
   27); OBI 129; PEI 513 (bw); (Neoaratus sp.)
   BIN cp 102; (Laphria sp.) ALE 2:394; OBI
   129; PEI 511 (bw).
Fly, St. Mark's —See **Hairfly.**
Fly, Saw —See **Sawfly.**
Fly, Scorpion (Panorpa sp.) ALE 2:302, 363;
   BIN pls 95, 96; CIB 148 (cp 15); GAL 387
   (bw); KIW 96 (cp 61); LAW 210; LEA 138
   (bw); OBI 37; PEI 302, 303 (bw); PWI 66, 73
   (bw).
Fly, Snake (Raphidia notata) ALE 2:301; CIB
   148 (cp 15); LEA 137 (bw); OBI 37; PEI 289
   (bw); PWI 66 (bw).
Fly, Snipe (Rhagio scolopacea) CIB 204 (cp
   31); LEA 117 (bw); OBI 127.
Fly, Soldier (Stratiomys sp.) ALE 2:394; OBI
   127; PEI 508 (bw).
Fly, Spongilla (Sisyra fuscata) ALE 2:301;
   OBI 37.
Fly, Stable (Biting) (Stomoxys calcitrans)
   ALE 2:403; CIB 240 (cp 35); GAL 398 (bw);
   OBI 141; PEI 524 (bw).
Fly, Stone (Perla sp.) ALE 2:74; KIW 53 (bw);
   OBI 9; PEI 37 (bw); ZWI 191; (Chloroperla
   torrentium) CIB 77 (cp 4); PEI 36 (bw);
   (Dinocras cephalotes) CIB 77 (cp 4); OBI 9;
   (Leuctra fusca) CIB 77 (cp 4); OBI 9.
Fly, Sun (Helophilus sp.) OBI 133; PEI 515
   (bw).
Fly, Tsetse (Glossina sp.) ALE 2:403; LAW
   222; PEI 524 (bw); ZWI 220.
Fly, Warble —See **Gadfly.**
Fly, White (Aleurodes sp.) GAL 447 (bw);
   LAW 233; OBI 35; (Aleurochiton complana-
   tus) ALE 2:197; (Trialeurodes vaporari-
   orum) EPD 129; OBI 35; PEI 128, 129 (bw).
Fly, Window (Scenopinus fenestralis) CIB
   204 (cp 31); OBI 129.

Flycatcher, African Paradise (Terpsiphone viridis) BBT 76; CDB 185 (cp 799); GPB 288; NBA 65; PSA 273 (cp 32).

Flycatcher, Asiatic Paradise (Terpsiphone paradisi) ABW 261; ALE 9:260; KBA 368 (cp 55); RBI 191, 192; WAB 179.

Flycatcher, Black or Japanese Paradise (Terpsiphone atrocaudata) ABW 261; ALE 9:260; DPB 353; KBA 368 (cp 55).

Flycatcher, Black-faced (Monarcha melanopsis) CDB 183 (cp 793); HAB 208; MBA 8, 9; RBU 224; SAB 42 (cp 20).

Flycatcher, Black-naped Blue (Hypothymis azurea) ABW 261; ALE 9:260; DPB 345; KBA 368 (cp 55); (Monarcha azurea) WBV 179.

Flycatcher, Blue (Cyornis sp.) ALE 9:260; CAS 146; DPB 343; KBA 369 (cp 56), 384 (pl 57); (Cyanoptila cyanomela) ALE 9:260.

Flycatcher, Boat-billed (Machaerirhynchus flaviventris) HAB 207; SAB 42 (cp 20); (Megarhynchus pitangua) SSA cp 38.

Flycatcher, Broad-billed (Myiagra ruficollis) HAB 211; SAB 40 (cp 19).

Flycatcher, Brown (Muscicapa latirostris) BBE 249; CDB 182 (cp 784); DPB 345; FBE 241; (Microeca leucophaea or M. fascinans) CDB 181 (cp 783); HAB 209; SAB 36 (cp 17).

Flycatcher, Cape or Puff-back (Batis capensis) ALE 9:260; PSA 273 (cp 32).

Flycatcher, Chinspot (Batis molitor) CDB 183 (cp 792); PSA 273 (cp 32).

Flycatcher, Collared (Ficedula albicollis) ALE 9:260; BBE 249; FBE 241; OBB 161; PBE 253 (cp 60).

Flycatcher, Fairy (Stenostira scita) CDB 184 (cp 797); PSA 273 (cp 32).

Flycatcher, Fiscal (Melaenornis silens) CDB 181 (cp 782); PSA 273 (cp 32); (Sigelus silens) NBA 67.

Flycatcher, Fork- or Swallow-tailed (Muscivora tyrannus) ALE 9:143; CDB 144 (cp 580); DTB 69 (cp 34); SSA pl 48.

Flycatcher, Frill-necked (Arses lorealis) HAB 207; SAB 42 (cp 20).

Flycatcher, Gray-headed (Culicicapa ceylonensis) DPB 343; KBA 337 (cp 52).

Flycatcher, Japanese Blue (Cyanoptila cyanomelana) ABW 260; CDB 179 (cp 776); KBA 384 (pl 57); LAW 508; (Ficedula cyanomelana) DPB 342.

Flycatcher, Jungle (Rhinomyias sp.) DPB 333; KBA 369 (cp 56).

Flycatcher, Leaden (Myiagra rubecula) CDB 184 (cp 796); HAB 211; SAB 40 (cp 19).

Flycatcher, Lemon-breasted (Microeca flavigaster) HAB 209; SAB 36 (cp 17).

Flycatcher, Little Pied (Ficedula westermanni) DPB 335; KBA 384 (pl 57).

Flycatcher, Narcissus (Ficedula narcissina) ABW 261; ALE 9:260; DPB 335.

Flycatcher, Pearly (Monarcha frater) HAB 208; SAB 42 (cp 20).

Flycatcher, Pied (Ficedula hypoleuca) ALE 9:260; BBB 139; BBE 249; CDB 180 (cp 780); FBE 241; GPB 284; LAW 439; OBB 161; PBE 253 (cp 60); WEA 152; (Arses kaupi) HAB 207; SAB 42 (cp 20).

Fly-catcher Plant (Cephalotus follicularis) EWF cp 133d; HPW 146; MLP 93.

Flycatcher, Red-breasted (Ficedula parva) ALE 9:260; BBB 275; BBE 249; FBE 241; GPB 287; KBA 368 (cp 55); OBB 161; PBE 253 (cp 60); (Muscicapa parva) FBW 83; RBE 275; WBV 179.

Flycatcher, Restless (Seisura inquieta or Myiagra inquieta) CDB 184 (cp 795); HAB 210; SAB 40 (cp 19).

Flycatcher, Royal (Onychorhynchus coronatus) DTB 71 (cp 35); SSA cp 38; (O. mexicanus) ABW 207; ALE 9:143.

Flycatcher, Satin (Myiagra cyanoleuca) CDB 183 (cp 794); HAB 211; RBU 243; SAB 40 (cp 19).

Flycatcher, Seychelles Paradise (Terpsiphone corvina) CDB 184 (cp 798); LVB 63.

Flycatcher, Shining (Piezorhynchus alecto) HAB 211; SAB 40 (cp 19).

Flycatcher, Snowy-browed or Thicket (Ficedula hyperythra) DPB 335; KBA 369 (cp 56).

Flycatcher, Spectacled (Monarcha trivirgata) HAB 208; SAB 42 (cp 20).

Flycatcher, Spotted (Muscicapa striata) ABW 261; ALE 9:260; 10:223; BBB 64; BBE 249; CDB 182 (cp 785); CEU 191; FBE 241; OBB 161; PBE 253 (cp 60); PSA 273 (cp 32).

Flycatcher, Streaked (Myiodynastes maculatus) CDB 144 (cp 581); SSA pl 48.

Flycatcher, Tody (Todirostrum sp.) ALE 9:143; SSA cp 38.

Flycatcher, White-eared (Monarcha leucotis) HAB 208; SAB 42 (cp 20).

Flying Fox —See **Bat.**

Flying Squirrel —See **Squirrel.**

Flying-fish, Sharpchin (Fodiator acutus) ALE 4:449; HFI 80.

Flying-fish, Tropical Two-wing (Exocoetus volitans) ALE 4:449; FEF 261; HFI 80; OBV 33.

Foalfoot —See **Coltsfoot.**

Fody (Foudia madagascariensis) ALE 10:283; (F. sechellarum) CDB 214 (cp 952).

Forget-me-not (Myosotis sp.) EWF cp 5h; SNZ 123 (cp 382).

Forget-me-not, Alpine (Myosotis alpestris) DFP 44 (cp 347); EWF cp 5d; MBF cp 60; MWF 205; NHE 281; OGF 33; PFE cp 103 (1067); PWF 160a.

Forget-me-not, Bur (Lappula myosotis) NHE 229; WFB 191 (8); (L. echinata) MCW 83.

Forget-me-not, Chatham Island (Myosotidium hortensia) EWF cp 5e; SNZ 165 (cp 524).

Forget-me-not, Common or Field (Myosotis arvensis) MBF cp 60; MFF cp 7 (658); OWF 173; PWF 31f; WFB 191 (4).

Forget-me-not, Creeping —See **Blue-eyed-Mary** (Omphalodes).

Forget-me-not, Marsh (Myosotis secunda) BFB cp 15 (5); MBF cp 60.

Forget-me-not, Summer (Anchusa capensis) DFP 30 (cp 235); EGA 94.

Forget-me-not, Water (Myosotis scorpiodes) EFC cp 4; EGC 141; EPF pl 797; EWF cp 20b; LFW 94 (cp 217); MBF cp 60; MFF cp 11 (656); PFE cp 103 (1065); PWF 91h; TEP 31; TGF 173; WFB 191 (6); (M. palustris) NHE 98; OWF 173.

Forget-me-not, Wood (Myosotis sylvatica) BWG 129; DEW 2:163 (cp 83); EGA 138; EGG 128; ESG 130; MBF cp 60; MFF cp 13 (657); OWF 173; PFF 286; PFW 49; PWF 30b; SPL cp 266; TGF 173; WFB 191 (3).

Forget-me-not, Yellow (Myosotis discolor) MBF cp 60; MFF cp 61 (659); OWF 173; WFB 191 (5); (M. australis) SNZ 153 (cp 483).

Forkbeard, Greater (Phycis blennoides) CSF 113; OBV 37.

Forkbeard, Lesser (Raniceps raninus) ALE 4:432; CSF 113; OBV 37.

Forktail, Leschenault's or White-crowned (Enicurus leschenaulti) ALE 9:280; KBA 385 (pl 58); WBV 191.

Forktail, Slaty-backed (Enicurus schistaceus) KBA 385 (pl 58); WBV 191.

Forktail, Spotted (Enicurus maculatus guttatus) RBI 224; WAB 162.

Forsythia or Golden Bells (Forsythia sp.) DEW 2:115; DFP 203 (cps 1611, 1612); EDP 115; EEG 95; EFS 21, 66, 112; EGC 133; EJG 117; ELG 116; EPF pl 666; EPG 121; GDS cps 196-199; HPW 226; HSC 54, 57; LFW 52 (cp 120); MLP 204; MWF 134; OGF 23; PFF 281; PFW 198; SPL cp 260; TGS 342, 344; TVG 225.

Fossa (Cryptoprocta ferox) ALE 10:283, 12:157; LVS 78; MLS 305 (bw); SMW 172 (bw).

Fountain Plant (Russelia equisetiformis) FHP 142; LFW 264 (cps 589, 590); MEP 133; SPL cp 347; (R. juncea) MWF 260.

Fountain Tree —See **Tulip Tree, African.**

Four O'Clock —See **Marvel of Peru.**

Fowl, Gray Jungle —See **Fowl, Sonnerat's Jungle.**

Fowl, Green Jungle (Gallus varius) ALE 8:57; WPW 71.

Fowl, Jungle or Scrub (Megapodius sp.) CDB 65 (cp 177); DPB 61; HAB 52; LVB 55; PWC 220 (bw); RBU 15; SNB 58 (pl 28); (M. laperouse) ALE 7:424.

Fowl, Mallee (Leipoa ocellata) ABW 85; ALE 7:424; CAU 103 (bw); CDB 65 (cp 175); GPB 110; HAB 52; RBU 12; SNB 58 (pl 28); WAB 194; WAW 14-15 (bw).

Fowl, Red Jungle (Gallus gallus) ABW 97; ALE 8:57; 13:285; CAS 215; CDB 69 (cp 201); DPB 61; KBA 81 (pl 8); WAB 165; WBV 67; WPW 60.

Fowl, Sonnerat's Jungle (Gallus sonneratii) ALE 8:57; GPB 120; RBI 36; WPW 62.

Fox, Arctic or White (Alopex lagopus) ALE 11:151, 12:215; DAE 112; HMA 53; HMW 142; JAW 97; MLS 258 (bw).

Fox, Bat-eared (Octocyon megalotis) ALE 12:274, 283; CAF 69 (bw); DDM 84 (cp 13); MLS 266 (bw); SMR 33 (cp 2); SMW 200 (bw); WEA 154.

Fox, Bengal (Vulpes bengalensis) ALE 12:215; DAA 73.

Fox, Cape (Vulpes chama) ALE 12:216; DDM 84 (cp 13).

Fox, Corsac (Alopex corsac) ALE 12:215; (Vulpes corsac) DAA 16.

Fox, Fennec (Fennecus zerda) ALE 12:216, 274; CAA 77; CAS 90 (bw); DAA 13; DDM 84 (cp 13); DSM 151; HMA 53; JAW 99; LEA 555 (bw); MLS 260 (bw); SMW 188 (cp 118); (Megalotis zerda) HMW 144.

Fox, Pale or Sand (Vulpes pallidus) ALE 12:216; DDM 84 (cp 13).

Fox, Red (Vulpes vulpes) ALE 11:321; 12:216, 273; 13:163; BMA 194-195; CAS 42 (bw); DAA 73; DAE 43; DDM 81 (cp 12); DSM 148; GAL 22 (bw); HMA 53; HMW 142; JAW 98; LAW 531; LEA 554 (bw); MLS 259 (bw); OBV 167; SMW 188 (cp 117); TWA 83.

Fox, Semien (Canis simensis) ALE 12:215; DDM 81 (cp 12); LVS 82.

Fox, Tibetan Sand (Vulpes ferrilatus) ALE 12:215; DAA 57.

Fox-and-cubs —See **Hawkweed, Orange.**

Fox-bat —See **Bat.**

Foxglove, Fairy —See **Erinus, Alpine.**

Foxglove, Purple or Common (Digitalis purpurea) AMP 75; BFB cp 1 (10); BWG 119; CTH 50; DEW 2:164 (cp 84); DFP 37 (cps 289, 290); EDP 134; EET 38, 39; EFC cp 60; EGA 115; EGH 112; EPF pl 846; ESG 108; KVP cp 125; LFW 48 (cps 112, 113), 49 (cp 114); MBF cp 63; MFF cp 21 (700); MWF 114; NHE 35; OWF 123; PFE cp 125 (1234); PFF 297; PFW 276; PWF 95f; SPL cp 390; TEP 255; TGF 221; TVG 87; WFB 212, 213 (2).

Foxglove, Rusty (Digitalis ferruginea) EPF pl 848; PFE cp 125 (1230); (D. obscura) HPW 244; PFE cp 125 (1233).

Foxglove, Small Yellow (Digitalis lutea) AMP 75; NHE 281; PFE cp 125 (1233).

Foxglove, Yellow (Digitalis grandiflora) EPF pl 845; EWF cp 21b; NHE 35; PFE cp 125 (1232); RWF 29; SPL cp 481; TEP 255; WFB 213 (3); (D. ambigua) AMP 75; EFC 60; OGF 119.

Foxtail —See also **Millet, Foxtail or Italian** and **Bristle-grass.**

Foxtail (Alopecurus sp.) EPF pl 1059; MBF pl 95; MFF pl 122 (1239-1241); NHE 138, 194, 245; PFE cp 181 (1780); TEP 153; TVG 137.

Frailea (Frailea sp.) LCS cp 55; EMB 107.

Francolin, Black (Francolinus francolinus) ALE 7:470; FBE 105; WAB 168.

Francolin, Chinese (Francolinus pintadeanus) ABW 91; DPB 61; KBA 81 (pl 8).

Francolin, Coqui (Francolinus coqui) CDB 69 (cp 198); PSA 113 (cp 12).

Francolin, Swainson's (Francolinus swainsoni) ALE 7:470; CDB 69 (cp 200); PSA 113 (cp 12).

Frangipani (Plumeria sp.) DEW 2:103 (cp 56); EET 220; EFS 126; EWF cp 180c; GFI 63 (cp 12); HPW 224; LFW 252 (cps 560, 561); MEP 111; MLP 202; MWF 237; PFW 29; SST cps 258, 259.

Franklinia (Franklinia alatamaha or Gordonia alatamaha) EEG 116; EFS 113; EGT 114; MLP 151; TGS 231; (G. axillaris) MWF 144.

Freckle Face (Hypoestes sanguinolenta) EFP 119; EGL 120; LHP 114, 115; MWF 160.

Freesia (Freesia sp.) BIP 87 cp 131; DFP 92 (cps 734, 735); EGB 114; EGG 110; EPF pl 950; LFW 52 (cp 122); MWF 136; PFF 387; PFW 149; SPL 261; TVG 195.

Friarbird, Helmeted (Philemon yorki) CDB 195 (cp 847); HAB 246; (P. novaeguineae) SAB 66 (cp 32).

Friarbird, Little (Philemon citreogularis) HAB 247; SAB 66 (cp 32).

Friarbird, Noisy (Philemon corniculatus) ABW 276; ALE 9:320; CDB 194 (cp 846); HAB 247; RBU 267; SAB 66 (cp 32).

Friarbird, Silver-crowned (Philemon argenticeps) HAB 247; SAB 66 (cp 32).

Friar's Cowl (Arisarum vulgare) NHE 246; PFE cp 183 (1821); PFM cp 221.

Fried Egg Tree (Oncoba spinosa) EWF cp
  59d; HPW 101; LFT 222 (cp 110); LFW 178
  (cp 401); MEP 88; MLP 157.
Friendly Neighbor (Bryophyllum tubi-
  florum) DEW 1:281 (cps 168, 169); MWF 63;
  RDP 110.
Friendship Plant —See **Queen's Tears.**
Frigatebird, Ascension (Fregata aquila)
  GPB 64; SAO 89.
Frigatebird, Greater (Fregata minor) ABW
  48; ALE 7:177, 178; BBT 114-115; CAU 225;
  DPB 9; GPB 64; HAB 18; HBG 65 (pl 2);
  SAO 89; SNB 22 (pl 10); WAN 118; WAB 25
  (bw).
Frigatebird, Lesser (Fregata ariel) DPB 9;
  GPB 64; HAB 19; LEA 368 (bw); RBU 95;
  SAO 89; SNB 22 (pl 10); WAN 118.
Frigatebird, Magnificent (Fregata magnif-
  icens) ABW 48-49; ALE 7:177; CDB 41 (cp
  53); GPB 64; HBG 65 (pl 2); OBB 11; SAO
  87, 89; SSA pl 21; WAB 24, 98.
Fringe Flower (Loropetalum chinense)
  MWF 190; TGS 279.
Fringe Tree (Chionanthus sp.) EGT 42, 105;
  ESG 100; MWF 82; TGS 374.
Frithia (Frithia pulchra) LFT cp 65; SCS cp
  84.
Fritillary (Fritillaria acmopetala) DFP 92
  (cp 736); PFM cp 229; (F. pallidiflora) DFP
  93 (cp 741); EWF cp 24g; OGF 31.
Fritillary, Snake's Head (Fritillaria
  meleagris) BFB cp 23 (10); DEW 2:233; DFP
  93 (cp 739); EFC cp 29; EGB 115; EMB 101;
  EPF pl 897; ESG 112; KVP cp 34; LFW 53 (cp
  123); MBF cp 85; MFF cp 34 (997); NHE 192;
  OWF 163; PFE cp 167 (1622); PFW 171;
  PWF cp 21m; TGF 28; TVG 195; WFB 262,
  263 (1).
Frog (Atelopus sp.) ALE 5:425, 440; CAW 87
  (cps 44, 45), 109 (bw); CSA 16; HWR 13, 17,
  18; LVS 145; MAR 45 (cp III); (Chiromantis
  sp.) ALE 5:425; LAW 373.
Frog, Arrow-poison (Dendrobates sp.) ALE
  5:399, 425; CAW 84-86 (cps 40-42); ERA 79;
  HRA 102; LAW 366; MAR 31 (pl 11), 55 (cp
  IV); VRA 149; (Phyllobates bicolor) ALE
  5:399; CAW 84 (cp 39).
Frog, Bull —See **Bullfrog.**

Frog, Catholic (Notaden bennetti) CAW 78
  (cp 28); WAW 100.
Frog, Chilean Water (Calyptocephalella
  gayi) CAW 92 (bw); MAR 20 (pl 6).
Frog, Clawed —See **Toad, Clawed.**
Frog, Corroboree —See **Toadlet,
  Corroboree.**
Frog, Darwin's Dwarf (Rhinoderma
  darwini) ALE 5:449; CAW 106 (bw); ERA
  103; HRA 101; MAR 31 (pl 11).
Frog, Edible (Rana esculenta) ALE 5:400;
  11:105, 257; ART 65; CAW 155, 157 (bw);
  ERA 90; LAW 382, 383; LEA 272; MAR 52
  (pl 20), 59 (pl 21); OBV 117; WEA 156.
Frog, European (Rana temporaria) ALE
  2:363; 5:378, 400, 423; ART 26, 57; CAW
  159, 160 (bw); HRA 96; LEA 272 (bw); MAR
  51 (pl 19); OBV 117.
Frog, Field (Rana arvalis) HRA 96; VRA 153.
Frog, Gliding (Rhacophorus sp.) ALE 5:409;
  CAW 134 (cp 73), 170, 171 (bw); ERA 94;
  HRA 100; MAR 60 (pl 22).
Frog, Golden (Mantella auriantiaca) ALE
  5:424; LAW 360.
Frog, Goliath (Gigantorana goliath) ALE
  5:426; (Rana goliath) HRA 98.
Frog, Hairy (Trichobatrachus robustus) ALE
  5:450; HRA 99.
Frog, Hochstetter's (Leiopelma hochstet-
  teri) ALE 5:361, 375; ERA 95; SLP 153.
Frog, Horned (Ceratophrys sp.) ART 67;
  CAW 89-91 (bw); ERA 96; HRA 105; HWR
  75; VRA 151; WEA 196.
Frog, Leaf (Phyllomedusa sp.) ALE 5:426;
  CAW 121 (cp 47), 122 (cp 49); HRA 105;
  LEA 270 (bw); MAR 41 (pl 15), 42 (pl 16);
  WEA 371.
Frog, Leptopelid (Leptopelis sp.) ALE 5:425;
  CAW 169 (bw).
Frog, Marsh (Rana ridibunda) ART 13; LEA
  279 (bw); OBV 117.
Frog, Marsupial or Pouched (Gastrotheca
  sp.) CAW 141, 143-145; cp 55; ERA 99; HRA
  104; LEA 282 (bw); MAR 49 (pl 17), 50 (pl
  18).
Frog, Paradoxical (Pseudis paradoxa) ALE
  5:410; CAW 146, 147 (bw); ERA 106; HRA
  106.

Frog, Pig-nosed (Hemisus marmoratus)
CAW 166 (bw); HWR 14; MAR 61 (pl 23).

Frog, Reed or Sedge (Hyperolius sp.) ALE
5:409, 426; CAA 122 (bw); CAW 134 (cp 72),
167 (bw); ERA 109, 110; MAR 45 (cp III).

Frog, Senegal (Kassina senegalensis) ALE
5:409; CAW 169 (bw).

Frog Shell (Biplex perca) CTS 35 (cp 50); GSS
67; (Tritonalia rubeta) CTS 38 (cp 58).

Frog, Short-headed (Breviceps sp.) ALE
5:410; CAW 136 (cps 76, 77), 174 (bw); ERA
107; HRA 101.

Frog, Slender-fingered Bladder (Leptodactylus sp.) ALE 5:449; CAW 77 (cp 26);
HRA 106; HWR 98.

Frog, Spotted (Hylambates maculatus) ALE
5:409; ART 20, 45; CAW 135 (cp 75).

Frog, Strawberry —See **Frog, Arrow-
poison.**

Frog, Tree —See **Treefrog.**

Frog, Water-holding (Cyclorana sp.) ERA
115; WAW 100.

Frog, West African (Hylarana albolabris)
ART 61; HWR 16; (Rana malabaricus) HRW
63; WEA 156.

Frog-bit (Hydrocharis morsus-ranae) EPF pl
875; EWF cp 23c; HPW 271; MBF cp 79;
MFF pl 69 (964); NHE 102; OWF 103; PFE
cp 161 (1566); PWF 140a; TEP 18; WFB 259
(8).

Frogfish (Antennarius sp.) ALE 4:422; HFI
10; HFW 271 (cps 142, 143), 283 (bw); WEA
47; WFW cps 182-184.

Froghopper (Cercopis sp.) ALE 2:182; CIB
129 (cp 12); KIW 81 (cps 30-32); OBI 33; PEI
97 (cp 12b).

Froghopper, Alder (Aphrophora alni) OBI
33; PEI 123 (bw).

Froghopper, Meadow (Philaenus spumarius) CIB 129 (cp 12); OBI 33; PEI 122 (bw);
(P. leucophtalmus) GAL 445 (bw); (Neo-
philaenus lineatus) BIN pl 36.

Frogmouth, Javanese (Batrachostomus
javensis) ALE 8:425; DPB 181; WBV 123.

Frogmouth, Marbled (Podargus ocellatus)
HAB 129; RBU 172; SNB 128 (cp 63).

Frogmouth, Owlet —See **Nightjar, Owlet.**

Frogmouth, Papuan (Podargus papuensis)
HAB 129; SNB 128 (cp 63).

Frogmouth, Plumed (Podargus plumiferus)
HAB 129; RBU 175.

Frogmouth, Tawny (Podargus strigoides)
ABW 157; CDB 119 (cp 455); GPB 195; HAB
127; LEA 430 (bw); NBB 43; SNB 128 (cp
63); WAB 188.

Frogwort —See **Buttercup, Bulbous.**

Fruit Salad Plant —See **Swiss Cheese
Plant.**

Fruit-bat —See **Bat.**

Fruit-eater, Green-and-Black (Euchlornis
riefferi) ALE 9:144; CDB 146 (cp 593);
(Pipreola riefferii) DTB 57 (cp 28).

Frullania (Frullania sp.) DEW 3:236; NFP
115; ONP 177.

Fuchsia (Fuchsia sp.) BIP 87 (cp 132); DEW
2:51 (cp 7); DFP 67, 68; EGG 110; EPF 719;
EWF cps 113c, f; 133a; FHP 121; GDS
203-208; HPW 163; HSC 55; LFW 199 (cp
450); MEP 101; MWF 136, 137; PFW 201,
203; RDP 216, 217; SNZ cps 25, 190, 191;
SPL cp 336.

Fuchsia, Australian (Correa sp.) BIP 61 (cp
84); EWF cp 138a, c.

Fuchsia, Magellan (Fuchsia magellanica)
DEW 2:52 (cp 11); DFP 203 (cps 1617-1618);
EFS 114; EPG 122; HSC 55; MFF cp 17
(438); MWF 136; OGF 169; PFW 201; PWF
147b; TGF 158.

Fuchsia, Tree —See **Tree Fuchsia.**

Fulmar (Fulmarus glacialis) ABW 31; ALE
7:61, 147; 8:229; BBB 255; BBE 25; CDB 37
(cp 28); DAE 32; FBE 25; GPB 46; LAW 502;
LEA 364 (bw); OBB 7; PBE 5 (pl 2); SAO 35;
WAB 73.

Fulmar, Antarctic (Fulmarus glacialoides)
SAO 33; SAR 111, 112; WAN 93.

Fulmar, Giant —See **Petrel, Giant.**

Fulvetta —See **Babbler, Nun.**

Fumewort or Fumitory —See also
Corydalis.

Fumewort (Corydalis sp.) NHE 29; PFE cp 30
(278); TEP 239.

Fumitory, Common (Fumaria officinalis)
AMP 143; BWG 104; EGH 115; MBF cp 6;
MFF cp 26 (45); NHE 214; OWF 137; PWF
521; WFB 79 (5).

Fumitory, Dense-flowered (Fumaria
densiflora) MBF cp 6; NHE 214.

Fumitory, Ramping (Fumaria capreolata) NHE 214; PFE cp 30 (280); PFM cp 37; PWF 79i; WFB 79 (7).

Fumitory, Small (Fumaria parviflora) MBF cp 6; NHE 214; PWF 104b; WFB 79 (8).

Fumitory, Wall (Fumaria muralis) HPW 53; NHE 214; WFB 79 (6).

Funeral Flower (Lisianthus sp.) EWF cp 172 c, d; PFW 122.

Fungus (Phaeolus schweinitzii) DLM 154, 155; LHM 67; ONP 111; SMF 67.

Fungus, Amethyst (Laccaria amethystina) CTM 34; DLM 203; LHM 99; NFP 64; PMU 89b; REM 133; (L. laccata) CTM 34; DEW 3: cp 60; DLM 204; KMF 107; LHM 99; NFP 64; ONP 133; PMU cp 89a; REM 133 (cp 101); RTM 119; SMG 156-159.

Fungus, Beefsteak (Fistulina hepatica) CTM 67; DEW 3:164; DLM 148; LHM 75; ONP 129; PFF 80; PMU cp 18; REM 171 (cp 139); RTM 213; SMF 21; SMG 68.

Fungus, Birch or Razor Strop (Piptoporus betulinus) DLM 156; KMF 47; LHM 67; ONP 117; SMF 69.

Fungus, Bird's-nest (Crucibulum sp.) DLM 131; KMF 107; LHM 223; NFP 89; PFF 87; SMG 53; (Cyathus sp.) DEW 3:190; DLM 33, 131, 132; LHM 223; NFP 89; RTM 227; SMF 58.

Fungus, Black-stud —See **Black Bulgar.**

Fungus, Carbon —See **Cramp Balls.**

Fungus, Cauliflower or Sponge (Sparassis crispa) DEW 3:160; DLM 159; KMF 79; LHM 57; NHE 47; PMU cp 11; REM 174 (cp 142); RTM 216; SMF 21; SMG 110, 111.

Fungus, Coral (Ramaria sp.) DLM 157; KMF 51; LHM 57; ONP 147; PUM cp 154b; TEP 213.

Fungus, Curled-edge (Paxillus involutus) DEW 3:179; DLM 233; KMF 71; LHM 185; ONP 115; PMU cp 47; REM 149 (cp 117); RTM 175, 267; SMG 204.

Fungus, Doughnut (Rhizina undulata) DEW 3:121; DLM 126; KMF 83; LHM 37; ONP 111.

Fungus, Dry Rot (Serpula lacrymans) DEW 3:cp 54; DLM 83, 159; LHM 53; (Merulius sp.) DLM 154; LHM 53; SMF 60.

Fungus, Earspoon (Auriscalpium vulgare) DEW 3:158; DLM 143; KMF 79; LHM 61; RTM 217.

Fungus, Edible Stump (Pholiota mutabilis) NHE 45; PMU cp 95; RTM 60; SMF 18.

Fungus, Field or Moor Club (Clavaria argillacea) DLM 145; KMF 127; LHM 55.

Fungus, Green Slime (Leotia lubrica) DLM 123; KMF 31; LHM 45; PFF 69; RTM 239.

Fungus, Hoax —See **Fungus, Amethyst.**

Fungus, Honey (Armillaria mellea) CTM 27; DEW 3:167; DLM 46, 170; KMF 99; LHM 93; NFP 56; NHE 45; ONP 141; PFF 81; PMU cp 82; SMF 19; SMG 136-138; TEP 221.

Fungus, Horsehair (Marasimius androsaceus) DLM 222; LHM 115; NFP 67; ONP 47.

Fungus, Insect (Cordyceps sp.) DLM 120; LHM 49; NFP 39; PFF 71; SMF 57.

Fungus, Magpie (Coprinus picaceus) CTM 49; DLM 184; LHM 137; RTM 52; SMF 42.

Fungus, Maze (Daedalea quercina) DLM 147; LHM 67; NFP 50; ONP 125; PFF 80; RTM 210, 245.

Fungus, Orange-vein (Phlebia radiata) KMF 26; LHM 53.

Fungus, Pine-cone (Strobilomyces strobilaceus) CTM 65; NFP 48; PFF 78; PMU cp 19.

Fungus, Porcelain —See **Beech Tuft.**

Fungus, Purple Knot (Coryne sarcoides) KMF 63; LHM 45; NFP 38; ONP 147.

Fungus, Quivering or Trembling (Tremella foliacea) KMF 35; ONP 149; PUM cp 157; RTM 245; (T. mesenterica) DEW 3: cp 49; DLM 142; LHM 225; NFP 45; ONP 149; RTM 301; SMF 56; (T. sp.) LHM 225; PFF 76; SMG 250.

Fungus, Red Cabbage —See **Fungus, Amethyst.**

Fungus, Rust Spot (Collybia maculata) DLM 181; KMF 67; LHM 101; ONP 103; REM 138; RTM 86, 244.

Fungus, Scaly Cluster (Pholiota squarrosa) DEW 3:176; DLM 237; KMF 59; LHM 149; PMU cp 94; RTM 58; SMG 208.

Fungus, Scaly Prickle (Sarcodon imbricatum) DEW 3:158; DLM 158; KMF 75; LHM 59; PMU cp 114.

Fungus, Sticky Coral or Staghorn (Calocera viscosa) CTM 71; DEW 3: cp 51; DLM 140; KMF 83; LHM 225; NFP 45; ONP 149; RTM 233; SMF 58.

Fungus, Stiptic (Panellus sp.) DLM 231; LHM 105; ONP 145.

Fungus, Thimble (Verpa digitaliformis) DEW 3:120; DLM 129; KMF 135; REM 183 (cp 151).

Fungus, Tinder —See **Rot, Yellowish Sapwood.**

Fungus, Yellow Knight —See **Man on Horseback.**

Funkia —See **Lily, Plantain.**

Furrow Shell, Peppery (Scrobicularia plana) ALE 3:168; OIB 79.

Furze —See **Gorse.**

Furze, Needle —See **Petty Whin.**

G

Gadfly (Hypoderma bovis) CIB 225 (cp 34); GAL 380 (cp 14); OBI 139; ZWI 125.

Gadwall (Anas strepera) ALE 7:255, 305; BBB 196; BBE 53; FBE 53; OBB 17; PBE 28 (cp 9), 40 (pl 13), 48 (pl 15); WAB 61.

Galactites (Galactites tomentosa) PFE cp 154 (1493); PFM cp 208.

Galago, Allen's (Galago alleni) ALE 10:252 (cp 6); DDM 52 (cp 5); LAW 610.

Galago, Common or Senegal (Galago senegalensis) ALE 10:252 (cp 6); CAA 86 (bw); CAF 171; DDM 52 (cp 5); DSM 70; SMR 16 (cp 1); WEA 6.

Galago, Demidoff's or Dwarf (Galago demidovii) ALE 10:252 (cp 6); DDM 52 (cp 5); DSM 70.

Galago, Needle-clawed (Euoticus elegantulus or Galago elegantulus) DDM 52 (cp 5); (G. inustus) ALE 10:252 (cp 6).

Galago, Thick- or Bush-tailed (Galago crassicaudatus) ALE 10:252 (cp 6); DDM 52 (cp 5); DSM 70; HMW 30; MLS 125 (bw); SMR 16 (cp 1); WEA 87.

Galah —See **Cockatoo, Roseate.**

Galax (Galax aphylla) EEG 142; EGC 133; ERW 116; ESG 113; EWG 110; HPW 131.

Galerina (Galerina sp.) DLM 192; LHM 175; ONP 47, 141; SMG 215.

Galingale (Cyperus longus) MBF cp 91; MFF pl 115 (1120); NHE 104; PFE 564.

Gall, Plant (Many genera and species) ALE 2:438; CIB 256 (cp 37); EPD 130, 138; GAL 464 (pl 15); OBI 149; PEI 139-141 (bw); ZWI 178, 179.

Gallant Soldier (Galinsoga parviflora) NHE 238; PFE cp 147 (1407); TEP 205; WUS 411.

Gallinule, Allen's or Lesser (Porphyrio alleni) BBE 109; FBE 117; PSA 160 (cp 17).

Gallinule, Common (Gallinula chloropus) ABW 109; ALE 8:97, 11:257; BBB 212; BBE 109; CDB 74 (cp 227);CEU 210; DPB 73; FBE 117; HBG 80 (pl 3); KBA 97 (cp 10); OBB 59; PBE 52 (cp 17); PSA 145 (pl 16); RBE 124; WBV 75; WEA 252.

Gallinule, Dusky (Gallinula tenebrosa) HAB 57; SNB 64 (cp 31).

Gallinule, Purple (Porphyrio porphyrio) BBE 109; CDB 75 (cp 231); DPB 73; FBE 117; KBA 97 (cp 10); PBE 52 (cp 17); SNB 64 (cp 31); WBV 75; WEA 304.

Gallito, Crested or Gray (Rhinocrypta lanceolata) ABW 202; SSA pl 26.

Gannet —See also **Booby.**

Gannet (Sula bassana or Morus bassanus) ABW 47; ALE 7:61, 177; BBB 248; BBE 31; CDB 39 (cp 42); DAE 31; FBE 31; GPB 55; LAW 502; OBB 11; PBE 5 (pl 2); PSA 32 (cp 1); RBE 63; SAO 73; SBB 23 (pl 22); SLP 52, 53; WAB 25, 215.

Gannet, Australian (Sula serrator) CAU 252-253; GPB 55; HAB 16.

Gannet, Cape (Sula capensis or Morus capensis) GPB 54, 55; SAO 77.

Gansies —See **Cancerbush.**

Gardener —See **Bowerbird.**

Gardenia (Gardenia sp.) BIP 89 (cp 133); DEW 2:133; EWF cp 79g; LFT cps 161, 162; MWF 138.

Gardenia, Butterfly —See **Jasmine, Crape.**

Gardenia, Common or Cape Jasmine (Gardenia jasminoides) BIP 89 (cp 134); EDP 124; EGE 71, 127; EGG 111; EGL 111; EJG 117; ELG 138; EMB 113; EPG 122; FHP 122; LHP 100, 101; MEP 149; MWF 138; PFW 266; RDP 218; SPL cp 47; TGS 359.

Gardenia, Tree (Rothmannia sp.) EWF cp 59c; LFT cp 162; MWF 258.

Garfish (Belone bellone) CSF 89; FEF 260 (bw); HFI 79; OBV 33.

Garganey (Anas querquedula) ABW 68; ALE 7:305, 385; BBB 194; BBE 55; DPB 33; FBE 55; GPB 81; KBA 80 (pl 7); OBB 19; PBE 28 (cp 9), 40 (pl 13), 48 (pl 15); SNB 39 (cp 18); WBV 49.

Garland Flower (Daphne cneorum) DFP 6 (cp 47); EGE 121; ERW 110; HSC 42, 44; NHE 267; PFE cp 74 (756); PFW 296; SPL cp 448; TGS 247.

Garland-flower (Hedychium) —See **Ginger-lily.**

Garlic (Allium sativum) AMP 33; CTH 23 (cp 14); EGH 95; EVF 143; OFP 169; PFF 372.

Garlic, Field (Allium oleraceum) MBF cp 85; MFF cp 35 (1012); NHE 191; WFB 265 (3a).

Garlic, Field or Crow (Allium vineale) EGC 71; MBF cp 84; MFF pl 65 (1011); NHE 191; OWF 163; PWF 130b; WFB 265 (3); WUS 107.

Garlic, Golden (Allium moly) DFP 83 (cp 664); EGB 92; ERW 93; EWF cp 24e; PFW 19; TGF 25.

Garlic, Keeled (Allium carinatum) MFF cp 32 (1013); NHE 241; PFE cp 165 (1607); PWF 122a; WFB 265 (5).

Garlic, Long-rooted (Allium victorialis) EFC cp 29; EPF pl 891; NHE 290.

Garlic, Mountain (Allium montanum) EPF pl 892; NHE 241.

Garlic, Naples (Allium neapolitanum) PFE cp 165 (1609); PFM cp 250.

Garlic, Rose (Allium roseum) MFF cp 32 (1015); PFE cp 166 (1611); PFM cp 251; PWF 49k.

Garlic, Society (Tulbaghia violacea) BIP 179 (cp 278); MWF 288.

Gas Plant —See **Burning Bush.**

Gasteria (Gasteria sp.) BIP 89 (cp 135); EWF cp 85c; LCS cp 257; SCS cp 85.

Gasteria, Lilliput (Gasteria liliputana) ECS 119; LCS cp 259; RDP 219.

Gasteria, Spotted (Gasteria maculata) EGL 112; SCS cp 86.

Gastromyzon (Gastromyzon borneensis) HFI 105; WFW cp 146.

Gaur (Bos gaurus) ALE 13:285, 342; CAS 174-175, 191, 205 (bw); DAA 111; DSM 250; HMW 91; JAW 187; LVS 128; MLS 401 (bw); SMW 257 (bw); TWA 158 (bw); VWA cp 28.

Gavial (Gavialis gangeticus) ALE 6:139; ERA 152, 153; HRA 74; LAW 434; LEA 297 (bw); LVS 168; VRA 163; WEA 160; (Gangetica gharial) MAR 83 (pl 33).

Gayal (Bos frontalis or Bibos frontalis) ALE 13:341; DAA 112; HMW 91; LAW 566.

Gazelle, Arabian (Gazella arabica) ALE 13:426, 438; CAA 90 (bw); DAA 12; SMW 284 (cp 179).

Gazelle, Clarke's —See **Dibatag.**

Gazelle, Dama (Gazella dama) ALE 13:426; CAF 67 (bw); DDM 225 (cp 38); DSM 266; HMW 77.

Gazelle, Dorcas (Gazella dorcas) ALE 13:426; DDM 240 (cp 39); LAW 570; LVA 17; LVS 121.

Gazelle, Goitered or Persian (Gazella subgutturosa) ALE 13:438; DAA 38.

Gazelle, Grant's (Gazella granti) ALE 13:426; DDM 225 (cp 38); DSM 267; HMA 53; HMW 77; SMW 284 (cp 180); TWA 60 (bw); WEA 161.

Gazelle, Mongolian (Procapra gutturosa) ALE 13:438; (Gazella gutturosa) DAA 37.

Gazelle, Mountain (Gazella gazella) CAS 167 (bw); LVS 132.

Gazelle, Petzeln's (Gazella pelzelni) ALE 13:426, 438; DDM 240 (cp 39).

Gazelle, Red-fronted (Gazella rufifrons)
ALE 13:426; CAF 65 (bw); DDM 240 (cp 39);
SMW 284 (cp 182).

Gazelle, Slender-horned (Gazella
leptoceros) ALE 13:426; DDM 240 (cp 39);
LVS 125.

Gazelle, Soemmering's (Gazella soemmeringi) ALE 13:426; DDM 225 (cp 38).

Gazelle, Speke's (Gazella spekei) ALE
13:426; DDM 240 (cp 39).

Gazelle, Thomson's (Gazella thomsoni) ALE
13:426, 435, 438; CAF 150-151 (bw); DDM
240 (cp 39); DSM 266; LAW 517; MLS 418
(bw).

Gazelle, Tibetan (Procapra picticaudata)
ALE 13:438; DAA 40.

Gazelle, Waller's —See **Gerenuk.**

Gean —See **Cherry, Sweet or Wild**

Gecko, African (Pachydactylus sp.) ALE
6:161, 163; MAR 96 (pl 38).

Gecko, Barking (Gymnodactylus sp.) ALE
6:161, 169; COG cp 8; HWR 67; MAR 95 (pl
37); MEA 26.

Gecko, European Leaf-fingered (Phyllodactylus sp.) ALE 6:164; ART 201; HRA 38.

Gecko, Fan-footed (Ptyodactylus hasselquisti) ALE 6:169; LEA 301 (bw); MAR 97
(pl 39); SIR 73 (bw).

Gecko, Fat-tailed or Robust (Oedura sp.)
ALE 6:163, 179; COG cp 11.

Gecko, Great House (Gekko gecko) ALE
6:161, 170; ART 158; HRA 39; HWR 70;
LEA 300 (bw), 304; MAR 98 (pl 40); SIR 60
(cps 22, 23); VRA 165; (G. lineatus) CAS
287.

Gecko, Knob-tailed (Nephrurus laevis) ALE
6:164; CAU 183; COG cp 5; (N. asper) WAW
72 (bw).

Gecko, Kuhl's or Flying (Ptychiozoon
kuhlii) ALE 6:170; HRA 38; LBR 41, 53;
MAR 101 (pl 41).

Gecko, Leaf-tailed (Phyllurus sp.) ALE
6:162; COG cp 6; MEA 27; SIR 75 (bw);
WAW 73 (bw); (Uroplatus fimbriatus) ALE
6:161, 170; ART 203; MAR 87 (cp VI).

Gecko, Madagascar or Malagasy
(Phelsuma madagascariensis) ALE 6:170;
10:283; CAA 109; HRA 38; HWR 69; LAW
409; MAR 87 (cp VI); (P. sp.) ALE 6:163,
170; ERA 169; HWR 68.

Gecko, Moorish or Wall (Tarentola
mauritanica) ALE 6:170; CAA 112; HRA
38; HWR 66; LAW 408; MAR 95 (pl 37); SIR
61 (cp 24).

Gecko, Panther (Eublepharis macularius)
ALE 6:169; MAR 95 (pl 37).

Gecko, Spiny-tailed (Diplodactylus sp.)
ALE 6:161, 163; COG cps 7, 10; MAR 97 (pl
39); SIR 76 (bw).

Gecko, Turkish (Hemidactylus turcicus)
ALE 6:170; HRA 38.

Gecko, Web-footed (Palmatogecko rangei)
ALE 6:169; MAR 105 (cp VII).

Geese —See **Goose.**

Gelada —See **Baboon, Gelada.**

Gemsbok (Oryx gazella) ALE 13:391, 402;
CAA 74 (bw); CAF 207; DDM 188 (cp 31);
DSM 258; HMA 25; JAW 192; MLS 409
(bw); SLP 231; SMR 97 (cp 8).

Genet, Abyssinian (Genetta abyssinica)
ALE 12:147; DDM 92 (cp 15).

Genet, Common or Small-spotted (Genetta
genetta) CAA 94 (bw); DAE 16; DDM 93 (cp
16); LAW 539; LEA 565 (bw); (G. servalina)
DDM 93 (cp 16).

Genet, Giant (Genetta victoriae) ALE
12:147; DDM 93 (cp 16).

Genet, Large-spotted (Genetta tigrina)
DDM 93 (cp 16); MLS 298 (bw); SMR 48 (cp
3).

Gentian (Gentiana sp.) EPF 681; EWF cps 4,
97; EWG 111; HPW 223; PFW 121; RWF 85;
SNZ 113 (cp 346), 151 (cp 476).

Gentian, Autumn (Gentianella amarella)
MBF cp 59; MFF cp 4 (643); NHE 185; OWF
139; PWF 157b; WFB 183 (6).

Gentian, Bladder (Gentiana utriculosa)
NHE 74; PFE cp 96 (990).

Gentian, Chiltern (Gentianella germanica)
MBF cp 59; NHE 185; WFB 183 (6a).

Gentian, China (Gentiana sino-ornata) DFP
10 (cp 75); EWF cp 97h; PFW 122; SPL cp
451.

Gentian, Closed or Bottle (Gentiana
andrewsi) ERW 142; TGF 157.

Gentian, Cross (Gentiana cruciata) NHE
185; PFE cp 96 (989); WFB 183 (3).

Gentian, Field (Gentianella campestris)
MBF cp 59; MFF cp 4 (642); OWF 139; PFE
cp 96 (999); PWF 157c; WFB 183 (5).

Gentian, Fringed (Gentiana crinita) DEW
  2:121; MLP 194; PFF 282.

Gentian, Fringed (Gentianella tenella)
  NHE 280; WFB 183 (7).

Gentian, Great Yellow (Gentiana lutea)
  AMP 25; CTH 44 (cp 61); DEW 2:120; EFC
  cp 13; EPF pl 685; KVP cp 133; NHE 279;
  PFE cp 95 (996); PFW 122; SPL cp 392; WFB
  181 (7).

Gentian, Marsh (Gentiana pneumonanthe)
  BFB cp 15 (4); KVP cp 74; MBF cp 59; MFF
  cp 12 (639); NHE 185; OWF 177; PFE cp 96
  (993); PWF 146f; WFB 183 (2).

Gentian, Purple (Gentiana purpurea) EFC
  cp 12; NHE 279; PFE cp 97 (998); WFB 183
  (4).

Gentian, Small Alpine (Genitiana nivalis)
  MBF cp 59; MFF cp 3 (641); NHE 280; PWF
  157a; WFB 183 (1a).

Gentian, Snow (Gentiana corymbifera) MLP
  206; SNZ 154 (cp 484).

Gentian, Spotted (Gentiana punctata) EFC
  cp 12; EPF pl 684; NHE 279; PFE cp 97
  (997); RWF 28; SPL cp 486; WFB 183 (4a).

Gentian, Spring (Gentiana verna) CTH 44
  (cp 60); EPF pl 682; KVP cp 129; MBF cp 59;
  MFF cp 12 (640); NHE 280; OWF 177; PFE
  cp 96 (991); PFW 121; PWF 47a; RWF 22;
  SPL cp 262; TEP 93; WFB 183 (1).

Gentian, Stemless or Trumpet (Gentiana
  acaulis) AMP 25; CTH 44 (cp 59); DFP 9 (cp
  71); EMB 138; ERW 117; MWF 141; OGF
  35; PFW 120; SPL cp 484.

Gentian, Stemless Trumpet (Gentiana
  clusii) EPF pl 680; NHE 280; PFE cp 96
  (992); TEP 93; TVG 151.

Gentian, Trumpet (Gentiana kochiana)
  DEW 2:106 (cp 62); NHE 280; PFE cp 96 (992).

Gentian, Willow (Gentiana asclepiadea)
  AMP 25; DFP 143 (cp 1142); EFC cp 13; EPF
  pl 683; ESG 84; OGF 111; PFE cp 97 (994);
  PFW 121; SPL cp 485; WFB 183 (2a).

Gentianella, Yellow (Cicendia filiformis)
  MBF cp 58; MFF cp 40 (632); NHE 74; WFB
  181 (5).

Geranium, Spotted or Wild (Geranium
  maculatum) ESG 83, 114; EWG 6, 112.

Gerbil (Gerbillus sp.) ALE 11:300; CAS 95;
  (Pachyuromys duprasi) MLS 213 (bw);
  (Taterillus gracilis) LEA 539 (bw).

Gerenuk (Lithocranius walleri) ALE
  13:423-426; CAA 81 (bw); CAF 82, 87 (bw);
  DDM 225 (cp 38); DSM 265; HMA 53; HMW
  78; JAW 199; LAW 570; LEA 612 (bw); MLS
  417 (bw); SLP 240; SMW 267 (bw).

German or Parlor Ivy (Senecio mikanioides) EFP 143; EGV 135; PFF 330.

Germander (Teucrium sp.) DFP 175 (cp
  1394); EWF cp 37c; PFE cp 107 (1096).

Germander, Cut-leaved (Teucrium botrys)
  MBF cp 70; MFF cp 28 (785); NHE 230;
  WFB 197 (9).

Germander, Mountain (Teucrium
  montanum) NHE 230; PFE cp 107 (1103);
  WFB 197 (6).

Germander, Tree (Teucrium fruticans)
  HPW 238; PFE cp 107 (1101); PFM cp 160.

Germander, Wall (Teucrium chamaedrys)
  AMP 19; EGC 147; EGH 144; MBF cp 70;
  MFF cp 28 (783); NHE 230; PFE cp 107
  (1099); PWF 121h; SPL cp 409; TGF 204;
  WFB 197 (7).

Germander, Water (Teucrium scordium)
  MBF cp 70; MFF cp 10 (784); NHE 98; PFE
  352; PWF 139j; WFB 197 (8).

Gesneria (Gesneria sp.) BIP 91 (cp 138); EGL
  112; EMB 126; HPW 247.

Gharial —See **Gavial.**

Ghostweed —See **Snow-on-the-mountain.**

Giant Velvet Rose (Aeonium canariense)
  MLP 106; SPL cp 125.

Gibbon, Capped or Pileated (Hylobates
  pileatus) DSM 95; LVS 42.

Gibbon, Dark-handed (Hylobates agilis)
  CAS 220; DAA 136.

Gibbon, Gray or Silvery (Hylobates
  cinereus) HMW 15; SMW 87 (cp 49), 101
  (bw); (H. leuciscus) DAA 136; (H. moloch)
  ALE 10:476.

Gibbon, Hoolock or White-browed
  (Hylobates hoolock) ALE 10:476; DAA 134;
  DSM 95; HMW 15; JAW 59.

Gibbon, Siamang —See **Siamang.**

Gibbon, White-handed (Hylobates lar) ALE
10:476; CAS 267 (bw); DAA 134; DSM
94-95; HMA 54; HMW 15; JAW 60; LEA 511
(bw); MLS 158 (bw); SMW 88 (cp 50); TWA
162.

Gilia, Blue (Gilia capitata) EGA 121; EWG
113; TGF 172; (G. rubra) TGF 185; (G. sp.)
DFP 38 (cp 298); HPW 232.

Gill-go-by-the-ground —See **Ground Ivy.**

Gillyflower —See **Stock.**

Gilthead (Sparus auratus) HFI 30; LMF 126;
OBV 29.

Ginger Blossom, Red (Alpinia purpurata)
LFW 179 (cp 403); MEP 26; MWF 37; (A.
caerulea) WAW 125.

Ginger, European (Asarum europaeum)
AMP 51; DEW 1:131; EGW 91; ERW 100;
HPW 41; MBF cp 76; MFF cp 35 (512); NHE
27; PFE cp 7 (74); PWF cp 20e; TEP 239;
WFB 39.

Ginger Plant (Zingiber officinale) AMP 21;
HPW 298; OFP 135; PFF 389; WWT 104; (Z.
spectabile) EWF cp 124d; PFW 311; (Z.
ottensii) EPF pl 1040.

Ginger, Torch (Nicolai sp.) BIP 131 (cp 196);
MEP 27.

Ginger, Western Wild (Asarum canadense)
ESG 93; EWG 88; MPL 78; PFF 151; (A.
caudatum) EGC 123; EWF cp 150b; PFW
36.

Ginger-lily, White (Hedychium coronarium)
MEP 27; MWF 148.

Ginger-lily, Yellow (Hedychium gardneria-
num) BIP cp 146; EPF cp XXXII; PFW 311;
SPL cp 502; (H. flavum or H. flavescens)
EWF cp 1240; MEP 27.

Gipsy-wort (Lycopus europaeus) MBF cp 67;
MFF pl 112 (751); NHE 98; OWF 97; PFE cp
116 (1166); PWF 135k; WFB 205 (5).

Giraffe (Giraffa camelopardalis) ALE
13:262; BMA 98; CAA 64 (bw); CAF 180
(bw), 198-199; DDM 176 (cp 27); DSM 238;
HMA 8, 55; JAW 179; LAW 572, 573; LEA
602 (bw); MLS 393 (bw); SLP 238, 239; SMR
85 (cp 6); TWA 40.

Giraffe, Masai (Giraffa camelopardalis
tippelskirchi) ALE 13:261, 262; CAF
140-141, 143 (bw); HMW 67; SMW 276 (cps
162, 163.

Giraffe, Reticulated (Giraffa camelopardalis reticulata) ALE 13:262; CAF 83; DDM
176 (cp 27); HMW 66.

Gladdon —See **Iris, Stinking.**

Gladiolus (Gladiolus sp.) EDP 134; EGB 117;
EGG 112; EMB 103; EPF pls 947, 949; EWF
cp 87d; HPW 311; LFT cp 40; LFW 58, 59
(cps 137-140); MBF cp 83; MFF cp 34 (1055);
MWF 142; NHE 193; OGF 117; PFM cp 256;
PWF 146, 147; SPL cp 210; TGF 44; TVG
197; WFB 273 (7).

Gladiolus, Eastern or Byzantine (Gladiolus byzantinus) DFP 94 (cp 748); EGB 117;
PFE cp 176 (1695).

Gladiolus, Field (Gladiolus segetum) NHE
246; PFE cp 176 (1695).

Glassfish, Indian (Chanda ranga) ALE
5:116; CTF 48 (cp 61); FEF 332 (bw); HFI 20;
HFW 209 (cp 91); SFW 525 (pl 142); WFW cp
257.

Glassfish, Wolff's (Chanda wolffi) FEF 332
(bw); SFW 524 (pl 141).

Glasswort (Salicornia sp.) DEW 1:112 (cp
70); EPF pl 12; OWF 57; SNZ 18 (cp 7).

Glasswort, Common (Salicornia europaea)
HPW 73; KVP cp 100; MBF cp 72; MFF pl
111 (220); MPL 84; NHE 135; PFF 158; PWF
164a; TEP 59; WFB 49 (2).

Glider, Feathertail —See **Phalanger,
Pygmy Flying.**

Glider, Greater —See **Phalanger, Greater
Flying.**

Glider, Honey or Sugar —See **Phalanger**
(Petaurus sp.).

Globe Flower (Trollius sp.) DFP 176 (cp
1406); ERW 149; EWF cp 8c; TVG 125.

Globe Flower Bush (Kerria japonica) DFP
209 (cp 1667); EFS 120; EGW 121; EJG 124;
EPF pl 399; ESG 123; GDS cp 268; HPW
142; HSC 71, 73; LFW 64 (cp 152); MWF
173; PFW 262; SPL cp 312; TGS 230.

Globe Flower, European (Trollius
europaeus) BFB cp 2 (3); DEW 1:116; DFP
176 (cps 1403, 1404); EPF pl 205; HPW 47;
MBF cp 2; MFF cp 38 (2); NHE 178; OGF 39;
OWF 5; PFE cp 19 (204); PFW 252, 253;
PWF 88d; SPL cp 496; TEP 139; TGF 61;
WFB 69 (1).

Globe Tulip (Calochortus sp.) DFP 85 (cp 679); EGB 100; EWF cp 167f; EWG 93; HPW 313; TGF 28.

Globe-daisy —See also **Globularia.**

Globe-daisy, Common (Globularia elongata) DEW 2:184; NHE 283; (G. sp.) HPW 245; TGF 236.

Globe-daisy, Heart-leaved (Globularia cordifolia) NHE 283; PFE cp 129 (1263); SPL cp 452; TVG 153.

Globefish —See Pufferfish.

Globe-thistle (Echinops ritro) DFP 140 (cp 1115); OGF 115; PFE cp 152 (1462); PFM cp 205; PFW 85; SPL cp 203; TGF 237; (E. sp.) EDP 131; EWF cp 106d; LFW 51 (cp 118).

Globe-thistle, Great (Echinops sphaero-cephalus) DEW 2:206; EPF pls 574, 575; MLP 248; NHE 239; TEP 205; WFB 245 (1).

Globularia —See also **Globe-daisy.**

Globularia, Common (Globularia vulgaris) PFE cp 129 (1264); WFB 219 (6).

Globularia, Shrubby (Globularia alypum) PFE cp 129 (1262); PFM cp 179.

Gloeophyllum (Gloeophyllum saepiarum) DLM 150; LHM 71; RTM 245.

Gloeoporus (Gloeoporus adustus) LHM 71; ONP 157; (G. fumosus) LHM 73; ONP 125.

Glory Bower (Clerodendrum splendens) EGV 52; EWF cp 60a; PFW 305; (C. capitatum) EWF cp 60c; PFW 304; (C. speciosis-simum) BIP 55 (cp 71); DFP 59 (cp 471); MWF 88.

Glory Bower (Clerodendrum thomsoniae) BIP 55 (cp 72); DEW 2:174 (cp 104); DFP 59 (cp 472); EGG 99; EGV 42, 53, 103; EPF pl 804; FHP 113; HPW 237; LFW 33 (cps 77, 78); MEP 121; PFW 305; RDP 147.

Glory Bower, Harlequin (Clerodendron trichotomum) BKT cp 184; DFP 190 (cps 1518, 1519); HSC 31; PFW 305; SPL cp 77; TGS 374; (C. fargessii) OGF 189.

Glory Bush (Tibouchina sp.) BIP 175 (cp 272); DFP 241 (cp 1928); EET 24; EGG 144; LFW 296 (cp 595); MEP 100; MWF 284; PFW 186; RDP 383; SPL 121.

Glory Flower (Eccremocarpus scaber) DFP 247 (cp 1976); EWF cp 181c; HSC 46, 48; PFW 46.

Glory of the Snow (Chionodoxa luciliae) DFP 86 (cp 683, 684); EGB 102; LFW 24 (cp 50); MWF 83; PFW 172; TGF 13; TVG 185; (C. sardensis) BIP 39 (cp 41); DFP 86 (cp 685); EGW 101; EMB 98; (C. gigantea rosea) OGF 9.

Glory of the Sun (Leucocoryne ixiodes) BIP 113 (cp 171); EWF cp 190d; LFW 206 (cp 462).

Glory Pea (Clianthus dampieri) MEP 55; WWT 47.

Glory Vine (Vitis coignetiae) DFP 250 (cps 1995, 1996); EGV 144; ESG 148; HSC 123, 125; MWF 295; TGS 71.

Glory-of-Texas (Thelocactus bicolor) ECS 31, 146; LCS cps 181, 182; SCS cp 64.

Glory-of-the-Seas —See **Cone Shell.**

Glow-worm (Lampyris noctiluca) ALE 2:363; CIB 304 (cp 51); GAL 288 (bw); OBI 173; PEI 209 (bw); PWI 150 (bw); WEA 167.

Gloxinera (Gloxinera hybrids) EGG 113; FHP 123.

Gloxinia (Gloxinia perennis) DEW 2:169 (cp 94); EGL 113; RDP 226.

Gloxinia (Sinningia sp.) BIP 165 (cp 252); DFP 81 (cp 648); EDP 120; EGB 138; EGG 137, 141; EGL 144; EMB 130; EPF pl 867; EWF cp 185c, d; FHP 144; LFW 266 (cp 594); LHP 168, 169; MEP 140; MLP 221; MWF 267; PFW 128; RDP 361-363; SPL cp 57.

Gloxinia, New Zealand —See **Taurepo.**

Glutton —See **Wolverine.**

Gnat, Fungus (Sciara sp.) CIB 197 (cp 30); EPD 117; (Dynatosoma fuscicorne) OBI 125.

Gnat, Winter (Trichocera sp.) CIB 197 (cp 30); OBI 123.

Gnat-eater (Conopophaga sp.) ABW 201; CDB 142 (cp 571); WAB 91.

Gnatwren, Long-billed (Ramphocaenas melanurus) CDB 170 (cp 723); SSA pl 29.

Gnu, Brindled —See **Wildebeest.**

Gnu, White-bearded (Connochaetes talboju-batus) HMA 55; SMW 283 (cp 178).

Gnu, White-tailed (Connochaetes gnu) ALE 13:61; DDM 224 (cp 37); HMW 75; LVA 17; WEA 168.

Goa —See **Antelope, Tibetan.**

Goanna —See **Monitor.**

Goat (Capra hircus) ALE 13:482, 488; BMA 100-101; DAA 20; DAE 76; HMW 86, 87; OBV 151; PFF 731.

Goat-antelope —See Goral and Serow.

Goatfish (Pseudupeneus sp.) FEF 363; HFW cps 92, 93; MOL 114; WFW cps 306, 307.

Goat's Beard, False —See Astilbe.

Goat's Rue (Galega officinalis) AMP 155; DFP 143 (cps 1140, 1141); NHE 181; PFE cp 53 (524); PWF 85j; SPL cp 391; TGF 146; TVG 93; WFB 119 (8).

Goatsbeard (Aruncus sylvester) DFP 123 (cp 979); EPF 394, 397; EWG 88; KVP cp 120; LFW 17 (cp 32); NHE 271; OGF 111; TGF 108; (A. dioicus) PFE cp 44 (420); SPL cp 327; TEP 241; WFB 107 (2).

Goatsbeard (Tragopogon sp.) DEW 2:209 (cp 112); KVP cp 173; MBF cp 54; MCW 130; NHE 76.

Goatsbeard, Meadow (Tragopogon pratensis) BWG 54; EFP pls 601, 602; MCW 131; MFF cp 58 (935); NHE 190; OWF 41; PWF 81e; TEP 148; WFB 251 (1).

Go-away Bird (Corythaixodes sp.) CDB 109 (cp 411), 110 (cp 412); GPB 181; PSA 192 (cp 21).

Goby, Black (Gobius niger) ALE 5:171; CSF 157; CSS cp 32; HFI 56; OBV 83.

Goby, Celebes (Stigmatogobius hoeveni) HFI 56; SFW 597 (pl 164).

Goby, Celebes Sleeper (Hypseleotris cyprinoides) ALE 5:171; HFI 57; SFW 608 (cp 167).

Goby, Crystal (Crystallogobius linearis or C. nilssoni) ALE 5:171; CSF 159; OBV 57.

Goby, Golden-banded or Wasp (Brachygobius xanthozona) CTF 58 (cp 76); FEF 496 (bw); HFI 56; SFW 597 (pl 164).

Goby, Rock (Gobius paganellus) CSS cp 32; HFI 56; OBV 83; WEA 169.

Goby, Sand or Painted (Pomatoschistus minutus or Gobius minutus) ALE 5:171; CSF 159; CSS cp 32; OBV 83; WFW cp 434.

Goby, Sleeper (Eleotris sp.) HFI 57; SFW 617 (pl 170).

Goby, Spotted (Chaparrudo flavescens) CSF 157; OBV 83.

Goby, Transparent (Aphya minuta) ALE 5:171; CSF 159.

God Almighty's Flower —See Trefoil, Common Bird's Foot.

Godwit, Bar-tailed (Limosa lapponica) BBB 228; BBE 121; BNZ 141; CEU 276; DPB 87; FBE 139; FNZ 129 (pl 12); GPB 149; HAB 68; KBA 128 (pl 13), 149 (pl 18); OBB 69; PBE 92 (pl 29), 101 (cp 32); RBE 143; SNB 84 (cp 41), 86 (pl 42); WBV 81.

Godwit, Black-tailed (Limosa limosa) ALE 7:385; 8:155; BBB 172; BBE 121; CDB 86 (cp 287); DPB 87; FBE 139; HAB 68; KBA 128 (pl 13), 149 (pl 18); OBB 69; PBE 92 (pl 29), 101 (cp 32); RBE 140; SNB 84 (cp 41), 86 (pl 42); WBV 81; (L. melanuroides) FNZ 129 (pl 12).

Golah —See Cockatoo, Roseate.

Gold Dust Tree —See Laurel, Spotted.

Gold of Pleasure (Camelina sativa) MFF cp 44 (111); NHE 219; PFE 128 (338); WFB 85 (5).

Gold Vine, Guinea —See Snake Vine.

Goldcrest (Regulus regulus) ALE 9:259; BBB 160; BBE 247; CDB 175 (cp 747); FBE 239; FBW 85; OBB 159; PBE 212 (cp 51); RBE 272.

Golden Bells —See Forsythia.

Golden Club (Orontium aquaticum) DEW 2:288 (cp 174); MLP 291; PFF 366.

Golden Cup (Adonis chrysocyanthus) EWF cp 91e; (A. amurensis) DFP 1 (cp 1), EJG 99.

Golden Cup —See also Chalice Vine (Solandra sp.) and Poppy, Mexican Tulip (Hunnemannia sp.).

Golden Dewdrop or Golden Tears —See Pigeonberry (Duranta).

Golden Drop (Onosma sp.) PFE cp 105 (1076), 341; PFM cp 146.

Golden Horn (Decabelone grandiflora or Tavaresia grandiflora) EWF cp 78d; LFT 270 (cp 134); MEP 113.

Golden Moss —See Stonecrop, Mossy (Sedum).

Golden Rain or Golden Chain Tree (Laburnum anagyroides) BKT cp 142; DFP 209 (cp 1669); EET 182, 183; EPF pls 447, 448; LFW 81 (cp 188); MLP 108; MWF 176; NHE 21; OBT 184; PFE cp 50 (496); PWF 34a; SPL cp 97; SST cp 248; TGS 137; WFB 121 (5).

Golden Rain or Golden Chain Tree (Laburnum watereri) EGT 41, 121; EPG 127; MTB 324 (cp 29); PFW 162; TGS 135; TVG 229.

Golden Rain Tree (Koelreuteria paniculata)
BKT cp 155; EEG 118; EGT 44, 121; EGW 122;
MEP 74; MTN 33 (cp 34); OBT 189; PFF 243;
PFW 270; SST cp 247; TGS 294.

Golden Trumpet (Allamanda cathartica) BIP
23 (cp 11); DFP 51 (cp 406); EGG 86; EGV 39,
92; FHP 104; GFI 55 (cp 10); HPW 224; LFW
228 (cp 510); MEP 108; MLP 201; PFW 29;
RDP 79.

Golden Vine (Stigmaphyllon ciliatum) EGV
138; EWF cp 178a; MEP 67; (S. heterophyllum) LFW 266 (cp 596).

Goldenback —See Woodpecker, Crimson-
backed.

Goldeneye (Bucephala clangula) ALE 7:255,
305, 385; BBB 197; BBE 61; FBE 65; FBW 51;
GPB 84; OBB 25; PBE 29 (cp 10), 41 (pl 14), 49
(pl 16).

Goldenrod, Canada (Solidago canadensis)
DFP 173 (cps 1381-1384); EPF pl 531; MCW
125; MWF 268; OGF 151; PFF 313; PFW 84;
PWF 122d; SPL cp 436; WFB 231 (3); WUS
431.

Goldenrod, European (Solidago virgaurea)
AMP 137; BFB cp 19 (7); CTH 56 (cp 87); EPF
pl 530; KVP cp 55; MBF cp 44; MFF pl 57
(874); NHE 37; OWF 33; PFE cp 143 (1358);
PWF 119e.

Goldfinch (Carduelis carduelis) ABW 301;
ALE 9:334, 391; BBB 60; BBE 283; CDB 208
(cp 918); DAE 133; FBE 287; FNZ 208 (cp 18);
GAL 118 (cp 1); HAB 256; LAW 438; NBB 32;
OBB 183; PBE 261 (cp 62); RBE 296; SAB 70
(cp 34); (C. caniceps) RBI 311.

Goldfish (Carassius auratus) ALE 4:334-335,
361; CTF 25 (cp 17), 27 (cps 20, 21); FEF 165,
166 (bw); GAL 239 (bw); HFI 94; HFW 42 (cp
21), 120 (bw); LAW 316; OBV 113; WEA 171;
WFW cps 121, 122.

Goldfish Plant (Hypocyrta nummularia) BIP
103 (cp 155); EGL 119; EWF cp 185c; FHP 126;
(Nematanthus wettsteinii) EGG 117; EGV
124.

Goldilocks (Ranunculus auricomus) MBF cp
3;MFF cp 39 (17); NHE 26; OWF 3; PWF 22d;
WFB 71 (1).

Goldsinny (Ctenolabrus rupestris) CSF 131;
CSS cp 30; FEF 460 (bw); OBV 63.

Golf Balls —See Cactus, Button.
Gomphidius sp. —See Pink Nail.
Gonolek (Laniarius barbarus) ABW 268; ALE
9:195; (L. atrococcineus) PSA 289 (cp 34);
WAB 139.

Goober —See Peanut.
Good Friday Grass —See Woodrush, Field.
Good King Henry (Chenopodium bonus-
henricus) EGH 107; MBF cp 72; MFF pl 110
(204); NHE 227; OFP 191; OWF 55; PFE cp
10 (103); PWF 117; TEP 190; WFB 47 (1).

Goosander —See Merganser, Common.
Goose, African Pygmy (Nettapus auritus)
ALE 7:319; CDB 53 (cp 123); GPB 83.

Goose, Andean (Chloephaga melanoptera)
ALE 7:295; SBB 42 (pl 49).

Goose, Ashy-headed (Chloephaga poliocephala) ALE 7:295; SSA cp 2; SBB 43 (pl 51).

Goose, Bar-headed (Anser indicus) ABW 66;
ALE 7:282; BBE 51; FBE 45; SBB 33 (pl 34).

Goose, Barnacle (Branta leucopsis) ABW 66;
ALE 7:284, 385; BBB 220; BBE 47; CDB 51
(cp 104);CEU 241 (bw); FBE 49; OBB 33;
PBE 16 (cp 5), 20 (pl 7).

Goose, Bean (Anser fabalis) ALE 7:281, 385;
BBE 51; FBE 47; OBB 31; PBE 17 (cp 6), 21
(pl 8); (A. arvensis) SBB 30 (pl 31).

Goose, Brant or Brent —See Brant.
Goose, Canada (Branta canadensis) ABW 66;
ALE 7:284; BBB 193; BBE 47; CDB 51 (cp
103); FBE 49; GPB 86; LEA 377 (bw); OBB
33; PBE 16 (cp 5), 20 (pl 7); WAB 65; WEA
162.

Goose, Cape Barren (Cereopsis novaehollandiae) ABW 64; ALE 7:295; CDB 51
(cp 109); HAB 29; LVB 39; LVS 192; PWC
218 (bw); RBU 52; SBB 44 (cp 52); SNB 34
(cp 16); WAW 18 (bw).

Goose, Cotton Pygmy —See Goose, Indian
Pygmy.

Goose, Egyptian (Alopochen aegyptiacus)
ABW 66; ALE 7:295; 13:112-113; BBB 280;
BBE 57; CAA 29 (bw); CAF 68; CDB 49 (cp
94); FBE 51; GPB 87; OBB 21; PSA 49 (cp
4), 64 (pl 5); SBB 42 (pl 48).

Goose, Graylag (Anser anser) ABW 66; ALE
7:281, 11:257; BBB 216; BBE 42, 49; CDB
50 (cp 99); FBE 47; GPB 86; OBB 31; PBE
17 (cp 6), 21 (pl 8); SBB 30 (pl 30); WBV 49.

Goose, Green Pygmy (Nettapus pulchellus) HAB 30; SNB 40 (cp 19).

Goose, Indian Pygmy (Nettapus coromandelianus) ALE 7:307; DPB 33; HAB 30; KBA 80 (pl 7); RBU 59; SNB 40 (cp 19); WBV 49.

Goose, Kelp (Chloephaga hybrida) ALE 7:295; CDB 51 (cp 110); WFI 102, 115 (bw).

Goose, Lesser White-fronted (Anser erythropus) ALE 7:281; BBE 49; FBE 47; OBB 31; PBE 17 (cp 6); SBB 31 (pl 32).

Goose, Magellan or Upland (Chloephaga picta) ALE 7:295; CDB 52 (cp 111); SBB 43 (pl 50); WFI 113 (bw); (C. leucoptera) ABW 66.

Goose, Magpie or Pied (Anseranas semipalmata) ALE 7:267; CDB 50 (cp 98); HAB 33; RBU 55; SNB 34 (cp 16); WAB 201.

Goose, Maned (Chenonetta jubata) ALE 7:319; 10:89; HAB 32; RBU 56; SNB 38 (cp 18).

Goose, Pink-footed (Anser brachyrhynchus) ALE 7:281; BBB 218; BBE 51; FBE 47; OBB 31; PBE 17 (cp 6), 21 (pl 8).

Goose, Red-breasted (Branta ruficollis) ABW 66; ALE 7:284; BBB 275; BBE 47; FBE 49; OBB 33; PEB 16 (cp 5), 21 (pl 8); RBE 111; SBB 34 (pl 37).

Goose, Snow (Anser caerulescens) ALE 7:283; 11:151; CDB 50 (cp 100); FBW 47; FBE 45; OBB 33; PBE 16 (cp 5), 20 (pl 7); (A. hyperboreus) BBE 51.

Goose, Spur-winged (Plectropterus gambiensis) ABW 64; ALE 7:318; CDB 54 (cp 125); PSA 49 (cp 4), 64 (pl 5).

Goose, Swan (Anser cygnoides) ALE 7:282; SBB 32 (pl 33).

Goose, Toulouse —See **Goose, Graylag.**

Goose, White-fronted (Anser albifrons) ABW 64; ALE 7:281, 385; BBB 217; BBE 49; FBE 47; OBB 31; PBE 16 (cp 6), 21 (pl 8); RBE 107.

Goose, White-quilled Pygmy —See **Goose, Indian Pygmy.**

Gooseberry (Ribes uva-crispa) EPF pls 372, 374; MBF cp 33; MFF pl 78 (414); NHE 19; PFE cp 43 (416); WFB 141 (21).

Gooseberry, Chinese (Actinidia chinensis) EGV 90; MWF 31; (A. arguta) TGS 71.

Gooseberry, English (Ribes grossularia) OFP 81; PFF 194.

Goosefoot, Fig-leaved (Chenopodium ficifolium) MBF cp 72; NHE 228; PWF 117k.

Goosefoot, Many-seeded (Chenopodium polyspermum) BWG 159; EWF cp 14e; MBF cp 71; MFF pl 110 (205); NHE 227; PWF 142d; WFB 47 (5).

Goosefoot, Maple-leaved (Chenopodium hybridum) MFF pl 110 (209); NHE 227; PWF 117i; WFB 47 (4).

Goosefoot, Nettle-leaved (Chenopodium murale) MBF cp 72; MFF pl 110 (208); NHE 227.

Goosefoot, Oak-leaved (Chenopodium glaucum) MBF cp 72; MFF pl 110 (211); NHE 227.

Goosefoot, Red (Chenopodium rubrum) BFB cp 12 (10); MBF cp 72; MFF pl 110 (210); NHE 227; OWF 55; PWF 166c; WFB 47 (3).

Goosefoot, Sticky (Chenopodium botrys) EGH 107; NHE 227; PFE 71.

Goosefoot, Stinking (Chenopodium vulvaria) MBF cp 71; MFF pl 110 (206); NHE 227; OWF 55; PFE 71.

Goosefoot, Strawberry (Chenopodium foliosum) DEW 1:202; KVP 45; PFE cp 10 (104).

Goosefoot, White —See **Fat Hen.**

Goosegrass (Eleusine indica) EGC 71; WUS 63; (E. coracana) OFP 11. Galium sp. — See **Cleavers.**

Goral, Gray or Short-tailed (Naemorhaedus goral) ALE 13:464; DAA 24; DSM 268; HMW 79; LAW 571; MLS 421 (bw).

Gordonia —See **Franklinia.**

Gorgonian (Hookerella sp.) VWS 82 (bw); (Iridogorgia sp.) MOL 130; (Melitoda sp.) LAW 59.

Gorilla (Gorilla gorilla) ALE 10:521-524; BMA 102-104; CAA 93 (bw), 100; CAF 126, 270 (bw), 277; DDM 80 (cp 11); DSM 96; HMA 57; HMW 11; JAW 63; LAW 624; LVA 9; LVS 41; MLS 162 (bw); PWC 234 (bw); SLS 201; SMW 89 (cp 53); TWA 13 (bw); VWA cp 1; WEA 172.

Gorse, Common (Ulex europaeus) BFB cp 7 (4); BWG 28; GDS cp 482; HSC 119, 121; MBF cp 21; MFF cp 47 (273); MWF 290; NHE 69; OBT 36; OGF 23; OWF 19; PFF 214; PWF cp 20c; TGS 135; WFB 121 (1).

Gorse, Dwarf (Ulex minor) BFB cp 7 (10); MBF cp 21; OWF 19; PFE cp 53 (517); PWF 155h; (U. gallii) MBF cp 21; PWF 152a.

Gorse, Spanish (Genista hispanica) DFP 203 (cp 1622); HSC 58, 61; PFE cp 52 (511); TGS 130.

Goshawk (Accipiter gentilis) ABW 76; ALE 7:332-333, 355; 13:163; BBB 269; BBE 81; DAE 116; FBE 75; GAL 153 (bw); GBP 238 (bw); KBA 65 (pl 6); LAW 503; LEA 388 (bw); OBB 41; PBE 53 (cp 18), 73 (pl 24); RBE 12; WAB 39.

Goshawk, Black (Accipiter melanoleucus) PSA 112 (cp 11); (A. melanochlamys) WAB 204.

Goshawk, Brown (Accipiter fasciatus) HAB 39; SNB 42 (cp 20), 50 (pl 24).

Goshawk, Chanting (Melierax metabates) FBE 75; PSA 81 (pl 8), 112 (cp 11); (M. canorus) CDB 59 (cp 157); (M. musicus) ALE 7:355; GBP 256 (bw); PSA 81 (pl 8), 112 (cp 11).

Goshawk, Chinese (Accipiter soloensis) DPB 45; KBA 65 (pl 6).

Goshawk, Crested (Accipiter trivirgatus) DPB 45; KBA 48 (pl 3), 65 (pl 6).

Goshawk, Gabar (Melierax gabar) PSA 112 (cp 11); (Micronisus gabar) WAB 151.

Goshawk, Gray or White (Accipiter novae-hollandiae) ALE 7:355; HAB 39; SNB 42 (cp 20), 50 (pl 24).

Goshawk, Red (Erythrotriorchis radiatus) GBP 253 (bw); HAB 40; RBU 80; SNB 44 (cp 21), 50 (pl 24).

Goura —See **Pigeon, Crowned.**

Gourami, Croaking or Talking (Trichopsis vittatus) FEF 493 (bw); SFW 633 (pl 174); WFW cp 459.

Gourami, Dwarf (Colisa lalia) ALE 5:229; CTF 43 (cp 52); FEF 480 (bw); HFI 54; SFW 668 (cp 183); WFW cp 457; (Trichopsis pumilus) FEF 528; SFW 632 (pl 173).

Gourami, Giant or Striped (Colisa fasciata) ALE 5:229; FEF 525; HFI 54; SFW 640 (pl 175); WFW cp 455.

Gourami, Kissing (Helostoma temmincki) ALE 5:220, 229; CTF 46 (cp 57); FEF 490 (bw); HFI 54; HFW 244 (bw); SFW 648 (cp 177); WFW cp 461.

Gourami, Pearl (Trichogaster leeri) ALE 5:220, 229; FEF 491 (bw); HFI 54; LAW 345; SFW 668 (cp 183); WEA 173.

Gourami, Three-spot (Trichogaster trichopterus) ALE 5:229; CTF 43 (cp 51); FEF 491 (bw); HFI 54; SFW 641 (pl 176).

Gourd (Cucurbita pepo ovifera) EGA 122; EVF 107; MWF 98; PFF 307; (C. foetidissima) EWF cp 159d.

Gourd, Bottle or Dipper (Lagenaria siceraria) DEW 1:251; EGA 122.

Gourd, Snake (Trichosanthes sp.) HPW 116; OFP 121.

Goutweed (Aegopodium podagraria) BWG 76; EGC 120; ESG 90; MBF cp 38; MFF pl 97 (488); NHE 32; OWF 91; PWF 65f; TEP 245; TGF 125; WFB 159 (4).

Grandala (Grandala coelicolor) ALE 12:103; RBI 219.

Granny's Bonnets —See **Columbine, Common.**

Grape Hyacinth (Muscari atlanticum) EPF pl 910; MBF cp 85; MFF cp 6 (1007); NHE 241; WFB 267 (5).

Grape Hyacinth (Muscari sp.) DFP 104 (cps 827, 828); EFC cp 30; EGB 129; EPF pl 912; ESG 129; EWF cp 41e; NHE 38, 241; OGF 9; PFE cps 169, 170 (1646, 1647); PFM cps 246, 247; SPL 265.

Grape Hyacinth, Armenian (Muscari armeniacum) DFP 103 (cp 823); EDP 122; EGB 129; FHP 133; MWF 204; TVG 205.

Grape Hyacinth, Common or Small (Muscari botryoides) DFP 103, 104 (cps 824, 825); ERW 127; LFW 68 (cp 158); NHE 241; PFE cp 170 (1648); PFF 376; PFW 173; TGF 13; WFB 267 (6).

Grape Hyacinth, Tassel (Muscari comosum) DFP 104 (cp 826); EPF pl 911; NHE 241; PFE cp 169 (1645); PFM cp 245; WFB 266.

Grape Ivy (Cissus rhombifolia) EGL 95; EGV 77, 83; MWF 85; RDP 27, 144.

Grape, Seaside (Coccoloba sp.) BIP 57 (cp 75); DFP 60 (cp 475); HPW 78.

Grape Vine (Vitis sp.) DFP 250 (cp 1997); EGV 50, 51; EPG 152; EPF pls 721, 722; EVF 124, 125; HPW 189; NHE 23; OFP 91, 93.

Grapefruit (Citrus paradisi) EVF 118; OFP 87; PFF 228; SST cp 195.

Grapevine, Evergreen (Rhoicissus rhomboidea) BIP 155 (cp 238); DFP 80 (cp 636); LHP 156, 157; RDP 340.

Grapple Plant (Harpagophytum sp.) DEW 2:180; EWF cp 79a.

Graptopetalum (Graptopetalum sp.) EMB 107; LCS cp 261; RDP 226.

Grass, Barnyard or Cockspur (Echinochloa crus-galli) MCW 18; MFF pl 133 (1252); NHE 243; PFE cp 182 (1804); PFF 348; WUS 61.

Grass, Bermuda (Cynodon dactylon) EGC 76, 116; HPW 289; MBF pl 96; MFF pl 126 (1251); NHE 244; PFF 354.

Grass, Cloud (Agrostis nebulosa) EDP 129; EGA 91.

Grass, Fern (Catapodium rigidum) MBF pl 98; MFF pl 128 (1185); PFE cp 180 (1759); PFM pl 22 (425).

Grass Nut (Brodiaea laxa or Triteleia laxa) DFP 85 (cp 678); EGB 99; OGF 71; SPL cp 282; TGF 18.

Grass, Pampas (Cortaderia sp.) DFP 133 (cp 1057); EDP 140; MLP 297; MWF 94; RWF 93; SNZ 102 (cp 311); SPL cp 198; TVG 127.

Grass, Poly (Lythrum hyssopifolia) MBF cp 34; MFF cp 26 (419); NHE 93; PFE 264; PWF 80c; WFB 151 (8).

Grass, Rabbit-tail —See **Hare's-tail Grass.**

Grass, Squirrel-tail (Hordeum jubatum) DFP 39 (cp 308); EDP 141; EGA 125; EPF pl 1050; MCW 19; PFF 356; TVG 53; WUS 67.

Grass, Timothy (Phleum pratense) BWG 178; DEW 2:260; EPF pl 1058; HPW 287; MBF pl 95; MFF pl 122 (1236); NHE 194; PFE cp 181 (1784); PFF 353; TEP 153.

Grass Tree (Dracophyllum sp.) HPW 128; SNZ 49 (cp 113), 101 (cp 307), 143 (cps 448, 449), 144 (cp 450).

Grass Tree or **Grass Spear** (Xanthorrhoea sp.) CAU 39; EWF cp 142c; MLP 260, 261; MWF 298; RWF 80; WAW 103.

Grassbird (Megalurus timoriensis) DPB 313; HAB 182; SAB 26 (cp 12); (M. gramineus) HAB 182; SAB 26 (cp 12); (M. palurus) DPB 313.

Grasshopper, Blue-winged Wasteland (Oedipoda caerulescens) ALE 2:116; PEI 60 (bw); PWI 39; ZWI 98.

Grasshopper, Field or Meadow (Chorthippus sp.) CIB 81 (cp 6); OBI 15, 17; PWI 46 (bw).

Grasshopper, Great Green (Tettigonia viridissima) ALE 2:115; CIB 80 (cp 5); GAL 314 (bw); KIW 17 (bw); LAW 194; OBI 13; PEI 53 (bw).

Grasshopper, Green (Omocestus sp.) GAL 312 (bw), 313 (bw); OBI 15.

Grasshopper, Mottled (Myrmeleotettrix maculatus) CIB 81 (cp 6); OBI 17.

Grasshopper, Steppe (Ephippiger sp.) ALE 2:115; LAW 176; PEI 56 (bw); PWI 101.

Grasshopper, Stripe-winged (Stenobothrus lineatus) CIB 81 (cp 6); OBI 15.

Grasshopper, Wart-biter (Decticus verrucivorus) ALE 2:115; GAL 314 (bw).

Grass-of-Parnassus (Parnassia palustris) EWF cp 116c; HPW 147; MBF cp 32; MFF pl 84 (411); NHE 179; OWF 83; PFE cp 42 (413); PWF 151h; WFB 101 (8).

Grassquit, Yellow-faced (Tiaris olivacea) ABW 299; ALE 9:342.

Grass-wrack —See **Eel-grass** (Zostera).

Grasswren —See **Wren.**

Graybeard or **Old Man's Beard** —See **Spanish Moss.**

Graybird —See **Cuckoo-shrike.**

Grayling, European (Thymallus thymallus) ALE 4:237, 255; FEF 59-61; FWF 55; GAL 248 (bw); HFI 123; OBV 103; SFW 56 (pl 5).

Grebe, Australian Little (Podiceps novaehollandiae) CAU 81 (bw); SNB 20 (pl 9).

Grebe, Black-necked (Podiceps nigricollis) ALE 7:255; BBB 210; DPB 7; FBE 23; LAW 501; OBB 5; PBE 4 (pl 1); PSA 32 (cp 1); RBE 56; (P. caspicus) BBE 23.

Grebe, Great Crested (Podiceps cristatus) ABW 22; ALE 7:255; 11:257; BBB 208; BBE 23; CAU 271; CDB 36 (cp 19); CEU 266 (bw); FBE 23; GPB 41; HAB 5; LEA 362 (bw); NBB 85 (bw); OBB 3; PBE 4 (pl 1); PSA 32 (cp 1); RBE 55; SNB 20 (pl 9); WAB 126.

Grebe, Hoary-headed (Podiceps poliocephalus) HAB 4; SNB 20 (pl 9).

Grebe, Horned —See **Grebe, Slavonian.**

Grebe, Little (Podiceps ruficollis or Tachybaptus ruficollis) ABW 23; ALE 7:385; BBB
209; BBE 23; CDB 36 (cp 21); DPB 7; FBE 23;
GPB 40; KBA 80 (pl 7); LEA 363 (bw); OBB 5;
PBE 4 (pl 1); PSA 32 (cp 1); RBE 59; WBV 41.

Grebe, Red-necked (Podiceps grisegena) ABW
22; BBB 275; BBE 23; FBE 23; OBB 3; PBE 4
(pl 1).

Grebe, Slavonian (Podiceps auritus) ABW 22;
BBB 211; BBE 23; CDB 36 (cp 18); FBE 23;
OBB 5; PBE 4 (pl 1); WAB 130.

Greek Valerian —See **Jacob's Ladder**
(Polemonium).

Green Ebony —See **Jacaranda.**

Greenbul —See **Bulbul** (Phyllastrephus sp.).

Greenfinch, Black-headed (Carduelis
ambigua) KBA 429 (bw); (C. monguilloti)
WBV 221.

Greenfinch, European (Carduelis chloris) ALE
9:391; BBB 60; BBE 281; CDB 208 (cp
919); FBE 287; LAW 469; LEA 335 (bw);
PBE 261 (cp 62); (Chloris chloris) FNZ 208 (cp
18); GAL 118 (cp 1); HAB 256; OBB 177; RBE
295; SAB 70 (cp 34).

Greenlet, Gray-headed (Hylophilus decurtatus) ABW 282; ALE 9:372.

Greenlet, Tawny-crowned (Hylophilus
ochraceiceps) CDB 206 (cp 905); SSA cp 40.

Greenovia (Greenovia aurea) ECS 120; LCS cp
322.

Greenshank (Tringa nebularia) ALE 8:170;
BBB 119; BBE 125; DPB 91; FBE 131; HAB 68;
KBA 129 (pl 14), 148 (pl 17); OBB 71; PBE 84
(pl 27), 116 (cp 33); PSA 144 (pl 15); SNB 70 (cp
34), 72 (pl 35); WBV 87.

Greenshank, Nordmann's or Spotted (Tringa
guttifera) DPB 91; KBA 148 (pl 17).

Greenweed, Dyer's (Genista tinctoria) DFP
203 (cp 1624); MBF cp 21; MFF cp 47 (270);
NHE 69; OWF 19; PWF 103i; TGS 135; WFB
121 (3).

Greenweed, Dyer's (Reseda luteola) MBF cp
11; MFF cp 50 (113); NHE 219; PWF 75i; WFB
101 (1).

Greenweed, German (Genista germanica)
NHE 69; PFE 185.

Greenweed, Hairy (Genista pilosa) MBF cp 21;
MFF cp 47 (272); NHE 69, 80; TGS 39.

Grenadier (Coryphaenoides rupestris) ALE
4:432; CSF 95.

Grevillea (Grevillea sp.) BIP 91 cp 140; DFP
204 (cp 1625); EPF pl 697; EWF cp 134e; HSC
59, 64; MEP 32; MWF 144, 145; PFW 245.

Griffon —See **Vulture.**

Grimmia (Grimmia sp.) DEW 3: cp 114; ONP
1, 75.

Grindelia (Grindelia robusta) AMP 97; EWG
114.

Grisette (Amanita vaginata) DEW 3:172; DLM
169; EPF pl 67; LHM 117;PMU cp 120;REM 56
(cp 24); (A. fulva) DLM 166; LHM 117; ONP
113; SMG 172, 173.

Grison (Galictis vittata) ALE 12:66; DSM 169;
LAW 535.

Grison, Lesser or Little (Galictis cuja) ALE
12:66; MLS 287 (bw).

Gromwell, Blue (Lithospermum purpurocaeruleum) EPF pl 800; MBF cp 61; MFF cp 13
(661); OWF 171; PFE cp 104 (1073); PFM
cp 147; WFB 189 (8).

Gromwell, Common (Lithospermum officinale)
MBF cp 61; MFF pl 112 (662); OWF 95; PFE
cp 105 (1072); PWF 128d; WFB 189 (7).

Gromwell, Corn (Lithospermum arvense) MBF
cp 61; MFF pl 112 (663); TEP 179; WFB 189
(6); WUS 305.

Gromwell, Scrambling (Lithospermum
diffusum) DFP 14 (cp 107); GDS cp 273;
PFE cp 104 (1074); PFM cp 150.

Grosbeak (Cyanocompsa sp.) ALE 9:352; SSA
cp 42; (Eophona sp.) ALE 9:392; RBI 312.

Grosbeak, Pine (Pinicola enucleator) ALE
9:382; BBB 271; BBE 285; FBE 297; PBE
260 (cp 61).

Grosbeak, Scarlet (Carpodacus erythrinus)
ALE 9:391; BBB 275; BBE 285; FBE 295; KBA
433 (cp 64); OBB 191; PBE 260 (cp 61); RBE
308; RBI 316.

Ground Elder —See **Goutweed.**

Ground Ivy (Glechoma hederacea) AMP 101;
BWG 149; EFC cp 53; EGC 71; MBF cp 68;
MCW 86; MFF cp 10 (779); NHE 230; OWF
145; PFE cp 109 (1118); PFF 288; PWF cp 22a;
WFB 199 (3); WUS 313.

Ground-cistus (Rhodothamnus chamaecistus)
NHE 266; OGF 67; SPL cp 462.

Ground-dove —See **Dove.**

Groundhopper (Tetrix sp.) CIB 81 (cp 6); OBI 17; PEI 64 (bw).

Ground-pine (Ajuga chamaepitys) AMP 149; MBF cp 70; MFF cp 61 (787); NHE 230; OWF 29; PFE cp 106 (1094); PFM cp 159; PWF 138d; WFB 197 (2).

Groundsel, Common (Senecio vulgaris) AMP 147; BWG 47; MBF cp 47; MFF cp 57 (848); OWF 43; PFP 329; PWF 15e; WFB 243 (6).

Groundsel, Gray Alpine (Senecio incanus) NHE 286; PFE cp 150 (1452).

Groundsel, Spring (Senecio vernalis) EPF pl 568; NHE 237; PFE cp 151 (1452).

Groundsel, Sticky or Stinking (Senecio viscosus) MBF cp 47; MFF cp 57 (847); NHE 237; PWF 101l.

Groundsel, Wood (Senecio sylvaticus) MBF cp 47; MFF cp 57 (846); OWF 43; PWF 153h.

Grouper (Epinephelus sp.) ALE 5:79; FEF 338, 340 (bw); HFI 21; HFW 87 (cp 43), 88 (cp 45); LAW 336.

Grouper, Giant (Promicrops lanceolatus) ALE 5:94; FEF 340.

Grouper, Six-lined (Grammistes sexlineatus) FEF 341 (bw); HFI 21; HFW 86 (cp 41); LAW 343.

Grouse, Black (Lyrurus tetrix) ABW 88; ALE 7:446, 467; BBB 153; BBE 97; CDB 67 (cp 186); DAE 124; FBE 99; OBB 53; PBE 164 (cp 43); RBE 43; WAB 25.

Grouse, Caucasian Black (Lyrurus mlokosiewiczi) FBE 99; RBI 3.

Grouse, Hazel (Tetrastes bonasia) ALE 7:446; BBE 97; CDB 67 (cp 189); DAE 125; FBE 97; PBE 164 (cp 43); RBE 47.

Grouse, Red or Willow (Lagopus lagopus) ABW 90; ALE 7:446; BBB 114; BBE 95; CDB 67 (cp 183); FBE 97; GPB 113, 115; OBB 51; PBE 157 (cp 42).

Grouse, Sand —See **Sandgrouse.**

Grysbok —See **Steinbok.**

Guacharo —See **Oilbird.**

Guan (Penelope sp.) ABW 87; CSA 129; WAB 79; (Chamaepetes goudotii) SSA cp 1.

Guan, Piping (Pipile pipile) ALE 7:436; GPB 112; SSA cp 1; WAB 90.

Guanaco (Lama glama huanacus) ALE 13:133; CSA 166-167 (bw); HMW 56; LEA 593 (bw); (L. guanacoe) JAW 169; MLS 377 (bw); SLP 109.

Guanay —See **Cormorant, Guanay.**

Guava (Psidium guajava) EET 223; MWF 246; OFP 99; SST cp 220; (P. cattleianum) MWF 246.

Guava, Pineapple (Feijoa sellowiana) EGE 126; EWF cp 179c; MWF 132; SPL cp 88.

Gudgeon (Gobio gobio) ALE 4:237, 341; FEF 151, 152 (bw); FWF 75; GAL 234 (bw); HFI 99; OBV 99; SFW 268 (pl 66).

Gudgeon, Australian or Purple-striped (Mogurnda mogurnda) FEF 494 (bw); HFI 57; SFW 616 (pl 169).

Guelder Rose (Viburnum opulus) AMP 149; BFB cp 11 (3); DFP 244 (cp 1945); EPF pl 770-772; GDS cps 491-493; HSC 122, 124; LFW 145 (cp 324); MBF cp 41; MFF pl 74 (822); NHE 25; OBT 33; OWF 193; PFE cp 133 (1298); PFF 304; PFW 69; PWF 58b; TEP 290; TGS 343; WFB 223 (6).

Guenon, Diademed (Cercopithecus mitis) ALE 10:392 (cp 7); DDM 64 (cp 9); SMR 16 (cp 1).

Guenon, Dwarf (Cercopithecus talapoin) ALE 10:391 (cp 6); MLS 153 (bw); (Miopithecus talapoin) DDM 61 (cp 8); DSM 80.

Guenon, Moustached (Cercopithecus cephus) ALE 10:391 (cp 6); DDM 61 (cp 8); DSM 80; HMW 23; LAW 618.

Guenon, Owl-faced (Cercopithecus hamlyni) ALE 10:392 (cp 7); DDM 61 (cp 8).

Guenon, Red-bellied (Cercopithecus erythrogaster) ALE 10:391 (cp 6); DDM 61 (cp 8).

Guereza —See **Colobus.**

Guillemot, Arctic or Thick-billed —See **Murre, Thick-billed.**

Guillemot, Black (Cepphus grylle) ABW 138; ALE 7:61; BBB 261; BBE 163; CDB 95 (cp 344); FBE 165; OBB 101; PBE 125 (pl 36); SAO 143; WAB 72.

Guillemot, Bridled —See **Murre, Common.**

Guinea-fowl, Helmeted or Tufted (Numida meleagris) ALE 8:40; CAA 27 (bw); CDB 71 (cp 210); FBE 101; GPB 122; WAB 147; WEA 178.

Guinea-fowl, Vulturine (Acryllium vulturinum) ABW 99; ALE 8:24; CAA 26 (bw); CAF 79; CDB 70 (cp 209); LEA 401 (bw)

Guinea-hen Flower —See **Fritillary, Snake's Head.**

Guinea-pig (Cavia porcellus) ALE 11:436; DSM 132; HMW 178; JAW 84; MLS 225 (bw); WEA 99; (C. aperea tschudii) ALE 11:436.

Guitarfish (Rhinobatus sp.) ALE 4:117; HFI 138; LAW 289.

Gularis, Blue (Aphyosemion gulare caeruleum) FEF 275 (bw); HFI 68; SFW 396 (cp 103). (cp 103).

Gull, Audouin's (Larus audouinii) BBE 147; FBE 147; LVB 49; PBE 132 (pl 37); SAO 105; SLS 97; WAB 29.

Gull, Black-backed or Dominican (Larus dominicanus) CAF 238; FNZ 144 (pl 13); HAB 79; PSA 176 (cp 19); SAO 105, 109; SAR 119, 120; SNB 92 (cp 45); SSA pl 20; WAN 121-123; WFI 165-168 (bw).

Gull, Black-billed or Buller's (Larus bulleri) BNZ 145; FNZ 144 (pl 13).

Gull, Black-headed (Larus ridibundus) ABW 131; ALE 7:255; 11:257; BBB 42; BBE 149; CDB 93 (cp 332); DAE 33; DPB 109; FBE 149; GPB 161; KBA 156 (pl 19); NBB 50-51 (bw); OBB 93; PBE 132 (pl 37), 133 (pl 38); RBE 175; SAO 113, 115; WAB 25; WBV 97; WEA 74, 180.

Gull, Common or Mew (Larus canus) ALE 8:187; BBB 240; BBE 147; CDB 93 (cp 327); FBE 153; OBB 91; PBE 132 (pl 37), 133 (pl 38); RBE 172; SAO 105, 109.

Gull, Dolphin (Larus scoresbii) SAO 113, 115; (Leucophaeus scoresbii) WFI 171.

Gull, Foul —See **Fulmar.**

Gull, Glaucous (Larus hyperboreus) ALE 8:210; BBB 276; BBE 143; DAE 33; FBE 151; OBB 89; PBE 132 (pl 37), 133 (pl 38); SAO 105, 109; WAB 73.

Gull, Gray-headed (Larus cirrhocephalus) CDB 93 (cp 328); PSA 176 (cp 19); SAO 121.

Gull, Great Black-backed (Larus marinus) ABW 131; ALE 7:61; BBB 253; BBE 145; DAE 33; FBE 151; OBB 87; PBE 132 (pl 37), 133 (pl 38); SAO 105, 109.

Gull, Great Black-headed (Larus ichthyaetus) ALE 8:210; BBE 151; FBE 147; SAO 117.

Gull, Herring (Larus argentatus) ABW 131; ALE 8:187, 206; BBB 241; BBE 145; CDB 92 (cp 325); DPB 109; FBE 151; GPB 163; KBA 156 (pl 19); LAW 473, 485; LEA 415 (bw); OBB 89; PBE 132 (pl 37), 133 (pl 38); SAO 101, 109; WAB 100, 132; WBV 97; WEA 180.

Gull, Iceland or Kumlien's (Larus glaucoides) BBB 279; BBE 143; FBE 151; OBB 89; PBE 132 (pl 37); SAO 105, 109.

Gull, Ivory (Pagophila eburnea) BBB 276; BBE 143; FBE 153; OBB 91; PBE 132 (pl 37); SAO 113, 115.

Gull, Kelp —See **Gull, Black-backed or Dominican.**

Gull, Lava (Larus fuliginosus) HBG 140 (pl 11); LAW 459.

Gull, Lesser Black-backed (Larus fuscus) ALE 8:210; BBB 252; BBE 145; FBE 151; OBB 87; PBE 132 (pl 37), 133 (pl 38); SAO 105, 109; WEA 180.

Gull, Little (Larus minutus) ALE 8:210; BBB 269; BBE 151; CDB 93 (cp 329); FBE 149; OBB 93; PBE 132 (pl 37), 133 (pl 38); SAO 113, 115.

Gull, Mediterranean (Larus melanocephalus) BBB 275; BBE 149; FBE 149; GPB 162; PBE 132 (pl 37), 133 (pl 38); SAO 117.

Gull, Pacific (Larus pacificus) CDB 93 (cp 330); GPB 163; HAB 79; SNB 92 (cp 45).

Gull, Red-billed or Silver (Larus novaehollandiae or L. scopulinus) BNZ 145; FNZ 144 (pl 13); HAB 79; MBA 54; SAO 121; SNB 92 (cp 45); WAN 123, 124.

Gull, Ross' (Rhodostethia rosea) BBE 151; FBE 153; PBE 285 (cp 66); (Larus roseus) SAO 113, 115.

Gull, Sabine's (Larus sabini) BBB 276; FBE 149; PBE 132 (pl 37), 133 (pl 38); SAO 113, 115; (Xema sabini) BBE 151; CDB 95 (cp 341); OBB 93.

Gull, Sea —See **Gull, Herring.**

Gull, Slender-billed (Larus genei) BBE 147; FBE 147; PBE 132 (pl 37); SAO 121.

Gull, Swallow-tailed (Creagus furcatus) ALE 8:210; CDB 92 (cp 321); CSA 240; HBG 140 (pl 11); LAW 455; (Larus furcatus) GPB 164.

Gulper-eel (Eurypharynx pelecanoides) ALE 4:176; HFI 114; MOL 99.

Gum, Blue (Eucalyptus globulus) AMP 101; BIP 77 (cp 114); DEW 2:19, 50 (cp 4); EPF pls 709, 710; MWF 128; PFF 262; SST cp 108.

Gum, Cider (Eucalyptus gunni) DFP 200 (cp 1597); GDS cp 183; OBT 137; RDP 195.

Gum, Red-flowering (Eucalyptus ficifolia) EET 238; EGE 123; ELG 127; EPG 119; EWF cp 137a; MEP 96; MLP 174; MWF 128; SST cp 107; WAW 119.

Gum Tree (Eucalyptus sp.) BIP 200 (cps 1598-1600); EET 206, 238-241; EWF cp 137c; HPW 161; MEP 96; MLP 170; MTB 368 (cp 37); MWF 128; OBT 137; SST cp 106; WAW 104, 105, 119.

Gum, Yellow (Eucalyptus leucoxylon) MLP 175; PFW 191.

Gumi (Elaeagnus multiflora) EFS 109; HPW 169; TGS 278.

Gum-succory (Chondrilla sp.) NHE 239, 289; WFB 253 (9).

Gunnell —See **Butterfish.**

Gunnera (Gunnera sp.) DEW 2:27; EWF cp 3g; HPW 155; SNZ 107 (cps 326, 327).

Guppy (Lebistes reticulatus or Poecilia reticulata) ALE 4:51, 458, 459; CTF 37 (cps 40-42), 38 (cp 43); FEF 255; HFI 70; SFW 492 (cp 133); WEA 181; WFW cp 209.

Gurnard, Flying (Dactylopterus volitans) ALE 5:46; FEF 509 (bw); HFI 66; HFW 218 (cp 107); LEA 209 (bw); VWS 25.

Gurnard, Gray (Trigla gurnardus or Eutrigla gurnardus) CSF 163; OBV 69; VWS 13 (bw).

Gurnard, Red (Aspitrigla cuculus) CSF 163; OBV 69.

Gurnard, River —See **Flathead.**

Gurnard, Streaked (Trigla lineata or T. lastovitza) ALE 5:46; LEA 212 (bw); OBV 69.

Gurnard, Tub or Yellow (Trigla lucerna) CSF 163; FEF 507 (bw); HFI 59; OBV 69; WFW cp 252; (T. hirundo) HFW 252 (bw).

Guzmania (Guzmania sp.) BIP 93 (cp 141); EGL 114; FHP 124; LHP 102, 103; PFW 55; RDP 228; SPL cps 48, 49.

Gymnogene —See **Hawk, Harrier.**

Gymnopilus (Gymnopilus sp.) DLM 193; LHM 175; ONP 109; REM 91 (cp 59).

Gymnure —See **Moon-rat.**

Gyrfalcon (Falco rusticolus) ALE 7:61, 404; BBB 277; BBE 87; FBE 91; GBP 109 (bw), cp after 112, 388 (bw); GPB 108; OBB 47; PBE 73 (pl 24); WAB 73.

H

Haberlea (Haberlea sp.) DFP 11 (cp 82); EPF pl 863; OGF 45.

Hackberry (Celtis sp.) HPW 96; LFT 110 (cp 54); SST cp 92.

Hacquetia (Hacquetica epipactis) DFP 11 (cp 83); EPF pl 755; NHE 32; OGF 13; PFE cp 83 (851); PFW 300.

Haddock (Melanogrammus aeglefinus) ALE 4:431; CSF 104-105; FEF 262 (bw); HFI 77; OBV 35.

Haddock, Norway (Sebastes viviparus) CSF 161; FEF 506; OBV 41; (S. marinus) ALE 5:67; CSF 161; HFI 59; OBV 41.

Hagfish (Myxine glutinosa) ALE 4:41; CSF 33; HFI 150; OBV 1.

Hairfly (Bibio sp.) ALE 2:394; GAL 401 (bw); OBI 125; PEI 500, 501 (bw); PWI 76 (bw).

Hair-grass (Deschampsia sp.) MBF pl 96; NHE 106; SAR 28.

Hair-grass (Koeleria sp.) DFP 154 (cp 1229); EPF pl 1063; MBF pl 97; MFF pl 122 (1216); NHE 194, 244.

Hair-grass, Early (Aira praecox) MBF pl 96; MFF pl 128 (1224); NHE 244.

Hair-grass, Gray (Corynephorus canescens) MBF pl 96; MFF pl 128 (1226); NHE 245.

Hair-grass, Silver (Aira carophyllea) MBF pl 96; MFF pl 128 (1225); NHE 244.

Hair-grass, Tufted (Deschampsia cespitosa) MBF pl 96; MFF pl 133 (1222); NHE 196; TEP 153; TVG 127.

Hair-grass, Wavy (Deschampsia flexuosa) MBF pl 96; MFF pl 131 (1223); NHE 40; PFE cp 180 (1772).

Hake (Merluccius merluccius) CSF 117; HFI 76; OBV 39.

Halfbeak (Hemirhamphus sp.) HFI 80; HFW 138 (bw).

Halfbeak, Wrestling (Dermogenys pusillus) ALE 4:459; CTF 35 (cp 36); FEF 261 (bw); HFI 80; HFW 137 (bw); SFW 485 (pl 132).

Half-insect (Acerentomon sp.) OBI 1; PEI 27 (bw).

Halibut (Hippoglossus hippoglossus) ALE 5:234; CSF 179; HFI 18; LMF 163 (bw); OBV 47.

Halibut, Greenland (Reinhardtius hippoglossoides) CSF 181; HFI 18.

Halimium (Halimium sp.) DFP 204 (cp 1629); HSC 59, 64; PFE cp 78 (796), 257.

Hamadryas (Papio hamadryas) ALE 10:389 (cp 4), 439; DDM 53 (cp 6); HMA 27; HMW 19; LEA 510 (bw); MLS 146 (bw); SMW 99 (bw).

Hammerhead or **Hamerkop** (Scopus umbretta) ABW 56; ALE 7:215; CAA 47 (bw); CDB 45 (cp 66); GPB 70; PSA 48 (cp 3); WAB 154.

Hamster, Common (Cricetus cricetus) ALE 11:300, 321; DAE 85; DSM 126; GAL 55 (bw); HMW 167; LAW 586; LEA 536 (bw); MLS 208 (bw); SMW 129 (cp 70); TWA 96.

Hamster, Golden (Mesocricetus auratus) ALE 11:298; DAE 85; HMW 167; JAW 79; WEA 183.

Hangnest —See **Troupial.**

Hangul —See **Deer, Red.**

Hanuman —See **Langur, Sacred.**

Hard-grass, Sea (Parapholis strigosa) MBF pl 99; MFF pl 121 (1248); (P. incurva) NHE 138; PFE 534.

Hardhead —See **Bluet** and **Knapweed** (Centaurea sp.).

Hardon —See **Lizard, Agama.**

Hare, Blue or Mountain (Lepus timidus) ALE 12:423, 446; 13:463; CEU 159; DAE 48; GAL 49 (bw); HMW 155; OBV 141.

Hare, Brown or European (Lepus europaeus or L. capensis) ALE 11:321, 12:418, 446; DAE 42; DSM 107; FWA 127; GAL 48 (bw); HMA 59; HMW 156; MLS 181 (bw); OBV 141; SMR 129 (cp 10); TWA 93.

Hare, Cape Jumping —See **Springhare.**

Hare, Red Rock (Pronolagus crassicaudatus) HMW 157; SMR 129 (cp 10); (P. rupestris) ALE 12:436.

Hare, Scrub or Bush (Lepus saxatilis) ALE 12:418; SMR 129 (cp 10).

Hare, Spring —See **Springhare.**

Harebell —See also **Bluebell** (Wahlenbergia).

Harebell (Campanula rotundifolia) EFC cp 5; EWG 94; MBF cp 54; MFF cp 4 (799); NHE 188; OWF 167; PFW 62; PWF 151f; WFB 227 (1).

Harebells —See **Wand Flower, South African.**

Hare's Ear —See **Lemon Peel.**

Hare's-ear, Shrubby (Bupleurum fruticosum) DFP 184 (cp 1469); HSC 17, 19.

Hare's-ear, Sickle (Bupleurum falcatum) MBF cp 36; MFF cp 51 (472); NHE 222; PWF 119f; WFB 167 (9).

Hare's-ear, Slender (Bupleurum tenuissimum) MBF cp 36; MFF cp 58 (471); NHE 135; WFB 164 (6).

Hare's-ear, Small (Bupleurum baldense) MBF cp 36; WFB 167 (8).

Hare's-tail Grass (Lagurus ovatus) DFP 40 (cp 315); EDP 140; EGA 129; EPF pl 1060; MFF pl 123 (1235); PFE cp 181 (1779); TVG 55.

Harlequin —See **Duck, Harlequin.**

Harlequin Fish (Rasbora heteromorpha) ALE 4:302; CTF 24 (cp 14); FEF 172, 203 (bw); HFI 101; SFW 276 (cp 69); WEA 185; WFW cp 134.

Harlequin Flower (Sparaxis tricolor) EGB 139; MWF 270.

Harp Shell (Harpa sp.) CTS 48 (cps 81, 82); VWS 93 (bw).

Harrier, African Marsh (Circus ranivorus) CDB 57 (cp 144); PSA 81 (pl 8), 112 (cp 11).

Harrier, Hen —See **Hen-harrier.**

Harrier, Marsh (Circus aeruginosus) ALE 7:398; 11:257; BBB 167; BBE 85; DPB 41; FBE 87; GAL 156 (bw); GBP 351 (bw); KBA 79 (bw); OBB 45; PBE 60 (cp 19), 72 (pl 23); RBE 20; WAB 125; WBV 61.

Harrier, Montagu's (Circus pygargus) ALE 7:398; BBB 92; BBE 85; DAE 116; FBE 87; GAL 157 (bw); NBB 47 (bw), 57 (bw); OBB 45; PBE 60 (cp 19), 72 (pl 23).

Harrier, Pallid (Circus macrourus) ALE 7:398; BBE 85; FBE 87; KBA 79 (bw); PBE 60 (cp 19), 72 (pl 23); PSA 81 (pl 8), 112 (cp 11).

Harrier, Pied (Circus melanoleucus) ALE 7:398; DPB 41; KBA 48 (pl 3), 79 (bw).

Harrier, Spotted (Circus assimilis) HAB 41; RBU 88; SNB 46 (cp 22), 54 (pl 26).

Harrier, Swamp (Circus approximans) BNZ 109; CAU 265 (bw); CDB 57 (cp 143); FNZ 98 (bw); HAB 41; SNB 46 (cp 22), 54 (pl 26).

Harrisia (Harrisia sp.) ECS 121; LCS cps 68, 69.

Hartebeest, Bastard —See **Sassaby.**

Hartebeest, Bubal or Red (Alcelaphus buselaphus) ALE 13:302; CAF 96 (bw); DDM 208 (cp 35); DSM 260; JAW 193; MLS 411 (bw); SMR 97 (cp 8).

Hartebeest, Coke's (Alcelaphus cokei) LAW 517; LEA 611 (bw); SMW 283 (cp 177); WEA 186.

Hartebeest, Hunter's (Damaliscus hunteri) DDM 209 (cp 36); LAW 569.

Hartebeest, Lichtenstein's (Alcelaphus lichtensteini) DDM 208 (cp 35); SMR 97 (cp 8); (Bubalus lichtensteini) HMW 74.

Hartwort (Tordylium apulum) PFE cp 86 (904); PFM cp 117.

Hartwort, Great (Tordylium maximum) MFF pl 95 (509); NHE 222; WFB 165 (4).

Harvestman (Phalangium opilio) ALE 1:400; GSP 133; LAW 138; LEA 179 (bw); OIB 153; (Trogulus tricarinatus) GSP 131; OIB 165; (Opilio parietinus) GAL 491 (bw).

Hatchet-cactus (Pelecyphora aselliformis) EHP 125; LCS cp 157.

Hatchetfish, Black-winged or Silver (Carnegiella marthae) ALE 4:291; FEF 138 (bw); HFI 89; HFW 36 (cp 8).

Hatchetfish, Common (Gasteropelecus sternicla) ALE 4:307; CTF 18 (cp 3); FEF 137 (bw); HFI 89; SFW 204 (pl 47); (G. levis) WFW cp 94.

Hatchetfish, Deep-sea (Argyropelecus sp.) ALE 4:266; CSF 73; HFI 127; HFW 97 (bw); (Sternoptyx diaphana) HFI 127.

Hatchetfish, Marbled (Carnegiella strigata) ALE 4:291; FEF 138 (bw); HFI 89; LAW 314; SFW 177 (cp 40).

Hawfinch (Coccothraustes coccothraustes) ALE 9:392; BBB 150; BBE 279; CDB 209 (cp 921); FBE 289; GAL 118 (cp 1); GPB 319; OBB 175; PBE 261 (cp 62); RBE 292; WAB 24, 108.

Hawk (Leucopternis sp.) ALE 7:356; GBP 282, 283 (bw); SSA pl 43.

Hawk, African Long-tailed (Urotriorchis macrourus) ALE 7:355; GBP 255 (bw).

Hawk, Brown —See **Falcon, Brown.**

Hawk, Crane (Geranospiza caerulescens) ALE 7:398; GBP cp after 160, 355 (bw); SSA pl 43.

Hawk, Crested (Aviceda subcristata) HAB 42; MEA 59; RBU 87; SNB 42 (cp 20), 50 (pl 24); (A. cuculoides) ALE 7:346; GPB 215 (bw); (A. jerdoni) DPB 37; KBA 64 (pl 5); (A. leuphotes) GBP 217 (bw); KBA 48 (pl 3), 64 (pl 5); WBV 61.

Hawk, Galapagos (Buteo galapagoensis) CDB 56 (cp 139); GPB 105; HBG 81 (pl 4); LVB 27; LVS 246; WAB 216.

Hawk, Gray Frog —See **Goshawk, Chinese.**

Hawk, Harrier (Polyboroides radiatus) ALE 7:398; PSA 81 (pl 8), 128 (pl 13); (P. typus) CDB 61 (cp 162); (Gymnogenys typicus) GBP 356 (bw).

Hawk, Marsh —See **Hen-harrier.**

Hawk, Roadside (Buteo magnirostris) GBP 275 (bw); SSA pl 22; WEA 188.

Hawk, Savannah (Heterospiza meridionalis) ALE 7:355; CDB 59 (cp 154); GBP cp after 160, 169 (bw), 258 (bw).

Hawk, Sparrow —See **Sparrow-hawk.**

Hawkbit, Autumn (Leontodon autumnalis) MCW 121; MFF cp 55 (930); NHE 190; OWF 37; PWF 137e; TEP 148.

Hawkbit, Lesser or Hairy (Leontodon taraxacoides) MBF cp 53; MFF cp 55 (932); WFB 256 (6).

Hawkbit, Rough or Greater (Leontodon hispidus) EPF pl 600; MBF cp 53; MFF cp 55 (931); NHE 288; OWF 37; PFE cp 158 (1524); PWF 61l; WFB 255 (5).

Hawk-cuckoo —See **Cuckoo.**

Hawk-eagle —See **Eagle.**

Hawkmoth, Bedstraw (Celerio galii or Hyles galii) ALE 2:363; CIB 177 (cp 26); LEA 141 (bw); SBM 194.

Hawkmoth, Bee (Hemaris fuciformis) ALE 2:347, 363; CIB 177 (cp 26); GAL 332 (cp 9); KIW 160 (bw); OBI 63; PEI 366 (bw); WDB cp 341; (H. tityus) CIB 177 (cp 26); OBI 63; PEI 366 (bw).

Hawkmoth, Convolvulus (Herse convolvuli) OBI 61; PEI 354 (bw); SBM 168 (bw), 171; (Agrius convolvuli) CIB 176 (cp 25).

Hawkmoth, Death's-head (Acherontia atropos) ALE 2:347; CIB 176 (cp 25); CTI 27 (cp 29); GAL 332 (cp 9); OBI 61; PEI 353 (cp 34), 354 (bw); SBM 180-188; WDB cp 342; ZWI 151.

Hawkmoth, Elephant (Deilephila elpenor) ALE 2:328; CIB 177 (cp 26); CTB 22 (cp 12); GAL 332 (cp 9); LEA 114 (bw); OBI 65; PEI 361 (bw); SBM 205-207; WDB cp 348; WEA 224; (D. porcellus) CIB 177 (cp 26); OBI 65; SBM 201-203; (D. euphorbiae) LEA 128.

Hawkmoth, Eyed (Smerinthus ocellata) ALE 2:327; BIN pls 75, 76; CIB 176 (cp 25); CTB 22 (cp 11); GAL 332 (cp 9); KIW 171 (cp 81); OBI 63; PEI 362 (bw), 368 (cp 35a); SBM 215-217; WDB cp 347.

Hawkmoth, Hummingbird (Macroglossum stellatarum) ALE 2:363; CIB 176 (cp 25); GAL 332 (cp 9); KBM 23 (pl 3); OBI 63; PEI 366 (bw); SBM 230; WDB cp 344d.

Hawkmoth, Lime (Mimas tiliae) ALE 2:363; CIB 176 (cp 25); GAL 332 (cp 9); OBI 63; PEI 363 (bw); SBM 224-226; WDB cp 346.

Hawkmoth, Oak (Marumba quercus) ALE 2:363; PEI 365 (bw); SBM 220 (bw), 223.

Hawkmoth, Oleander (Daphnis nerii) CIB 177 (cp 26); PEI 352 (cp 33), 360 (bw); SBM 172-177; WDB cp 344a.

Hawkmoth, Pine (Hyloicus pinastri) ALE 2:363; CIB 176 (cp 25); (Sphinx pinastri) WDB cp 350.

Hawkmoth, Poplar (Laothoe populi) CIB 176 (cp 25); GAL 332 (cp 9); OBI 63; PEI 364 (bw); SBM 211; WDB cp 349.

Hawkmoth, Privet (Sphinx ligustri) CIB 176 (cp 25); GAL 332 (cp 9); OBI 61; PEI 358 (bw); SBM 166; WDB cp 344j.

Hawkmoth, Silver-striped or Vine (Hippotion celerio) CIB 177 (cp 26); KBM 71 (pl 21); WDB cp 344q.

Hawkmoth, Spurge (Celerio euphorbiae or Hyles euphorbiae) CIB 177 (cp 26); GAL 332 (cp 9); LAW 216; PEI 289 (cp 32b); SBM 196-199.

Hawkmoth, Striped (Celerio livornica or Hyles livornica) CIB 177 (cp 26); SBM 192 (bw).

Hawkmoth, Willow-herb (Proserpinus proserpina) CIB 176 (cp 25); PEI 367 (bw); SBM 227-229.

Hawk-owl —See **Owl.**

Hawk's-beard, Beaked (Crepis taraxacifolia) BFB cp 21 (8); MBF cp 50; OWF 37; (C. vesicaria) MFF cp 54 (951); NHE 190; PFE cp 160 (1546); PWF 551; WFB 257 (7).

Hawk's-beard, Golden (Crepis aurea) KVP cp 144; NHE 289; PFE cp 160 (1549).

Hawk's-beard, Greater or Rough (Crepis biennis) MBF cp 50; MFF cp 54 (953); NHE 190.

Hawk's-beard, Marsh (Crepis paludosa) MBF cp 50; MFF cp 54 (955); NHE 101; PWF 83g; WFB 257 (8).

Hawk's-beard, Pink (Crepis rubra) DFP 35 (cp 278); EGA 111; PFM cp 216; (C. incana) EWF cp 40d.

Hawk's-beard, Smooth (Crepis capillaris) BFB cp 21 (3); MBF cp 50; MCW 109; MFF cp 55 (954); OWF 37; PWF 68c; WFB 255 (8).

Hawk's-beard, Soft (Crepis mollis) MBF cp 50; MFF cp 54 (952); WFB 257 (9).

Hawkweed (Hieracium brittannicum) MBF cp 51; PWF 67l.

Hawkweed, Brownish-orange (Hieracium brunneocroceum) EWF cp 22e; EWG 76; NHE 289; PWF 152d.

Hawkweed, Leafy (Hieracium umbellatum) MBF cp 52; PWF 130d; WFB 257 (2).

Hawkweed, Mouse-ear (Hieracium pilosella) EPF pl 598; EWF cp 22a; MBF cp 51; MFF cp 55 (948); NHE 240; PFE cp 160 (1550); PWF 55h; TEP 125; (Pilosella officinarium) OWF 37; WFB 255 (9).

Hawkweed, Orange (Hieracium aurantiacum) EGC 72; MCW 134; MFF cp 58 (949); OWF 105; PFF 334; (Pilosella auranticum) WFB 257 (4).

Hawkweed, Purple —See **Saw-wort, Alpine.**

Hawkweed, Woolly or Shaggy (Hieracium villosum) DFP 147 (cp 1174); NHE 289.

Haworthia (Haworthia sp.) BIP 95 cp (144;) ECS 11, 12, 122; EDP 138; EGL 117; EFP 117; LCS cps 263-265; RDP 233; SCS cps 87-90.

Hawthorn, Chinese (Photinia serrulata) DFP 217 (cp 1734); MEP 46; MWF 231.

Hawthorn, Cockspur (Crataegus crus-galli) BKT cp 124; EET 167; EGT 108; EWF cp 152b; MTB 288 (cp 25); OBT 172; TGS 198.

Hawthorn, Common (Crataegus monogyna) BKT cps 119, 120; EET 166, 167; MBF cp 31; MFF pl 75 (380); NHE 20; OBT 172; OWF 181; PWF 39g, 163i; TEP 130; WFB 115 (4).

Hawthorn, English (Crataegus oxyacantha) AMP 77; BFB cp 10 (3); DFP 193 (cp 1544); EET 167; MWF 96; SST cp 98; TGS 198.

Hawthorn, Indian (Raphiolepis indica) EGW 65; MEP 47; MWF 250; (R. delacourii) GDS cp 341.

Hawthorn, Midland (Crataegus oxyacanthoides) EPF pl 431; MBF cp 31; NHE 20; (C. laevigata) OBT 172; TEP 129; WFB 114.

Hawthorn, Water —See **Pondweed, Cape.**

Hawthorn, Yeddo (Raphiolepis umbellata) EJG 141; MEP 47; MWF 250; SPL cp 322.

Hayrattle —See **Yellow-rattle** (Rhinanthus).

Hazel or **Hazelnut** —See also **Filbert.**

Hazel, Beaked (Corylus cornuta) EET 144; PFF 136.

Hazel, Corkscrew or Curly (Corylus avellana contorta) DFP 192 (cp 1533); EFS 104; EGW 104; GDS 114.

Hazel, European or Common (Corylus avellana) DEW 1:188; EET 144, 145; EPF pl 338; EVF 135; MBF cp 77; MFF pl 71 (562); NHE 23; OBT 32, front end paper; OFP 27; OWF 189; PWF 13m, 165e; SST cp 198; TEP 275; TGS 87; WFB 33 (2).

Hazel, Turkish (Corylus colurna) EEG 115; OBT 152.

Hazel-grouse or **Hazel-hen** —See **Grouse, Hazel.**

Hazelmouse —See **Dormouse, Hazel.**

Headache —See **Poppy, Corn** (Papaver).

Head-and-tail-light Fish (Hemigrammus ocellifer) ALE 4:286; FEF 84 (bw); SFW 136 (pl 27); WFW cp 57.

Headstander (Abramites microcephalus) HFI 89; HFW 112 (bw); SFW 161 (cp 35).

Headstander, Spotted (Chilodus punctatus) ALE 4:291; FEF 118 (bw); SFW 161 (pl 36), 165 (cp 38); WFW cp 95.

Heal-all —See **Self-heal.**

Heart Urchin —See **Sea Urchin, Sand.**

Hearts Entangled —See **Rosary Vine.**

Heartsease —See **Pansy, Wild.**

Heath (Epacris sp.) BIP 75 (cp 108); EWF cp 136a, b, e; HPW 128; MWF 122; PFW 105; SNZ cp 71; WAW 118.

Heath, Christmas (Erica canaliculata) EDP 129; EGG 108; MEP 103.

Heath, Cornish (Erica vagans) DFP 199 (cps 1588-1589); GDS cps 178, 179; HSC 47, 52; MBF cp 55; MFF cp 24 (595); TGS 375; WFB 171 (4).

Heath, Cross-leaved —See **Heather, Bog.**

Heath, Dorset (Erica ciliaris) BFB cp 13 (5); MBF cp 55; MFF cp 24 (592); OWF 119; PFE cp 90 (929); PWF 147f; WFB 171 (3a).

Heath, Irish or Biscay (Erica mediterranea) HSC 47, 52; MBF cp 55; MFF cp 24 (594); TGS 393; (E. erigena) GDS cp 172; WFB 171 (5).

Heath, Irish or St. Dabeoc's (Daboecia cantabrica) DEW 1:173 (cp 101); EGE 120; GDS cps 143-145; HSC 42, 44; MBF cp 55; MFF cp 24 (585); MWF 103; PFE cp 88 (922); PWF 115a; TGF 156; WFB 171 (6).

Heath, Portugal or Spring (Erica lusitanica) DFP 199 (cp 1585); GDS cp 176; LFW 50 (cp 116); MWF 123; PFE cp 89 (927); PFW 107.

Heath, Sea —See **Sea Heath.**

Heath, Spike (Bruckenthalia spiculifolia) ERW 103; TGF 156.

Heath, Spring or Flesh-colored (Erica herbacea) EGE 122; EPG 118; GDS cps 173-175; NHE 266; PFE cp 89 (933).

Heath, Tree (Erica arborea) DFP 198 (cp 1577); GDS cps 164, 165; HSC 47, 49; PFE cp 89 (927); PFM cp 121; RWF 33; (E. veitchii) DFP 199 (cp 1590); GDS cp 180.

Heath, Winter-flowering (Erica darleyensis) DFP 198 (cp 1584); GDS cp 169; HSC 47; OGF 3.

Heather (Calluna vulgaris) BFB cp 13 (8); BWG 116; DFP 185 (cps 1474-1478); EGC 125; EGE 116; EGW 13, 96-97; EPF pl 485; ERW 104; GDS cps 30-38; HSC 20, 21; MBF cp 55; MFF cp 24 (590); NHE 70; OWF 119; PFW 106; PWF 155; TGF 156; TVG 219; WFB 171 (1).

Heather (Cassiope sp.) DFP 4 (cp 29); EWF cp 161d; GDS cp 54; HPW 126; NHE 70, 266; PFW 108; WFB 173 (7).

Heather, Bell or Scotch (Erica cinerea) BFB cp 13 (4); BWG 116; GDS cps 166, 168; MBF cp 55; MFF cp 24 (593); NHE 70; OWF 119; PFW 106; PWF 155; TGS 393; WFG 171 (2).

Heather, Blue Mountain (Phyllodoce caerulea) MBF cp 55; MFF cp 24 (584); NHE 70; PFE cp 88 (922); WFB 171 (7).

Heather, Bog (Erica tetralix) BWG 116; DFP 199 (cp 1586); ERW 12, 114; GDS cp 177; KVP cp 72; MBF cp 55; MFF cp 24 (590); MLP 186; NHE 70; OWF 119; PWF 155n; TGS 375; WFB 171 (3).

Heath-grass (Sieglingia decumbens) MBF pl 96; MFF pl 121 (1170); NHE 196.

Heath-rose (Fumana procumbens) HPW 108; NHE 72; PFE 257; WFB 149 (3).

Heath-wren, Chestnut-tailed (Hylacola pyrrhopygia) HAB 181; SAB 32 (cp 15).

Hebe (Hebe sp.) DFP 205 (cps 1636-1638); GDS cps 219, 221, 223; HSC 62, 65; SNZ 31, 45, 92, 122, 124, 125, 140, 141, 148, 149; TVG 227.

Hebe, Showy (Hebe speciosa) DFP 205-206 (cps 1639-1641); EGE 127; GDS cps 222, 224; MWF 147; PFW 279; SNZ 27 (cps 35, 36).

Hedgehog (Paraechinus sp.) ALE 10:180; DAA 128; (Neotetracus sinensis) ALE 10:179.

Hedgehog, Algerian (Aethechinus algirus) ALE 10:179; DAE 11.

Hedgehog, Common or European (Erinaceus europaeus) ALE 2:363; 10:179, 223; CEU 200, 221 (bw);DAE 26; DSM 47; GAL 64 (bw); HMA 59; HMW 192; LEA 490 (bw); MLS 81 (bw); OBV 127; SMW 42 (cp 18); TWA 101 (bw); WEA 188.

Hedgehog, Desert (Hemiechinus auritus) ALE 10:180; CAS 72 (bw); DAA 128; DSM 47; (Parachinus deserti) LAW 596-597.

Hedgehog, Hairy —See **Moon-rat.**

Hedgehog, Southern African (Erinaceus frontalis) DDM 32 (cp 1); SMR 129 (cp 10).

Hedgehog, Tropical (Atelerix sp.) ALE 10:179; DDM 32 (cp 1).

Hedge-hyssop (Gratiola officinalis) NHE 98; PFE cp 124 (1218); WFB 213 (4).

Hedge-parsley, Knotted (Torilis nodosa) MBF cp 40; MFF pl 96 (463); NHE 222; OWF 91; WFB 157 (5).

Hedge-parsley, Spreading (Torilis arvensis) MBF cp 40; MFF pl 96 (462); NHE 221.

Hedge-parsley, Upright (Torilis japonica) MBF cp 40; MFF pl 99 (461); NHE 221; OWF 91; PWF 135g, 148a; WFB 157 (4).

Hedge-sparrow —See **Sparrow, Hedge.**

Hedysarum (Hedysarum sp.) HSC 63, 68; PFE cp 62 (634).

Heliconia, Hanging (Heliconia collinsiana) MEP 25; MWF 151.

Heliophila (Heliophila sp.) DFP 39 (cp 306); HPW 121.

Heliotrope (Heliotropium europaeum) NHE 229; PFE 101 (1045); PFW 49; (H. nelsonii) LFT cp 139.

Heliotrope, Common (Heliotropium arborescens) FHP 124; LFW 201 (cp 455); PFF 285; TGF 173; (H. peruvianum) BIP 97 (cp 147); EGA 124; EGH 116; MWF 151; OGF 125.

Heliotrope, Winter (Petasites fragrans) BWG 125; MBF cp 46; MFF cp 21 (857); PWF 26a; WFB 239 (1a).

Hellebore (Helleborus sp.) DFP 146 (cp 1163); EGW 115; EWF cp 8b; OGF 5, 13; PFW 250.

Hellebore, Green (Helleborus cyclophyllus) HPW 47; PFE cp 18 (200); PFM cp 19; (H. viridis) AMP 51; EFC cp 48; MBF cp 4; MFF pl 65 (4); NHE 268; OWF 51; PWF 36d; WFB 67 (2).

Hellebore, Stinking (Helleborus foetidus) BFB cp 2 (1); EFC cp 47; MBF cp 4; MFF pl 65 (3); NHE 178; PFE cp 19 (199); PFW 251; PWF cp 16k; WFB 67 (1).

Helleborine (Epipactis sp.) EGO 17; NHE 41; OWF 161; PWF 132b, 156a; WFB 284, 285.

Helleborine, Broad-leaved (Epipactis helleborine) MBF cp 80; MFF pl 65 (1062); NHE 41; OWF 161; PWF 94h; WFB 284, 285 (3).

Helleborine, Dark-red (Epipactis atrorubens) MBF cp 80; MFF cp 34 (1063); NHE 41; OWF 161; PWF 71b; TEP 263; WFB 284, 285 (2).

Helleborine, Long-leaved (Cephalanthera longifolia) BFB cp 22; EPF pl 972; KVP cp 176; MBF cp 80; MFF pl 83 (1059); NHE 41; PFE cp 191 (1919); PWF 27e; WFB 283 (4).

Helleborine, Marsh (Epipactis palustris)
BFB cp 22 (1); KVP cp 25; MBF cp 81; MFF cp
35 (1061); NHE 197; OWF 161; PFE cp 191
(1916); PWF 132a; WFB 284, 285 (1).

Helleborine, Narrow-lipped (Epipactis
leptochila) MBF cp 80; PWF 89c; WFB 285
(4).

Helleborine, Red (Cephalanthera rubra)
MBF cp 80; MFF cp 34 (1060); NHE 41; PFE
cp 191 (1920).

Helleborine, White (Cephalanthera damasonium) BFB cp 22; EWF cp 26f; MBF cp 80;
MFF pl 83 (1058); NHE 41; OWF 101; PWF
96a; TEP 263; WFB 283 (3).

Helmet Shell (Cassis sp.) ALE 3:55; CTS 33
(cp 45), 34 (cp 47); GSS 63; (Cassidaria echinophora) GSS 20, 66.

Helmet Shell, Bull-mouth (Cypraecassis
rufa) ALE 3:71; CTS 35 (cp 49); GSS 65.

Hemipode, Andalusian —See Quail,
Kurrichane Button.

Hemipode, Collared —See Plains
Wanderer.

Hemispingus (Hemispingus verticalis) SSA pl
30; WAB 97; (H. atropileus) DTB 131 (cp 65).

Hemlock, Poison (Conium maculatum) AMP
163; CTH 40 (cp 53); MBF cp 36; MFF pl 97
(468); NHE 221; OWF 87; PFE cp 85 (879);
PFF 264; PWF 65h; WFB 161 (3); WUS 281.

Hemp, African (Sparmannia africana) DEW
1:212 (cps 114, 115); EWF cp 76a; LHP 172,
173; PFW 297; RDP 366, 367.

Hemp or Marijuana (Cannabis sativa) AMP
123; DEW 1:179; EHP 31-33, 37; EPF pl 364;
PFE cp 6 (64); PFF 146; PWF 154d; WFB 39
(6); WUS 108; (C. indica). MLP 72.

Hemp-nettle, Common (Galeopsis tetrahit)
BFB cp 17 (8); MBF cp 69; MCW 85; MFF cp
28 (776); NHE 230; OWF 149; PFE cp 111
(1126); PWF 134e; WUS 311.

Hemp-nettle, Downy (Galeopsis segetum)
MBF cp 69; NHE 230; PFE 360; PWF 139k;
WFB 201 (8).

Hemp-nettle, Large-flowered (Galeopsis
speciosa) BFB cp 17 (3); EPF pl 808; KVP cp
37; MBF cp 69; MFF cp 60 (777); NHE 230;
PFE cp 111 (1126); PWF 141f; TEP 253; WFB
201 (7).

Hemp-nettle, Narrow-leaved (Galeopsis
angustifolia) BFB cp 17 (10); MBF cp 69;
MFF cp 28 (774); NHE 231; OWF 149; WFB
201 (5).

Hen of the Woods (Grifolia frondosa) DLM
150, 151; LHM 75; SMF overleaf.

Hen-and-Chicks —See Houseleek,
Common.

Henbane (Hyoscyamus niger) AMP 121; BFB
cp 15 (10); CTH 48 (cp 70); DEW 2:151; EFC
cp 23; EHP 49; EPF pls 826, 827; MBF cp 61;
MFF cp 58 (672); NHE 232; OWF 29; PFE cp
117 (1176); PFF 294; PWF 114d; TEP 195;
WFB 207 (5).

Henbane, Golden or Yellow (Hyoscyamus
aureus) EWF cp 38a; PFE cp 117 (1177); PFM
cp 166.

Henbane, White (Hyoscyamus albus) PFE cp
117 (1177); PFM cp 165.

Henbit —See Dead-nettle, Henbit.

Hen-harrier (Circus cyaneus) ABW 81; ALE
7:398; BBB 111; BBE 85; FBE 87; GAL 157
(bw); GBP cp after 144, 347 (bw), 353 (bw);
GPB 100; KBA 79 (bw); LEA 387 (bw); OBB
45; PBE 60 (cp 19), 72 (pl 23); RBE 23; WAB
63.

Hepatica, European (Hepatica nobilis) AMP
115; EPF pl 215; NHE 26; PFE cp 20 (219);
PFW 253; TEP 236; TVG 153.

Herald's Trumpet (Beaumontia grandiflora)
EGV 95; MEP 108; PFW 30.

Herb Bennet —See Avens, Wood.

Herb Christopher —See Baneberry,
Common.

Herb of Grace —See Rue, Common.

Herb Paris (Paris quadrifolia) BFB 309; EPF
pl 922; KVP cp 50; MBF cp 86; MFF pl 65
(1020); NHE 38; OWF 53; PFE cp 170 (1658);
PWF 32c; TEP 261; WFB 269 (3).

Herb Robert (Geranium robertianum) AMP
165; BFB cp 6 (2); BWG 106; EFC cp 14; EGH
115; MBF cp 19; MFF cp 16 (247); NHE 31;
OWF 129; PFF 225; PWF 33h, 44i; WFB 135
(6).

Herb Robert, Lesser or Little Robin
(Geranium purpureum) MBF cp 19; MFF cp
16 (248).

Hermit —See Hummingbird, Hermit.

Heron, Black-crowned Night (Nicticorax nicticorax) ABW 51; ALE 7:198; BBB 269; BBE 39; CDB 45 (cp 64); CEU 58 (bw); DPB 17; FBE 39; KBA 32 (pl 1); LAW 478; OBB 13; PBE 8 (cp 3); PSA 33 (cp 2); SBB 11 (pl 6); WAB 101; WBV 41; WFI 104 (bw), 106 (bw).

Heron, Black-headed (Ardea melano-cephala) ALE 7:184-185; BBT 78-79; LAW 502; PSA 33 (cp 2).

Heron, Boat-billed (Cochlearius cochlearius) ABW 55; ALE 7:198; CDB 43 (cp 57); GPB 69; SBB 11 (pl 7); SSA pl 22; WAB 94.

Heron, Chestnut-bellied (Agamia agami) ABW 54; ALE 7:197; SSA cp 3.

Heron, Chinese Pond (Ardeola bacchus) CAS 146; DPB 15; KBA 112 (cp 11).

Heron, Common or Gray (Ardea cinerea) ALE 7:196, 255; 11:257; BBB 163; BBE 39; CDB 42 (cp 54); CEU 155; DPB 17; FBE 35; KBA 112 (cp 11); NBB 25, 63 (bw), 87 (bw); OBB 13; PBE 8 (cp 3); PSA 33 (cp 2); SBB 7 (pl 1); WBV 41.

Heron, Goliath (Ardea goliath) ALE 7:195; CDB 42 (cp 55); FBE 35; PSA 33 (cp 2).

Heron, Great White —See **Egret, Common or Great White.**

Heron, Great-billed or Giant (Ardea sumatrana) DPB 17; KBA 112 (cp 11); SNB 30 (cp 14).

Heron, Green or Lava (Butorides sundevalli) CSA 242 (bw); HBG 96 (cp 5).

Heron, Green-backed —See **Heron, Mangrove or Striated.**

Heron, Indian Pond (Ardeola grayii) KBA 112 (cp 11); WEA 190.

Heron, Little Green —See **Heron, Mangrove or Striated.**

Heron, Malayan Night —See **Bittern, Tiger.**

Heron, Mangrove or Striated (Butorides striatus) ALE 7:196; BBT 79; DPB 15; HAB 26; HBG 96 (cp 5); KBA 112 (cp 11); RBU 75; PSA 33 (cp 2); SNB 32 (cp 15); WBV 41.

Heron, Nankeen or Rufous Night (Nycti-corax caledonicus) DPB 17; HAB 25; SNB 32 (cp 15).

Heron, Pied (Ardea picata or Notophoyx picata) ALE 7:196; RBU 72; SNB 30 (cp 14).

Heron, Purple (Ardea purpurea) ABW 51; ALE 7:195; BBB 269; BBE 39; DPB 17; FBE 35; KBA 112 (cp 11); LEA 369 (bw); NBB 37 (bw); OBB 15; PBE 8 (cp 3); RBE 67; SBB 8 (pl 2); WBV 41.

Heron, Reef —See **Egret, Pacific Reef.**

Heron, Squacco (Ardeola ralloides) ALE 7:197; BBB 273; BBE 37; CEU 33; FBE 37; GPB 66; LEA 370 (bw); NBB 38 (bw); PBE 9 (cp 4); SBB 9 (pl 4).

Heron, Tiger (Tigrisoma lineatum) ABW 52; CSA 51.

Heron, Whistling (Syrigma sibilatrix) ALE 7:197; SSA cp 2.

Heron, White-faced (Ardea novaehollandiae or Notophoyx novaehollandiae) ALE 7:196; 10:89; CDB 45 (cp 65); FNZ 113 (pl 10); MBA 53; MWA 111; SNB 30 (cp 14); WAW 28 (bw).

Heron, White-necked Pacific (Ardea pacifica) FNZ 113 (pl 10); SNB 30 (cp 14).

Heron's-bill —See **Storksbill.**

Herring, Atlantic (Clupea harengus) ALE 4:197; CSF 64; FEF 45 (bw); HFI 119; HFW 74 (bw); OBV 21.

Herring, Ox-eye (Megalops cyprinoides) ALE 4:158; LMF 30; WFW cp 20.

Herschelia (Herschelia sp.) EWF cp 89c, h; LFT cp 48.

Heurnia —See **Huernia.**

Hiba —See **Arborvitae** (Thujopsis).

Hibiscus (Hibiscus rosa-sinensis) DEW 1:220 (cp 134); DFP 69 (cp 548); EGE 128; EGG 114; EPF pl 615; EPG 124; FHP 125; GFI 71 (cp 14), 75 (cp 15); LFW 166, 167 (cps 369-374); LHP 108, 109; MEP 78; MLP 146; MWF 154; PFW 183; RDP 238; SPL cp 91.

Hibiscus, Fringed (Hibiscus schizopetalus) DEW 1:221 (cp 137); EWF cp 57b; GFI 67 (cp 13); HPW 94; MEP 78; MWF 155; PFW 183; RDP 238.

Hickory, Bitternut (Carya cordiformis) BKT cp 85; EET 138; PFF 136.

Hickory, Mockernut (Carya tomentosa) BKT cp 91; DEW 1:191; EET 139; PFF 135.

Hickory, Shagbark or Shellbark (Carya ovata) EET 139; EGW 98; MLP 76; PFF 134; SST cp 88.

Hillstar —See **Hummingbird, Hillstar.**

Hilo Holly —See **Ardisia.**

Himalayan Whorlflower of Nepal (Morina longifolia) DFP 160 (cp 1277); EWF cp 99f; PFW 189; (M. betonicoides) HPW 262.

Hippopotamus (Hippopotamus amphibius) ALE 13:111-114; BMA 113; CAA 66 (bw), 69; CAF 108-109, 111 (bw); DDM 160 (cp 25); DSM 222; HMA 60; HMW 52; JAW 166; LAW 562; LEA 591 (bw), 593; MLS 374 (bw); SMR 85 (cp 6); SMW 235 (cp 147); TWA 36.

Hippopotamus, Pygmy (Choeropsis liberiensis) ALE 13:111; BMA 113; DDM 160 (cp 25); DSM 223; HMW 53; LAW 561; LVA 21; LVS 126; MLS 375 (bw); PWC 243 (bw); SMW 235 (cp 148).

Hoatzin (Opisthocomus hoazin) ABW 100; ALE 8:64; CDB 71 (cp 212); GPB 123; SSA cp 3; WAB 95.

Hobby (Falco subbuteo) ALE 7:418; BBB 93; BBE 89; DAE 117; FBE 95; FBW 69; GAL 155 (bw); GBP 395 (bw); GPB 107; KBA 65 (pl 6); OBB 47; PBE 61 (cp 20), 73 (pl 24); PSA 97 (cp 10); RBE 27; WAB 119.

Hobby, Oriental (Falco severus) DPB 51; KBA 65 (pl 6).

Hog, Giant Forest (Hylochoerus meinertzhageni) ALE 13:93; DDM 161 (cp 26); DSM 220; MLS 371 (bw).

Hog, Red River —See **Bushpig.**

Hog, Wart —See **Warthog.**

Hog's Fennel (Peucedanum officinale) MBF cp 40; MFF cp 51 (504); NHE 182; PWF 119h; WFB 165 (1).

Hogweed (Heracleum sphondylium) AMP 117; BWG 78; CTH 41 (cp 54); DEW 2:87; EFC cp 42; EPF pl 765; HPW 220; MBF cp 40; MFF pl 95 (508); NHE 182; OWF 87; PWF 63f; TEP 145; TVG 101; WFB 161 (1).

Hogweed, Giant (Heracleum mantegazzianum) BWG 80; PFE cp 86 (902); PWF 63e.

Holly, American (Ilex opaca) EEG 125; EGE 130; ELG 139; EPG 125; PFF cp 131; SST cp 131; TGS 199.

Holly, Chinese (Ilex cornuta) EEG 106; EGE 129; EGW 117; ELG 139; MWF 162; TGS 199.

Holly, English (Ilex aquifolium) AMP 17; BFB cp 7 (3); DEW 2:75; DFP 208 (cps 1659-1662); EDP 118; EET 186, 187; EGE 129; ELG 138; EPF pls 663, 664; ESG 119; HPW 182; MBF cp 20; MFF pl 76 (263); MTB 325 (cp 30); MWF 162; NHE 21; OBT 36; OWF 183; PFE cp 71 (717); PWF 39f, 174c; SPL cp 94; SST cp 130; TGS 199; TVG 229; WFB 36, 37a.

Holly, False —See **Osmanthus, Holly.**

Holly, Japanese (Ilex crenata) EEG 107; EGE 130; EGW 118; EJG 121; EMB 114; ERW 123; TGS 199.

Hollyhock (Alcea sp.) PFE cp 72 (748); WFB 142 (7A).

Hollyhock (Althaea rosea) AMP 87; CTH 36 (cp 44) DEW 1:220 (cp 135); DFP 29 (cps 228-229); EGA 92; EGP 94; EPF pl 617; LFW 10 (cp 20); MLP 146; MWF 38; PFF 248; PFW 182; SPL cp 380; TGF 124; TVG 33.

Holy-grass (Hierochloe odorata) MFF pl 129 (1244); NHE 107; PFF 353; (H. redolens) SNZ 103 (cp 312).

Hondapara (Dillenia indica) HPW 80; MEP 85; PFW 101; (D. triquetra) DEW 1:233.

Honesty (Lunaria annua) BWG 137; DEW 1:141; DFP 42 (cp 333); EWF cp 11b; NHE 217; OGF 191; PFE cp 35 (322); PFF 189; PFW 99; PWF 59e; SPL cp 225; TGF 588; (L. biennis) MWF 191.

Honesty, Perennial (Lunaria rediviva) DFP 156(cp 1245); NHE 30; TEP 239; WFB 91 (7).

Honewort (Trinia glauca) MBF cp 37; MFF pl 99(473); NHE 223; WFB 159 (9).

Honey Plant —See **Wax Plant.**

Honey-badger (Mellivora capensis) ALE 12:42; CAA 85 (bw); CAF 210 (bw); DAA 81; DDM 85 (cp 14); DSM 171; JAW 116; LAW 537; LEA 561 (bw); MLS 290 (bw); SMR 33 (cp 2); TWA 53; WEA 305; (M. ratel) HMW 138.

Honeycreeper, Green (Chlorophanes spiza) ALE 9:362; SSA cp 14.

Honeycreeper, Purple (Cyanerpes caeruleus) ALE 9:335; CDB 203 (cp 887); DTB 83 (cp 41); WEA 192.

Honeycreeper, Red-legged or Yellowwinged (Cyanerpes cyaneus) ABW 295; ALE 9:335, 362; DTB 85 (cp 42); LEA 464; SSA cp 39.

Honey-eater (Myzomela sp.) ABW 277; CDB
194 (cp 845); HAB 230-232; RBU 268; SAB 56
(cp 27); WAB 212; WEA 193.

Honey-eater, Blue-faced (Entomyzon
cyanotis) CDB 193 (cp 838); HAB 246; RBU
271; SAB 62 (cp 30); WAB 181.

Honey-eater, Brown (Lichmera indistincta)
SAB 54 (cp 26); (Gliciphila indistincta)
MWA 35, 36.

Honey-eater, Brown-headed (Melithreptes
brevirostris) HAB 228; SAB 62 (cp 30).

Honey-eater, Crescent (Phylidonyris pyrrhoptera) ALE 9:320; HAB 240; SAB 64 (cp 31).

Honey-eater, Golden-backed (Melithreptes
laetior) SAB 62 (cp 30); WAB 190.

Honey-eater, Graceful (Meliphaga gracilis)
HAB 223; SAB 58 (cp 28).

Honey-eater, Lewin (Meliphaga lewinii)
HAB 223; SAB 58 (cp 28); WAW 29 (bw).

Honey-eater, Mallee —See **Honey-eater,
Yellow-plumed.**

Honey-eater, New Holland (Meliornis
novaehollandiae) HAB 241; MBA 19; MWA
31, 32; (Phylidonyris novaehollandiae) SAB
64 (cp 31); WAB 187.

Honey-eater, Painted (Grantiella picta) HAB
237; (Conopophila picta) SAB 64 (cp 31).

Honey-eater, Pied (Certhionyx variegatus)
HAB 236; SAB 56 (cp 27).

Honey-eater, Regent (Zanthomiza phrygia)
HAB 237; RBU 263; SAB 64 (cp 31).

Honey-eater, Rufous-throated (Conopophila rufogularis) HAB 236; SAB 64 (cp 31).

Honey-eater, Spiny-cheeked (Acanthagenys
rufogularis) CDB 192 (cp 834); HAB 244;
(Anthochaera rufogularis) SAB 66 (cp 32).

Honey-eater, Striped (Plectorhyncha
lanceolata) CDB 195 (cp 848); HAB 229; SAB
62 (cp 30).

Honey-eater, Tawny-breasted (Xanthotis
flaviventer) HAB 240; SAB 58 (cp 28).

Honey-eater, Tawny-crowned (Gliciphila
melanops) ALE 10:89; CDB 194 (cp 840);
HAB 234; SAB 64 (cp 31); (Phylidonyris
melanops) MEA 60.

Honey-eater, White-cheeked (Meliornis
niger) CDB 194 (cp 841); HAB 241; (Phylidonyris niger) MEA 60; SAB 64 (cp 31).

Honey-eater, White-eared (Meliphaga
leucotis) CDB 194 (cp 843); HAB 223; SAB 58
(cp 28).

Honey-eater, White-fronted (Gliciphila
albifrons) CDB 193 (cp 839); HAB 234;
(Phylidonyris albifrons) SAB 64 (cp 31).

Honey-eater, White-lined (Meliphaga albilineata) HAB 223; SAB 58 (cp 28).

Honey-eater, White-naped (Melithreptes
lunatus) ALE 9:320; HAB 227; RBU 260;
SAB 62 (cp 30).

Honey-eater, White-throated (Melithreptes
albogularis) HAB 119, 227; SAB 62 (cp 30).

Honey-eater, Yellow-faced (Meliphaga
chrysops) CDB 194 (cp 842); HAB 226; MBA
18.

Honey-eater, Yellow-plumed (Meliphaga
ornata) HAB 225; MBA 14, 15.

Honey-eater, Yellow-tufted (Meliphaga
melanops) CDB 194 (cp 844); GPB 299; SAB
58 (cp 28).

Honey-flower (Melianthus sp.) DFP 214 (cp
1705); HPW 191; LFT 202 (cp 100).

Honey-guide (Indicator archipelagicus) KBA
240 (cp 31); (I. xanthonotus) RBI 139; (Prodotiscus regulus) PSA 225 (cp 26).

Honey-guide, Black-throated or Greater
(Indicator indicator) ABW 185; ALE 9:67;
PSA 225 (cp 26); WAB 144.

Honey-guide, Lesser (Indicator minor) ALE
9:997; CDB 134 (cp 533); GPB 223.

Honeysuckle (Lonicera americana) DEW
2:110 (cp 72); DFP 248 (cp 1980); GDS cp 244;
HSC 75; (L. etrusca) PFE cp 135 (1305).

Honeysuckle (Lonicera sp.) DEW 2:108 (cp
69); EGW 66; EWF cp 155d; GDS cp 281;
HPW 259; HSC 75; LFW 90 (cp 209); MLP
226; NHE 267; SPL cp 173; TGS 359, 385,
386.

Honeysuckle, Alpine (Lonicera alpigena)
NHE 267; PFE 399.

Honeysuckle, Amur (Lonicera maackii) EFS
23; GDS cps 276, 277; TGS 386.

Honeysuckle, Australian (Banksia sp.) MLP
73; MWF 52; PWF 247; WAW 119.

Honeysuckle, Blue (Lonicera caerulea) NHE
25; PFE 135 (1303).

Honeysuckle, Cape (Tecomaria capensis) EGV 139; EWF cp 79c; GFI 35 (cp 5); LFT cp 150; LFW 134 (cp 303); MEP 139.

Honeysuckle, Cape (Protea mellifera) MEP 33; MWF 243.

Honeysuckle, Fly (Lonicera xylosteum) MBF cp 41; MFF pl 64 (825); NHE 25; PFE cp 135 (1302); WFB 223 (10).

Honeysuckle, French (Hedysarum coronarium) PFE cp 62 (634); PFM cp 74.

Honeysuckle, Giant (Lonicera hildebrandiana) EGV 119; EWF cp 95f; MEP 150; MWF 189.

Honeysuckle, Himalayan (Leycesteria formosa) DEW 2:111 (cp 76); DFP 210 (cp 1675); GDS cp 271; HSC 74, 77; MWF 183; OGF 189; PFW 68.

Honeysuckle, Japanese (Lonicera japonica) EEG 144; ESG 127; GDS cp 275; HSC 75; SPL cp 174; TGS 63; WUS 359.

Honeysuckle or **Woodbine** (Lonicera periclymenum) BFB cp 11; DFP 248 (cp 1982); GDS cp 278; HSC 80; MBF cp 41; MFF cp 64 (826); MWF 189; NHE 25; OWF 15; PFE cp 135 (1304); PFW 67; PWF 86b, 163k; WFB 223 (9).

Honeysuckle, Perfoliate (Lonicera caprifolium) EPF pl 777; NHE 25; TGS 63.

Honeysuckle, Scarlet Trumpet (Lonicera fuchsiodes) DFP 248 (cp 1980); OGF 63.

Honeysuckle, Tartarian (Lonicera tatarica) EEG 99; EFS 124; LFW 90 (cp 208); NHE 25; TGS 374; TVG 231.

Honeysuckle, Trumpet or Coral (Lonicera sempervirens) DFP 248 (cp 1983); EGV 120; ELG 151; EPG 130; GDS cp 280; SPL cp 175; TGS 70.

Honeysuckle, Winter or Fragrant (Lonicera fragrantissma) EFS 123; HSC 80; MWF 189; TGS 359; (L. purpusii) GDS cp 279; HSC 75.

Honeywort (Cerinthe sp.) DEW 2:147; EGA 105; EWF cp 37f; HPW 236; NHE 35, 281; PFE cp 105 (1078, 1079); PFM cp 148, 149, 151; TEP 179.

Hoodia (Hoodia bainii) DEW 2:130; EWF cp 78c; LCS cp 269; (H. gordonii) DEW 2:129; LCS cp 323.

Hoopoe, African (Upupa africana) BBT 55; CAF 244; NBA 41; PSA 240 (cp 27).

Hoopoe, Eurasian (Upupa epops) ABW 181; ALE 9:29, 57; BBB 66; BBE 185; CAA 52 (bw); CDB 130 (cp 509); CEU 66 (bw); DAE 138; DPB 209; FBE 191; FBW 73; GAL 148 (cp 4); GPB 216; KBA 217 (bw); LEA 437 (bw); NBB 8, 77 (bw); OBB 113; PEB 181 (cp 46); RBE 204; WAB 101, 118; WBV 135.

Hoopoe, Wood (Phoeniculus purpureus) ALE 9:57; CDB 130 (cp 510); GPB 217; PSA 240 (cp 27); (P. bollei) WAB 140.

Hop, Common (Humulus lupulus) AMP 107; BFB cp 14 (10); DEW 1:179; EGA 125; EGV 114; EPF pl 363; MBF pl 76; MFF pl 76 (555); NHE 96; OFP 137; OWF 65; PFE cp 6 (63); PFF 147; PWF 147c, 173e; WFB 39 (5).

Hop, Japanese (Humulus scandens or H. japonica) DFP 39 (cp 309); EGA 125.

Hop-trefoil (Trifolium campestre) MBF cp 23; MFF cp 48 (303); NHE 180; OWF 21; PWF 49e; WFB 129 (8); (T. aureum) MBF cp 23; NHE 180.

Horehound (Marrubium vulgare) AMP 175; EGH 122; MBF cp 69; MFF pl 112 (780); NHE 230; OWF 97; PFE cp 109 (1112); PFF 287; TGF 204; WFB 199 (4).

Horehound, Black (Ballotta nigra) BFB cp 17 (5); MBF cp 70; MFF cp 29 (768); NHE 231; OWF 149; PFE cp 112 (1134); PWF 85l; TEP 195; WFB 199 (8); (B. pseudo-dictamnus) DFP 125 (cp 1000).

Horn of Plenty (Craterellus sp.) CTM 52; DEW 3:160; DLM 147; KMF 27; LHM 59; NFP 47; NHE 46; ONP 123; PFF 77; PMU cp 13; PUM cp 151; RTM 223, 245; SMF 17; SMG 123.

Hornbeam, European (Carpinus betulus) BKT cp 68; EET 143; EFP pl 337; EGT 101; EPG 111; MBF cp 77; MFF pl 71 (561); MTB 208 (cp 19); NHE 23; OBT 13, 17; OWF 189; PFE cp 5 (42); PWF 20a, 109h; SST cp 87; TEP 275; TGS 87; WFB 33 (1).

Hornbeam, European Hop (Ostrya carpinifolia) BKT cp 69; EET 143; EPF pl 334; NHE 42; OBT 152; PFE 51; SST cp 138.

Hornbill (Aceros sp. or Rhyticeros sp.) BBT 17; CDB 130 (cp 511); DPB 215; KBA 209 (pl 28); WAB 171; (Ceratogymna atrata) ALE 9:58.

Hornbill, Asian White-crested (Berenicornis comatus) CDB 130 (cp 513); KBA 209 (pl 28).

Hornbill, Brown or Tickell's (Ptilolaemus tickelli) KBA 209 (pl 28); WBV 135.

Hornbill, Crowned (Tockus alboterminatus) CDB 131 (cp 517); PSA 224 (pl 25).

Hornbill, Gray (Tockus nasutus) CDB 132 (cp 521); PSA 224 (cp 25).

Hornbill, Great Indian (Buceros bicornis) ABW 181; ALE 9:47; BBT 45; CDB 131 (cp 514); GPB 218; KBA 209 (pl 28); LVS 229; SBB 63 (pl 80); (B. hydrocorax) DPB 215.

Hornbill, Ground (Bucorvus leadbeateri) CAA 52 (bw); CDB 131 (cp 516); GPB 219; WAB 139; (B. abyssinicus) ALE 9:31, 58; CDB 131 (cp 515).

Hornbill, Helmeted (Rhinoplax vigil) ALE 9:58; KBA 209 (pl 28).

Hornbill, Indian Pied (Anthracoceros malabaricus) CDB 130 (cp 512); WBV 135; (A. albirostris) KBA 209 (pl 28).

Hornbill, Palawan (Anthracoceros marchei) DPB 215; (A. undulatus) CAS 254 (bw).

Hornbill, Red-billed (Tockus erythrorhynchus) ALE 9:32, 58; CDB 132 (cp 519); PSA 224 (cp 25); SBB 61 (pl 77); (T. camurus) CAA 42 (bw); GPB 218; WAB 142.

Hornbill, Van der Decken's (Tockus deckeni) ALE 9:32, 58; CDB 131 (cp 518); WEA 195.

Hornbill, Yellow-billed (Tockus flavirostris) BBT 62; CAF 169; CDB 132 (cp 520); GPB 218; PSA 224 (cp 25); (Lophoceros flavirostris) LEA 437 (bw).

Horned-poppy, Red (Glaucium corniculatum) NHE 139; PFE cp 29 (273); PFM cp 35; PWF 83i; WFB 80.

Horned-poppy, Violet (Roemeria hybrida) MBF cp 5; PFE cp 28 (271); (R. refracta) EWF cp 46c.

Horned-poppy, Yellow (Glaucium flavum) DEW 1:140; DFP 38 (cp 299); EPF pl 231; EWF cp 30g; HPW 52 (2); MBF cp 5; MFF cp 38 (41); NHE 139; OWF 7; PFE cp 29 (272); PWF 114b; SPL cp 487; WFB 81 (7).

Hornero (Furnarius rufus) ABW 198; ALE 9:125; WAB 89; (F. leucopus) SSA pl 25.

Hornet, European (Vespa crabro) ALE 2:476; CIB 269 (cp 42); CTI 62 (cp 85); GAL 380 (cp 14); OBI 155; PEI 167, 168 (bw); ZWI 154.

Hornet, Yellow-jacket —See **Yellow-jacket.**

Horn-fern, Elk's —See **Fern, Stag's-horn.**

Horntail —See **Wasp, Wood.**

Hornwort, Common (Ceratophyllum demersum) DEW 1:130; HPW 43; MBF cp 34; MFF pl 68 (34); NHE 92; OWF 53; PFF 167; WFB 293; WUS 171.

Horse, Wild (Equus caballus var.) ALE 12:554, 571; DAA 26-27; DAE 72; DSM 209; HMA 61; HMW 43; JAW 155; LEA 584 (bw); LVA 45; LVS 118; MLS 348 (bw); PWC 95; SLP 46; SMW 230 (cp 136); VWA cp 19.

Horse-chestnut, Common (Aesculus hippocastanum) AMP 175; BKT cps 165-167; DEW 2:73; EET 192; EGT 39, 96; EPF pls 735-737; EWF cp 19a; HPW 195; MLP 134; MTB 337 (cp 34); NHE 294; OBT 189; OWF 189; PFE cp 69 (713); PFF 243; PWF 39h, 175l; SST cp 226; TGS 145; WFB 36, 37 (1).

Horse-chestnut, Dwarf (Aesculus parviflora) DEW 2:72; EFS 95; HSC 11; SPL cp 64; TGS 145.

Horse-chestnut, Indian (Aesculus indica) EET 192, 193; MTB 352 (cp 35); OBT 189.

Horse-chestnut, Red (Aesculus carnea) BKT cp 168; DFP 181 (cp 1441); EET 193; EGT 41, 95; MTB 337 (cp 34); OBT 189; PFE cp 69 (713); PFW 138; PWF 38i, 175m; SST cp 225.

Horsefly —See **Fly, Horse.**

Horsehoof —See **Coltsfoot.**

Horsemint (Mentha longifolia) MBF cp 67; MFF cp 11 (749); NHE 98; OWF 143; PFE cp 116 (1171); PWF 139l.

Horseradish (Armoracia rusticana) DEW 1:144; EGH 100; OFP 135; PFF 188; PWF 65g; TEP 41; WFB 93 (5).

Horseradish Tree (Moringa oleifera) DEW 1:149; MEP 41; SST cp 289.

Horse's Hoof —See **Clam, Giant.**

Horsetail, Common or Field (Equisetum arvense) BWG 156; CGF 205; DEW 3: Fig. 342; EPF pl 97; KVP cp 41; MCW 13; MFF cp 36, pl 137 (1273); NFP 120; NHE 109; ONP 191; PFF 107; WUS 7.

Horsetail, Dwarf (Equisetum scirpoides) CGF 199; EJG 115; EMB 123; NFP 119.

Horsetail, Great (Equisetum telmateia) AMP 141; CTH 19 (cps 4, 5); DEW 3: Fig. 343; EPF pl 98; NFP 119; NHE 43; ONP 191.

Horsetail, Marsh (Equisetum palustre) CGF 211; DEW 3: cps 126, 127; MFF pl 137 (1271); NFP 121; NHE 109; ONP 95; PWF 64d.

Horsetail, Rough (Equisetum hyemale) CGF 203; EGF 115; MFF pl 136 (1268); NFP 120; NHE 43; ONP 95; PFF 107.

Horsetail, Water (Equisetum fluviatile) CGF 213; EPF pl 101; NFP 121; NHE 109; ONP 95.

Horsetail, Wood (Equisetum sylvaticum) CGF 209; DEW 3: cps 124, 125, 128; EPF pls 99, 100; MFF pl 137 (1272); NFP 121; NHE 43; TEP 231.

Horseweed —See **Fleabane, Canadian.**

Hotwater Plant —See **Achimenes, Trumpet.**

Hound, Smooth (Mustelus mustelus) CSF 41; HFI 149; HFW 26 (bw); OBV 7; WFW cp 7.

Hound's-tongue (Cynoglossum officinale) EPF pl 791; MBF cp 59; MFF cp 37 (647); NHE 229; OWF 105; PFE cp 102 (1049); PWF 89a; WFB 189 (2); (C. creticum) PFE cp 101 (1049); (C. germanicum) MBF cp 59; (C. nervosum) DFP 133 (cp 1059).

Housefly —See **Fly, House.**

Houseleek, Cobweb (Sempervivum arachnoideum) BIP 163 (cp 250); CTH 31 (cp 31); DEW 1:280 (cp 164); DFP 26 (cp 205); ECS 145; KVP 152, 152a; NHE 270; OGF 91; PFE cp 39 (381).

Houseleek, Common or Roof (Sempervivum tectorum) EGH 140; ERW 136; EWF cp 16b; NHE 210, 270; PFW 96; SPL cp 465; TEP 90; TGF 92; TVG 175.

Houseleek, Hen-and-chickens (Jovibara sobolifera) TEP 91; WFB 103 (10); (J. hirta) EPF pl 378.

Houseleek, Mountain (Sempervivum montanum) CTH 30 (cp 29); DFP 26 (cp 207); EPF pl 382; NHE 270; PFE cp 39 (382).

Houseleek, Wulfen's (Sempervivum wulfenii) NHE 270; SPL cp 466.

Houting (Coregonus lavaretus oxyrhynchus) FEF 59 (bw); FWF 53; HFI 122.

Houttuynia (Houttuynia cordata) DEW 1:101 (cp 48); HPW 39; PFW 271.

Hoverfly —See **Fly, Flower or Hover.**

Howler —See **Monkey, Howler.**

Huchen (Hucho hucho) FWF 47; HFI 121; SFW 57 (pl 6).

Huernia (Huernia sp.) CAF 201; ECS 123; EMB 108; EWF cp 78a; LCS cp 267, 268; LFT cp 129; RWF 60; SPL cp 142.

Hug-Me-Tight —See **Cat's Claw.**

Humantin (Oxynotus centrina) FEF 19; OBV 7.

Humble Plant —See **Sensitive Plant.**

Humbug, Banded —See **Damselfish** (Dascyllus sp.).

Hummingbird —See also specific names (i.e., **Emerald, Sabre-wing, Sylph, Wedgebill,** etc.).

Hummingbird, Bearded Helmet-crest (Oxypogon guerinii) ALE 8:453; SSA cp 32.

Hummingbird, Black-throated Train-bearer (Lesbia victoriae) ALE 8:443; SHU 42, 76, 108; SSA cp 33; WEA 201.

Hummingbird, Brilliant (Heliodoxa sp.) SHU 42, 137; SSA cp 32.

Hummingbird, Coquette (Discosura longi-cauda) ALE 8:454; SSA cp 5; (Lophornis sp.) ABW 170; SSA cp 33; WAB 79.

Hummingbird, Coronet (Boissonneaua flavescens) SHU 74; (B. jardini) SHU 90, 140; SSA cp 32.

Hummingbird, Fairy (Heliothryx barroti) SHU 76; SSA cp 33; (H. aurita) WAB 90.

Hummingbird, Giant (Patagona gigas) ABW 171; ALE 8:453.

Hummingbird, Hermit (Phaetornis sp.) ABW 171; ALE 8:453; CDB 123 (cp 474); LAW 479; SSA cp 32; WAB 90; (Ramphodon naevius) SSA cp 5; (Glaucis hirsuta) SSA cp 6.

Hummingbird, Herran's or Rainbow-bearded Thornbill (Chalcostigma herrani) ABW 171; SSA cp 33.

Hummingbird, Hillstar (Oreotrochilus estella) ALE 8:454; WAB 92; (O. melano-gaster) SSA cp 6.

Hummingbird, Inca (Coeligena sp.) ABW 170; ALE 8:454; SHU 57, 58, 73, 107; SSA cps 6, 32.

Hummingbird, Marvelous Spatule-tail (Loddigesia mirabilis) ABW 170; ALE 8:454; SSA cp 6.

Hummingbird, Peruvian Sheartail (Thaumastura cora) ALE 8:454; SSA cp 7.

Hummingbird, Puff-leg (Eriocnemis mirabilis) ALE 8:454; CDB 122 (cp 472); DTB 5 (cp 2); SSA cp 6; (E. luciani) SHU 26; SSA cp 33.

Hummingbird, Purple-backed Thornbill (Ramphomicron microrhynchum) ALE 8:454; SSA cp 33.

Hummingbird, Ruby-topaz (Chrysolampis mosquitus) ABW 171; ALE 8:453, 463; SHU 25, 28, 41; SSA cp 5.

Hummingbird, Sappho Comet (Sappho sparganura) ABW 171; SSA cp 7.

Hummingbird, Star-throat (Heliomaster sp.) ALE 8:454; SHU 43; SSA cp 7.

Hummingbird, Streamer-tail (Trochilus polytmus) ABW 171; ALE 8:453, 463; BBT 86.

Hummingbird, Sun Gem (Hellaetin cornuta) NBB 33; (Heliactin cornuta) SSA cp 6.

Hummingbird, Sun-angel (Heliangelus exortis) ALE 8:453; SHU 27; (H. amethysticollis) SSA cp 33.

Hummingbird, Swallow-tailed (Eupetomena macroura) ALE 8:453; SSA cp 6.

Hummingbird, Sword-billed (Ensifera ensifera) ABW 171; ALE 8:454, 463; SHU 76; SSA cp 33.

Hummingbird, Thorntail (Popelairia popelarii) ABW 170; (P. conversii) SHU 104; SSA cp 33.

Hummingbird, Topaz (Topaza pella) ABW 170; ALE 8:454; WAB 79; (T. pyra) SSA cp 32.

Hummingbird, White-necked Jacobin (Florisuga mellivora) ABW 170; ALE 8:454; SHU 58, 75; SSA cp 5.

Hunter's Robe —See **Ivy, Devil's.**

Huntsman's Horn (Sarracenia flava) MLP 92; WWT 92.

Hutchinsia (Hornungia petraea) MBF cp 10; MFF pl 90 (73); NHE 216; WFB 97 (8).

Hutchinsia, Alpine (Hutchinsia alpina) ERW 121; NHE 274; SPL cp 454; TVG 153.

Hutia (Capromys prehensilis) ALE 11:438; (Plagiodontia aedium) LVA 33.

Hyacinth, Giant Summer (Galtonia candicans) DFP 94 (cp 747); EGB 116; MWF 138; SPL cp 285; TGF 13.

Hyacinth Shrub —See **Yellow Horn.**

Hyacinth, Wild —See **Bluebell.**

Hydnora (Hydnora sp.) EWF cp 53c; LFT cp 63.

Hydrangea (Hydrangea sp.) DEW 1:293, 297 (cp 161); DFP 207 (cp 1653); ESG 25; EWF cp 99b; GDS cps 237, 238, 247-249; HSC 66, 67; PFW 139, 140; TGS 181; (H. villosa) DFP 207 (cp 1654); HSC 67, 69; OGF 145.

Hydrangea, Climbing (Hydrangea petiolaris) EGV 48, 115; EGW 116; ELG 151; HSC 67, 72; LFW 71 (cp 164); TGS 70.

Hydrangea, House (Hydrangea macrophylla) BIP 101 (cp 153); DFP 207 (cps 1649-1651); EDP 122; EEG 97; EFS 68, 69, 74, 75, 117; EGG 117; ELG 116; EPF pl 369; EPG 124; ESG 119; FHP 126; GDS cps 239-246; HSC 66, 69; LHP 112, 113; MWF 158; OGF 145; RDP 244; SPL cp 213; TGS 182; TVG 227.

Hydrangea, Oak-leaved (Hydrangea quercifolia) EFS 25, 30; LFW 71 (cp 166); SPL cp 93; TGS 181.

Hydrangea, Panicle or Plumed (Hydrangea paniculata) DFP 207 (cp 1652); GDS cps 250-252; HSC 66, 69; MWF 158; PFF 193; OGF 145; TGS 182; TVG 227.

Hydroid (Antennularia antennina) CSS pl XXV; (Tubularia larynx) CSS cp 2; VWS 36 (bw), 44, 69; (Branchiocerianthus imperator) MOL 85.

Hydroid or **Cerith** (Clava sp.) CSS cp 2; CTS 20 (cp 11); VWS 42, 45.

Hyena, Brown (Hyaena brunnea) ALE 12:178, 188; DDM 113 (cp 20); HMW 129; LVA 19; LVS 83; SMR 65 (cp 4).

Hyena, Spotted (Crocuta crocuta) ALE 12:178, 187, 232; CAA 73; CAF 157 (bw); DDM 113 (cp 20); DSM 154; HMA 62; JAW 124; LAW 540; LEA 567 (bw); MLS 307 (bw); SMR 65 (cp 4); SMW 184 (cp 111); TWA 47; WEA 202; (Hyaena crocuta) HMW 129.

Hyena, Striped (Hyaena hyaena) ALE 12:178, 188; DAA 72; DDM 113 (cp 20); DSM 154; HMA 62; HMW 129; JAW 125; MLS 308 (bw).

Hygrocybe (Hygrocybe sp.) DEW 3: cp 59; DLM 65, 195-197; RTM 170, 172.

Hygrophorus (Hygrophorus sp.) CTM 35; DLM 197-199; KMF 119; LFO 26, 27; LHM 77, 79; NFP 69; ONP 33; PFF 83; PMU cp 50; PUM cp 17; REM 143-146 (cps 111-114); RTM 165-171; SMF 22, 44; SMG 229.

Hypecoum (Hypecoum sp.) PFE cp 30 (276); PFM cp 36.

Hypholoma (Hypholoma sp.) CTM 43; DLM 200; LHM 147; PMU cp 101; PUM cps 128b, 138a; RTM 57.

Hypocolius, Gray (Hypocolius ampelinus) ABW 266; FBE 217.

Hyrax, Abyssinian (Procavia habessinica)
ALE 12:509; HMW 104.

Hyrax, Cape or Rock (Procavia capensis)
ALE 12:509, 519; CAF 170 (bw); DAA 13;
DDM 224 (cp 37); DSM 207; HMA 62; LAW
576; LEA 582 (bw); MLS 341 (bw); SMR 80 (cp
5); SMW 291 (bw).

Hyrax, Rock (Heterohyrax brucei) DDM 224

(cp 37); JAW 151; (H. syriacus) ALE 12:509.

Hyrax, Tree (Dendrohyrax arboreus or D.
dorsalis) CAF 292 (bw); DDM 224 (cp 37);
DSM 206; HMW 104; LAW 576; SMR 80 (cp
5); SMW 287 (cp 188).

Hyssop (Hyssop officinalis) AMP 43; DFP 148
(cp 1181); EGH 116; NHE 231; PFE cp 115
(1159); WFB 199 (10).

I

Ibex, Alpine (Capra ibex) ALE 13:463; CEV 19
(bw), 148, 151; HMA 56; HMW 85; JAW 204;
LEA 609 (bw); MLS 428 (bw); PWC 114 (bw);
SLP 27; SMW 272 (bw); WEA 203; (C. hircus
ibex) DAE 50.

Ibex, Nubian (Capra nubiana) CAS 112, 116
(bw); DAA 15; DSM 273; TWA 61 (bw).

Ibex, Siberian (Capra sibirica) DAA 25; WEA
203.

Ibex, Walia (Capra walia) LVS 122; SLS 215;
VWA cp 7.

Ibis, Bald or Hermit (Geronticus eremita)
ALE 7:237; FBE 41; SLP 42, 43; WAB 29; (G.
calvus) CDB 48 (cp 82); PSA 48 (cp 3).

Ibis, Giant (Pseudibis gigantea or Thaumatibis gigantea) KBA 33 (pl 2); LVB 37; PWC
218 (bw).

Ibis, Glossy (Plegadis falcinellus) ABW 59;
ALE 7:237; BBB 274; BBE 39; CDB 48 (cp
85); DPB 25; FBE 41; HAB 22; KBA 33 (pl 2);
LEA 375 (bw); OBB 15; PBE 9 (cp 4); PSA 48
(cp 3); RBE 83; SBB 19 (pl 16); SNB 30 (cp 14).

Ibis, Hadeda (Hagedashia hagedash) ALE
7:237; CDB 48 (cp 83); (Bostrychia hagedash)
PSA 48 (cp 3).

Ibis, Japanese Crested (Nipponia nippon)
GPB 74; LVB 29; LVS 199; PWC 218 (bw);
SLS 174; WAB 29.

Ibis, Sacred or White (Threskiornis
aethiopica) ABW 59; BBE 39; CAA 38 (bw);
CDB 49 (cp 87); DPB 25; FBE 41; GPB 73;
LEA 375 (bw); PSA 48 (cp 3); SBB 19 (pl 17);
(T. molucca) CAU 100 (bw); SNB 30 (cp 14).

Ibis, Scarlet (Eudocimus ruber) ABW 59; ALE
7:237; 8:359; BBT 93; CDB 48 (cp 81); SBB 20
(cp 18); WAB 79, 94; (Guara ruber) GPB 74.

Ibis, Straw-necked (Threskiornis spinicollis)
ALE 10:89; CAU 100 (bw); CDB 49 (cp 88);
HAB 22; RBU 71; SNB 30 (cp 14); WAW 28
(bw).

Ibis, Wood or Yellow-billed (Ibis ibis or
Mycteria ibis) ALE 7:222; CAA 41; CAF 113;
CDB 46 (cp 75); FBE 43; GPB 71; PSA 48 (cp3).

Ibis-bill (Ibidorhyncha struthersii) ALE
8:170; KBA 149 (pl 18).

Ice Plant (Lampranthus sp.) HPW 66; LCS cp
275; MEP 37; MWF 178, 179; SPL cp 143.

Ice Plant, Pink (Lampranthus spectabilis)
EGC 137; LFW 181 (cp 408); MWF 179; PFW

Ichneumon (insect) —See **Wasp,
Ichneumon.**

Ichneumon (mammal) —See **Mongoose,
Egyptian.**

Ide (Leuciscus idus or Idus idus) ALE 4:341;
FBW 35; FEF 144 (bw), 154; FWF 65; GAL
232; HFI 97; OBV 113; WEA 170; WFW cp
130.

Iguana (Corythophanes sp.) CSA 21 (bw);
MAR pl 46.

Iguana, Common or Green (Iguana iguana)
ALE 6:33, 180, 188; 8:359; ART 117; ERA
179, 181; HRA 27; HWR 78, 79; LAW 411;
LEA 305 (bw); SIR 99 (cps 35, 36); VRA 185.

Iguana, Fiji Banded (Brachylophus
fasciatus) ALE 6:180; LVS 180.

Iguana, Galapagos Land (Conolophus subcristatus) ALE 6:180, 185; CSA 245 (bw),
247; HRA 28; LVS 179; (C. pallidus) LAW 414.

Iguana, Marine (Amblyrhynchus cristatus)
ALE 6:180, 187; ART 152, 192, 193; CSA 236,
244 (bw); ERA 184, 185; HRA 28; HWR 80;
LAW 413; LBR 119; LEA 306 (bw); LVS 173;
MAR pl 49, cp IX; SLP 107.

Iigiri Tree (Idesia polycarpia) EGW 117; HPW 101; SST cp 129; TGS 199.

Illawarra Flame Tree —See Flame Tree of Australia.

Illecebrum (Illecebrum verticillatum) MBF cp 71; MFF pl 104 (196); NHE 97; PFE cp 13 (152); PWF 46a; WFB 57 (8).

Illicium, Japanese (Illicium anisatum) HPW 29; PFW 142.

Immortelle —See Pink Everlasting (Xeranthemum).

Immortelle, Mountain —See Bucare Tree.

Impala (Aepyceros melampus) ALE 13:426, 436; CAF 195, 197 (bw); DDM 225 (cp 38); DSM 265; JAW 198; LAW 567; LEA 613 (bw); MLS 416 (bw); SLP 245; SMR 97 (cp 8); SMW 284 (cp 181); TWA 33; WEA 206.

Impala Lily (Adenium obesum) ECS 92; EWF cp 61d; LFT cp 123; RWF 63.

Incense Plant (Humea elegans) BIP cp 152; MWF 157.

Inch Plant (Callisia elegans) EDP 145; EFP 96; RDP 126.

Inch Plant, Flowering (Tradescantia blossfeldiana) LHP 183; RDP 386.

India Rubber Plant (Ficus elastica) BIP 85 (cp 125); DFP 67 (cps 529, 530); EDP 142; EFP 113; LHP 95; MWF 133; PFF 149; RDP 12, 211-213; SPL cp 19; SST cp 282.

Indian Blanket (Gaillardia pulchella) DFP 37 (296); EGA 120; EPF pl 550; EWF cp 165e; LFW 198 (cp 449); TGF 269; TVG 51.

Indian Cress—See Nasturium.

Indian Currant—See Coralberry (Symphoricarpos).

Indian Mallow—See Velvetleaf.

Indian Shot —See Canna.

Indigo (Baptisia australis or B. tinctoria) DFP 126 (cp 1001); EGH 103; EWF 91; LFW 186 (cp 417); OGF 77; TGF 140.

Indigo (Indigofera sp.) EFS 119; EJG 122; HSC 70, 72; LFT cp 85; MWF 163; SPL cp 95.

Indigo-bird —See Widow-finch.

Indris (Indri indri) ALE 10:251 (cp 5), 283; DSM 67; HMW 31; LVA 25; MLS 119 (bw); PWC 232 (bw); SLS 191; VWA cp 3.

Ink-cap, Bonfire (Coprinus lagopus) DLM 184; LHM 139; ONP 139; SMF 62.

Ink-cap, Common (Coprinus atramentarius) DLM 68, 182; LHM 137; NFP 84; ONP 35; PFF 85; REM 81 (cp 49); RTM 50, 267; SMF 27; SMG 223.

Ink-cap, Furrowed (Coprinus plicatilis) DLM 185; KMF 115; LHM 139; ONP 35.

Ink-cap, Glistening (Coprinus micaceus) CTM 49; DEW 3:176; DLM 184; LHM 141; NFP 84; ONP 139; PFF 86; REM 82 (cp 50); RTM 52; SMF 43; SMG 226.

Ink-cap, Shaggy or Shaggy-mane (Coprinus comatus) CTM 50; DEW 3: cp 64; DLM 53, 69, 183; EPF pl 63; KMF 115; LHM 137; NFP 84; ONP 35; PFF 86; PUM cp 134; REM 80 (cp 48); RTM 51; SMF 15; SMG 224, 225.

Innocence —See Bluets (Houstonia).

Inocybe (Inocybe sp.) DEW 3:178; DLM 201-203; LHM 155-159; NFP 79; NHE 48; ONP 119, 139; PMU cp 63h; PUM cps 126, 127; REM 105-107 (cps 73-75); RTM 79-81, 267; SMF 34.

Inonotus (Inonotus sp.) DLM 153; LHM 63; ONP 157; RTM 212.

Inula (Inula sp.) DFP 149 (cps 1185, 1186); NHE 76; PFE 446; TGF 285.

Iora, Common (Aegithina tiphia) ABW 245; ALE 9:190; DPB 289; KBA 273 (cp 40); WAB 161; WBV 179.

Iris (Iris histrio) EWF cp 51h; PFM cp 267; (I. laevigata) DFP 150 (cps 1194, 1195); HPW 311; PFW 145; (I. pumila) EWF cp 42g; PFE cp 176 (1691); PFM cps 260, 261; TVG 155.

Iris, Algerian (Iris unguicularis) DFP 151 (cps 1203, 1204); OGF 5; PFW 146.

Iris, Baker (Iris bakeriana) DFP 97 (cp 775); EWF cp 51j.

Iris, Bokhara (Iris bucharica) DEW 2:224 (cp 143); DFP 97 (cp 776); TVG 105.

Iris, Butterfly (Iris spuria) DFP 151 (cp 1201); LFW 76 (cp 173); MBF cp 83; MFF cp 6 (1048); NHE 77; WFB 273 (5).

Iris, Common (Iris germanica) DEW 2:251; HPW 311; LFW 76 (cps 171, 174, 175), 76 (cps 177, 178); PFE cp 175 (1693); PFF 385; PFW 145; SPL cp 215, 216; WFB 273 (4).

Iris, Crested (Iris cristata) ESG 120; EWG 23; LFW 202 (cp 457); TGF 36.

Iris, Danford (Iris danfordiae) DFP 98 (cp 777); EGW 79, 119; EMB 102; OGF 11; PFW 142; TVG 199.

Iris, English (Iris xiphioides) MWF 167; PFE cp 175 (1685); PFW 143.

Iris, Grass-leaved (Iris graminea) EPF pl 94; NHE 193; PFE cp 175 (1686).

Iris, Harput (Iris histrioides) EGW 79, 119; OGF 11; PFW 142.

Iris, Leafless (Iris aphylla) EPF pl 946; NHE 193.

Iris, Lesser (Iris chamaeiris) EWF cp 42b; KVP cp 160; PFE cp 175 (1692); PFM cps 265, 266.

Iris, Netted (Iris reticulata) BIP 39 (cp 43); DFP 98 (cps 780-783); EGB 122; EPF pl 939; OGF 11; PFW 142.

Iris, Peacock (Moraea pavonia) EGB 129; MEP 23; MWF 203; (M. villosa) EWF cp 87c.

Iris, Siberian (Iris sibirica) DFP 150 (cps 1199, 1200); EFC cp 15; LFW 76 (cp 176); OGF 75; PFW 144; TVG 105; WFB 273 (1).

Iris, Snake's Head (Hermodactylus tuberosus) DFP 96 (cp 764); EWF cp 42a; MWF 153; PFE cp 174 (1683); PFM cp 262; PFW 146; SPL cp 488.

Iris, Spanish (Iris xiphium) DFP 99 (cps 785-787); MWF 166; PFE cp 175 (1685); PFM cp 263; SPL cp 218.

Iris, Stinking (Iris foetidissima) BFB cp 23 (9); BWG 153; DFP 149 (cp 1188); HPW 311; MBF cp 83; MFF cp 6 (1049); OWF 163; PFW 144; PWF 19i, 107e; WFB 273 (6).

Irish Daisy —See **Dandelion** (Taraxacum).

Iron Root —See **Orache, Common.**

Iron Tree (Parrotia persica) BKT cp 104; DFP 216 (cp 1721); EGT 131; OBT 148; PFW 136, 137; TGS 199.

Ironbark, White —See **Gum, Yellow.**

Isoloma (Isoloma sp.) DEW 1:84; MWF 167.

Isopyrum (Isopyrum sp.) EPF pl 207; NHE 178.

Ithuriel's Spear —See **Grass Nut.**

Ivy, Boston or Japanese (Parthenocissus tricuspidata) DFP 249 (cp 1985); EGV 25, 125; EJG 135; EPF pl 723, 724; HSC 86; MWF 222; SPL cp 178; TGS 71.

Ivy, Canary Island (Hedera canariensis) BIP 95 (cp 145); DFP 68 (cp 543); EFP 55, 117; EGC 24, 134; EGV 66, 112; LHP 107; PFW 36; RDP 235.

Ivy, Common English or Evergreen (Hedera helix) AMP 105; BIP 79 (cp 145); BWG 164; CTH 40 (cp 52); DEW 2:83; DFP 68 (cp 544), 69 (cps 545-546); EEG 17, 143; EFP 25, 118; EGC 37, 134; EGL 118; EGV 59, 61, 113; EGW 114; ELG 146; EMB 127; EPF pls 751, 752; ESG 115; GDS cp 225; HPW 218; HSC 63; LHP 106, 107; MBF cp 41; MFF pl 76 (450); MWF 148; NHE 21; OWF 65; PFE cp 83 (848); PFF 264; PFW 36; PWF 18c, 173l; RDP 234; SPL cp 167; TGS 39; TVG 227; WFB 155 (4).

Ivy, Devil's (Scindapsus aureus) BIP 161 (cp 246); DFP 81 (cp 645); EFP 142; EGV 134; LHP 166, 167; RDP 355; SPL 33.

Ivy, German or Parlor —See **German or Parlor Ivy.**

Ivy, Japanese or Tree (Fatshedera lizei) BIP 81 (cp 121); DFP 66 (cp 527); EFP 112; EGV 108; LHP 90, 91; MWF 132; RDP 201; TGS 246.

Ivy, Kenilworth —See **Toadflax, Ivy-leaved.**

Ivy, Persian (Hedera colchica) HSC 63, 65; MWF 148.

Ivy, Violet —See **Cup and Saucer Vine.**

Ixora (Ixora sp.) DEW 2:108 (cps 67, 68); GFI 83 (cp 17); HPW 258; LFW 170 (cp 383); MEP 149; MWF 168.

J

Jabiru (Jabiru mycteria) ABW 57; ALE 7:222; CDB 46 (cp 72); CSA 54 (bw); SBB 17 (cp 14); (Xenorhynchus asiaticus) HAB 24; KBA 33 (pl 2); SNB 30 (cp 14); WAW 7 (bw).

Jacamar (Jacamerops aurea) ABW 188; ALE 9:68; SSA cp 34; WAB 90; (Galbalcyrhynchus leucotis) SSA cp 9.

Jacamar, Paradise (Galbula dea) ABW 188; ALE 9:68.

Jacamar, Rufous-tailed (Galbula ruficauda)
BBT 91; CDB 132 (cp 522); CSA 67; DTB 20
(cp 10); SSA cp 34; WAB 83.

Jacana, African (Actophilornis africanus)
ABW 116; CDB 78 (cp 248); GPB 136; LEA
409 (bw); PSA 160 (cp 17).

Jacana, Comb-crested (Irediparra gallinacea) CDB 78 (cp 249); DPB 81; HAB 74; RBU
32; (Jacana gallinacea) SNB 64 (cp 31).

Jacana, Pheasant-tailed (Hydrophasianus
chirurgus) ABW 116; ALE 8:170; CAS 161;
DPB 81; KBA 97 (cp 10); RBI 60; WAB 176.

Jacaranda (Jacaranda acutifolia or J.
mimosifolia) BIP 107 (cps 161, 162); EET
225; EGT 119; MEP 137; MWF 169; PFW 44;
RDP 248; SST cp 246.

Jack Dempsey (Cichlasoma biocellatus) ALE
5:132; FEF 417 (bw); SFW 549 (pl 148), 561
(cp 152).

Jack, Yellow (Gnathodon speciosus) HFI 26;
HFW 167 (cp 72).

Jackal, Abyssinian —See **Fox, Semien.**

Jackal, Black-backed (Canis mesomelas)
ALE 12:187, 206, 232; CAA 85 (bw); CAF 157
(bw); DDM 81 (cp 12); DSM 144; HMA 45;
JAW 96; MLS 257 (bw); SMR 33 (cp 2); TWA
48.

Jackal, Common or Golden (Canis aureus)
ALE 12:187, 206, 232; CAS 109; DAE 75;
DDM 81 (cp 12); HMW 141; (Thos aureus)
DAA 73.

Jackal, Side-striped (Canis adustus) ALE
12:206; DDM 81 (cp 12); HMW 141; LEA 553
(bw); SMR 33 (cp 2); WEA 208

Jack-by-the-hedge —See **Mustard, Garlic.**

Jackdaw (Corvus monedula) ALE 9:498; BBB
81; BBE 217; CDB 223 (cp 1000); FBE 311;
LEA 465; OBB 127; PBE 309 (bw); WAB 25.

Jackdaw, Alpine —See **Chough, Alpine.**

Jack-in-the-green —See **Primrose, English.**

Jack-in-the-pulpit (Arisaema triphyllum)
ESG 84, 92; EWG 26; LFW 186 (cp 419); MLP
287; PFF 364; TGF 38.

Jack-in-the-pulpit (Arisaema ringens) DEW
2:314; EPF pl 1008; (A. candidissimum) EWF
cp 108a; (A. fimbriatum) EWF cp 125a; (A.
sikokiamum) EWF cp 108e.

Jacky Winter —See **Flycatcher, Australian
Brown.**

Jacobin —See **Hummingbird.**

Jacobinia (Jacobinia sp.) BIP 107 (cp 163);
FHP 128; LFW 253 (cps 562, 563); RDP 249;
SPL cp 220.

Jacob's Coat —See **Snow Bush.**

Jacob's Ladder (Pedilanthus) —See **Redbird
Cactus.**

Jacob's Ladder (Polemonium caeruleum)
BFB cp 15 (3); DEW 2:142; EPF pl 787; EWF
cp 14c; KVP cp 141; LFW 112 (cp 246); MBF
cp 59; MFF cp 12 (646); NHE 186; PFE cp 100
(1032); PWF 56c; TGF 172; WFB 185 (5); (P.
carneum) OGF 37; (P. reptans) EWG 133.

Jade Plant (Crassula argentea) EDP 136; EFP
103; LCS cps 232, 233; LHP 188; RDP 158.

Jade Plant (Portulacaria) —See **Elephant
Bush.**

Jade Vine (Strongylodon macrobotrys) MEP
58; PFW 164; (S. lucidus) EWF cp 147b.

Jaeger, Pomarine (Stercorarius pomarinus)
ABW 129; BBB 277; BBE 141; DAE 32; DPB
107; HAB 76; HBG 110 (bw); KBA 156 (pl 19);
OBB 85; PBE 124 (pl 35); RBE 160; SAO 97.

Jaguar (Panthera onca or Felis onca) ALE
12:324, 339; BMA 121, 122; CSA 118, 119
(bw); DSM 190; HMA 36, 37; HMW 123; JAW
133; LEA 573 (bw); LVS 89; MLS 321; SMW
180 (cp 103); TWA 136; WEA 209.

Jaguarundi —See **Cat, Jaguarundi.**

Jasmine —See also **Jessamine** and **Cestrum.**

Jasmine or Jessamine (Jasminum sp.) GDS
cp 259; HSC 70; LFT cp 120; MWF 170; PFE
cp 95 (983); TGS 327, 329.

Jasmine, Arabian (Jasminum sambac) EGL
121; EGV 117; MEP 105.

Jasmine, Cape —See **Gardenia, Common.**

Jasmine, Chilean (Mandevilla suaveolens)
MWF 196; PFW 29.

Jasmine, Chinese (Jasminum polyanthum)
BIP 107 (cp 164); DFP 248 (cp 1978); EWF cp
95d; FHP 128; LHP 118, 119; MWF 170; SPL
cp 170.

Jasmine, Common (Jasminum officinale)
EGH 118; EGV 116; ESG 121; OGF 165; PFE
cp 95 (984); PFW 199; RDP 250; SPL cp 169;
TGS 39.

Jasmine, Confederate Star (Trachelospermum jasminoides) BIP 177 (cp 276);
EGC 148; EGV 140; FHP 148; HSC 119; MEP
107; MWF 286; PFW 29; SPL cp 183; TGS 70.

Jasmine, Crape (Ervatamia coronaria) FHP 119; MWF 125; (E. divaricata) MEP 109.

Jasmine, King (Jasiminum rex) EWF cp 121b; MWF 170; PFW 199.

Jasmine, Madagascar —See **Wax Flower** (Stephanotis).

Jasmine, Orange (Murraya sp.) MEP 65; MWF 203.

Jasmine, Poet's (Jasminum officinale grandiflorum) EDP 129; EGG 120.

Jasmine, Primrose (Jasminum mesnyi) EGE 132; MEP 105; MWF 170; RDP 250.

Jasmine, Winter (Jasminum nudiflorum) DFP 208 (cp 1664); EFS 119; EGW 120; EJG 123; EPF cp XVII; EPG 125; HSC 73; MWF 170; OGF 3; PFW 199; SPL cp 264; TGS 327.

Jasmine, Yellow (Jasminum revolutum) DFP 248 (cp 1979); SPL cp 171.

Jatropha (Jatropha sp.) EWF cp 178c; LFT cp 94; MEP 72.

Jay (Cyanocorax sp.) ALE 9:497; SSA cp 16; WAB 91; (Cyanolyca sp.) ABW 225; ALE 9:497.

Jay, Common or Eurasian (Garrulous glandarius) ABW 225; ALE 9:497; 13:163; BBB 149; BBE 215; CDB 224 (cp 1005); FBE 305; KBA 277 (pl 42); LAW 510; NBB 91 (bw); OBB 127; PBE 197 (cp 50); RBE 235; WAB 109; WEA 210.

Jay, Ground (Podoces sp.) ALE 9:469; RBI 283; WAB 120; (Pseudopodoces sp.) ALE 9:469.

Jay, Siberian (Perisoreus infaustus) ALE 9:497; BBE 215; FBE 305; PBE 197 (cp 50); WAB 104.

Jelly Babies —See **Fungus, Green Slime.**

Jelly Beans —See **Christmas Cheer.**

Jellyfish (Chrysaora sp.) ALE 1:183; CSS cp 4; OIB 13; VWS 61 (bw); (Cyanea sp.) ALE 1:183; CSS cp 3; OIB 13.

Jellyfish (Rhizostoma sp.) ALE 1:183, 272; CSS cp 4; (Physophora sp.) ALE 1:271; OIB 11; MOL 53; OIB 13.

Jellyfish, Comb —See **Comb-jelly.**

Jellyfish, Moon (Aurelia sp.) ALE 1:183; CSS cp 4; MOL 134 (bw); OIB 13.

Jellyfish, Stalked (Haliclystus sp.) CSS cp 2; MOL 53; OIB 13; VWS 26 (bw).

Jerboa (Allactaga sp.) ALE 11:399; CAS 78 (bw); MLS 221 (bw); (Dipus sagitta) ALE 11:399.

Jerboa, African (Jaculus orientalis) LAW 593; SMW 137 (cp 84).

Jerboa, Lesser Egyptian (Jaculus jaculus) ALE 11:399; HMW 164; TWA 63 (bw); WEA 211.

Jerusalem Sage (Phlomis fruticosa) DFP 217 (cp 1731); EWF cp 37b; GDS cp 215; HSC 87, 93; MWF 230; OGF 105; PFE cp 110 (1124); PFM cp 163; SPL cp 239; TGS 342; (P. sp.) PFE cp 110 (1122, 1123), 360; PFM cps 162, 164.

Jessamine, Carolina Yellow (Gelsemium sempervirens) BIP 91 (cp 136); EGV 110; ELG 150; FHP 123; MEP 105; MWF 140; TGS 70.

Jessamine, Night-flowering (Cestrum nocturnum) FHP 111; MEP 127; MWF 79.

Jessamine, Orange —See **Jasmine, Orange.**

Jessamine, Purple (Cestrum purpureum) DEW 2:163 (cp 82); MEP 127; SPL cp 72.

Jet Bead —See **Kerria, White.**

Jewelfish (Hemichromis bimaculatus) ALE 5:122, 131; CTF 57 (cp 74); FEF 440, 441 (bw); HFW 176 (cp 87).

Jewel-weed (Impatiens capensis) LFW 202 (cp 459); MBF cp 20; MFF pl 62 (257); OWF 13; PFE cp 70 (714); PWF 134c; WFB 139 (2a).

Jewfish, Australian —See **Grouper, Giant.**

Jew's Ear (Auricularia auricula) DEW 3:152; DLM 140; LHM 225; NFP 45; ONP 149; RTM 231; SMF 24; (A. mesenterica) KMF 35; ONP 149; RTM 231.

Jew's Mallow —See **Globe Flower Bush.**

Jigger —See **Flea, Chigoe.**

Jimson Weed —See **Thorn-apple.**

John Dory —See **Dory, John.**

Joint-pine or **Joint-fir** (Ephedra sp.) AMP 97; DEW 1:65, 66; 98 (cp 40); NFP 150; PFE cp 2 (18); PFF 113.

Jonquil (Narcissus jonquilla) DFP 105 (cp 1834); EGF 132; PFF 384; SPL cp 269; TGF 18.

Joseph's Coat (Amaranthus tricolor) DFP 30 (cp 233); EGA 93; MWF 39.

Juanulloa (Juanulloa sp.) EWF cp 183b; PFW 287.

Judas Tree (Cercis siliquastrum) BKT cps
134, 135; DFP 189 (cp 1507); EET 180, 181;
EPF pl 444; EWF cp 28c; GDS cp 64; MTB 324
(cp 29); MWF 79; NHE 294; OBT 184; PFE cp
50 (485); PFM cp 54; PFW 160; SST cp 236.

Judas Tree, Chinese (Cercis chinensis) EFS
100; EJG 106; TGS 130.

Jujube (Zizyphus jujuba) EGT 147; HPW 188;
(Z. mucronata) LFT cp 102.

June Berry —See **Serviceberry.**

Jungle Fowl —See **Fowl.**

Juniper, Chinese (Juniperus chinensis)
DEW 1:55 (cp 15); EEG 107, 126; EET 93;
EGB 20; EGE 96, 97; EJG 123; ELG 132; EMB
115; MTB 65 (cp 4); OBT 104, 132; TGS 12.

Juniper, Common (Juniperus communis)
AMP 137; BFB cp 14 (9); DEW 1:45; DFP 252
(cps 2015, 2017); EET 92; EPF pl 162; GDS

cps 260, 261; MFB pl 100; MFF cp 64 (1258);
NHE 69, 266; OBT 37; OFP 137; OWF 185;
PFE cp 1 (12); PFF 124; PFM cp 1; PWF 174b;
SST cp 21; TGS 22; TVG 253; WFB 25 (6).

Juniper, Creeping (Juniperus horizantalis)
DFP 253 (cp 2018); GDS cp 264; MWF 171; (J.
h. wiltonii) EEG 143; EGC 38, 137; EGE 98;
ELG 146; EPG 126.

Juniper, Phoenician (Juniperus phoenicea)
PFE cp 2 (14); PFM cp 2.

Juniper, Prickly (Juniperus oxycedrus) CTH
21 (cp 9); PFE cp 2 (13); SST cp 22.

Jupiter's Distaff (Salvia glutinosa) EFC cp
49; KVP cp 111; NHE 35; OGF 153; PFE cp
112 (1146); WFB 202; (S. argentea) PFE cp
113 (1144); (S. occinea) EFC cp 50.

Jurupari —See **Cichlid, Pearl.**

Justicia (Justicia sp.) EHP 83; LFT 158; MWF 171.

K

Kaffir Broom —See **Coral Tree.**

Kaffir Fig —See **Fig, Hottentot.**

Kagu (Rhynochetus jubatus) ABW 112; ALE
8:100; CDB 76 (cp 239); GPB 133; LVB 57;
LVS 216; PWC 188; WAB 205.

Kaka (Nestor meridionalis) ALE 8:299; BNZ
89; CDB 106 (cp 397); FNZ 48 (cp 2).

Kakapo (Strigops habroptilus) ALE 8:299;
BNZ 97; CAU 282; CDB 109 (cp 408); FNZ 48
(cp 2); GPB 177; LVB 43; RBU 120; SLP 148;
SLS 235; WAB 33, 208.

Kakariki —See **Parakeet, Yellow-crowned.**

Kaki, Japanese —See **Persimmon,
Oriental.**

Kalanchoe (Kalanchoe blossfeldiana) BIP 109
(cp 166); EGG 121; FHP 129; LCS cp 271;
LHP 120, 121; MWF 172; RDP 251; SPL cp 51.

Kalanchoe (Kalanchoe sp.) EWF cp 70d; HPW
145; LFT cp 75; LFW 170 (cp 381); MWF 172;
PFW 95; RDP 251.

Kalanchoe, Rainbow (Kalanchoe fedtschenkoi) ECS 124; LCS 272; LFW 170 (cp 382);
MWF 172.

Kangaroo, Great Gray (Macropus giganteus)
ALE 10:89, 138 (cp 8), 143; CAU 20 (bw), 23;

JAW 30; LVS 35; MEA 113; TWA 186; (M.
canguru) DSM 37; HMA 63; LAW 530; (M.
major) MWA 77; WAW 35, 36, 39 (bw).

Kangaroo Ivy or **Kangeroo Vine** (Cissus
antarctica) BIP 55 (cp 68); DFP 59 (cp 469);
EGV 101; LHP 59; RDP 143.

Kangaroo, Rat —See **Rat-kangaroo.**

Kangaroo, Red (Macropus rufus) ALE 10:138
(cp 8), 144, 145; CAU 91; DSM 36; HMA 63;
HMW 205; LEA 486 (bw); MLS 64 (bw); MWA
57, 112 (bw); SMW 39 (cp 13); TWA 187;
WAW 34 (bw), 38; WEA 212; (Megaleia rufa)
LAW 528; MEA 114.

Kangaroo, Tree —See **Tree-kangaroo.**

Kangaroo-paw or **Kangaroo Flower**
(Anigozanthos sp.) EWF 143c, 144a; HPW
317; MLP 257; PFW 135; RWF 77-78.

Kankerbos —See **Cancerbush.**

Kapok Tree (Ceiba pentandra) EPF pl 609;
MLP 149; PFF 251; SST cp 271; WWT 101.

Karo (Pittosporum crassifolium) BIP 147 (cp
221); DFP 218 (cp 1740); EWF cp 131e; HWF
138; SNZ 53 (cps 128, 129).

Karoo Rose (Lapidaria margaretae) ECS 125;
LCS cp 276.

Kingfisher, Red-backed (Halcyon pyrrhopygia) HAB 135; MBA 32, 33; SNB 126 (cp 62); WAB 191.

Kingfisher, Ruddy (Halcyon coromanda) DPB 205; KBA 177 (cp 24).

Kingfisher, Sacred (Halcyon sancta) BNZ 67; CAU 263 (bw); CDB 126 (cp 489); FNZ 48 (cp 2); HAB 134; MBA 62; MWA 106; RBU 184; SNB 126 (cp 62).

Kingfisher, Smyrna —See **Kingfisher, White-breasted or -throated.**

Kingfisher, Stork-billed (Pelargopsis capensis) ALE 9:38; DPB 205; KBA 177 (cp 24); WBV 131.

Kingfisher, White-breasted or -throated (Halcyon smyrnensis) ABW 176; ALE 9:32, 38; CAS 207; CDB 127 (cp 491); DAA 150; DPB 205; FBE 191; GPB 209; KBA 177 (cp 24); RBI 120; WAB 175; WBV 131.

Kingfisher, White-collared (Halcyon chloris) ABW 177; ALE 9:37; DPB 205; HAB 133; KBA 177 (cp 24); SNB 126 (cp 62); WBV 131.

Kingfisher, White-tailed (Tanysiptera sylvia) RBU 191; SNB 126 (cp 62); WAB 181.

King's Crown —See **Plume Flower, Brazilian.**

King's Spear (Asphodeline lutea) EWF cp 41D; NHE 42; OGF 73; PFE cp 162 (1593); PFM cp 235; TGF 29.

Kinkajou (Potos flavus) ALE 12:94, 113; HMW 147; MLS 277 (bw); SMW 194 (bw).

Kinnikinnick —See **Bearberry.**

Kiss-me-at-the-garden-gate —See **Pansy, Wild.**

Kite, African Swallow-tailed (Chelictinia riocourii) ALE 7:346; CDB 57 (cp 141); GBP 211 (bw).

Kite, Black (Milvus migrans) ALE 7:346, 385; BBE 83; CAS 219 (bw); DPB 37; FBE 73; GBP 230 (bw); HAB 38; KBA 48 (pl 3), 64 (pl 5); PBE 60 (cp 19), 72 (pl 23); PSA 81 (pl 8), 128 (pl 13); RBE 15; SNB 44 (cp 21), 52 (pl 25); WAB 179; WBV 57.

Kite, Black-shouldered or -winged (Elanus caeruleus) ALE 7:346; BBE 83; CDB 57 (cp 145); DAE 123; DPB 37; FBE 73; GBP cp after 112, 168 (bw), 207 (bw); GPB 102; KBA 48 (pl 3), 65 (pl 6); PBE 60 (cp 19); PSA 81 (pl 8), 128 (pl 13); (E. notatus) SNB 42 (cp 20), 50 (pl 24).

Kite, Brahminy (Haliastur indus) ABW 76; BBT 43; DPB 41; GBP 235, 237 (bw); HAB 46; KBA 48 (pl 3), 64 (pl 5); SNB 44 (cp 21), 52 (pl 25); WAB 175.

Kite, Cayenne or Gray-headed (Leptodon cayanensis) ALE 7:346; GBP 222 (bw).

Kite, Hook-billed (Chondrohierax uncinatus) GBP 223 (bw); SSA pl 22.

Kite, Letter-winged (Elanus scriptus) HAB 38; RBU 84; SNB 42 (cp 20), 50 (pl 24).

Kite, Pearl (Gampsonyx swainsonii) ALE 7:346; GBP 210 (bw); SSA pl 43.

Kite, Plumbeous (Ictinia plumbea) ALE 7:346; GBP cp before 161, 226 (bw); SSA pl 43.

Kite, Red (Milvus milvus) ALE 7:346; BBB 130; BBE 83; FBE 73; GAL 156 (bw); GBP 232 (bw); GPB 101; LEA 385 (bw); OBB 43; PBE 60 (cp 19), 72 (pl 23).

Kite, Red-backed —See **Eagle, Red-backed Sea.**

Kite, Square-tailed (Lophoictinia isura) ALE 7:346; GBP 33 (bw); HAB 38; SNB 44 (cp 21), 52 (pl 25).

Kite, Whistling (Haliastur sphenurus) CDB 59 (cp 153); HAB 34; SNB 44 (cp 21), 52 (pl 25).

Kite, Yellow-billed (Milvus aegyptius) CDB 59 (cp 158); PSA 81 (pl 8), 128 (pl 13).

Kittiwake, Black-legged (Rissa tridactyla or Larus tridactyla) ALE 7:61; 8:207, 229; BBB 254; BBE 147; CDB 93 (cp 333); DAE 32; FBE 153; GPB 162; LAW 504; LEA 416 (bw); OBB 91; PBE 132 (pl 37), 133 (pl 38); SAO 119; WAB 134; WEA 180.

Kiwi Berry or Fruit —See **Gooseberry, Chinese** (Actinidia).

Kiwi, Brown or Common (Apteryx australis) ABW 15; ALE 7:88; BNZ 243; CAU 261; CDB 34 (cp 5); FNZ 18 (bw); GPB 34; LAW 446, 501; LEA 359 (bw); RBU 7; SLP 146; WAB 211.

Kiwi, Little Spotted or Owen's (Apteryx oweni) ALE 7:88; BNZ 253; FNZ 19 (bw); RBU 8.

Kleinia (Kleinia sp.) DFP 70 (cp 553); ECS 124, 125; EWF cp 83c; RDP 252.

Klipspringer (Oreotragus oreotragus) ALE 13:426; DSM 262; HMA 54; HMW 76; JAW 195; SMR 92 (cp 7); SMW 266 (bw); TWA 31.

Knapweed, Black or Lesser (Centaurea
nigra) BFB cp 20 (9); MBF cp 49; MFF cp 31
(919); OWF 155; PWF 85h; WFB 249 (1).

Knapweed, Brown (Centaurea jacea) EFC cp
38; MBF cp 49; NHE 240; PWF 143e; TEP 125.

Knapweed, Diffuse (Centaurea diffusa)
MCW 103; WUS 381.

Knapweed, Greater (Centaurea scabiosa)
BFB cp 20 (5); LFW 23 (cp 45); MBF cp 49; MFF
cp 31 (917); NHE 240; OWF 155; PWF
101h; WFB 249 (2).

Knapweed, Spotted (Centaurea maculosa)
MCW 104; NHE 240; WUS 383.

Knapweed, Yellow (Centaurea salonitana)
EWF cp 39f; PFE cp 156 (1503).

Knawel, Annual (Scleranthus annus) MBF cp
71; MFF pl 89 (191); NHE 184; OWF 49; PFE cp
13 (148); WFB 57 (5); WUS 163.

Knawel, Perennial (Scleranthus perennis)
MBF cp 71; NHE 74; PFE 81; PWF 81a; (S.
uniflorus) SNZ 117 (cp 359), 118 (cp 362).

Knifefish, Banded (Gymnotus carapo) HFI
91; SFW 328 (pl 82).

Knifefish, Black African (Xenomystis nigri)
ALE 4:208; CAA 19 (bw); HFI 128; HFW 101;
WFW cp 32.

Knifefish, Green (Eigenmannia virescens)
ALE 4:317; FEF 139 (bw); HFI 90; HFW 115
(bw).

Knit-bone —See **Comfrey, Common.**

Knot, European or Gray-crowned (Calidris
canutus) ABW 121; ALE 11:151; BBB
226-227; BBE 129; CDB 85 (cp 278); DPB 97;
FBE 129; FNZ 129 (pl 12); HAB 67; KBA 144
(pl 15), 148 (pl 17); OBB 75; PBE 85 (pl 28),
117 (cp 34); SNB 70 (cp 34), 72 (pl 35); WAB
131; WBV 91.

Knot, Great or Stripe-crowned (Calidris
tenuirostris) DPB 97; HAB 67; KBA 144 (pl
15), 148 (pl 17); SNB 70 (cp 34), 72 (pl 35).

Knot, Red —See **Knot, European or**

Knot-grass (Illecebrum sp.) —See **Illecebrum.**

Knot-grass, Common—See **Knotweed,
Prostrate.**

Knotweed, Himalayan (Polygonum campanulatum) DFP 166 (cp 1326); EWF cp 96a;
OGF 149; PFW 236; TVG 119.

Knotweed, Japanese (Polygonum cuspidatum) BWG 81; EPF pl 323; MFF pl 107 (535);
PWF 147a; (Reynoutria japonica) PFE cp 8
(88); WFB 43 (2).

Knotweed, Prostrate (Polygonum aviculare)
BWG 115; EGC 72; MBF cp 73; MCW 24;
MFF pl 104 (527); NHE 224; OWF 127; PFF
153; PWF 135l; WFB 41 (6); WUS 117; (P.
salicifolium) LFT cp 64.

Koala (Phascolarctos cinereus) ALE 10:51, 101
(cp 3); BMA 128-130; CAU 15; DSM 34; HMA
63; HMW 201; JAW 27; LAW 528; LEA 486
(bw), 496; LVA 55; MEA 115; MLS 60 (bw);
MWA 37 (bw); SLP 141; SMW 38 (cp 11); TWA
194; VWA cp 43.

Kob (Kobbus kob) CAA 80; CAF 99, 100 (bw);
DDM 189 (cp 32); DSM 255; (Adenota kob)
ALE 13:402.

Koel (Eudynamis scolopacea) ALE 8:369; DPB
157; KBA 192 (pl 25); SNB 122 (cp 60).

Koellikeria (Koellikeria erinoides) EGL 122;
EMB 127.

Kohleria (Kohleria sp.) BIP 109 (cp 167); EGG
121; FHP 129; PFW 130; RDP 253.

Kohlrabi (Brassica oleracea caulorapa) EVF
95; PFF 187.

Kohuhu (Pittosporum tenuifolium) GDS cp
324; HSC   90, 91, 96; SNZ 53 (cp 127).

Kokako (Callaeas cinerea) ALE 9:469; BNZ 5;
FNZ 176 (cp 16); WAB 210.

Koklas (Pucrasia macrolopha) ALE 8:54; RBI
37; WPW 49.

Kolomikta (Actinidia kolomikta) DFP 245
(cp 1957); EGV 90; HSC 11, 12 (bw).

Kongoni —See **Hartebeest, Coke's.**

Kookaburra, Blue-winged (Dacelo leachii)
CAU 157 (bw); HAB 133; RBU 180; SNB 122
(cp 60).

Kookaburra, Laughing (Dacelo gigas or D.
novaeguineae) ABW 176; ALE 9:38; 10:89;
CAU 32 (bw); CDB 126 (cp 487); GPB 209;
LAW 477; MBA 64, 65 (bw); RBU 180; SNB
122 (cp 60); WAB 189; WEA 216.

Korhaan, Black (Afrotis afra) ALE 8:134;
(Eupodotis afra) PSA 161 (cp 18).

Korikori —See **Buttercup, Hairy Alpine.**

Korrigum —See **Topi.**

Kouprey (Bos sauveli) ALE 13:341; DAA 110.

Kowhai, Yellow (Sophora microphylla) EET
245; MWF 269; SNZ cp 132 ; (S. tetraptera)
DFP 238 (cp 1904); EET 181; EWF cp 129a;
(S. macrocarpa) PFW 165.

Kowharawhara —See **Lily, Perching.**

Krait (Bungarus sp.) ALE 6:430; CAS 200; LEA 326 (bw); SSW 107.

Krait, Black-banded (Laticauda laticauda) ALE 6:430; CAS 282; HRA 51; LAW 430; SSW 115.

Kudu, Greater (Tragelaphus strepsiceros) ALE 13:299; CAF 196 (bw); DDM 180 (cp 29); DSM 242; HMA 25; HMW 69; JAW 181; MLS 396 (bw); SMR 112 (cp 9); SMW 281 (cp 174); WEA 216; (Strepsiceros strepsiceros) LEA 611 (bw); TWA 22.

Kudu, Lesser (Tragelaphus imberbis) ALE 13:302; DDM 180 (cp 29); DSM 242.

Kudzu Vine (Pueraria lobata) EGA 148; EGV 132; WUS 237; (P. thunbergiana) PFF 219; TGS 74.

Kulan —See **Ass, Asian or Indian Wild.**

Kumquat, Marumi (Fortunella japonica) BIP 87 (cp 130); MWF 135; OFP 89.

Kumquat, Nagami (Fortunella margarita) EVF 120; FHP 121; PFF 230; RDP 215; SST cp 204.

Kurrajong —See **Flame Tree of Australia.**

L

Labeo (Labeo sp.) ALE 4:352; FEF 186 (bw); SFW 293 (pl 74), 336 (cp 85).

Lablab —See **Bean, Hyacinth.**

Labrador Tea —See also **Ledum.**

Labrador Tea (Ledum groenlandicum) EWG 54; MFF pl 77 (582); PFF 272; TGS 406.

Laburnum, East African (Calpurnia sp.) EWF cp 54d; MEP 53.

Lace Flower Vine (Episcia dianthiflora) EMB 125; RDP 192.

Lace Plant, Madagascar (Aponogeton fenestralis or A. madagascariensis) HPW 272; PFW 31.

Lacebark (Hoheria sp.) DFP 206 (cp 1648); EWF cp 135e; MEP 80; OGF 159; PFW 184; SNZ 57 (cp 141), 81 (cps 229-233).

Lace-flower, Blue (Trachymene caerulea or Didiscus caeruleus) DFP 36 (cp 288); EDP 122; EGA 158; EGG 145; LFW 176 (cp 395); TGF 125.

Lacertid, Plated —See **Lizard, Sand.**

Lacewing, Brown (Hemerobius sp.) ALE 2:301; CIB 148 (cp 15).

Lacewing, Giant (Osmylus fulvicephalus) CIB 148 (cp 15); OBI 37.

Lacewing, Green (Chrysopa sp.) ALE 2:292; BIN pls 53-55; CIB 148 (cp 15); EPD 55; KIW 94 (cp 58); LAW 209; OBI 37; PEI 293 (bw); PWI 33; ZWI 120.

Lacewing, Thread (Nemoptera sp.) ALE 2:302; CIB 145 (cp 14); PEI 295 (bw).

Ladies' Fingers —See **Kidney-vetch.**

Ladies' Purses —See **Pocketbook Flower.**

Ladybird or Ladybug —See **Beetle, Ladybird.**

Ladyfish (Elop saurus) ALE 4:158; LMF 30.

Lady's Mantle (Alchemilla sp.) EWF cp 9e; MBF cp 26; OGF 35; PWF 32j, 61i, j.

Lady's Mantle, Alpine (Alchemilla alpina) MBF cp 26; MFF pl 88 (364); NHE 271; PWF 93k; WFB 107 (7).

Lady's Mantle, Common (Alchemilla vulgaris) DEW 1:299; EGH 94; EPF pl 411; MFF pl 88 (365); NHE 179; OWF 49; TEP 141; WFB 107 (6).

Lady's Slipper —See **Trefoil, Common Bird's Foot.**

Lady's Tresses, American or Irish (Spiranthes romanzoffiana) MFF pl 82 (1066); WFB 283 (7).

Lady's Tresses, Autumn (Spiranthes spiralis) EWF cp 26h; MBF cp 80; MFF pl 82 (1065); NHE 197; OWF 101; PFE cp 192 (1922); PWF 156e.

Lady's Tresses, Creeping (Goodyera repens) MBF cp 80; MFF pl 82 (1070); NHE 40; OWF 101; PFE cp 192 (1925); PWF 132d; WFB 283 (8).

Lady's Tresses, Summer (Spiranthes aestivalis) MBF cp 80; NHE 197; WFB 282.

Lady's-slipper (Cypripedium calceolus) DEW 2:279 (cp 157); DFP 6 (cp 44); EFC cp 55; EGO 107; EPF pls 961, 962; EWG 28, 82; HPW 324; MBF cp 82; MFF cp 63 (1057); MLP 323; NHE 40; PWF 27a; SPL cp 518; TEP 263; TGF 44; TVG 147; WFB 275 (1).

Lady's-slipper, Pink (Cypripedium acaule)
EGO 18; ESG 83; EWG 29; LFO 83.

Lady's-slipper, Showy (Cypripedium
reginae) DEW 2:297; DFP 62 (cps 493, 494);
ESG 106; EWG 101; PFF 391; PFW 206.

Lady's-smock —See **Cuckoo-flower.**

Lady's-thumb (Polygonum persicaria) BWG
116; DEW 1:162 (cp 76); MBF cp 73; MCW
27; MFF cp 26 (531); NHE 224; OWF 127;
PFF 154; PWF 135h, 170d; WFB 41 (3); WUS
127.

Lamb's Ears (Stachys lanata) DFP 174 (cp
1386); EGC 146; EGH 143; LFW 134 (cp
304); MWF 272; TVG 175.

Lamb's Lettuce (Valerianella locusta) BWG
131; MBF cp 43; MFF cp 7 (828); NHE 188;
OWF 177; PFE cp 136 (1309); PWF 59g;
WFB 223 (4).

Lamb's Quarters —See **Fat Hen.**

Lamb's Tails (Lachnostachys verbascifolia)
EWF cp 132c; MWF 177.

Lamb's-tails (Ptilotus sp.) EWF cp 132 e, f;
MLP 85; PFW 22.

Lamb's-tongue —See **Plantain, Hoary.**

Lammergeier —See **Vulture, Bearded.**

Lamprey, Brook (Lampetra planeri) FEF 10,
11 (bw); HFI 151; OBV 1; SFW 49 (pl 4);
WFW cp 1.

Lamprey, River (Petromyzon fluvialis or
Lampetra fluvialis) ALE 4:41; CSF 33; GAL
224 (bw); HFI 151; LEA 213 (bw); OBV 1.

Lamprey, Sea (Petromyzon marinus) ALE
4:41; CSF 33; FEF 12 (bw); GAL 225 (bw);
HFI 151; HFW 14, 15 (bw); OBV 1.

Lampshell (Terebratula sp.) ALE 3:253;
LAW 106; OIB 171.

Lancelet —See **Amphioxus.**

Langur, Capped (Presbytis pileatus) ALE
10:454; DAA 132; DSM 92; SMW 78 (bw).

Langur, Douc (Pygathrix nemaeus) ALE
10:454; DSM 93; LVS 53; MLS 155 (bw).

Langur, Sacred (Presbytis entellus) ALE
10:453, 13:285; CAA 93 (bw); CAS 159, 188
(bw); DAA 132; JAW 56; LAW 619;
(Semnopithecus entellus) HMW 16; TWA
156 (bw).

Lanner —See **Falcon, Lanner.**

Lantana, Common (Lantana camara) BIP
111 (cp 169); DEW 2:172 (cp 102); EGA 129;

EPF pl 803; FHP 130; LFW 254 (cps 564,
566); MWF 179; PFF 287; RDP 256; SPL cp
339; TGF 189.

Lantana, Trailing (Lantana montevidensis
or L. sellowiana) EGC 137; EGV 44, 45, 117;
FHP 130; MWF 179; (L. rugosa) LFT cp 140.

Lantern Tree (Crinodendron hookerianum)
DFP 194 (cp 1545); GDS cp 128; PFW 104;
SST cp 241.

Lantern-eye Fish (Photoblepharon palpebratus) ALE 5:35; HFI 64.

Lanternfish (Bathypterois ventralis) —See
Spiderfish.

Lanternfish (Myctophum sp.) ALE 4:266;
HFI 116; LAW 310.

Lanternfly (Fulgora sp.) LAW 233; PEI 124
(bw).

Lanternfly, European (Dictyophora
europaea) LAW 233; PEI 125 (bw).

Lapeyrousia (Lapeyrousia sp.) DFP 100 (cp
793); EGB 124; LFT cps 36, 37.

Laportea (Laportea sp.) DEW 1:105 (cp 55);
EPF pl 362.

Lapwing (Vanellus vanellus) ABW 120; ALE
7:385; BBB 68; BBE 113; CDB 84 (cp 274);
FBE 125; GPB 142; NBB 69 (bw); OBB 61;
PBE 93 (pl 30), 100 (cp 31); RBE 131; (V.
duvaucelii) KBA 128 (pl 13), 149 (pl 18).

Lapwing, Gray-headed (Vanellus cinereus)
DPB 81; KBA 128 (pl 13), 149 (pl 18).

Lapwing, Red-wattled (Vanellus indicus)
FBE 125; KBA 128 (pl 13), 149 (pl 18); WBV
75; (Lobivanellus indicus) ABW 120.

Larch, American or Tamarack (Larix
laricina) EET 107; MTB 128 (cp 9); PFF 120.

Larch, European (Larix decidua) BKT cps
35, 37, 38; DFP 253 (cps 2022, 2023); EET
13, 106; EPF pls 131-133; MTB 128 (cp 9);
NHE 18; OBT 100, 112; OWF 185; PFE cp 1
(3); PWF 169i; SST cp 24; TEP 268; TGS 22;
WFB cp 3.

Larch, Golden (Pseudolarix amabilis) BKT
cp 36; DEW 1:35; EEG 128; EET 106, 107;
EGE 105; EPF pl 130; MTB 128 (cp 9); TGS
22.

Larch, Hybrid or Dunkeld (Larix eurolepis)
EET 107; MTB 128 cp 9; OBT 112.

Larch, Japanese (Larix kaempferi) EET 107;
NHE 18; OBT 112; PWF 25n, 169h; (L.
leptolepis) EGE 99.

Lark, Bifasciated or Hoopoe (Alaemon alaudipes) ALE 9:179; FBE 199; WAB 153.

Lark, Black (Melanocorypha yeltoniensis) BBE 197; FBE 203; PBE 189 (cp 48).

Lark, Bush (Mirafra sp.) ALE 9:179; CDB 149 (cp 608); DPB 231; HAB 142; KBA 400 (pl 59); PSA 240 (cp 27); SAB 10 (cp 4); WAB 193.

Lark, Calandra (Melanocorypha calandra) ALE 9:179; BBE 195; DAE 78; FBE 203; PBE 189 (cp 48); RBI 235.

Lark, Crested (Galerida cristata) ALE 9:179; BBE 197; CAS 75; CDB 148 (cp 603); FBE 205; OBB 123; PBE 189 (cp 48); (G. modesta) CDB 148 (cp 605).

Lark, Desert (Ammomanes deserti) FBE 201; GPB 247; WAB 153; (A. cincturus) FBE 201.

Lark, Du Pont's (Chersophilus duponti) BBE 195; FBE 201; PBE 189 (cp 48).

Lark, Finch —See **Finch-lark.**

Lark, Horned or Shore (Eremophila alpestris) ABW 215; ALE 9:179; BBB 270; BBE 199; CDB 148 (cp 601); FBE 199; OBB 123; PBE 189 (cp 48); RBE 219; SSA pl 29; WAB 130.

Lark, Lesser Short-toed (Calandrella rufescens) BBE 195; FBE 201; PBE 189 (cp 48); WEA 219.

Lark, Magpie —See **Magpie-lark.**

Lark, Meadow —See **Meadowlark.**

Lark, Mud —See **Magpie-lark.**

Lark, Shore —See **Lark, Horned.**

Lark, Short-toed (Calandrella cinerea) BBB 272; BBE 195; CDB 148 (cp 600); FBE 201; PBE 189 (cp 48); PSA 240 (cp 27); (C. brachydactyla) ALE 9:179; OBB 123.

Lark, Sky —See **Skylark.**

Lark, Song —See **Songlark.**

Lark, Temminck's Horned (Eremophila bilopha) CAS 92 (bw); CDB 148 (cp 602); FBE 199; GPB 249; WEA 219.

Lark, Thekla (Galerida theklae) ALE 9:179; BBE 197; CDB 148 (cp 606); FBE 205; PBE 189 (cp 48).

Lark, Thick-billed (Rhamphocorys clotbey) ALE 9:179; FBE 203; (Galerida magnirostris) CDB 148 (cp 604).

Lark, White-winged (Melanocorypha leucoptera) ALE 9:179; BBE 197; FBE 203; PBE 189 (cp 48); RBE 215.

Lark, Wood —See **Woodlark.**

Larkspur (Delphinium sp.) DFP 136 (cps 1084-1089), 137 (cps 1090, 1091); EGA 109; EGG 100; EPF pl 208; EWF cps 27d, 52a, 91a; EWG 102; LFW 40 (cps 94, 95); MBF cp 4; MWF 107; NHE 210; OGF 81, 85; PFE cp 20 (213); PFM cp 20; PFW 249; PWF 99f; SPL cp 201; TGF 141; TVG 87; WFB 75 (4); WUS 179, 181, 183.

Larkspur, Candle or Alpine (Delphinium elatum) EDP 123; EFC cp 46; EGG 104; LFW 40 (cp 93), 42 (cp 96), 43 (cps 98, 99); MWF 107; NHE 268; PFE cp 20 (211).

Lasiandra —See **Glory Bush** (Tibouchina).

Latiaxis Shell (Latiaxis sp.) CTS 36 (cp 53), 39 (cp 59); GSS 77.

Laticauda —See **Krait, Black-banded.**

Laurel, Alexandrian (Danae racemosa) HSC 42; OGF 51.

Laurel, Australian —See **Tobira**

Laurel, Cherry (Prunus laurocerasus) AMP 113; EEG 111; EET 176, 177; EGE 141; EGW 135; EPF pl 416; GDS cp 333; HSC 95, 100; MWF 244; OBT 140; PFE cp 47 (483); TVG 236.

Laurel, Portugal (Prunus lusitanica) EET 177; GDS cp 334; HSC 95; OBT 140; PFE cp 48 (483); TGS 283.

Laurel, Spotted or Japanese (Aucuba japonica) DFP 182 (cp 1453); EDP 146; EEG 104; EFP 91; EGE 113; EGW 92; EJG 67, 102; ELG 134; EPF pl 749; HPW 168; HSC 14, 15; MWF 49; PFE cp 83 (847); PFW 93; SPL cp 5; TGS 182.

Laurel, Spurge —See **Spurge-laurel.**

Laurustinus (Viburnum tinus) DFP 244 (cp 1948); EGW 67; GDS cps 494, 499; HPW 259; HSC 122; OBT 141; PFE cp 134 (1300); PFM cp 186; PFW 69; TGS 343.

Lavender Cotton (Santolina chamaecyparissus) AMP 57; DFP 238 (cp 1899); EGC 145; EGE 146; EGH 139; EGP 145; EGW 38, 140; MWF 262; SPL cp 371; TGF 237; TGS 327; TVG 169.

Lavender Cotton (Santolina virens) GDS cp 440; HSC 111; (S. caespitosa) EPF pl 552.

Lavender, English or Common (Lavandula officinalis or L. spica or L. vera) AMP 109; CTH 51 (cp 76); DFP 209 (cp 1671); EDP 134; EGH 120; GDS cp 270; HSC 71, 76; OGF 105; SPL cp 393; TGF 189.

Lavender, French or Spanish (Lavandula stoechas) EGH 120; EWF cp 37g; NHE 81; PFE 108 (1110); PFM cp 153.

Lavender, Fringed or Toothed (Lavandula dentata) EGH 120; EGL 124; MWF 180.

Lawyer (Rubus sp.) EWF cp 129g; SNZ 82 (cps 235-237), 121 (cps 372, 373).

Lawyer's Wig —See **Ink-cap, Shaggy.**

Leadwort (Plumbago auriculata or P. capensis) BIP 147 (cp 222); DEW 1:164 (cp 78); EGE 140; EGG 135; EGV 43, 131; EWF cp 80e; HPW 79; LFW 174 (cp 390); MEP 104; MWF 237; PFW 232; RDP 324; SPL cp 180; (P. indica) DEW 1:162 (cp 75); MEP 104; (P. rosea) EWF cp 114a.

Leadwort, Blue (Ceratostigma plumbaginoides) DFP 189 (cp 1505); EGC 33, 127; ERW 105; EWF cp 105b; SPL cp 421; TGS 157.

Leadwort, Willmott Blue (Ceratostigma willmottianum) DFP 189 (cp 1506); GDS cp 63; HSC 26, 28; OGF 161; PFW 232.

Leaf Flower —See **Snow Bush.**

Leafbird, Gold-fronted (Chloropsis aurifrons) ABW 245; ALE 9:184, 190; CDB 156 (cp 643); KBA 273 (cp 40); RBI 291.

Leafbird, Hardwicke's or Orange-bellied (Chloropsis hardwickii) KBA 273 (cp 40); RBI 288.

Leaf-fish (Nandus nandus) ALE 5:116; SFW 541 (pl 146); (N. nebulosus) WEA 220.

Leaf-fish, African (Polycentropsis abbreviata) ALE 5:116; FEF 400 (bw); HFI 34; WFW cp 353.

Leaf-fish, Schomburgk's (Polycentrus schomburgkii) ALE 5:116; FEF 398, 399 (bw); HFI 34; SFW 541 (pl 146).

Leaf-fish, South American (Monocirrhus polyacanthus) ALE 5:116; FEF 397 (bw); HFI 34; SFW 541 (pl 146).

Leafhopper (Cicadella viridus) OBI 33; (Cicadula sexnotata) GAL 445 (bw); (Penthimia nigra) CIB 129 (cp 12).

Leaf-insect (Phyllium sp.) ALE 2:125; CTI 18 (cp 18); (Extatosoma tiaratum) BIN pls 39, 41; (Pulchriphyllium sp.) CAS 264 (bw); LEA 132 (bw).

Leaf-scraper, Gray-throated (Sclerurus albigularis) SSA cp 35; WAB 91.

Leatherjacket —See **Talang Talang.**

Leatherleaf (Chamaedaphne calyculata) NHE 70; WFB 173 (8).

Leatherwood (Cyrilla sp.) HPW 125; SST cp 100.

Lecanora (Lecanora sp.) DEW 3: cp 89; ONP 3, 53, 77, 79, 171, 173.

Lechwe —See **Waterbuck, Lechwe.**

Lecidea (Lecidea sp.) DEW 3: cp 97, 220; NFP 95; ONP 53, 71, 79, 171.

Ledum (Ledum palustre) EPF pl 476; MBF cp 55; NHE 70; TEP 79; WFB 173 (5).

Leech, Dog or Worm (Erpobdella octoculata) ALE 1:381; OIB 117.

Leech, Fish (Piscicola geometra) ALE 1:376, 381; OIB 117.

Leech, Horse (Haemopis sanguisuga) ALE 1:381; GAL 521 (bw); OIB 117.

Leech, Medicinal (Hirudo medicinalis) ALE 1:376, 381; GAL 521 (bw); LAW 103; OIB 117.

Leech, Snail (Glossiphonia sp.) ALE 1:381; OIB 117.

Leek (Allium porrum) EGH 95; EVF 96.

Leek, Few-flowered (Allium paradoxum) EPF pl 894; MFF pl 81 (1017).

Leek, Round-headed (Allium sphaerocephalon) MFF cp 32 (1010); NHE 191; PFE cp 165 (1604); PWF 129i; WFB 265 (10).

Leek, Sand (Allium scorodoprasum) MBF cp 84; MFF cp 32 (1009); NHE 241; WFB 265 (7).

Leek, Three-cornered (Allium triquetrum) MBF cp 85; MFF pl 81 (1016); PFE cp 165 (1610); PFM cp 248; WFB 265 (2).

Leek, Wild (Allium ampeloprasum) MFF pl 65 (1008); OFP 169; PFE cp 166 (1615); PWF 131g; WFB 265 (8).

Lelwel —See **Hartebeest, Bubal or Red.**

Lemmaphyllum (Lemmaphyllum microphyllum) EGF 119; EMB 121.

Lemming (Dicrostonyx torquatus) SMW 131 (cp 73); (Ellobius talpinus) HMW 168; (Myopus schisticolor) DAE 104.

Lemming, Norwegian (Lemmus lemmus)
ALE 11:300; BMA 131, 132; CEU 246; DAE
104; DSM 126; GAL 56 (bw); HMA 64; HMW
168; LAW 587; LEA 538 (bw); TWA 94; WEA
221.

Lemming, Steppe (Lagurus lagurus) MLS
212 (bw); WEA 221.

Lemon (Citrus limon) AMP 29; DEW 2:41;
EET 222; EPF pl 658; EVF 119; HPW 203;
MWF 86; OFP 85; PFE cp 67 (693); PFF 228;
SST cp 192.

Lemon Peel (Otidea onotica) DEW 3:121;
DLM 125; ONP 151.

Lemon Vine —See **Barbados Gooseberry**
(Pereskia).

Lemur, Avahi (Avahi laniger) ALE 10:251
(cp 5); LVS 54-55.

Lemur, Black (Lemur macaco) ALE 10:250
(cp 4); DSM 65; HMW 31; LAW 607; LVA 25;
LVS 57; SMW 47 (cp 35).

Lemur, Brown (Lemur fulvus) ALE 10:250
(cp 4), 283; MLS 116 (bw).

Lemur, Fat-tailed Dwarf (Cheirogaleus
medius) ALE 10:248 (cp 2); WEA 222.

Lemur, Flying —See **Colugo**.

Lemur, Fork-marked Mouse (Phaner
furcifer) ALE 10:248 (cp 2); PWC 232 (bw).

Lemur, Gray Gentle (Hapalemur griseus)
ALE 10:249 (cp 3); DSM 66; LAW 606.

Lemur, Greater Dwarf (Cheirogaleus major)
ALE 10:248 (cp 2); DSM 66.

Lemur, Lesser Mouse (Microcebus murinus)
ALE 10:248 (cp 2), 263, 283; DSM 66; LAW
609; LVS 60-61; MLS 117 (bw); SMW 67
(bw).

Lemur, Mongoose (Lemur mongoz) DSM 64;
LVS 56; PWC 232 (bw).

Lemur, Ring-tailed (Lemur catta) ALE
10:250 (cp 4), 283; CAA 88 (bw); DSM 64;
HMW 31; LEA 501 (bw); MLS 114 (bw);
TWA 58 (bw).

Lemur, Ruffed (Lemur variegatus) ALE
10:250 (cp 4); DSM 65; HMW 31; MLS 115
(bw); (L. v. ruber) LVS 62.

Lemur, Sportive or Weasel (Lepilemur
mustelinus) ALE 10:249 (cp 3); LAW 608;
LVS 58; (L. ruficaudatus) ALE 10:249 (cp 3).

Lenten Rose (Helleborus orientalis) DFP 146
(cp 1166); MWF 152; PFW 250.

Lentil (Lens culinaris) NHE 212; OFP 43; PFF
221.

Lentil, Wild —See **Milk-vetch, Yellow**.

Lentinellus (Lentinellus cochleatus) DLM
214; LHM 103; REM 142; RTM 128.

Lentinus (Lentinus sp.) DLM 214, 215; LHM
107; NFP 71; PUM cps 44, 45; REM 142 (cp
110); RTM 126, 128; SMG 144.

Lentisc —See **Mastic Tree**.

Lenzites (Lenzites sp.) DEW 3:164; DLM 154;
NFP 76; PFF 80; RTM 209.

Leopard (Panthera pardus or Felis pardus)
ALE 10:31; 12:337, 367; BMA 133-135; CAA
102 (bw), 104; CAF 127 (bw), 163; CAS 180
(bw); DAA 58; DDM 140 (cp 23); HMA 37;
HMW 120; JAW 134; LAW 545; LEA 570
(bw), 576; LVA 23; LVS 88; MLS 320 (bw);
SLP 221; SMR 80 (cp 5); SMW 178 (cps 99,
100); TWA 45.

Leopard, Clouded (Felis nebulosa or
Neofelis nebulosa) ALE 12:355; DAA 116;
HMW 123; JAW 139; LVS 82; MLS 317 (bw);
TWA 172.

Leopard Flower —See **Lily, Blackberry**.

Leopard, Hunting —See **Cheetah**.

Leopard, Snow (Felis uncia or Panthera
uncia) DAA 52; HMA 37; HMW 121; LVA
39; LVS 86; MLS 322 (bw); SLS 129; (Uncia
uncia) ALE 12:339; JAW 140.

Leopard's Spots (Andromischus maculatus)
EDP 136; EFP 86; (A. saxicola) LCS cp 214;
(A. trigynus) LCS cp 215; (A. umbraticola)
LFT 151 (cp 75).

Leopard's-bane (Doronicum sp.) DFP 140
(cps 1113, 1114); EEG 133; EPF pls 565, 567;
MWF 116; NHE 286; OGF 39; TGF 284; TVG
89.

Leopard's-bane, Clusius' (Doronicum
clusii) EPF pl 564; NHE 286.

Leopard's-bane, Great (Doronicum
pardalianches) LFW 51 (cp 119); MBF cp 46;
MFF cp 56 (853); NHE 286; PFE cp 151
(1447); PWF 31i; WFB 237 (9).

Leopard's-bane, Large-flowered (Doronicum grandiflora) KVP cp 137; NHE 286;
PFE cp 150 (1448).

Lepiota, Crested (Lepiota cristata) DLM
216; LHM 127; NFP 58; REM 78 (cp 46);
RTM 38; (L. excoriata) CTM 25; DLM 217;
LHM 123; PUM cp 146; RTM 37.

Leporinus, Banded (Leporinus fasciatus)
ALE 4:291; FEF 119; HFW 113.

Leschenaultia, Blue (Leschenaultia biloba)
EWF cp 139b; MWF 182.

Lettuce, Acrid or Greater Prickly (Lactuca
virosa) MBF cp 53; NHE 241.

Lettuce, Blue (Lactuca perennis) NHE 240;
PFE cp 160 (1541); WFB 251 (7).

Lettuce, Blue (Lactuca pulchella) MCW 119;
WUS 425.

Lettuce, Garden or Common (Lactuca
sativa) EMB 149; EVF 96, 97; OFP 151; PFF
333.

Lettuce, Hare's or Purple (Prenanthes
purpurea) KVP cp 117; NHE 37; PFE cp 160
(1545); WFB 251 (8).

Lettuce, Least (Lactuca saligna) MFF cp 53
(939); NHE 240; WFB 253 (6).

Lettuce, Prickly (Lactuca serriola) BFB cp
21; MBF cp 53; MFF cp 53 (938); NHE 240;
PWF 142c; TEP 202; WFB 253 (5); WUS 427.

Lettuce, Wall (Mycelis muralis) MBF cp 53;
MFF cp 53 (940); NHE 237; PWF 95e; WFB
258 (4).

Lettuce, Wild or Prickly (Lactuca scariola)
MCW 120; PFF 334.

Leucopaxillus (Leucopaxillus sp.) LFO 86,
87; PUM cp 49; SMG 148.

Liberty Cap (Psilocybe semilanceata) DLM
242; LHM 149; ONP 33; (P. sp.) DLM 241,
242; EHP 66, 67; LHM 149.

Lichen (Parmelia sp.) DEW 3: cps 92, 94; 222;
EPF pl 78; NFP 97; NHE 44; ONP 63, 161,
163, 164; PFF 74; TEP 227.

Lichen (Umbilicaria sp.) DEW 3: cp 86; NFP
97, 98; ONP 67; PFF 74; (Verrucaria sp.)
DEW 3:216; NFP 95; ONP 7, 77.

Lichen (Usnea sp.) DEW 3: cps 91, 93; NFP 98;
NHE 44; ONP 159; PFF 73, 74; SAR 28;
(Xanthoria sp.) DEW 3: cps 90, 95; NHE 44;
ONP 3, 77.

Lichen, Dog (Peltigera canina) DEW 3: cp 82;
219; NFP 95; ONP 29; TEP 227.

Lichen, Haematomma (Haematomma sp.)
DEW 3: cp 88; NFP 95; ONP 71.

Lichen, Horsehair (Alectoria sp.) NFP 98;
ONP 159.

Lichen, Map (Rhizocarpon geographicum)
DEW 3: cp 81; NFP 96; ONP 71; (R. sp.) ONP
53, 71.

Lichen, Orange (Caloplaca sp.) DEW 3: cp 97;
NFP 96; ONP 3, 77.

Lichen, Pink Earth (Baeomyces roseus)
DEW 3:220; NFP 96; ONP 53; (B. rufus)
ONP 53.

Lichen, Sticta (Sticta sp.) AMP 89; DEW
3:219; ONP 166.

Licorice —See **Liquorice.**

Lifeplant (Bryophyllum sp.) DEW 1:28 (cp
166); RDP 110.

Lignon-berry —See **Cowberry.**

Lignum Vitae (Guaiacum officinale) EWF cp
178e; MEP 64; SST cp 124.

Ligularia (Ligularia sp.) DFP 155 (cps
1236-1240), 172 (cp 1376); EWF cp 106 a, b;
OGF 121; TGF 285.

Lilac, Amur or Japanese (Syringa
amurensis) EGT 145; TGS 374.

Lilac, Chinese or Rouen (Syringa chinensis)
EPF pl 669; TGS 372.

Lilac, Common (Syringa vulgaris) DFP 241
(cps 1922-1926); EDP 120; EFS 22, 72, 73,
140; ELG 119; EPF pl 667; EPG 147; GDS
cps 473, 477; HPW 226; HSC 118; LFW 82
(cps 189, 190); MWF 278; OBT 196; PFF 281;
PFW 200; PWF 39j; SPL cp 120; TGS 374;
TVG 247.

Lilac, Littleleaf (Syringa microphylla) GDS
cp 470; HSC 118, 120; MWF 278; TGS 373.

Lilac, Persian (Syringa persica) EFS 140;
TGS 374; (S. laciniata) EWF cp 95e.

Lily, African Corn (Ixia sp.) BIP 105 (cp 159);
EGB 123; EWF cp 87; MWF 168; SPL cp 219.

Lily, Alpine —See **Lily, St. Bruno's.**

Lily, Arum —See **Calla Lily** (Zantedeschia).

Lily, Amazon (Eucharis grandiflora) BIP 77
(cp 116); DFP 66 (cp 525); EGB 113; EPF pl
930; FHP 120; MEP 21; PFW 26.

Lily, Barbados (Hippeastrum equestre) LFW
67 (cp 155); MEP 19; MWF 155; SPL cp 290.

Lily, Belladonna (Amaryllis belladonna)
BIP 25 (cp 15); DEW 2:219 (cp 134); DFP 84
(cp 670); EGB 94; EWF cp 86c; MWF 39; OGF
175; PFW 24; SPL cp 281.

Lily, Berg (Galtonia) —See **Hyacinth, Giant
Summer.**

Lily, Blackberry (Belamcanda chinensis)
EGB 97; MEP 23; SPL cp 191; TGF 46.

Lily, Black-stick (Vellozia retinervis) EWF
cp 88b; LFT cp 35.

Lily, Blood (Haemanthus coccineus) EGB
119; EPF pl 926, cp XXII; MWF 146; PFW
25; (H. multiflorus) FHP 124; MEP 19;
MWF 146; SPL cp 286.

Lily, Blood (Haemanthus katherinae) BIP 93
(cp 143); DEW 2:217 (cp 130); DFP 68 (cp
542); EGL 116; EPF pl 927; (H. magnificus)
LFT cps 26, 27; (H. albiflos) EWF cp 86d;
RDP 231.

Lily, Blue African (Agapanthus africanus)
DEW 2:235; EGB 91; MEP 15; MWF 32; OGF
175; SPL cp 37; TGF 3.

Lily, Blue African (Agapanthus sp.) DFP 120
(cp 953); EPF pl 888; FHP 103; LFT 26 (cp
12); 27 (cp 13); PFW 20; TVG 77.

Lily, Bugle (Watsonia sp.) DFP 118 (cp
942); EGB 146; LFT cp 39; MEP 23; MWF
296.

Lily, Calla —See **Calla Lily.**

Lily, Chatham Island—See **Forget-me-not,
Chatham Island.**

Lily, Corsican or Spirit (Pancratium
illyricum) DFP 109 (cp 872); SPL cp 304.

Lily, Crinum (Crinum sp.) DEW 2:247; DFP
87 (cp 693); EGB 106; EWF cp 66c, cp 143b;
FHP 114; LFT 59 (cp 29); MEP 18; MWF 96;
OGF 175; PFW 24; RDP 160; SPL cp 284.

Lily, Cuban—See **Squill, Peruvian.**

Lily, Cunjevoi or Spoon (Alocasia macrorhiza) MWF 36; WAW 124.

Lily, Day—See **Day Lily.**

Lily, Easter or Bermuda (Lilium longiflorum) EDP 125; EGB 126; FHP 131; LFW 84
(cp 196); MLP 258; SPL cp 298.

Lily, Fire or Ifafa (Cyrtanthus sp.) HPW 315;
LFT 66 (cp 32); MWF 102.

Lily, Foxtail (Eremurus sp.) DFP 140 (cps
1116, 1117); EGB 112; EPF pl 882; EWF cp
50a, b; MWF 123; PFW 171; TGF 29; TVG
193.

Lily, Glory or Climbing (Glorisa superba)
BIP 91 (cp 139); CAF 186; GFI 103 (cp 22);
LFT cp 9; LFW 165 (cp 368); MEP 16; MLP
261; (G. rothschildiana) DEW 2:212 (cp
120); EGB 119; FHP 123; PFW 176; SPL cp
166; (G. sp.) EWF cp 66 b; LFT cp 9.

Lily, Golden Spider or Hurricane (Lycoris
aurea) MEP 20; MWF 191.

Lily, Golden-rayed (Lilium auratum) DFP
100 (cp 797); LHP 190; MWF 184; PFW 169;
SPL cp 295.

Lily, Guernsey (Nerine sarniensis) BIP 129
(cp 193); DFP 71 (cp 564); EGB 133; MEP
18; MWF 212; PFW 26.

Lily, Hawaiian—See **Chalice Vine.**

Lily, Henry's (Lilium henryi) DFP 101 (cp
808); LFW 87 (cp 200), 88 (cp 203).

Lily, Kaffir (Clivia miniata) BIP 57 (cp 74);
DEW 2:246; DFP 60 (cp 474); EGB 63, 103;
EPF pl 929; FHP 113; HPW 315; LHP 62, 63;
MEP 19; MWF 88; PFW 24; RDP 149, 150;
SPL cp 283; (C. caulescens) LFT 67 (cp 33).

Lily, Kaffir or River (Schizostylis
coccinea) DFP 110 (cps 876, 877); EGB 137;
LFT cp 40; OGF 173; PFW 149.

Lily, Kerry (Simethis planifolia) MBF cp 84;
MFF pl 80 (988); WFB 261 (4).

Lily, Leopard—See **Cowslip, Cape.**

Lily, Madonna (Lilium candidum) AMP 169;
DFP 101 (cp 803); EGB 128; EPF pl 898;
EWF pl 41b; MWF 184; PFF 373; PFW 167;
SPL cp 297; TGF 29.

Lily, Margaton or Turk's-cap (Lilium
margaton) DFP 102 (cp 812); EFC cp 28;
EGB 126; EPF pl 901, 902; ESG 124; HPW
313; KVP cp 122; MFF cp 32 (995); NHE 38;
OWF 163; PFE cp 167 (1618); PFW 169;
PWF 157e; SPL cp 299; TEP 259; WFB 263
(3).

Lily, Mariposa—See **Globe Tulip.**

Lily, May (Maianthemum bifolium) MBF cp
84; MFF pl 80 (992); NHE 38; TEP 261; WFB
260, 261 (2).

Lily, Mount Cook—See **Buttercup, Giant
or Mountain** (Ranunculus).

Lily of the Field—See **Daffodil, Winter.**

Lily, Orange (Lilium bulbiferum) EFC cp 28;
PFE cp 166 (1620); RWF 26; SPL cp 296.

Lily, Perching (Astelia solandri) EWF cp
141d; SNZ 72 (cps 196-199).

Lily, Peruvian (Alstroemeria aurantiaca)
DFP 120 (cp 957); EGB 93; MWF 38; OGF
173; PFW 21; TGF 29; (A. ligtu) DFP 120 (cp
958); PFW 21; (A. pelegrina) SPL cp 280; (A.
violacea) RWF 83.

Lily, Plantain (Hosta sp.) DEW 2:238; DFP 147 (cps 1175, 1176), 148 (cps 1177, 1179); EEG 135; EJG 120; EPF pl 883; ESG 117; LFW 83 (cp 191); MWF 156; OGF 73; PFF 378; SPL cp 292; TGF 28; TVG 101.

Lily, Poor Knight's (Xeronema callistemon) EWF cp 141c; SNZ 163 (cp 518).

Lily, Red (Lilium pomponium) NHE 294; PFE cp 166 (1619).

Lily, Red Spider (Lycoris radiata) MWF 191; SPL cp 432.

Lily, St. Bernard's (Anthericum lilago) DFP 122 (cp 971); EGB 96; EPF pl 884; KVP cp 171; NHE 38; PFE cp 164 (1596); PFF 379; WFB 261 (5).

Lily, St. Bruno's (Paradisea liliastrum) EFC cp 27; EGB 135; EPF pl 881; PFE cp 163 (1595).

Lily, Scarborough (Vallota speciosa) BIP 179 (cp 279); DFP 118 (cp 941); EGB 146; FHP 150; MWF 291; PFW 26; RDP 390.

Lily, Sea (Pancratium maritimum) EGB 135; EWF cp 42c; PFE cp 174 (1673); SPL cp 305; (P. trianthum) EHP 29.

Lily, Snowdon (Lloydia serotina) EWF cp 7d; MBF cp 86; MFF pl 84 (999); NHE 290; PFE cp 164 (1629); PWF 51 b; WFB 261 (3); (L. graeca) PFM cp 239.

Lily, Spice or Galanga (Kaempferia galanga) DEW 2:274 (cp 146); EHP 29.

Lily, Spider —See **Spider-lily.**

Lily, Spire (Galonia)—See **Hyacinth, Giant Summer.**

Lily, Spirit—See **Lily, Corsican.**

Lily, Toad (Tricyrtis stolonifera) DFP 176 (cp 1402); EWF cp 107b; (T. sp.) ESG 84, 144; TGF 47.

Lily Torch—See **Red Hot Poker.**

Lily Tree (Magnolia denudata) DEW 1:74, 75; DFP 211 (cp 1683); EET 161; EGT 124; EPF pl 177; MWF 193; OBT 144; TGS 278.

Lily, Triplet —See **Grass Nut.**

Lily, Yellow Turk's-cap (Lilium pyrenaicum) DFP 102 (cp 816); EWF cp 41c; MBF cp 85; MFF pl 63 (996); PFE 167 (1619); PWF 50e.

Lily-of-the-Nile —See **Calla Lily** (Zantedeschia).

Lily-of-the-valley (Convallaria majalis) AMP 77; BFB cp 23 (8); BIP 61 (cp 81); CTH 24 (cp 16); DEW 2:236; EDP 124; EEG 140; EGB 105; EGC 127; EPF pl 921; ESG 104; HPW 313; KVP cp 63; LFW 36 (cp 87); MBF cp 84; MFF pl 80 (989); MWF 92; NHE 38; OGF 51; PFE cp 171 (1657); PFF 377; PFW 171; PWF 31k, 168c; SPL cp 256; TEP 259; TGF 13; TVG 147; WFB 261 (1).

Lily-of-the-valley Shrub —See **Andromeda, Japanese.**

Lily-trotter —See **Jacana, Comb-crested.**

Lily-turf (Ophiopogon jaburan) MWF 218; RDP 282; (O. japonicus) EEG 145; EGC 141; EJG 133; ESG 132; TGF 36.

Lily-turf, Blue (Liriope muscari) EEG 144; MWF 187; RDP 257; (L. graminifolia) OGF 173.

Lily-turf, Creeping (Liriope spicata) EGC 138; EGW 124.

Limacella (Limacella sp.) LHM 121; PUM cp 147; RTM 32, 261, 262; SMG 183, 184.

Lime (Citrus aurantifolia) EWF cp 119d; OFP 87; SST cp 191.

Lime (Tilia) —See also **Linden.**

Lime, Caucasian (Tilia euchlora) MTB 352 (cp 35); OBT 192.

Lime, Common (Tilia europaea) BKT cp 147; CTH 36 (cp 45); DFP 242 (cp 1929); EET 35, 194; MBF cp 18; MFF pl 71 (222); MTB 353 (cp 36); TGS 192.

Lime, Silver Pendant (Tilia petiolaris) EET 195; HPW 91; MTB 353 (cp 36); OBT 192.

Limodore (Limodorum abortivum) KVP cp 178; NHE 41; PFE cp 191 (1921); PFM cp 288; WFB 287 (7).

Limpet (Patella sp.) ALE 3:45, 74; CSS cp 17; GSS 27; LAW 119; LEA 87 (bw); MOL 66; OIB 35; (Patiniger polaris) SAR 51.

Limpet, Blue-rayed (Patina pellucida) CSS cp 17; OIB 35.

Limpet, Keyhole (Diodora apertura) CSS cp 17; OIB 35; (D. italica) ALE 3:45.

Limpet, River (Ancylastrum fluviatile or Ancylus fluviatilis) ALE 3:80; GAL 513 (bw); OIB 71.

Limpet, Slipper —See **Slipper Shell.**

Limpet, Slit (Emarginula reticulata) CSS cp 17; OIB 35.

Limpet, Tortoise-shell (Acmaea sp.) CSS cp 17; GSS 27; MOL 66; OIB 35.

Limpkin (Aramus guarauna) ABW 106, 109; ALE 8:100; CDB 73 (cp 223); GPB 128; SSA pl 23; WAB 68.

Linanthus (Linanthus sp.) DEW 2:111 (cp 71); EWF cp 163e; HPW 232.

Linden (Tilia) —See also **Lime.**

Linden, Common (Tilia vulgaris) AMP 91; OBT 192; OWF 189; PWF 109p, 159k.

Linden, Large-leaved (Tilia platyphyllos) BKT cps 145, 146; EET 195; EPF pl 604; HPW 91; MBF cp 18; MTB 353 (cp 36); NHE 21; OBT 13, 16; PWF 159n; SST cp 178; TEP 285; TGS 192.

Linden, Silver (Tilia tomentosa) EET 195; EGT 146; EPF pl 603; TGS 183.

Linden, Small-leaved (Tilia cordata) EEG 123; EET 195; EGT 145; ELG 112; EPG 150; MBF cp 18; MFF pl 70 (221); MTB 353 (cp 36); NHE 21; OBT 13, 16; OWF 189; PFE 239; PWF 159l; SST cp 177; TEP 285; TGS 183; WFB 36, 37 (3).

Lindernia (Lindernia sp.) NHE 99; WFB 303.

Ling (fish) (Molva molva) CSF 115; HFI 78; OBV 39.

Ling (plant) —See **Heather.**

Linnet (Acanthis cannabina or Carduelis cannabina) ALE 9:391; BBB 108; BBE 283; CDB 207 (cp 915); FBE 291; GAL 118 (cp 1); LAW 471; OBB 179; PBE 260 (cp 61); RBE 300.

Linsang, African (Poiana richardsoni) ALE 12:147; DDM 92 (cp 15).

Linsang, Banded (Prionodon linsang) ALE 12:138; DAA 120; DSM 180.

Linsang, Spotted (Prionodon pardicolor) ALE 12:138; DAA 76; DSM 180.

Lion (Panthera leo or Felis leo) ALE 12:337; BMA 136-140; CAA 98-99, 103 (bw), 104, 105; CAF 146-147, 154-155 (bw), 286-287; CAS 165 (bw); DAA 60-61; DDM 140 (cp 23); DSM 194; HMA 8, 65; HMW 118; JAW 138; LAW 543; LEA 572 (bw), 577; LVA 39; MLS 318 (bw); PWC 136-137, 152 (bw); SLP 184, 185, 217-219; SMW 177 (cp 98); TWA 56; VWA cp 24.

Lion, Mountain —See **Cougar.**

Lionfish (Pterois sp.) ALE 5:59; CAA 16 (bw); FEF 501 (bw); HFI 59; HFW 224 (cps 117, 118), 249 (bw); LAW 335; LEA 223 (bw); VWS 7; WFW cps 244-247.

Lion's Ear (Leonotis leonurus) DEW 2:174 (cp 106); EWF cp 63b; MEP 124; MWF 181; OGF 105; PFW 154; SPL cp 221; (L. microphylla) LFT cp 141.

Lion's Foot —See **Edelweiss.**

Lion's Tooth —See **Dandelion** (Taraxacum).

Lion's-eye (Wormskioldia sp.) HPW 104; LFT cp 111.

Lipfern (Cheilanthes sp.) CGF 151, 153, 155; EGF 103; EPF 108; NFP 144, 145.

Lippia (Lippia sp.) LFT cp 140; PFE 341 (1086).

Lipstick Plant (Aeschynanthus lobbianus) BIP 21 (cp 8); EGV 91; (A. longiflorus) EWF cp 122a; (A. parviflorus) FHP 103; (A. pulcher) EGG 86; HPW 247; (A. specious) DEW 2:170 (cp 96); EGL 82; PFW 128; RDP 74; (A. tricolor) EWF cp 122c.

Liquorice (Glycyrrhiza glabra) AMP 105; CTH 34; OFP 135; PFE 192.

Liquorice, Wild (Astragalus glycphyllos) BFB cp 8 (8); MBF cp 24; MFF pl 79 (315); NHE 28; OWF 23; PFE cp 54 (527); PWF 99g; TEP 243; WFB 119 (4).

Litchi or Lychee (Litchi chinensis) BIP 113 (cp 173); HPW 193; OFP 105; SST cp 207.

Little Darling —See **Mignonette.**

Little Robin —See **Herb Robert, Lesser.**

Live-forever —See **Stonecrop.**

Livelong —See **Orpine.**

Liverwort (Pellia sp.) EPF pl 83; NFP 114; ONP 101.

Liverwort, Common (Marchantia polymorpha) DEW 3: cp 99; 231, 232; EPF pl 85; LFL 25; NFP 113; NHE 109; ONP 99; PFF 90.

Liverwort, Great Scented (Conocephalum conicum) DEW 3: cp 100; 233; NFP 113; ONP 101; PFF 90.

Living Stones (Conophytum sp.) DEW 1:196; ECS 108, 109; EPF cp VIIIa; EWF cp 73j, k; LCS cp 220-227; PFW 18; SCS cp 74.

Living Stones (Lithops sp.) DFP 70 (cp 555); ECS 12, 127; EWF cp 73a, b, c; HPW 66; LCS cps 277-279; LFT cp 65; PFW 18; RDP 257; RWF 59; SCS cps 92, 93.

Lizard (Diploglossus sp.) ALE 6:300; HWR
   97; MAR 139 (cp X).
Lizard, Agama (Agama agama) ALE 6:237;
   ERA 155; HRA 32; HWR 5; LAW 415; (A.
   atricollis) SIR 81 (bw).
Lizard, Agama (Phrynocephalus sp.) ALE
   6:210, 237; CAS 87 (bw), 107; HRA 32; MAR
   102 (pl 42); SIR 83, 84 (bw).
Lizard, Angle-headed (Gonyocephalus sp.)
   COG cp 17; HRW 97; MEA 16; SIR 87 (bw).
Lizard, Armadillo (Cordylus cataphractus)
   ALE 6:261; MAR 124 (pl 50); SIR 106 (cp 47).
Lizard, Bearded (Amphibolurus barbatus)
   ALE 6:209, 237; ART 204; COG cp 20; ERA
   160, 161; HRA 36; HWR 6; LAW 414; LBR
   44; LEA 304 (bw); MEA 15, 24; SIR 80 (bw);
   WAW 76 (bw); WEA 38.
Lizard, Brazilian Tree (Enyalius catenatus)
   ALE 6:193; SIR 121 (bw).
Lizard, Burrowing —See Worm-lizard,
   Wiegmanni's.
Lizard, Caiman (Dracaena guianensis) ALE
   6:277; MAR 136 (pl 56).
Lizard, Cape Snake (Chamaesaura anguina)
   ALE 6:268; FSA 68 (pl 1).
Lizard, Changeable (Calotes sp.) ALE 6:211,
   237; ERA 154; HRA 34; LEA 284 (bw).
Lizard, Common Flying (Draco volans) ALE
   6:237; HRA 34.
Lizard, Common or Viviparous (Lacerta
   vivipara) ALE 6:278, 291; 13:463; ART 106,
   186; DAE 147; ERA 200, 201; HRA 11; LBR
   77; LEA 309 (bw); OBV 123; SIR 112 (cp 58);
   (L. pityuensis) HRW 99; WEA 381.
Lizard, Corfu (Algyroides nigropunctatus)
   ALE 6:299; HRA 14.
Lizard, Eyed or Jeweled (Lacerta lepida)
   ALE 6:299; CAA 115 (bw); DAE 147; HRA
   13; LBR 80, 81; MAR 142 (pl 58); SIR
   110-111 (cp 56); VRA 199.
Lizard, Fin-tailed or Sail (Hydrosaurus
   amboinensis) ALE 6:237; HRA 35; MAR 109
   (pl 43); SIR 78 (bw).
Lizard, Frilled (Chlamydosaurus kingi) ALE
   6:204, 209; ART 205; CAU 147; COG cp 15;
   ERA 168; HRA 36; LBR 41; LEA 288 (bw);
   MAR 110 (pl 44); MEA 21; MWA 61 (bw);
   SIR 82, 88 (bw); VRA 169.
Lizard, Fringe-toed (Acanthodactylus sp.)
   ALE 6:291, 299; HRA 14; SIR 142 (bw).

Lizard, Glass (Ophisaurus sp.) ALE 6:300;
   CAS 74 (bw); ERA 174, 175; HRA 25; LAW
   421; LEA 315 (bw); MAR 125 (pl 51); SIR 108
   (cp 51); VRA 191.
Lizard, Green (Lacerta viridis) ALE 6:278;
   DAE 145; ERA 176, 177; HRA 13; LBR 120,
   121 (cp 78); LEA 308 (bw); MAR 141 (pl 57);
   SIR 144 (bw); VRA 197.
Lizard, Horned Agama (Ceratophora sp.)
   ALE 6:210; ART 207.
Lizard, Legless (Lialis burtonis) ALE 6:179;
   COG cp 12; ERA 183; HRW 122; MEA 30;
   WAW 73 (bw); (Delma sp.) COG cp 14; MWA
   56 (bw).
Lizard, LeSueur's Water (Physignathus
   lesueuri) ALE 6:237; COG cp 19; HRA 35;
   (Hydrosaurus lesueuri) SIR 67 (cp 27).
Lizard, Monitor —See Monitor.
Lizard, Patagonian (Diplolaemus darwinii)
   ALE 6:194; SIR 121 (bw).
Lizard Plant —See Chestnut Vine.
Lizard, Plated Rock (Gerrhosaurus validus)
   ALE 6:260, 268; HWR 51; (G. major) SIR 112
   (cp 59); (G. nigrolineatus) MAR 124 (pl 50).
Lizard, Ruin (Lacerta sicula) ALE 6:278;
   HRA 12.
Lizard, Sand (Lacerta agilis) ALE 6:278, 291;
   10:223; DAE 147; ERA 192, 193; GAL 208
   (bw); HRA 11; LAW 419; LBR 80; LEA 306
   (bw); OBV 123.
Lizard, Sand (Psammodromus sp.) ALE
   6:299; DAE 147; HRA 14; SIR 143 (bw).
Lizard, Sharp-headed (Lacerta oxycephala)
   ALE 6:293; HRA 12.
Lizard, Shingle-back —See Skink,
   Shingle-back.
Lizard, Spiny-tailed or Palm (Uromastyx
   sp.) ALE 6:237; ART 153; CAS 89, 95; HRA
   33; LEA 303; SIR 62 (cp 26); VRA 167.
Lizard, Starred Agama (Agama stellio) ALE
   6:210; ART 110; CAF 98 (bw), 169; HRA 31;
   LEA 302 (bw); SIR 63 (cp 28).
Lizard, Wall (Lacerta muralis) ALE 6:278;
   DAE 147; ERA 202; HRA 11; LEA 307 (bw);
   SIR 144 (bw); (Podarcis muralis) ART 155.
Lizard, Worm —See Worm-lizard.
Llama (Lama glama) ALE 13:133; CSA 208;
   DSM 225; HMW 57; MLS 376 (bw); SMW 238
   (cp 151); TWA 135; (L. peruana) JAW 170.

Loach, Blue (Botia modesta) FEF 219 (bw); HFI 106; SFW 285 (cp 72).

Loach, Clown (Botia macracantha) CTF 31 (cp 28); FEF 189; HFI 106; HFW 39 (cp 15); LEA 246 (bw); SFW 284 (cp 71).

Loach, Coolie (Acanthophthalmus kuhlii) ALE 4:372; CTF 31 (cp 29); FEF 214, 215 (bw); HFI 106; (A. semicinctus) SFW 284 (cp 71).

Loach, Hora's Clown (Botia horae) ALE 4:372; SFW 324 (pl 81); WFW cp 149.

Loach, Pond (Misgurnus fossilis) ALE 4:327, 371; FEF 212 (bw); FWF 105; GAL 226 (bw); HFI 105; SFW 320 (pl 79).

Loach, Spined (Cobitis taenia) ALE 4:237, 371; CTF 32 (cps 30, 31); FEF 213 (bw); FWF 103; GAL 227 (bw); HFI 105; OBV 93; SFW 320 (pl 79).

Loach, Stone (Nemacheilus barbatulus) ALE 4:237, 371; FEF 211 (bw); FWF 103; GAL 244 (bw); HFI 105; OBV 101; SFW 320 (pl 79); WFW cp 151.

Loach, Sucker (Gyrinocheilus aymonieri) ALE 4:362; CTF 30 (cps 26, 27); FEF 210; HFI 105; SFW 292 (pl 73).

Loach, Tiger (Botia hymenophysa) ALE 4:372; HFI 106; SFW 285 (cp 72); WEA 230.

Lobelia (Lobelia deckenii) DEW 2:196; EWF cp 56a; RWF 51.

Lobelia (Lobelia sp.) AMP 97; DEW 2:195; EHP 153; EWF cp 55b; LFT cps 166, 167; MLP 241; MWF 188; OGF 121; PFW 63; SNZ 33 (cp 58); WFB 291 (7).

Lobelia, Blue (Lobelia urens) MBF cp 54; MFF cp 7 (806); WFB 227 (10).

Lobelia, Common (cultivated) (Lobelia erinus) DFP 42 (cps 330-332); EGA 132; LFW 90 (cp 207); MWF 188; SPL cp 456; TGF 236.

Lobelia, Water (Lobelia dortmanna) MBF cp 54; MFF pl 66 (807); NHE 100; PWF 115c; WFB 291 (8).

Lobster Claw —See also **Flaming Sword.**

Lobster Claw (Heliconia humilis) GFI 95 (cp 20); MEP 25; MWF 151.

Lobster, European (Homarus vulgaris) ALE 1:452, 460; CSF 198; CSS cp 14; LEA 174 (bw); MOL 82; OIB 123.

Lobster, Norway or King (Nephrops norvegicus) ALE 1:451, 460; CSF 200.

Lobster, Spiny (Palinurus sp. or Panulirus sp.) ALE 1:454, 460, 473; CSF 201; CSS cp 14; LAW 146; LEA 174; MOL 82; OIB 123; VWS 65.

Locust, Black (tree) (Robinia pseudoacacia) BKT cps 140, 141, 143, 144; EET 182; EGT 142; EPF pl 454; EPG 142; GDS cp 405; MTB 324 (cp 29); MWF 255; NHE 212; OBT 185; PFE cp 53 (523); PFF 216; SPL cp 401; SST cp 262; TEP 275; TGS 135; WFB 121 (9).

Locust, Clammy (tree) (Robinia viscosa) EPF pl 455; TGS 141; (R. kelseyi) EGT 142; HSC 104.

Locust, Desert (insect) (Schistocerca gregaria) LAW 170, 171, 173; LEA 118 (bw), 131 (bw); (S. peregrina) ALE 2:116.

Locust, Honey (tree) (Gleditsia triacanthos) BKT cp 52; EET 183; GDS cp 213; PFF 208; SST cp 122; TGS 135; (G. sp.) BKT cp 139; DEW 1:306.

Locust, Migratory (insect) (Locusta migratoria) ALE 2:26; BIN pl 29; CIB 81 (cp 6); LEA 130 (bw); PEI 61 (bw); ZWI 110.

Locust, Thornless Honey (tree) (Gleditsia triacanthos inermis) EEG 117; EGT 117; ELG 105; EPG 123.

Locust Tree —See **Carob.**

Log-runner, Northern (Orthonyx spaldingi) CDB 170 (cp 719); SAB 14 (cp 6).

Log-runner, Southern (Orthonyx temmincki) ALE 9:222; RBU 227; SAB 14 (cp 6).

Lomatogonium (Lomatogonium sp.) NHE 279; WFB 182.

London Pride —See **Aaron's Beard** (Saxifraga sp.).

Long-claw (Macronyx capensis) ALE 9:189; CDB 153 (cp 629); PSA 288 (cp 33); (M. ameliae) CDB 153 (cp 628); (M. croceus) ABW 264.

Longfin (Pterolebias sp.) ALE 4:450; SFW 500 (cp 135); WFW cp 204.

Longleaf (Falcaria vulgaris) MFF pl 95 (481); NHE 221; PFE 287; PWF 119g; TEP 115; WFB 161 (8).

Longspur —See **Bunting.**

Loon, Black-throated or Arctic (Gavia arctica) ABW 25; BBB 206; BBE 21; CDB 35 (cp 14); FBE 21; GPB 40; OBB 1; PBE 4 (pl 1); WAB 36.

Loon, Common or Great Northern (Gavia immer) ABW 25; BBB 235; BBE 21; CDB 35 (cp 15); FBE 21; LEA 361 (bw); OBB 1; PBE 4 (pl 1).

Loon, Red-throated (Gavia stellata) ABW 25; ALE 7:61; BBB 207; BBE 21; CDB 35 (cp 16); FBE 21; OBB 1; PBE 4 (pl 1); RBE 52.

Loon, White-billed (Gavia adamsii) BBE 21; FBE 21; PBE 4 (pl 1).

Loosestrife (Lysimachia sp.) DFP 157 (cp 1256); EWF cp 114d; LFW 92 (cp 212); OGF 113; TGF 174.

Loosestrife, Fringed (Lysimachia ciliata) MBF cp 57; PFF 278.

Loosestrife, Hyssop-leaved —See **Grass Poly.**

Loosestrife, Large Yellow or Dotted (Lysimachia punctata) DFP 158 (cp 1257); EPF pl 268; HPW 135; OGF 113; PFE cp 93 (963); PWF 105k.

Loosestrife, Purple (Lythrum salicaria) AMP 67; BFB cp 10; DFP 158 (cp 1258); EPF pl 714; HPW 157; KVP cp 17; MBF cp 34; MCW 74; MFF cp 20 (418); MLP 170; MWF 192; NHE 93; OWF 123; PFE cp 79 (820); PFW 178; PWF 97g; SPL cp 228; TEP 43; TGF 124; WFB 151 (7).

Loosestrife, Slender (Lythrum virgatum) DFP 158 (cp 1259); PFE cp 80 (821).

Loosestrife, Tufted (Lysimachia or Naumburgia thrysiflora) MBF cp 57; MFF cp 59 (621); NHE 97; PFE 304; WFB 175 (7).

Loosestrife, Yellow (Lysimachia vulgaris) BFB cp 13; EFC cp 21 (3); KVP cp 15; MBF cp 57; MFF cp 40 (620); NHE 97; OWF 27; PFW 243; PWF 141e; TEP 43; WFB 175 (6).

Lophocolea (Lophocolea sp.) DEW 3:235; EPF pl 84; ONP 39, 175; PFF 92.

Loquat (Eriobotrya japonica) BIP 75 (cp 112); DEW 1:295; DFP 199 (cp 1591); EGE 122; ELG 137; EPG 119; MWF 124; OFP 105; PFE cp 48 (470); PFM cp 47; RDP 194; SPL cp 429; SST cp 202.

Lords-and-ladies —See **Cuckoo-pint.**

Lorikeet (Psitteuteles versicolor) HAB 90; SNB 106 (cp 52); (P. goldiei) CDB 109 (cp 407); (Domicella domicella) CDB 105 (cp 391); SBB 52 (cp 65); (D. garrula) ALE 8:299; (Eos squamata) CDB 106 (cp 392); (Lorius domicella) ABW 146.

Lorikeet, Blue-browed —See **Fig-parrot, Blue-browed.**

Lorikeet, Little (Glossopsitta pusilla) CDB 106 (cp 393); HAB 89; SNB 106 (cp 52).

Lorikeet, Malay (Loriculus gulgulus) ALE 8:325; KBA 160 (cp 21); (L. philippensis) DPB 147.

Lorikeet, Musk (Glossopsitta concinna) HAB 89; SNB 106 (cp 52).

Lorikeet, Purple-crowned (Glossopsitta porphyrocephala) CAU 46; HAB 89; MBA 36; MWA 87; SNB 106 (cp 52); WAB 194.

Lorikeet, Rainbow (Trichoglossus haematodus or T. moluccanus) ABW 146; ALE 8:293, 299; CDB 109 (cp 409); GPB 178; HAB 88; MBA 37; RBU 124; SBB 52 (cp 64); SNB 106 (cp 52); WAW 12.

Lorikeet, Scaly-breasted (Trichoglossus chlorolepidotus) CAU 108; HAB 90; MBA 34; SNB 106 (cp 52).

Loris, Slender (Loris tardigradus) ALE 10:253 (cp 7); DAA 143; DSM 69; HMW 30; LAW 611; MLS 121 (bw); TWA 155 (bw).

Loris, Slow (Nycticebus coucang) ALE 10:253 (cp 7), 266; CAS 218, 256 (bw); DAA 142; DSM 69; HMA 66; HMW 30; JAW 40; LAW 611; LEA 502 (bw); MLS 122 (bw); SMW 48 (cp 36); WEA 232.

Lotus, Blue (Nymphaea caerulea) LFT cp 68; MWF 215; (N. lotus) DFP 72 (cp 569); EPF pls 188, 190.

Lotus, Sacred (Nelumbo nucifera) DEW 1:101 (cp 49), 129; EJG 52, 131; EPF cp Va; ERW 144; HPW 44; OFP 33; PFW 192, 193; SPL cp 504.

Lotus-bird —See **Jacana, Comb-crested.**

Lourie —See **Turaco.**

Louse, Bee (Braula sp.) ALE 2:403; GAL 461 (bw); OBI 137; PEI 520 (bw); ZWI 122.

Louse, Book (Liposcelis divinatorius) ALE 2:157; GAL 474 (bw); ZWI 148; (L. silvarum) PEI 86 (bw); Trogium pulsatorium) ALE 2:157; GAL 474 (bw); OBI 19.

Louse, Crab (Phthirus pubis) ALE 2:157; GAL 460 (bw); LAW 227; PEI 95 (bw).

Louse, Dog (Trichodectes canis) ALE 2:157; GAL 461 (bw); PEI 91 (bw).

Louse, Fish (Argulus foliaceus) ALE 1:443; GAL 466 (bw); LEA 164 (bw); OIB 141.

Louse, Head or Body (Pediculus humanus var.) ALE 2:157; GAL 460 (bw); LAW 227; LEA 134 (bw); OBI 19; PEI 94 (bw).

Louse, Hog (Haematopinus suis) ALE 2:157; PEI 95 (bw), 96 (cp 11b).

Louse, Pigeon (Columbicola columbae) ALE 2:157; OBI 19.

Louse, Water or Wood —See **Sowbug.**

Lousewort (Pedicularis sp.) EWF cp 5c; EWG 131; NHE 282; PFE cp 127 (1250), cp 128 (1255).

Lousewort, Common (Pedicularis sylvatica) BFB cp 16 (10); MBF cp 65; MFF cp 28 (722); NHE 75; OWF 141; PFE cp 128 (1256); PWF 57j; WFB 217 (4).

Lousewort, Leafy (Pedicularis foliosa) KVP cp 138; NHE 75; PFE cp 127 (1252); WFB 217 (6).

Lousewort, Marsh —See **Red Rattle.**

Lousewort, Truncate or Alpine (Pedicularis recutita) NHE 282; PFE cp 128 (1254).

Lousewort, Yellow (Pedicularis oederi) EPF pl 850; NHE 282.

Lovage (Levisticum officinale) EGH 121; EVF 144; OFP 147; WFB 167 (6).

Lovage, Northern or Scots (Ligusticum scoticum) MBF cp 40; MFF pl 97 (502); PFE cp 85 (892); PWF 153e; WFB 165 (8); (L. mutellina) NHE 275.

Love Apple —See **Mandrake.**

Love Entangled —See **Trefoil, Common Bird's Foot.**

Lovebird, Black-cheeked (Agapornis nigrigensis) ALE 8:325; GPB 179.

Lovebird, Black-collared (Agapornis swinderniana) ALE 8:325; GPB 179.

Lovebird, Fischer's (Agapornis fischeri) ALE 8:325; GPB 179; LAW 505; WEA 233.

Lovebird, Rosy-faced (Agapornis roseicollis) ALE 8:325; PSA 177 (cp 20).

Lovebird, Yellow-collared (Agapornis personata) ABW 146; ALE 8:325; CDB 104 (cp 384).

Love-in-a-mist (Nigella damascena) DEW 1:118; DFP 44 (cp 352), 45 (cp 353); EGA 140; EPF pl 206; EWF cp 27b; HPW 47; LFW 104 (cp 231); MWF 214; NHE 199; OGF 131; PFE cp 18 (203); TGF 61; TVG 63; WFB 67 (3).

Love-in-idleness —See **Pansy, Wild.**

Love-lies-bleeding (Amaranthus caudatus) DFP 29 (cp 232); EGA 93; EPF pl 314; HPW 71; LFW 158 (cp 348); PFW 22; SPL cp 414; TGF 45; TVG 33; WFB 48.

Lucernarian —See **Jellyfish, Stalked** (Haliclystus).

Lucerne —See **Alfalfa.**

Lucine (Codakia sp.) CTS 61 (cp 117); GSS 143.

Luculia (Luculia sp.) BIP 175; EWF cp 104b; MWF 190.

Lugworm (Arenicola marina) CSS cp 12, pl V; LAW 96; OIB 103.

Lumpsucker (Cyclopterus lumpus) ALE 5:66; CSF 168; CSS cp 30; FEF 509 (bw); HFI 60; HFW 253 (bw); MOL 106; OBV 81; WEA 233; WFW cp 256.

Lungfish, African (Protopterus sp.) ALE 5:258, 272; HFI 134; HFW 289 (bw); LEA 267 (bw); SFW 692 (pl 191), 693 (pl 192).

Lungfish, Australian (Neoceratodus forsteri) ALE 5:272; HFI 134; SFW 847 (bw).

Lungfish, South American (Lepidosiren paradoxa) HFI 135; LAW 294; SFW 847 (bw).

Lungwort (Pulmonaria sp.) DEW 2:146; EPF pl 798; NHE 35; OGF 17; PFW 50.

Lungwort, Common (Pulmonaria officinalis) AMP 91; EFC cp 4; MBF cp 60; MFF cp 13 (655); NHE 34; PWF cp 17j; TEP 251; WFB 191 (1).

Lungwort, Mountain (Pulmonaria montana) EFC cp 4; NHE 281.

Lungwort, Narrow-leaved (Pulmonaria angustifolia) DEW 2:162 (cp 81); DFP 21 (cp 166); OGF 17; PFW 50; PWF cp 17i; TGF 173; (P. longifolia) KVP cp 124; MBF cp 60; NHE 281; PFE cp 104 (1064); WFB 190.

Lungwort, Sea —See **Shorewort, Northern.**

Lupin or Lupine (Lupinus sp.) EWF cp 153d, e; MLP 110; PFE cp 54 (521).

Lupin (Lupinus hybrids) DEW 1:302; DFP 42 (cp 334), 156 (cps 1246-1248); EEG 136; EGA 133; EGG 125; EPF pl 446, cp XIIA; LFW 208 (cps 467-469); MLP 109; MWF 191; OGF 77; PFW 163; SPL cp 226; TEP 243; TGF 140; TVG 109.

Lupin, Blue (Lupinus hirsutus) PFF 210; PFM cp 53.

Lupin, Narrow-leaved (Lupinus angustifolius) PFE cp 54 (520); PMF cp 52.

Lupin, Scottish (Lupinus nootkatensis) MBF cp 20; MFF cp 12 (268); WFB 121 (8).

Lupin, Tree (Lupinus arboreus) BWG 26; HSC 75, 81; MBF cp 20; MFF cp 47 (269); OGF 77; PWF 77f; WFB 121 (7).

Lupin, Yellow (Lupinus luteus) LFW 209 (cp 470); NHE 180; PFE cp 53 (519); (L. sulphureus) MWF 191.

Lychee Nut (Nephelium sp.) MWF 211; OFP 101.

Lyme-grass (Elymus arenarius) MBF pl 99; MFF pl 125 (1211); MLP 298; NHE 138; PFE cp 179 (1736); TEP 65.

Lynx, African or Desert —See **Caracal.**

Lynx, European or Northern (Lynx lynx or Felis lynx) ALE 12:284, 298; 13:163; BMA 142-143; CEU 137 (bw); DAE 108; DSM 188; GAL 24 (bw); HMW 127; MLS 310 (bw); TWA 79; WEA 234.

Lynx, Spanish (Felis lynx pardina or Lynx lynx pardina) ALE 12:298; DAE 12; LVA 59; LVS 74-75; PWC 19; SLS 81; VWA cp 14.

Lyophyllum (Lyophyllum sp.) DLM 221; LHM 81, 83; PUM cp 50; REM 113 (cp 81); RTM 105, 107.

Lyrebird, Prince Albert's (Menura alberti) HAB 139; RBU 196; SAB 6 (cp 2).

Lyrebird, Superb (Menura novaehollandiae or M. superba) ABW 213; ALE 9:134-135; BBT 18, 19; CAU 75; CDB 147 (cp 597); GPB 245; HAB 139; MWA 83, 84; RBU 195; SAB 6 (cp 2); SLP 143; WAB 185; WAW 31 (bw); WEA 235.

Lyretail (Aphyosemion christyi) FEF 207; LAW 325; (A. walkeri) FEF 225; WFW cp199.

Lyretail, Banner (Aphyosemion calliurum) CTF 36 (cp 39); FEF 225, 278 (bw).

Lyretail, Calabar (Aphyosemion calabaricus) SFW 445 (cp 118); (Roloffia liberiense) FEF 190; WFW cp 205.

Lyretail, Cape Lopez (Aphyosemion australe) ALE 4:450, 452; FEF 190, 274 (bw); SFW 397 (cp 104); WFW cp 195.

Lyretail, Red-striped (Aphyosemion lujae) FEF 208; WFW cp 198.

Lyretail, Steel-blue (Aphyosemion gardneri) FEF 207, 278 (bw); SFW 429 (pl 112); WFW cp 197.

M

Macaque, Bonnet (Macaca radiata) CAS 186-187 (bw); DAA 132; HMW 22; LAW 617.

Macaque, Crab-eating or Long-tailed (Macaca irus) ALE 10:386 (cp 1); CAS 261; HMW 22; LEA 508 (bw); SMW 97 (bw); (M. fascicularis) DSM 83.

Macaque, Japanese (Macaca fuscata) ALE 10:387 (cp 2); CAS 57, 63 (bw); DSM 82.

Macaque, Lion-tailed (Macaca silenus) ALE 10:386 (cp 1); CAS 196 (bw); DAA 132; DSM 83; LVS 52; MLS 140 (bw).

Macaque, Moor (Macaca maurus) ALE 10:387 (cp 2); HMW 20.

Macaque, Pig-tailed (Macaca nemestrina) ALE 10:368 (cp 1); DSM 82; JAW 50.

Macaque, Rhesus (Macaca mulatta) ALE 10:386 (cp 1); DSM 83; HMA 71; HMW 22; JAW 48; LAW 618; LEA 507 (bw); MLS 141 (bw); TWA 163; WEA 236, 237.

Macaque, Stump-tailed (Macaca arctoides) ALE 10:387 (cp 2); DSM 83; HMW 20; (M. speciosa) MLS 142 (bw).

Macaw, Blue or Hyacinthine (Anodorhynchus hyacinthinus) ABW 149; ALE 8:336; CDB 104 (cp 385); WAB 90; WEA 278; (A. leari) LVS 224.

Macaw, Blue-and-yellow (Ara ararauna) ABW 149; ALE 8:336; BBT 101; SBB 55 (cp 70); WAB 79, 83.

Macaw, Military Green (Ara militaris) ABW 149; ALE 8:336.

Macaw, Red-and-blue (Ara chloroptera) ALE 8:336, 359; SBB 49 (cp 59).

Macaw, Scarlet (Ara macao) ABW 149; ALE 8:336; LAW 505; NBB 9; (A. rubrogenys) SSA cp 4.

Mace —See **Nutmeg.**

Mackerel, Common (Scomber scombrus) ALE 5:191; CSF 140; HFI 50; LAW 349; LEA 225 (bw); LMF 49; OBV 25; WFW cp 451.

Mackerel, Frigate (Auxis thazard) CSF 148; LMF 43 (bw).

Mackerel, Horse —See **Scad.**

Mackerel, Spanish (Scomber colias) CSF 142; HFI 50; OBV 25; (Scomberomorus maculatus) LMF 53; MOL 95.

Madder, Field (Sherardia arvensis) MBF cp 42; MFF cp 26 (808); NHE 234; OWF 139; PFE cp 99 (1014); PWF 115b; TEP 179; WFB 185 (6).

Madder, Wild (Rubia peregrina) MBF cp 42; MFF pl 102 (818); OWF 49; PFE cp 98 (1031); PWF 143i; WFB 187 (1); (R. cordifolia) AMP 69.

Madrona Tree (Arbutus menziesii) BKT cp 177; DFP 182 (cp 1449); MTB 368 (cp 37); PFF 276.

Madwort (Asperugo procumbens) MFF cp 7 (649); NHE 229; WFB 191 (7).

Magic Flower (Achimenes hybrids) BIP 19 (cp 4); EGB 90; EGG 85; EGL 80; FHP 102; RDP 68; (Achimenes erecta) EPF pl 866.

Magnolia, Oyama (Magnolia sieboldii) DFP 212 (cp 1689); SPL cp 102; SST cp 253; TGS 276.

Magnolia, Saucer (Magnolia soulangeana) DEW 1:56 (cp 18); DFP 212 (cps 1690-1693); EET 161; EGT 125; ELG 106; GDS cps 286-289; LFW 210 (cp 471); MWF 193; OBT 145; OGF 23; PFW 180; SPL cp 103; SST cp 254; TGS 278; TVG 231.

Magnolia, Southern (Magnolia grandiflora) DFP 212 (cp 1684); EDP 142; EEG 126; EET 160, 161; EGE 71, 135; ELG 129; EPG 131; HPW 27; MLP 48; MWF 193; OBT 137; PFW 179; SPL cp 101; SST cp 251; TGS 278.

Magnolia, Star (Magnolia stellata) DFP 212 (cp 1694); EET 161; EJG 128; EWF cp 90b; GDS cp 290; HPW 27; HSC 79, 81; MLP 48; MWF 193; OBT 145; PFW 181; SPL cp 104; TGS 278; TVG 231.

Magnolia, Whiteleaf Japanese (Magnolia obovata) BKT cp 110; DFP 211 (cp 1687); EPF pl 179; SST cp 252.

Magpie, Azure-winged (Cyanopica cyanus) ABW 225; ALE 9:497; BBE 215; CDB 223 (cp 1004); DAE 127; FBE 307; PBE 197 (cp 50).

Magpie, Black-backed (Gymnorhina tibicen) CDB 220 (cp 984); HAB 155; SAB 76 (cp 37).

Magpie, Black-billed or Common (Pica pica) ABW 225; ALE 9:497, 501; 11:321; BBB 80; BBE 215; CDB 224 (cp 1007); FBE 307; KBA 277 (pl 42); LAW 510; OBB 127; PBE 197 (cp 50); WBV 155.

Magpie, Ceylon Blue (Cissa ornata or Urocissa ornata) ABW 225; ALE 9:497.

Magpie, Green (Cissa chinensis) ABW 225; ALE 9:497; CDB 222 (cp 995); DAA 147; KBA 277 (pl 42); RBI 279; WBV 155.

Magpie, Red-billed Blue (Cissa erythrorhyncha or Urocissa erythrorhyncha) ALE 9:497; DAA 147; KBA 277 (pl 42); RBI 272; WAB 163.

Magpie, Short-tailed Green (Cissa thalassina) KBA 277 (pl 42); WBV 165.

Magpie, Western (Gymnorhina dorsalis) ALE 10:89; HAB 155; SAB 76 (cp 37).

Magpie, White-backed (Gymnorhina hypoleuca) ALE 9:469; HAB 154; RBU 284; SAB 76 (cp 37); WEA 239.

Magpie, Whitehead's Blue (Cissa whiteheadi or Urocissa whiteheadi) KBA 277 (pl 42); WBV 165.

Magpie-lark (Grallina cyanoleuca) ABW 228; ALE 9:469; CDB 219 (cp 981); GPB 333; HAB 153; SAB 10 (cp 4); WAB 202.

Magpie-robin (Copsychus saularis) DPB 293; KBA 385 (pl 58); RBI 203; WAB 178; WBV 191.

Magpie-robin, Seychelles (Copsychus seychallarum) LVB 63; PWC 147; WAB 30.

Mahogany (Swietenia sp.) EET 218; HPW 201; SST cp 172.

Mahogany, Cape (Trichilia sp.) DEW 2:56 (cp 20); MEP 66.

Mahonia (Mahonia japonica) DFP 213 (cp 1697); GDS cp 292; HSC 78; OGF 1; PFW 43; (M. lomariifolia) DFP 213 (cp 1698); HSC 78, 84; MWF 194; (M. repens) DEW 1:115; EGC 139; ESG 128.

Maidenhair Tree (Gingko bilboa) BKT cp 234; DEW 1:50 (cp 3); DFP 252 (cp 2014); EEG 117; EET 29, 86; EGT 38, 70, 71, 117; EJG 119; EPF pls 167-171; MLP 26, 27; MTB 48 (cp 1); NFP 150; OBT 105, 133; PFF 112; SST cp 121; TGS 22.

Mai-take —See **Hen of the Woods.**

Maize —See **Corn** (Zea).

Mako —See **Shark, Mako.**

Malcoha or Malkoha (Phaenicophaeus sp.)
DAA 145; DPB 157; KBA 161 (cp 22); WAB
31.

Maleo (Macrocephalon maleo) ALE 7:424;
(Megacephalon maleo) CDB 65 (cp 176).

Mallard (Anas platyrhynchos) ABW 68; ALE
7:255, 293, 306; 11:257; BBB 192; BBE 42,
53; CDB 50 (cp 97); FBE 53; FNZ 161 (cp 15);
GPB 79; LAW 442, 491; LEA 334 (bw); OBB
17; PBE 28 (cp 9), 40 (pl 13), 48 (pl 15); SBB
38 (pl 42); SNB 39 (cp 18); WAB 60.

Mallee Fowl —See **Fowl, Mallee.**

Mallow (Lavatera sp.) DFP 155 (cp 1233);
EPF pl 618; HSC 71, 76; KVP cp 103; NHE
220.

Mallow, Annual (Lavatera trimestris) DEW
1:258; DFP 41 (cp 322); EGA 131; EWF cp
35f; PFE cp 72 (742); PFW 185; SPL cp 340;
TGF 128; TVG 57.

Mallow, Common (Malva sylvestris) AMP
87; BWG 104; CTH 35 (cp 43); EFC cp 17;
EWF cp 18b; HPW 94; KVP cp 75; MBF cp
18; MFF cp 15 (224); NHE 220; OWF 131;
PFM cp 98; PWF 75f; SPL cp 395; WFB 142,
143 (4).

Mallow, Dwarf (Malva neglecta) BFB cp 6;
MBF cp 18; MFF cp 15 (225); NHE 220; PFF
248; PWF 123h; TEP 195; WFB 143 (2); WUS
263.

Mallow, Lesser or Smaller Tree (Lavatera
cretica) MBF cp 18; WFB 143 (6).

Mallow, Marsh (Althaea officinalis) AMP 87;
BFB cp 6 (3); EGH 96; EPF pl 616; KVP cp
97; MBF cp 18; MFF cp 15 (227); MLP 142;
NHE 94; PFE cp 72 (747); PFF 249; PWF
145f; WFB 143 (7).

Mallow, Musk (Malva moschata) BFB cp 6;
MBF cp 18; MFF cp 15 (223); MWF 195; NHE
220; OWF 131; PFF 248; PWF 91f; TGF 128;
WFB 143 (1).

Mallow, Poppy (Callirhoe involucrata) ERW
104; EWG 92; TGF 128.

Mallow, Rough (Althaea hirsuta) MBF cp 18;
MFF cp 15 (228); NHE 220; WFB 143 (8).

Mallow, Small (Malva pusilla) MBF cp 18;
NHE 220.

Mallow, Tree (Lavatera arborca) BFB cp 6
(8); DFP 209 (cp 1672); MBF cp 18; MFF cp
15 (226); PFE cp 72 (740); PWF 157d; WFB
142, 143 (5).

Mallow, Vervain (Malva alcea) DFP 158 (cp
1262); EFC cp 17; NHE 73; PFE cp 72 (735).

Malope (Malope trifida) DFP 43 (cp 337);
EGA 134; EWF cp 35e; HPW 94; OGF 131.

Maltese Cross (Lychnis chalcedonica) DFP
157 (cp 1249); LFW 92 (cp 213); TVG 111.

Maltese Cross (Tribulus terrestris) NHE
139; PFE cp 64 (655); WUS 243; (T. cistoides)
MEP 64.

Mamba, Black (Dendroaspis polylepis) CAA
110 (bw); ERA 228; FSA 160 (pl 23), 164 (cp
25); LBR 14; LEA 327 (bw); SIR cp 118; SSW
104.

Mamba, Green (Dendroaspis angusticeps)
ALE 6:406, 430; ART 102; CAA 111 (bw);
ERA 227; FSA 160 (pl 23), 164 (cp 25); HWR
55, 100; LAW 430; SSW 105; VRA 235; WEA
240; (D. viridis) HRA 50.

Man on Horseback (Tricholoma flavovirens)
DLM 259; LHM 87; TEP 221.

Manakin (Chiroxiphia sp.) ALE 9:143; CDB
145 (cp 587); GPB 241; SSA cp 37; WAB 79;
WEA 241; (Pipra sp.) ABW 206; ALE 9:143;
DTB 65 (cp 32); SSA cps 37, 45.

Manakin, Golden-collared (Manacus
vitellinus) SSA cp 37; WAB 84.

Manakin, Striped (Machaeropterus regulus)
CDB 146 (cp 588); DTB 67 (cp 33); SSA cp 37.

Manakin, White-bearded (Manacus
manacus) ALE 9:143; CDB 146 (cp 589); SSA
pl 27.

Manakin, Wire-tailed (Teleonema filicauda)
ABW 206; ALE 9:143; DTB 63 (cp 31); SSA
cp 37.

Manatee, African (Trichechus senegalensis)
ALE 12:510; DDM 148 (bw).

Manatu (Plagianthus betulinus) GDS cp 325;
SNZ 89 (cps 260, 261).

Mandarin —See **Duck, Mandarin.**

Mandrake (Mandragora officinarum) EHP
50, 51; EWF cp 38d; PFE cp 118 (1185); PFM
cps 169, 170; PFW 286.

Mandrill (Mandrillus sphinx) ALE 10:385, 389 (cp 4); BMA 145, 146; HMA 71; HMW 18; JAW 54; LEA 510 (bw), 513; MLS 147 (bw); TWA 15 (bw); (Papio leucophaeus) DDM 53 (cp 6); DSM 86; MLS 148 (bw); (P. mandrillus) DDM 53 (cp 6); SMW 86 (cp 47); (P. sphinx) DSM 86.

Mangabey, Black (Cercocebus aterrimus) ALE 10:390 (cp 5); DDM 60 (cp 7); DSM 89.

Mangabey, Crested or Agile (Cercocebus galeritus) ALE 10:390 (cp 5); DDM 60 (cp 7).

Mangabey, Gray-cheeked (Cercocebus albigena) ALE 10:390 (cp 5); DDM 60 (cp 7); DSM 88.

Mangabey, Sooty or White-collared (Cercocebus torquatus) ALE 10:390 (cp 5); DDM 60 (cp 7); DSM 88; JAW 51; MLS 145 (bw); SMW 80 (bw).

Mango (tree) (Mangifera indica) DEW 2:52 (cp 10), 53 (cps 12, 13); EET 221; EWF cp 120d; LFW 160 (cp 358); MWF 197; OFP 101; PFF 237; SST cp 209; WWT 118.

Mango, Black-throated (bird) (Anthracothorax nigricollis) ALE 8:453; SHU 41; SSA cp 5.

Mangosteen (Garcinia mangostana) EWF cp 118d; OFP 101.

Mangrove (Avicennia sp.) DEW 2:187; EET 226; SNZ 23 (cps 22-23); (Rhizophora sp.) DEW 2:12, 13; EET 227.

Manilagrass —See **Zoysia Grass.**

Manioc —See **Cassava.**

Mannikin (Lonchura cucullata) NBA 95; PSA 316 (cp 39); (L. fuscans) DPB 413; (L. atricapilla) SAB 70 (cp 34); L. striata) KBA 401 (pl 60).

Mannikin, Chestnut (Lonchura malacca) ALE 9:439; DPB 413; KBA 401 (pl 60); (L. ferruginosa) ABW 302.

Man-of-War Fish (Nomeus gronovi) ALE 5:101; HFI 57.

Man-o'-war Bird —See **Frigatebird, Magnificent.**

Manta, Atlantic (Manta birostris) ALE 4:118, 127; HFI 140; LEA 207 (bw); LMF 26; WFW cp 12.

Mantis (Ameles sp.) BIN pls 56, 57; CIB 96 (cp 7); PEI 81 (cp 10); (Iris oratoria) CIB 96 (cp 7); LAW 241.

Mantis, African Praying (Sphodromantis lineola) PEI 65 (cp 8), 74 (bw), 80 (cp 9); PWI 24 (bw).

Mantis, Mediterranean Praying (Empusa sp.) BIN pls 63, 64; LAW 192, 193; PEI 75 (bw).

Mantis, Praying (Mantis religiosa) ALE 2:125, 140; CIB 96 (cp 7); EPD 53; KBM 83 (pl 27); LAW 192; LEA 120 (bw); PEI 72, 73 (bw); PWI 24 (bw); (Orthodera ministralis) BIN cp 59, pls 60, 61.

Manucode (Phonygammus keraudreni) ALE 9:479; HAB 165; RBU 299; SAB 80 (cp 39).

Manuka —See **Tea Tree, Manuka.**

Maple, Amur (Acer ginnala) EGT 44, 69, 90; ELG 100; MTB 338 (bw); TGS 169 (bw).

Maple, Ash-leaved —See **Box-elder.**

Maple, Big-leaf or Oregon (Acer macrophyllum) EET 189; MTB 336 (cp 33); PFF 240.

Maple, Cappadocian (Acer cappadocicum) BKT cp 160; MTB 336 (cp 33).

Maple, Common or Field (Acer campestre) DFP 179 (cp 1430); EET 189; MBF cp 20; MFF pl 70 (262); MTB 332 (cp 31); NHE 22; OBT 21, 25; OWF 197 (bw); PWF 351; 109g; SST cp 70; WFB 34, 35 (4).

Maple, Flowering (Abutilon hybridum) BIP 17; EGA 90; EGG 84; LFW 265 (cp 592); PFF 249; PFW 184; RDP 66; TGF 130.

Maple, Flowering (Abutilon megapotamicum) DFP 179 (cp 1427); EWF cp 177f; FHP 102; HSC 9 (bw) 10; MEP 77; PFW 184; RDP 66.

Maple, Flowering (Abutilon sp.) DEW 1:220 (cp 133); EWF cp 177a; GDS cp 2; LFT 214 (cp 106); MWF 26, 27; RDP 66.

Maple, Flowering Grape-leaved (Abutilon vitifolium) DFP 179 (cp 1428); GDS cp 3; OGF 103.

Maple, Italian (Acer opalus) NHE 22; PFE 239; SST cp 73.

Maple, Japanese (Acer palmatum) DFP 180 (cps 1434-1436); EEG 112; EET 189; EGT 92; EGW 89; EJG 96, 98; EMB 122; ESG 28, 88; GDS cps 4, 5; HSC 9, 10; MTB 336 (cp 33); MWF 29; OBT 188; SST cp 74; TGS 151; (A. japonicum) DFP 179 (cp 1432); EGT 38-39; SPL cp 63.

Maple, Montpelier or French (Acer monspessulanum) NHE 22; PFM pl 5 (359); SST cp 71; WFB 38 (4a).

Maple, Norway (Acer platanoides) BKT cps 158, 163; DEW 2:69; DFP 180 (cp 1438); EEG 113; EET 189, 191; EGT 92-93; ELG 100; EFP 729-730; EWF cp 19b; HPW 196; MTB 332 (cp 31); NHE 22; OBT 188; PFE cp 69 (710); PFF 239; PWF 35k, 109f; SST cp 75; TEP 283; TGS 151.

Maple, Paperbark (Acer griseum) BKT cp 162; DFP 179 (cp 1431); EGT 46, 91; EGW 88; MTB 325 (cp 30); OBT 188.

Maple, Red or Scarlet or Swamp (Acer rubrum) BKT cp 156; DEW 2:55 (cp 18); DFP 180 (cp 1440); EET 189; EGT 67, 93; ELG 101; MTB 332 (cp 31); PFF 242; SST cp 77; TGS 172 (bw).

Maple, Silver or Soft (Acer saccharinum) EGT 94; MTB 332 (cp 31); PFF 241.

Maple, Striped (Acer pensylvanicum) DFP 180 (cp 1437); EET 189; PFF 241.

Maple, Sugar or Rock (Acer saccharum) BKT cp 159; EET 128, 188, 190-191; EGT 65, 74-75, 94, 95; ELG 101; EPG 103; MLP cp 128, 132, 134 (bw); MTB 336 (cp 33); OFP 17; PFF 243; SST cp 267; TGS 172 (bw).

Maple, Sycamore (Acer pseudoplatnus) BFB cp 7 (5); BKT cp 157, 161, 164; BWG 161 (bw); DFP 180 (cp 1439); EET 188, 191; EPF pls 728, 731; EWF cp 19c; MBF cp 20; MFF pl 70 (261); MTB 333 (cp 32); NHE 22; OBT 21, 25, (bw), rear fly; OWF 197 (bw); PFE cp 69 (708); PFF 241; PWF 35j, 109l; SST cp 76; TEP 283; TGS 151; WFB 34-35 (3).

Maple, Tartar (Acer tataricum) MTB 339 (bw); NHE 294.

Mara (Dolichotis patagonum) ALE 11:425, 438; CSA 162 (bw); DSM 134; LEA 526 (bw); MLS 226 (bw); SMW 140 (cp 90).

Marabou, African (Leptoptilos crumeniferus) ABW 57; ALE 7:220-221; CDB 46 (cp 73); GPB 72; LAW 450; LEA 373 (bw); PSA 161 (cp 18); SBB 15 (pl 12); WAB 145; WEA 243.

Marabou, Greater (Leptoptilos dubius) ALE 7:217; KBA 33 (pl 2); SBB 15 (pl 12), 16 (cp 13).

Marabou, Lesser (Leptoptilus javanicus) ALE 7:217; KBA 33 (pl 2).

Maral (Cervus elaphus maral) ALE 13:216; DAA 18.

Marasimius, Garlic (Marasimius scorodonius) EPF pl 60; NFP 67; PUM cp 78; RTM 92.

Marasimius, Little Wheel (Marasimius rotula) DEW 3:169; DLM 224; LHM 115; NFP 66; RTM 92.

Marcgravia (Marcgravia sp.) EWF cp 174d; HPW 88.

Mare's-tail (Hippuris vulgaris) DEW 2:28; MBF cp 34; MFF pl 137 (443); NHE 93; OWF 53; PWF 102a; TVG 131; WFB 293 (3).

Margay —See **Cat, Margay.**

Marguerite (Chrysanthemum frutescens) FHP 112; LFW 35 (cp 83); MWF 84.

Marguerite (Chrysanthemum leucanthemum) EGC 70; EPF pl 556; EWG 77; MBF 46; MCW 105; MFF pl 100 (892); NHE 190; OWF 99; PFE cp 148 (1427); PFF 326; PWF 31f; SPL cp 195; TEP 147; WUS 389.

Marguerite, Blue —See **Daisy, Blue.**

Marguerite, Golden —See **Chamomile, Yellow.**

Marguerite Othonna (Euryops sp.) BIP 81; DFP 9 (cp 70); MWF 131; OGF 67.

Marianthus (Marianthus candidus) EWF cp 131a; HPW 138.

Marica (Neomarica caerulea) BIP 125 (cp 190); MEP 23; (N. gracilis) FHP 134.

Marigold, African (Tagetes erecta) DFP 48 (cps 378, 379); EDP 84; EGA 32-35, 73, 76, 155; EGG 144; EPF pl 551; LFW 233 (cps 501, 502); MWF 278; SPL cp 248; TGF 253; TVG 71.

Marigold, Cape (Dimorphotheca sp.) DFP 7 (cp 54), 37 (cp 291); EGA 115; MWF 114, 115; OGF 139; SPL cp 202; TGF 269; TVG 49.

Marigold, Corn (Chrysanthemum segetum) BFB cp 19 (4); BWG 49; DFP 34 (cp 266); EGA 106; EWF cp 39b; MBF cp 46; MFF cp 56 (891); NHE 237; OWF 39; PFM cp 198; PWF 112g; TVG 41; WFB 241 (4).

Marigold, Field (Calendula arvensis) KVP cp 167; NHE 237; PFE cp 151 (1460); PFM cp 204.

Marigold, French (Tagetes patula) DFP 48 (cps 380, 381); EGA 155; EMB 141; LFW 222 (cps 489-500); MLP 237; MWF 279; OGF 143; PFF 324; SPL cp 376.

Marigold, Pot (Calendula officinalis) AMP
159; BWG 59, 60; CTH 64 (cp 103); DFP 31
(cp 248), 32 (cps 249, 250); EDP 116; EGA
101; EGG 94; EGH 105; EPF pl 573; LFW 18
(cp 35); MWF 66; OGF 137; PFF 325; SPL cps
255, 387; TGF 269; TVG 35; WFB 241 (7).

Marjoram, Alpine —See **Adenostyles,
Gray.**

Marjoram, Pot or Wild (Origanum vulgare)
DFP 162 (cp 1293); EGH 130; MBF cp 67;
MFF cp 26 (752); NHE 186; OFP 141; OWF
147; PFE cp 115 (1160); PWF 108b; TEP 119;
TGF 204; TVG 159; WFB 205 (8); (O. onites)
OFP 141.

Marjoram, Sweet (Marjorana hortensis)
EGH 129; EPF pl 814; EVF 144; PFE 365;
TGF 204.

Markhor (Capra falconeri) DAA 55; DSM
272; HMA 56; JAW 205; LEA 608 (bw); LVA
53; LVS 130; MLS 429 (bw); SMW 271 (bw);
VWA cp 30.

Marlin, Black or Blue (Makaira nigricans)
ALE 5:192; LMF 63 (bw); MOL 95.

Marlin, Striped (Makaira sp.) HFW 233, 234
(bw); LMF 63 (bw).

Marmoset (Hapale sp.) DSM 73; HMA 71;
HMW 28; TWA 133.

Marmoset, Black-plumed (Callithrix
penicillata) ALE 10:375; MLS 136 (bw);
SMW 82 (cp 39).

Marmoset, Cotton-topped or Pinche
(Oedipomidas oedipus) ALE 10:350, 375;
BMA 148; DSM 73; HMW 28; SMW 71 (bw);
(Saguinus oedipus) JAW 47.

Marmoset, Golden Lion or Silky (Leontideus rosalia or Leontocebus rosalia) ALE
10:375; BMA 149; DSM 73; HMW 28; JAW
46; LVA 29; MLS 139 (bw); SMW 81 (cp 38);
TWA 132; (Leontopithecus rosalia) LVS 48.

Marmoset, Pygmy (Cebuella pygmaea) BMA
149 (bw); DSM 73; LEA 506 (bw); MLS 138
(bw); SMW 70 (bw); (Callithrix pygmaea)
ALE 10:375.

Marmoset, Silvery (Callithrix argentata)
DSM 73; MLS 137 (bw); (Mico argentata)
BMA 150 (bw); (Saguinus argentata) ALE
10:376.

Marmoset, White-plumed (Callithrix
aurita) BMA 150; SMW 82 (cp 40).

Marmot (Marmota camtschatica) CAS 118
(bw); (M. caudata) DAA 128.

Marmot, Alpine (Marmota marmota) ALE
11:216, 239; 13:463; CEU 159; DAE 47; GAL
59 (bw); HMW 160; LAW 585; PWC 145 (bw);
TWA 89; WEA 244.

Marmot, Bobak or Steppe (Marmota bobak)
ALE 11:239; CAS 73; DAE 99; DSM 118;
HMW 161.

Marram-grass (Ammophila arenaria) DEW
2:265; MBF pl 96; MFF pl 123 (1227); NHE
138; PFE cp 181 (1775); TEP 65.

Marshbird —See **Grassbird** (Megalurus sp.)

Marsh-marigold, Common (Caltha palustris) BFB cp 2 (4); DFP 127 (cp 1009); EFC cp
41 (1); EPF pl 203; ERW 139; EWG 93; MBF
cp 4; MFF cp 38 (1); MPL 53; NHE 91; OWF
5; PFF 172; PWF 31j; SPL cp 497; TEP 29;
TGF 61; TVG 131; WFB 69 (2).

Marshwort (Apium inundatum) MBF cp 37;
MFF pl 104 (476); NHE 95; OWF 89; PWF
89e.

Marten, Beech or Stone (Martes foina) ALE
11:321; 12:39, 65; DAA 84; DAE 45; DSM
168; GAL 26 (bw); HMW 133.

Marten, Pine (Martes martes) ALE 12:40, 65;
13:163; CEU 160 (bw); DAE 44; DSM 168;
FBW 29; GAL 25; HMW 133; LAW 535; LEA
559 (bw); OBV 171; TWA 85.

Marten, Yellow-throated (Martes flavigula)
ALE 12:65; DAA 82.

Martin (Progne modesta) HBG 141 (pl 12); (P.
tapera) ABW 217.

Martin, Asian House (Delichon dasypus)
DPB 231; KBA 245 (pl 34).

Martin, Crag (Hirundo rupestris) BBE 201;
CDB 150 (cp 616); FBE 207; PBE 196 (cp 49);
PSA 208 (cp 23); (H. concolor) KBA 245 (pl
34); (H. obsoleta) FBE 207; (Ptyonoprogne
rupestris) ALE 9:180.

Martin, Fairy (Hylochelidon ariel) HAB 143;
(Petrochelidon ariel) RBU 280; SAB 8 (cp 8);
SNB 130 (cp 64).

Martin, House (Delichon urbica) ALE 9:180;
BBB 47; BBE 201; CDB 149 (cp 609); FBE
207; KBA 245 (pl 34); OBB 121; PBE 196 (cp
49); RBE 223.

Martin, River (Pseudochelidon sp.) KBA 245
(pl 34); WAB 157.

Martin, Sand —See **Swallow, Bank.**

Martin, Tree (Hylochelidon nigricans) HAB 143; MBA 60-61; (Petrochelidon nigricans) SAB 8 (cp 8); SNB 130 (cp 64).

Marvel of Peru (Mirabilis jalapa) DFP 44 (cp 345); EGA 137; HPW 69; MEP 36; MWF 201; PFW 194; SPL cp 229; TGF 66; TVG 61.

Mary's Flower —See **Rose of Jericho.**

Mask Flower (Alonsoa warscewiczii) DFP 29 (cp 227); EGA 92.

Masterwort (Peucedanum ostruthium) MFF pl 97 (506); PFE 280; WFB 165 (3).

Masterwort, Great (Astrantia major or maxima) DFP 125 (cp 999); EPF pl 756; EWF cp 17e; KVP cp 146; MBF cp 36; MFF cp 25 (453); NHE 32; OGF 119; PFE cp 83 (852); PFW 301; PWF 72b; SPL cp 472; TEP 247; TGF 125; WFB 155 (7).

Mastic Tree (Pistacia lentiscus) EWF cp 31c; HPW 198; NHE 246; PFE cp 68 (703); PFM cp 97.

Mastic-tree —See **Pepper-tree.**

Matamata —See **Terrapin, Matamata.**

Match-me-if-you-can —See **Beefsteak Plant.**

Mate (Ilex paraguariensis) HPW 182; SST cp 285.

Mat-grass (Nardus stricta) MBF pl 99; MFF pl 121 (1249); NHE 195.

Matipo —See **Myrsine.**

Matrimony Vine (Lycium chinense) HSC 75; MBF cp 61; PWF 113h; TGS 375; (L. halimifolium) EGV 120; MFF cp 8 (670); OWF 131; (L. europaeum) PFE 384.

May or **Maytree** —See **Hawthorn, Common.**

Mayflower (Epigaea sp.) —See **Trailing Arbutus.**

Mayflower, Himalayan (Podophyllum emodii) DEW 1:99 (cp 44); DFP 163 (cp 1320); EWF cp 91c.

Mayfly —See **Fly, May.**

Mayweed —See **Chamomile, Stinking.**

Mayweed, Scentless (Matricaria inodora) OWF 99; SPL cp 364; TVG 59.

Mayweed, Scentless (Tripleurospermum inodorum) BWG 96; EPF pl 555; MBF cp 46; WFB 233 (2); (T. maritimum) EPF pl 555; MFF pl 100 (886); NHE 236.

Mazus (Mazus reptans) EGC 139; OGF 89; TGF 234; (M. radicans) SNZ 104 (cp 318).

Meadow-grass (Puccinellia sp.) —See **Salt-marsh Grass.**

Meadow-grass (Poa sp.) MBF pl 97; NHE 106; SNZ 130 (cp 402), 131 (cp 405).

Meadow-grass, Alpine (Poa alpinus) MBF pl 97; MFF pl 129 (1188); NHE 291; PFE cp 179 (1751).

Meadow-grass, Annual (Poa annua) BWG 174; HPW 287; MBF pl 97; MFF pl 128 (1187); NHE 244; TEP 205.

Meadow-grass, Common or Smooth (Poa pratensis) EGC 23, 75, 80, 81, 119; MBF pl 97; MFF pl 30 (1191); NHE 195; PFF 359; TEP 151.

Meadow-grass, Flattened or Flat-stalked (Poa compressa) MBF pl 97; MFF pl 128 (1190); NHE 244; PFF 359.

Meadow-grass, Rough (Poa trivialis) EGC 118; MBF pl 97; MFF pl 130 (1192); NHE 195.

Meadow-grass, Wood or Woodland (Poa nemoralis) MBF pl 97; MFF pl 130 (1189); NHE 244.

Meadow-rue, Alpine (Thalictrum alpinum) MBF cp 1; MFF cp 50 (27); NHE 268.

Meadow-rue, Common (Thalictrum flavum) MBF cp 1; MFF cp 50 (26); NHE 91; OWF 5; PFE cp 26 (257); PWF 129g; WFB 73 (1).

Meadow-rue, Great (Thalictrum aquilegifolium) DEW 1:125; DFP 175 (cp 1395); EPF pl 210; KVP cp 115; LFW 224 (cp 504); NHE 178; PFE 109; TEP 237; TGF 86; TVG 123; WFB 73 (2).

Meadow-rue, Lesser (Thalictrum minus) HPW 48; MBF cp 1; MFF cp 50 (28); NHE 178; PFE 109; PWF 120a; WFB 73 (3).

Meadowsweet —See also **Queen-of-the-Meadow** and **Dropwort.**

Meadowsweet (Filipendula sp.) DFP 142 (cp 1136); TVG 91.

Mealworm —See **Beetle, Mealworm.**

Medick (Medicago sp.) EWF cp 29a; PFE cp 58 (584); PFM cp 64.

Medick, Black (Medicago lupulina) BWG 28; EGC 72; EWF cp 10d; MBF cp 21; MCW 63; MFF cp 48 (279); NHE 179; OWF 21; PFF 213; PWF 38b, 53e; WFB 129 (6); WUS 231.

Medick, Hairy or Toothed (Medicago polymorpha) MBF cp 22; NHE 212; PFE cp 58 (588); PWF 61g; (M. hispida) MFF pl 48 (281); OWF 21.

Medick, Sea (Medicago marina) NHE 139; PFE cp 58 (589); PFM cp 62.

Medick, Small (Medicago minima) MBF cp 22; MFF cp 48 (280); NHE 179.

Medick, Spotted (Medicago arabica) BWG 29; MBF cp 22; MFF cp 48 (282); OWF 21; PWF 53f; WFB 129 (7).

Medick, Yellow or Sickle (Medicago falcata) MBF cp 21; MFF cp 47 (277); NHE 179.

Medinilla (Medinilla magnifica) BIP 119 (cp 182); DEW 2:21; EWF cp 121c; MEP 101; MWF 198; PFW 187; RDP 266; SPL cp 52; (M. sp.) EPF pl 711; HPW 164.

Mediolobivia (Mediolobivia sp.) EPF cp VI b; SCS cps 46, 47.

Medlar (Mespilus germanica) BKT cps 121, 123; DEW 1:94; EPF pls 429, 430; MFF pl 75 (381); MWF 200; NHE 20; OBT 172; OFP 63; PFW 264.

Meerkat, Gray or Suricate (Suricata suricatta) ALE 12:168, 177; CAF 208 (bw); DDM 97 (cp 18); DSM 183; LAW 541; LEA 566 (bw); MLS 303 (bw); SMW 174 (bw).

Meerkat, Red or Bushy-tailed (Cynictis penicillata) ALE 12:177; DDM 97 (cp 18); TWA 52.

Megapode —See **Fowl, Jungle or Scrub.**

Megrim (Lepidorhombus whiffiagonis) CSF 177; FEF 511 (bw); OBV 55.

Melanoleuca (Melanoleuca sp.) DLM 224, 225; LHM 103; PUM cp 72; PMU cp 70; RTM 106, 109.

Melick, Mountain or Nodding (Melica nutans) DEW 2:264; MBF pl 97; MFF pl 126 (1199); NHE 40.

Melick, Wood (Melica uniflora) MBF pl 97; MFF pl 126 (1198); NHE 40; (M. ciliata) PFE cp 179 (1745).

Melilot, Common (Melilotus officinalis) BFB cp 8; MBF cp 22; MCW 65; MFF cp 47 (283); NHE 134, 212; OWF 23; PFE 206; PFW 73f; TEP 193; WFB 127 (6).

Melilot, Tall or Yellow (Melilotus altissima) BWG 30; MBF cp 22; NHE 212; PFE cp 57 (576).

Melilot, White (Melilotus alba) MBF cp 22; MCW 64; MFF pl 107 (284); NHE 212; PFF 213; PWF 55g; WFB 127 (7).

Membrillo (Gustavia sp.) HPW 99; MEP 94.

Menziesia, Scottish —See **Heather, Blue Mountain.**

Mercury —See **Good King Henry.**

Mercury, Annual (Mercurialis annua) AMP 139; NHE 223; PWF 117d.

Mercury, Dog's (Mercurialis perennis) BWG 165; MBF cp 75; MFF pl 110 (514); NHE 33; OWF 63; PFE cp 65 (666); PWF cp 12d; TEP 245; WFB 137 (10).

Merganser, Common (Mergus merganser) ALE 7:61, 308; BBB 204; BBE 67; FBW 51; FBE 67; LEA 380 (bw); OBB 29; PBE 37 (cp 12), 40 (pl 13), 48 (pl 15); RBE 100.

Merganser, Red-breasted (Mergus serrator) ABW 71; ALE 7:309; 8:229; BBB 205; BBE 43, 67; FBE 67; OBB 29 PBE 37 (cp 12), 40 (pl 13), 48 (pl 15); WAB 130.

Merlin (Falco columbarius) ALE 7:418; BBB 110; BBE 89; FBE 95; OBB 49; PBE 61 (cp 20), 73 (pl 24); RBE 28; WAB 122; WBV 61.

Merou —See **Grouper** (Epinephelus sp.).

Merry Widow (Phalichthys amates) ALE 4:459; SFW 477 (pl 128).

Mescal Bean (Sophora secundiflora) EHP 95; MEP 61.

Mescal Buttons —See **Peyote.**

Mesia (Leiothrix argentaurus) ABW 240; ALE 9:221; CDB 168 (cp 711); KBA 288 (cp 45); RBI 184; WBV 199.

Mesite, Brown (Mesoenas unicolor) ABW 101; ALE 10:283.

Mespil or Mespilus —See **Serviceberry.**

Mexican Bamboo —See **Knotweed, Japanese** (Polygonum sp.).

Mezereon (Daphne mezereum) AMP 163; CTH 37 (cp 46); DFP 195 (cp 1554); EFC cp 25; EFS 108; EPF pl 701, cp XVIII; EWF cp 18g; GDS cp 147; HPW 160; HSC 42, 44; MBF cp 76; MFF cp 17 (421); NHE 267; OGF 3; PFE cp 74 (759); PWF cp 16b; SPL cp 480; TEP 286; TGS 247; TVG 223; WFB 139 (6).

Mezereon, Mediterranean (Daphne gnidium) DEW 1:276 (cp 154); PFM cp 113.

Michauxia (Michauxia sp.) EWF cp 47b; PFW 63.

Mickey Mouse Plant —See **Carnival Plant.**

Midge, Feather (Chironomus plumosus) GAL 400 (bw); OBI 125; PEI 504 (bw); (C. annularis) CIB 197 (cp 30).

Midge, Owl (Psychoda sp.) ALE 2:394; OBI 125.

Midshipman (Porichthys sp.) ALE 4:418; HFI 12.

Midsummer Men —See **Roseroot** (Rhodiola sp.).

Midsummer-men —See **Roseroot** (Sedum sp.).

Mieri (Leporinus sp.) FEF 120; WFW cp 100.

Mignonette (Reseda odorata) DFP 46 (cp 365); EDP 141; EGA 150; MWF 252; PFW 256; TGF 92; (R. villosa) HPW 123.

Mignonette Tree (Lawsonia inermis) DEW 2:11; HPW 157.

Mignonette, White or Upright (Reseda alba) DEW 1:147; PFE cp 38 (373); PFM cp 45; PWF 97; (R. phyteuma) PFE cp 38 (374).

Mignonette, Wild (Reseda lutea) EPF pl 240; EWF cp 11d; MBF cp 11; MFF cp 50 (114); NHE 219; OWF 49; PFE cp 38 (372); PFM cp 44; PWF 37k; WFB 101 (2).

Mildew, Downy (Phytophthora infestans) DLM 84; NFP 33.

Milfoil —See also **Yarrow.**

Milfoil, Dwarf (Achillea nana) CTH 58 (cp 92); PFE cp 147 (1419).

Milkbush (Euphorbia tirucalli) EDP 27; RDP 199.

Milk-cap, Oak (Lactarius quietus) DLM 208; LHM 211; ONP 127; PUM cp 40; RTM 143.

Milk-cap, Orange-brown (Lactarius volemus) DEW 3: cp 69; DLM 212, 213; LHM 207; NFP 73; PMU cp 62; REM 38 (cp 6); RTM 135; SMG 241; TEP 219.

Milk-cap, Red (Lactarius rufus) DLM 209; LHM 209; ONP 105; PMU cp 64; REM 37 (cp 7); RTM 140; SMF 52; SMG 245.

Milk-cap, Sulphur (Lactarius chrysorrheus) DLM 61, 203; KMF 47; REM 37 (cp 5); RTM 136; SMG 236.

Milk-cap, Woolly or Shaggy (Lactarius torminosus) CTM 38; DEW 3:181; DLM 210; KMF 47; LHM 211; NFP 73; ONP 113; PMU cp 67; REM 34 (cp 2); RTM 133; SMF 30, 31.

Milkfish (Chanos chanos) ALE 4:158; LMF 31.

Milkmaids —See **Cuckoo-flower.**

Milk-parsley, Cambridge —See **Parsley, Cambridge.**

Milk-thistle (Silybum marianum) DFP 47 (cp 376); EGH 141; EPF pl 587; MFF cp 30 (914); OWF 151; PFE cp 155 (1492); PFM cp 207; PWF 77h; WFB 245 (7).

Milk-vetch (Astragalus sp.) CTH 34 (cp 39); NHE 212, 272, 273, 294.

Milk-vetch (Oxytropis sp.) EWF cp 2h; NHE 272; WFB 118 (6a).

Milk-vetch, Alpine (Astragalus alpinus) MBF cp 24; MFF cp 3 (314); NHE 273; PFE cp 54 (530); (A. frigidus) NHE 273; PFE 192; WFB 119 (3).

Milk-vetch, Drooping (Astragalus penduliflorus) EPF pl 460; NHE 273.

Milk-vetch, Hairy Meadow Beaked (Oxytropis pilosa) EPF pl 459; NHE 181; TEP 111.

Milk-vetch, Purple (Astragalus danicus) MBF cp 24; MFF cp 3 (313); NHE 181; PWF 68b; WFB 119 (1).

Milk-vetch, Purple Mountain Beaked (Oxytropis halleri) EWF cp 10e; MBF cp 23; MFF cp 3 (316); PWF 81f; WFB 119 (6).

Milk-vetch, Yellow (Astragalus cicer) NHE 181; PFE cp 54 (528); WFB 119 (2).

Milk-vetch, Yellow Beaked (Oxytropis campestris) MBF cp 23; MFF pl 79 (317); NHE 272; PFE 192; PWF 103l.

Milkweed (Asclepias fruticosa) LFT 250 (cp 124); MWF 47; (A. sp.) MLP 202, 205.

Milkweed, Common (Asclepias syriaca) DEW 2: (cp 60); MCW 79; PFE cp 99 (1011); PFF 283; WUS 287.

Milkweed, Giant (Calotropis procera) DEW 2:128; EWF cp 59a; MEP 113; WWT 63.

Milkwort (Polygala sp.) EPF pl 650; HPW 216; LFT cp 93; MBF cp 11; MEP 68; MFF cp 5 (129); NHE 183, 273; PFE cp 68 (700), 220.

Milkwort, Box-leaved or Shrubby (Polygala chamaebuxus) NHE 268; PFE cp 68 (696); SPL cp 461; WFB 139 (5).

Milkwort, Chalk (Polygala calcarea) MBF cp 11; MFF cp 5 (128); NHE 183; OWF 169; PWF 29k.

Milkwort, Common (Polygala vulgaris) AMP 103; MBF cp 11; MFF cp 5 (126); NHE 73; OWF 113, 169; PFE cp 68 (698); PWF 57f; WFB 139 (4).

Milkwort, Heath (Polygala serpyllifolia)
EWF cp 18d; MBF cp 11; MFF cp 5 (127);
NHE 73; OWF 169; PWF 571.

Milkwort, Sea (Glaux maritima) MBF cp 58;
MFF cp 27 (626); NHE 136; OWF 125; PFE
cp 92 (968); PWF 114f; TEP 61.

Miller's Thumb —See **Bullhead.**

Millet, Common or Broom-corn (Panicum
miliaceum) MFF pl 133 (1256); OFP 13; PFE
cp 182 (1803); PFF 348; (P. miliare) OFP 13;
(P. repens) PFE 553.

Millet, Dwarf (Milium scabrum) EWF cp 25e;
MFF pl 121 (1243).

Millet, Foxtail or Italian (Setaria italica)
MFF pl 123 (1255); OFP 13; PFE cp 182
(1805); PFF 349; TVG 71.

Millet, Wood (Milium effusum) MBF pl 95;
MFF pl 130 (1242); NHE 40; PFE cp 181
(1791); PFW 132.

Millipede (Polydesmus sp.) ALE 1:509, 2:51;
GAL 478 (bw); LAW 158; OIB 153; PEI 22
(bw); (Julus sp.) ALE 2:51; GAL 477, 478
(bw); LEA 109 (bw).

Millipede (Polyxenus lagurus) ALE 1:509;
GSP 147; OIB 155.

Millipede, Greenhouse (Oxidus gracilis)
EPD 119; GSP 149.

Millipede, Pill (Glomeris sp.) ALE 1:496, 509;
2:51; GAL 478 (bw); GSP 147; LEA 109 (bw);
OIB 165.

Millipede, Spotted (Blaniulus guttulatus)
OIB 153; PEI 48 (cp 5b).

Mimicry Plant (Pleiospilow sp.) DEW 1:197;
ECS 138; HPW 66; LCS cp 282.

Mind-your-own-business —See **Baby's
Tears.**

Miner, Bell (Manorina melanophrys) CAU
75; HAB 242; RBU 272.

Miner, Noisy (Manorina melanocephala or
Myzantha melanocephala) HAB 243; SAB
66 (cp 32).

Miner, Yellow-throated (Manorina
flavigula or Myzantha flavigula) ALE
9:333; HAB 243; MBA 17; SAB 66 (cp 32).

Minivet, Ashy (Pericrocotus divaricatus)
ABW 219; ALE 9:450; DPB 243; KBA 272
(cp 39).

Minivet, Flame or Scarlet (Pericrocotus
flammeus) ABW 219; ALE 9:450; DPB 243;
KBA 272 (cp 39); LEA 448; RBI 292; WAB 167.

Minivet, Gray-chinned or Mountain
(Pericrocotus solaris) KBA 272 (cp 39);
WBV 165.

Minivet, Long-tailed (Pericrocotus
ethologus) KBA 272 (cp 39); WBV 165.

Minivet, Rosy (Pericrocotus roseus) KBA 272
(cp 39); WBV 165.

Minivet, Small (Pericrocotus cinnamomeus)
DPB 243; KBA 272 (cp 39).

Mink (Mustela lutreola) ALE 11:105; 12:55;
DAE 21; HMW 136; SMW 226 (cp 129).

Minnow, European (Phoxinus phoxinus)
ALE 4:371; 11:105; FEF 146 (bw); FWF 67;
GAS 242; (bw); HFI 100; LEA 245 (bw); OBV
101; SFW 244 (pl 57).

Minnow, White-cloud Mountain (Tanichthys albonubes) ALE 4:318; CTF 24 (cp 15);
FEF 175 (bw); SFW 164 (cp 37); WFW cp 142.

Mint, Corn or Field (Mentha arvensis) MBF
cp 67; MFF cp 10 (745); NHE 231; OWF 143;
PWF 135i; WFB 205 (2).

Mint, Corsican or Creeping (Mentha
requienii) EGC 140; MFF cp 11 (743); OGF
89; TGF 204.

Mint, Round-leaved or Apple-scented
(Mentha rotundifolia) EGH 125; EVF 144;
MBF cp 67; MFF cp 11 (750); NHE 98; OWF
143; PWF 125i.

Mint, Water (Mentha aquatica) BFB cp 16;
DEW 2:188; EFC cp 52; KVP cp 13; MBF cp
67; MFF cp 11 (747); NHE 98; OWF 143; PFE
cp 116 (1169); PWF 141d; WFB 205 (1).

Mintbush (Prostanthera ovalifolia) BIP 149
(cp 227); DFP 219 (cp 1751); MEP 124; MWF
242; (P. lasianthos) EWF cp 132c; (P. nivea)
WAW 118.

Mist Flower (Eupatorium coelestinum) EWG
108; TGF 237; (E. sordidum) MWF 129; SPL
cp 85.

Mistle-thrush —See **Thrush, Mistle.**

Mistletoe (Loranthus sp.) DEW 2:99 (cps 46,
47); HPW 174; LFT cp 61; NHE 26; SNZ 92
(cp 271).

Mistletoe, Eurasian (Viscum album) AMP
85; DEW 2:98 (cps 43-45); EFC cp 16; EPF pl
699; HPW 174; MBF cp 75; MFF pl 73 (446);
NHE 26; OWF 51; PFE cp 7 (73); PFW 177;
PWF 25o, 174d; SPL 412; WFB 39; (V. sp.)
DEW 2:93; EWF cp 35a; LFT cp 61.

Mistletoebird (Dicaeum hirundinaceum)
ABW 280; ALE 9:372; CAU 26; CDB 187 (cp
815); HAB 254; MBA 61; MEA 54; MWA
71-76; RBU 255; SAB 52 (cp 25).

Mite (Spinturnix sp.) GSP 134; OIB 147.

Mite, Fruit Tree Spider (Panonychus ulmi)
ALE 1:437; OIB 153.

Mite, Itch (Sarcoptes scabei) ALE 1:437; GSP
137; OIB 145.

Mite, Round Water (Hydrachna globosa)
GAL 463 (bw); OIB 139.

Mite, Spider (Tetranychus sp.) EPD 121; GSP
135.

Mite, Velvet Spider (Trombidium sp.) ALE
1:426; GSP 135.

Mite, Water (Arrhenurus superior) GSP 136;
(A. caudatus) OIB 139.

Miter Shell (Mitra sp.) CTS 49 (cps 85, 86), 50
(cp 87); GSS 92-95; MOL 67.

Mocassin-flower—See Lady's-slipper.

Mock Orange —See also Box, Victorian.

Mock Orange (Philadelphus sp.) DEW 1:
290; DFP 216, 217 (cps 1725-1729); EEG 100;
EFS 124, 125; ELG 117; GDS cp 313; HSC 87,
92; OGF 61; PFW 228; TGS 158, 160; TVG
235.

Mock Orange, Sweet (Philadelphus coronarius) DFP 216 (cp 1728); EPF pl 368; GDS cp
312; HSC 92; MWF 229; PFF 193; PFW 228;
SPL cp 108; TGS 151.

Mockingbird, Galapagos (Nesomimus parvulus) ALE 9:221; HBG 97 (cp 6); (N. melanotis); CSA 256 (bw); HBG 97 (cp 6); (N. trifasciatus); CDB 161 (cp 671); GPB 267; HBG 97
(cp 6).

Mock-privet (Phillyrea decora) GDS cp 314;
HSC 87; TGS 359; (P. latifolia); MTB 385 (cp
40); NHE 246.

Moderlieschen (Leucaspius delineatus)
FEF 143 (bw); GAL 240 (bw); HFI 99; SFW
205 (pl 48).

Moepel (Mimusops zeyheri) HPW 132; LFT
cp 119.

Mole, European (Talpa europaea) ALE 10:
223; DAE 62; DSM 52; GAL 63 (bw); HMW
191; JAW 33; LAW 599; LEA 491 (bw); MLS
86 (bw); OBV 127; TWA 103 (bw); (T. caeca)
ALE 10:190.

Mole, Golden (Chrysochloris sp.) ALE 10:178;
DSM 52; LAF 15 (bw); LAW 595; (Amblysomus sp.) ALE 10:178.

Mole, Marsupial or Pouched (Notoryctes
typhlops) ALE 10:71; HMW 200; SMW 37 (cp
9); TWA 205 (bw).

Mole Plant—See Spurge, Caper.

Mole, Shrew—See Shrew-mole.

Mole-rat (Spalax microphthalmus) ALE 11:
300; CAS 109; DAE 82; HMW 166.

Mole-rat, African (Tachyoryctes splendens)
LAW 522; LEA 539 (bw); (T. daemon) ALE
11:300.

Mole-rat, Cape (Georychus capensis) ALE
11:426; HMA 82.

Mole-rat, Naked (Heterocephalus sp.) ALE
11:426; HMW 166; LEA 522 (bw); SMW 127(bw).

Mole-rat, Zech's (Cryptomys zechi) ALE
11:426; MLS 236 (bw).

Moloch, Australian—See Devil, Mountain
or Thorny.

Moltkia (Moltkia sp.) DFP 14 (cp 111); GDS cp
297.

Monal—See Pheasant.

Monarch (bird)—See Flycatcher.

Monarch-of-the-Veldt (Venidium fastuosum) DFP 49 (cp 391); EGA 160; TGF 268;
TVG 73.

Mondo Grass—See Lily-turf.

Money Plant—See Honesty.

Moneywort (Lysimachia nummularia) BFB
cp 13; BWG 44; DFP 14; (cp 109); EGC 139;
EGV 121; ESG 127; MBF cp 57; MFF cp 40
(619); NHE 186; OWF 27; PFE cp 93 (961);
PFF 278; PFW 243; PWF 93m; SPL cp 364;
TVG 157; WFB 175 (8).

Moneywort, Cornish (Sibthorpia europaea)
HPW 244; MBF cp 63; MFF pl 105 (698); PWF
150b; WFB 217 (9).

Mongoose, Banded or Zebra (Mungos
mungo) ALE 12:177; DDM 97 (cp 18); DSM
182; SMR 48 (cp 3).

Mongoose, Black-legged (Bdeogale nigripes)
ALE 12:177; DDM 96 (cp 17); SMW 174 (bw).

Mongoose, Crab-eating (Herpestes urva)
DAA 102; DSM 182; HMW 132.

Mongoose, Dwarf (Helogale parvula) ALE
12:168, 177; DDM 112 (cp 19); DSM 183; SMR
48 (cp 3).

Mongoose, Egyptian (Herpestes ichneumon) ALE 12:177; DAE 11; DDM 96 (cp 17); HMW 132; JAW 123; SMR 48 (cp 3).

Mongoose, Gray (Herpestes edwardsii) CAS 196 (bw) LEA 566 (bw); SMW 183 (cp 110); TWA 174; (H. pulverulentus); DDM 112 (cp 19).

Mongoose, Marsh or Water (Atilax paludinosus) ALE 12:177; DDM 96 (cp 17); MLS 304 (bw); SMR 48 (cp 3).

Mongoose, Meller's (Rhynchogale melleri) ALE 12:177; DDM 96 (cp 17); SMR 65 (cp 4).

Mongoose, Pousargues' (Dologale dybowshii) ALE 12:177; DDM 112 (cp 19).

Mongoose, Selous' (Paracynictus selousi) ALE 12:177; DDM 97 (cp 18); SMR 65 (cp 4).

Mongoose, Slender (Herpestes sanguineus) DDM 112 (cp 19); SMR 48 (cp 3); (H. gracilis) TWA 51.

Mongoose, Tropical Savannah—See **Mongoose, Pousargues'.**

Mongoose, White-tailed (Ichneumia albicauda) ALE 12:177; DDM 96 (cp 17); SMR 65 (cp 4).

Mongoose, Yellow—See **Meerkat, Red or Bushy-tailed.**

Monitor, Bosc's or Savannah (Varanus exanthematicus) ERA 188, 189; LEA 310 (bw); SIR 108 (cp 52).

Monitor, Desert or Gray (Varanus griseus) ART 176, 206; CAS 107; HRA 22; SIR 171 (bw).

Monitor, Earless (Lanthanotus borneensis) ALE 6:341; ERA 165.

Monitor, Gould's or Sand (Varanus gouldii) CAU 147; COG cp 24; ERA 190; MAR 133 (pl 53) SIR 172 (bw).

Monitor, Komodo Dragon (Varanus komodoensis) ALE 6:318, 334, 335; ART 101; CAS 287; DAA 153; ERA 182; HRA 23; HWR 89, 90 (bw); LBR 29, 34; LVS 176-177; MAR 134 (pl 54), 135 (pl 55); SIR 169 (bw); SLP 190, 191.

Monitor, Lace or Variegated (Varanus varius) ALE 6:318; COG cp 23; ERA 190; WAW 78 (bw).

Monitor, New Guinea or Salvador's (Varanus salvadori) VRA 203; WAW 80 (bw).

Monitor, Nile (Varanus niloticus) ALE 6:318, 333; ERA 190; HRA 20; HRW 106; LAW 420; LEA 288 (bw); MAR 134 (pl 54).

Monitor, Perentie (Varanus giganteus) COG cp 22; MEA 27; WAW 81 (bw).

Monitor, Spiny-tailed (Varanus acanthurus) ALE 6:317, 333; COG cp 25.

Monitor, Two-banded (Varanus salvator) ALE 6:318; ART 175; HRA 21.

Monkey, Black Spider (Ateles paniscus) ALE 8:359, 10:340; DSM 78; TWA 131.

Monkey, Black-cheeked White-nosed (Cercopithecus ascanius) DDM 61 (cp 8); SMR 16 (cp 1).

Monkey, Blue—See **Guenon, Diademed.**

Monkey, Brazza's (Cercopithecus neglectus) ALE 10:392 (cp 7); DDM 64 (cp 9) DSM 81; LAW 619; LEA 509 (bw) MLS 152 (bw); (C. brazzae) HMW 24.

Monkey, Capuchin—See **Capuchin.**

Monkey, Colobus—See **Colobus.**

Monkey, Crab-eating—See **Macaque, Crab-eating or Long-tailed.**

Monkey Cups—See **Pitcher Plant.**

Monkey, Diana (Cercopithecus diana) ALE 10:393 (cp 8); DDM 64 (cp 9); DSM 81; HMA 71; HMW 24; SMW 85 (cp 45).

Monkey, Grass—See **Monkey, Vervet.**

Monkey, Greater White-nosed (Ceropithecus nicticans) DDM 61 (cp 8); LAW 619.

Monkey, Green (Cercopithecus sabaeus) HMW 23; TWA 16 (bw).

Monkey, Guereza—See **Colobus, Black and White.**

Monkey, Howler (Alouatta seniculus) ALE 8:359; 10:328; DSM 75; HMW 27; JAW 42; LEA 505 (bw); MLS 131 (bw); TWA 130; (A. palliata) LAW 613; (A. villosa) HMA 71.

Monkey, Langur or Leaf—See **Langur.**

Monkey, Lesser White-nosed (Cercopithecus petaurista) ALE 10:391 (cp 6); DDM 61 (cp 8).

Monkey, L'Hoest's (Cercopithecus l'hoesti) ALE 10:392 (cp 7); DDM 64 (cp 9).

Monkey, Macaque—See **Macaque.**

Monkey, Mangabey—See **Mangabey.**

Monkey, Marmoset—See **Marmoset.**

Monkey, Mona (Cercopithecus mona) ALE 10:393 (cp 8); DDM 64 (cp 9); DSM 81; HMW 23; MLS 151 (bw).

Monkey, Moor—See **Macaque, Moor.**

Monkey, Night—See **Ape, Night.**

Monkey, Patas (Erythrocebus patas) ALE 10:394 (cp 9); CAF 97 (bw); DDM 60 (cp 7); DSM 88; HMW 22; LEA 509 (bw); MLS 154 (bw); SMW 80 (bw).

Monkey, Proboscis (Nasalis larvatus) ALE 10:446; CAS 232, 233 (bw); DAA 137; HMA 71; HMW 17; JAW 57; MLS 156 (bw).

Monkey Puzzle Tree (Araucaria araucana) BKT cp 15; DEW 1:41; EET 246; EPF pl 153; MTB 48 (cp 1); OBT 105; 132; SST cp 7; TGS 23; (A. imbricata) NFP 152.

Monkey, Red-eared Nose-spotted (Cercopithecus erythrotis) DDM 61 (cp 8); HMW 23.

Monkey, Rhesus—See **Macaque, Rhesus.**

Monkey, Saki—See **Saki.**

Monkey, Snub-nosed (Rhinopithecus roxellanae) ALE 10:455, 12:103; DAA 49; SLS 103; (R. bieti) ALE 10:455.

Monkey, Spider (Ateles geoffroyi) ALE 10:340; HMA 71; HMW 25; JAW 44; LAW 614, 615; MLS 134 (bw); SMW 84 (cp 43); (A. ater) LEA 505 (bw); (Brachyteles arachnoides) LVS 46.

Monkey, Squirrel (Saimiri sp.) ALE 10:317, 337; DSM 79; HMW 26; LEA 505 (bw); MLS 133 (bw); SMW 84 (cp 44).

Monkey, Talapoin—See **Guenon, Dwarf.**

Monkey, Tamarin—See **Tamarin.**

Monkey, Titi—See **Titi.**

Monkey, Uakari—See **Uakari.**

Monkey, Vervet (Cercopithecus aethiops) ALE 10:394 (cp 9); CAA 97, 101; CAF 92, 133 (bw); DDM 64 (cp 9); DSM 80; HMA 8; JAW 55; LEA 475 (bw); MLS 150 (bw); SMR 16 (cp 1); SMW 85 (cp 46); TWA 17; (C. pygerythrus) LAW 619.

Monkey, Woolly (Lagothrix sp.) ALE 10:337; DSM 79; HMW 25; JAW 45; MLS 135 (bw); SMW 75 (bw).

Monkey-faces—See **Pansy, Wild.**

Monkey-flower (Mimulus cupreus) DFP 159 (cps 1271, 1272); EGA 137; ESG 129; LFW 102 (cps 228, 229) 225 (cp 506); OGF 75; TVG 175; (M. moschatus) PFE cp 123 (1217).

Monkey-flower (Mimulus sp.) DEW 2:168 (cp 92); DFP 14, 43, 159, 160, 214; EDP 146; EGA 137; ESG 129; EWG 126; LFT cp 146; OGF 103; PFW 279; SNZ 23 (cp 21); TGF 221; TVG 59; WFB 213 (1a).

Monkey-flower, Common (Mimulus guttatus) BFB cp 16(4); EPF pl 842; MBF cp 63; MFF cp 59 (695); OWF 25; PFE cp 123 (1217); PWF 107j; WFB 213 (1).

Monkfish (Squatina squatina) CSF 51; HFI 143; LEA 232 (bw); OBV 11.

Monk's Pepper Tree—See **Chaste Tree.**

Monk's Rhubarb (Rumex alpinus) MFF pl 108 (542); NHE 276; PFE cp 9 (94).

Monkshood (Aconitum anglicum) BWF 128; MBF cp 4; MFF cp 12 (6).

Monkshood, Common (Aconitum napellus) AMP 83; CTH 28 (cp 24); DEW 1:120; DFP 119 (cp 952); HPW 47 (7); NHE 268; PFE cp 19 (210); PWF 77e, 123e; SPL cp 379; TEP 237; WFB 75 (2).

Monkshood, Lesser (Aconitum variegatum) AMP 83; NHE 268; WFB 74.

Monkshood, Panicled (Aconitum paniculatum) EFC cp 45; NHE 268.

Monkshood, Spark's Variety (Aconitum cammarum or A. henryi) EGP 93; PFW 251.

Monkshood, Yellow (Aconitum vulparia) EPF pl 209; KVP cp 114; PFE cp 19 (208); TEP 237; WFB 75 (2).

Montbretia, Garden (Crocosmia crocosmiflora) DEW 2:222 (cp 139); DFP 87 (cp 694); EGB 106; MFF cp 63 (1054); PWF 148 d; SPL cp 249; (C. masonorum) DFP 87 (cp 695); TVG 185.

Monvillea (Monvillea sp.) ECS 15; LCS cps 111, 112.

Moon Carrot (Seseli sp.) MBF cp 39; MFF pl 95 (492); NHE 182, 222; WFB 159 (5).

Moonfish (Lampris sp.)—See **Opah**

Moonfish (Monodactylus argenteus)—See **Fingerfish, Common.**

Moon-rat (Echinosorex gymnurus) ALE 10:179; DSM 46; MLS 80 (bw); (Hylomis suillus) ALE 10:179.

Moonseed (Menispermum sp.) DEW 1:113; EGV 123.

Moonwort—See **Honesty.**

Moonstones (Pachyphytum sp.) ECS 136; EDP 142; EFP 129; EGL 132; HPW 145; RDP 291.

Moonwort (Botrychium lunaria) CGF 189; DEW 3: cp 183; KVP cp 112; MFF pl 136 (1305); NFP 128; NHE 198; ONP 57.

Moor-grass, Blue (Sesleria caerulea) MBF pl 96; MFF pl 123 (1200); NHE 292; PFE cp 179 (1737); TEP 95; (S. sp.) EPF pls 1061, 1062; NHE 292.

Moor-grass, Purple (Molinia caerulea) MBF pl 97; MFF pl 134 (1169); NHE 79; PFE cp 179 (1742).

Moorhen—See **Gallinule.**

Moorish Idol (Zanclus cornutus) ALE 5:202; CAA 15; HFI 49; VWS 6(bw); (Z. canescens) HFW 209 (cp 90); LEA 225.

Moor-king (Pedicularis sceptrum-carolinum) MLP 214; NHE 75; WFB 217 (5).

Moose, European—See **Elk, European.**

Moosewood—See **Maple, Striped.**

Morel, Common (Morchella vulgaris) CTM 78; DLM 124; LHM 41; REM 181 (cp 149); RTM 236.

Morel, Conical (Morchella conica) DEW 3: cp 43; DLM 123; LHM 41; NHE 47; PMU cp 3a; REM 181 (cp 149); RTM 237.

Morel, Early (Verpa bohemica) DLM 129; LHM 41; SMG 25.

Morel, Edible (Morchella esculenta) DLM 94, 124; EPF pl 45; KMF 127; LHM 40; NFP 8, 36; NHE 47; OFP 189; ONP 41; PFF 70; PMU 2b; SMG 28.

Morel, False (Gyromitra esculenta) CTM 78; DEW 3:120; DLM 121; LHM 39; NFP 36; NHE 48; PMU cp 2a; RTM 234; SMF 37.

Morepork—See **Owl, Boobook.**

Morisia (Morisia monantha) DFP 14 (cp 112); OGF 21.

Mormyrid (Gymnarchus niloticus) ALE 4:208; (Hyperopisus bebe) ALE 4:208.

Mormyrid, Elephant-snout (Gnathonemus sp.) FEF 71; HFI 114; HFW 103 (bw); LAW 306; SFW 72 (cp 9); WFW cp 34.

Mormyrid, Peters' or Ubangi (Gnathonemus petersi) ALE 4:208; HFI 114; HFW 102 (bw); SFW 61 (pl 8); WFW cp 33.

Morning Glory, Ground (Convolvulus mauritanicus) DFP 5 (cp 38); EGC 128; MWF 92.

Morning-glory (Ipomoea sp.) DFP 39 (cp 312); EHP 128, 132, 133, 136; EWF cp 79d; LFT cps 136, 137; MEP 116, 117; MWF 165; OGF 163; PFW 91.

Morning-glory, Common or Tall (Ipomea purpurea) EGA 127; EGV 116; FHP 127; HPW 230; LFW cp 2; PFF 283; TVG 55; WUS 303.

Morning-glory, Ivy-leaved (Ipomoea hederacea) PFE cp 100 (1033); WUS 301.

Morning-glory, Woolly (Argyreia nervosa) EHP 136; MEP 115.

Mosaic Plant—See **Nerve Plant.**

Moschatel (Adoxa moschatellina) EWF cp 16a; MBF cp 41; MFF pl 78 (827); NHE 36; OWF 49; PFE cp 137 (1308); PWF cp 22e; WFB 221 (8).

Moschosma—See **Nutmeg Bush.**

Moses in a Basket or **Boat**—See **Boat Lily.**

Mosquito (Culex sp.) ALE 2:385, 393; CIB 197 (cp 30); GAL 399 (bw); LAW 220; LEA 143 (bw); OBI 123; PEI 507 (bw); PWI 87 (bw); WEA 253; ZWI 187; (Aedes sp.) ALE 2:393; KIW 263 (cp 135); PWI 85 (bw); ZWI 220.

Mosquito, Malaria (Anopheles sp.) ALE 2:385, 393; GAL 399 (bw); OBI 123; ZWI 186.

Moss, Broom (Dicranum scoparium) DEW 3:244; NFP 106; ONP 183; TEP 227; (D. sp.) EPF pl 90; ONP 83.

Moss, Catherine's (Atrichum undulatum) DEW 3:243; NFP 105; ONP 181.

Moss, Cord (Funaria hygrometrica) DEW 3:246; NFP 106; ONP 41.

Moss, Forest-star (Mnium hornum) DEW 3:227; NFP 109; ONP 181; (M. sp.) DEW 3: cp 111; 247; NFP 109; NHE 44; ONP 179; PFF 94; TEP 228.

Moss, Hair-cap (Polytrichum commune) DEW 3:cps 107-109; 243; EGC 37; EMB 64; EPF pl 87; NFP 105; NHE 44; ONF 81; PFF 95; TEP 228; (P. sp.) DEW 3:228; EPF pl 91; ONP 29, 81; PFF 96.

Moss, Iceland (Cetraria sp.) AMP 89; DEW 3: Fig. 278; EPF pl 79; NHE 44; ONP 67, 163; TEP 277.

Moss, Mountain Fern (Hylocomium splendens) DEW 3: cp 116; NFP 110; ONP 179; TEP 228.

Moss, Nut (Diphyscium foliosum) DEW 3: cp 112; NFP 105.

Moss, Purple Horn-tooth or Burned Ground (Ceratodon purpureus) NFP 106; ONP 49; PFF 93.

Moss, Reindeer (Cladonia rangiferina)
DEW 3: cp 79; EPF pl 81; NFP 99; TEP 227.

Moss, Shaggy (Rhytidiadelphus sp.) NFP 109;
ONP 43, 179; TEP 228.

Moss, Slender (Plagiothecium sp.) DEW 3: cp
110; 249; NFP 110; ONP 183.

Moss, Tree (Climacium dendroides) NFP 110;
ONP 178.

Moss, Twisted (Tortella tortuosa) NFP 108;
ONP 61.

Moss, Water (Fontinalis sp.) DEW 3: 248; NFP
111.

Moth (Brahmaea wallichii) ALE 2:327; PEI
394 (bw); SBM 246, 247; WDB cp 340d;
(Acanthobrahmaea europaea) SBM 255;
WDB cp 340b; (Urania leilus) KBM 157, 181;
SBM 334; WDB cp 315; (Charagia sp.) ALE
2:359; WDB cp 7b; (Thysania agrippina) PEI
407 (bw); SBM 304-305 (bw); WDB cp 389; (T.
zenobia) KBM 181.

Moth, Ailanthus (Samia sp.) KBM 64 (pl 20);
WDB cp 333b.

Moth, Alder (Apatele alni) CIB 172 (cp 23);
SBM 298; (A. psi); CIB 172 (cp 23).

Moth, Atlas (Attacus atlas) CAS 265; KBM 55;
PEI 379 (bw); SBM 264-265 (bw); WDB cp
327; (A. edwardsii); PEI 369 cp (36); SBM
266, 267.

Moth, Australian Day-flying (Alcidis
metaurus) BIN cp 68; SBM 334; (A. agathyrsus) WDB cp 314b.

Moth, Bag-worm (Psyche sp.) ALE 2:341;
WDB cp 24; (Sterrhopteryx fusca) CIB 164
(cp 21).

Moth, Bee (Aphomia sociella) OBI 117; WDB
cp 60.

Moth, Black Arches (Lymantria monacha)
ALE 2:342; CIB 173 (cp 24); PEI 402 (bw);
SBM 312 (bw); WDB cp 357t.

Moth, Bordered White Beauty (Bupalus
piniaria) PEI 352 (bw); WDB cp 312.

Moth, Brimstone (Opisthograptis luteolata)
OBI 109; WDB cp 309.

Moth, Brindled Beauty (Lycia hirtaria) LEA
128; OBI 111.

Moth, Brown China Mark (Hydrocampa
nympheata or Nymphula nympheata) ALE
2:341; WDB cp 68.

Moth, Brown Tiger (Hyphoraia aulica) PEI
428 (bw); SBM 295.

Moth, Buff Arches (Habrosyne pyritoides)
CIB 192 (cp 27); OBI 69; SBM 296 (bw); WDB
cp 269.

Moth, Buff-tip (Phalera bucephala) CIB 165
(cp 22); CTI 27 (cp 28); OBI 67; PEI 343 (bw);
SBM 331; WDB cp 355.

Moth, Bulrush Wainscot (Nonagria typhae)
OBI 85; PEI 419 (bw); WDB cp 404.

Moth, Burnet (Zygaena sp.) ALE 2:348; CIB
160 (cp 19); CTB 19 (cp 4), 20 (cp 5); GAL 332
(cp 9); LAW 177; OBI 113; PEI 288 (cp 31a),
336 (bw); SBM 335; WDB cps 46a, 49, 50; ZWI
156; (Erasmia sanguiflua) KBM 77; PEI 337
(bw); WDB cp 45g.

Moth, Burnished Brass (Plusia sp.) CIB 172
(cp 23); WDB cps 396, 398.

Moth, Burren Green (Calamia tridens) SBM
311; WDB cp 391d.

Moth, Cabbage (Mamestra brassicae) OBI 75;
WDB cp 391h.

Moth, Chinese Character (Cilix glaucata)
CIB 192 (cp 27); OBI 69; WDB cp 270.

Moth, Chocolate-tip (Clostera sp.) OBI 67;
SBM 321 (bw).

Moth, Clearwing—See **Clearwing.**

Moth, Clifden Nonpareil (Catocala fraxini)
PEI 410 (bw); SBM 309 (bw); WDB cp 392e.

Moth, Clothes (Case-making) (Tinea pellionella) ALE 2:341; CIB 164 (cp 21); GAL 473
(bw).

Moth, Clothes (Webbing) (Tineola bisselliella) CIB 164 (cp 21); GAL 472 (bw); LAW 213;
PEI 317 (bw); WDB cp 5d.

Moth, Clouded Buff (Diacrisia sannio) SBM
292 (bw); WDB cp 371.

Moth, Codlin (Cydia pomonella) CIB 161 (cp
20); WDB cps 14e, 19; (C. saltitans) WDB cp
14b.

Moth, Common Footman (Lithosia lurideola)
OBI 93; (Eilema lurideola) CIB 165 (cp 22).

Moth, Common Heath (Ematurga atomaria)
OBI 111; WDB cp 299.

Moth, Coxcomb Prominent (Lophopteryx
capucina) OBI 67; SBM 325 (bw).

Moth, Crimson Speckled (Utethesia pulchella) PEI 426 (bw); WDB cp 366.

Moth, Dagger (Acronycta sp.) ALE 2:327; PEI
425 (bw); SBM 297 (bw); WDB cp 393.

Moth, Drinker (Philudoria potatoria) CIB 173
(cp 24); OBI 91; PEI 389 (bw); WDB cp 324.

Moth, Emperor (Saturnia pavonia) ALE 2:342; CIB 173 (cp 24); GAL 332 (cp 9); KBM 55, 67; LEA 141 (bw); OBI 65; SBM 257, 258; WDB cp 334; ZWI 128, 129.

Moth, Ermine (Spilosoma sp.) CIB 165 (cp 22); CTB 21 (cp 9); OBI 93; WDB cp 377; WEA 226; (Yponomeuta sp.) ALE 2:341; CIB 164 (cp 21); CTB 17 (cp 1); PEI 320, 321 (bw); WDB cp 31.

Moth, Flour or Meal (Pyralis farinalis) CIB 161 (cp 20); OBI 119; WDB cp 62.

Moth, Fox (Macrothylacia rubi) OBI 91; PEI 387 (bw); WDB cp 323.

Moth, Garden Tiger (Arctia caja) ALE 2:342; 11:105; CIB 165 (cp 22); GAL 332 (cp 9); KBM 67; KIW 174 (cp 86); OBI 95; PEI 429 (bw), 432 (cp 41a); SBM 291; (A. villica) CTB 21 (cp 10).

Moth, Geoffroy's Tubic (Alabonia geoffrella) OBI 121; WDB cp 40.

Moth, Goat (Cossus cossus) ALE 2:347; CIB 160 (cp 19); OBI 113; PEI 314 (bw); SBM 339; WDB cp 25.

Moth, Gold Fringe (Hypsopygia costalis) OBI 119; WDB cp 61.

Moth, Gold- or Yellow-tail (Euproctis similis) CIB 173 (cp 24); OBI 69; WDB cp 358.

Moth, Great Peacock (Saturnia pyri) CTB 23 (cp 14); PEI 369 (bw); SBM 258; WDB cp 340c.

Moth, Green Carpet (Colostygia pectinataria) CIB 192 (cp 27); OBI 99; WDB cp 274.

Moth, Green-silver Lines (Bena prasinana or B. fagana) CIB 172 (cp 23); GAL 332 (cp 9); OBI 87; PEI 422 (bw); WDB cp 403.

Moth, Gypsy (Lymantria dispar) CIB 173 (cp 24); PEI 403 (bw); WDB cp 357s; ZWI 82.

Moth, Hawk—See **Hawkmoth.**

Moth, Heart and Dart (Agrotis exclamationis) CIB 172 (cp 23); OBI 71.

Moth, Herald or Scalloped (Scoliopteryx libatrix) CIB 172 (cp 23); OBI 89; PEI 421; SBM 311; WDB cp 395.

Moth, Honeycomb (Galleria mellonella) GAL 473 (bw); OBI 117; WDB cp 56a.

Moth, Hornet—See **Clearwing.**

Moth, Ilia Underwing (Catocala ilia) KBM 87, KIW 163 (cp 66).

Moth, Indian Meal (Plodia interpunctella) PEI 330 (bw); WDB cp 56k.

Moth, Indian or Luna (Actias selene) ALE 2:327; PEI 385 (cp 38); SBM 282; WDB cp 336; (A. isis) PEI 377 (bw); SBM 276 (bw).

Moth, Iron Prominent (Notodonta dromedarius); CIB 165 (cp 22); PEI 342 (bw); SBM 322.

Moth, Kentish Glory (Endromis versicolora) CIB 173 (cp 24); PEI 392, 393 (bw); SBM 242-244; WDB cp 320.

Moth, Kitten (Harpyia furcula) CIB 165 (cp 22); (H. bifida) SBM 326.

Moth, Lackey (Malacosoma neustria) ALE 2:342; CIB 173 (cp 24); OBI 89; PEI 385 (bw); SBM 241 (bw).

Moth, Lappet (Gastropacha quercifolia) CIB 173 (cp 24); GAL 332 (cp 9); LAW 214; OBI 91; PEI 390 (bw); SBM 233-235; WDB cp 321.

Moth, Large Emerald (Geometra papilionaria) CIB 192 (cp 27); OBI 97.

Moth, Large Yellow Underwing (Noctua pronuba) CIB 172 (cp 23); OBI 73.

Moth, Leopard (Zeuzera pyrina) CIB 160 (cp 19); CTB 18 (cp 3); OBI 113; PEI 314 (bw).

Moth, Light Emerald (Campaea margaritata) OBI 107; WDB cp 300.

Moth, Lobster (Stauropus fagi) OBI 65; WDB cp 353.

Moth, Magpie (Abraxas grossulariata) ALE 2:348; CIB 192 (cp 27); KBM 113 (pl 39); OBI 107; PEI 351 (bw); WDB cp 301; (A. marginata) ALE 2:363; (A. sylvata) WDB cp 304.

Moth, Malagasy Silk (Argema mittrei) ALE 2:359; KBM 29; PEI 376 (bw), 400 (cp 39); SBM 270; (A. mimosae) WDB cp 332.

Moth, Many-plumed (Alucita pentadactyla) ALE 2:341; PEI 323 (bw); SBM 341 (bw); (A. hexadactyla) CIB 161 (cp 20); OBI 121; (Orneodes grammodactyla) ALE 2:341; SBM 340 (bw).

Moth, Mediterrranean Flour (Ephestia kuhniella) PEI 331 (bw); WDB cp 56e, s.

Moth, Mottled Umber (Erannis defoliaria) CIB 192 (cp 27); OBI 109; WDB cps 294d, 308.

Moth, Nail-mark—See **Moth, Tau Emperor.**

Moth, Oak Eggar (Lasiocampa quercus) ALE 2:342; CIB 173 (cp 24); CTI 30 (cp 33); OBI 91; SBM 236 (bw); WDB cp 325.

Moth, Oak or Poplar Beauty (Biston stratarius) OBI 111; PEI 344 (bw); WDB cp 280.

Moth, Oak Silk (Antherea sp.) PEI 372-374 (bw); SBM 277 (bw), 278; WDB cp 337a.

Moth, Oak-roller (Tortrix viridana) CIB 161 (cp 20); OBI 121; PEI 326 (bw); SBM 336 (bw); WDB cp 12.

Moth, Owlet (Agarista agricola) KBM 131; PEI 401 (cp 40a); SBM 303; WDB cp 399n; (Lobocraspis griseifusca)ALE 2:328.

Moth, Pale Prominent (Pterostoma palpina) CIB 165 (cp 22); OBI 67; WDB cp 354.

Moth, Pale Tussock (Dasychira pudibunda) CIB 173 (cp 24); OBI 69; PEI 400 (bw); SBM 317 (bw); WDB cp 360.

Moth, Peach-blossom (Thyatira batis) CIB 192 (cp 27); OBI 69; WDB cp 271.

Moth, Pebble Hook-tip (Drepana falcataria) CIB 192 (cp 27); OBI 69; WDB cp 272r.

Moth, Pebble Prominent (Notodonta ziczac) OBI 67; SBM 324 (bw).

Moth, Peppered (Biston betularia) CIB 192 (cp 27); KBM 145 (pl 49); WDB cps 302, 303.

Moth, Pine Lappet (Dendrolimus pini) PEI 388 (bw); SBM 236 (bw); WDB cp 318n.

Moth, Plume (Platyptilia sp.) OBI 121; WDB cp 78.

Moth, Primitive (Micropteryx calthella) ALE 2:341; SBM 343; WDB cp 2.

Moth, Processionary (Thaumetopoea processionea) CIB 165 (cp 22); PEI 398 (bw); WDB cp 357d.

Moth, Pug (Eupithecia sp.) CIB 192 (cp 27); OBI 105; WDB cps 288, 289.

Moth, Puss (Cerura sp.) ALE 2:363; CIB 165 (cp 22); KIW 173 (cp 84); OBI 65; PEI 341 (bw); SBM 328, 329 (bw); WDB cp 356; (Dicranura vinula) CTI 27 (cp 27); PEI 338, 339 (bw).

Moth, Red Underwing (Catocala nupta) ALE 2:347, 363; CIB 172 (cp 23); GAL 332 (cp 9); OBI 87; WDB cp 394; (C. elocata) PEI 401 (cp 40b), 408, 409 (bw).

Moth, Rosy Footman (Miltochrista miniata) CIB 165 (cp 22); WDB cp 382c.

Moth, Scalloped Oak (Crocallis elinguaria) OBI 107; WDB cp 297.

Moth, Scarlet Tiger (Panaxia dominula) GAL 332 (cp 9); PEI 429 (bw); SBM 293 (bw).

Moth, Shark (Cucullia umbratica) CIB 172 (cp 23); OBI 77; PEI 414 (bw); (C. campanulae) PEI 415 (bw).

Moth, Silk (Graellsia isabellae) KBM 29; PEI 375 (bw); SBM 269 (bw); WDB cp 337b; (Rothcschildia sp.) ALE 2:327; PEI 380 (bw); SBM 279; WDB cp 337c.

Moth, Silver-Y (Plusia gamma or Phytometra gamma) CIB 172 (cp 23); KBM 23 (pl 3); OBI 89; PEI 423 (bw).

Moth, Small Eggar (Eriogaster lanestris) PEI 386 (bw); SBM 237 (bw).

Moth, Small Magpie (Eurrhypara hortulata) CIB 161 (cp 20); OBI 119; WDB cp 71.

Moth, Speckled Yellow (Pseudopanthera macularia) CIB 192 (cp 27); OBI 109; WDB cp 310.

Moth, Sphinx—See **Hawkmoth.**

Moth, Square-spot Rustic (Amathes xanthographa) OBI 71; (A. c-nigrum); WDB cp 391j.

Moth, Swallow Prominent (Pheosia tremula) CIB 165 (cp 22); OBI 67.

Moth, Swallow-tailed (Ourapteryx sambucaria) CIB 192 (cp 27); OBI 109; WDB cp 306.

Moth, Swift (Hepialus sp.) CIB 160 (cp 19); OBI 115; PEI 311 (bw); SBM 343, 344 (bw); WDB cp 10; (Leto staceyi) KBM 131; (L. venus) WDB cp 7c.

Moth, Sycamore (Apatele aceris) OBI 81; SMB 296 (bw).

Moth, Tapestry (Trichophaga tapetzella) CIB 164 (cp 21); PEI 317 (bw); WDB cp 5b.

Moth, Tau Emperor (Aglia tau) ALE 2:327; GAL 332 (cp 9); PEI 381 (bw); SBM 259-263; WDB cp 331k.

Moth, Tiger (Callimorpha sp.) ALE 2:363; CIB 165 (cp 22); GAL 332 (cp 9); KIW 213 (cp 99); OBI 95; WDB cp 378.

Moth, True Silk (Bombyx mori) CTI 28 (cp 30); SBM 284-287; WDB cp 317e; ZWI 226.

Moth, Vapourer (Orgyia antiqua) CIB 173 (cp 24); LAW 179; OBI 69; SBM 316, 317 (bw); (O. recens) ALE 2:342.

Moth, Wainscot (Leucania pallens) OBI 77; PEI 418 (bw); (L. impura) CIB 172 (cp 23).

Moth, White Plume (Pterophorus pentadactyla) CTB 20 (cp 6); LAW 213; WDB cp 55.

Moth, White Prominent (Leucodonta bicoloria) CIB 165 (cp 22); WDB cp 352.

Moth, Winter (Operophtera brumata) OBI 103; WDB cp 282; ZWI 82.

Moth, Wood Tiger (Parasemia plantaginis) OBI 93; WDB cp 375.

Moth, Yellow Underwing (Triphaena pronuba) ALE 2:347; PEI 424 (bw).

Mother-in-law's Armchair—See **Cactus, Barrel.**

Mother-in-law's Tongue (Sansevieria trifasciata) BIP 159 (cp 242); DEW 2:211 (cp 118); DFP 81 (cp 643); ECS 141; EDP 134; EFP 141; EPF pl 918; LCS cp 295; LHP 161; MWF 262; RDP 27, 348; (S. sp.); DFP 81 (cps 641, 642); LCS cp 292; LHP 160, 161; PFF 376.

Mother-of-thousands—See **Aaron's Beard, Baby's Tears, Youth-on-Age.**

Mother-of-thyme—See **Thyme, Wild.**

Motherwort (Leonurus cardiaca) AMP 83; MFF cp 29 (773); NHE 230; OWF 149; PFE cp 112 (1133); PFF 289; WFB 201 (4).

Motmot, Blue-crowned (Momotus momota) ALE 9:32, 48; CDB 127 (cp 496); SSA cp 34; WAB 96; WEA 253.

Motmot, Rufous (Baryphthengus ruficapillus) ALE 9:32; CDB 127 (cp 495).

Motmot, Turquoise-browed (Eumomota superciliosa) ABW 180; ALE 9:48; GPB 211.

Mottlecah—See **Blue Bush.**

Mouflon (Ovis musimon) ALE 13:497; CEU 78 (bw); DSM 274; JAW 208; LEA 608 (bw); (O. mouflon) HMW 82; SMW 286 (cp 186).

Mountain Beauty (Hovea longifolia) EWF cp 129c; MWF 157.

Mountain Lion—See **Cougar.**

Mountain-ash, Korean (Sorbus alnifolia) EEG 122; EJG 145.

Mountain-tanager (Anisognathus sp.) ALE 9:335, 361; DTB 113 (cp 56), 115 (cp 57); SSA cp 40; (Dubusia taeniata) SSA cp 17; (Buthraupis montana) DTB 117 (cp 58).

Mournful Widow—See **Scabious, Sweet.**

Mourning Bride—See **Scabious, Sweet.**

Mourning Widow—See **Cranesbill, Dusky.**

Mouse, African Climbing (Dendromus insignis) ALE 11:372; CAA 77.

Mouse, Birch (Sicista betulina) ALE 11:399; DAE 101; LEA 532 (bw); (S. subtilis) DAE 101.

Mouse, European Harvest (Micromys minutus) ALE 11:257, 371; DAE 65; GAL 52 (bw); HMW 172; LEA 534 (bw); OBV 133; SMW 136 (cp 83).

Mouse, Fat-tailed Pouched (Sminthopsis crassicaudata) ALE 10:62; DSM 28; HMW 198; MEA 92; MLS 51 (bw); MWA 42; (S. murina) CAU 30 (bw); WAW 46 (bw).

Mouse, Hopping or Kangaroo (Notomys sp.) ALE 10:89; MEA 114; WAW 65 (bw).

Mouse, House (Mus musculus) ALE 11:371; DAE 64; DSM 122; GAL 51 (bw); HMA 68; HMW 170; JAW 82; LAW 590; OBV 135; SMW 134 (cp 78); WEA 199.

Mouse, Jerboa-like Pouched (Antechinomys laniger) DSM 28; HMW 198; (A. spenseri) ALE 10:62.

Mouse, Long-tailed Field (Apodemus sylvaticus) ALE 2:363; 11:299, 371; DAE 64; GAL 52 (bw); HMW 172; LAW 589; LEA 535 (bw); MLS 214 (bw); OBV 133; TWA 95; WEA 379.

Mouse, Marsupial (Antechinus sp.) ALE 10:62; MEA 91; SMW 37 (cp 7); WAW 46 (bw); (Dasycercus cristicauda) DSM 28; (Saccostomus sp.) SMW 130 (cp 71).

Mouse, Spiny (Acomys sp.) ALE 11:371; CAS 95; DAE 65; HMA 68; MLS 217 (bw); SMW 134 (cp 79).

Mouse, Striped Field (Apodemus agrarius) ALE 11:371; GAL 53 (bw); HMW 172.

Mouse, Wood—See **Mouse, Long-tailed Field.**

Mouse, Yellow-necked (Apodemus flavicollis) ALE 11:321, 371; LAW 58; OBV 133.

Mouse, Zebra (Rhabdomys pumilio) ALE 11:371; LEA 539 (bw).

Mousebird, Red-faced (Colius indicus) CDB 124 (cp 479); NBA 45; PSA 209 (cp 24).

Mousebird, Speckled (Colius striatus) CDB 124 (cp 480); NBA 47.

Mousebird, White-backed (Colius colius) GPB 206; NBA 47.

Mouse-ear, Clustered—See **Chickweed, Sticky.**

Mouse-ear, Common (Cerastium holosteoides) BWG 66; MFF pl 86 (167); NHE 184.

Mouse-deer—See **Chevrotain.**

Mouse, Honey—See **Possum, Honey.**

Mouse-lemur, Lesser—See **Lemur, Lesser Mouse.**

Mouse-tail (Myosurus minimus) EWF cp 8g; HPW 48; MBF cp 1; MFF pl 106 (24); NHE 210; OWF 5; PFE cp 26 (252); PWF 30a; WFB 75 (8).

Moutan—See **Peony, Tree.**

Mouthbrooder (Haplochromis sp.) ALE 5:128; FEF 410, 438, 439 (bw); HFI 36; SFW 564 (pl 153); WFW cps 362-365.

Mouthbrooder (Tilapia sp.) ALE 5:122; FEF 454, 455 (bw); HFI 36; HFW 176 (cp 88); SFW 565 (pl 154), 572 (pl 155); 596 (pl 162), 597 (pl 163).

Mrs. Robb's Bonnet (Euphorbia robbiae) DFP 142 (cp 1133); EWF cp 35c.

Mucuna (Mucuna sp.) EET 8; EWF cp 112b.

Mudfish—See **Loach, Pond.**

Mudlark—See **Magpie-lark.**

Mudminnow, European (Umbra krameri) FEF 63 (bw); HFI 124; LEA 241 (bw).

Mudskipper (Periophthalmus sp.) ALE 5:172; FEF 496, 497 (bw); HFI 55; HFW 236 (bw); LEA 256; SFW 617 (pl 170); WEA 255.

Mudwort (Limosella sp.) LFT cp 149; MBF cp 63; MFF cp 66 (696, 697); NHE 99; WFB 291 (5).

Mugger—See **Crocodile, Mugger.**

Mugwort (Artemisia vulgaris) BWG 51; MBF cp 46; MFF pl 101 (896); NHE 238; OWF 61; PWF 142a; TEP 201; TGF 237; WFB 239 (4); WUS 375.

Mugwort, Breckland (Artemisia campestris) MBF cp 46; Mff pl 101 (899); NHE 238; PWF 162 a.

Mulberry, Black (Morus nigra) AMP 59; BKT cp 96; CTH 25 (cp 19); DEW 1:158; EET 159; EPF pl 350; HPW 97; MFB 256 (cp 23); MWF 203; OBT 160, 161; OFP 95; SST cp 137.

Mulberry, Paper (Broussonetia papyrifera) EGT 101; EPF pl 360; PFF 148; TGS 134.

Mulberry, Red (Morus rubra) EET 159; PFF 149.

Mulberry, White (Morus alba) BKT cp 95; EET 159; EFP pl 349; EGT 128; EGW 128; MTB 256 (cp 23); PFE cp 6 (61); PFF 148; SST cp 136; TGS 87.

Mulla Mulla—See also **Lamb's-tails** (Ptilotus sp.).

Mulla Mulla (Trichinium sp.) DFP 27 (cp 215); MWF 287.

Mullein (Verbascum sp.) DFP 28 (cp 217), 176 (cps 1407, 1408) 177 (cps 1409, 1410); HPW 244; NHE 199; PFE cp 120 (1195, 1196); PFM cps 176, 177.

Mullein, Common or Great (Verbascum thapsus) AMP 61; BFB cp 16 (9); BWG 44; EFC cp 59; MBF cp 62; MFF cp 62 (676); OWF 29; PFF 295; PWF 95k; TEP 119; WFB 209 (1).

Mullein, Dark (Verbascum nigrum) BFB cp 1 (6); MBF cp 62; MFF cp 62 (679); NHE 233; OGF 113; PFE cp 120 (1190); PWF 99l; WFB 209 (2).

Mullein, Hoary (Verbascum pulverulentum) MBF cp 62; MFF cp 62 (678); NHE 233.

Mullein, Large-flowered (Verbascum thapsiforme) DEW 2:169 (cp 95); EPF pl 836; NHE 233; PFE cp 120 (1193); TEP 120.

Mullein, Moth (Verbascum blattaria) EWG 145; NHE 98; PFE cp 120 (1192); WFB 209 (4); WUS 333.

Mullein, Orange (Verbascum phlomoides) CTH 50 (cp 73); KVP cp 80; NHE 233; SPL cp 411.

Mullein, Purple (Verbascum phoeniceum) EWF cp 21a; NHE 187; PFE cp 121 (1192); TEP 120; TGF 221.

Mullein, White (Verbascum lychnitis) MBF cp 62; MFF pl 101 (677); NHE 233; TEP 120; WFB 209 (3).

Mullet (Mugil sp.) ALE 5:138; FEF 326; HFI 63; WFW cp 395.

Mullet, Golden Gray (Liza auratus) CSF 138; OBV 75.

Mullet, Red (Mullus surmuletus) ALE 5:102; CSF 125; FEF 362 (bw); HFI 31; LAW 343; LEA 240; OBV 31.

Mullet, Red (Pseudupeneus sp.) —See **Goatfish.**

Mullet, Thick-lipped Gray (Crenimugil labrosus) CSF 139; FEF 326 (bw); OBV 75.

Munia, Chestnut —See **Mannikin, Chestnut.**

Munia, Scaly-breasted —See **Finch, Spice.**

Muntjac (Muntiacus sp.) ALE 13:209; CAS 153 (bw); DAA 39, 108; DAE 58; DSM 228; FBW 23; GAL 43 (bw); HMA 43; LEA 600 (bw); MLS 384 (bw); OBV 149; SMW 238 (cp 152); WEA 136.

Murex Shell (Murex sp.) ALE 3:71; CTS 42 (cp 68), 43 (cp 70); GSS 18, 20, 71-75; LAW 111; VWS 95, 96 (bw).

Murre, Common (Uria aalge) ABW 139; ALE
  7:61, 8:208; BBB 260; BBE 161; CDB 96 (cp
  346); DAE 30; FBE 165; OBB 101; PBE 125
  (pl 36); RBE 180; SAO 143.
Murre, Thick-billed (Uria lomvia) ALE
  8:229; BBE 161; CAS 22 (bw); FBE 165; PBE
  125 (pl 36); SAO 143; WAB 72.
Mushroom (Agaricus bisporus) DLM 101,
  162; LHM 133; ONP 31; PUM cp 140.
Mushroom, Amanitopsis (Amanitopsis sp.)
  CTM 23; KMF 11; NFP 55; RTM 31-33; SMF
  8.
Mushroom, Big or Shaggy (Agaricus
  augustus) DLM 161; EPF cp IIa; LHM 131;
  PMU cp 107; SMF 15; SMG 194, 195.
Mushroom, Caesar's or Orange (Amanita
  caesarea) CTM 17; DLM 72, 165; KMF 11;
  NFP 54; PFF 81; REM 57 (cp 25); RTM 17.
Mushroom, Deer (Pluteus cervinus) CTM
  47; DEW 3:172; DLM 239; LHM 121; NFP
  77; ONP 135; PMU cp 108; REM 68 (cp 36);
  RTM 41.
Mushroom, Fairy Cake (Hebeloma crustu-
  liniforme) DLM 194; LHM 161; ONP 47;
  REM 103 (cp 71); RTM 65.
Mushroom, Fairy Ring (Marasmius
  oreades) CTM 33; DLM 25, 91, 223; LHM
  115; NFP 67; OFP 189; ONP 35; PFF 85;
  PMU cp 83a; REM 134 (cp 102); RTM 91;
  SMF 20; SMG 155.
Mushroom, Field (Agaricus campestris)
  DLM 64, 163; KMF 115; LHM 133; NFP 81;
  OFP 189; ONP 31; PFF 87; PMU cp 104;
  SMF 10; SMG 190.
Mushroom, Fly (Amanita muscaria) CTM 20;
  DEW 3:171; DLM 45, 55, 76, 167; EET 34;
  EHP 23, 24, 26; EPF pl 64, cp IIb; LHM 117;
  NFP 53; NHE 48; ONP 113; PFF 81; PMU cp
  116; REM 58 (cp 26); RTM 18, 19, 269; SMF
  33; SMG 177, 178; TEP 221.
Mushroom, Garden (Agaricus hortensis)
  PMU cp 105; PUM cp 139.
Mushroom, Green-lined Parasol (Lepiota
  morgani) DLM 219; NFP 58; RTM 267.
Mushroom, Gypsy (Rozites caperata) DLM
  242; LHM 173; PMU cp 97; REM 104 (cp 72);
  RTM 62; SMG 214.
Mushroom, Horse (Agaricus arvensis) CTM
  48; DEW 3: cp 66, 175; DLM 161; EPF pl 61;

LHM 135; NFP 81; PMU 106; SMF 10.
Mushroom, Oyster (Pleurotus ostreatus)
  DEW 3:168; DLM 97, 238; LHM 107; NFP
  70; OFP 189; ONP 125; PFF 83; PMU cp 68;
  REM 140 (cp 108); RTM 124; SMF 25.
Mushroom, Parasol (Lepiota procera) CTM
  24; DEW 3: cp 65; DLM 96, 219; EPF pls 65,
  68; KMF 111; LHM 125; NFP 58; NHE 45;
  ONP 31; PFF 81; PMU cp 112; REM 75 (cp
  43); RTM 35; SMF 111; TEP 225.
Mushroom, Pavement (Agaricus bitorquis)
  DLM 162; LHM 133.
Mushroom, Psalliota (Psalliota sp.) NHE 45;
  REM 60-73 (cps 37-41); RTM 43-49, 267.
Mushroom, St. George's (Tricholoma
  gambosum or T. georgii) LHM 83; PMU cp
  72; SMF 9.
Mushroom, Shaggy Parasol (Lepiota
  rhacodes) DLM 64, 219; LHM 125; NFP 58;
  NHE 45; OFP 189; PMU cp 110; REM 76 (cp
  44); RTM 36; SMF 1.
Mushroom, Sweetbread (Clitopilus sp.)
  CTM 33; DLM 180; LHM 179; NFP 78; PMU
  cp 90; REM 111; RTM 122.
Mushroom, Winter or Velvet Foot
  (Flammulina velutipes) DLM 191; LHM
  109; ONP 141; SMG 154.
Mushroom, Wood (Agaricus silvaticus) DLM
  163; LHM 131; ONP 103; PMU cp 102; SMG
  197; (A. silvicola) DLM 59, 164; NFP 82;
  ONP 103; SMG 199.
Mushroom, Yellow Cow-pat (Bolbitius
  vitellinus) DLM 171; LHM 153.
Musk —See **Monkey-flower.**
Mussel (Musculus discors) CSS cp 22; OIB 73;
  (M. marmoratus) ALE 3:140; CSS cp 22.
Mussel, Blue or Edible (Mytilus edulis) ALE
  3:140; CSF 207; CSS cp 22; GSS 133; LEA 92
  (bw); OIB 73.
Mussel, Honey Orb (Sphaerium corneum)
  ALE 3:142; GAL 517 (bw); OIB 91.
Mussel, Horse (Modiolus modiolus) CSS cp
  22; OIB 73; (M. barbatus) ALE 3:140; CSS
  cp 22.
Mussel, Painter's (Unio pictorum) ALE
  3:142; CTS 60 (cp 113); GAL 515 (bw); OIB
  91.
Mussel, Pea (Pisidium sp.) ALE 3:142; GAL
  516, 517 (bw); OIB 91.

Mussel, Pearl (Margaritifera margaritifera) ALE 3:142; GAL 516 (bw); (Unio margaritifera) OIB 91.

Mussel, Swan (Anodonta cygnea) ALE 3:142; GAL 516 (bw); OIB 91.

Mussel, Zebra (Dreissena polymorpha) ALE 3:142; OIB 91.

Mustard, Ball (Neslia paniculata) MCW 54; NHE 217; PFE 135; WFB 85 (6).

Mustard, Black (Brassica nigra) BFB cp 3 (10); MBF cp 9; MFF cp 45 (48); NHE 215; OFP 133; OWF 9; PFF 185; WFB 89 (3); WUS 195.

Mustard, Buckler (Biscutella laevigata) EWF cp 30; NHE 217; PFM pl 5 (338); WFB 85 (7); (B. didyma) HPW 121.

Mustard, Field or Wild (Brassica kaber) EWG 79; PFF 184; WUS 193.

Mustard, Garlic (Alliaria petiolata) BFB cp 4 (8); BWG 65; EPF pl 251; EWG 78; MBF cp 8; MFF pl 93 (106); OWF 69; PFE cp 31 (289); PWF cp 23i; TEP 239; WFB 91 (5); (A. officinalis) NHE 30.

Mustard, Hare's-ear (Conringia orientalis) MCW 49; NHE 216; PFE 135; WUS 203.

Mustard, Hedge (Sisymbrium officinale) BWG 24; MBF cp 8; MFF cp 44 (107); NHE 219; OWF 9; PWF 36b; WFB 87 (2).

Mustard, Hoary (Hirschfieldia incana) MFF cp 45 (53); PFE 128.

Mustard, Tower (Arabis glabra) NHE 30; PWF 73i; WFB 95 (6); (Turritis glabra) MBF cp 7; MFF cp 43 (97).

Mustard, Tumbling —See **Rocket, Tall.**

Mustard, White (Sinapis alba) AMP 23; DEW 1:145; MBF cp 9; MFF cp 45 (52); NHE 215; OFP 133, 153; OWF 9; PFE cp 37 (362); (Brassica alba) EGH 104; EPF pl 241; PFF 184.

Mustard, Wild —See **Charlock.**

Mutisia (Mutisia olidodon) BIP 123 (cp 188); DFP 248 (cp 1); HPW 264; HSC 84; (M. sp.) EWF cps 186 c, e; HSC 79; MEP 153; PFW 89.

Myall —See **Thorn-tree** (Acacia sp.).

Mycena, Amethyst or Lilac (Mycena pura) DLM 227; LHM 113; NFP 68; ONP 121; PMU cp 83b; REM 139 (cp 107); RTM 92.

Mycena, Bleeding (Mycena haematopus) DLM 226; KMF 23; LHM 111; NFP 68; RTM 93.

Mycena, Yellow-stemmed (Mycena epipterygia) DLM 225; LHM 111; ONP 105; RTM 93.

Mynah, Bali or Rothschild's (Leucopsar rothschildi) ALE 9:449; BBT 42; CDB 217 (cp 971); LVS 233.

Mynah, Coleto (Sarcops calvus) ALE 9:449; DPB 371.

Mynah, Common or Indian (Acridotheres tristis) ALE 9:449; CDB 216 (cp 963); FNZ 208 (cp 18); GPB 328; HAB 222; KBA 416 (cp 61); NBA 77; SAB 72 (cp 35); (Sturnus tristis) WBV 165.

Mynah, Crested (Acridotheres cristatellus) DPB 371; KBA 411 (bw).

Mynah, Golden-crested (Ampeliceps coronatus or Mino coronatus) ABW 273; KBA 416 (cp 61).

Mynah, Hill or Talking (Gracula religiosa) ABW 273; ALE 9:449; CDB 216 (cp 968); DPB 371; KBA 416 (cp 61); WBV 165; WEA 256.

Myrmecodia (Myrmecodia sp.) DEW 2:135; EPF pl 766.

Myrobalan (Terminalia sp.) HPW 165; LFT cp 115; SST cp 298.

Myrsine (Myrsine sp.) HPW 136; PFW 190; SNZ 71 (cp 195), 97 (cp 289).

Myrtle, Common or True (Myrtus communis) AMP 159; BIP 125 (cp 189); EFP 125; EGE 136; EGH 126; EPF pl 704; GDS cp 298; HSC 79, 84; MWF 206; NHE 246; OGF 103; PFE 81 (824); PFW 191; RDP 271; SPL cp 397.

Myrtle, Creeping —See **Periwinkle, Lesser.**

Myrtle, Downy or Rose (Rhodomyrtus tomentosa) MEP 99; MWF 253.

Myrtle, Wax —See **Bayberry.**

Mysore-thorn (Caesalpinia sepiaria) MEP 52; SPL cp 157.

N

Nimblewill (Muhlenbergia schreberi) EGC 72; WUS 71.

Nipplewort (Lapsana communis) BWG 52; MBF cp 50; MFF cp 53 (925); NHE 239; OWF 41; PWF 87h; WFB 253 (7).

Noah's Ark (Arca sp.) ALE 3:139; OIB 85.

Noddy —See **Tern, Noddy.**

Nolanea (Nolanea sp.) DLM 228; ONP 31, 103; RTM 83.

Nonpareil —See **Parrot-finch.**

Norfolk Island Pine (Araucaria heterophylla or A. excelsa) BIP 27 (cp 18); EDP 136; EET 246; EFP 89; EGE 90; LHP 31, 32; PFF 122; RDP 86; SST cp 9.

Nothobranchius (Nothobranchius sp.) ALE 4:450, 453; FEF 253, 254, 286, 287; WFW cp 202.

Nudibranch (Acanthodoris pilosa) CSS cp 20; MOL 70; OIB 47; (Dendronotus sp.) ALE 3:113; VWS 85.

Nudibranch (Coryphella sp.) LAW 125; LEA 89 (bw); WEA 328; (Hexabranchus sp.) ALE 3:103, 119; MOL 70.

Nudibranch (Facelina sp.) ALE 3:104, 114, 120; CSS cp 21; (Glaucus sp.) ALE 3:103, 120, 121; MOL 60.

Nudibranch (Glossodoris sp.) ALE 3:104, 119; CAU 136; LAW 113; MOL 70; (Polycera quadrilineata) ALE 3:103; CSS cp 20; OIB 47; (Onchidoris sp.) CSS cp 19; OIB 47.

Nudibranch, Gray (Aeolidia papillosa) CSS cp 21; OIB 49.

Nudibranch, Spotted (Peltodoris atromaculata) ALE 3:104, 119; LAW 47.

Numbat —See **Anteater, Banded or Marsupial.**

Nunlet (Nonnula sp.) ALE 9:68; SSA cp 34.

Nursehound —See **Dogfish, Greater Spotted.**

Nut Orchid —See **Magic Flower.**

Nutcracker (Nucifraga caryocatactes) ALE 9:497; BBB 270; BBE 217; FBE 305; PBE 197 (cp 50); RBE 232.

Nuthatch, Chestnut-bellied (Sitta castanea) ALE 9:319; KBA 320 (cp 49).

Nuthatch, Coral-billed or Madagascar (Hypositta corallirostris) ABW 239; ALE 9:319.

Nuthatch, Corsican (Sitta whiteheadi) ALE 9:319; BBE 273; DAE 59; FBE 271; PBE 212 (cp 51).

Nuthatch, European (Sitta europaea) ALE 9:301, 319, 13:163; BBB 147; BBE 273; CDB 186 (cp 811); DAE 128; FBE 271; GAL 148 (cp 4); LEA 451 (bw); OBB 133; PBE 212 (cp 51); RBE 244; WBV 221; WEA 261.

Nuthatch, Rock (Sitta neumayer) ALE 9:319; BBE 273; CDB 187 (cp 812); FBE 271; PBE 212 (cp 51); (S. tephronota) FBE 271.

Nuthatch, Turkish (Sitta krueperi) ALE 9:319; FBE 271.

Nuthatch, Velvet-fronted (Sitta frontalis) ABW 238; ALE 9:319; DPB 261; KBA 320 (cp 49).

Nutmeg (Myristica fragrans) EET 223; EHP 20; HPW 31; OFP 131; SST cp 290.

Nutmeg Bush (Iboza riparia) LFT 290; MEP 125; MWF 161.

Nutmeg, Calabash (Monodora myristica) EWF cp 53b; HPW 30; MEP 39; SST cp 288; (M. crispata) DEW 1:59 (cp 25).

Nutmeg, California (Torreya californica) EET 87; MTB 49 (cp 2); OBT 105, 132; PFF 113; SST cp 45; (T. sp.) DEW 1:47, 48; NFP 152; TGS 6.

Nutmeg, Flowering —See **Honeysuckle, Himalayan.**

Nutria (Myocastor coypu) ALE 11:438; AWW 52; CSA 169 (bw); DAE 54; DSM 135; FBW 49; GAL 62 (bw); HMA 73; HMW 175; LEA 523 (bw); MLS 233 (bw); OBV 131; SMW 143 (cp 94); WEA 115.

Nutsedge or **Nutgrass** (Cyperus esculentus) EGC 72; PFF 361; WUS 97.

Nyala, Mountain (Tragelaphus buxtoni) ALE 13:302; DDM 180 (cp 29); LVA 15; VWA cp 8.

Nyala, Zululand (Tragelaphus angasii) ALE 13:302; CAF 288 (bw); DDM 180 (cp 29); DSM 241; HMW 69; SLP 241; SMR 112 (cp 9); TWA 23.

O

Oak, Caucasian (Quercus castaneifolia) DEW 1:183; MTB 240 (cp 21); OBT 157.

Oak, Common or English (Quercus robur) BFB cp 14 (4); BKT cps 72, 78; DEW 1:105 (cp 57); EET 150, 152, 182; EPF pl 341; HPW 61; MBF cp 77; MFF pl 70 (564); MTB 209 (cp 20); NHE 23; OBT 12, 16; OWF 20, 202, 203; PFE cp 5 (52); PWF 35f, 175j; SST cp 156; TEP 279; TGS 102; WFB 32 (6), 33,(6).

Oak, Cork (Quercus suber) BKT cp 70; DEW 106 (cp 58), 183; EET 82; MLP 68; MTB 209 (cp 20); NHE 42; PFF 142; PFM cp 6, pl 2 (319); SST cp 293.

Oak, Holm or Holly (Quercus ilex) BKT cp 76; EET 151; EWF cp 31f; HPW 61; MLP 67; MTB 209 (cp 20); NHE 42; OBT 136; PWF 76d, 175h; SST cp 153; WFB 33 (5).

Oak, Hungarian (Quercus frainetto) MTB 240 (cp 21); OBT 157; SST cp 152.

Oak, Indoor (Nicodemia diversiflora) EDP 112; EFP 126; RDP 277.

Oak, Japanese (Quercus acuta) DEW 1:106 (cp 59); MTB 240 (cp 21).

Oak, Kermes or Holly (Quercus coccifera) EWF cp 31e; NHE 81; PFE cp 5 (47).

Oak, Pedunculate —See **Oak, Common or English.**

Oak, Sessile or Durmast (Quercus petraea) BKT cp 71; EET 151, 152; EPF pls 340, 342; MBF cp 77; MTB 209 (cp 20); NHE 23; OBT 12; PWF 35e, 175i; SST cp 155; TEP 280; WFB 33 (6b).

Oak, Turkey (Quercus cerris) BKT cp 74; EET 151; EFP pl 343; EWF cp 12b; MTB 209 (cp 20); OBT 157; PFE 54; SST cp 151.

Oat (Avena sativa) DEW 2:267; EPF pl 1044; HPW 287; MLP 306; OPF 5; PFF 353; TEP 170.

Oat, Animated (Avena sterilis) EGA 97; PFE cp 180 (1765); PFM pl 22 (426).

Oat Grass (Arrhenatherum elatius) BWG 178; DEW 2:261; MBF pl 96; MFF pl 127 (1220); NHE 195; TVG 127.

Oat, Wild (Avena fatua) MBF pl 96; MCW 15; MFF pl 127 (1218); NHE 244; PFF 354; WUS 39; (A. sempervirens) TVG 127.

Oat-grass (Helictotrichon sp.) DFP 145 (cp 1160); MBF pl 96; MFF pl 127 (1219); NHE 196, 292.

Oat-grass, Yellow (Trisetum flavescens) EWF cp 25d; MBF pl 96; MFF pl 131 (1217); NHE 196, 292; PFE cp 180 (1764).

Ocelot (Felis pardalis or Panthera pardalis) ALE 8:359; 12:300, 323; DSM 184; HMA 37; HMW 124; JAW 130; LVS 73; MLS 314 (bw); TWA 137; WEA 262.

October Plant or **October Daphne** (Sedum sieboldii) DFP 81 (cp 646); ERW 135; LFW 131 (cp 297); RDP 356; TGF 92.

Octopus, Common (Octopus vulgaris) ALE 3:198, 223; CSS pl XXI; LAW 129; LEA 93 (bw); OIB 93; VWS 66 (bw); WEA 263.

Octopus, Curled or Lesser (Eledone cirrhosa) CSF 205; CSS pl XXI; MOL 75; OIB 93.

Oilbird (Steatornis caripensis) ABW 157; ALE 8:425; CDB 119 (cp 454); GPB 194; SSA pl 24; WAB 82.

Okapi (Okapia johnstoni) ALE 13:262; CAA 74 (bw); CAF 136 (bw); DDM 176 (cp 27); DSM 239; HMW 65; JAW 178; LEA 603 (bw); LVA 23; MLS 392 (bw); PWC 172 (bw); SMW 255 (bw); TWA 62 (bw); WEA 264.

Old Man of the Woods (Strobilomyces floccopus) DEW 3:184; DLM 253; LHM 195; SMG 108.

Old Man's Beard—See **Traveler's Joy.**

Old Man's Beard or **Graybeard**—See **Spanish Moss.**

Old Woman—See **Dusty Miller.**

Old-squaw (Clangula hyemalis) ABW 69; ALE 7:309; BBB 233; BBE 63; CDB 52 (cp 112); FBE 65; OBB 27; PBE 36 (cp 11), 41 (pl 14), 49 (pl 16).

Oleander (Nerium oleander) AMP 77; BIP 129 (cp 194); CTH 45 (cps 62-64); DFP 214 (cp 1710); EEG 109; EGE 74, 137; EGG 130; ELG 141; EPG 133; EWF cp 36b; GFI 99 (cp 21); HPW 224; LFW 178 (cp 400); LHP 134, 135; MEP 110; MWF 213; NHE 109; PFE cp 98 (1007); PFM cp 129; PFW 30; RDP 275; SPL cp 106; TGS 359; TVG 233.

Oleander, Yellow (Thevetia peruviana) MEP 110; MWF 282.

Oleaster (Elaeagnus angustifolia) EGT 111; EPF pl 698; EPG 118; NHE 134; SST cp 105; TGS 278.

Oleaster (Elaeagnus pungens) DFP 197 (cp 1571) EEG 105; EGE 121; EGW 109; ELG 137; GDS cps 158, 159; HSC 46, 49; PFW 104; TGS 268.

Olingo (Bassaricyon sp.) ALE 12:94; MLS 278 (bw).

Olive (Olea europaea) AMP 85; DEW 2:116, 117; DFP 215 (cp 1713); EET 202, 203; EFP 127; EGE 137; ELG 129; EPF pls 675, 676; EPG 134; MLP 201; NHE 246, 294; OFP 23; PFE cp 94 (985); PFW 200; SST cp 211.

Olive Shell (Oliva sp.) ALE 3:56; CTS 50 (cp 88), 51 (cp 90); GSS 18, 88, 89; (Olivancillaria sp.) CTS 51 (cp 89); MOL 67; GSS 88.

Olive, Tea or Sweet—See **Tea Olive.**

Olm (Proteus anguineus) ALE 5:351; ART 18; CAW cp 19; ERA 77; HRA 119; LAW 378; LEA 275 (bw).

Omphalina (Omphalina sp.) DLM 229; LHM 99; ONP 47, 85, 89.

Onager —See **Ass, Asian or Indian Wild.**

One-day Flower—See **Spiderwort.**

Onion, Common (Allium cepa) AMP 153; CTH 23 (cp 15); EGF 100; EGH 94; OFP 167; PFF 371.

Onion, Nodding or Flowering (Allium cernuum) DFP 83 (cp 660); LFW 8 (cp 16).

Onion, Ornamental (Allium giganteum) MWF 36; TVG 181.

Onion, Turkestan (Allium karataviense) DFP 83 (cp 663); EPF pl 893; SPL cp 251; TVG 181.

Onion, Welsh (Allium fistulosum) NHE 241; OFP 167; WFB 265 (6).

Onion, Yellow (Allium flavum) DFP 83 (cp 662); LFW 8 (cp 15); PFE cp 165 (1606).

Opah (Lampris sp.) ALE 5:36; CSF 121; HFI 71; OBV 41.

Opossum, Australian or Brush-tailed—See **Phalanger, Brush-tailed.**

Opossum, Common (Didelphis marsupialis) ALE 10:62; HMA 74; JAW 20; MLS 50 (bw); SMW 34 (cp 4).

Opossum, Four-eyed (Philander opossum) DSM 24; HMA 74; (Metachirus nudicaudatus) ALE 8:359; (Metachirops opossum) ALE 10:62.

Opossum, Mouse or Murine (Marmosa murina) BMA 160, 161; DSM 23; HMA 74; MLS 49 (bw).

Opossum, Thick-tailed (Lutreolina crassicaudata) ALE 10:62; DSM 25; (Metachirus crassicaudatus) HMW 196.

Opossum, Water (Chironectes minimus) ALE 10:62; DSM 25; HMA 74; HMW 197; SMW 18 (bw).

Opossum, Woolly (Caluromys sp.) ALE 10:61; MLS 48 (bw); SMW 35 (cp 5).

Orache (Atriplex hortensis) DFP 31 (cp 241); EGA 97; EGH 102.

Orache, Common (Atriplex patula) MBF cp 72; MFF pl 111 (214); NHE 227; OWF 55; PWF 158a; WFB 47 (6).

Orache, Frosted (Atriplex laciniata) MBF cp 72; MFF pl 111 (215); NHE 136; PWF 164b; WFB 47 (7).

Orache, Grass-leaved or Shore (Atriplex littoralis) MBF cp 72; MFF pl 111 (213); NHE 136; OWF 54; PWF 166b.

Orache, Halberd-leaved (Atriplex hastata) KVP cp 105; MBF cp 72; NHE 227; PWF 166d.

Orache, Shining (Atriplex nitens) EPF pl 313; NHE 227; TEP 190.

Orache, Shrubby (Atriplex halimus) GDS cp 13; HSC 15.

Orange Ball Tree —See **Butterfly Bush, Globe.**

Orange, Calamondium (Citrus mitis) DFP 59 (cp 470); EDP 116; EGG 98; EGH 108; LHP 60, 61; RDP 23, 145; (C. calamondia) BIP 55 (cp 69).

Orange, Chinese or Hardy (Poncirus trifoliata) DFP 218 (cps 1742, 1743); EGW 134; HSC 91; PFW 269; SPL cp 112; TGS 326.

Orange, Mandarin—See **Tangerine.**

Orange Marmalade—See **Browallia, Orange.**

Orange, Osage (Maclura pomifera) DEW 1: 159; EGT 123; EPF pls 351; 352; PFF 147; SST cp 134; TGS 87; (M. aurantica) WWT 107.

Orange Peel or **Orange Cup** (Aleuria laurantia) DEW 3:cp 45, 122; DLM 118; NFP 37; ONP 151; (A. sp.) RTM 245, 301.

Orange, Seville or Sour (Citrus aurantium) AMP 41; BKT cp 150; EWF cp 119c; HPW 203; MWF 86; OFP 85; PFF 229; SST cp 275.

Orange, Sweet (Citrus sinensis) DEW 2:57 (cps 21, 22); EGE 119; EPG 115; EVF 118; OFP 85; PFE cp 67 (690); PFF 229; PFW 268; SST cp 196.

Orangutan (Pongo pygmaneus) ALE 10:509-512; BMA 162-165; CAS 263; DAA 131; DSM 98-99; HMA 75; HMW 14; JAW 61; LAW 622; LVA 41; LVS 43; MLS 160 (bw); PWC 234 (bw); SLP 188, 189; SLS 107; SMW 89 (cp 52); TWA 161; VWA cp 23; WEA 265.

Orchid (Ascocenda hybrid) EGO 46, 94; EGG 59; (Angraecum sp.) EWF cp 69b; FHP 55; LFT 87 (cp 43); (Brassocattleya sp.) DFP 55 (cp 434); EGO 98; EPF cp XXVII; MWF 62; (Brassolaelio cattleya sp.) EGO 98; PFW 214.

Orchid (Bulbophyllum sp.) HPW 322; LFT 87 (cp 43); (Caladenia sp.) EWF cp 145g; SNZ 155 (cp 490); (Cirrhopetalum sp.) PFW 209; RWF 76.

Orchid (Coelogyne sp.) BIP 133 (cp 202); DEW 2:300; DFP 60 (cp 477); HPW 322; PFW 210; RWF 75.

Orchid (Comparettia sp.) EGL 97; EGO 105; EMB 132; (Corybas sp.) EWF cp 145f; HPW 324; SNZ 108 (cp 331).

Orchid (Dactylorchis sp.) EPF pls 966, 967; KVP cp 73; MFF cp 33 (1091-1093); PFM cps 303, 304.

Orchid (Dispersis sp.) EWF cp 89a; LFT cps 43, 44; (Disa sp.) HPW 324; LFT cps 49-51.

Orchid (Masdevallia sp.) EGG 61-63; EGO 88, 121; EMB 133; PFW 209; (Maxillaria sp.) EGO 41, 122; FHP 58; RDP 265.

Orchid (Paphiopedilum sp.) BIP 135 (cp 205); DEW 2:297; EDP 146; EGG 132; EGO 42, 81, 83, 129, 130; EPF cp XXIX b; EWF cp 1092; FHP 55; HPW 322; LFW 162 (cp 359); MWF 221; PFW 207; RDP 297; RWF 75; SPL cp 520.

Orchid (Phragmipedium sp.) EGO 36, 133; PFW 208; (Pleurothallis sp.) EGG 63; EGO 45, 134; (Polystachya sp.) EGG 62, 63; EGO 134; PFW 212; (Restrepia sp.) EGG 62; EGO 137.

Orchid (Stanhopea so.) EGO 19, 142; EPF pls 982-985; PFW 212; SPL cp 522; WWT 46.

Orchid, Angel (Coelogyne cristata) EGG 99; MWF 90; RWF 152.

Orchid, Bee (Ophrys apidera) BFB cp 24(3); KVP cp 177; MBF cp 82; MFF cp 34 (1080); NHE 197; OWF 161; PFE cp 188 (1878); PFM cp 290; PWF 94d; WFB 274, 275 (2).

Orchid, Bertoloni's Bee (Ophrys bertolonii) HPW 324 (9); NHE 42; PFE cp 188 (1881).

Orchid, Bird's-nest (Neottia nidus-avis) HPW 325; MBF cp 80; MFF cp 36 (1069); NHE 41; OWF 45; PFE cp 192 (1924); PFW 206; PWF 96c; TEP 264; WFB 286, 287 (1).

Orchid, Black (Coelogyne pandurata) EGO 104; EWF cp 127b.

Orchid, Black Vanilla (Nigritella nigra) NHE 293; PFE cp 189 (1901); WFB 275 (6); (N. miniata) NHE 293.

Orchid, Bog (Hammarbya paludosa or Malaxis paludosa) EWF cp 26c; MBF cp 80; MFF pl 83 (1071); NHE 79; OWF 161; WFB 287 (4).

Orchid, Broad-leaved Marsh (Dactylorhiza majalis) NHE 198; PFE 187 (1898); WFB 279 (4).

Orchid, Brown Bee (Ophrys fusca) EWF cp 26e; PFE cp 188 (1873).

Orchid, Bug (Orchis coriophora) NHE 198; PFE cp 187 (1886); WFB 277 (7).

Orchid, Bumble Bee (Ophrys bombyliflora) PFE cp 188 (1879); PFM cp 282.

Orchid, Butterfly (Oncidium papilio) EGO 127; MWF 217.

Orchid, Calanthe (Calanthe vestita) EGG 93; EGO 101; PFW 214.

Orchid, Cattleya (Cattleya hybrids) BIP 135 (cp 206); DFP 56 (cp 442); EDP 124; EGG 95; EGO 46, 63, 102, 103; EPF cp XXVI; MWF 76; PFW 205; RDP 132; WWT 44.

Orchid, Cattleya (Cattleya sp.) EPF 979, cp XXVI; EGL 93; EGO 44, 102, 103; EMB 132; FHP 55; LFW 260 (cps 573, 574), 261 (cps 580, 582, 584); PFF 390; PFW 209.

Orchid, Chain (Dendrochilum sp.) CAU 244; EGO 110; EPF pl 976.

Orchid, Chinese Ground (Bletilla striata or B. hyacinthina) DFB 126 (cp 1004); EGB 98; MWF 58; PFW 206; SPL cp 515.

Orchid, Clamshell (Epidendrum cochleatum) EGL 106; MLP 328.

Orchid, Coral (Rodriguezia secunda) EGL
140; EGO 138; (R. venusta) FHP 58.

Orchid, Coralroot (Corallorhiza
trifida) EWF cp 7; MBF cp 80; MFF pl 83
(1073); NHE 41; OWF 45; PFE cp 192 (1926);
PWF 51c; WFB 287 (2).

Orchid, Crucifix (Epidendrum ibaguense)
DFP 66 (cp 522); EGO 113; MWF 122.

Orchid, Cymbidium (Cymbidium sp.) BIP
133 (cp 200); DFP 61 (cps 487, 488), 62 (cps
489-491); EGG 103; EGL 101; EGO 25, 63;
EPF pls 989, XXIX A; EWF cps 109d, 126c,
145a; FHP 55; MEP 30; MWF 101; PFW 211;
RDP 168; SPL cp 517; WWT 42.

Orchid, Dancing Lady (Oncidium varicosum) EGG 131; EPF pl 987; FHP 56; (O.
ornithorhynchum) RDP 282; (O. crispum)
MWF 217.

Orchid, Dark-winged or Burnt (Orchis
ustulata) EWF cp 26a; MBF cp 81; MFF cp
33 (1087); NHE 198; PFE cp 187 (1888);
WFB 277 (5).

Orchid, Dendrobium (Dendrobium nobile)
BIP 133 (cp 201); DEW 2:301; DFP 62 (cp
496), 63 (cps 497, 498); EGG 105; EPF cp
XXVII, pl 978; MWF 108; RDP 174.

Orchid, Dendrobium (Dendrobium sp.) EGG
59; EGO 37, 44, 109; EMB 133; EPF 980;
EWF cp 109a, c, cp 127e, cp 145b; FHP 57;
HPW 322; MLP 327; MWF 108, 109; RWF
75; SNZ 94 (cps 278, 279); SPL cp 519; WWT 46.

Orchid, Dense-flowered (Neotinea intacta)
MFF pl 82 (1079); WFB 281 (3).

Orchid, Early Marsh (Dactylorhiza
incarnata) MBF cp 81; NHE 79; PWF 27f;
TEP 154; WFB 279 (2); (Orchis strictifolia)
BFB cp 24 (7); OWF 159.

Orchid, Early Purple (Orchis mascula) BFB
cp 24 (2); CTH 25 (cp 18); MBF cp 81; MFF cp
33 (1090); NHE 197; OWF 159; PWF 28c;
WFB 277 (1).

Orchid, Early Spider (Ophrys sphegodes)
EWF cp 26k; MBF cp 82; MFF 34 (1081);
NHE 197; PFM cps 279, 280; WFB 274, 275(4).

Orchid, Elder-flowered (Dactylorhiza
sambucina) NHE 198; PFE cp 189 (1899);
WFB 279 (6).

Orchid, Epidendrum (Epidendrum sp.) EGG
107; EGO 36, 45, 114; EPF pl 977; MWF 122;
RDP 190; WWT 42.

Orchid, Eyelash (Epidendrum ciliare) EGO
113; EWF cp 192c.

Orchid, Fen (Liparis loeselii) MBF cp 80;
MFF pl 83 (1072); NHE 79; PWF 110a; WFB
287 (6).

Orchid, Flor de San Miguel (Laelia anceps)
EGO 118; EWF cp 192d; MLP 327.

Orchid, Fly (Ophrys insectifera) BFB cp 23
(5); EPF pls 963, 965; MBF cp 82; MFF cp 34
(1082); NHE 197; OWF 161; PWF 27d; RWF
70; WFB 275 (5).

Orchid, Four-spotted (Orchis quadipunctata) PFE cp 189 (1895); PFM cp 302.

Orchid, Fragrant (Gymnadenia conopsea)
BFB cp 24 (10); MBF cp 82; MFF cp 32
(1076); NHE 198, 293; OWF 159; PFE cp 190
(1912); PWF 92b; TEP 154; WFB 279 (7).

Orchid, Frog (Coeloglossum viride) BFB cp
24 (1); MBF cp 82; MFF pl 83 (1075); NHE
197; OWF 45; PWF 132c; WFB 280, 281 (6).

Orchid, Ghost (Epipogium aphyllum) EPF pl
973; MFF cp 36 (1064); NHE 41; RWF 70;
WFB 286 (3), 287 (3).

Orchid, Giant (Himantoglossum longibracteatum) PFE cp 190 (1906); PFM cp 291.

Orchid, Greater Butterfly (Platanthera
chlorantha) KVP cp 59; MFF pl 83 (1078);
NHE 197; OWF 101; PFE cp 191 (1915);
PWF 96b; WFB 283 (1).

Orchid, Green-winged (Orchis morio) AMP
47; BFB cp 24 (4); EPF pl 968; MBF cp 81;
MFF cp 33 (1088); NHE 198; OWF 159; PWF
27b; TEP 153; WFB 277 (2).

Orchid, Hooded (Pterostylis sp.) EWF cp
145e; PFW 211; SNZ 44 (cps 95, 96), 95 (cps
282, 284).

Orchid (hybrid) (Laeliocattleya) EGG 58;
EGO 119; FHP 57; RDP 255.

Orchid, Jersey or Loose-flowered (Orchis
laxiflora) EWF cp 26j; MBF cp 81; MFF cp 33
(1089); PFE cp 187 (1893); PFM cp 299; WFB
277 (8).

Orchid, Jewel (Anoectochilus sp. or Macodes
sp.) EWF cp 126A; HPW 324; PFW 213;
(Haemaria discolor) EGL 116; EGO 115.

Orchid, Lady (Orchis purpurea) AMP 47;
EPF pl 969; HPW 324; MBF cp 81; MFF cp 33
(1084); NHE 41; PFE cp 187 (1892); PFM cp
293; WFB 277 (3), 280.

Orchid, Lady-of-the-night (Brassavola
nodosa) EGG 92; EGO 88; FHP 55; (B.
cordata) EGO 45; (B. digbyana) EGO 41, 96.

Orchid, Lady's-slipper —See **Lady's-
slipper.**

Orchid, Laelia (Laelia sp.) EGG 122; EGL
123; EGO 88, 118, 119; FHP 56; PFW 214;
RDP 254; WWT 45.

Orchid, Late Spider (Ophrys fuciflora) DEW
2:280 (cp 159); MBF cp 82; NHE 197; PFM cp
283; WFB 274, 275 (3).

Orchid, Leopard (Ansellia gigantea or A.
africana) EGO 93; EWF cp 69a; LFT 86 (cp 42).

Orchid, Leptotes (Leptotes sp.) EGG 62;
EGO 120.

Orchid, Lesser Butterfly (Platanthera
bifolia) BFB cp 24 (9); MBF cp 82; NHE 197;
PFE cp 191 (1914); PWF 27c; TEP 264.

Orchid, Lizard (Himantoglossum hircinum)
EFC 54; EPF pl 971; MBF cp 81; MFF pl 65
(1083); NHE 197; PFE cp 190 (1907); WFB
280, 281 (2).

Orchid, Lycaste (Lycaste sp.) EGG 126; EGO
121; PFW 212; RDP 260; RWF 75.

Orchid, Man (Aceras anthropophorum) EWF
cp 26G; MBF cp 82; MFF cp 58 (1094); NHE
40; OWF 45; PFE cp 189 (1905); PFM cp 306;
PWF 94C; WFB 280 (1), 281 (1).

Orchid, Marsh (Orchis palustris) NHE 79;
WFB 276.

Orchid, Medusa's-head (Bulbophyllum
medusae) EGO 100; EWF cp 127d.

Orchid, Mirror (Ophrys speculum) PFM cp
277; RWF 70.

Orchid, Monkey (Orchis simia) MBF cp 81;
MFF cp 33 (1086); NHE 245; PFM cp 295;
WFB 277 (6), 280.

Orchid, Moorland Spotted (Dactylorhiza
maculata) MBF cp 81; PWF 96d.

Orchid, Moth (Phalaenopsis amabilis) DEW
2:277 (cp 153); EWF cp 126b; LFW 260 (cp
577); MWF 228.

Orchid, Moth (Phalaenopsis sp.) EGG 134;
EGL 136; EGO 22, 43, 63, 88, 131, 132; PFW
209; RDP 310; SPL cp 521; WWT 44.

Orchid, Musk (Herminium monorchis) MBF
cp 82; MFF pl 83 (1074); NHE 197; OWF 45;
PFE cp 190 (1909); PWF 70e; WFB 281 (5).

Orchid, Northern Marsh (Dactylorhiza
purpurella) MBF cp 81; PWF 71d.

Orchid, Oncidium (Oncidium sp.) EGG 60,
62; EGL 129; EGO 21, 39, 74, 126-128; HPW
322; LFW 26 (cp 575).

Orchid, One-leaved Bog (Malaxis
monophyllos) NHE 79; WFB 287 (5).

Orchid, Pansy (Miltonia sp.) DFP 70 (cps
558-559); EGG 127; EGO 122, 123; MWF
201; PFW 213; RDP 268.

Orchid, Pink Butterfly (Orchis papilionacea) NHE 42; PFE cp 187 (1884); PFM cp
292.

Orchid, Pleione (Pleione sp.) DFP 18 (cps
139-141); EGO 133; PFW 206; TVG 209.

Orchid, Poor Man's —See **Butterfly
Flower.**

Orchid, Pyramidal (Anacamptis
pyramidalis) BFB cp 24 (8); EWF cp 26D;
MBF cp 81; MFF cp 32 (1095); NHE 197;
OWF 159; PFE cp 190 (1908); PFM cp 307;
PWF 92D; WFB 279 (1).

Orchid, Samurai (Neofinetia falcata) EGO
125; EMB 134.

Orchid, Sawfly (Ophrys tenthredinifera)
PFE cp 188 (1882); PFM cp 289; RWF 70.

Orchid, Scorpion (Arachnis flos-aeris) EGO
93; EWF cp 127A.

Orchid, Small White (Leucorchis albida)
MBF cp 82; MFF pl 82 (1077); NHE 293; PFE
cp 192 (1913).

Orchid, Soldier or Military (Orchis
militaris) AMP 47; EFC cp 55; EPF pl 970;
MBF cp 81; MFF cp 33 (1085); NHE 198; PFE
cp 189 (1891); WFB 277 (4), 280.

Orchid, Sophronitis (Sophronitis sp.) EGG
62, 142; EGO 142.

Orchid, Southern Marsh (Dactylorhiza
praetermissa) MBF cp 81; PWF 89h; WFB
279 (3).

Orchid, Spice (Epidendrum atropurpureum)
EGO 41, 112; LFW 260 (cp 578).

Orchid, Spider (Brassia sp.) EGL 90; EGO
88, 97; EWF cp 192b; FHP 57; LFW 261 (cp
583); RDP 104.

Orchid, Spotted (Dactylorhiza fuchsii) HPW
324; MBF cp 81; NHE 198; PFE cp 189
(1900); PFW 212; PWF 92a; WFB 279 (5);
(Orchis fuchsii) BFB cp 24 (5); OWF 159.

Orchid, Star-of-Bethlehem (Angraecum
sesquipedale) DEW 2:301; EGO 91; PFW
208.

Orchid, Swan (Cycnoches chlorochilon) EGG
102; EGO 105.

Orchid, Tiger (Odontoglossum grande) DFP
73 (cps 577-579); EGG 130; EGO 6, 126; EPF
pl 990, cp XXVIII; PFW 211; RDP 281; WWT
43.

Orchid, Tongue (Serapias lingua) NHE 42;
PFM cp 311.

Orchid, Toothed (Orchis tridentata) NHE
198; PFE cp 187 (1889); PFM cp 296; WFB
276.

Orchid Tree (Bauhinia sp.) BIP 31 (cp 26);
EGE 113; EGT 99; ELG 125; EWF cp 54a;
GFI 139 (cp 31); LFW 232 (cp 518); MWF 54;
PFW 158, 159; SST cps 229, 230.

Orchid, Vanda (Vanda sp.) EDP 149; EGG
147; EGO 88, 146; EPF pl 991; EWF cp 127c;
LFW 260 (cp 576), 261 cp (579); MLP 327;
MWF 291; PFW 215; RDP 391; RWF 72, 73.

Orchid, Vanilla (Vanilla planifolia) AMP 39;
EGO 146; EGV 142; OFP 131; PFF 391; (V.
imperialis) PFW 213.

Orchid, Virgin (Diacrium bicornutum) DEW
2:297; EGO 111.

Orchid, Woodcock (Ophrys scolopax) PFE cp
188 (1876); PFM cps 278, 284, 285.

Orchid, Yellow Bee (Ophrys lutea) PFE cp
188 (1874); PFM cp 273.

Orchid, Zygopetalum (Zygopetalum
mackaii) BIP 135 (cp 203); MWF 303; PFW
208; (Z. intermedium) EGO 147; EWF cp
192e.

Oregon Grape (Mahonia aquifolium) DFP
212 (cp 1695); EEG 109; EGE 135; EGW 126;
ELG 140; EPF pl 194; EPG 131; HPW 45;
HSC 78, 81; MFF cp 64 (31); PFE cp 27 (262);
PFF 179; SPL cp 105; TGS 298; TVG 231;
WFB 78.

Orfe —See **Ide.**

Oribi (Ourebia ourebi) ALE 13:426; DSM 262;
MLS 413 (bw); SMR 92 (cp 7).

Oriole, Black-headed (Oriolus larvatus)
CAA 49 (bw); CDB 218 (cp 976); NBA 53;
PSA 241 (cp 28).

Oriole, Black-hooded (Oriolus xanthornus)
KBA 272 (cp 39); WBV 155.

Oriole, Black-naped (Oriolus chinensis)
ABW 222; ALE 9:450; DPB 253; KBA 272
(cp 39); RBI 299; WBV 155.

Oriole, Dark-throated (Oriolus xanthonotus) DPB 253; KBA 272 (cp 39).

Oriole, Golden (Oriolus oriolus) ALE 9:450;
13:163; BBB 268; BBE 213; CDB 218 (cp
977); DAE 136; FBE 303; FBW 67; GAL 126
(cp 2); GPB 331; LAW 510; NBB 81; OBB
173; PBE 197 (cp 50); PSA 241 (cp 28); RBE
224; (O. auratus) PSA 241 (cp 28).

Oriole, Maroon (Oriolus traillii) ABW 222;
ALE 9:450; KBA 272 (cp 39); RBI 300; WBV
155.

Oriole, Olive-backed (Oriolus sagittatus)
CDB 218 (cp 978); HAB 148; MBA 58, 59;
SAB 72 (cp 35).

Oriole, Yellow (Oriolus flavocinctus) HAB
148; SAB 72 (cp 35).

Orlaya (Orlaya grandiflora) DEW 2:86; EPF
pl 761; NHE 221.

Ormer —See **Abalone.**

Oropendola, Montezuma (Gymnostinops
montezuma) ABW 288; ALE 9:381; WAB 97.

Oropendola, Wagler's (Psarocolius wagleri
or Zarhynchus wagleri) ABW 288; ALE
9:381.

Orpine (Sedum telephium) BFB cp 10 (3);
DEW 1:272; EPF pl 386; MBF cp 33; MFF cp
26 (389); NHE 27; PFE cp 40 (395); PWF
152c; TEP 91; TVG 173; WFG 103 (1).

Ortolan —See **Bunting, Ortolan.**

Oryx, Arabian (Oryx leucoryx) ALE 13:402;
DAA 10-11; LVA 11; PWC 247 (bw); SLP
230; SLS 163; VWA cp 13; WEA 268.

Oryx, Beisa (Oryx beisa) ALE 13:402; DDM
188 (cp 31); DSM 258; HMW 73.

Oryx, East African —See **Gemsbok.**

Oryx, Scimitar-horned or White (Oryx tao)
DSM 259; LVA 13; PWC 247 (bw); (O.
algazel) CAF 66 (bw); HMW 73; (O.
dammah) ALE 13:402; DDM 188 (cp 31);
LVS 135.

Oscularia (Oscularia deltoides) DFP 73 (cp 584); HPW 67.

Osier, Common (Salix viminalis) EET 135; MBF cp 77; MFF pl 72 (572); OBT 89; OWF 187; PWF 167g; SST cp 165; WFB 27 (5); (S. woodii) LFT cp 54.

Osier, Purple —See Willow, Purple.

Osmanthus (Osmanthus delavayi) BIP 137 (cp 207); DFP 215 (cp 1716); GDS cp 301; OGF 23; TGS 334.

Osmanthus, Holly (Osmanthus hetero-phyllus) EEG 110; EFP 128; EGE 36, 138; EGW 130; GDS cp 302; RDP 290; TVG 233; (O. ilicifolius) HSC 82; TGS 334.

Osprey (Pandion haliaetus) ABW 82; ALE 7:385, 398; BBB 166; BBE 83; CDB 60 (cp 160); CEU 259 (bw); DPB 37; FBE 71; FBW 37; HAB 43; HBG 81 (pl 4); KBA 48 (pl 3); LEA 389 (bw); OBB 43; PBE 68 (pl 21); PSA 81 (pl 8), 128 (pl 13); RBE 24; RBU 76; SNB 46 (cp 22), 54 (pl 26); WAB 219; WBV 57.

Ostrich (Struthio camelus) ABW 17; ALE 7:88, 98; 13:61; BBT 52-53; CAA 22-25 (bw); CAF 158 (bw); CDB 33 (cp 1); GPB 31; LAW 443, 500; LEA 356 (bw); WAB 149; WEA 269.

Ostrowskia (Ostrowskia magnifica) EWF cp 47a; PFW 63.

Otter, Asian Short-clawed (Amblonyx cinerea) CAS 231; DSM 177.

Otter, Brazilian or Giant (Pteronura brasiliensis) ALE 12:41, 93; LAW 536; LVS 96; MLS 295 (bw); PWC 238 (bw); SLS 41.

Otter, Clawless (Aonyx sp.) ALE 12:93; BMA 167 (bw); DAA 82; DDM 85 (cp 14); MLS 296 (bw); SMR 48 (cp 3).

Otter, European or Common River (Lutra lutra) ALE 11:257; 12:41, 93; BMA 167 (bw); DAE 21; DSM 176; GAL 30 (bw); HMW 139; LAW 536, 538; LEA 563 (bw); LVA 59; OBV 175; SLP 100; SMW 227 (cp 131); TWA 71 (bw).

Otter, Indian Smooth-coated (Lutrogale perspicillata) ALE 12:41; DSM 177.

Otter Shell —See Clam, Otter Shell.

Otter, Spotted-necked (Lutra maculicollis) ALE 12:93; DDM 85 (cp 14); SMR 48 (cp 3).

Otter-civet —See Civet, Otter.

Otter-shrew (Potamogale velox) ALE 10:178; DDM 32 (cp 1); DSM 43; HMW 190; LAW 594; SMR 16 (cp 1); (Micropotamogale lamottei) ALE 10:178.

Ouratea (Ouratea sp.) DEW 1:164 (cp 79), 165 (cp 80); HPW 84.

Ourisia (Ourisia sp.) DFP 16 (cp 122); EWF cp 5f; SNZ 114 (cp 351), 115 (cps 352-354).

Ouzel, Ring (Turdus torquatus) ALE 9:290; 13:463; BBB 126; BBE 263; FBE 257; OBB 139; PBE 220 (cp 53); RBE 256.

Ovenfish —See Firebrat.

Owl, African Marsh (Asio capensis) BOW 146; CDB 114 (cp 433); FBE 179; PSA 193 (cp 22).

Owl, Barking (Ninox connivens) BOW 149; GBP 440 (bw); HAB 124; RBU 167; SNB 124 (cp 61); WAB 191.

Owl, Barn (Tyto alba) ABW 156; ALE 8:380, 393, 394, 396; BBB 77; BBE 177; BOW 43-48; CDB 113 (cp 429); DAE 120; FBE 177; GAL 160 (bw); GBP cp after 144; HAB 125; HGB 81 (pl 4); KBA 193 (pl 26); LEA 426 (bw); NBB 35 (bw), 65, 96 (bw); OBB 107; PBE 165 (cp 44); PSA 193 (cp 22); RBE 163; SSA pl 22; SNB 124 (cp 61); WAB 219; WBV 117, WEA 271.

Owl, Barred Eagle (Bubo sumatranus) BOW 90; KBA 195 (bw).

Owl, Bay (Phodilus badius) ALE 8:380; BOW 59; DPB 171; GBP 412 (bw); KBA 193 (pl 26); RBI 107; WBV 117; (P. prigoginei) BOW 60.

Owl, Black-banded (Ciccaba huhula) BOW 123; SSA pl 24.

Owl, Boobook (Ninox novaeseelandiae) BNZ 103; BOW 148; CAU 256 (bw); CDB 116 (cp 444); GPB 190; MBA 44, 45; MWA 43; RBU 164; SNB 124 (cp 61); (N. boobook) HAB 125; RBU 163.

Owl, Boreal (Aegolius funereus) ALE 8:380, 396; BBB 271; BBE 173; BOW 170, 171; CDB 114 (cp 432); FBE 181; GAL 161 (bw); OBB 109; PBE 180 (cp 45); (A. harrisii) BOW 174.

Owl, Brown Fish (Ketupa zeylonensis) BOW 66; FBE 177; KBA 195 (bw); (K. blakistoni) BOW 62-63.

Owl, Brown or Oriental Hawk (Ninox scutulata) BOW 157; DPB 171; KBA 193 (pl 26); WEA 272.

Owl, Brown Wood (Strix leptogrammica) BOW 136; KBA 193 (pl 26).

Owl, Buffy Fish —See **Owl, Malay Fish.**

Owl, Collared Scops (Otus bakkamoena or Scops bakkamoena) BOW 100; CAS 209 (bw); DPB 171; KBA 208 (pl 27); WBV 117.

Owl, Crested or Maned (Lophostrix cristata) BOW 115; GBP 420 (bw); (L. lettii) BOW 114; (Jubula lettii) GBP 419 (bw).

Owl, Dusky Eagle (Bubo coromandus) BOW 89; KBA 195 (bw).

Owl, Eagle (Bubo bubo) ALE 8:380, 13:163; BBE 175; BOW 72-75; CDB 115 (cp 438); DAE 122; FBE 177; GAL 162 (bw); GBP 423 (bw); LAW 445, 471; NBB 42 (bw); PBE 165 (cp 44); RBE 164; SLP 30; (B. capensis) BOW 78, 79.

Owl, European Scops (Otus scops) ALE 8:380, 396; BBB 273; BBE 173; BOW 104; CDB 117 (cp 448); CEU 214; DAE 122; DPB 171; FBE 179; KBA 208 (pl 27); OBB 109; PBE 165 (cp 44); PSA 193 (cp 22); WAB 101.

Owl, Fearful (Nesasio solomonensis) BOW 160; GBP 463 (bw).

Owl, Ferruginous Pygmy (Glaucidium brasilianum) ABW 157; BOW 177, 178; SSA pl 24.

Owl, Fishing (Scotopelia peli) ABW 158; ALE 8:380; BOW 70-71; GBP 427 (bw); WAB 156; (S. bouvieri) BOW 71; (S. ussheri) BOW 69.

Owl, Forest Eagle (Bubo nipalensis) BOW 87; KBA 195 (bw).

Owl, Giant Scops (Otus gurneyi or Mimizuku gurneyi) BOW 109; DPB 171.

Owl, Grass (Tyto capensis) BOW 52, 53; DPB 173; KBA 193 (pl 26); (T. longimembris) HAB 125; RBI 108; SNB 124 (cp 61); (T. soumagnei) BOW 58.

Owl, Great Gray or Lapland (Strix nebulosa) ALE 8:380, 393; BBE 175; BOW 129; CDB 118 (cp 451); CEU 262 (bw); DAE 105; FBE 183; GPB 188; PBE 180 (cp 45); WAB 105.

Owl, Great Hawk (Ninox strenua) ALE 8:380; BOW 151; GBP 440 (bw); HAB 124; RBU 168; SNB 124 (cp 61).

Owl, Hawk (Surnia ulula) ALE 8:380; BBE 177; BOW 162, 163; CDB 118 (cp 453); DAE 105; FBE 181; GBP 433 (bw); OBB 107; PBE 180 (cp 45); PFF 615; RBE 171; WAB 39.

Owl, Hume's Tawny (Strix butleri) BOW 135; FBE 183.

Owl, Laughing (Sceloglaux albifacies) BNZ 103; GBP 445 (bw).

Owl, Little (Athene noctua) ALE 2:363; 8:394; BBB 76; BBE 173; BOW 165; CDB 115 (cp 436); CEU 190; DAE 120; FBE 181; FBW 43; GAL 161 (bw); GBP 446 (bw); LEA 428 (bw); NBB 64 (bw); OBB 109; PBE 180 (cp 45); WAB 112.

Owl, Long-eared (Asio otus) ALE 8:379; BBB 154; BBE 177; BOW 139, 140; CDB 115 (cp 435); DAE 122; FBE 179; GAL 161 (bw); GBP cp after 64, 458 (bw); GPB 192-193; LAW 504; LEA 427 (bw); OBB 111; PBE 165 (cp 44); RBE 168; (A. madagascariensis) BOW 142.

Owl, Malay Fish (Ketupa ketupa) ALE 8:380; BOW 64; GBP cp after 144, 426 (bw); KBA 195 (bw); WBV 117; (Bubo ketupa) CDB 115 (cp 439).

Owl, Masked (Tyto novaehollandiae) BOW 55; HAB 125; SNB 124 (cp 61); (T. castanops) HAB 125.

Owl, Milky Eagle (Bubo lacteus) ABW 155; ALE 8:380; BOW 83; CDB 115 (cp 440); GBP 424 (bw).

Owl, Mottled (Ciccaba virgata) BOW 120; GBP 449 (bw).

Owl, Mountain or Spotted Scops (Otus spilocephalus) BOW 110; KBA 208 (pl 27); WBV 117.

Owl, New Guinea Hawk (Uroglaux dimorpha) BOW 161; GBP 439 (bw); WAB 204.

Owl, Oriental Screech —See **Owl, Collared Scops.**

Owl, Philippine Boobook (Ninox philippensis) BOW 159; DPB 171.

Owl, Philippine Eagle (Bubo philippensis) BOW 86; DPB 173.

Owl, Philippine Hawk —See **Owl, Brown or Oriental Hawk.**

Owl, Powerful —See **Owl, Great Hawk.**

Owl, Pygmy (Glaucidium passerinum) ALE 8:380; BBE 173; BOW 175; FBE 181; PBE 180 (cp 45); (G. minutissimum) BOW 179.

Owl, Rufous (Ninox rufa) BOW 151; HAB 124; SNB 124 (cp 61).

Owl, Rufous Scops (Otus rufescens) BOW 100; DPB 171.

Owl, Seychelles (Otus insularis) BOW 189; LVB 61; PWC 224 (bw).

Owl, Short-eared (Asio flammeus) ALE 8:380; 11:105; BBB 113; BBE 175, 177; BOW 143; CDB 114 (cp 434); DAE 122; DPB 173; FBE 179; FBW 47; GAL 162 (bw); GBP 460 (bw); HBG 81 (pl 4); KBA 193 (pl 26); OBB 111; PBE 165 (cp 44); WBV 117.

Owl, Snowy (Nyctea scandiaca) ABW 156; BBB 267; BBE 175; BOW 91; CDB 116 (cp 445); DAE 100; FBE 177; FBW 45; GBP 124 (bw), 431-432 (bw); GPB 187-187; LEA 430 (bw); OBB 107; PBE 165 (cp 44); SLP 56; WAB 37.

Owl, Sooty (Tyto tenebricosa) BOW 54; CDB 113 (cp 430); HAB 124; SNB 124 (cp 61).

Owl, Spectacled (Pulsatrix perspicillata) ABW 156; ALE 8:380; BOW 117; GBP cp before 145, 429 (bw); WAB 91.

Owl, Spotted —See **Owl, Boobook.**

Owl, Spotted Eagle (Bubo africanus) BOW 35, 80, 81; CDB 115 (cp 437); PSA 193 (cp 22).

Owl, Spotted Little (Athene brama) BOW 166; KBA 193 (pl 26).

Owl, Spotted Wood (Strix seloputo) BOW 137; DPB 173; KBA 193 (pl 26).

Owl, Tawny (Strix aluco) ALE 8:380, 394, 396; BBB 50; BBE 177; BOW 133, 134; CDB 118 (cp 450); DAE 120; FBE 183; GAL 161 (bw); GBP cp after 144, 146-147, 456 (bw); LEA 429 (bw); NBB 11, 67 (bw); OBB 109; PBE 180 (cp 45); RBE 167; WAB 137.

Owl, Tengmalm's —See **Owl, Boreal.**

Owl, Tropical Screech (Otus choliba) BOW 104; SSA pl 24.

Owl, Ural (Strix uralensis) ALE 8:380; BBE 175; BOW 131; CAS 47 (bw); FBE 183; GPB 191; PBE 180 (cp 45).

Owl, White-faced Scops (Otus leucotis) BOW 107; CDB 117 (cp 447); GBP 418 (bw); WEA 272.

Owl, White-fronted Scops (Otus sagittatus) BOW 109; KBA 208 (pl 27).

Owl, Winking—See **Owl, Barking.**

Owl, Wood (Ciccaba woodfordi) BOW 125; CDB 116 (cp 442); GBP 451 (bw); PSA 193 (cp 22).

Owlet, Asian Barred or Cuckoo (Glaucidium cuculoides) BOW 185; KBA 193 (pl 26)

Owlet, Collared (Glaucidium brodiei) BOW 184; KBA 193 (pl 26).

Owlet, Pearl-spotted (Glaucidium perlatum) BOW 180; PSA 193 (cp 22).

Owlet-nightjar —See **Nightjar.**

Oxalis (Oxalis sp.) DFP 16 (cps 124, 125); EGL 131; EPF pls 639, 641; EWF cps 3h, 171b-f; LFW 256 (cp 569); MWF 219; OFP 179; OGF 45; PFW 216; SNZ 116 (cp 357); SPL cp 368.

Oxalis, Pink (Oxalis articulata) PFM cp 81; WFB 133 (2); (O. floribunda) MFF cp 23 (256).

Ox-eye, Yellow (Buphthalmum salicifolium) DFP 126 (cp 1006); NHE 287; OGF 121, 147; PFE 446 (1396); TGF 284; WFB 237 (4).

Oxlip (Primula elatior) EFC cp 21; EPF pl 256; MBF cp 57; MFF cp 40 (614); NHE 33; OFW 27; PFE cp 90 (940); PWF cp 24d; TEP 249; TGF 171; TVG 165; WFB 175 (2).

Oxpecker, Red-billed (Buphagus erythrorhynchus) CDB 216 (cp 965); GPB 330; PSA 288 (cp 33); WAB 147 (bw).

Oxpecker, Yellow-billed (Buphagus africanus) ABW 274; ALE 9:449; CDB 216 (cp 964); LAW 510.

Ox-tongue (Gasteria verrucosa) EDP 135; EFP 115; RDP 219; SPL cp 139.

Ox-tongue, Bristly (Picris echioides) BFB cp 20 (7); BWG 53; MBF cp 50; MFF cp 53 (933); OWF 35; PWF 95i; WFB 257 (6).

Ox-tongue, Hawkweed (Picris hieracioides) BFB cp 20 (10); MBF cp 50; MFF cp 54 (934); NHE 239; PWF 145j; WFB 257 (5).

Oyster, Hammerhead (Malleus malleus) ALE 3:140; GSS 132; (M. albus) CTS 59 (cp 112); GSS 132.

Oyster, Native (Ostrea sp.) CSF 207; CSS pl XVII; OIB 85; WEA 274.

Oyster, Pearl (Pinctada sp.) ALE 3:140; GSS 131.

Oyster Plant (Mertensia sp.) —See also **Shorewort, Sea.**

Oyster Plant, Spanish (Scolymus hispanicus) EPF pl 596; PFM cp 213; (S. maculatus) PFE cp 157 (1510).

Oyster, Saddle (Anomia ephippium) ALE 3:141; CSS pl XVII; CTS 59 (cp 111); OIB 85.

Oyster, Spiny or Thorny (Spondylus sp.) ALE 3:141; CTS 57 (cp 107), 58 (cp 110); GSS 139; MOL 71; VWS 84, 89 (bw).

Oyster, Wing (Pteria sp.) CTS 57 (cp 105); GSS 131; OIB 73.

Oystercatcher, Black (Haematopus moquini) CDB 79 (cp 253); FBE 119; PSA 129 (pl 14); (H. ater) WFI 145.

Oystercatcher, European or Pied (Haematopus ostralegus) ALE 7:385; 8:160, 187; BBB 246; BBE 113; CDB 80 (cp 254); FBE 119; GPB 140; HAB 61; HBG 132 (pl 9); KBA 146 (bw); LEA 347 (bw); OBB 61; PBE 92 (pl 29), 101 (cp 32); RBE 128; RBU 36; SNB 88 (cp 43); WAB 132.

Oystercatcher, Sooty (Haematopus unicolor) HAB 61; (H. fuliginosus) SNB 88 (cp 43).

P

Paca, Spotted (Cuniculus paca) ALE 11:438; CSA 99; DSM 133; LEA 525 (bw); MLS 228 (bw); (Coelogenys paca) HMW 177; SMW 150 (bw).

Pacarana (Dinomys branickii) ALE 11:438; HMW 177; LAW 590; LEA 525 (bw); SMW 141 (cp 91).

Pachypodium (Pachypodium sp.) DEW 2:124-126; LCS cp 326; LFT cp 123; RWF 66.

Pachystachys (Pachystachys lutea) BIP 139 (cp 209); LHP 136, 137; RDP 292.

Pacu (Colossoma nigripinnis) FEF 75 (bw); SFW 112 (cp 19); WFW cp 54.

Pademelon, Red-bellied (Thylogale billard-ierii) ALE 10:136 (cp 6); CAU 61; WAW 37.

Pagoda Tree —See also **Frangipani.**

Pagoda Tree (Sophora japonica) BKT cp 138; EEG 122; EET 180, 181; EGT 43, 144; EJG 144; EPF pl 445; EPG 147; OBT 185; SST cp 170; TGS 150.

Paigle —See **Oxlip.**

Paint Brush —See **Flaming Sword.**

Paintbrush, Cupid's —See **Cupid's Paintbrush.**

Painted Drop-tongue (Aglaonema crispum or A. robelinii) BIP 21 (cp 10); DFP 51 (cp 404); LHP 20, 21; RDP 76, 77.

Painted Feather —See **Flaming Sword.**

Painted Nettle —See **Coleus.**

Painted Tongue (Salpiglossis sinuata) DFP 47 (cp 369); EDP 149; EGA 62, 151; EGG 140; EWF cp 183c; HPW 228; OGF 129; PFW 285; TGF 223; TVG 69.

Pak-choi (Brassica chinensis) EVF 89; OFP 155; (B. pekinensis) OFP 155.

Pallenis (Pallenis spinosa) PFE cp 145 (1397); PFM cp 202.

Palm, Betel (Areca catechu) EET 236; SST cp 47.

Palm, Blue (Erythea armata) BKT cp 191; SST cp 54.

Palm, Bread Fern (Encephalartos sp.) DEW 1:21; LFT cp 1; RWF 55.

Palm, Butterfly (Chrysalifocarpus lutescens) EFP 50, 100; ELG 153; RDP 140.

Palm, Cabbage —See **Palmetto.**

Palm, Canary Islands (Phoenix canariensis) BKT cp 193; LHP 149; PFE cp 176 (1718); RDP 314; SPL cp 28; SST cp 61.

Palm, Chinese Fan (Livistona chinensis) EDP 133; EFP 121; MLP 282; RDP 258; (L. sp.) CAU 171; HPW 302; RWF 66; SST cp 59; WAW 112.

Palm, Chusan or Windmill (Trachycarpus fortunei) DFP 242 (cp 1930); MTB 385 (cp 40); OBT 200; RDP 385, 386; SST cp 67; (T. excelsa) EPF pl 1011.

Palm, Coconut (Cocos nucifera) DEW 2:303, 304; EET 232, 233; EPF pl 1016, 1017; EWF cp 191b; LHP 64, 65; MEP 10; MLP 278, 280, 299; OFP 19; PFF 363; SST cp 51; WWT 111, 115, 121; (C. weddeliana) LHP 65; SPL cp 8.

Palm, Curly (Howea belmoreana) EDP 27; EFP 119; LHP 122; RDP 38, 39, 241; (H. forsteriana) EFP 17; LHP 123; SST cp 56.

Palm, Date (Phoenix dactylifera) DEW 2:282 (cps 162-164); EET 25, 237; MLP 274; MWF 231; OFP 107; PFF 363; RDP 315; RWF 49; SST cp 62; WWT 112.

Palm, Doum (Hyphaene thebaica) DEW 2:305; EET 232; EWF cp 19f; HPW 302; MLP 280; SST cp 57.

Palm, Dwarf (Phoenix robelinii) DFP 79 (cp 626); EFP 133; ELG 153; LHP 148; RDP 315.

Palm, Dwarf Fan (Chamaerops humilis) BIP 139 (cp 210-2); DFP 189 (cp 1510); EFP 99; NHE 81; PFE cp 176 (1719); PFM cp 219; RDP 11, 138; SPL cp 73.

Palm, Feather (Chamaedorea sp.) EPF pl 1014; EWF cp 191d; HPW 302.

Palm, Fishtail (Caryota sp.) EDP 24, 136; EFP 97; HPW 302; MEP 10; RDP 130.

Palm, Lady (Rhapis excelsa) EFP 140; EJG 141; ELG 153; MWF 252; RDP 334.

Palm Lily (Cordyline stricta) EWF cp 142a; SPL cp 11.

Palm, Natal (Carissa grandiflora or C. macrocarpa) BIP 49 (cp 58); EDP 125; EGC 125; EGE 117; ELG 136; FHP 110; MEP 109; (C. bispinosa) LFT 247 (cp 123).

Palm, Nikau (Rhopalostylis sapida) MLP 274; SNZ 57 (cps 138-140.

Palm, Nipa (Nypa fruticans) DEW 2:309; EWF cp 191a.

Palm, Oil (Elaeis guineensis) DEW 2:283 (cp 165); EET 236; EPF pl 1015; HPW 302; OFP 21; SST cp 53.

Palm, Palmyra (Borassus sp.) CAF 281; DEW 2:306; EPF pl 1010; OFP 17, 107.

Palm, Paradise or Sentry —See **Palm, Curly.**

Palm, Parlor (Chamaedorea elegans) EFP 98; EGL 93; LHP 54-55; RDP 137; SPL cp 7.

Palm, Petticoat or Desert (Washingtonia filifera) EWF cp 191j; MLP 273, 282; RDP 394; SST cp 68.

Palm, Pot-bellied or "Barrel" (Colpothrinax wrightii) EWF cp 191h; RWF 97.

Palm, Royal (Roystonea regia) EET 234; EWF cp 191e; HPW 302; SST cp 64.

Palm, Sago (Cycas revoluta) CAS 241; DEW 1:18; EDP 132; EFP 104; EPF pl 172; LFW 160 (cp 356); MWF 100; NFP 149; PFF 111; RDP 166; SPL cp 13; SST cp 20.

Palm, Sago (Metroxylon sago) EET 234; OFP 185; SST cp 60.

Palm, Senegal (Phoenix reclinata) SST cp 63; WWT 35; (P. sp.) BIP 139 (cp 210); OFP 17.

Palm, Sugar (Arenga sp.) EET 235; HPW 302; OFP 17, 185.

Palm-civet (Paradoxurus sp.) ALE 12:148; DAA 77; SMW 170 (bw); TWA 180 (bw); (Hemigalus derbyanus) ALE 12:158, 167.

Palm-civet, African or Two-spotted (Nandinia binotata) ALE 12:148; DDM 92 (cp 15); DSM 178; LAW 540; SMR 48 (cp 3); SMW 171 (bw).

Palm-civet, Masked (Paguma larvata) ALE 12:148; DAA 122; DSM 178; MLS 300 (bw).

Palm-civet, Small-toothed (Arctogalidia trivirgata) ALE 12:148; DAA 122; DSM 178; SMW 170 (bw).

Palmetto (Sabal palmetto) PFF 364; SST cp 65; (S. texana) SST cp 66.

Pamianthe (Pamianthe peruviana) DFP 74 (cp 585); PFW 26.

Panaeolus (Panaeolus sp.) DLM 52, 231; LHM 145; NFP 85; ONP 41; RTM 53, 267; SMF 50.

Panamiga or Pan-American Friendship Plant (Pilea involucarta) EDP 138; EFP 134; EGL 136; PFW 302; RDP 317; (P. sp.) EFP 21; EGV 129; HPW 98; LHP 150; RDP 316.

Panchax (Aplocheilus sp.) ALE 4:450; FEF 272, 273 (bw); HFI 68; SFW 421 (pl 110), 444 (cp 117).

Panchax (Epiplatys sp.) —See **Epiplatys.**

Panchax, Playfair's (Pachypanchax playfairi) ALE 4:450; FEF 288, 289 (bw); HFI 68; SFW 465 (pl 124); WFW cp 203.

Panda, Giant (Ailuropoda melanoleuca) ALE 12:103, 114, 137; BMA 94-97; CAS 133 (bw); DAA 51; DSM 165; HMA 76; HMW 152; JAW 110; LAW 534; LEA 559 (bw); LVA 43; LVS 99; MLS 280 (bw); PWC frontis., 79 (bw); SLS 111; SMW 195 (bw); VWA cp 17; WEA 275.

Panda, Lesser or Red (Ailurus fulgens) ALE 12:94, 103; CAS 124, 128 (bw); DAA 51; DSM 164; HMW 147; MLS 279 (bw); SMW 192 (cp 125); TWA 175; WEA 276.

Panda Plant (Kalanchoe tomentosa) EDP 113; EFP 120; EWF cp 70a; LCS cp 273; RDP 251; SCS cp 91.

Pandora (Pagellus erythrinus) CSF 127; OBV 29.

Pangolin, Giant (Manis gigantea) ALE 11:199; DDM 32 (cp 1); HMA 24; JAW 69; SMW 111 (bw).

Pangolin, Indian (Manis crassicaudata) ALE 11:199; CAS 190 (bw); DAA 82; DSM 105; LAW 519.

Pangolin, Long-tailed Tree (Manis
  tetradactyla) ALE 11:199; DDM 32 (cp 1);
  DSM 105; SMW 111 (bw); (M. longicaudata)
  BMA 172.
Pangolin, Malayan (Manis javanica) ALE
  11:199; CAS 255; LAW 580.
Pangolin, Small-scaled Tree (Manis
  tricuspis) ALE 11:199; BMA 170 (bw), 171;
  CAF 137 (bw); DDM 32 (cp 1); DSM 105; LEA
  518 (bw).
Pangolin, Temminck's or Cape Giant
  (Manis temmincki) ALE 2:135; 11:199, 200;
  BMA 172 (bw); HMW 182; MLS 176 (bw);
  SMR 33 (cp 2); TWA 42.
Pansy, Dwarf (Viola kitaibeliana) MBF cp
  12; WFB 147 (9a).
Pansy, Field (Viola arvensis) BFB cp 2 (3);
  MFF pl 87 (125); PWF cp 211; WFB 147 (9).
Pansy, Long-spurred (Viola calcarata) CTH
  37 (cp 47); EWF cp 18f; NHE 275; PFE cp 76
  (786); SPL cp 468.
Pansy, Mountain (Viola lutea) EPF pl 507;
  MBF cp 12; MFF cp 42 (123); NHE 275; PFE
  cp 76 (785); PWF 67g; TVG 179; WFB 147
  (8a).
Pansy, Tufted (Viola cornuta) DFP 178 (cps
  1421-1423); EWF cp 34a; LFW 146 (cps 327,
  328); MWF 294; PFW 306; TGF 140; TVG
  125.
Pansy, Wild (Viola tricolor) AMP 93; BWG
  140; DEW 1:166 (cp 86); DFP 50 (cps 395,
  396); EGA 161; EGH 149; EMB 142; EPF pl
  506; KVP cp 78; LFW 146 (cp 326); MBF cp
  12; MFF cp 1 (124); MLP 152; MWF 294;
  NHE 220; OWF 7, 165; PFE 75 (783); PFF
  252; PFW 306; PWF 61h; TEP 175; TGF 140;
  WFB 147 (8).
Panther —See **Leopard**.
Panther Cap (Amanita pantherina) CTM 22;
  DLM 167; KMF 11; LHM 119; NFP 53; NHE
  48; PMU cp 115; REM 60 (cp 28); RTM 18;
  SMG 180; TEP 223.
Papaw or **Pawpaw** (Asimina triloba) EGT
  99; HPW 30; PFF 177; TGS 246.
Papaya (Carica papaya) DEW 1:250; EPF pl
  517; MLP 152, 153; MWF 73; OFP 115; SST
  cp 187.
Paper Bush (Edgeworthia papyrifera) DFP
  196 (cp 1568); EFC cp 25; GDS cp 157; PFW
  296.

Paper Flower —See **Bougainvillea**.
Paper Nautilus (Argonauta argo) ALE 3:198;
  GSS 156; MOL 60, 72; (A. hians) CTS 64 (cp
  125); GSS 156.
Paper Reed —See **Egyptian Paper Plant**.
Paradise-fish (Macropodus sp.) ALE 5:229;
  CTF 45 (cps 55, 56); FEF 483 (bw); HFI 53;
  SFW 640 (pl 175), 661 (cp 182), 663 (pl 174);
  WFW cp 458.
Parakeelya —See **Purslane**.
Parakeet (Pyrrhura sp.) ALE 8:335; SBB 53
  (pl 66); SSA cp 31.
Parakeet, Black-headed (Psittacula
  himalayana) ABW 146; ALE 8:320; WBV
  135.
Parakeet, Blossom- or Red-headed
  (Psittacula cyanocephala rosa) ALE 8:320;
  DAA 147; RBI 99; (P. roseata) KBA 160 (cp
  21).
Parakeet, Chinese or Derbian (Psittacula
  derbyana) ALE 12:103; RBI 103; SBB 58 (cp
  73).
Parakeet, Golden (Aratinga guarouba)
  ABW 146; ALE 8:335; LAW 505; LVS 224;
  (A. mitrata) SSA cp 4.
Parakeet, Grass —See **Budgerigar**.
Parakeet, Green or Quaker (Myiopsitta
  monachus) ALE 8:335; SBB 58 (cp 74).
Parakeet, Hanging —See **Lorikeet**.
Parakeet, Long-tailed (Psittacula
  longicauda) KBA 160 (cp 21); RBI 104.
Parakeet, Orange-chinned (Brotogeris
  jugularis) ALE 8:335; SSA cp 31.
Parakeet, Orange-fronted (Cyanoramphus
  malherbi) BNZ 83; FNZ 48 (cp 2).
Parakeet, Rainbow —See **Lorikeet,
  Rainbow**.
Parakeet, Red-breasted (Psittacula
  alexandri) ALE 8:320; KBA 160 (cp 21).
Parakeet, Red-crowned (Cyanoramphus
  novaezeelandiae) BNZ 83; CAU 261; CDB
  105 (cp 390); FNZ 48 (cp 2).
Parakeet, Rose-ringed (Psittacula krameri)
  ALE 8:320; CDB 108 (cp 405); FBE 191;
  WAB 161.
Parakeet, Yellow-crowned (Cyanoramphus
  auriceps) BNZ 83; FNZ 48 (cp 2); WAN 132.
Parakeet, Yellow-naped —See **Parrot,
  Twenty-eight**.

Paramecium (Paramecium sp.) ALE 1:108, 113, 305; LEA 30 (bw).

Pardalote, Black-headed (Pardalotus melanocephalus) CDB 188 (cp 816); HAB 253; SAB 52 (cp 25); WAB 189.

Pardalote, Forty-spotted (Pardalotus quadragintus) HAB 253; SAB 52 (cp 25).

Pardalote, Red-browed (Pardalotus rubricatus) HAB 253; MBA 64; SAB 52 (cp 25).

Pardalote, Spotted (Pardalotus punctatus) ALE 9:372; 10:89; CDB 188 (cp 817); GPB 296; HAB 253; MBA 60-61; MWA 69, 70; RBU 256; SAB 52 (cp 25).

Pardalote, Striated (Pardalotus substriatus) HAB 254; MBA 61; SAB 52 (cp 25); (P. ornatus) SAB 52 (cp 25).

Pardalote, Yellow-tailed (Pardalotus xanthopygus) ABW 280; HAB 253; SAB 52 (cp 25).

Pardalote, Yellow-tipped (Pardalotus striatus) CDB 188 (cp 818); HAB 254; SAB 52 (cp 25).

Parrot (Pionus sp.) ALE 8:326; CDB 107 (cp 401); SSA cp 31; (Eclectus roratus) SNB 110 (cp 54); (Psittinus cyanurus) KBA 160 (cp 21); (Tanygnathus sp.) DPB 147; (Triclaria malachitacea) SSA cp 4.

Parrot, Amazon (Amazon aestiva) ALE 8:326; SBB 56 (pl 71); SSA cp 4; (A. amazonica) ALE 8:326, 359; (A. ochrocephala) ABW 146; WAB 17.

Parrot, Black-headed (Nandayus nenday) ALE 8:335; (Pionites melanocephala) SSA cp 4.

Parrot, Blue-bonnet (Psephotus haematogaster) HAB 110, 111; SNB 114 (cp 56); (P. narathae) HAB 11.

Parrot, Bourke's (Neophema bourkii) ALE 8:309; HAB 113; SNB 116 (cp 57); WAB 195.

Parrot, Brown-necked or Cape (Poicephalus robustus) ALE 8:326; PSA 177 (cp 20).

Parrot, Cloncurry (Barnardius macgillivrayi) HAB 104; SNB 112 (cp 55).

Parrot, Elegant (Neophema elegans) ALE 8:309; CDB 106 (cp 396); HAB 112; SNB 116 (cp 57).

Parrot, Fig (Opopsitta coxeni) HAB 91; RBU 127; (O. diophthalma) SNB 106 (cp 52); (O. leadbeateri) CDB 107 (cp 399).

Parrot Flower (Heliconia psittacorum) EWF cp 189a; LFW 236 (cp 524).

Parrot, Golden-shouldered (Psephotus chrysopterygius) HAB 109; LVS 224; RBU 159; SNB 114 (cp 56).

Parrot, Gray (Psittacus erithacus) ABW 146; ALE 8:326; CAA 42 (bw); CDB 108 (cp 406); SBB 50 (pl 60).

Parrot, Ground (Pezoporus wallicus) ALE 8:309; CDB 107 (cp 400); HAB 114; SNB 110 (cp 54); WAB 187.

Parrot, Hanging —See **Lorikeet, Malay.**

Parrot, King (Alisterus scapularis or Aprosmictus scapularis) ALE 8:320; CDB 104 (cp 386); HAB 116; RBU 144; SNB 110 (cp 54).

Parrot, Meyer's (Poicephalus meyeri) CDB 108 (cp 403); PSA 177 (cp 20).

Parrot, Mulga (Psephotus varius) ALE 8:309; CAU 97; HAB 111; MBA 40; MWA 50; SNB 114 (cp 56).

Parrot, Night (Geopsittacus occidentalis) HAB 14; SNB 110 (cp 54); WAB 33, 196.

Parrot, Owl —See **Kakapo.**

Parrot, Paradise (Psephotus pulcherrimus) ALE 8:309; HAB 109; SNB 114 (cp 56).

Parrot, Port Lincoln —See **Parrot, Twenty-eight.**

Parrot, Princess (Polytelis alexandrae) ALE 8:320; HAB 101; RBU 143; SNB 110 (cp 54); WAB 195.

Parrot, Racket-tailed (Prioniturus sp.) DPB 145; RBI 92, 95.

Parrot, Red-backed (Psephotus haematonotus) ALE 8:309; HAB 111; SNB 114 (cp 56); (P. haematodus) CDB 108 (cp 404).

Parrot, Red-capped (Purpureicephalus spurius) ALE 8:309; CAU 46; HAB 106; MBA 40; MWA 50; SNB 112 (cp 55).

Parrot, Red-fan (Deroptyus accipitrinus) ALE 8:292, 326; BBT 100.

Parrot, Red-winged (Aprosmictus erythropterus) ALE 8:320; HAB 116; SNB 110 (cp 54).

Parrot, Regent (Polytelis anthopeplus) ALE 8:320; HAB 101; SNB 110 (cp 54).

Parrot, Ringnecked (Barnardius barnardi) CDB 104 (cp 387); HAB 104; SNB 112 (cp 55).

Parrot, Rosella —See **Rosella.**

Parrot, Scarlet-chested (Neophema splendida) HAB 112; LVB 45; LVS 225; RBU 160; SNB 116 (cp 57).

Parrot, Superb (Polytelis swainsonii) ALE 8:320; HAB 100; SNB 110 (cp 54).

Parrot, Swift (Lathamus discolor) ALE 8:299; CDB 106 (cp 394); HAB 107; SNB 106 (cp 52).

Parrot, Twenty-eight (Barnardius zonarius semitorquatus) HAB 105; MBA 39, 41; MWA 46, 50; RBU 156; SNB 112 (cp 55).

Parrot, Vasa —See **Vasa.**

Parrot, Wax-billed (Lorius roratus) ALE 8:326; (L. pectoralis) HAB 99.

Parrotbill, Black-throated (Paradoxornis nipalensis) KBA 320 (cp 49); RBI 187; (P. webbiana) ABW 241.

Parrotbill, Gray-headed (Paradoxornis gularis) KBA 320 (cp 49); RBI 188; WBV 221.

Parrot-finch (Erythrura psittacea) ALE 9:439; (E. viridifacies) DPB 413.

Parrot-finch, Blue-faced (Erythrura trichroa) ABW 302; HAB 257; SAB 68 (cp 33).

Parrot-finch, Pin-tailed (Erythrura prasina) ALE 9:439; KBA 433 (cp 64).

Parrot-fish (Scarus sp.) ALE 5:144; HFW 170-172 (cps 81-84); LAW 343; MOL 118; WFW cp 413; (Pseudoscarus guacamaia) HFW 205 (bw).

Parrotlet (Forpus sp.) ALE 8:335; SSA cp 31; (Touit huetii) SSA cp 4.

Parrot's Beak —See **Coral Gem.**

Parrot's Bill (Clianthus puniceus) DFP 247 (cp 1974); EGV 103; HSC 31, 33; SNZ 33 (cps 56, 57).

Parsley (Petroselinum crispum) EDP 41; EGH 133; EVF 145; MBF cp 37; MFF cp 51 (477); OFP 147; PFF 266; WFB 162; (P. felicinum) EGH 133.

Parsley, Cambridge (Selinum carvifolia) MFF pl 98 (501); NHE 182; WFB 161 (9).

Parsley, Corn (Petroselinum segetum) MBF cp 37; MFF pl 95 (478); WFB 163 (7).

Parsley, Fool's (Aethusa cynapium) BWG 77; MBF cp 39; MFF pl 96 (497); NHE 222; OWF 87; PWF 119J; WFB 159 (6).

Parsley, Milk (Peucedanum palustre) MBF cp 40; MFF pl 98 (505); NHE 182; PWF 145h; WFB 165 (2); (P. oreoselinum) NHE 32.

Parsley Piert (Aphanes arvensis) BWG 162

(bw); MBF cp 26; MFF pl 89 (366); NHE 211; PWF 50B; WFB 107 (8).

Parsley, Stone (Sison amomum) MBF cp 38; MFF pl 99 (479); WFB 163 (8).

Parsnip (Pastinaca sativa) BWG 43; EVF 101; MBF cp 40; MCW 78; MFF cp 51 (507); NHE 222; OFP 175; OWF 13; PFE cp 86 (900); PFF 267; PWF 95g; WFB 167 (1).

Parsnip, Water (Sium latifolium) MBF cp 37; MFF pl 97 (489); NHE 95; PWF 127g; WFB 161 (5).

Partridge (Francolin sp.) —See **Francolin.**

Partridge, Barbary (Alectoris barbara) ALE 7:481; BBE 99; FBE 103; PBE 157 (cp 42).

Partridge, Bearded or Daurian (Perdix barbata) DPB 61; RBI 8.

Partridge, Chukar or Rock —See **Chukar.**

Partridge, Crested or Crowned Wood —See **Roulroul.**

Partridge, French —See **Partridge, Red-legged.**

Partridge, Gray or Hungarian (Perdix perdix) ALE 7:467-469, 481; 11:321; BBB 71; BBE 99; CDB 70 (cp 205); DAE 124; FBE 103; NBB 66; OBB 57; PBE 157 (cp 42); RBE 51; WAB 112.

Partridge, Green Wood —See **Roulroul.**

Partridge, Red-legged (Alectoris rufa) ABW 91; ALE 7:481; BBB 70; BBE 99; CDB 68 (cp 191); FBE 103; FBW 57; LAW 498; LEA 396 (bw); OBB 57; PBE 157 (cp 42); RBE 48.

Partridge, See-see (Ammoperdix griseo-gularis) ALE 7:481; FBE 105.

Partridge, Snow (Lerwa lerwa) ALE 7:480; 12:103; RBI 4.

Partridge, Tree (Arborophila sp.) KBA 96 (cp 9); RBI 15.

Partridge-berry (Mitchella repens) ERW 126; ESG 63; EWG 126; TGF 188.

Pasang —See **Goat.**

Pastor, Rosy —See **Starling, Rose-colored or Rosy.**

Pasque Flower (Pulsatilla vulgaris or Anemone pulsatilla) BFB cp 1 (7); CTH 26 (cp 22); DEW 1:97 (cp 38); DFP 22 (cps 172, 173); EGH 97; EPF pl 221; ERW 95; EWF cp 8e; LFW 13 (cp 26); MBF cp 1; MFF cp 4 (9); NHE 71; OGF 35; OWF 139; PFE cp 22 (223); PFW 252; SPL 253; TGF 61; TVG 167; WFB 77 (1); (P. halleri) DFP 21 (cp 168); TVG 167.

Pasque Flower, Small (Pulsatilla pratensis) NHE 71; PFE cp 22 (222); PWF cp 20k; TEP 105; WFB 77 (3).

Passion Vine, Purple —See **Velvet Plant, Purple.**

Patience Plant or **Patient Lucy** (Impatiens sultanii) FHP 127; LFW 168 (cp 379); LHP 116, 117; MWF 163.

Pavonia (Pavonia multiflora) DEW 1:219 (cp 132); EPF pl 613; EWF cp 177b; MEP 80; PFW 185; (P. columella) LFT cp 106.

Pea, Asparagus (Tetragonolobus purpureus) EWF cp 29f; PFE cp 61 (617); PFM cp 72, pl 7 (345).

Pea, Beach or Sea (Lathyrus japonica) MBF cp 25; MFF cp 18 (338); MLP 100; PFE cp 56 (565); PWF 54e; WFB 125 (5); (L. maritimus) NHE 134; OWF 135.

Pea, Black (Lathyrus niger) EWF cp 10b; MBF cp 25; PFE cp 56 (563); PWF 103g.

Pea, Butterfly (Clitoria ternatea) EWF cp 53e; MEP 55.

Pea, Darling (Swainsona galegifolia) EGG 143; EWF cp 130d; LFW 115 (cp 251); MEP 62.

Pea, Earth-nut (Lathyrus tuberous) KVP cp 35; MBF cp 25; MFF cp 18 (335); PFE cp 56 (559); PWF 103k; TEP 177; WFB 125 (7).

Pea, Everlasting or Perennial (Lathyrus latifolius) EGV 118; MBF cp 25; MWF 180; OGF 163; PFE cp 56 (560); PWF 82d; TGF 140; TVG 107; WFB 125 (1).

Pea, Garden (Pisum sativum) EPF pls 466, 467; EVF 102; OFP 43; PFE cp 56 (566); PFF 221; TEP 163; (P. elatius) PFM cp 76.

Pea, Large Wedge (Gompholobium sp.) EWF cp 129f; MWF 143.

Pea, Marsh (Lathyrus palustris) MFF cp 18 (337); NHE 92; PWF 74c; WFB 125 (6).

Pea, Spring (Lathyrus vernus) DFP 154 (cp 1232); NHE 72; PFE cp 56 (564); TEP 243; WFB 125 (3).

Pea, Sweet (Lathyrus odoratus) DFP 40 (cps 316-320), 41 (cp 321); EGA 71, 130; EGG 123; EWF 29b; LFW 80 (cp 185); MWF 180; NHE 213; PFF 221; PFW 160; SPL cp 172; TGF 140; TVG 55; (L. clymenum) KVP cp 180; PFE cp 56 (561).

Pea, Wild or Everlasting (Lathyrus sylvestris) DEW 1:311; HPW 150; MBF cp 25; MFF cp 18 (336); OWF 133; PWF 139b; TEP 243.

Pea, Winged —See **Dragon's Teeth.**

Peach, Guinea or African (Nauclea latifolia) EWF cp 60f; HPW 258; WAW 110, 111.

Peacock —See **Peafowl.**

Peacock Flower —See **Iris, Peacock.**

Peacock Plant (Calathea makoyana) EFP 23, 96; LHP 125; MWF 65; PFW 186; RDP 15, 124; SPL cp 6; (Kaempferia sp.) EWF cps 124c, 124e; MEP 28.

Peacock-pheasant, Germain's (Polyplectron germaini) WBV 67; WPW 126.

Peacock-pheasant, Gray (Polyplectron bicalcaratum) ALE 8:39; KBA 81 (pl 8); WBV 67; WPW 126.

Peacock-pheasant, Malay (Polyplectron malacense) ALE 8:39; RBI 52; WPW 143.

Peacock-pheasant, Palawan (Polyplectron emphanum) BBT 33; CDB 70 (cp 206); DPB 63; LVS 208; WPW 144.

Peafowl, African or Congo (Afropavo congensis) ALE 8:30; WAB 141; WPW 144.

Peafowl, Blue or Indian (Pavo cristatus) ABW 94; ALE 8:21, 23, 29; 13:285; BBT 40; CDB 70 (cp 203); GPB 121; LEA 400; NBB 18 (bw), 44; WAB 164-165; WEA 280; WPW 144.

Peafowl, Green or Javan (Pavo muticus) ALE 7:29; CAS 270 (bw); WPW 144.

Peanut (Arachis hypogaea) EFP pl 458, 461; OFP 23; PFF 218.

Pear (Pyrus calleryana) EEG 121; EGT 73, 139; EPG 140; (P. cordata) MBF cp 31; (P. kawakamii) EGE 142; EGW 61; (P. nivalis) BKT cp 133; (P. pyraste) WFB 115 (2).

Pear, Common (Pyrus communis) EET 174, 175; EVF 127; MBF cp 31; MFF pl 75 (386); NHE 20; OFP 57, 59; PFF 197; SST cp 222.

Pear, Willow-leaved (Pyrus salicifolia) DFP 222 (cp 1771); EET 174, 175; MTB 289 (cp 26); OBT 180.

Pearl Bush (Exochorda sp.) DFP 201 (cp 1607); EFS 111; EPF pl 396; GDS cp 192; HSC 50, 56; PFW 265; TGS 279.

Pearlfish (Carapus acus) ALE 4:432; HFI 47; (Cynopoecilus ladigesi) ALE 4:450; FEF 294 (bw); SFW 445 (cp 118).

Pearlfish, Argentine (Cynolebias bellotti)
   ALE 4:453; FEF 291 (bw); SFW 433 (pl 114),
   440 (pl 115), 444 (cp 117).
Pearlfish, Black-finned (Cynolebias
   nigripinnis) FEF 292 (bw); SFW 460 (pl
   121), 461 (pl 122), 464 (pl 123).
Pearl-side (Maurolicus muelleri) ALE 4:266;
   CSF 73; OBV 23.
Pearlwort, Alpine (Sagina saginoides) MBF
   cp 16; MFF pl 89 (178); NHE 277.
Pearlwort, Annual or Common (Sagina
   apetala) MBF cp 16; MFF pl 89 (175); NHE
   97; (S. caespitosa) EWF cp 3c.
Pearlwort, Heath or Corsican (Sagina
   sublata) EEG 147; EGC 145; MBF cp 16;
   MFF pl 90 (179); TGF 91.
Pearlwort, Knotted (Sagina nodosa) MBF cp
   16; MFF pl 90 (180); NHE 185; OWF 77; PFE
   81; WFB 57 (3).
Pearlwort, Procumbent or Mossy (Sagina
   procumbens) BWG 69; MBF cp 16; MFF pl 89
   (177); NHE 226; OWF 77; PFE cp 13 (146);
   PWF 47f; WFB 57 (4).
Pearlwort, Sea (Sagina maritima) MBF cp
   16; MFF pl 89 (176); NHE 135.
Pearly Everlasting (Anaphalis sp.) DFP 120
   (cp 960), 121 (cps 961-962); EGP 94; MFF pl
   94 (872); OGF 151; PFF 317; TGF 45.
Pea-tree (Caragana arborescens) EFS 22, 99;
   HSC 23; TGS 137.
Pecan (Carya illinoiensis) EET 139; EGT 102;
   EVF 137; OFP 29; PFF 134; SST cp 188; TGS
   86.
Peccary (Tayassu tajacu or Pecari tajacu)
   ALE 13:92; DSM 218; JAW 165; MLS 373
   (bw); SMW 234 (cp 145); TWA 134; WEA
   281; (T. albirostris) ALE 8:359; 13:92.
Pediunker —See **Petrel, Gray.**
Peepul Tree (Ficus religiosa) HPW 97; SST cp
   115.
Pelican, Australian (Pelecanus conspicil-
   latus) ALE 7:158; CAU 84-85 (bw); CDB 38
   (cp 39); GPB 52; HAB 14; LEA 365 (bw);
   MBA 55; RBU 91; SNB 20 (pl 9); WAW 22-23
   (bw).
Pelican, Brown (Pelecanus occidentalis)
   ABW 41, 42; ALE 7:157; CSA 225; CDB 38
   (cp 40); GPB 52; HBG 68 (bw); WAB 67;
   WEA 281.

Pelican, Dalmatian (Pelecanus crispus) ALE
   7:158; BBE 31; FBE 31; GPB 52; LVS 201;
   SBB 23 (pl 21).
Pelican Flower (Aristolochia grandiflora)
   DEW 1:64(cp 36), 133; MEP 34; MWF 46;
   RWF 35; (A. gigantea) EWF cp 170b.
Pelican, Pink-backed (Pelecanus rufescens)
   ALE 7:158, 183, 220-221; BBT 58, 59; CAA
   40; GPB 52; LAW 450; SAO 71.
Pelican, White (Pelecanus onocrotalus) ABW
   40; BBE 31; CAA 30 (bw); CAF 108-109;
   CDB 39 (cp 41); FBE 31; GAL 193 (bw); GPB
   51, 52; LAW 441; SBB 22 (pl 20); WEA 283.
Pelican-ibis —See **Stork, Painted.**
Pelican's-foot Shell (Aporrhais pespelicani)
   ALE 3:55; CSS pl XV; CTS 21 (cp 14); GSS
   20; OIB 41; (A. serresianus) CTS 21 (cp 15);
   MOL 69.
Pellionia (Pellionia sp.) BIP 69 (cp 98); PFW
   302; RDP 305.
Pellitory-of-the-wall (Parietaria diffusa)
   MBF cp 76; MFF cp 37 (551); NHE 223; OWF
   65; PWF 67k; (P. officinalis) AMP 131; PFE
   cp 7 (68); (P. judaica) HPW 98; WFB 39 (8).
Pen Shell, Mediterranean (Pinna sp.) ALE
   3:140; CTS 58 (cp 108); GSS 130; OIB 73.
Pencilfish (Nannostomus sp.) ALE 4:286;
   FEF 123, 124, 129; HFI 88; SFW 164 (cp 37),
   184 (pl 4), 185 (pl 42); WFW cps 91, 92.
Pencilfish, Tube-mouthed (Nannobrycon
   eques) FEF 133 (bw); SFW 192 (pl 43).
Penguin, Adelie (Pygoscelis adeliae) ABW
   29; ALE 7:121; CDB 35 (cp 11); GPB 36, 38;
   LAW 501; NBB 15, 94 (bw); SAO 21; SAR 32,
   41, 87, 93-97; SLP 116; WAB 25; WAN
   62-74; WEA 282.
Penguin, Bearded or Chinstrap (Pygoscelis
   antarctica) ABW 28; ALE 7:121; GPB 38;
   SAO 21; SAR 98.
Penguin, Black-footed —See **Penguin,
   Jackass.**
Penguin, Crested (Eudyptes pachyrhynchus)
   BNZ 239; GPB 39; (E. sclateri) GPB 39.
Penguin, Emperor (Aptenodytes forsteri or
   Aptenoides forsteri) ABW 26; ALE 7:121-
   123; CDB 34 (cp 8); GPB 37, 38; SAO 21; SAR
   88-91; SLP 118-120; WAB 221; WAN 41-59.
Penguin, Galapagos (Spheniscus mendicul-
   us) ALE 7:124; GPB 39; HBG 52 (bw); LVB
   27; LVS 203; PWC 218 (bw); SLS 67; WAB 221.

Penguin, Gentoo (Pygoscelis papua) ABW
28; ALE 7:121; CSA 288-289; GPB 38; LAW
466; LEA 361 (bw); SAO 21; SAR 99; WEA
282; WFI 65.

Penguin, Jackass (Spheniscus demersus)
ABW 28; ALE 7:124; CAF 238-239; CDB 35
(cp 12); NBB 14 (bw); PSA 32 (cp 1); SAO 21.

Penguin, King (Aptenodytes patagonica or
Aptenoides patagonicus) ABW 26; ALE
7:121; GPB 38; HAB 267 (bw); LAW 451;
LEA 360 (bw); PWC 46 (bw); SAO 21; SAR
103-105; WAN 39, 40.

Penguin, Little (Eudyptula minor) ABW 28;
ALE 7:124; CAU 70-71 (bw); GPB 39; HAB 7;
WAB 213; (Eudyptes minor) CDB 34 (cp 10).

Penguin, Macaroni (Endyptes chrysolophus) ALE 7:121; CDB 34 (cp 9); CSA 288;
GPB 39; SAO 21; SAR 100; WAB 221; WFI
72 (bw).

Penguin, Magellanic (Spheniscus
magellanicus) ABW 28; ALE 7:124; CSA
282 (bw); GPB 39; SAO 21; SSA pl 20; WAB
79, 221; WFI 65.

Penguin, Peruvian (Spheniscus humboldti)
ALE 7:124; CDB 35 (cp 13); GPB 38.

Penguin, Rockhopper (Eudyptes crestatus)
ABW 29; ALE 7:121; GPB 39; LAW 457, 465;
LEA 361 (bw); RBU 103; SAO 21; SAR 101;
WFI 70 (bw).

Penguin, Royal (Eudyptes schlegeli) GPB
39; WAB 221; WAN 75.

Penguin, Yellow-eyed (Megadyptes
antipodes) BNZ 239; GPB 39; WAN 75.

Penguin-fish (Thayeria sp.) ALE 4:286; CTF
22 (cp 11); FEF 78 (bw); SFW 133 (pl 26);
WFW cps 87, 88.

Penicillium (Penicillium sp.) DEW 3: cp 77;
210; DLM 86; NFP 24, 35; PFF 68.

Pennycress, Alpine (Thlaspi alpestre) MBF
cp 10; MFF pl 92 (70); NHE 217; WFB 99 (6).

Pennycress, Field (Thlaspi arvense) BWG
62; EPF pl 245; MBF cp 10; MCW 58; MFF pl
92 (67); NHE 217; OWF 71; PFE cp 35 (343);
PWF 59f; TEP 173; WFB 99 (5); WUS 215.

Pennycress, Mountain (Thlaspi montanum)
NHE 217; PFE 139; WFB 99 (8).

Pennycress, Perfoliate (Thlaspi perfoliatum) MBF cp 10; MFF pl 92 (69); NHE 217;
WFB 99 (7).

Pennycress, Round-leaved (Thlaspi
rotundifolium) DFP 27 (cp 213); NHE 274;
PFE cp 36 (345).

Penny-royal (Mentha pulegium) EGH 124;
MBF 67; MFF cp 11 (744); NHE 98; OWF 143;
PFE cp 116 (1167); PWF 150c; WFB 205 (4).

Pennywort, Marsh (Hydrocotyle vulgaris)
MBF cp 36; MFF pl 105 (451); NHE 95; OWF
47; PWF 98d; WFB 155 (5); (H. rotundifolia)
EGC 73.

Pennywort, Wall (Umbilicus rupestris) BWG
74; MBF cp 33; MFF pl 88 (397); NHE 246;
OWF 53; PFE cp 39 (379); PWF 73k; WFB 101
(5); (U. spinosus) TVG 177.

Peony (Paeonia lutea) DFP 215 (cp 1718);
EWF cp 90a; HSC 82, 88; OGF 53.

Peony (Paeonia sp.) DFP 162, 163 (cps
1295-1298); GDS cp 304, 305; HPW 81; HSC
83; LFW 106 (cp 235); MLP 54; OGF 53; PFF
173; PFM cp 28; PFW 218; TGF 61.

Peony, Chinese (Paeonia lactiflora) DFP 163
(cp 1297); EDP 130; EEG 136; LFW 108 (cps
236-239); MWF 220; SPL cp 235; TVG 113.

Peony, Common (Paeonia officinalis) DEW
1:61 (cp 30); DFP 163 (cp 1299); NHE 268;
PFE cp 27 (258); PFW 217; SPL cp 234.

Peony, Fine-leaved (Paeonia tenuifolia)
EFC cp 41; HPW 81.

Peony, Tree (Paeonia suffruticosa) DFP 215
(cps 1719, 1720); EFS 125; EJG 67, 134; ELG
117; EPF pl 199, cp VI a; EWF 90c; GDS cp
306; HSC 83, 88; MWF 220; PFW 219; SPL cp
236; TVG 233.

Peony, Wild (Paeonia mascula) HPW 81;
MBF cp 4; MFF cp 15 (29); PFE cp 27 (259).

Pepper, Black (Piper nigrum) AMP 21; OFP
129.

Pepper, Celebes (Piper ornatum) BIP 147 (cp
220); PFW 231; (P. tiliifolium) EPF pl 182.

Pepper, Red (Capsicum annuum) BIP 49 (cp
56); DFP 32 (cp 256); EDP 142; EGL 92; EPF
pls 820, 821; FHP 110; LHP 52, 53; MWF 73;
OFP 129; PFF 292; PFW 284; RDP 129; (C.
frutescens) EVF 103; OFP 129; (C.
minimum) AMP 29.

Pepper, Saffron (Piper crocatum) EFP 135;
EGV 130; RDP 318.

Pepperbush (Clethra sp.) DFP 190 (cp 1520);
EEG 93; EFS 101; ESG 103; EWF cp 161f;
HSC 31, 33; TGS 407.

Pepper-elder or **Pepper-face** (Peperomia
sp.) BIP 143 (cp 216); DEW 1:150, 151; DFP
78 (cps 618-621); EDP 145; EFP 21, 50, 130,
131; EMB 109; EPF 131; HPW 40; LHP 142,
143; MWF 226; RDP 306, 308; SPL sp 25.

Peppermint (Mentha piperita) AMP 43; EGH
124; MBF cp 67; OFP 141; OWF 143; PFF
291; PWF 139i.

Peppershrike, Rufous-browed (Cyclarhis
gujanensis) ALE 9:372; CDB 206 (cp 904);
DTB 73 (cp 36); SSA cp 16; WAB 91.

Pepper-tree (Schinus molle) DEW 2:68; EGE
147; EPG 145; MEP 74; MWF 264; PFF 237;
SST cp 168; (S. terebintifolius) SST cp 169.

Pepperwort, Broad-leaved —See
Dittander.

Pepperwort, Downy (Lepidium heterophyllum) MBF cp 10; PWF 118b.

Pepperwort, Field or Common (Lepidium
campestre) MBF cp 10; MFF pl 92 (60); NHE
216; OWF 71; PWF 100b; WFB 99 (2); WUS
207.

Pepperwort, Hoary —See **Cress, Hoary.**

Pepperwort, Narrow-leaved (Lepidium
ruderale) MBF cp 10; MFF pl 92 (61); NHE
216; PFE 139; WFB 99 (4).

Perch, Ansorge's Climbing (Ctenopoma
ansorgei) HFI 53; SFW 656 (pl 179).

Perch, Climbing (Anabas testudineus) HFI
53; SFW 624 (pl 171).

Perch, Common (Perca fluviatilis) ALE 5:85;
FEF 346 (bw); FWF 121; GAL 235 (bw); HFI
24; OBV 95; SFW 532 (pl 143); WEA 284.

Perch, Ocean —See **Haddock, Norway.**

Perch, Sea —See **Sea-perch.**

Perch, Sharp-nosed Climbing (Ctenopoma
oxyrhynchus) LAW 346; SFW 656 (pl 179).

Perch, Spotted Climbing (Ctenopoma
acutirostre) HFI 53; HFW 243 (bw); WFW cp
453.

Peregrine —See **Falcon, Peregrine.**

Perentie —See **Monitor, Perentie.**

Periwinkle (mollusk) (Littorina sp.) ALE
3:46, 71, 74; CSS pl XIV; GSS 37; OIB 39;
WEA 285; (Tectarius pagodus) GSS 37.

Periwinkle, Greater (plant) (Vinca major)
CTH 46 (cp 66); DFP 244 (cp 1951); EFC cp 2;
KVP cp 181; MFF cp 6 (631); MWF 293; PFM
cp 130; PWF cp 24e; SPL 275.

Periwinkle, Lesser or Common (plant)
(Vinca minor) AMP 165; BFB cp 13 (10);
BWG 141; DFP 178 (cp 1420); EEG 147; EFC
cp 2; EGC 149; EGV 143; EGW 146; EJG 149;
ELG 148; ESG 147; EWF cp 21c; HPW 224;
LFW 144 (cp 323); MBF cp 58; NHE 34; OWF
137; PFE cp 97 (1005); PFF 282; PFW 28;
PWF cp 17f; SPL cp 276; TEP 251; TGF 157;
TGS 358; TVG 179; WFB 179 (8).

Pernettya (Pernettya mucronata) DFP 216
(cp 1724); GDS cps 307-310; HSC 87, 89; PFW
109; (P. macrostigma) SNZ 141 (cps 439-441);
(P. furens and parviflora) EHP 126, 127.

Persian Shield (Strobilanthes dyerianus)
BIP 173 (cp 268); MWF 277; RDP 374; (S.
isophyllus) BIP 173 (cp 267).

Persimmon, Common (Diospyros virginiana) EET 202, 203; EGT 111; EGW 108;
PFF 279; SST cp 201; TGS 375.

Persimmon, Oriental (Diospyros kaki) EET
203; EVF 128; MWF 115; OFP 105; SST cp
200.

Peruvian Daffodil —See **Spider-lily.**

Petrel, Antarctic (Thalassoica antarctica)
SAO 33; SAR 111; WAN 99-103.

Petrel, Band-rumped Storm —See **Petrel,
Madeiran Storm.**

Petrel, Black-bellied Storm (Fregetta
tropica) HAB 12; SAO 61; SAR 117; SNB 18
(pl 8).

Petrel, Blue (Halobaena caerulea) HAB 13;
SAO 33; SAR 114; SNB 14 (cp 6); WAN 94,
95.

Petrel, Bulwer's (Bulweria bulwerii) BBE
29; FBE 29; SAO 49; WAN 113.

Petrel, Cape —See **Petrel, Pintado.**

Petrel, Common Diving (Pelecanoides
urinatrix) ABW 38; CDB 38 (cp 37); FNZ 57
(bw); SAO 67; WAB 212.

Petrel, Giant (Macronectes giganteus) CDB
37 (cp 29); HAB 13; PSA 65 (pl 6); SAO 33;
SAR 110; WAB 220; WAN 91-93; WFI 101.

Petrel, Gould —See **Petrel, White-winged.**

Petrel, Gray (Procellaria cinerea) SAO 49;
SAR 115; WAN 104, 105.

Petrel, Gray-backed Storm (Garrodia
nereis) FNZ 54 (bw); RBU 51; SAO 61; SAR
117; SNB 18 (pl 8).

Petrel, Great-winged (Pterodroma
macroptera) HAB 13; SAO 41; SAR 114.

Petrel, Herald or Trinidade (Pterodroma arminjoniana) SAO 41; WAN 108-110.

Petrel, Kerguelen (Pterodroma brevirostris) HAB 13; SAO 41; SAR 114.

Petrel, Leach's (Oceanodroma leucorrhoa) ABW 31; BBB 266; BBE 29; FBE 29; HBG 64 (pl); OBB 7; PBE 5 (pl 2); SAO 57; SNB 18 (pl 8).

Petrel, Madeiran Storm (Oceanodroma castro) FBE 29; HBG 64 (pl 1); SAO 61; WAN 114.

Petrel, Magellan Diving (Pelecanoides magellani) ALE 7:148; SAO 67; SSA pl 20; (P. georgicus) SAR 116.

Petrel, Pintado (Daption capensis) ABW 31; ALE 7:147; BBE 29; CDB 37 (cp 27); FBE 25; HAB 13; HBG 54 (bw); PSA 65 (pl 6); SAO 33; SAR 72, 111, 112; SSA pl 20; WAN 94.

Petrel, Snow (Pagodroma nivea) SAO 37; SAR 111, 112; WAN 110-112.

Petrel, Soft-plumaged (Pterodroma mollis) FBE 25; HAB 13; SAO 41; WAN 108.

Petrel, Storm (Hydrobates pelagicus) ABW 31; ALE 7:148; BBB 266; BBE 29; FBE 29; OBB 7; PBE 5 (pl 2); PSA 65 (pl 6); SAO 61; (Oceanodroma tethys) HBG 64 (pl 1).

Petrel, White-bellied Storm (Fregetta grallaria) HAB 12; HBG 60 (bw); SNB 18 (pl 8).

Petrel, White-chinned (Procellaria aequinoctialis) ALE 7:147; PSA 65 (pl 6); SAO 49; SAR 115; WFI 87 (bw).

Petrel, White-faced Storm (Pelagodroma marina) ALE 7:148; CDB 37 (cp 26); FBE 29; FNZ 56 (bw); HAB 12; HBG 64 (pl 1); SAO 61; SNB 18 (pl 8).

Petrel, White-headed (Pterodroma lessonii) HAB 13; SAR 114; WAN 107.

Petrel, White-vented Storm (Oceanites gracilis) HBG 64 (pl 1); SSA pl 20.

Petrel, White-winged (Pterodroma leucoptera) ABW 31; HAB 13; WAW 27 (bw).

Petrel, Wilson's Storm (Oceanites oceanicus) ABW 31; ALE 7:148; BBE 29; FBE 29; FNZ 54 (bw); HAB 12; PBE 5 (pl 2); PSA 65 (pl 6); SAO 61; SNB 18 (pl 8); WAB 220; WAN 114, 115.

Petty Whin (Genista anglica) MBF cp 21; MFF cp 47 (271); NHE 69; PWF 57k; WFB 121 (4).

Peyote (Lophophora williamsii) EHP 115; MLP 161; PFW 58; SCS cp 38; (L. weddellii) HPW 176.

Phaeolepiota (Phaeolepiota aurea) DLM 233; LHM 131; RTM 62.

Phalanger —See also **Cuscus.**

Phalanger (Distoechurus pennatus) ALE 10:100 (cp 2); (Dromicia nana) HMW 203.

Phalanger, Brush-tailed (Trichosurus vulpecula) ALE 10:99 (cp 1); BMA 36, 37; CAU 83 (bw); DSM 32; LEA 482 (bw); MLS 58 (bw); SMW 38 (cp 12); TWA 192; WAW 50 (bw); (T. caninus) HMW 204.

Phalanger, Flying (Petaurus australis) ALE 10:100 (cp 2); BMA 91-93; DSM 33; WEA 153; (P. norfolcensis) CAU 116; MLS 59 (bw); (P. sciureus) TWA 202 (bw).

Phalanger, Greater Flying (Schoinobates volans) ALE 10:101 (cp 3), 104, 105; DSM 33; WAW 49 (bw); (Petauroides volans) TWA 193.

Phalanger, Honey —See **Possum, Honey.**

Phalanger, Leadbeater's —See **Possum, Leadbeater's.**

Phalanger, Pygmy Flying (Acrobates pygmaeus) ALE 10:100 (cp 2), 103; WAW 37.

Phalanger, Scaly-tailed (Wyulda squami-caudata) ALE 10:99 (cp 1); PWC 230 (bw).

Phalanger, Short-headed Flying (Petaurus breviceps) ALE 10:100 (cp 2); HMA 77; HMW 202; JAW 26; LEA 485 (bw); MEA 89; SMW 25 (bw).

Phalanger, Striped (Dactylopsila trivirgata) ALE 10:99 (cp 1); (D. picata) LEA 485 (bw).

Phalarope, Northern or Red-necked (Phalaropus lobatus or Lobipes lobatus) ABW 124; BBB 175; BBE 135; CDB 89 (cp 304); DPB 107; FBE 129; GPB 153; HBG 133 (pl 10); KBA 150 (bw); OBB 79; PBE 85 (pl 28), 100 (cp 31), 117 (cp 34); RBE 155; SAO 93; SNB 70 (cp 34), 74 (cp 36); WAB 214.

Phalarope, Red or Gray (Phalaropus fulicarius) ABW 124; ALE 8:170; BBB 277; BBE 135; CDB 89 (cp 303); FBE 129; GPB 152; HBG 133 (pl 10); OBB 79; PBE 85 (pl 28), 100 (cp 31), 117 (cp 34); SAO 93; SNB 70 (cp 34), 74 (cp 36).

Phalarope, Wilson's (Phalaropus tricolor or Steganopus tricolor) ABW 124; BBE 135; CAA 43 (bw); FBE 129; HBG 108 (bw), 132 (pl 9); PBE 285 (cp 66); SAO 93; SNB 70 (cp 34), 74 (cp 36); SSA pl 23.

Phascogale, Brush-tailed (Phascogale penicillata) HMW 198; TWA 206 (bw); (P. calura) MWA 42; (P. tapoatafa) WAW 47 (bw).

Pheasant, Blood (Ithaginis cruentus) ALE 7:480; 12:103; RBI 19; WPW 18.

Pheasant, Blue Eared (Crossoptilon auritum) ALE 8:46; BBT 30-31; WPW 106.

Pheasant, Blyth's (Tragopan blythii blythii) ALE 7:482; WPW 35.

Pheasant, Brown Eared (Crossoptilon mantchuricum) ABW 93; ALE 8:46; LVB 33; LVS 209; RBI 27; SLS 174; WAB 29; WPW 106.

Pheasant, Bulwer's Wattled (Lophura bulweri) RBI 35; WPW 83; (Lobiophasis bulweri) ALE 8:53.

Pheasant, Cheer (Catreus wallichi) ALE 8:54; WPW 123.

Pheasant, Chinese Monal (Lophophorus lhuysi) ALE 12:103; RBI 23; WAB 29; WPW 59.

Pheasant, Crested Fireback (Lophura ignita) ALE 8:53; KBA 81 (pl 8); RBI 28; WPW 86.

Pheasant, Crimson Horned—See **Pheasant, Satyr.**

Pheasant, Diard's or Siamese Fireback (Lophura diardi) ABW 93; KBA 81 (pl 8); RBI 32; WBV 67.

Pheasant, Edward's (Lophura edwardsi) WAB 31; WBV 67; WPW 84.

Pheasant, Elliott's (Syrmaticus ellioti) ALE 8:60; BBT 36-37; LVB 35; PWC 220 (bw); RBI 47; WAB 31; WPW 123.

Pheasant, Elwes' or Tibetan Eared (Crossoptilon crossoptilon) ALE 8:46, 12:103; CDB 68 (cp 196); WPW 106.

Pheasant, Golden (Chrysolophus pictus) ABW 92; ALE 8:36, 73; BBB 280; BBT 26, 27; CDB 68 (cp 193); FBE 108; FBW 55; LAW 441; OBB 55; RBI 48; WAB 101; WEA 287; WPW 124.

Pheasant, Great Argus (Argusianus argus) ABW 95; ALE 8:39; RBI 55; WPW 144.

Pheasant, Green or Japanese (Phasianus versicolor) ALE 8:63; FBE 107; FBW 53; RBI 40; WPW 125.

Pheasant, Himalayan or Impeyan Monal (Lophophorus impejanus) ABW 92; ALE 8:45; DAA 145; WAB 162; WPW 50.

Pheasant, Hume's (Syrmaticus humiae) KBA 81 (pl 8); WPW 123.

Pheasant, Imperial (Lophura imperialis) KBA 81 (pl 8); WBV 67.

Pheasant, Kalij (Lophura leucomelana) FBW 53; KBA 81 (pl 8).

Pheasant, Lady Amherst's (Chrysolophus amherstiae) ABW 92; ALE 8:73; 12:103; BBB 280; BBT 28; CDB 68 (cp 192); FBE 108; FBW 53; OBB 55; RBI 51; WPW 126.

Pheasant, Malay Crestless Fireback (Lophura erythrophthalma) WPW 85; (Houppifer erythrophthalmus) ALE 8:53.

Pheasant, Mikado (Syrmaticus mikado) ALE 8:60; BBT 34-35; LVB 35; LVS 207; PWC 86, 220 (bw); WPW 124.

Pheasant, Ocellated (Rheinartia ocellata) ALE 8:39; WPW 144.

Pheasant, Peacock —See **Peacock-pheasant.**

Pheasant, Reeves' (Syrmaticus reevesi) ABW 93; ALE 8:60; BBB 280; DAA 148; FBE 108; FBW 53; RBI 43; WPW 124.

Pheasant, Ring-necked (Phasianus colchicus) ABW 92; ALE 8:36, 64; BBB 69; BBE 101; CDB 70 (cp 204); DAE 125; FBE 107; FBW 55; GPB 121; LAW 465; LEA 399 (bw); OBB 55; PBE 164 (cp 43); WEA 286; WPW 125.

Pheasant, Satyr (Tragopan satyra) ALE 7:482; BBT 29; CDB 70 (cp 208).

Pheasant, Sclater's Monal (Lophophorus sclateri) LVB 33; RBI 24; WPW 48.

Pheasant, Silver (Lophura nycthemera) BBT 34-35; FBE 108; GPB 119; KBA 81 (pl 8); RBI 31; WPW 74; (Gennaeus nycthemerus) ABW 92; ALE 8:47.

Pheasant, Soemmering's Copper (Syrmaticus soemmeringii) ABW 93; ALE 8:63; RBI 44; WPW 125.

Pheasant, Swinhoe's (Lophura swinhoii) LVB 35; LVS 208; PWC 220 (bw); WAB 31; WPW 103; (Gennaeus swinhoii) ABW 93; (Hierophasis swinhoii) ALE 8:47.

Pheasant, Temminck's (Tragopan temmincki) ABW 92; ALE 12:103; RBI 20; WAB 161.

Pheasant, Western (Tragopan melano-cephalus) ALE 7:482; LVB 33; WPW 38.

Pheasant's Eye, Scarlet or Large (Adonis flammea) NHE 210; WFB 75 (6).

Pheasant's Eye, Spring or Yellow (Adonis vernalis) DEW 126; EGP 93; EPF pl 214; NHE 210; PFE cp 24 (231); TEP 105; TGF 86 (bw); TVG 137; WFB 75 (5).

Pheasant's Eye, Summer (Adonis aestivalis) DEW 1:98 (cp 41); EGA 90; EPF pl 211; NHE 210; TEP 172; WFB 75 (7).

Pheasant's Eye, Winter (Adonis annua) AMP 79; EWF cp 27E; MBF cp 1; MFF cp 14 (23); OWF 105; PFE cp 24 (230); PFM cp 32; PWF 520.

Phebalium (Phebalium sp.) EWF cp 138 e, g; SNZ 63 (cp 162).

Philesia (Philesia magellanica or P. buxifolia) DFP 217 (cp 1730); EWF cp 190a; PFW 230.

Philodendron (Philodendron sp.) BIP 145 (cp 219); DFP 78, 79 (cps 622-625); EDP 112, 127; EFP 56, 132, 133; EGV 129; HPW 308; LHP 144-147; MWF 229; RDP 311-313; SPL cps 26, 27.

Photinia, Fraser (Photinia fraseri) EGE 139; EPG 135.

Photinia, Oriental (Photinia villosa) EFS 126; EJG 135; TGS 252; (P. glabra) TGS 246.

Phyllanthus (Phyllanthus pulcher) EWF cp 119a; HPW 186.

Pick-a-back Plant or **Piggyback Plant** —See **Youth-on-age.**

Piculet (Picumnus sp.) ALE 9:108; CDB 138 (cp 552); KBA 240 (cp 31); SSA cps 34, 45; WAB 80; (Sasia sp.) KBA 240 (cp 31).

Piddock, Common (Pholas dactylus) CSS pl XX; CTS 62 (cp 121); OIB 87.

Piddock, Oval or Frilled (Zirphaea crispata) ALE 3:177; CSS pl XX; OIB 87.

Piddock, White (Barnea candida) ALE 3:177; CSS pl XX; OIB 87.

Piddock, Wood (Xylophaga dorsalis) CSS pl XX; OIB 87.

Piemarker or **Pieprint** —See **Velvetleaf.**

Pieris, Chinese (Pieris forrestii) DFP 218 (cp 1737); GDS cp 318; HSC 90, 93; (P. taiwanensis) DFP 218 (cp 1739); SPL cp 109.

Pig, Bush —See Bushpig.

Pig Squeak (Bergenia crassifolia) EGC 124; EPF pl 391; ESG 97; HPW 147; SPL cp 254; TGF 93.

Pigeon, Bleeding Heart (Gallicolumba luzonica) ABW 143; ALE 8:246; CDB 360 (cp 360); DPB 137; RBI 87; WAB 173.

Pigeon, Brush Bronze-wing (Phaps elegans) ALE 10:89; HAB 84; SNB 102 (cp 50); WAB 186.

Pigeon, Cape —See **Petrel, Pintado.**

Pigeon, Chestnut-quilled Rock (Petrophassa rufipennis) HAB 87; SNB 102 (cp 50).

Pigeon, Common (Columba livia) ABW 144; ALE 7:35, 36; 8:244; BBB 48, 265; BBE 167; CDB 98 (cp 356); FBE 171; OBB 103; PBE 80 (pl 25); RBE 187; WEA 289; (C. l. intermedia) RBI 84.

Pigeon, Crested (Ocyphaps lophotes) CAU 181; CDB 101 (cp 368); HAB 84; LEA 421 (bw); MEA 67; RBU 116; SNB 102 (cp 50); WAB 193.

Pigeon, Crowned (Goura cristata) ABW 143; BBT 8; CAU 241.

Pigeon, Flock (Histriophaps histrionica) SNB 102 (cp 50); WAB 181; (Phaps histrionica) CDB 102 (cp 372).

Pigeon, Forest Bronze-wing (Phaps chalcoptera) HAB 84; SNB 102 (cp 50).

Pigeon, Green (Treron australis) PSA 177 (cp 20); WAB 140; (T. capellei) ABW 143; (T. formosae) DPB 117; (T. seimundi) WBV 103.

Pigeon, Green Imperial (Ducula aenea) BBT 8; DPB 129; KBA 113 (cp 12); WAB 172; WBV 109.

Pigeon, Green-winged —See **Dove, Emerald.**

Pigeon, Imperial Fruit (Ducula sp.) ABW 143; ALE 8:243; DPB 129.

Pigeon, Lesser Thick-billed Green (Treron curvirostra) DPB 117; KBA 113 (cp 12); WBV 103.

Pigeon, Magnificent Fruit (Megaloprepia magnifica) ABW 143; SNB 104 (cp 50); WAW 9.

Pigeon, Magnificent Ground (Otidiphaps nobilis) RBI 91; WAB 173.

Pigeon, Mindoro Imperial (Ducula mindorensis) DPB 129; LVB 57.

Pigeon, Mountain Imperial (Ducula badia) KBA 113 (cp 12); WBV 109.

Pigeon, New Zealand (Hemiphaga novaeseelandiae) BNZ 121; FNZ 176 (cp 16); WAB 209.

Pigeon, Nicobar (Caloenas nicobarica) ALE 8:246; CDB 97 (cp 350); DPB 137; KBA 175 (bw); WAB 172; WEA 289.

Pigeon, Orange-breasted Green (Treron bicincta) KBA 113 (cp 12); WBV 103.

Pigeon, Pale-capped or Purple Wood (Columba punicea) KBA 160 (cp 21); WBV 109.

Pigeon, Partridge (Geophaps smithi) HAB 85; SNB 102 (cp 50).

Pigeon, Pink-necked Green (Treron vernans) DPB 117; KBA 113 (cp 12); WBV 103.

Pigeon, Pin-tailed Green (Treron apicauda) KBA 113 (cp 12); WBV 103.

Pigeon, Plumed (Lophophaps plumifera) ALE 8:259; HAB 85; SNB 102 (cp 50); WAB 172; (L. ferruginea) HAB 84; RBU 119.

Pigeon, Red-crowned (Ptilinopus regina) RBU 111; SNB 104 (cp 51); WAW 9.

Pigeon, Speckled or Rock (Columba guinea) ALE 8:443; CDB 98 (cp 355); NBA 35; PSA 177 (cp 20).

Pigeon, Squatter (Petrophassa scripta or Geophaps scripta) CDB 102 (cp 371); SNB 102 (cp 50).

Pigeon, Tooth-billed (Didunculus strigirostris) ALE 8:260; WAB 173.

Pigeon, Top-knot (Lopholaimus antarcticus) CDB 101 (cp 366); RBU 115; SNB 104 (cp 50); WAB 183.

Pigeon, Victoria Crowned (Goura victoria) ALE 8:269; LVS 227; SLP 156; WAB 205.

Pigeon, Wedge-tailed Green (Treron sphenura) KBA 113 (cp 12); WBV 103.

Pigeon, White-headed (Columba norfolciensis) RBU 112; SNB 104 (cp 50).

Pigeon, White-quilled Rock (Petrophassa albipennis) CDB 102 (cp 370); HAB 87; SNB 102 (cp 50).

Pigeon, Wompoo —See **Pigeon, Magnificent Fruit.**

Pigeon, Wonga (Leucosarcia melanoleuca) CDB 101 (cp 365); HAB 84; SNB 102 (cp 50); WAB 173.

Pigeon, Wood (Columba palumbus) ALE 8:246; BBB 74; BBE 167; CDB 99 (cp 358); FBE 171; OBB 103; PBE 80 (pl 25); WAB 113; (C. vitiensis) DPB 135.

Pigeon, Yellow-bellied Fruit (Leucotreron cincta) ABW 143; ALE 8:243; (L. alligator) HAB 83.

Pigeon, Yellow-footed Green (Treron phoenicoptera) KBA 113 (cp 12); WBV 103.

Pigeonberry (Duranta repens) BIP 73(cp 104); MEP 122; MWF 118.

Pigeonwood (Hedycarya arborea) HPW 34; SNZ 91 (cps 267, 268).

Pigface —See **Rose-moss.**

Pignut (Conopodium majus) MBF cp 38; MFF pl 96 (485); OWF 87; PWF 37e; WFB 159 (1).

Pig's Ear —See **Silver Crown.**

Pigweed, Redroot or Common (Amaranthus retroflexus) HPW 71; MCW 35; MFF pl 106 (202); NHE 228; PFE cp 10 (117); PFF 159; WUS 148.

Pigweed, White (Amaranthus albus) MFF pl 104 (203); WUS 143.

Piha (Lipaugus strephophorus) ALE 9:144; (L. vociferans) SSA pl 45.

Pika, Alpine or Siberian (Ochotona alpinus) ALE 12:463; DSM 106; HMW 154; SMW 91 (cp 56).

Pika, Northern or Japanese (Ochotona hyperborea) ALE 12:463; CAS 66 (bw).

Pika, Royle's (Ochotona roylei) ALE 12:463; DAA 59.

Pika, Steppe (Ochotona pusilla) ALE 12:463; WEA 290.

Pike, Northern (Esox lucius) ALE 4:223, 265, 327; 11:105; FEF 62, 64 (bw), 65 (bw); FWF 57; GAL 227 (bw); HFI 124; HFW 36 (cp 7); LAW 311; LEA 217 (bw); OBV 111; SFW 73 (cp 10).

Pikehead (Luciocephalus pulcher) ALE 5:229; FEF 494 (bw); HFI 55; SFW 660 (cp 181).

Pike-perch (Stizostedion lucioperca or Lucioperca lucioperca) ALE 5:85; FEF 349 (bw); FWF 117; GAL 236 (bw); HFI 24; OBV 95; SFW 532 (pl 143); WEA 290; (S. volgense) FWF 119.

Pilchard —See **Sardine.**

Pilewort —See **Celandine, Lesser.**

Pilgrim's Tree —See **Traveler's Tree.**

Pill-bug —See **Sea Slater** (Ligia sp.) and **Sowbug.**

Pillwort (Pilularia globulifera) DEW 3: cp 154; MFF pl 136 (1303); NHE 108; ONP 97.

Pilot-bird (Pycnoptilus floccosus) CDB 178 (cp 770); HAB 194; SAB 32 (cp 15).

Pilotfish (Naucrates ductor) ALE 5:101; HFI 26; MOL 93; OBV 31.

Pimpernel, Blue (Anagallis foemina) EWF cp 33b; MBF cp 58; WFB 177 (5a); (A. linifolia) DFP 1 (cp 6); EGA 94; PFE cp 93 (967); TGF 174; (A. monelli) EWF cp 33c; PFM cp 127.

Pimpernel, Bog (Anagallis tenella) MBF cp 58; MFF cp 23 (623); NHE 74; OWF 125; PWF 98A; WFB 117 (6).

Pimpernel, Scarlet (Anagallis arvensis) BWG 121; DEW 1:231; EGA 94; KVP cp 46; MBF cp 58; MFF cp 26 (624); NHE 228; OWF 105; PFM cp 128; PFW 243; PWF 46d; TEP 177; WFB 177 (5).

Pimpernel, Yellow or Wood (Lysimachia nemorum) EPF pl 267; MBF cp 57; MFF cp 40 (618); NHE 34; OWF 27; PFE cp 93 (962); PWF 41k; WFB 175 (8a).

Pin Cushion Plant —See **Brass Buttons.**

Pinatoro —See **Rice Flower.**

Pinche —See **Marmoset, Cotton-topped or Pinche.**

Pincushion (Leucospermum sp.) HPW 170; LFT cp 59; MWF 183; PFW 245; RWF 56.

Pincushion Flower —See **Scabious.**

Pine, Aleppo or Jerusalem (Pinus halepensis) AMP 95; DEW 1:52 (cp 7); EGE 102; NHE 42; PFE cp 1 (6); PFM cp 3; SST cp 31.

Pine, Arolla (Pinus cembra) BKT cp 29, 42; CTH 20 (cp 7); EEG 127; EET 125; EGE 102; EPF pls 136, 139; MLP 29; NHE 18; OBT 120; SST cp 29; TGS 28.

Pine, Beach or Lodge-pole (Pinus contorta) EET 118; OBT 113; PFF 115; PWF 169g.

Pine, Bhutan (Pinus wallichiana) BKT cp 50; DFP 255 (cp 2036); EET 124, 125; MTB 160 (cp 13); OBT 121; (P. excelsa) SST cp 30.

Pine, Black or Austrian (Pinus nigra) DEW 1:28; DFP 255 (cp 2033); EPF pl 142; NHE 18; OBT 121; PFE cp 1 (7); PFF 117; PWF 60d, 171f; SST cp 33; TEP 268; TGS 23; WFB cp 25 (4a).

Pine, Bosnian (Pinus leucodermis) MTB 161 (cp 14); TVG 255.

Pine, Buddhist —See **Yew, Japanese.**

Pine, Chile —See **Monkey Puzzle Tree.**

Pine, Corsican (Pinus nigra var. maritima) EET 122, 123; MTB 160 (cp 13); OBT 100, 112.

Pine, Cow's-tail (Cephalotaxus harringtonia) BKT cp 7; EGE 92; EJG 105; MTB 49 (cp 2); (C. drupacea) TGS 6; (C. fortuni) MTB 49 (cp 2).

Pine, Japanese Black (Pinus thunbergii) EGE 41, 104; EJG 70, 137; ELG 123; MTB 161 (cp 14).

Pine, Japanese Umbrella (Sciadopitys verticillata) BKT cp 17; DEW 1:40; DFP 255 (cp 2037); EEG 129; EET 127; EGE 106; EJG 144; GDS cp 443; MTB 85 (cp 6); OBT 101, 125; TGS 23.

Pine, Japanese White (Pinus parviflora) EET 127; EGE 83; MWF 234.

Pine, Kauri (Agathis australis) CAU 255; MLP 28; SNZ 54 (cps 130-131); SST cp 6; (A. palmerstonii) WAW 108, 109.

Pine, Lace Bark (Pinus bungeana) BKT cp 46; EET 127.

Pine, Macedonian (Pinus peuce) DEW 1:54 (cp 14); EET 126.

Pine, Maritime (Pinus pinaster) BKT cps 47-49; DEW 1:28; EET 117, 119; MTB 160 (cp 13); NHE 42; OBT 121; PFE cp 1 (4); PFM cp 4, pl 1 (317); WFB cp 25 (4b).

Pine, Monterey (Pinus radiata) DFP 255 (cp 2034); EET 122, 123; EGE 44, 45; MTB 160 (cp 13); OBT 112, 121; PWF 171g.

Pine, Mugo or Mountain (Pinus mugo) DEW 1:54 (cp 11); EGE 103; EGW 6; ELG 133; EMB 116; EPF pls 143, 145; NHE 18; PFF 116; SST cp 32; TEP 78, 268.

Pine, Norway or Red (Pinus resinosa) EDP 139; EEG 128; EGE 103; PFF 117.

Pine, Pitch (Pinus rigida) BKT cp 45; EET 118; EPF pl 144; PFF 115.

Pine, Red (Dacrydium cupressinum) EET 247; SNZ 67 (cp 179); (D. bidwillii) SNZ 150 (cp 473).

Pine, Scots (Pinus sylvestris) AMP 95; DFP 255 (cp 2035); EET 18, 116, 117, 120, 121; EPF 140, 141; GDS cp 323; MBF pl 100; MTB 161 (cp 14); NHE 18; OBT 21; OWF 185; PFF 116; PWF 62d, 168f; TEP 266; WFB cp 25 (4).

Pine, Stone (Pinus pinea) BKT cp 43; DEW 1:29, 31; EET 41, 117, 124; NHE 42; OBT 121; OFP 31; SST cp 34.

Pine, Weymouth or White (Pinus strobus)
EET 122, 123; EGE 104; ELG 122; EPF pl
138; EPG 137; NHE 18; OBT 121; PFF 114;
SST cp 35; TGS 23.

Pine, White (Podocarpus dacrydoides) EET
247; EPF pl 146; SNZ 68 (cp 183).

Pineapple (Ananas comosus) EGL 85; EPF pl
1027-1028, cp XXXI; FHP 104; HPW 295;
LHP 22-23; OFP 97; PFF 368; PFW 55; RDP
26, 81; WWT 1; (A. sp.) DEW 2:221 (cp 137);
LHP 22; MLP 294.

Pineapple Flower (Eucomis sp.) DFP 92 (cp
733); EGB 114; EPF pl 907; EWF cp 85b;
LFT cp 16; MWF 128; TVG 193.

Pineapple Weed (Matricaria matricaroides)
BWG 171; CTH 59 (cp 94); MBF cp 46; MCW
122; MFF pl 101 (887); NHE 236; OWF 39;
PFE cp 148 (1432); PWF 78c; TEP 203; WFB
233 (3).

Pine-cone Fish (Monocentrus japonicus)
ALE 5:35; HFI 64.

Pine-marten —See **Marten, Pine.**

Pinesap —See **Bird's-nest, Yellow.**

Pink, Alpine (Dianthus alpinus) DFP 7 (cp
49); EMB 136; SPL cp 449; TVG 147.

Pink Bottoms —See **Mushroom, Field.**

Pink, Carthusian (Dianthus carthusianorum) DEW 1:110 (cp 66); KVP cp 77; NHE
184; PFE cp 17 (187); TEP 107; WFB 65 (8).

Pink, Cheddar (Dianthus gratianopolitanus) DFP 7 (cp 51); MFF cp 22 (160); NHE
74; PFE 90; PWF 80b; TEP 90; TVG 149;
WFB 65 (1).

Pink, Childing (Kohlrauschia prolifera)
MBF cp 13; MFF cp 22 (163); (K. glumacea)
PFM cp 18.

Pink, Childing or Proliferous —See **Tunic
Flower** (Petrorhagia).

Pink, Clove —See **Carnation.**

Pink, Deptford (Dianthus armeria) EWG 80;
MBF cp 13; MFF cp 22 (159); NHE 33; PFE
cp 17 (186); PWF 136d; WFB 65 (7).

Pink Everlasting (Xeranthemum annum)
EGA 162; EPF pl 576; OGF 191; PFE cp 151
(1464); TGF 45.

Pink, Fringed (Dianthus monspessulanus)
NHE 276; PFE cp 17 (190).

Pink, Garden or Common or Chelsea
(Dianthus plumarius) EWF cp 13d; LFW 46

(cps 104-106), 47 (cps 107-110); MWF 110;
NHE 224; PFW 71; TGF 60; WFB 65 (3).

Pink, Gravel or Mountain or Winter —See
Trailing Arbutus.

Pink, Maiden (Dianthus deltoides) EGC 129;
HPW 68; MBF cp 13; MFF cp 22 (161); NHE
184; PWF 80a; TEP 107; TGF 60; TVG 149;
WFB 65 (6).

Pink Nail (Gomphidius roseus) KMF 83; RTM
174; (G. sp.) CTM 42; DLM 192, 193; LHM
185; PMU cps 48, 59; REM 147, 148 (cps 115,
116); RTM 173, 174, 245; SMG 227, 228.

Pink Polka-Dot Plant —See **Freckle Face.**

Pink, Sand (Dianthus arenarius) MWF 110;
NHE 184.

Pink, Superb (Dianthus superbus) EPF pl
308; KVP cp 139; NHE 184; PFE cp 17 (188);
WFB 65 (5).

Pink, Wood (Dianthus sylvestris) NHE 276;
PFE cp 17 (194).

Pinkweed —See **Knotweed, Prostrate.**

Pintado —See **Petrel, Pintado.**

Pintail, Bahama or White-cheeked —See
Duck, Bahama.

Pintail, Common (Anas acuta) ABW 68; ALE
7:255, 304; BBB 198; BBE 53; CDB 50 (cp
95); DPB 29; FBE 57; KBA 80 (pl 7); OBB 17;
PBE 28 (cp 9), 40 (pl 13), 48 (pl 15); SAR 123;
WAB 25, 61; WBV 49.

Pinwheel (Aeonium haworthii) ECS 93; RDP
73.

Pipefish (Nerophis sp.) CSF 94, 95; CSS pl
XXII; HFI 72; OBV 65; (Dunckerocampus
caulleryi) HFW 147 (bw); LAW 331; (Siphonostoma sp.) LEA 222 (bw).

Pipefish, Broad-nosed (Syngnathus typhle)
ALE 5:40; CSF 93; HFI 72; OBV 65.

Pipefish, Greater (Syngnathus acus) CSF 93;
CSS pl XXII; FEF 267 (bw); MOL 111; OBV
65; (S. pelagicus) MOL 92.

Pipefish, Lesser Freshwater (Syngnathus
pulchellus) SFW 493 (cp 134); WEA 291;
WFW cp 236.

Pipefish, Ocean or Snake (Entelurus
aequoreus) ALE 5:40; CSF 94; CSS pl XXII;
HFI 72; HFW 148 (bw); OBV 65.

Pipes and Matches —See **Lily, Glory.**

Pipewort (Eriocaulon aquaticum) HPW 281;
WFB 261 (9); (E. septangulare) MBF cp 90;
MFF pl 66 (985).

Pipistrelle —See Bat, Pipistrelle.

Pipit, Indian or Olive Tree (Anthus hodgsoni) BBE 203; DPB 357; FBE 209; KBA 400 (pl 59).

Pipit, Meadow (Anthus pratensis) ALE 9:189; BBB 124; BBE 205; CDB 152 (cp 626); FBE 209; OBB 163; PBE 245 (cp 58); RBE 279; (A. caffer) PSA 288 (cp 33).

Pipit, New Zealand or Richard's (Anthus novaeseelandiae) ALE 9:181; BBB 275; BBE 203; BNZ 37; CDB 152 (cp 625); DPB 357; FBE 211; KBA 400 (pl 59); PBE 245 (cp 58); PSA 288 (cp 33); SAB 10 (cp 4); WAN 132; WBV 155; (A. australis) HAB 142; (A. richardi) OBB 165.

Pipit, Pechora (Anthus gustavi) BBB 277; BBE 203; DPB 357; FBE 209; PBE 284 (cp 65).

Pipit, Red-throated (Anthus cervinus) ALE 9:189; BBB 277; BBE 205; CDB 151 (cp 623); DPB 357; FBE 209; KBA 400 (pl 59); OBB 165; PBE 245 (cp 58).

Pipit, Rock or Water (Anthus spinoletta) ABW 264; ALE 9:184, 189; BBB 264; BBE 205; FBE 211; OBB 163, 165; PBE 245 (cp 58); WAB 52.

Pipit, Tawny (Anthus campestris) ALE 9:189; BBB 268; BBE 205; FBE 211; OBB 165; PBE 245 (cp 58).

Pipit, Tree (Anthus trivialis) ALE 9:189; BBB 106; BBE 203; CDB 152 (cp 627); FBE 209; GPB 254; OBB 163; PBE 245 (cp 58).

Pipsissewa —See Wintergreen (Chimaphila sp.).

Piranha (Serrasalmus sp.) ALE 4:292, 299; FEF 72 (bw); HFI 86; HFW 38 (cps 12-14); LAW 315; (Pygocentrus piraya) FEF 73 (bw), 79; WFW cp 85.

Piranha, Natterer's or Red (Rooseveltiella nattereri) CTF 23 (cps 12, 13); SFW 72 (cp 9); WEA 292; WFW cp 86.

Pirri-pirri-bur (Acaena anserinifolia) MFF pl 88 (369); WFB 106; (A. adscendens) SAR 80-81.

Pistachio (Pistacia vera) OFP 29; PFF 236; SST cp 213.

Pistachio or Pistache, Chinese (Pistacia chinensis) EGT 132; ELG 108; MWF 234.

Pitcairnia (Pitcairnia sp.) EWF cp 188e; HPW 295; MEP 13; MWF 235.

Pitcher Plant —See also Fly Catcher Plant.

Pitcher Plant (Nepenthes sp.) BIP 127 (cp 192); CAS 259, 261; DEW 1:136, 161 (cp 72); EPF pl 181; EWF cp 111 a, b; HPW 42; MLP 94; MWF 210; PFW 194.

Pitcher Plant (Sarracenia purpurea) EPF pls 237, 238; EWF cp 158c; EWG 22, 137; HPW 54; MFF cp 25 (417); PFF 190; PFW 271; WWT 89; (S. leucophylla syn. drummondii) DEW 1:101 (cp 47); EWF cp 158b.

Pitcher-plant, California (Darlingtonia california) DEW 1:135; EWG 102; HPW 54; PFW 271; WWT 94.

Pitta, Ant —See Antpitta.

Pitta, Banded or Blue-tailed (Pitta guajana) ALE 9:133, 149; CDB 142 (cp 573); KBA 243 (bw).

Pitta, Black-headed —See Pitta, Hooded.

Pitta, Blue (Pitta cyanea) KBA 224 (cp 29), 239 (bw); WBV 147.

Pitta, Blue-breasted (Pitta mackloti) HAB 141; RBU 192; SAB 4 (cp 1).

Pitta, Blue-winged (Pitta moluccensis) KBA 224 (cp 29), 239 (bw); SAB 4 (cp 1); WEA 293.

Pitta, Buff-breasted or Noisy (Pitta versicolor) GPB 236; HAB 141; SAB 4 (cp 1); WAW 9.

Pitta, Elliot's (Pitta ellioti) ALE 9:149; KBA 224 (cp 29); WBV 147.

Pitta, Garnet or Red-headed (Pitta granatina) ABW 211; ALE 9:149; KBA 224 (cp 29); RBI 151.

Pitta, Great Blue or Giant (Pitta caerulea) ALE 9:149; KBA 224 (cp 29).

Pitta, Hooded (Pitta sordida) ALE 9:149; DPB 229; KBA 224 (cp 29); RBI 156.

Pitta, Indian (Pitta brachyura) CDB 142 (cp 572); DPB 229; RBI 148; WAB 161.

Pitta, Rainbow (Pitta iris) HAB 140; SAB 4 (cp 1).

Pitta, Steer's (Pitta steerii) ABW 211; ALE 9:149; DPB 229.

Pittosporum (Pittosporum sp.) MEP 44; SNZ 67 (cp 177), 77 (cp 218), 84 (cps 242, 243).

Pittosporum, Japanese (Pittosporum tobira) EDP 142; EEG 110; EFP 20, 135; EGE 40; EJG 138; ELG 142; EPG 137; HSC 91, 96; PFW 231; RDP 319; SPL cp 110.

Pixie Cup (Cladonia pyxidata) DEW 3:221; NFP 99; PFF 74.

Plaice (Pleuronectes platessa) ALE 5:233, 239; CSF 185; CSS pl XXIV; FEF 513-514 (bw); HFI 18; HFW 274 (bw); LAW 349; LEA 261 (bw); OBV 45.

Plains Wanderer (Pedionomus torquatus) ABW 102; ALE 8:100; HAB 50; SNB 60 (cp 29).

Plane, London (Platanus acerifolia) DEW 1:155; EGT 133; EPF pls 347, 348; EPG 138; MWF 236; OWF 197; SST cp 142; TGS 183; (P. hybrida or P. hispanica) BKT cp 105; EET 164, 165; ELG 109; MTB 257 (cp 24); OBT 192; PWF 25p, 161j; WFB 35 (2).

Plane, Oriental (Platanus oriental) BKT cp 106; DFP 218 (cp 1741); EET 164; MTB 257 (cp 24); OBT 193; PFE cp 43 (418); PFM cp 46; SST cp 144.

Plantain, Alpine (Plantago rigida) DEW 2:185; MLP 238; (P. alpina) NHE 283.

Plantain, Branched (Plantago indica) EPF pl 872; NHE 234; PFE cp 133 (1283); PFM cp 182; WFB 221 (4).

Plantain, Buck's-horn (Plantago coronopus) MBF cp 71; MFF pl 106 (793); NHE 137; OWF 61; PFE cp 133 (1285); PWF 54b; TEP 61; WFB 221 (5).

Plantain, Common or Great (Plantago major) BWG 87; DEW 2:184; EGC 73; MBF cp 71; MCW 91; MFF pl 106 (789); NHE 187; OWF 61; PFF 300; PWF 70c; TEP 197; WFB 221 (1); WUS 349b.

Plantain, Hoary (Plantago media) AMP 129; DEW 2:184; EPF pl 871; EWF cp 15j; MBF cp 71; MCW 92; MFF cp 26 (790); NHE 187; OWF 61; PFE cp 133 (1292); PWF 105i, 69k; WFB 221 (1a).

Plantain, Ribort or Narrow-leaved (Plantago lanceolata) AMP 129; BWG 89; EGC 73; EPF pl 870; MBF cp 71; MCW 90; MFF cp 37 (791); NHE 187; OWF 61; PFF 300; PWF 55k; TEP 145; WFB 221; WUS 349.

Plantain, Sea (Plantago maritima) MBF cp 71; MFF pl 106 (792); NHE 137; PFE cp 132 (1287); PWF 90a; TEP 63; WFB 221 (3).

Plantain, Shrubby (Plantago sempervirens) NHE 187; PFE 399.

Plantain, Wild (Heliconia bihai) MLP 316, 319; PFW 137; (H. wagneriana) MEP 25.

Plant-cutter (Phytotoma rutila) ABW 210; ALE 9:150; SSA cp 16.

Platypus (Ornithorhynchus anatinus) ALE 10:41, 42; BMA 175, 176 (bw); DSM 20; HMA 11, 72; HMW 208; JAW 19; LAW 523; MEA 114; MLS 39 (bw); MWA 109 (bw); SLP 138; SMW 33 (cp 1); TWA 199; WAW 68 (bw); WEA 295.

Plectranthus (Plectranthus sp.) EWF cp 80f; LFT cp 144; MWF 237.

Ploughman's Spikenard (Inula conyza) BFB cp 19 (10); MBF cp 45; MFF cp 52 (860); NHE 237; OWF 35; PFE cp 144 (1387); PWF 145g; WFB 231 (6).

Plover, Australian Spur-winged (Lobibyx novaehollandiae) CDB 82 (cp 264); FNZ 121 (bw); HAB 64; RBU 40.

Plover, Banded (Vanellus tricolor or Zonifer tricolor) CDB 84 (cp 273); HAB 65; SNB 80 (cp 39), 82 (pl 40); WAB 181.

Plover, Blackhead (Vanellus tectus or Sarciphorus tectus) ALE 8:443; CDB 84 (cp 272).

Plover, Blacksmith (Vanellus armatus or Hoplopterus armatus) ABW 120; CDB 82 (cp 268); PSA 129 (pl 14).

Plover, Crab (Dromas ardeola) ABW 125; ALE 8:170; CDB 89 (cp 305); FBE 119; KBA 151 (bw).

Plover, Crowned (Vanellus coronatus or Stephanibyx coronatus) CAA 47 (bw); CDB 83 (cp 269); LEA 410 (bw); PSA 129 (pl 14).

Plover, Egyptian (Pluvianus aegypticus) ALE 8:170; CDB 90 (cp 313); FBE 143; GPB 156; RBI 64.

Plover, Golden (Pluvialis apricarius) BBB 120; BBE 117; CDB 82 (cp 265); FBE 123; OBB 63; PBE 93 (pl 30), 100 (cp 31); (P. dominica) ABW 120; BBE 117; CDB 82 (cp 266); DPB 81; HAB 69; KBA 128 (pl 13), 145 (pl 16); SNB 76 (cp 37), 78 (pl 38); (Charadrius dominica) WBV 81.

Plover, Gray (Pluvialis squatarola) ALE 8:170; BBB 225; BBE 117; CDB 82 (cp 267); DPB 81; FBE 123; HAB 69; KBA 128 (pl 13); OBB 63; PBE 93 (pl 30), 100 (cp 31); SNB 76 (cp 37), 78 (pl 38); WAB 101; (Charadrius squatarola) WBV 81.

Plover, Kentish —See **Plover, Snowy.**

Plover, Kittlitz's (Charadrius pecuarius)
CDB 81 (cp 260); FBE 121; GPB 143; PSA
129 (pl 14).

Plover, Little Ringed (Charadrius dubius)
ALE 7:385; 8:151, 169; BBB 174; BBE 115;
DPB 83; FBE 121; GPB 141; KBA 129 (pl
14), 145 (pl 16); OBB 65; PBE 93 (pl 30), 100
(cp 31); WBV 81.

Plover, Malaysian (Charadrius peronii)
DPB 83; KBA 145 (pl 16).

Plover, Masked (Lobibyx miles) HAB 64;
(Vanellus miles) SNB 80 (cp 39), 82 (pl 40).

Plover, Mongolian (Charadrius mongolus)
FBE 121; HAB 67; KBA 129 (pl 14), 145 (pl
16); SNB 76 (cp 37), 78 (pl 38).

Plover, New Zealand or Shore (Thinornis
novaeseelandiae) ALE 8:170; BNZ 131; LVB
41.

Plover, Oriental (Charadrius asiaticus) ALE
8:169; BBE 115; FBE 123; HAB 67; SNB 76
(cp 37), 78 (pl 38); (C. veredus) DPB 83.

Plover, Ringed (Charadrius hiaticula) ALE
8:154, 169, 187; BBB 238; BBE 115; CDB 80
(cp 258); FBE 121; KBA 129 (pl 14); LAW
496; OBB 65; PBE 93 (pl 30), 100 (cp 31);
PSA 129 (pl 14); SNB 76 (cp 37), 78 (pl 38);
WAB 133.

Plover, Sand —See **Sandplover.**

Plover, Snowy (Charadrius alexandrinus or
C. ruficapillus) ALE 8:169, 187; BBB 269;
BBE 115; CDB 80 (cp 256); DPB 83; FBE
121; HAB 67; KBA 129 (pl 14), 145 (pl 16);
OBB 65; PBE 93 (pl 30), 100 (cp 31); RBU 43;
SNB 80 (cp 39), 82 (pl 40); WBV 81.

Plover, Sociable (Vanellus gregarius) BBE
113; FBE 125; (Chettusia gregaria) ALE
8:170.

Plover, Spur-winged (Vanellus spinosus or
Hoplopterus spinosus) ALE 8:170; BBE 113;
CDB 84 (cp 271); FBE 125; PBE 285 (cp 66).

Plover, Three-banded (Charadrius
tricollaris) CDB 81 (cp 261); PSA 129 (pl 14).

Plover, Wattled (Vanellus senegallus or
Afribyx senegallus) ALE 8:170; CDB 83 (cp
270); PSA 129 (pl 14); WAB 139.

Plover, Wrybill (Anarhynchus frontalis)
ALE 8:170; BNZ 131; CAU 272 (bw); FNZ
129 (pl 12); GPB 142; LEA 346 (bw); WAB
181.

Plum, Common or European (Prunus

domestica) EPF pl 421; EVF 129; MBF cp 26;
OFP 69; SST cp 218.

Plum, Japanese —See **Loquat.**

Plum, Kaffir or Sour (Ximenia caffra) HPW
173; LFT cp 62.

Plume Flower, Brazilian (Jacobinia carnea)
DEW 2:171 (cps 98, 99); EPF pl 859; MEP
145; MWF 169; RDP 249; SPL cp 50.

Plum-yew, Japanese —See **Pine,
Cow's-tail.**

Plush Plant (Echeveria pulvinata) LCS cp
243; SCS cp 77; (E. dactylifera) EFP 109;
LCS 242.

Poa, Hard —See **Grass, Fern.**

Poacher or **Pogge** (Agonus cataphractus)
ALE 5:66; CSF 167; HFI 60; OBV 81.

Pochard, African and South American
(Netta erythrophthalma) ALE 7:307; CDB
53 (cp 122); PSA 49 (cp 4), 64 (pl 5).

Pochard, Common (Aythya ferina) ALE
7:255, 308, 385; BBB 201; BBE 43, 59; DPB
33; FBE 59; OBB 23; PBE 29 (cp 10), 41 (pl
14), 49 (pl 16); RBE 92; SBB 44 (cp 53); WAB
25.

Pochard, Red-crested (Netta rufina) ALE
7:255, 308; BBB 269; BBE 59; FBE 57; OBB
23; PBE 29 (cp 10), 41 (pl 14), 49 (pl 16); RBE
88.

Pocketbook Flower (Calceolaria sp.) BIP 45
(cp 51); DFP 3 (cp 22), 31 (cp 247), 55 (cps
438, 439), 184 (cp 1472); EDP 147; EGG 93;
EWF cp 5g, cp 184d; EPF pl 837; FHP 108;
LFW 241 (cp 536); LHP 193; MWF 66; PFW
278; RDP 125; SPL cp 41.

Pocket-handkerchief Tree —See **Dove
Tree.**

Podocarpus (Podocarpus sp.) DFP 79 (cp
629); LFT cp 1; SNZ cps 144, 145, 150, 151,
153, 461, 462.

Pohuehue (Muehlenbeckia sp.) BIP 123 (cp
186); SNZ 20 (cp 13), 83 (cps 238, 239), 136
(cps 423, 424).

Poinciana (Caesalpinia gilliesii or Poinciana
gilliesii) HPW 151; MEP 52; SPL cp 111;
TGS 135; (P. regia) RWF 112; (Caesalpinia
japonica) BIP 43 (cp 49); DFP 184 (cp 1471).

Poinciana, Royal (Delonix regia) EET 224,
225; EGT 110; EWF cp 113b; GFI 59 (cp 11);
LFW 162 (cps 360, 362); MEP 57; MWF 107;
PFW 158; SST cp 243.

Poinsettia (Euphorbia pulcherrima) BIP 79 (cp 118); DEW 1:224 (cps 142-145); EGE 124; EGG 109; EPF cp XV; EPG 120; FHP 120; GFI 107 (cp 23); LFW 250 (cp 557); LHP 88, 89; MEP 70; MWF 130; PFF 232; PFW 114; RDP 198; SPL cp 46.

Pokeberry or **Pokeweed** (Phytolacca americana) DEW 1:162 (cp 74); EPF pls 288, 289; PFE cp 11 (119); PFF 160; PFW 230; TVG 117; WUS 149; (P. sp.) AMP 71; DEW 1:195; HPW 72; SST cp 141.

Polecat, European (Mustela putorius) ALE 2:363; 11:257, 321; 12:40, 56; DAE 86; DSM 167; FBW 31; GAL 29 (bw); HMA 51; HMW 136; JAW 113; LAW 536; MLS 284 (bw); OBV 169; SMW 225 (cp 126); TWA 84.

Polecat, Marbled (Vormela peregusna) ALE 12:56; DAA 78; DAE 86.

Polecat, Striped —See **Zorilla.**

Polecat Tree or **Poison Bay Tree** (Illicium floridanum) EWF cp 150c; HPW 29.

Policeman's Helmet (Impatiens glandulifera) BWG 108; HPW 211; MBF cp 20; MFF cp 20 (260); OWF 109; PFE cp 70 (716); PWF 116b.

Pollack or **Pollock** (Pollachius sp.) ALE 4:431; CSF 111; HFI 77; LEA 251; LMF 157 (bw); OBV 39; WFW cp 189.

Pollan (Coregonus albula) HFI 122; OBV 111.

Polycnemum (Polycnemum majus) NHE 226; WFB 47 (9).

Polypody, Common (Polypodium vulgare) AMP 35; CGF 131; DEW 3: (cps 151, 152), 292; EGF 135; EPF pl 116; ESG 137; MFF pl 140 (1302); NHE 80; TEP 89; (P. sp.) NFP 131; ONP 189; RDP 326.

Polypore (Polystictus sp.) EPF pl 52; KMF 67; NFP 49; PUM 152.

Pomegranate (Punica granatum) AMP 55; BIP 151 (cp 229); CTH 38 (cp 49), 39 (cp 50); DEW 1:288 (cp 186); DFP 22 (cp 174), 221 (cp 1768); EFS 128; EGL 138; EPF cp XIX; EPG 139; EWF cp 28b; FHP 140; GPI 111 (cp 241); LFW 172 (cp 389); MEP 93; MLP 168; MWF 247; OFP 95; PFE cp 80 (827); PFW 248; RDP 331; SPL cps 113, 114; SST cp 221; TGS 151.

Pomme de Terre —See **Potato, Irish** (Solanum).

Pompadour —See **Discus, Brown or Green.**

Pond Skater —See **Water Strider.**

Pondweed, Broad-leaved or Floating (Potamogeton natans) KVP cp 3; MBF cp 89; MFF pl 67 (972); NHE 103; PFE cp 161 (1570); PFF 336; PWF 102b; TEP 20; WFB 295 (1).

Pondweed, Canadian (Elodea canadensis) EPF pl 877; HPW 271; MBF cp 79; MFF pl 68 (966); NHE 102; OWF 53; PWF 140b; TEP 20; WFB 293 (5); WUS 29.

Pondweed, Cape (Aponogeton distachys) DEW 2:231; DFP 122 (cp 272); EWF cp 84b; HPW 272; PFW 31; (A. natans) DEW 2:211 (cp 117).

Pondweed, Curled or Crisp (Potamogeton crispus) MBF cp 90; MFF pl 67 (979); NHE 103; PFF 337; TEP 22; WFB 295 (4); WUS 13.

Pondweed, Dense-leaved (Egeria densa) NHE 102; WFB 293 (5c); WUS 31.

Pondweed, Fen (Potamogeton coloratus) MBF cp 89; NHE 103; PFE 484; WFB 295 (1b)

Pondweed, Fennel-like (Potamogeton pectinatus) MBF cp 90; MFF pl 66 (980); NHE 103; PFE 484; PFF 336; WFB 295 (6).

Pondweed, Grassy or Blunt-leaved (Potamogeton obtusifolius) MBF cp 90; MFF pl 67 (978); PFE 484.

Pondweed, Horned (Zannichellia palustris) HPW 275; MBF cp 90; MFF pl 67 (983); NHE 104; PFE 480; PFF 337; WFB 295 (10).

Pondweed, Loddon (Potamogeton nodosus) EPF pl 1022; MBF cp 89; NHE 103; WUS 17.

Pondweed, Opposite-leaved (Groenlandia densa) MBF cp 89; MFF pl 67 (981); NHE 103; WFB 295 (7).

Pondweed, Perfoliate (Potamogeton perfoliatus) EWF cp 23f; MBF cp 89; MFF pl 67 (975); NHE 103; TEP 21; WFB 295 (3).

Pondweed, Sharp-leaved (Potamogeton acutifolius) EWF cp 23e; MBF cp 90; MFF pl 67 (977); OWF 53.

Pondweed, Shining (Potamogeton lucens) EWF cp 23b; MBF cp 89; MFF pl 67 (973); NHE 103; PFE 484; TEP 21; WFB 295 (2).

Pondweed, Small or Lesser (Potamogeton berchtoldii) MBF cp 90; MFF pl 66 (976); WFB 295 (5); (P. pusillus) MBF cp 90; NHE 103; WUS 19.

Pondweed, Tassel (Ruppia sp.) MBF cp 90; MFF pl 68 (982); NHE 137; PFF 337; WFB 295 (9).

Pondweed, Various-leaved (Potamogeton gramineus) MBF cp 89; MFF pl 67 (974); NHE 103.

Pony Tail —See **Elephant-foot Tree.**

Poor Man's Weather Glass —See **Pimpernel, Scarlet.**

Poor-cod (Trisopterus minutus) CSF 107; OBV 37.

Pope —See **Ruffe.**

Poplar, Berlin (Populus berolinensis) MTB 176 (cp 15); OBT 168.

Poplar, Black (Populus nigra) AMP 173; BFB cp 14 (5); BKT cp 92; DEW 1:192; EET 130; EGE 37; EGT 134; EPF pl 503; HPW 117; MBF cp 78; MTB 176 (cp 15); MWF 239; NHE 24; OBT 168; OWF 187; PWF 13i; 161e, i; SST cp 147; TGS 92.

Poplar, Gray (Populus canescens) EET 131; MBF cp 78; MTB 176 (cp 15); OBT 92; PWF 13e, 161l; WFB 31 (1).

Poplar, White (Populus alba) EET 130; EGT 134; MBF cp 78; MFF pl 70 (565); MTB 176 (cp 15); NHE 24; OBT 169; OWF 187; PWF 13h, 161h; SST cp 145; TEP 50; TGS 86; WFB 31 (1a).

Poppy (Papaver sp.) EWF cp 1c; LFT cp 72; OGF 83; PFE cp 28 (267), 29 (267).

Poppy, Arctic (Papaver radicatum) EWF cp 1d; WFB 81 (6); (P. alboroseum) MLP 90.

Poppy, Bristly or Prickly Round-headed (Papaver hybridum) MBF cp 5; MFF cp 14 (37); WFB 81 (3).

Poppy, California (Eschscholtzia californica) DFP 37 (cp 294); EGA 117; EPF pl 229; EWF cp 154a; EWG 108; HPW 52; LFW 252 (cp 559); MLP 89; MWF 127; OGF 83; PFE cp 29 (275); PFF 181; PFW 220, 224; RWF 108; SPL cp 205; TGF 76; TVG 49.

Poppy, California Tree (Dendromecon rigida) BIP 69 (cp 96); DFP 195 (cp 1560); GDS cp 152; MWF 109.

Poppy, Corn or Field or Shirley (Papaver rhoeas) AMP 93; BFB cp 3 (4); BWG 100; CTH 29 (cp 26); DEW 1:137; DFP 45 (cp 355); EFC cp 19; EGA 141; EPF pl 232; KVP cp 38; MBF cp 5; MFF cp 14 (35); MWF 221; NHE 214; OGF 83; OWF 105; PFE cp 29 (265); PFM cp 34; PFW 220; PWF 64b; RWF 32; TEP 173, TVG 63, WFB 81 (1).

Poppy, Long-headed (Papaver dubium) BFB cp 3 (1); HPW 52; MBF cp 5; MFF cp 14 (36); MLP 90; NHE 214; PWF 71a; WFB 81 (4).

Poppy, Long Rough-headed (Papaver argemone) MBF cp 5; MFF cp 14 (38); WFB 81 (2).

Poppy, Mexican Tulip (Hunnemannia fumariaefolia) EGA 126; EWG 118; OGF 83; TGF 76.

Poppy, Opium (Papaver somniferum) AMP 125; BWG 102; CTH 29 (cps 27, 28); DEW 1:138; EPF pl 233; LFW 108 (cp 237), 110 (cp 243); MBF cp 5; MFF cp 14 (39); MWF 221; NHE 214; PFE cp 28 (264); PFW 220; PWF 82c; TEP 161; TVG 63; WFB 81 (5).

Poppy, Oriental (Papaver orientale) DFP 163 (cps 1300-1302); EEG 137; EGP 20; EPF pl 234; EWF cp 46a; LFW 110 (cps 240-242), 111 (cp 244); MWF 221; PFF 180; PFW 221; SPL cp 237; TGF 76; TVG 115.

Poppy, Plume (Macleaya cordata) DFP 158 (cp 1260); EDP 128; EJG 127; HPW 52; MEP 40; TGF 76; (M. microcarpa) DFP 158 (cp 1261).

Poppy, Prickly (Argemone mexicana) CAU 162; DFP 30 (cp 240); EWF cp 154b; MEP 40; MWF 46; OGF 83; (A. grandiflora) EGA 96; TGF 76; (A. hispida) EWG 87.

Poppy, Welsh (Meconopsis cambrica) BFB cp 3; DFP 158 (cp 1264); EWF cp 11f; MBF cp 5; MFF cp 38 (40); OWF 7; PFE cp 28 (269); PFW 220; PWF 73g; TGF 76; WFB 81 (8).

Porbeagle —See **Shark, Mackerel.**

Porcelain Shell —See **Cowry.**

Porcupine (Hystrix afrae-australis) TWA 41; (H. leucura) ALE 11:410; 13:285.

Porcupine, Brazilian Tree (Coendou prehensilis) ALE 11:409, 423; BMA 182-183; HMW 174; LEA 524 (bw); MLS 224 (bw); (C. villosus) HMW 174.

Porcupine, Brush-tailed (Atherurus africanus) ALE 11:409; DDM 49 (cp 4); DSM 131; LAW 593.

Porcupine, Crested (Hystrix cristata) ALE 11:400; DAE 24; DDM 49 (cp 4); DSM 131; HMA 78; HMW 173; LAW 593; LEA 524 (bw); SMR 129 (cp 10); SMW 139 (cp 88).

Porcupine, Indian Crested (Hystrix indica) CAS 94 (bw); DAA 128; MLS 222 (bw).

Porcupine Plant (Hymenanthera sp.) HPW 100; SNZ 19 (cp 9), 96 (cp 288), 122 (cps 376-378).

Porcupine-fish (Diodon sp.) ALE 5:240, 258; HFI 16; HFW 262 (cp 126), 279, 280 (bw); LAW 353; LEA 257; WEA 296; (Lophodiodon calori) WFW 500.

Porpoise, Common or Harbor (Phocaena phocaena) ALE 11:480; HMA 78; HMW 107; LEA 545 (bw); OBV 183.

Porpoise, Spectacled (Phocaena dioptrica) ALE 11:480; SAR 131.

Portuguese Man-of-War (Physalia physalis) ALE 1:198, 271; CSS cp 3; LAW 55; MOL 58; OBV 11.

Possum —See also **Phalanger** and **Opossum**.

Possum, Brush-tailed —See **Phalanger, Brush-tailed.**

Possum, Honey (Tarsipes spenserae) ALE 10:100 (cp 2); HMA 77; MEA 101; MWA 41 (bw); (T. rostratus) HMW 204.

Possum, Leadbeater's (Gymnobelideus leadbeateri) ALE 10:100 (cp 2); DSM 32; MWA 40 (bw); PWC 177, 230 (bw); SLP 142; VWA cp 34; WAW 51 (bw).

Possum, Pygmy (Eudromecia caudata) ALE 10:99 (cp 1); (Burramys parvus) LVS 27; (Cercartetus concinnus) MWA 38, 39.

Possum, Ringtail (Pseudocheirus sp.) ALE 10:100; (cp 2); CAU 26; DSM 31; MWA 45 (bw).

Potato, Irish (Solanum tuberosum) EPF pl 817; EVF 104; PFF 292; TEP 166.

Potato Tree (Solanum macranthum) MEP 131; MWF 268.

Potato Tree, Chilean (Solanum crispum) DFP 250 (cp 1993); GDS cp 448; PFW 282.

Potato Vine (Solanum jasminoides) GDS cp 449; HSC 113, 114; OGF 165.

Pothos —See **Ivy, Devil's.**

Potoo, Common (Nyctibius griseus) ALE 8:425; CDB 119 (cp 457); SSA pl 24; WAB 81.

Potoo, Great (Nyctibius grandis) ABW 161; CDB 119 (cp 456).

Potoroo —See **Rat-kangaroo, Long-nosed.**

Potto, Bosman's (Perodicticus potto) ALE 10:252 (cp 6); CAA 89 (bw); DDM 52 (cp 5); DSM 71; LEA 512; MLS 124 (bw); SMW 48 (cp 37); WEA 298.

Potto, Golden (Arctocebus calabarensis) ALE 10:252 (cp 6); DDM 52 (cp 5); DSM 71; LAW 610; MLS 123 (bw); SMW 65 (bw); WEA 48.

Poui —See **Trumpet Tree.**

Pout, Norway (Trisopterus esmarkii) CSF 109; OBV 37.

Pouting (Trisopterus luscus) CSF 105; OBV 37; WFW cp 190.

Powan —See **Houting.**

Powder Puff, Pink (Calliandra inaequilatera) EGE 115; FHP 109; LFW 4 (cp 8); MWF 67; (C. surinamensis) MLP 107; MWF 67.

Powder Puff, Red (Calliandra haematocephala) EWF cp 170d; MEP 53; PFW 159.

Pratia (Pratia sp.) HPW 254; SNZ 114 (cps 349, 350), 164 (cp 523).

Pratincole, Australian (Stiltia isabella) CDB 90 (cp 314); HAB 72; RBU 39; SNB 88 (cp 43).

Pratincole, Black-winged (Glareola nordmanni) BBE 137; FBE 143; RBI 67; PBE 101 (cp 32); PSA 145 (pl 16).

Pratincole, European (Glareola pratincola) ABW 125; ALE 8:170; BBB 273; BBE 137; CDB 90 (cp 311); CEU 54; FBE 143; GPB 157; HAB 72; OBB 81; PBE 84 (pl 27), 101 (cp 32); RBE 159; SNB 88 (cp 43); WEA 298.

Pratincole, Oriental (Glareola maldivarum) DPB 107; KBA 144 (pl 15).

Prawn, Common (Palaemon serratus or Leander serratus) CSS pl XI; LEA 172 (bw); OIB 125.

Prawn, Deep-sea (Pandalus sp.) CSF 197; CSS pl XI.

Prayer Plant (Maranta leuconeura) BIP 119 (cp 181); DFP 70 (cp 556); EDP 148; EFP 123; EGL 126; LHP 124; MWF 197; PFW 186; RDP 264; SPL cp 22; (M. bicolor) DEW 2:296; EPF pl 1039.

Pride of Barbados (Caesalpinia pulcherrima or Poinciana pulcherrima) DEW 1:282 (cp 172); EFS 98; GFI 115 (cp 25); MEP 52; MWF 65; PFW 159.

Pride of Burma (Amherstia nobilis) EWF cp 113a; MEP 48; SST cp 228.

Pride of China —See **Trumpet Flower.**

Pride of De Kaap (Bauhinia galpinii) HPW 151; LFT 162 (cp 80); MEP 51.

Pride of India (Koelreuteria) —See **Golden Rain Tree.**

Pride of India (Lagerstroemia) —See **Crape Myrtle.**

Pride of Table Mountain (Disa uniflora) EWF cp 89b; RWF 48.

Primrose (Primula sp.) BIP 149 (cp 226); DFP 19, 20, 21, 167, 168; EMB 140; EWF cps 4, 98, 114; HPW 135; MWF 241; NHE 278; OGF 15, 21; PFE cp 89 (945); PFW 241, 242; RDP 328; TVG 163, 165.

Primrose, Bird's-eye (Primula farinosa) DEW 1:213 (cp 117); DFP 19 (cp 152); KVP cp 135; MBF cp 57; MFF cp 22 (610); NHE 278; OWF 113; PFE cp 91 (943); PWF cp 21j; WFB 175 (4).

Primrose, Bog (Primula rosea) DFP 20 (cp 159); EWF cp 98e; PFW 241.

Primrose, Cape (Streptocarpus hybrids) BIP 91 (cp 138), 173 (cp 265); DFP 82 (cp 652); EGL 146; EMB 131; EWF cps 64d, 82 a-d; FHP 147; HPW 247; LHP 178, 179; LFT cps 153-155; MWF 276; PFW 127; RDP 373.

Primrose, Chinese (Primula sinensis) EPF pl 259; RDP 328.

Primrose, English or Common (Primula vulgaris) BFB cp 13 (1); CTH 42 (cp 55); DFP 21 (cp 163); EWF cp 15g; KVP cp 66; LFW 115 (cp 255); MBF cp 57; MFF cp 40 (615); MWF 241; NHE 33; OGF 15; OWF 27; PFE cp 90 (942); PFW 239; PWF cp 17h; SPL cps 369, 491; TGF 156; TVG 165; WFB 175 (1).

Primrose, Entire-leaved (Primula integrifolia) NHE 278; PFE cp 91 (946).

Primrose, Fairy or Baby (Primula malacoides) BIP 149 (cp 224); DFP 79 (cp 631); FHP 139; LFW 115 (cp 252); LHP 192; MWF 241; PFW 242; RDP 329.

Primrose, Ganges (Asystasia gangetica) MEP 146; PFW 14.

Primrose, Gold (Vitaliana primuliflora) NHE 278; PFE 304.

Primrose, Himalayan (Primula denticulata) DFP 19 (cp 150); EPF pl 258; ERW 131; OGF 15; PFW 241; TVG 163.

Primrose, Japanese or Candelabra (Primula japonica) DFP 167 (cps 1334, 1335); ERW 148; MWF 241.

Primrose, Least (Primula minima) EPF pl 257; NHE 278; PFE cp 91 (944).

Primrose, Long-flowered or -leaved (Primula halleri) EWF cp 15c; NHE 278.

Primrose Peerless (Narcissus biflorus) MBF cp 83; WFB 271 (2).

Primrose, Poison (Primula obconica) BIP 149 (cp 225); DFP 79 (cp 632); EGG 137; LHP 192; MWF 241; PFW 242; RDP 328.

Primrose, Polyanthus (Primula polyantha) DFP 46 (cp 364); EEG 137; EGA 147; LFW 115 (cps 253, 254, 256); MWF 241; PFF 277.

Primrose, Red Alpine or Hairy-leaved (Primula hirsuta) NHE 278; PFE cp 91 (945).

Primrose, Scottish (Primula scotia) MBF cp 57; MFF cp 22 (611); PWF 88b; WFB 175 (5).

Primrose, Showy (Primula spectabilis) DFP 21 (cp 161); SPL cp 493.

Primrose, Siebold's (Primula sieboldii) DFP 20 (cp 160); ESG 83, 138; TVG 165.

Princess Flower—See **Glory Bush.**

Prinia, Black-chested (Prinia flavicans) NBA 63; PSA 257 (cp 30).

Prinia, Tawny-flanked (Prinia subflava) KBA 353 (cp 54); NBA 63; PSA 257 (cp 30).

Prion, Broad-billed (Pachyptila vittata) SAO 41; WAN 97.

Prion, Dove (Pachyptila desolata) CDB 37 (cp 30); HAB 11; SAO 41; SAR 113, 114; WAN 96.

Prion, Fairy (Pachyptila turtur) HAB 11; SAO 33; SNB 14 (cp 6).

Prion, Thin-billed (Pachyptila belcheri) SAO 33; SAR 113.

Privet (Ligustrum sp.); EEG 99; EFS 33, 76, 77, 122; MTB 385 (cp 40); TGS 379.

Privet, California (Ligustrum ovalifolium) HSC 74, 80; OBT 141; TGS 374.

Privet, Chinese (Ligustrum sinensis) GDS cp 272; TGS 374.

Privet, Common (Ligustrum vulgare) BFB cp 13; BWF 81; DEW 2:115; EPF pls 670, 671; MBF cp 58; MFF pl 74 (630); NHE 24; OBT 37; OWF 183; PFF 279; PWF 86c, 172d; SPL cp 100; TGS 379; TVG 229; WFB 179 (5).

Privet, Vicary Golden (Ligustrum vicaryi) EEG 99; EFS 122.

Privet, Wax-leaved (Ligustrum japonicum) EDP 143; EEG 108; EFP 121; EGE 134; EJG 126; EPG 130; TGS 350.

Propeller Plant—See **Scarlet Paintbrush.**

Prophet Flower (Arnebia echioides or Aipyanthus echioides) DFP 122 (cp 975); EWF cp 49d; PFW 51.

Protea (Protea sp.) DEW 2:100 (cp 49); EWF cp 75 d-f; LFT cps 57, 58; MWF 243; PFW 244; WWT 15.

Protea, Giant or King (Protea cynaroides) DEW 2:95, 113; MEP 33; MWF 242; PFW 244; RWF 57.

Prunella —See **Accentor.**

Pseudoscorpion—See **Scorpion, Book or False.**

Ptarmigan, Rock (Lagopus mutus) BBB 115-117; BBE 95; CDB 67 (cp 185); DAE 49; FBE 97; FBW 27; LEA 393 (bw); OBB 51; PBE 157 (cp 42); WAB 103.

Ptarmigan, Willow —See **Grouse, Red or Willow.**

Pterodiscus (Pterodiscus sp.) DEW 2:180; LFT cp 151.

Puapilo—See **Caper Bush.**

Pudu—See **Deer, Pudu.**

Puffball (Pisolithus sp.) DLM 137; NFP 88; PMU cp 5b; SMG 51; (Tulostoma sp.) DLM 139; LHM 223; NFP 88.

Puffball, Common or Warted (Lycoperdon perlatum) CTM 72; DEW 3: cp 75; DLM 30, 135; EPF pl 99; KMF 123; LHM 217; NHE 47; ONP 155; SMG 47; TEP 225.

Puffball, Gemmed (Lycoperdon gemmatum) NFP 87; PFF 88; PMU cp 6b; REM 179 (cp 147); RTM 230.

Puffball, Giant (Lycoperdon giganteum) LHM 217; SMF 21.

Puffball, Mosaic (Calvatia uteriformis) DLM 130; KMF 119; (C. sp.) DLM 130; ONP 37; RTM 303.

Puffball, Pear-shaped or Wood (Lycoperdon pyriforme); DLM 135; LHM 219; NFP 87; NHE 47; ONP 155; RTM 231; SMF 41; SMG 48.

Puffball, Spiny (Lycoperdon echinatum) CTM 73; DLM 134; LHM 219; NFP 87; RTM 231.

Puffbird (Malacoptila panamensis) CDB 133 (cp 524); DTB 25 (cp 12); (M. fulvogularis) SSA cp 9; (M. striata) ALE 9:68.

Puffbird, Barred (Nystalus radiatus) ALE 9:68; SSA cp 34.

Puffbird, Chestnup-capped (Bucco macrodactylus) CDB 133 (cp 523); DTB 23 (cp 11).

Puffbird, Collared (Bucco capensis) ABW 188; ALE 9:68.

Puffbird, Large-billed (Notharchus macrorhynchus) ABW 188; ALE 9:68; SSA pl 45.

Pufferfish (Arothron)—See **Blowfish.**

Pufferfish (Canthigaster sp.) ALE 5:257; HFI 15; HFW 266 (cp 132); WFW cps 492, 493; (Sphaeroides sp.) ALE 5:265; FEF 529 (bw); HFI 15; HFW 264-265 (cps 129-131); MOL 112.

Pufferfish (Tetraodon sp.)ALE 5:265; CTF 47 (cps 58, 59); FEF 524, 529 (bw); LAW 350; OBV 43; SFW 669 (cp 184), 680 (pl 187) 681 (pl 188), 688 (pl 189); WFW cps 494, 495.

Puffin, Atlantic or Common (Fratercula arctica) ABW 139; ALE 8:209; BBB 259; BBE 163; CDB 95 (cp 345); DAE 30; FBE 165; GPB 171; LAW 476; LEA 417 (bw); NBB 13 (bw); OBB 101; RBE 183; SAO 143; SLP 48, 49; WAB 135; WEA 300.

Puff-leg—See **Hummingbird, Puff-leg.**

Puku (Kobus vardoni) DDM 189 (cp 32); SMR 97 (cp 8).

Puller, Banded—See **Damselfish.**

Puma—See **Cougar.**

Pumpkin, Field (Cucurbita pepo) EGA 122; EMB 147; EVF 104; OFP 123; PFF 307; TVG 45.

Puncture Vine—See **Maltese Cross.**

Punk-tree—See **Cajeput Tree.**

Purple Heart (Setcreasea purpurea) BIP 29 (cp 22); EDP 120; EFP 143; EGV 135; EWF cp 187c; PFW 77; RDP 360.

Purslane (Calandrinia sp.) CAU 181; DFP 31 (cp 246); EGA 100; EWF cp 132b; PFW 238; TGF 68; WAW 120.

Purslane (Portulaca oleracea) MCW 36; PFE cp 11 (122); PFF 165; WUS 153; (P. kermesina) ; LFT cp 66.

Purslane, Hampshire or Water (Lugwigia palustris) DEW 2:25; EWF cp 16g; MBF cp 35; MFF pl 104 (424); NHE 93; WFB 291 (4).

Purslane, Iceland (Koenigia islandica) MFF pl 105 (550); WFB 43.

Purslane, Pink (Claytonia alsinoides) MFF cp 23 (200); OWF 113; (C. arctica) EWF cp 3d.

Purslane, Pink (Montia sp.)—See **Claytonia, Pink.**

Purslane, Sea (Honkenya sp.)—See **Sandwort, Sea.**

Purslane, Sea (Halimione portulascoides)
DEW 1:203; MBF cp 72; MFF pl 111 (216);
NHE 136; OWF 57; PFE 73; PFW 164d; WFW
49i; (H. pedunculata) MBF cp 72; NHE 136.

Purslane, Tree —See **Orache, Shrubby.**

Purslane, Water (Peplis portula) HPW 157;
MBF cp 34; MFF pl 104 (420); NHE 93; OWF
113; (Lythrum portula); PFE cp 81 (819);
WFB 291 (3).

Pussy Ears (Kalanchoe sp.)—See **Panda
Plant.**

Pussy Ears (Cyanotis somaliensis) EGL 100;
RDP 165.

Pussy Willow—See **Willow, Pussy.**

Putoria (Putoria calabrica) EWF cp 36c; PFE
cp 98 (1013); PFM cp 183.

Puya (Puya raimondii) DEW 2:253; MLP 295;
RWF 89; (P. alpestris) DFP 168 (cp 1339); (P.
beteroniana) RWF 94.

Pygmy-tyrant—See **Tyrant.**

Pyrrhulina (Pyrrhulina sp.) ALE 4:291; FEF
106, 107 (bw); SFW 157 (pl 32).

Python (Liasis sp.) ALE 6:369; COG cp 36;
HWR 128; WAW 82-84.

Python, African (Python sebae) ALE 6:369;
ART 141; FSA 76 (pl 3), 84 (cp 5); HRA 70;
LAW 429; LBR cp 19, 60-61; LEA 322 (bw);
SIR 156 (cp 78).

Python, Angola or Dwarf (Python anchietae) ALE 6:386; FSA 76 (pl 3).

Python, Ball or Royal (Python regius) ALE
6:370; HRA 70; LBR 89; LEA 322 (bw); SIR
157 (cp 79); SSW 58; WEA 76.

Python, Carpet or Diamond (Morelia argus)
ALE 6:370, 386; HRA 71; LBR 36, 104; SIR
156 (cp 77), 182 (bw); SSW 61; (M. spilotes)
COG cp 35; ERA 237; WAW 85 (bw); (M. variegata) MEA 28.

Python, Green Tree (Chondropython viridis)
ALE 6:370, 385; ART 200; CAU 244; LEA
320; MAR 152 (pl 62); SSW 59.

Python, Indian or Rock (Python molurus)
ALE 6:369; DAA 151; HRA 68; LBR 87; LVS
186; MAR 150 (pl 60), 151 (pl 61); SIR 155 (cp
76); SSW 56; VRA 205, 207, 209.

Python, Reticulated (Python reticulatus)
ALE 6:369; HRA 69; SSW 55.

Pytilia (Pytilia afra) ALE 9:429; (P. phoenicoptera) CDB 212 (cp 943).

Q

Quackgrass (Agropyron repens) AMP 135;
BWG 177 (bw); EGC 73; EPF pl 1065; MBF
pl 99; MCW 14; MFF pl 125 (1208); NHE
242; PFE cp 178 (1728); PFF 356; TEP 182;
WUS 35.

Quail, Asiatic—See **Quail, Common or
Eurasian.**

Quail, Barred Button (Turnix suscitator)
ABW 102; ALE 8:100; DPB 65; KBA 96 (cp
9); WBV 67.

Quail, Black-breasted (Turnix melanogaster) HAB 51; RBU 19; SNB 60 (cp 29).

Quail, Blue-breasted or Painted (Coturnix
chinensis) DPB 61; KBA 96 (cp 9); WBV 67.

Quail, Brown (Synoicus australis) HAB 48;
SNB 60 (cp 29).

Quail, Bustard—See **Quail, Barred Button.**

Quail, Button (Turnix nana) ALE 8:443; (T.
ocellata) DPB 65.

Quail, Chinese Painted or King (Excalfactoria chinensis) ABW 91; ALE 7:480;
HAB 48; RBI 11; SNB 60 (cp 29).

Quail, Common or Eurasian (Coturnix
coturnix) ABW 96; ALE 7:480; 11:321; BBB
72; BBE 101; DAE 124; DPB 61; FBE 105;
GPB 118; OBB 57; PBE 157 (cp 42); WBV 67.

Quail, Kurrichane Button (Turnix sylvatica) BBE 101; DPB 65; FBE 110; KBA 96
(cp 9); PBE 157 (cp 42); PSA 113 (cp 12); RBE
115; WBV 67.

Quail, Little (Turnix velox) HAB 49; SNB 60
(cp 29).

Quail, Painted Button (Turnix varia) CDB
72 (cp 213); HAB 49; SNB 60 (cp 29).

Quail, Red-backed (Turnix maculosa) HAB
49; SNB 60 (cp 29).

Quail, Stubble (Coturnix pectoralis) CDB 68
(cp 195); HAB 48; SNB 60 (cp 29)

Quail-finch (Ortygospiza atricollis) ALE 9:440; (O. fuscocrissa); PSA 316 (cp 39).

Quail-thrush, Chestnut (Cinclosoma castanotum) ALE 9:222; CDB 169 (cp 717); SAB 16 (cp 7).

Quail-thrush, Chestnut-breasted (Cinclosoma castaneothorax) HAB 173; SAB 16 (cp 7).

Quail-thrush, Cinnamon (Cinclosoma cinnamomeum) ABW 240; CDB 169 (cp 718); HAB 172; SAB 16 (cp 7); WAB 195.

Quail-thrush, Spotted (Cinclosoma punctatum) GPB 275; HAB 173; SAB 16 (cp 7).

Quaker Ladies—See Bluets.

Quaking-grass, Common (Briza media) MBF pl 97; NHE 195; TEP 151.

Quaking-grass, Large (Briza maxima) DFP 31 (cp 245); EGA 99; MFF pl 126 (1197); NHE 199; PFE cp 179 (1746); PFM pl 21 (420); PFW 131.

Quarrion—See Cockatiel.

Queen Anne's Lace (Daucus carota) DEW 2:86; EPF pl 764; MBF cp 40; MCW 77; MFF pl 95 (510); NHE 182; OWF 91; PFE cp 87 (910); PFF 268; PWF 75h; TEP 15; WFB 157 (10); WUS 283.

Queen-of-the-Meadow (Filipendula ulmaria) AMP 141; EPF pl 410; KVP cp 11; MBF cp 26; MFF pl 94 (342); NHE 92; OWF 81; PFE cp 44 (421); PWF 112d; TEP 41; TGF 108; TVG 93; WFB 107 (1).

Queen's Tears (Billbergia nutans) BIP 33 (cp 33); DFP 54 (cp 432); EPF pl 1032; EWF cp 188b; MWG 57; RDP 100.

Queensland Lacebark—See Flame Tree of Australia.

Queensland Nut (Macadamia ternifolia) MLP 74; MWF 192; OFP 31.

Queensland Wheel Tree—See Firewheel Tree.

Quelea, Red-billed (Quelea quelea) ABW 304; ALE 9:430; CAF 171; CDB 215 (cp 962); GPB 325; LAW 499; PSA 308 (cp 37); WAB 149.

Quillwort (Isoetes sp.) CGF 243; DEW 3:255; cps 122, 130; MFF pl 136 (1266, 1267); NPF 122; NHE 108; ONP 97; PFF 110; WFB 291 (10).

Quince (Cydonia oblonga) EPF pl 432; OBT 172; PFE cp 47 (463); PFF 198; (C. vulgaris) OFP 63; SST cp 199.

Quince, Flowering (Chaenomeles speciosa) DFP 189 (cp 1508); EPF pl 435; EWF cp 92a; GDS cp 65; HPW 143; HSC 27, 28; SPL cp 309; (C. superba) DFP 189 (cp 1509); GDS cps 66, 67.

Quince, Japanese (Chaenomeles japonica) EEG 92; EFS 21, 101; EJG 106; ELG 114; EPG 114; LFW 24 (cp 46); PFF 197; TGS 254; (C. lagenaria) EFC cp 57; MWF 80; OGF 27; TGS 246; TVG 219.

Quinine Tree (Cinchona calisaya) CTH 53 (cp 81); DEW 2:131; SST cp 272.

Quinine-bush—See Silk Tassel.

Quiver (Aloe dichotoma) DEW 2:238; EET 231; LCS cp 316; RWF 66.

Quokka (Setonix brachyurus) ALE 10:89, 136 (cp 6); CAU 41 (bw); LVS 29.

R

Rabbit, African (Poelagus marjorita) ALE 12:436; DDM 49 (cp 4).

Rabbit, European Wild (Oryctolagus cuniculus) ALE 11:321, 12:424, 435, 446; BMA 189 (bw), 190; DAE 13; DSM 109; FBW 57; GAL 49 (bw); HMW 157; JAW 73; LAW 581; LEA 520 (bw); MLS 183 (bw); OBV 139; SMW 91 (cp 57); TWA 92.

Rabbit, Ryukyu (Pentalagus furnessi) ALE 12:436; PWC 235 (bw).

Rabbit Tracks—See Prayer Plant.

Rabbitfish (Chimaera monstrosa) ALE 4:104; CSF 59; FEF 28 (bw); HFI 136; OBV 19; (C. mirabilis) MOL 102; (Siganus sp.) HFI 49; HFW 226 (bw); MOL 117, 119; WFW cps 449, 450; (Lo vulpinus) LAW 342.

Raccoon, Crab-eating (Procyon cancrivorus) ALE 12:94; MLS 275 (bw).

Race, Indian—See Squirrel, Indian Giant.

Radish, Wild (Raphanus raphanistrum) BFB cp 4 (4); MBF cp 11; MCW 55; MFF cp 45 (56); NHE 215; OWF 69; PWF 87e; TEP 173; WFB 89 (8).

Rafflesia (Rafflesia sp.) DEW 1:134; EWF cp 117c; HPW 177; RWF 79.

Ragged Robin (Lychnis flos-cuculi) BFB cp 5; EFC cp 6; KVP cp 30; MBF cp 14; MFF cp 14 (157); NHE 184; OWF 107; PFE cp 15 (160); PWF 91f; TEP 141; WFB 61 (4).

Ragged Sailors—See **Cornflower.**

Ragweed—See **Ragwort.**

Rag-worm—See **Worm, Clam.**

Ragwort, Broad-leaved (Senecio fluviatilis) MFF pl 59 (849); NHE 101; PWF 113j.

Ragwort, Chamois or Leopardsbane (Senecio doronicum) DFP 172 (cp 1375); NHE 286; PFE cp 151 (1455).

Ragwort, Common or Tansy (Senecio jacobaea) BFB cp 18 (2); BWG 45; MBF cp 47; MCW 124; MFF pl 57 (842); NHE 189; OWF 43; PWF 105 g; WFB 243 (1); WUS 429.

Ragwort, Great Fen (Senecio paludosus) MBF cp 47; NHE 101; PFE 446; WFB 243 (4).

Ragwort, Hoary (Senecio erucifolius) BFB cp 18 (1); MBF cp 47; MFF cp 57 (844).

Ragwort, Marsh (Senecio aquaticus) BFB cp 18 (8); MBF cp 47; MFF cp 57 (843); NHE 101; OWF 43; PWF 148b, 153f.

Ragwort, Moorland—See **Fleawort, Marsh.**

Ragwort, Oxford (Senecio squalidus) BFB 18 (7); MBF cp 47; MFF cp 57 (845); OWF 43; PWF 23f; WFB 243 (2).

Ragwort, Silver (Senecio cineraria) DFP 34 (cp 267); EEG 147; EGA 153; MBF cp 47; MFF cp 57 (851); MWF 266; WFB 243 (7).

Rail, Banded (Rallus philippensis) BBT 120-121; BNZ 169; DPB 71; FNZ 105 (bw); SNB 62 (cp 30); (R. mirificus) DPB 71; (Hypotaenidia philippensis) HAB 56.

Rail, Chestnut (Eulabeornis castaneoventris) HAB 55; SNB 62 (cp 30).

Rail, Red-necked (Rallina tricolor) CDB 75 (cp 235); HAB 55; RBU 28; SNB 62 (cp 30).

Rail, Slaty-breasted (Rallus striatus) DPB 71; KBA 97 (cp 10); WBV 75.

Rail, Sooty—See **Crake, Spotless.**

Rail, Water (Rallus aquaticus) ALE 8:83; 11:105; BBB 168; BBE 107; FBE 115; GPB 130; KBA 97 (cp 10); PBE 52 (cp 17); RBE 120; (R. pectoralis); HAB 55; SNB 62 (cp 30).

Rail, Weka (Gallirallus australis) ABW 111; ALE 8:83; BNZ 173, 179; CDB 74 (cp 228); FNZ 103 (bw); LVB 43.

Rail, Wood (Aramides cajanea) ALE 8:83; SSA cp 3; (A. ypecaha) ABW 107.

Rainbow-bird—See **Bee-eater, Rainbow.**

Rainbow-fish, Australian Red-tailed (Melanotaenia nigrans) ALE 4:451, 454; FEF 328 (bw); SFW 676 (pl 185).

Rainbow-fish, Celebes (Telmatherina ladigesi) ALE 4:451; FEF 329 (bw); HFI 63; SFW 609 (cp 168); WFW cp 215.

Rainbow-fish, Dwarf (Melanotaenia macullochi) ALE 4:451; FEF 325 (bw); HFI 63; SFW 609 (cp 168); (Nematocentrus macullochi) WFW cp 212.

Rainbow-runner (Elegatis bipinnulatus) LMF 66; MOL 94.

Rambutan—See **Lychee Nut.**

Ramonda (Ramonda myconi) DFP 22 (cp 175); EWF cp 38c; HPW 247; OGF 45; PFE cp 131 (1268); (R. nathaliae) EPF pl 862.

Rampion (Campanula rapunculus) CTH 54 (cp 82); HPW 255; MBF cp 54.

Rampion, Black (Phyteuma nigrum) DEW 1:62 (cp 32); NHE 36.

Rampion, Clustered—See **Devil's Claw.**

Rampion, Horned (Phyteuma scheuchzeri) EWF cp 33g; TGF 201.

Rampion, Round-headed (Phyteuma orbiculare) HPW 254; NHE 188; PFE cp 141 (1352); (P. tenerum) MBF cp 54; MFF cp 8 (803); OWF 179; PWF 136c; WFB 225 (6).

Rampion, Spiked (Phyteuma spicatum) DEW 2:106 (cp 63), 194; EPF pl 525; KVP cp 49; MBF cp 54; MFF pl 106 (804); NHE 36; PFE cp 141 (1350); WFB 225 (7); (P. betonicifolium NHE 284; PFE cp 141 (1350).

Rangoon Creeper (Quisqualis indica) HPW 165; MEP 95; MWF 249; PFW 76.

Ransons (Allium ursinum) BFB cp 23; BWG 98; KVP cp 48; MBF cp 85; MFF pl 95 (1018); NHE 38; OWF 103; PFE cp 166 (1616); PWF cp 24c; TEP 261; WFB 265 (1).

Rape (Brassica napus) MBF 9; OFP 25, 173; PFF 185; PWF cp 23h, 123k; TEP 163.

Rapfen —See **Asp (fish).**

Rasbora (Rasbora sp.) FEF 203, 204, 206; HFI 101; SFW 261 (pl 64), 276 (cp 69), 308 (pl 75), 309 (pl 76); WFW cps 133, 135, 137.

Rasbora, Harlequin or Red —See **Harlequin Fish.**

Raspberry, European (Rubus idaeus) AMP 41; EPF pls 400, 401, cp X; EVF 131; MBF cp 28; MFF pl 78 (345); NHE 20; OFP 77; OWF 79; PFF 202; PWF 62b, 109j; TEP 271; WFB 109 (6).

Rat, Bamboo or Root (Rhizomys sp.) ALE 11:300; SMW 133 (cp 77).

Rat, Black (Rattus rattus) ALE 11:371; DAE 65; DSM 124; GAL 54 (bw); HMA 82; HMW 171; MLS 216 (bw); OBV 135; SMW 135 (cp 80); TWA 98 (bw); (R. tunneyi) MEA 114.

Rat, Brown or Common –See **Rat, Norway.**

Rat, Cane —See **Cane-rat.**

Rat, Cloud (Phloemys sp.) ALE 11:372; SMW 136 (cp 82); (Crateromys schadenbergi) DSM 125.

Rat, Crested or Maned (Lophiomys imhausi) DDM 49 (cp 4); LEA 533 (bw); MLS 209 (bw).

Rat, Giant Pouched (Cricetomys gambianus) ALE 11:371; DSM 124; LEA 536 (bw); MLS 218 (bw); SMR 129 (cp 10); (C. emini) DDM 49 (cp 4).

Rat, Mole —See **Mole-rat.**

Rat, Norway (Rattus norvegicus) ALE 11:371; DAE 65; DSM 124; GAL 54 (bw); HMA 82; HMW 171; JAW 81; LAW 589; LEA 534 (bw); OBV 135; SMW 135 (cp 81).

Rat, Rind —See **Rat, Cloud** (Phloemys sp.)

Rat, Sand —See **Sand-rat.**

Rata (Metrosideros sp.) CAU 283; EWF cp 137d; MWF 200; RWF frontispiece; SNZ 52 (cp 122), 64 (cps 165-168), 65 (cps 169-170), 166 (cp 527); SST cp 135.

Ratel —See **Honey-badger.**

Rat-kangaroo (Bettongia sp.) ALE 10:131 (cp 1); DSM 40; LEA 489 (bw); LVS 33; MLS 66 (bw); WAW 45 (bw).

Rat-kangaroo, Long-nosed (Potorous tridactylus) ALE 10:131 (cp 1); DSM 40; PWC 231 (bw); TWA 184 (bw).

Rat's Tail —See **Plantain, Great.**

Ratstripper (Pachistima canbyi or Paxistima canbyi) EEG 146; EGC 142; EGE 138; ELG 147; ESG 135.

Rattan, Ground —See **Palm, Lady** (Rhapis).

Rattlesnake, South American (Crotalus durissus) ALE 6:468; LAW 394; MAR 197 (pl 78); SIR 254 (cp 141); VRA 247.

Raven (Corvus corax) ABW 224; ALE 9:498; BBB 123; BBE 219; CDB 222 (cp 997); FBE 309; LAW 507; OBB 125; PBE 309 (bw); WAB 37; (C. coronoides) HAB 150.

Raven, Cape or White-necked (Corvus albicollis) CDB 222 (cp 996); PSA 161 (cp 18); WEA 306.

Ray (Raja sp.) CSF 55, 57; HFI 139; LEA 233; OBV 13, 15, 17.

Ray, Eagle (Myliobatis aquila) ALE 4:118; HFI 140; OBV 19.

Ray, Electric (Torpedo marmorata) ALE 4:117; CSF 51; FEF 20 (bw); HFI 137; LAW 289; OBV 11.

Ray, Manta —See **Manta, Atlantic.**

Ray, Starry (Raja radiata) CSF 53; HFI 139.

Ray, Sting —See **Stingray.**

Ray, Thornback (Raja clavata) ALE 4:117; CSF 53; FEF 21-23 (bw); HFI 139; HFW 57, 58 (bw); LAW 293; OBV 13; WEA 339; WFW cp 8.

Razorbill —See **Auk, Razorbill.**

Razorfish (Aeoliscus strigatus) ALE 5:40; HFI 73; HFW 151; WFW cp 229; (A. punctulatus) MOL 110.

Red Hot Poker (Kniphofia sp.) DFP 153 (cp 1224); HPW 313; LFT cps 18, 19, 20; LFW 73 (cp 169); OGF 173; TVG 107.

Red Hot Poker (Kniphofia uvaria) CAF 51; DFP 154 (cps 1225-1228); MEP 17; MWF 174; PFW 174; RWF 53; SPL cp 294; TGF 29.

Red Rattle (Pedicularis palustris) EPF pl 851; MBF cp 65; NHE 75; OWF 141; PWF 128a; WFB 217 (3).

Red Spider Flower (Grevillea punicea) EWF cp 134f; MWF 145; WAW 119.

Redbird Cactus (Pedilanthus tithymaloides) DEW 1:275 (cp 153); ECS 137; EFP 130; GFI 91 (cp 19); LFW 263 (cp 588); MWF 223; RDP 300.

Redbush —See **Christmas Bush.**

Redfish —See **Haddock, Norway.**

Red-flag Bush (Mussaenda erythrophylla) EWF cp 60e; MEP 149; MWF 204; PFW 267; (M. sp.) HPW 258; MLP 231; MWF 204.

Redondo Creeper (Drosanthemum floribunda) MEP 37; MWF 118; (D. hispidum) EGC 130; (D. splendens) EWF cp 73h.

Redpoll, Arctic (Acanthis hornemanni or Carduelis hornemanni) BBE 285; FBE 291; OBB 179; PBE 260 (cp 61).

Redpoll, Common or Lesser (Acanthis flammea or Carduelis flammea) ABW 302; ALE 9:391; BBB 156; BBE 285; CDB 207 (cp 916); FBE 291; FNZ 208 (cp 18); OBB 179; PBE 260 (cp 61); RBE 303.

Redshank, Common (Tringa totanus) ALE 8:170; BBB 179; BBE 125; CDB 88 (cp 299); DPB 87; FBE 131; GPB 148; KBA 129 (pl 14), 148(pl 17); OBB 71; PBE 84 (pl 27), 116 (cp 33); RBE 147; WAB 129; WBV 87.

Redshank, Spotted (Tringa erythropus) BBB 271; BBE 125; CDB 88 (cp 295); CEU 272 (bw); FBE 131; KBA 129 (pl 14); OBB 71; PBE 84 (pl 27), 116 (cp 33); WBV 81.

Redstart, Black (Phoenicurus ochrurus) ALE 9:279; BBB 44; BBE 257; FBE 251; FBW 81; GAL 126 (cp 2); LAW 467; OBB 143; PEB 221 (cp 54).

Redstart, Collared (Myioborus torquatus) ABW 285; ALE 9:371.

Redstart, Common (Phoenicurus phoenicurus) ABW 253; ALE 8:346; 9:279; BBB 138; BBE 257; CDB 166 (cp 698); FBE 251; GAL 126 (cp 2); GPB 272; LAW 481; OBB 143; PEB 221 (cp 54); RBE 260.

Redstart, Golden-fronted (Myioborus ornatus) CDB 205 (cp 898); SSA cp 39.

Redstart, Guldenstadt's (Phoenicurus erythrogaster) FBE 251; RBI 220.

Redstart, Water (Rhyacornis sp.) DPB 293; KBA 321 (cp 50).

Redthroat (Pyrrholaemus brunneus) HAB 181; SAB 32 (cp 15).

Redtop (Agrostis alba) EGC 114; PFF 352.

Redwing (Turdus iliacus) ALE 9:290; BBB 83; BBE 265; CDB 167 (cp 704); FBE 259; GPB 271; OBB 137; PBE 220 (cp 53).

Redwood, Dawn (Metasequoia glyptostroboides) BKT cp 8; DFP 254 (cp 2025); EET 94, 95; EGE 100; EPF pl 151; MTB 85 (cp 6); OBT 101, 125; SST cp 25; TGS 22.

Reed (Arundo sp.) MLP 302; MWF 47; NHE 109.

Reed, Common (Phragmites communis) EPF pl 1045; KVP cp 19; MFF pl 134 (1168); NHE 106; PFE cp 179 (1740); PFF 358; TEP 35; (P. australis) WUS 81.

Reedbuck, Bohor (Redunca redunca) ALE 13:402; CAA 79; DDM 193 (cp 34); DSM 256.

Reedbuck, Common or Southern (Redunca arundinum) DDM 193 (cp 34); DSM 256; HMW 71; SMR 92 (cp 7); TWA 26.

Reedbuck, Mountain (Redunca fulvorufula) ALE 13:402; DDM 193 (cp 34).

Reedfish (Calamoichthys calabaricus) ALE 4:140; HFI 132; SFW 44 (pl 1).

Reedling, Bearded —See **Tit, Bearded.**

Reedmace, Great (Typha latifolia) BFB cp 22 (7); DEW 2:317; EPF pl 994; EWF cp 23a; KVP cp 20; MBF cp 88; MFF cp 35 (1105); NHE 108; PFE cp 184 (1827); PFF 335; PWF 102c, 141h; WFB 297 (7); WUS 11.

Reedmace, Lesser or Narrow-leaved (Typha angustifolia) ERW 149; MBF cp 88; MFF cp 35 (1106); NHE 108; PFF 335; PWF 141g; TEP 38; WFB 297 (8); WUS 11; (T. sp.) MLP 270; SNZ 101 (cp 306).

Rehmannia (Rehmannia sp.) BIP 151 (cp 231); EWF cp 104 c, e; MWF 251.

Reindeer (Rangifer tarandus) ALE 13:192, 224; BMA 196-199; CAS 28-29, 34-35 (bw); CEU 276, 279 (bw), 281; DAE 102; DSM 234, 235; FBW 21; HMA 35; HMW 62; JAW 177; LEA 601 (bw); MLS 389 (bw); SLP 74; SMW 274 (cp 158); TWA 73; WEA 94.

Remember Me —See **Forget-me-not, Wood.**

Remora —See **Shark-sucker.**

Restharrow (Ononis sp.) AMP 129; NHE 211; PFE cp 57 (569, 570); PFM cp 61.

Restharrow, Common (Ononis repens) BFB cp 7 (1); KVP cp 95; MBF cp 21; MFF cp 18 (275); NHE 211; OWF 133; PWF 93i; WFB 127 (2).

Restharrow, Large Yellow (Ononis natrix) DEW 1:288 (cp 187); NHE 211; PFE cp 57 (571); WFB 127 (3).

Restharrow, Prickly (Ononis spinosa) MBF cp 21; NHE 211; PWF 106a; TEP 131.

Restharrow, Small (Ononis reclinata) MBF cp 21; WFB 127 (4).

Restio (Restio sp.) EWF cp 88f; HPW 284.

Resurrection Flower —See **Rose of Jericho.**

Rhamphichthyid —See **Knifefish, Green.**

Rhea (Rhea americana) ABW 19; ALE 7:88; CDB 33 (cp 2); CSA 163 (bw); LAW 500; NBB 92 (bw); SSA cp 1; WAB 79, 87; WEA 310.

Rhea, Darwin's (Pterocnemia pennata) ABW 19; ALE 7:88; LVS 205.

Rhebuck, Gray or Vaal (Pelea capreolus) ALE 13:402; DDM 193 (cp 34).

Rhinoceros, Black (Diceros bicornis) ALE 13:37-40; BMA 200-201, 203 (bw); CAA 69; CAF 164 (bw), 165 (bw); DDM 160 (cp 25); DSM 216; HMA 84; HMW 35; JAW 162; LAW 559; LEA 588 (bw), 592; LVA 21; LVS 113; MLS 358 (bw); PWC 167, 243 (bw); SLP 228; SMR 80 (cp 5); SMW 233 (cp 144); TWA 37; WEA 310.

Rhinoceros, Great Indian (Rhinoceros unicornis) ALE 13:37; BMA 202-203; CAS 168 (bw); DAA 100, 101; DSM 217; HMA 85; HMW 34; JAW 161; LAW 557, 558; LVA 47; LVS 114; MLS 356 (bw); PWC 85; SLP 182, 183; SMW 232 (cp 142); TWA 169; VWA cp 27.

Rhinoceros, Javan (Rhinoceros sondaicus) ALE 13:37; CAS 239 (bw); LVA 49; SLP 193; SLS 148-149; VWA cp 25.

Rhinoceros, Square-lipped or White (Diceros simus or Ceratotherium simus) ALE 13:37; CAA 67 (bw); CAF 285 (bw); DDM 160 (cp 25); DSM 216; HMA 85; LEA 589 (bw); LVA 21; MLS 357 (bw); PWC 243 (bw); SLP 229; SMR 80 (cp 5); SMW 233 (cp 143); VWA cp 6; WEA 311.

Rhinoceros, Sumatran (Didermocerus sumatrensis or Dicerorhinus sumatrensis) ALE 13:37; DAA 129; DSM 217; PWC 243 (bw); VWA cp 26.

Rhododendron, Miniature —See **Ground-cistus.**

Rhododendron, Wild (Rhododendron ponticum) OBT 140; PFE cp 88 (920); PWF 62c; SPL cp 54; WFB 170.

Rhodohypoxis (Rhodohypoxis sp.) DFP 23 (cps 181, 182); LFT cp 35.

Rhodophyllus (Rhodophyllus sp.) LHM 179-183; REM 108-109 (cps 76-78); SMG 201.

Rhubarb (Rheum rhaponticum) EVF 106; OFP 163; PFF 155; (R. sp.) DFP 169 (cp 1347); EPF pls 313, 317; EWF cp 96b; SPL cp 400.

Rhubarb, Sorrel (Rheum palmatum) AMP 27; CTH 27 (cp 23); DFP 169 (cp 1348).

Ribbon Grass —See **Basket Grass.**

Ribbon Plant —See **Spider Plant.**

Ribbon-bush —See **Pohuehue.**

Ribbonwood —See **Lacebark.**

Ribbon-worm —See **Worm, Bootlace** (Lineus).

Rib-seed, Austrian (Pleurospermum austriacum) NHE 275; PFE 280; WFB 161 (6).

Rice (Oryza sativa) DEW 2:269; EPF pl 1046; MLP 315; OFP 9; PFF 350.

Rice Flower (Pimelea sp.) EWF cp 134 a, b; HPW 160; MEP 92; MWF 234; SNZ 19 (cps 10, 11), 118 (cp 364), 162 (cp 516).

Rice Grass —See **Cord-grass** (Spartina).

Rice Paper Plant (Tetrapanax papyriferum) EFP 145; MEP 102; MWF 281; SST cp 299.

Ricebird —See **Finch, Spice.**

Rifle-bird, Magnificent (Ptiloris magnificus) ALE 9:479; HAB 165; RBU 304; SAB 80 (cp 39); WAB 207; (Craspedophora magnifica) ABW 233.

Rifle-bird, Paradise (Ptiloris paradiseus) HAB 166; RBU 300; SAB 80 (cp 39).

Rifle-bird, Victoria (Ptiloris victoriae) HAB 165; RBU 303; SAB 80 (cp 39).

Rifleman (Acanthisitta chloris) ALE 9:150; BNZ 61; CDB 143 (cp 575); FNZ 193 (cp 17); LEA 441 (bw); WAB 209.

Rimu —See **Pine, Red.**

Ringhals —See **Cobra, Ring-necked Spitting.**

River Horse —See **Hippopotamus.**

River Jack —See **Viper, Rhinoceros.**

River Rose (Bauera rubioides) BIP 29 (cp 25); MWF 53.

Rivulus (Rivulus sp.) ALE 4:453; FEF 297, 298, 301 (bw); HFI 68.

Roach (fish) (Rutilus rutilus) FEF 140, 142 (bw); FWF 59; GAL 232 (bw); HFI 95; OBV 87; SFW 244 (pl 57); WFW cp 140.

Roach (insect)—See **Cockroach.**

Roan —See **Antelope, Roan.**

Roast Beef Plant —See **Iris, Stinking.**

Robin, Buff-sided (Poecilodryas cerviniventris) HAB 204; SAB 38 (cp 18).

Robin, Cape (Cossypha caffra) ALE 9:279; CDB 163 (cp 678); GPB 182; NBA 59; PSA 257 (cp 30).

Robin, Dusky (Petroica vittata) HAB 203; SAB 38 (cp 18).

Robin, European or Eurasian (Erithacus rubecula) ABW 253; ALE 8:363; 9:269, 279; BBB 55; BBE 257; CDB 163 (cp 681); DAE 132; FBE 253; GAL 126 (cp 2); NBB 75; OBB 145; PBE 221 (cp 54); WAB 24; WEA 312.

Robin, Flame (Petroica phoenicea) GPB 287; HAB 203; MBA 20; RBU 247; SAB 36 (cp 17).

Robin, Gray-headed (Heteromyias cinereifrons) HAB 206; SAB 38 (cp 18).

Robin, Hooded (Petroica cucullata) CDB 182 (cp 786); HAB 203; SAB 38 (cp 18).

Robin, Magpie —See Magpie-robin.

Robin, Mangrove (Peneoenanthe pulverulentus) HAB 206; SAB 38 (cp 18).

Robin, Northern Yellow (Eopsaltria chrysorrhoa) HAB 205; MBA 25; SAB 38 (cp 18).

Robin, Pale Yellow (Eopsaltria capito) HAB 205; MEA 54; SAB 38 (cp 18).

Robin, Pekin (Leiothrix lutea) ABW 240; ALE 9:222; CAS 154; CDB 168 (cp 712); LAW 508.

Robin, Pink (Petroica rodinogaster) CDB 183 (cp 789); HAB 203; SAB 36 (cp 17); WAB 185.

Robin, Red-capped (Petroica goodenovii) CDB 182 (cp 787); HAB 203; MBA 22, 23; MEA 54; SAB 36 (cp 17).

Robin, Rose (Petroica rosea) HAB 203; SAB 36 (cp 17).

Robin, Scarlet (Petroica multicolor) CDB 182 (cp 788); HAB 203; MBA 21; SAB 36 (cp 17).

Robin, Scrub —See Scrub-robin.

Robin, Southern Yellow (Eopsaltria australis) ALE 9:241; CAU 75; CDB 180 (cp 779); HAB 205; RBU 248; SAB 38 (cp 18).

Robin, Western Yellow (Eopsaltria griseogularis) HAB 205; MBA 24; MWA 82; SAB 38 (cp 18).

Robin, White-faced (Eopsaltria leucops) HAB 205; SAB 38 (cp 18).

Robin, White-throated (Irania gutturalis) FBE 255; (Cossypha humeralis) CDB 163 (cp 679).

Rochea (Rochea coccinea) BIP 155 (cp 239); ERW 26; RDP 340.

Rock Cook (Centrolabrus exoletus) CSF 131; CSS cp 30; OBV 63.

Rock Cress, Purple (Aubrieta deltoidea) DFP 3 (cps 19, 20); ERW 101; LFW 16 (cp 31); MWF 48; OGF 41; PFW 98; SPL cp 442; TGF 77; TVG 143.

Rock-beauty, Pyrenean (Petrocallis pyrenaica) EFP pl 249; NHE 274.

Rock-borer, Wrinkled —See Clam, Red-nose.

Rock-cress, Alpine (Arabis alpina) MBF cp 7; NHE 274; PFE cp 34 (318); WFB 95 (7).

Rock-cress, Bristol (Arabis stricta) MBF cp 7; MFF pl 91 (96).

Rock-cress, Hairy (Arabis hirsuta) MBF cp 7; MFF pl 91 (95); OWF 71; PWF 67f; WFB 95 (5).

Rock-cress, Mountain or Northern (Cardaminopsis petraea) MBF cp 7; MFF pl 91 (93); WFB 95 (2).

Rock-cress, Tall or Sand (Cardaminopsis arenosa) NHE 218; PFE cp 35 (314); WFB 95 (3).

Rock-cress, Upright (Arabis recta) NHE 218; WFB 95 (9).

Rocket, Annual Wall (Diplotaxis muralis) MBF cp 10; MFF cp 43 (54); NHE 215; PWF 111b; WFB 85 (2).

Rocket, Dame's —See Dame's-rocket or Dame's-violet.

Rocket, Dyer's —See Greenweed, Dyer's.

Rocket, Eastern (Sisymbrium orientale) MFF cp 44 (108); PFE cp 32 (287); PWF 55j.

Rocket, Hairy (Erucastrum gallicum) MFF cp 45 (49); NHE 215; PFE 128; WFB 89 (5).

Rocket, London (Sisymbrium irio) MBF cp 8; NHE 219.

Rocket, Perennial Wall (Diplotaxis tenuifolia) AMP 99; MBF cp 9; MFF cp 43 (55); NHE 215; OWF 9; PWF 82b.

Rocket Salad (Eruca sativa) EWF cp 30a; OFP 153; (E. vesicaria) PFE cp 37 (363).

Rocket, Tall (Sisymbrium altissimum) MCW 57; MFF cp 44 (109); NHE 218; PFF 187; PWF 95j; WFB 87 (3); WUS 213.

Rocket, Yellow —See Cress, Winter.

Rockfoil (Saxifraga aizoon) DFP 24 (cp 188); TGF 93.

Rockfowl (Picathartes gymnocephalus or P. oreas) ABW 241; ALE 9:222; LAW 510; LEA 450 (bw); LVS 235; WAB 140; WEA 57.

Rockhopper or **Rock-penguin** —See
Penguin, Rockhopper.

Rock-jasmine (Androsace sp.) DFP 1 (cp 8), 2
(cp 9, 10); ERW 94; EWF cp 4b, cp 98a; NHE
279.

Rock-jasmine, Alpine (Androsace alpina)
EPF pl 261; KVP cp 128; NHE 279; PFE cp
91 (949); TVG 139.

Rock-jasmine, Flesh-red (Androsace
carnea) NHE 279; PFE cp 90 (953).

Rock-jasmine, Great (Androsace maxima)
NHE 228; WFB 176.

Rock-jasmine, Long-stalked (Androsace
elongata) NHE 228; WFB 176.

Rock-jasmine, Milk (Androsace lactea) NHE
279; WFB 176.

Rock-jasmine, Northern (Androsace septentrionalis) NHE 228; TGF 157; WFB 177 (8).

Rock-jasmine, Swiss (Androsace helvetica)
DFP 1 (cp 7); NHE 279.

Rock-jasmine, Woolly (Androsace villosa)
DFP 2 (cp 11); NHE 279.

Rockling, Five-bearded (Ciliata mustela or
Onos mustela) CSF 119; HFI 78; OBV 77;
WEA 313.

Rockling, Three-bearded (Gaidropsarus
sp.) CSF 119; HFI 78; OBV 77; (Onos
mediterraneus) CSS pl XXIII.

Rockrose (Cistus) —See also **Cistus.**

Rockrose (Helianthemum sp.) DEW 2:212 (cp
123); EPF pl 493; GDS cp 226; KVP cp 174;
NHE 273; OGF 65.

Rockrose, Annual or **Spotted** (Tuberaria
guttata) HPW 108; MBF cp 11; MFF cp 46
(139); NHE 72; PFE cp 78 (798); PFM cp 109;
WFB 149 (2).

Rockrose, Common (Helianthemum
chamaecistus) BFB cp 3 (8); MBF cp 11; MFF
cp 38 (140); NHE 72; OWF 15.

Rockrose, Common (Helianthemum
nummularium) DFP 11 (cps 84, 85); EGC
135; ERW 120; GDS cps 227-229 ; LFW 55
(cp 127); MWF 149; PFE cp 78 (802); PWF
47c; SPL cp 453; TEP 107; TGF 124; WFB
149 (1).

Rockrose, Hoary (Helianthemum canum)
MBF cp 11; MFF cp 38 (142); NHE 73; WFB
149 (1c).

Rockrose, Pink (Cistus incanus) PFE cp 77

(787); PFW 75; SPL cp 477; (C. incanus
creticus) EWF cp 34d.

Rockrose, Purple (Cistus purpureus) DFP
190 (cp 1515); HSC 29; MWF 85.

Rockrose, White (Helianthemum
apenninum) MBF cp 11; MFF pl 84 (141);
NHE 72; PFE cp 78 (803); PWF 42d; WFB
149 (1a).

Rockspray (Cotoneaster microphyllus) MFF
pl 71 (379); TGS 279; TVG 221.

Rock-wallaby, Brush-tailed (Petrogale
penicillata) ALE 10:133 (cp 3); TWA 201
(bw).

Rock-wallaby, Ring-tailed or **Yellow-
footed** (Petrogale xanthopus) ALE 10:133
(cp 3); HMA 63; HMW 205; LVS 31; PWC
187.

Rockweed —See **Wrack, Knotted.**

Rodgersia (Rodgersia sp.) DFP 169 (cps
1349-1351); OGF 75; TVG 119.

Roker —See **Ray, Thornback.**

Roller, Bengal or **Indian** (Coracias bengalensis) CAS 207; CDB 129 (cp 506); DAA 147;
KBA 177 (cp 24); RBI 127; WBV 135.

Roller, Broad-billed —See **Dollarbird.**

Roller, Common or **European** (Coracias
garrulus) ALE 9:32, 57; BBB 272; BBE 185;
CAA 54 (bw); CAS 89; CDB 130 (cp 508);
DAE 138; FBW 41; GAL 148 (cp 4); GPB 215;
OBB 113; PBE 181 (cp 46); PSA 224 (cp 25);
RBE 203; WAB 117.

Roller, Cuckoo —See **Cuckoo-roller.**

Roller, Ground (Atelornis sp.) ALE 9:57;
10:283; (Brachypteracias sp.) ALE 9:57;
10:283; WAB 158.

Roller (insect) —See **Weevil.**

Roller, Lilac-breasted (Coracias caudatus)
ABW 180; ALE 9:57; CAA 36, 41; CDB 129
(cp 507); GPB 215; LAW 474; PSA 224 (cp
25); WEA 314.

Roller, Long-tailed Ground (Uratelornis
chimaera) ALE 10:283; LVB 13; SLS 221;
WAB 139.

Rondeletia (Rondeletia amoena) MWF 255;
(R. odorata) EWF cp 182b; MEP 149.

Roof Nail (Mycena polygramma) DEW 3: cp
58; DLM 227; LHM 111; ONP 133; RTM 93.

Rook (Corvus frugilegus) ALE 9:498; BBB 78;
BBE 219; CDB 222 (cp 999); FBE 311; OBB
125; PBE 309 (bw); RBE 231; WAB 113.

Rorqual (Balaenoptera acutorostrata) ALE 11:464; OBV 191; (B. borealis) ALE 11:464; HMA 99.

Rosary Vine (Ceropegia woodii) BIP 51 (cp 61); ECS 106; EGV 100; PFW 38; RDP 135.

Roscoea (Roscoea cautleoides) DFP 23 (cp 183); (R. humeana) DFP 23 (cp 184); OGF 49; PFW 310.

Rose, Alpine or Drooping (Rosa pendulina) NHE 266; PFE cp 45 (435); TEP 274.

Rose, Burnet (Rosa pimpinellifolia) DEW 1:301; MBF cp 30; MFF pl 75 (371); PFE cp 45 (434); PWF 50d, 165k; WFB 109 (2); (R. spinosissima) BFB cp 9 (7); HSC 108; NHE 134; OGF 93; OWF 77; TGS 303.

Rose, China (Rosa chinensis) EPF pl 415; GDS cp 407; OGF 95; RDP 342; TGS 308.

Rose, Dog (Rosa canina) AMP 49; BFB cp 9 (9); EPF pl 413; MBF cp 30; MFF cp 17 (372); NHE 20; OFP 63; OWF 117; PFW 257; PWF 76b, 173k; TEP 127; WFB 109 (1).

Rose, Downy (Rosa tomentosa) EPF pl 414; MBF cp 30; WFB 109 (4); (R. villosa) MBF cp 30; MFF cp 17 (373).

Rose, Field or Trailing (Rosa arvensis) BFB cp 1 (1); MBF cp 30; MFF pl 75 (370); OWF 79; PFE cp 45 (430); PWF 70d; WFB 109 (3).

Rose, Multiflora (Rosa multiflora) EGW 138; LFW 128 (cp 291); SPL cp 325; TGS 306; WUS 223.

Rose of Jericho (Anastatica hierochuntica) EWF cp 46d.

Rose of Sharon —See also **Cochlospermum**.

Rose of Sharon (Hypericum calycinum) DFP 207 (cp 1655); EEG 143; EGC 136; ERW 121; EWF cp 45d; HPW 86; MBF cp 17; MFF cp 41 (131); PFE cp 75 (763); PWF 109k; SPL cp 361; TGF 124; WFB 145 (2).

Rose of Sharon (Hibiscus syriacus) DFP 206 (cps 1643-1646); EEG 97; EFS 27, 115; EJG 119; ELG 116; EPF cp XIV; GDS cps 231-236; HSC 63, 68; LFW 62 (cps 146-148), 63 (cp 150); MWF 155; OGF 171; PFF 250; PFW 183; SPL cp 92; TGS 231.

Rose, Ramanus (Rosa rugosa) EEG 101; EFS 134; ELG 118; EPF pl 412; HSC 106, 107; MWF 257; OGF 93, 187; TGS 295; TVG 243.

Rose, Sage —See **Sage Rose**.

Rosebay —See **Willow-herb, Rosebay**.

Rosefinch, Common or Scarlet —See **Grosbeak, Scarlet**.

Rosefinch, Long-billed (Uragus sibiricus) ALE 9:382; CDB 211 (cp 934).

Rosella, Adelaide (Platycercus adelaidae) HAB 103; SNB 112 (cp 55).

Rosella, Crimson (Platycercus elegans) CDB 107 (cp 402); HAB 102; MBA 36-37; RBU 147; SNB 112 (cp 55); WAB 203.

Rosella, Eastern (Platycercus eximius) FNZ 48 (cp 2); HAB 102; RBU 152; SNB 112 (cp 55).

Rosella, Green (Platycercus caledonicus) ALE 8:319; HAB 102; RBU 148; SNB 112 (cp 55).

Rosella, Northern (Platycercus venustus) ALE 8:319; HAB 103; RBU 155; SNB 112 (cp 55).

Rosella, Pale-headed (Platycercus adscitus) ALE 8:319; HAB 103; SNB 112 (cp 55).

Rosella, Western (Platycercus icterotis) ALE 8:319; 10:89; HAB 103; MBA 38; MWA 47; SNB 112 (cp 55).

Rosella, Yellow (Platycercus flaveolus) HAB 103; RBU 151; SNB 112 (cp 55).

Rose-mallow (Hibiscus sp.) BIP 97 (cp 148); EWF cps 135d, 147a; EWG 117; LFT cps 105, 106; LFW 63 (cp 149), 64 (cp 153); MWF 155; SPL cp 212; TGF 124.

Rosemary (Rosmarinus officinalis) AMP 30; CTH 51 (cp 77); EGC 145; EGE 146; EGH 84, 136; EPG 144; EVF 145; FHP 141; HPW 238; HSC 109, 110; MWF 258; NHE 81; OFP 143; OGF 105; PFE cp 107 (1105); PFM cp 161; SPL cp 402; TGF 189; TGS 358.

Rosemary, Bog (Andromeda polifolia) EPF pl 480; HSC 14; KVP cp 71; MBF cp 55; MFF pl 24 (586); NHE 70; OWF 119; PFE cp 88 (923); PWF 41F; TEP 79; WFB 171 (8); (A. glaucophylla) PFF 273; TGS 406.

Rosemary, Wild —See **Ledum**.

Rose-moss (Portulaca grandiflora) DFP 46 (cp 363); EGA 147; EPF cp VIII b; HPW 75; LFW 262 (cps 586, 587); MWF 239; PFF 165; PFW 238; SPL cp 150; TGF 68; TVG 67.

Rose-of-heaven (Lychnis coeli-rosa) DFP 42 (cp 335); TGF 71.

Roseroot (Rhodiola rosea) NHE 270; PFE cp 40 (397); PWF 51e; WFB 101 (6); (R. kirilowii) EPF pl 385.

Roseroot (Sedum rosea) BFB cp 10 (1); MBF cp 33; MFF cp 46 (388); OWF 15.

Rosy Morn —See **Trefoil, Common Bird's-foot.**

Rot, Yellowish Sapwood (Fomes fomentarius) DEW 3: cp 57, 162; DLM 40, 149; EPF pl 55; LHM 65; NFP 50; NHE 46; ONP 117; PFF 80; SMF 67; (F. sp.) LHM 65, 67; ONP 111; PFF 79.

Roulroul (Rollulus roulroul) ALE 7:480; CDB 70 (cp 207); KBA 96 (cp 9).

Roving Sailor —See **Aaron's Beard.**

Rowan (Sorbus aucuparia) BFB cp 1 (3); BKT cp 129; DEW 1:295; DFP 239 (cp 1905); EET 168; EGT 44, 144; ELG 111; EPF pls 436, 437; EPG 146; MBF cp 31; MFF pl 74 (383); MTB 288 (cp 25); MWF 270; NHE 19; OBT 20, 24; OWF 195; PFF 197; PWF 34d, 163f; SPL cp 326; SST cp 171; TEP 273; TGS 326.

Rowan, Hupeh (Sorbus hupehensis) DFP 239 (cp 1907); GDS cp 452; MTB 288 (cp 25).

Rubber Vine (Cryptostegia sp.) EGV 105; MEP 113.

Rubythroat, Siberian (Luscinia calliope) ALE 9:279; BBE 259; CDB 164 (cp 685); FBE 253; (Erithacus calliope) DPB 293; KBA 321 (cp 50).

Rudd (Scardinius erythrophthalmus) ALE 4:327, 341; FEF 147 (bw); FWF 69; GAL 238 (bw); HFI 97; LEA 245 (bw); OBV 97; SFW 244 (pl 57); WFW cp 141.

Rue, Common (Ruta graveolens) AMP 119; CTH 35 (cp 42); DEW 2:39; DFP 237 (cp 1891); EFC cp 22 (2); EGH 137; EWF cp 35b; GDS cps 434, 435; HPW 203; PFE cp 66 (686); SPL cp 403.

Rue, Fringed or Sicilian (Ruta chalepensis) NHE 199; PFM cp 80.

Rue-anemone (Anemonella thalictroides) ERW 96; EWG 25, 86; PFF 170; TGF 86 (bw).

Ruff (Philomachus pugnax) ABW 122; ALE 7:255; 8:156, 157, 170; BBB 173; BBE 125; CDB 87 (cp 292); CEU 180 (bw); DPB 99; FBE 133; FBW 63; GPB 146; KBA 128 (pl 13), 148 (pl 17); LAW 458; OBB 75; PBE 85 (pl 28), 116 (cp 33); PSA 144 (pl 15); RBE 148; SNB 66 (cp 32), 68 (pl 33); WAB 25, 125.

Ruffe (Gymnocephalus cernua or Acerina cernua) FEF 349 (bw); FWF 125; GAL 236 (bw); HFI 24; LAW 344; LEA 254 (bw); OBV 95; SFW 533 (pl 144); (G. schraetzer) ALE 5:86.

Rugby Football Plant —See **Pepper-elder.**

Runch —See **Charlock** (Sinapis sp.).

Runner, Rainbow —See **Rainbow-runner.**

Rupture-wort (Herniaria sp.) HPW 68; MBF cp 71; MFF pl 105 (195); NHE 226; PFE 81; WFB 57 (7).

Ruschia (Ruschia sp.) EPF pl 294; HPW 66.

Rush, Bulbous (Juncus bulbosus) HPW 291; MBF cp 87; MFF pl 113 (1032); NHE 104; PFE 519.

Rush, Capitate or Dwarf (Juncus capitatus) HPW 291; MBF cp 87; MFF pl 113 (1030); NHE 79; PFE 519.

Rush, Common or Compact (Juncus conglomeratus) EPF pl 951; MBF cp 86; MFF pl 114 (1028).

Rush, Dutch —See **Horsetail, Rough.**

Rush, Flowering (Butomus umbellatus) BFB cp 22 (4); DEW 2:227; DFP 126 (cp 1007); EPF pl 873; EWF cp 23d; KVP cp 10; MBF cp 79; MFF cp 21 (963); NHE 102; OWF 109; PFE cp 161 (1564); PFW 56; PWF 133e; TEP 32; WFB 259 (7).

Rush, Hard (Juncus inflexus) MBF cp 86; MFF pl 114 (1026); NHE 104; PFE cp 177 (1698).

Rush, Heath (Juncus squarrosus) MBF cp 86; MFF pl 114 (1021); NHE 79; PFE cp 177 (1702).

Rush, Jointed (Juncus articulatus) MBF cp 87; MFF cp 114 (1031); NHE 104; PFE cp 177 (1707); WFB 297 (2); (J. lampocarpus) SNZ 102 (cps 308, 309).

Rush, Rannoch (Scheuchzeria palustris) MBF cp 79; MFF pl 106 (969); NHE 102; WFB 220.

Rush, Sea (Juncus maritimus) MBF cp 87; MFF pl 114 (1029); NHE 137.

Rush, Soft (Juncus effusus) BWG 172; MBF cp 86; MFF pl 114 (1027); NHE 104; PFE cp 177 (1699); PFF 370; TEP 35; WFB 297 (1).

Rush, Toad (Juncus bufonius) HPW 291; MBF cp 86; MFF pl 113 (1025); NHE 104.

Rushbird, Wren-like (Phleocryptes melanops) ALE 9:125; SSA pl 25.

Russian Olive —See Oleaster (Elaeagnus angustifolia).

Russian Vine (Polygonum baldschuanicum) EPF pl 322; HSC 91, 97; TVG 235.

Russula (Russula sp.) CTM 34-41; DEW 3: cp 73; DLM 243-253; LHM 197-205; NFP 74, 75; NHE 45; ONP 107, 115, 121, 127, 135, 137; PMU cps 53-61; PUM cps 18-30; REM 41-55 (cps 9-23); RTM 144-164; SMG 247, 248.

Russula, Pungent (Russula emetica) CTM 36; DLM 44, 45, 246; KMF 91; LHM 203; NFP 74; NHE 48; ONP 137; PFF 84; REM 44

(cp 12); RTM 160; SMF 53; SMG 249.

Rust, White (Albugo sp.) NFP 33; PFF 67.

Rutabaga (Brassica napobrassica) EVF 110; PFF 184.

Rye (Secale cereale) EPF pls 1042, 1051, 1052; MLP 298; OFP 5; PFF 355; TEP 169.

Rye-grass, Italian (Lolium multiflorum) MFF pl 125 (1179); NHE 194; PFE cp 180 (1760).

Rye-grass, Perennial (Lolium perenne) BWG 174; EGC 79, 117; EPF pl 1064; HPW 287; MBF pl 99; MFF pl 125 (1178); NHE 194; PFF 357; TEP 151.

S

Sable (antelope)—See Antelope, Sable.

Sable, Eurasian (Martes zibellina) ALE 12:65; DAA 17; HMW 134; JAW 114.

Sabre-wing (Campylopterus sp.) DTB 3 (cp 1); SSA cp 5.

Saccobranch —See Catfish, Stinging.

Saddleback (Creadion carunculatus) ALE 9:469; BNZ 15; CDB 219 (cp 980); GPB 332; (Philesturnus carunculatus) FNZ 176 (cp 16).

Safflower or False Saffron (Carthamus tinctoris) DEW 2:210 (cp 116); EGH 105; WFB 248 (5c); (C. arborescens) PFM cp 210; (C. lanatus) PFE cp 157 (1508); PFM cp 211.

Saffron (Crocus sativus) AMP 21; EGB 108; EGH 110; NHE 241; OFP 132; PFW 148; TGF 14; (C. tomasinianus) DFP 89 (cp 708); EGB 107; MWF 97.

Saffron, Meadow (Colchium autumnale) AMP 177; BFB cp 23 (6); CTH 22; DFP 86 (cp 686); EGB 103; EPF pl 880; KVP cp 32; MBF cp 86; MFF cp 34 (1019); MWF 90; NHE 191; OWF 16; PFE cp 163 (1588); PFM cp 230; PWF 36a, 168a; TEP 148; TGF 12; WFB 266, 267 (7).

Saffron, Spring Meadow (Bulbocodium vernum) EGB 99; EGW 95; PFE cp 163 (1589); TGF 14.

Sage (Salvia) —See also Clary.

Sage, Blue (Eranthemum nervosum) FHP 119; PFW 13; (E. pulchellum) BIP 75 (cp 110); MEP 144.

Sage, Gentian (Salvia patens) EGA 151; PFW 154.

Sage, Red-topped (Salvia horminum) DFP 47 (cp 370); EGA 151; PFE cp 115 (1149).

Sage Rose (Turnera sp.) HPW 104 (2, 3); MEP 88.

Sage, Sticky —See Jupiter's Distaff (Salvia).

Sage, Three-lobed (Salvia triloba) PFE cp 113 (1143); PFM cp 155.

Sage, Wood (Teucrium scorodonia) BFB cp 12 (8); MBF cp 70; MFF pl 112 (786); OWF 31; PWF 108d; WFB 197 (5).

Saiga (Saiga tatarica) ALE 13:438; CAS 79 (bw); CEU 126 (bw); DAE 98-99; DSM 268; HMA 85; HMW 78; MLS 420 (bw); SMW 269 (bw); WEA 316.

Sailfish (Istiophorus sp.) ALE 5:192; HFW 232 (bw); LMF 60.

Sainfoin (Onobrychis viciifolia) BFB cp 8 (6); MBF cp 24; MFF cp 25 (322); NHE 212; OWF 135; PWF 54a; TEP 109; WFB 127 (1); (O. montana) PFE cp 59 (636); (O. radiata) HPW 150.

Sainfoin, Alpine (Hedysarum hedysaroides) NHE 273; PFE cp 61 (635).

St. Anthony's Turnip —See Buttercup, Bulbous.

St. Augustine Grass (Stenotaphrum secundatum) EGC 76, 77, 82, 83, 119; PFE 553; RDP 26, 371.

St. John's Bread —See Carob.

St. John's Wort (Hypericum sp.) DFP 12 (cps 90, 91), 208 (cp 1658); EFS 118; EGE 68, 128; EMB 64; EWF cp 121d; LFT cp 109; MBF cp 17; MWF 160; NHE 220; PFW 134, 135; SNZ 134 (cp 418); TGS 155.

St. John's Wort, Beautiful or Slender (Hypericum pulchrum) MBF cp 17; MFF cp 41 (135); NHE 31, 273; OWF 13; WFB 145 (6).

St. John's Wort, Common (Hypericum perforatum) AMP 171; BFB cp 5 (10); EWG 81; MBF cp 17; MCW 71; MFF cp 41 (132); NHE 220; OWF 13; PFE cp 75 (768); PFF 252; PWF 105e; TEP 113; WFB 145 (3); WUS 267.

St. John's Wort, Hairy (Hypericum hirsutum) BFB cp 5 (2); MBF cp 17; MFF cp 41 (136); NHE 31; PFE cp 75 (765); PWF 105f; WFB 145 (4).

St. John's Wort, Imperforate (Hypericum maculatum or H. dubium) BFB cp 5 (5); MBF cp 17; NHE 94; PWF 129k.

St. John's Wort, Marsh (Hypericum elodes) KVP cp 67; MBF cp 17; MFF cp 46 (138); NHE 73; PFE cp 75 (766); PWF 98b; WFB 145 (9).

St. John's Wort, Mountain (Hypericum montanum) MBF cp 17; MFF cp 41 (137); NHE 31; PFE cp 75 (764).

St. John's Wort, Square-stemmed (Hypericum tetrapterum) BFB cp 5 (1); MBF cp 17; MFF cp 41 (133); NHE 94; PWF 126a; WFB 145 (5).

St. John's Wort, Trailing (Hypericum humifusum) MBF 17; MFF cp 41 (134); NHE 220; OWF 13; PWF 98c; WFB 145 (7).

St. Patrick's Cabbage —See also **Houseleek, Common.**

St. Patrick's Cabbage (Saxifraga spathularis) MBF cp 32; MFF pl 82 (401); PWF 67j; WFB 105 (2); (S. umbrosa) DFP 25 (cp 200); PFE 153; TGF 93.

St. Peter's Fish —See **Dory, John.**

St. Thomas Tree (Bauhinia tomentosa) BIP cp 27; MEP 51; MWF 54.

Saithe —See **Pollack.**

Saker —See **Falcon, Saker.**

Saki (Chiropotes sp.) ALE 10:327; LVS 49.

Saki, Hairy or Monk (Pithecia monachus) ALE 10:527; BMA 205; MLS 130 (bw).

Saki, White-faced (Pithecia pithecia) ALE 10:327; BMA 206; DSM 75; SMW 73 (bw).

Salamander —See also **Newt.**

Salamander (Mertensiella sp.) ALE 5:306; MAR 13 (cp 1).

Salamander, Alpine or Black (Salamandra atra) ALE 5:296-307; 13:463; HRA 115.

Salamander, Brown Cave (Hydromantes geneii) HRA 117; LAW 374.

Salamander, Fire or Spotted (Salamandra salamandra) ALE 5:296, 306; 11:105; CAW 28; ERA 70-71; HRA 114; HWR 117, 118; LAW 371, 372; LEA 277 (bw); MAR 9 (pl 1); VRA 131.

Salamander, Giant (Megalobatrachus japonicus) HRA 118; HWR 71; LAW 369; WEA 165; (Andrias japonicus) ALE 5:305; ART 19.

Salamander, Gorman's (Hydromantes gormani) ALE 5:351; CAW 72 (cp 15).

Salamander, Spanish or Gold-striped (Chioglossa lusitanica) ALE 5:307; HRA 116.

Sallow, Common (Salix atrocinera) MBF cp 78; MFF pl 72 (574); PWF 167j.

Sallow, Common (Salix cinerea) EWF cp 12e; OBT 32; WFB 27 (3b).

Sallow, Eared (Salix aurita) MBF cp 77; MFF pl 72 (575); PFE 45; WFB 27 (3b).

Sallow, Great (Salix caprea) DEW 1:193, 210 (cp 110, 111); DFP 237 (cp 1893); EET 132, 135; EFS 19, 135; HPW 117; MBF cp 78; MFF pl 72; MWF 261; NHE 242; OBT 32; OWF 187; PFE cp 3 (29); PWF 13o, 167k, l; TEP 281; TGS 95; WFB 27 (3).

Sallow, White —See **Wattle, Sydney.**

Salmon, Atlantic (Salmo salar) ALE 4:224; CSF 74-76; FWF 21, 23; GAL 247 (bw); HFI 119; HFW 76 (bw); OBV 105.

Salsify (Tragopogon porrifolius) MFF cp 9 (936); OFP 173; PFE cp 158 (1528); PFF 332; PFM cp 215; WFB 251 (2).

Saltator (Saltator sp.) ABW 297; ALE 9:352; CDB 201 (cp 874); CSA 55; SSA cp 42.

Salt-marsh Grass, Common (Puccinellia maritima) MBF pl 98; MFF pl 129 (1182); NHE 138; TEP 65.

Salt-marsh Grass, Reflexed (Puccinellia distans) MBF pl 98; MFF pl 129 (1183); NHE 138.

Saltwort (Salsola kali) HPW 73; MBF cp 73; MFF pl 111 (218); NHE 135; OWF 57; PFE 73; PFF 158; PWF 137c; TEP 59; WFB 49 (5); WUS 139; (S. pestifera) MCW 34; MFF pl 111 (219).

Sambar (Cervus unicolor) ALE 13:212, 285; CAS 192 (bw), 214; DAA 94; DSM 231; TWA 179 (bw); (C. mariannus) ALE 13:213.

Samphire, Golden (Inula crithmoides) BFB cp 19 (2); MBF cp 45; MFF cp 56 (861); OWF 39; PFE cp 145 (1390); PWF 158d; WFB 231 (7).

Samphire, Marsh —See **Glasswort, Common.**

Samphire, Rock (Crithmum maritimum) DEW 2:90; EGH 110; MBF cp 39; MFF cp 51 (491); OFP 147; OWF 47; PFE cp 84 (871); PWF 158c; WFB 167 (3).

Sanchezia (Sanchezia nobilis) MEP 146; RDP 347.

Sand Dollar (Astrophytum asterias) ECS 100; LCS cp 4; SCS cp 2.

Sand-dotterel, Large —See **Sandplover, Greater.**

Sand-dotterel, Mongolian —See **Plover, Mongolian.**

Sand-eel, Greater (Ammodytes lanceolatus or Hyperoplus lanceolatus) ALE 5:161; CSF 133; HFI 47; OBV 61.

Sand-eel, Lesser (Ammodytes tobianus) CSF 133; OBV 61; (A. lancea) CSS pl XXIII.

Sand-eel, Smooth (Gymnammodytes semi-squamatus) CSF 133; OBV 61.

Sanderling (Calidris alba or Crocethia alba) ALE 8:170; BBB 236; BBE 131; CDB 85 (cp 276); DPB 99; FBE 129; HAB 69; HBG 104 (bw), 133 (pl 10); KBA 144 (pl 15), 145 (pl 16); OBB 77; PBE 85 (pl 28), 117 (cp 34); SNB 66 (cp 32), 68 (pl 33); WBV 91; WFI 154.

Sand-fish —See **Skink, Sand.**

Sand-grass, Early (Mibora minima) MBF pl 96; MFF pl 121 (1234); NHE 245.

Sandgrouse, Black-bellied or Imperial (Pterocles orientalis) ALE 8:260; BBE 165; FBE 169; PBE 81 (cp 26).

Sandgrouse, Chestnut-bellied (Pterocles exustus) ALE 8:260; CDB 97 (cp 349); FBE 169.

Sandgrouse, Four-banded (Eremialector quadricinctus) BBT 50-51; CDB 97 (cp 348).

Sandgrouse, Indian or Painted (Pterocles indicus) ALE 8:260; RBI 83.

Sandgrouse, Lichtenstein's (Pterocles lichtensteinii) FBE 167; WAB 153.

Sandgrouse, Pallas' (Syrrhaptes paradoxus) ALE 8:260; BBE 165; FBE 169; LAW 504; OBB 105; PBE 81 (pl 26); RBE 184; WAB 120.

Sandgrouse, Pin-tailed (Pterocles alchata) ABW 140; ALE 8:260; BBE 165; FBE 169; PBE 81 (pl 26); WEA 317.

Sandgrouse, Spotted (Pterocles senegallus) ALE 8:270; FBE 167.

Sandhopper —See **Flea, Sand.**

Sandpiper, Broad-billed (Limicola falcinellus) ALE 8:170; BBB 271; BBE 131; CDB 86 (cp 285); DPB 99; FBE 127; HAB 69; KBA 145 (pl 16); OBB 73; PBE 117 (cp 32); SNB 66 (cp 32), 68 (pl 33); WBV 87.

Sandpiper, Buff-breasted (Tryngites subruficollis) ALE 8:170; CDB 88 (cp 300); PBE 117 (cp 32); SNB 66 (cp 32), 68 (pl 33).

Sandpiper, Common (Tringa hypoleucos or Actitis hypoleucos) ALE 7:385; BBB 177; BBE 123; CDB 88 (cp 297); DPB 91; FBE 133; HAB 68; KBA 144 (pl 15), 148 (pl 17); OBB 73; PBE 85 (pl 28), 117 (cp 34); PSA 144 (pl 15); SNB 70 (cp 34); WAB 127; WBV 81.

Sandpiper, Curlew (Calidris ferruginea) BBB 277; BBE 129; CDB 85 (cp 279); DPB 99; FBE 127; FNZ 129 (pl 12); KBA 144 (pl 15), 148 (pl 17); OBB 77; PBE 85 (pl 28), 100 (cp 31), 117 (cp 34); PSA 144 (pl 15); SNB 66 (cp 32), 68 (pl 33); WBV 87 (Erolia ferruginea) HAB 69.

Sandpiper, Green (Tringa ochropus) BBB 176; BBE 123; DPB 91; FBE 133; KBA 129 (pl 14); OBB 73; PBE 84 (pl 27), 116 (cp 33); RBE 144; WBV 87.

Sandpiper, Marsh (Tringa stagnatilis) BBE 125; CDB 88 (cp 298); DPB 87; FBE 131; HAB 66; KBA 148 (pl 17); PSA 144 (pl 15); SNB 70 (cp 34), 72 (pl 35).

Sandpiper, Purple (Calidris maritima) BBB 277; BBE 129; CDB 85 (cp 281); FBE 129; OBB 79; PBE 85 (pl 28), 117 (cp 34).

Sandpiper, Red-backed —See **Dunlin.**

Sandpiper, Sharp-tailed (Calidris acuminata) BBE 133; DPB 99; FNZ 129 (pl 12); HAB 69; KBA 144 (pl 15), 149 (pl 18); SNB 66 (cp 32), 68 (pl 33).

Sandpiper, Terek (Xenus cinereus) ALE 8:170; BBE 123; DPB 91; HAB 68; KBA 144 (pl 15), 148 (pl 17); PBE 117 (cp 34); SNB 70 (cp 34), 72 (pl 35); (Tringa cinereus) FBE 131; FNZ 129 (pl 12); (T. terek) CAS 31 (bw).

Sandpiper, White-rumped (Calidris fuscicollis) BBE 133; CDB 85 (cp 280); FBE 136; PBE 117 (cp 32); WFI 154.

Sandpiper, Wood (Tringa glareola) ALE 8:170; BBB 267; BBE 123; CDB 88 (cp 296); DPB 91; FBE 133; HAB 66; KBA 129 (pl 14), 148 (pl 17); OBB 73; PBE 84 (pl 27), 116 (cp 33); PSA 144 (pl 15); SNB 70 (cp 34), 72 (pl 35); WBV 87.

Sandplover (Charadrius leschenaultii) BBE 115; DPB 83; FBE 121; HAB 67; KBA 145 (pl 16); SNB 76 (cp 37), 78 (pl 38).

Sand-puppy or **Sand-rat** —See **Mole-rat.**

Sand-smelt (Atherina sp.) CSF 137; HFI 63; OBV 75; WFW cp 213.

Sand-verbena (Abronia fragrans or A. villosa) HPW 69; RWF 106, 109.

Sand-verbena, Pink (Abronia umbellata) EGA 90; EWG 84; TGF 66 (bw).

Sandwort, Arctic (Arenaria norvegica) MBF cp 15; MFF pl 85 (189); WFB 53 (2).

Sandwort, Bog (Minuartia stricta) MFF pl 87 (183); NHE 185.

Sandwort, Fine-leaved (Minuartia tenuifolia) MBF cp 15; MFF pl 87 (184); NHE 225; WFB 53 (5).

Sandwort, Irish (Arenaria ciliata) MBF cp 15; NHE 277.

Sandwort, Mossy (Moehringia muscosa) NHE 277; PFE 81.

Sandwort, Mountain (Arenaria montana) ERW 98; OGF 43; PFE cp 12 (126); TGF 71.

Sandwort, Mountain (Minuartia rubella) MBF cp 15; MFF pl 87 (182).

Sandwort, Sea (Honkenya peploides) MBF cp 16; MFF pl 89 (186); NHE 135; OWF 76; PFE cp 11 (133); PWF 53g; WFB 53 (8).

Sandwort, Slender (Arenaria leptoclados) MBF cp 15; NHE 225.

Sandwort, Spring or Vernal (Minuartia verna) MBF cp 15; MFF pl 90 (181); NHE 277; OWF 77; PWF 47b; WFB 53 (4).

Sandwort, Three-nerved or Three-veined (Moehringia trinervia) MBF cp 15; MFF pl 86 (187); NHE 33; OWF 77; PWF 55m; WFB 53 (3).

Sandwort, Thyme-leaved (Arenaria serpyllifolia) MBF cp 15; MFF pl 86 (188); OWF 76; PWF 110c; WFB 53 (1).

Sanicle, Wood (Sanicula europaea) AMP 63; HPW 220; MBF cp 36; MFF pl 97 (452); NHE 32; OWF 89; PWF 31g; TEP 247; WFB 155 (6).

Sapodilla (Achras sapota) EPF pl 498; HPW 132 (4); OFP 99.

Sapphire Flower —See **Browallia.**

Sapphireberry (Symplocos paniculata) EFS 30, 139; GDS cp 468; TGS 407.

Sarcocaulon (Sarcocaulon sp.) DEW 2:35; HPW 208.

Sardine (Sardina pilchardus) CSF 71; HFI 119; OBV 23.

Sargassofish (Histrio histrio) HFW 272 (cp 144); MOL 92; WEA 318; WFW cp 184.

Saro —See **Otter, Brazilian or Giant.**

Sassaby (Damaliscus lunatus) ALE 13:302; DDM 209 (cp 36); SMR 97 (cp 8); SMW 264 (bw); TWA 32.

Satin Flower —See also **Spring Bell** (Sisyrinchium sp.).

Satin Flower (Godetia grandiflora) DFP 38 (cps 300, 301); EGA 27, 121; EGG 113; MWF 143; OGF 131; SPL cp 359; TGF 124; TVG 51.

Satyrium (Satyrium sp.) EWF cp 89e; LFT 106 (cp 52).

Saupe (Sarpa salpa) FEF 360 (bw); WFW cp 303.

Saury (Scomberesox saurus) CSF 90; HFI 79; OBV 33; (Cololabis saira) MOL 96.

Sausage Tree (Kigelia africana or K. pinnata) DEW 2:177; EWF cp 62a; MEP 136; MLP 216; PFW 44; SST cp 132; WWT 97.

Savory, Summer (Satureia hortensis) EGH 139; EVF 146; OFP 143; (S. thymbra) PFM cp 157.

Savory, Winter (Satureia montana) DFP 171 (cp 1365); EGH 140; EGL 142; OFP 143; PFE 365; PWF 138a; TGS 358; WFB 199 (9).

Sawfish, Small-tooth (Pristis pectinatus) ALE 4:118; HFI 137; LMF 23.

Sawfly (Rhogogaster viridis) ALE 2:437; (Teuthredo mesomelas) WEA 319.

Sawfly, Apple (Hoplocampa testudinea) GAL 417 (bw); OBI 145.

Sawfly, Birch-leaf (Cimbex femorata) GAL
379 (bw); OBI 143; PEI 150, 151 (bw);
(Croesus septentrionalis) ALE 2:437.

Sawfly, Currant or Gooseberry (Nematus
ribesii) ALE 2:437; CIB 241 (cp 36); OBI 145.

Sawfly, Hawthorne (Trichiosoma sp.) CIB
241 (cp 36); OBI 143.

Sawfly, Pine (Diprion pini) ALE 2:437; CIB
141 (cp 36); OBI 145.

Sawfly, Rose (Arge sp.) CIB 241 (cp 36); OBI
143; PEI 149 (bw).

Sawfly, Wheat-stem (Cephus pygmaeus)
ALE 2:437; CIB 241 (cp 36); OBI 143.

Saw-wort (Serratula tinctoria) BFB cp 18
(10); MBF cp 49; MFF cp 31 (923); NHE 190;
OWF 155; PFE cp 155 (1497); PWF 121g;
WFB 249 (7); (S. shawii) DFP 173 (cp 1378);
OGF 177.

Saw-wort (Saussurea sp.) EWF cp 97d; NHE
287; PFE 446.

Saw-wort, Alpine (Saussurea alpina) EPF pl
588; MBF cp 49; MFF cp 3 (916); NHE 287;
PFE cp 152 (1475); PWF 149g; WFB 245 (4).

Saxicave —See **Clam, Red-nose.**

Saxifrage, Alternate-leaved Golden
(Chrysosplenium alternifolium) MBF cp 32;
NHE 27.

Saxifrage, Arctic or Clustered Alpine
(Saxifraga nivalis) MBF cp 32; MFF pl 82
(398); WFB 105 (3).

Saxifrage, Bell (Saxifraga grisebachii) DFP
25 (cp 194); OGF 21; PFW 274; TVG 171.

Saxifrage, Bluish (Saxifraga caesia) NHE
271; PFE 153.

Saxifrage, Burser's (Saxifraga burseriana)
DEW 1:290; DFP 24 (cps 190, 191); PFW
274.

Saxifrage, Elephant-leaved (Bergenia
cordifolia) EGW 93; MWF 56; OGF 17; TVG
145.

Saxifrage, Fortune (Saxifraga fortunei)
DFP 25 (cp 193); EWF cp 99c; PFW 273.

Saxifrage, Highland or Alpine (Saxifraga
rivularis) MBF cp 32; MFF pl 87 (406); WFB
105 (7).

Saxifrage, Irish (Saxifraga rosacea) MBF cp
32; NHE 210.

Saxifrage, Kidney (Saxifraga hirsuta) MBF
cp 32; MFF pl 82 (402); PFE cp 42 (405).

Saxifrage, Lesser Bulbous or Drooping
(Saxifraga cernua) MBF cp 32; MFF 87
(405).

Saxifrage, Livelong or Silver (Saxifraga
paniculata) EPF pl 387; NHE 271; PFE cp 42
(400); WFB 105 (10).

Saxifrage, Meadow or Bulbous (Saxifraga
granulata) MBF cp 32; MFF pl 84 (404);
NHE 27; PWF 31l; TEP 109; WFB 105 (6).

Saxifrage, Mossy (Saxifraga hypnoides)
MBF cp 32; MFF pl 84 (407); PFE 153; PWF
42b; WFB 105 (8).

Saxifrage, Musky (Saxifraga moschata)
NHE 271; PFE cp 42 (411), 153.

Saxifrage, Opposite-leaved Golden
(Chrysosplenium oppositifolium) MBF cp
32; MFF cp 46 (410); NHE 27; OWF 15; PFE
cp 42 (412); PWF 30c; WFB 101 (7).

Saxifrage, Pepper (Silaum silaus) MBF cp
39; MFF cp 51 (499); NHE 182; OWF 13; PFE
287; PWF 127h; WFB 167 (5).

Saxifrage, Purple (Saxifraga oppositifolia)
DFP 25 (cp 197); EWF cp 2g; KVP cp 147;
MBF cp 32; MFF cp 23 (409); NHE 271; OGF
21; PFE cp 41 (398); PFW 274; PWF 149k;
TVG 171; WFB 105 (9).

Saxifrage, Pyrenean (Saxifraga longifolia)
DEW 1:290; DFP 25 (cp 196); PFE cp 41
(399); TVG 171.

Saxifrage, Rock Jasmine (Saxifraga
androsacea) NHE 271; PFE cp 41 (409).

Saxifrage, Round-leaved (Saxifraga
rotundifolia) NHE 271; PFE cp 42 (403).

Saxifrage, Rue-leaved or Fingered
(Saxifraga tridactylites) MBF cp 32; MFF pl
87 (403); NHE 210; PFE cp 41 (411); WFB
105 (4).

Saxifrage, Starry (Saxifraga stellaris) MBF
cp 31; MFF pl 82 (399); NHE 271; PFE cp 41
(407); PWF 149e; WFB 105 (1).

Saxifrage, Yellow Marsh (Saxifraga
hirculus) MBF cp 32; MFF cp 46 (400); NHE
179; PFE 153.

Saxifrage, Yellow Mountain (Saxifraga
aizoides) KVP cp 127; MBF cp 32; MFF cp 46
(408); NHE 271; PFE cp 41 (402); PWF 149h;
WFB 105 (5).

Scabies —See **Mite, Itch.**

Scabious (Scabiosa sp.) HPW 262; OGF 89;
PFE 411; PFM cp 187.

Scabious, Caucasian (Scabiosa caucasica) DFP 171 (cp 1366); OGF 155; PFW 101; SPL cp 246; TVG 121.

Scabious, Devil's-bit (Succisa pratensis) MBF cp 43; MFF cp 8 (837); NHE 235; OWF 179; PWF 139g, 151e; WFB 225 (3).

Scabious, Field (Knautia arvensis) EFC cp 9; MBF cp 43; MFF cp 8 (835); NHE 235; PFE cp 138 (1322); PWF 85e; WFB 225 (1).

Scabious, Hoary (Scabiosa canescens) NHE 76; WFB 225 (2b).

Scabious, Mountain or Wood (Knautia sylvatica) EFC cp 9; NHE 36; PFE cp 138 (1323).

Scabious, Shining (Scabiosa lucida) NHE 284; PFE cp 138 (1325).

Scabious, Small (Scabiosa columbaria) BFB cp 11 (1); EFC cp 9; LFT cp 163; MFF cp 8 (836); NHE 76; PWF 69g, 121j; WFB 225 (2).

Scabious, Sweet (Scabiosa atropurpurea) DFP 47 (cp 373); EGA 152; LFW 129 (cps 292, 293); MWF 263; OGF 87; PFE cp 138 (1326); TGF 236.

Scabious, Yellow (Scabiosa ochroleuca) EPF pl 782; NHE 76; PFE cp 137 (1327); WFB 225 (2a).

Scabweed —See **Daisy, Mat**.

Scad (Trachurus trachurus) CSF 123; HFI 26; OBV 31.

Scaldfish (Arnoglossus laterna) CSF 175; OBV 53.

Scale, Cottony-cushion (Icerya purchasi) CIB 144 (cp 13); ZWI 102.

Scale, Mussel or Oystershell (Lepidosaphes ulmi) CIB 144 (cp 13); EPD 124; KIW 82 (cp 34); OBI 35; PEI 147 (bw).

Scallop, Great (Pecten maximus) ALE 3:159; CSS cp 22; OIB 85.

Scallop, Hunchbank (Chlamys distorta) CSS cp 22; OIB 85.

Scallop, Jacob's or Pilgrim's (Pecten jacobaeus) ALE 3:141; GSS 20.

Scallop, Moon (Amusium japonicum) CTS 58 (cp 109); (A. pleuronectes) GSS 137.

Scallop, Noble (Chlamys nobilis) CTS 57 (cp 106); GSS 135; VWS 84.

Scallop, Queen (Chlamys opercularis) CSS cp 22; LEA 82, 191 (bw); OIB 85.

Scallop, Swift's (Chlamys swifti) GSS 135; VWS 84.

Scallop, Variegated (Chlamys varia) ALE 3:141; CSS cp 22; OIB 85.

Scaly-foot —See **Slowworm** (Pygopus sp.).

Scapania (Scapania sp.) DEW 3:235; NFP 115; ONP 59, 183.

Scarlet Paintbrush (Crassula falcata) EGL 98; MEP 42; RDP 159; SPL cp 152.

Scarlet Plume (Euphorbia fulgens) BIP 79 (cp 117); DFP 66 (cp 526); EGG 109; MEP 70; PFW 115; SPL cp 45.

Scat —See **Argus-fish**.

Scaup, Greater (Aythya marila) ABW 69; ALE 7:61, 308; BBB 231; BBE 59; DPB 33; FBE 59; OBB 25; PBE 29 (cp 10), 41 (pl 14), 49 (pl 16); RBE 91.

Scaup, New Zealand (Aythya novaeseelandiae) ALE 7:307; BNZ 233; FNZ 161 (cp 15).

Schefflera —See also **Umbrella Tree, Queensland**.

Schefflera (Schefflera digitata) BIP 159 (cp 245); SNZ 75 (cps 210, 211); (S. polybotrya) MWF 264.

Scheltopusik —See **Lizard, Glass**.

Schizocodon (Schizocodon soldanelloides) DFP 26 (cp 201); HPW 131.

Schneider (Alburnus bipunctatus) HFI 95; (Alburnoides bipunctatus) FWF 67.

Schoolmaster —See **Snapper**.

Schraetzer (Acerina schraetzer) FWF 125; HFI 24; SFW 533 (pl 144).

Scolopender or **Scolopendra** —See **Centipede**.

Scopolia (Scopolia carniolica) DEW 2:152; EPF pl 830; PFE cp 116 (1175).

Scorpion (Buthus sp.) ALE 1:407; GSP 126; LAW 140; LEA 178 (bw); PWI 10 (bw); (Euscorpius sp.) ALE 1:393; GSP 124; PEI 10 (bw); WEA 320.

Scorpion, African (Pandinus sp.) ALE 1:393, 407; GSP 123.

Scorpion, Book or False (Chelifer cancroides) ALE 1:416; GAL 492 (bw); GSP 121; OIB 149; (Allochernes panzeri) OIB 155; (Chthonius sp.) LEA 178 (bw); (Neobisium sp.) ALE 2:51, 97; OIB 131.

Scorpion Senna (Coronilla emerus) PFE cp 62 (624); PFM cp 68; WFB 121 (10).

Scorpion, Water —See **Water-scorpion**.

Scorpion, Whip —See **Whip-scorpion**.

Scorpionfish (Dendrochirus sp.) ALE 5:59,
60; WFW cps 241, 242; (Scorpaena sp.) HFI
59; HFW 223 (cp 115); LAW 332; WFW cps
248-250; (Scorpaeniopsis sp.) ALE 5:60.

Scorpion-grass —See **Forget-me-not.**

Scorpion-vetch (Coronilla sp.) NHE 212,
273; PFM cp 71.

Scorpion-vetch, Shrubby (Coronilla
glauca) DFP 192 (cp 1529); GDS cp 111; HSC
35, 37.

Scorpiurus (Scorpiurus muricatus) EWF cp
29e; PFE 185.

Scoter, Black or Common (Melanitta nigra)
ALE 7:310; BBB 232; BBE 63; CDB 53 (cp
119); FBE 63; OBB 27; PBE 36 (cp 11), 41 (pl
14), 49 (pl 16).

Scoter, Surf (Melanitta perspicillata) ABW
69; FBE 63; OBB 27; PBE 36 (cp 11), 41 (pl
14), 49 (pl 16).

Scoter, Velvet (Melanitta fusca) ALE 7:310;
BBB 232; BBE 63; FBE 63; OBB 27; PBE 36
(cp 11), 41 (pl 14), 49 (pl 16).

Scouring-rush —See **Horsetail.**

Scrambled Eggs —See **Buttercup Tree.**

Screamer, Black-necked (Chauna chavaria)
ABW 63; LEA 376 (bw).

Screamer, Crested or Southern (Chauna
torquata) ALE 7:267; WAB 87.

Screamer, Horned (Anhima cornuta) ABW
63; ALE 7:267; CDB 49 (cp 92); SBB 45 (cp
54).

Screwpine (Pandanus sp.) BIP 139 (cp 211);
CAU 11; DEW 2:288 (cp 175), 315, 316; EFP
129; HPW 306; MLP 270; RDP 296; WAW
122, 123; WWT 3, 119.

Scrub-bird, Noisy (Atrichornis clamosus)
HAB 137; LVB 45; RBU 199; SAB 4 (cp 1);
WAB 33, 187.

Scrub-bird, Rufous (Atrichornis rufescens)
ABW 214; ALE 9:150; CDB 147 (cp 598);
GPB 246; HAB 137; LVS 234; MBA 10, 11;
SAB 4 (cp 1).

Scrub-robin (Erythropygia sp.) CDB 163 (cps
682, 683); PSA 257 (cp 30).

Scrub-robin, Northern (Drymodes superciliaris) HAB 177; RBU 251; SAB 14 (cp 6).

Scrub-robin, Southern (Drymodes brunneopygia) CDB 180 (cp 777); HAB 177; SAB 14
(cp 6).

Scrub-wren, Little (Sericornis minimus)
HAB 188; (S. beccarii) SAB 34 (cp 16).

Scrub-wren, Spotted (Sericornis maculatus)
HAB 189; RBU 212; SAB 34 (cp 16).

Scrub-wren, White-browed (Sericornis
frontalis) CDB 179 (cp 771); HAB 188; SAB
34 (cp 16).

Scrub-wren, Yellow-throated (Sericornis
lathami) HAB 188; MBA 12; SAB 34 (cp 16).

Scurvy-grass, Common (Cochlearia offcin-
alis) BFB cp 4 (7); MBF cp 8; MFF pl 93 (74);
NHE 134; OWF 69; PWF 158b; WFB 93 (7).

Scurvy-grass, Danish or Early (Cochlearia
danica) MBF cp 8; MFF pl 90 (75); NHE 134;
PFE cp 36 (337); PWF 26b; WFB 93 (8).

Scurvy-grass, English or Long-leaved
(Cochlearia anglica) MBF cp 8; NHE 134;
PWF 53j.

Scurvy-pea —See **Trefoil, Pitch.**

Scythebill (Campylorhamphus trochilirostris) ABW 197; ALE 9:120; DTB 35 (cp 17);
SSA cp 35; WAB 90; (C. falcularius) CDB
139 (cp 558).

Sea Anemone (Cerianthus sp.) ALE 1:203,
271; LAW 59; MOL 57; OIB 17; VWS 69;
(Sagartia sp.) CSS cp 7; LAW 63, 157; MOL
55, 57; OIB 15; (Epizoanthus arenaceus)
ALE 1:204; OIB 19.

Sea Anemone, Beadlet (Actinia sp.) CSS cp
5; LAW 59; OIB 15; VWS 38 (bw).

Sea Anemone, Burrowing (Edwardsia
callimorpha) MOL 56; (Peachia sp.) CSS cp
5; MOL 56; OIB 17.

Sea Anemone, Dahlia (Tealia sp.) ALE
1:200, 271; CSS cp 6; OIB 15; VWS 26 (bw).

Sea Anemone, Daisy (Cereus pedunculatus)
ALE 1:202; CSS cp 7; LEA 50 (bw); MOL 57;
OIB 17.

Sea Anemone, Gem (Bunodactis sp.) CSS cp
6; OIB 15.

Sea Anemone, Hermit Crab (Calliactis
parasitica) ALE 1:202; CSS cp 6; CTF 61 (cp
81); OIB 19.

Sea Anemone, Jewel (Corynactis viridis)
OIB 19; VWS 42 (bw).

Sea Anemone, Plumose (Metridium sp.)
ALE 1:199, 271; CSS cp 6; LEA 51 (bw); OBV
15; VWS 38 (bw).

Sea Anemone, Snakelocks (Anemonia sp.) ALE 1:271; CSS cp 5; LAW 63; LEA 50 (bw); OIB 15; WEA 321.

Sea Bass —See **Sea-perch, Banded.**

Sea Cow —See **Dugong** and **Manatee.**

Sea Cucumber (Cucumaria sp.) CSS cp 28; OIB 187; WEA 322; (Leptosynapta inhaerens) ALE 3:302; OIB 187.

Sea Cucumber (Holothuria sp.) ALE 3:301, 312; CSS cp 28; LAW 261; LEA 196 (bw); OIB 187.

Sea Dahlia —See **Sea Anemone, Dahlia.**

Sea Dragon —See **Seahorse, Australian.**

Sea Elephant —See **Seal, Elephant (Southern).**

Sea Fan (Eunicella sp.) ALE 1:262-263; LAW 53; LEA 49 (bw); OIB 19; VWS 47 (bw).

Sea Fig —See **Fig, Hottentot.**

Sea Gooseberry —See **Comb-jelly.**

Sea Hare (Aplysia sp.) ALE 3:90, 122; CSS cp 18; LAW 124; LEA 49, 89 (bw); OIB 45.

Sea Heath (Frankenia laevis) HPW 111; MBF cp 13; MFF cp 27 (144); OGF 89; OWF 113; PWF 160c; WFB 149 (4).

Sea Holly —See also **Eryngo.**

Sea Holly (Eryngium giganteum) DEW 2:89; DFP 141 (cp 1125); MLP 180; OGF 155; PFW 301; TVG 91.

Sea Holly (Eryngium maritimum) BFB cp 11 (8); DEW 2:88; EPF pl 758; KVP cp 106; MBF cp 36; MFF cp 9 (454); NHE 135; OWF 179; PFE cp 83 (853); PWF 124a; TEP 61; WFB 155 (8).

Sea Lace (Chorda sp.) NHE 140; ONP 19; PFF 56.

Sea Leopard —See **Seal, Leopard.**

Sea Lettuce (Ulva lactuca) NFP 16; NHE 140; ONP 9; PFF 53.

Sea Lion, Australian (Neophoca cinerea) ALE 12:373; MEA 116; WAW 55.

Sea Lion, Patagonian or South American (Otaria byronia) ALE 12:373, 400; CSA 289; HMW 114.

Sea Moth —See **Dragonfish.**

Sea Mouse (Aphrodite aculeata) ALE 1:373; CSS cp 9; OIB 95.

Sea Nettle —See **Jellyfish** (Chrysaora sp.).

Sea Pen (Veretillum cynomorium) ALE 1:263; LAW 61.

Sea Pen, Gray (Pteroides griseum) ALE 1:271; LAW 62.

Sea Pen, Phosphorescent (Pennatula phosphorea) ALE 1:263; MOL 54.

Sea Pink or **Thrift** (Armeria maritima) BFB cp 12 (6); DFP 2 (cp 15); EGC 122; EPF pl 311; ERW 99; KVP cp 108; MBF cp 56; MFF cp 27 (609); MWF 46; NHE 136; OGF 81; OWF 127; PFE cp 94 (975); PFW 232; PWF 66c; TGF 59; WFB 179 (1).

Sea Rocket (Cakile maritima) BFB cp 3 (7); KVP cp 101; MBF cp 11; MFF pl 93 (59); NHE 134; OWF 69; PFE cp 37 (366); PWF 100d; TEP 57; WFB 93 (1).

Sea Scorpion (Cottus bubalis) CSF 167; CSS cp 32; HFI 58; OBV 67.

Sea Serpent —See **Krait, Black-banded.**

Sea Slater (Ligia sp.) ALE 1:496; CSS pl VII; GSP 152; WEA 395.

Sea Slug —See **Nudibranch.**

Sea Snail (fish) —See **Snailfish.**

Sea Spider (Pycnogonum littorale) ALE 1:426, 437; CSS pl XIII; OIB 127; (Nymphon sp.) LEA 185 (bw); OIB 127.

Sea Squirt (Ciona intestinalis) ALE 3:428; CSS cp 29; LEA 193; MOL 91; OIB 189.

Sea Squirt (Clavellina lepadiformis) ALE 3:450; CSS cp 29; LAW 273; OIB 191; (Ascidia mentula) CSS cp 29; OIB 189; (A. mamillata) MOL 91.

Sea Squirt (Phallusia mammillata) ALE 3:438; LAW 275; OIB 189; (Salpa sp.) LAW 268; MOL 129 (bw); OIB 189.

Sea Squirt, Red (Halocynthia papillosa) ALE 3:403, 428, 449; LAW 279.

Sea Squirt, Star (Botryllus schlosseri) ALE 3:428, 455; CSS cp 29; LAW 274; LEA 202 (bw); OIB 191; VWS 34 (bw); (B. violaceus) MOL 91.

Sea Star —See **Starfish.**

Sea Trout —See **Trout, Brown.**

Sea Urchin (Heterocentrotus mammillatus) ALE 3:336; LAW 252; VWS 64 (bw).

Sea Urchin, Edible or European (Echinus esculentus) ALE 3:336; CSS cp 27; MOL 87; OIB 179; VWS 62 (bw); WEA 329; (E. melo) ALE 3:333.

Sea Urchin, Pencil (Cidaris sp.) LAW 251; MOL 87.

Sea Urchin, Purple (Psammechinus miliaris) CSS cp 27; OIB 179; (Sphaerechinus granularis) ALE 3:345; LEA 193.

Sea Urchin, Sand (Echinocardium cordatum) ALE 3:348; OIB 179.

Sea-bat —See **Batfish.**

Sea-blite, Annual (Suaeda maritima) MBF cp 73; MFF pl 111 (217); NHE 135; OWF 57; PFE 73; PWF 164c; TEP 58; WFB 49 (3).

Seablite, Hairy (Bassia hirsuta) NHE 135; WFB 49 (6).

Sea-blite, Shrubby (Suaeda vera) PWF 118d; WFB 49 (4).

Sea-eagle —See **Eagle.**

Seagull —See **Gull, Herring.**

Seahorse (Hippocampus sp.) ALE 5:26, 40; FEF 270, 271 (bw); HFI 74; HFW 81 (cp 33), 82 (cp 34), 149 (bw); LAW 332, 333; LEA 211 (bw); MOL 110; OBV 65; VWS 21 (bw); WEA 323; WFW cps 233-235.

Seahorse, Australian (Phyllopteryx eques) ALE 5:40; HFI 75.

Seakale (Crambe maritima) MBF cp 11; MFF pl 93 (57); NHE 134; OFP 163; PFE cp 37 (368); PWF 54d; WFB 93 (2); (C. cordifolia) DFP 133 (cp 1058).

Seal, Baikal (Pusa sibirica) ALE 12:385; LEA 576 (bw); MLS 329 (bw).

Seal, Crab-eating (Lobodon carcinophaga) ALE 12:386; HMW 113; SAR 137, 146, 147.

Seal, Elephant (Southern) (Mirounga leonina) ALE 12:401, 402, 417; BMA 86; DSM 198; LAW 549; LEA 577 (bw); MLS 332 (bw); SAR 75, 137, 138, 143-145; TWA 148 (bw); WEA 326; WFI 42 (bw).

Seal, Fur (Arctocephalus sp.) ALE 12:368; LVS 108; SLS 265; WEA 327.

Seal, Gray (Halichoerus grypus) ALE 12:386; DAE 143; DSM 200; HMW 112; LEA 576; OBV 179; TWA 149 (bw); WEA 325.

Seal, Hair or Harbor (Phoca vitulina) ALE 8:187; 12:385; DAE 142; HMA 87; HMW 112; JAW 145; OBV 177; SMW 158 (bw).

Seal, Leopard (Hydrurga leptonyx) ALE 12:386; HMW 113; LAW 548; LEA 577 (bw); SAR 137, 138, 148; WAN 61.

Seal, Mediterranean Monk (Monachus monachus) ALE 12:386; DAE 143; SLS 259; (M. schauinslandi) WEA 325.

Seal, Ringed (Phoca hispida or Pusa hispida) ALE 12:385; CAS 18 (bw); HMW 112; OBV 181; (P. h. saimensis) PWC 198-199.

Seal, Ross (Ommatophoca rossi) ALE 12:386; SAR 137, 139, 148.

Seal, Weddell's (Leptonychotes weddelli) ALE 12:386; LAW 546; LEA 545; SAR 136, 137, 139, 146, 149; SLP 114, 115.

Sea-lavender, Common (Limonium vulgare) BFB cp 12; KVP cp 99; MBF cp 56; MFF cp 8 (606); NHE 136; OWF 137; PWF 106c; WFB 179 (2).

Sea-lavender, Matted (Limonium bellidifolium) MBF cp 56; MFF cp 8 (608); PFE 315; PWF 146c; WFB 179 (3).

Sea-lavender, Rock (Limonium binervosum) MBF cp 56; MFF cp 8 (607); PWF 121l; WFB 179 (4).

Sea-lavender, Wide-leaved (Limonium latifolium) EDP 129; TGF 62; TVG 109.

Sea-lavender, Winged (Limonium sinuatum) DEW 1:227; DFP 41 (cp 326); EGA 131; EGG 125; EWF cp 33h; LFW 135 (cps 305, 306); MWF 186; OGF 191; PFE cp 94 (972); PFM cp 132; PFW 232; SPL cp 363; TVG 57.

Sea-perch (Anthias anthias) LAW 334; WFW cp 262.

Sea-perch, Banded (Serranus scriba) ALE 5:79; FEF 334; HFI 22.

Sea-snake (Microcephalophis gracilis) SSW 116; (Aipysurus laevis) ALE 6:446.

Sea-snake, Banded (Laticauda colubrina) COG cp 50; LEA 330 (bw); MAR 173 (pl 70); (Hydrophis sp.) COG cp 49; HWR 120.

Sea-snake, Black-and-yellow (Pelamis platurus) FSA 172 (cp 27); HRA 51; SSW 117; WAW 86 (bw).

Secretary-bird (Sagittarius serpentarius) ABW 74; ALE 7:345; CAA 59 (bw); CAF 140; CDB 64 (cp 174); GBP 29 (bw), 205 (bw); GPB 109; LEA 392 (bw); PSA 80 (pl 7), 96 (cp 9); WAB 146; WEA 330.

Sedge, Bottle or Beaked (Carex rostrata) DEW 2:290; MBF cp 94; MFF pl 120 (1133); NHE 137; PFE cp 186 (1857); WFB 297 (12).

Sedge, Dark or Black (Carex atrata) EWF cp 7m; MBF cp 93; MFF pl 117 (1149); NHE 291; PFE cp 186 (1869).

Sedge, Fen (Cladium mariscus) EPF pl 953; MBF cp 91; MFF pl 115 (1125); NHE 105; (C. tetraquetrum) HPW 292.

Sedge, Great Pond (Carex riparia) EWF cp 25f; MBF cp 94; NHE 106; PFE cp 186 (1858).

Sedge, Lesser Pond (Carex acutiformis) DEW 2:290; MBF cp 94; MFF pl 120 (1135); NHE 106; WFB 297 (10).

Sedge, Rock (Carex rupestris) EWF cp 7h; MBF cp 92; MFF pl 117 (1163).

Sedge, Sand (Carex arenaria) MBF cp 92; MFF pl 118 (1154); NHE 137; TEP 63.

Sedge, Yellow (Carex flava) EPF pl 960; MBF cp 94; PFE 569.

Sedum, Golden (Sedum adolphi) RDP 356; SCS cp 94.

Seedeater, Band-tailed (Catamenia analis) SSA cp 42; WAB 97.

"Seeds of Paradise" —See **Cardamon.**

Seedsnipe, D'Orbigny's or **Gray-breasted** (Thinocorus orbignyanus) ALE 8:170; SSA pl 20.

Seedsnipe, Least or **Patagonian** (Thinocorus rumicivorus) ABW 127; WAB 86.

Seersucker Plant (Geogenanthus undatus) EDP 144; EFP 115.

Seladang —See **Gaur.**

Self-heal (Prunella vulgaris) BFB cp 17 (7); BWG 148; EGC 72; EPF pl 807; EWG 56; MBF cp 69; MFF cp 10 (761); NHE 186; OWF 145; PFF 289; PWF 85f; TEP 119; WFB 199 (1); WUS 317.

Self-heal, Cut-leaved (Prunella laciniata) MBF cp 69; MFF pl 112 (762); NHE 186; OWF 97; PFE cp 110 (1120); WFB 199 (2).

Self-heal, Large (Prunella grandiflora) NHE 186; OGF 81; PFE cp 110 (1120); (P. webbiana) DFP 168 (cp 1338).

Senna, Wild (Cassia marilandica) EGH 106; TGF 148.

Sensitive Plant (Mimosa pudica) DEW 1:283 (cp 175); EGA 136; EPF pl 442; HPW 151; RDP 269; WWT 93; (M. sensitiva) BIP 121 (cp 184).

Seps —See **Skink.**

Serapias, Heart-flowered (Serapias cordigera) KVP cp 175; PFM cp 308; (S. sp.) PFE cp 189 (1903, 1904); PFM cps 309, 310.

Sergeant-major (Abudefduf saxatilis) HFI 37; WFW cp 384.

Seriema, Crested or **Red-legged** (Cariama cristata) ABW 112; ALE 8:133; CDB 76 (cp 241); GPB 134; SSA cp 1; WAB 88.

Serin (Serinus serinus) ALE 9:382; BBB 267; BBE 281; CDB 210 (cp 931); FBE 293; FBW 75; PBE 261 (cp 62); (S. syriacus) FBE 293; (S. tristiatus) ALE 9:382.

Serin, Gold- or **Red-fronted** (Serinus pusillus) ALE 9:382; BBE 281; FBE 293.

Serow (Capricornis sumatraensis) ALE 13:464; DAA 54; HMW 80; MLS 422 (bw); SMW 269 (bw); (C. crispus) CAS 56 (bw).

Serpent-eagle —See **Eagle.**

Serpent-star —See **Brittle-star.**

Serradella (Ornithopus sativus) EWF cp 29c; NHE 212.

Serval —See **Cat, Serval.**

Service Tree (Sorbus domestica) BKT cps 125, 127; NHE 19; OBT 172; PFE 171.

Service Tree, Wild (Sorbus torminalis) BKT cp 128; EET 169; MBF cp 31; MFF pl 74 (385); NHE 19; OBT 173; OWF 195.

Serviceberry (Amelanchier sp.) DFP 181 (cps 1444, 1445); EFS 95; EGT 67, 99; EPF pl 441; ESG 91; EWF cp 152c; GDS cp 8; MFF pl 74 (382); MTB 289 (cp 26); NHE 267; OBT 181; PFE cp 47 (471); PFF 198; TGS 246, 255; WFB 115 (9).

Sesame or **Benne** (Sesamum indicum or S. orientale) EGH 141; OFP 23.

Shad —See also **Nase** or **Nasling.**

Shad, Allis (Alosa alosa) ALE 4:198; CSF 73; OBV 23.

Shad, Twaite (Alosa fallax) ALE 4:198; CSF 73; OBV 23.

Shadbush —See **Serviceberry.**

Shag (Phalacrocorax aristotelis) ALE 7:167; BBB 251; BBE 33; FBE 33; OBB 11; SAO 83; SLP 51; WAB 25, 134.

Shag, Little Black (Phalacrocorax sulcirostris) FNZ 97 (pl 8); HAB 14; SNB 26 (cp 12).

Shag, Long-tailed —See **Cormorant, Reed.**

Shag, New Zealand King (Phalacrocorax carunculatus) ABW 40; ALE 7:167; BNZ 191; FNZ 112 (pl 9).

Shag, Pied (Phalacrocorax varius) BNZ 191; CDB 41 (cp 51); FNZ 97 (pl 8); HAB 15; SNB 26 (cp 12); WAB 20 (bw).

Shag, Rock —See **Cormorant, Rock.**

Shag, Spotted (Phalacrocorax punctatus) ALE 7:167; BNZ 197; CAU 275; LEA 366 (bw); (Stictocarbo punctatus) FNZ 112 (pl 9).

Shaggy Soldier (Galinsoga ciliata) MCW 113; MFF pl 100 (840); NHE 238; PFF 323; PWF 108c; WFB 233 (8).

Shaggy-mane —See **Ink-cap, Shaggy.**

Shaketail (Cinclodes sp.) ALE 9:125; CDB 140 (cp 562); SSA cp 35; WFI 180 (bw).

Shallot (Allium ascalonicum) EGH 94; EVF 146.

Shama (Copsychus malabaricus) ALE 9:279; KBA 336 (cp 51); RBI 204; WAB 166; WBV 191; (C. luzoniensis) DPB 293; (Kittacincla malabarica) ABW 253.

Shanny —See **Blenny, Common.**

Shark (Carcharinus sp.) ALE 4:107; HFI 147; LMF 18; WFW cps 3, 4.

Shark, Angel —See **Monkfish.**

Shark, Basking (Cetorhinus maximus) ALE 4:93; CSF 39; HFI 144; OBV 5.

"Shark," Black (Morulius chrysophekadion) HFI 102; SFW 292 (pl 73).

Shark, Blue (Prionace glauca) ALE 4:103, 108-109; CSF 42; HFI 147; LMF 18; OBV 5.

Shark, Frilled (Chalamydoselachus anguineus) ALE 4:104; MOL 101; OBV 3.

Shark, Greenland (Somniosus micro-cephalus) CSF 49; HFI 141; OBV 5.

Shark, Hammerhead (Sphyrna zygaena) ALE 4:103; HFI 148; LMF 22; OBV 9; WFW cp 6.

Shark, Horn (Heterodontus sp.) FEF 14 (bw); MOL 109.

Shark, Mackerel (Lamna nasus) ALE 4:104; CSF 37; OBV 3; (L. cornubica) HFI 144.

Shark, Mako (Isurus oxyrhynchus) HFI 144; LMF 23; OBV 3.

Shark, Nurse (Ginglymostoma cirratum) HFW 34 (cp 2); MOL 109.

Shark, Porbeagle —See **Shark, Mackerel.**

"Shark," Red-tailed Black (Labeo bicolor) CTF 26 (cp 18); FEF 185 (bw); HFI 102; SFW 292 (pl 73); WFW cp 127.

Shark, Six-gilled (Hexanchus griseus) HFI 148; OBV 3.

Shark, Smooth —See **Hound, Smooth.**

Shark, Thresher (Alopias vulpinus) ALE 4:103; CSF 41; FEF 15 (bw); HFI 144; LMF 19; OBV 5.

Shark, Tiger (Galeocerdo cuvieri) HFI 146; LMF 19.

Shark, Whale (Rhincodon typus) ALE 4:93; HFI 145.

Shark, White (Carcharodon carcharias) ALE 4:93; HFI 144; LMF 22.

Shark-sucker (Echeneis naucrates) ALE 5:138; FEF 517; HFI 17; MOL 93; WFW cp 290; (Remora remora) LEA 262 (bw); OBV 43.

Sharpbill, Crested (Oxyruncus cristatus) ABW 210; ALE 9:150; CDB 144 (cp 583); SSA cp 16.

Shaving Brush (Penicillus sp.) NFP 16; PFF 52.

Shearwater, Buller's (Puffinus bulleri) BNZ 217; SNB 16 (pl 7).

Shearwater, Cory's (Procellaria diomedea) BBB 279; BBE 27; FBE 27; OBB 9; PBE 5 (pl 2); SAO 49.

Shearwater, Flesh-footed (Puffinus carneipes) HAB 11; SNB 16 (pl 7).

Shearwater, Greater (Puffinus gravis) BBB 279; BBE 27; CDB 37 (cp 32); FBE 27; OBB 9; PBE 5 (pl 2); PSA 65 (pl 6); SAO 53; WAB 215.

Shearwater, Little (Puffinus assimilis) BBE 27; FBE 27; HAB 10; SAO 49.

Shearwater, Manx (Puffinus puffinus) ABW 34; ALE 7:147; BBB 249; BBE 27; CDB 38 (cp 34); FBE 27; OBB 9; PBE 5 (pl 2); SAO 29, 49; (Procellaria puffinus) RBE 60.

Shearwater, Sooty (Puffinus griseus) BBB 279; BBE 27; FBE 27; HBG 57 (bw); OBB 9; PBE 5 (cp 2); PSA 65 (pl 6); SAO 49; SNB 16 (pl 7); WFI 90, 92 (bw).

Shearwater, Wedge-tailed (Puffinus pacificus) CDB 38 (cp 33); DPB 7; HAB 10; HBG 56 (bw); SNB 16 (pl 7).

Sheathbill, Snowy or Yellow-billed (Chionis alba) ABW 128; ALE 8:170; CDB 91 (cp 315); GPB 158; SSA pl 20; WAB 223; WEA 332; WFI 159 (bw).

Sheep, Barbary or Maned (Ammotragus lervia) ALE 13:488; DSM 274; HMA 88; HMW 84; JAW 206; MLS 430 (bw); SMW 286 (cp 185); WEA 61.

Sheep, Blue (Pseudois nayaur) ALE 13:487; DAA 22; SMW 285 (cp 184).

Sheep Ked —See **Tick, Sheep.**

Sheep, Wild (Ovis aries) OBV 151; (O. ammon var.) ALE 13:497, 507; DAA 32; (O. orientalis) DAA 14; (O. vignel) DAA 56.

Sheep's Bit (Jasione montana) KVP cp 68; MBF cp 54; MFF cp 8 (805); NHE 188; OWF 179; PFE cp 141 (1355); PWF 120d; WFB 225 (8); (J. perennis) NHE 188; TGF 200.

Shelduck, Australian (Tadorna tadornoides) ALE 7:295; SNB 34 (cp 16); (Casarca tadornoides) HAB 28, 32; WAW 18 (bw).

Shelduck, Common (Tadorna tadorna) ABW 69; ALE 7:296; 8:187; BBB 234; BBE 57; CDB 54 (cp 129); DPB 29; FBE 51; OBB 29; PBE 37 (cp 12), 40 (pl 13), 48 (pl 15); RBE 103; SBB 41 (cp 47); WAB 128.

Shelduck, Paradise (Tadorna variegata) ALE 7:295; BNZ 223; FNZ 161 (cp 15).

Shelduck, Rajah (Tadorna radjah) ALE 7:295; HAB 28; RBU 63; SNB 34 (cp 16).

Shelduck, Ruddy (Tadorna ferruginea) ALE 7:295; FBE 51; KBA 80 (pl 7); OBB 21; PBE 37 (cp 12), 40 (pl 13), 48 (pl 15); (Casarca ferruginea) ABW 69; BBB 273; BBE 57; RBE 104.

Shelduck, South African (Tadorna cana) ALE 7:295; FBW 25; PSA 49 (cp 4), 64 (pl 5).

Shell Flower (Alpinia speciosa) BIP 23 (cp 14); DEW 2:275 (cp 148), 294; HPW 298; LFW 154 (cp 340); MEP 26; MWF 37; PFW 309, 311.

Shell, Tower or Turret (Turritella sp.) CSS pl XV; GSS 37; OIB 41.

Shellflower —See **Tiger Flower.**

She-oak —See **Beefwood.**

Shepherd's Needle (Scandix pecten-veneris) MBF cp 38; MFF pl 96 (459); NHE 222; OWF 89; PWF 83e; WFB 159 (3).

Shepherd's Purse (Capsella bursa-pastoris) AMP 145; BWG 63; DEW 1:142; EPF pl 246; MBF cp 10; MCW 48; MFF pl 92 (72); NHE 217; OWF 71; PFF 182; PWF 14b, 55i; TEP 175; WFB 97 (2); WUS 198.

Shepherd's Rod —See **Teasel, Small.**

Shield-bug, Green (Palomena prasina) GAL 303 (bw); OBI 21; PEI 97 (cp 12a).

Shield-bug, Parent (Elasmucha grisea) ALE 2:179; CIB 97 (cp 8); OBI 21.

Shield-bug, Pied (Sehirus bicolor) CIB 97 (cp 8); OBI 21; (S. dubius) LAW 178.

Shipworm (Teredo sp.) ALE 3:177; CSS pl XX; GSS 155; OIB 87; WEA 334.

Shoebill —See **Stork, Shoebill or Whale-headed.**

Shoeflower —See **Hibiscus.**

Shoo Fly Plant —See **Apple-of-Peru.**

Shooting Star (Dodecatheon sp.) DFP 7 (cp 56), 139 (cp 1112); EPF pls 262, 266; EWF cp 156d; EWG 25, 104; HPG 135; MLP 192; OGF 45; PFW 243; TGF 157; TVG 149.

Shore-weed (Littorella uniflora) MBF cp 71; MFF pl 66 (794); NHE 100; WFB 291 (7).

Shorewort, Northern (Mertensia maritima) MBF cp 60; MFF cp 13 (644); OWF 171; PFE cp 103 (1062); PWF 71f; WFB 193 (6).

Shortwing (Brachypteryx sp.) DPB 293; KBA 385 (pl 58); RBI 208.

Shoveler (Anas clypeata) ABW 68; CDB 50 (cp 96); DPB 33; FBE 57; KBA 80 (pl 7); LEA 379 (bw); OBB 17; PBE 28 (cp 9), 40 (pl 13), 48 (pl 15); RBE 87; SBB 38 (pl 43); SNB 39 (cp 18); WBV 49; WEA 135; (Spatula clypeata) ALE 7:255; BBB 200; BBE 55.

Shoveler, Blue-winged (Anas rhynchotis) HAB 32; SNB 39 (cp 18); (Spatula rhynchotis) RBU 64.

Shoveler, New Zealand (Anas rhynchotis variegata) ALE 7:307; BNZ 227; FNZ 161 (cp 15).

Shower, Pink-and-white (Cassia javanica) MEP 54; SST cp 234.

Shrew, Alpine (Sorex alpinus) ALE 10:189; HMW 191.

Shrew, Common European (Sorex araneus) ALE 10:189; 13:163; BMA 211; DSM 51; HMW 191; LEA 490 (bw); MLS 83 (bw); OBV 129; TWA 102 (bw).

Shrew, Elephant —See **Elephant-shrew.**

Shrew, European Water (Neomys fodiens) ALE 10:189, 11:105; DAE 62; DSM 51; HMW 192; MLS 84 (bw); OBV 129; (N. anomalus) ALE 10:189.

Shrew, Laxman's (Sorex caecutiens) ALE 10:189; CAS 55 (bw).

Shrew, Lesser White-toothed (Crocidura suaveolens) ALE 10:189; DAE 62; OBV 129.

Shrew, Otter —See Otter-shrew.

Shrew, Pygmy (Sorex minutus) ALE 10:189; CEU 268 (bw); DSM 51; HMW 192; LEA 490 (bw); OBV 129.

Shrew, Savi's Pygmy (Suncus etruscus) ALE 10:189; DAE 62.

Shrew, Tree —See Tree-shrew.

Shrew, White-toothed (Crocidura russula) ALE 10:223; HMW 191; LAW 597; OBV 129; (C. leucodon) ALE 10:189, 223.

Shrike, Bou-bou (Laniarius ferrugineus) NBA 71; PSA 289 (cp 34); (L. aethiopicus) CDB 157 (cp 647).

Shrike, Brown (Lanius cristatus) ALE 9:195; BBB 98; DPB 363; KBA 416 (cp 61); LAW 507; OBB 171.

Shrike, Bush (Tehagra senegala) CDB 158 (cp 657); FBE 217; PSA 289 (cp 34); (T. tchagra) ALE 9:195; (Malaconotus sp.) CDB 158 (cp 655); PSA 289 (cp 34); (Telophorus sp.) ABW 268; NBA 73; WAB 148.

Shrike, Fiscal (Lanius collaris) CDB 157 (cp 649); NBA 69; PSA 289 (cp 34); (L. cabanisi) CDB 157 (cp 648).

Shrike, Great Gray —See Shrike, Northern.

Shrike, Lesser Gray (Lanius minor) ALE 9:195; BBB 268; BBE 211; CDB 157 (cp 652); FBE 219; OBB 171; PBE 253 (cp 60).

Shrike, Long-tailed or Rufous-backed (Lanius schach) ABW 268; ALE 9:195; CDB 157 (cp 653); DPB 363; KBA 416 (cp 61); WBV 147.

Shrike, Magpie (Urolestes melanoleucus) ALE 9:195; (Corvinella melanoleuca) PSA 289 (cp 34).

Shrike, Masked (Lanius nubicus) ALE 9:195; BBE 211; FBE 219; PBE 253 (cp 60).

Shrike, Northern (Lanius excubitor) ABW 268; ALE 9:195, 201, 202; BBB 268; BBE 211; CDB 157 (cp 651); FBE 219; GPB 259; OBB 171; PBE 253 (cp 60); WAB 119 (bw).

Shrike, Puff-backed (Dryoscopus cubla) ALE 9:195; CDB 156 (cp 645); PSA 289 (cp 34).

Shrike, Red-backed (Lanius collurio) ABW 268; ALE 9:195; 11:321; BBE 211; CDB 157 (cp 650); FBE 219; GPB 258; NBB 84; PBE 253 (cp 60); PSA 289 (cp 34); WAB 119 (bw); WEA 336.

Shrike, Schach —See Shrike, Long-tailed or Rufous-backed.

Shrike, Tiger (Lanius tigrinus) DPB 363; KBA 416 (cp 61).

Shrike, Vanga —See Vanga.

Shrike, White-crowned (Eurocephalus anguitimens) ALE 9:196; CDB 156 (cp 646).

Shrike, White-helmeted (Prionops plumata) ABW 270; ALE 9:196; CDB 158 (cp 656); PSA 289 (cp 34).

Shrike, Wood —See Wood-shrike.

Shrike, Woodchat (Lanius senator) ALE 9:195; BBB 268; BBE 211; CDB 158 (cp 654); FBE 219; GPB 260; OBB 171; PBE 253 (cp 60); RBE 287.

Shrike-babbler —See Babbler.

Shrike-thrush (Colluricincla harmonica) HAB 175; RBU 236; SAB 46 (cp 22); (C. rufiventris) HAB 177; SAB 46 (cp 22).

Shrike-tit, Eastern (Falcunculus frontatus) HAB 248; SAB 46 (cp 22).

Shrike-tit, Western (Falcunculus leucogaster) HAB 248; MBA 64; SAB 46 (cp 22).

Shrike-vireo (Vireolanius sp.) ABW 282; ALE 9:372.

Shrike-vireo, Slaty-capped (Smaragdolanius leucotis) DTB 75 (cp 37); SSA cp 16.

Shrimp, Common (Crangon vulgaris) CSF 197; CSS pl XI; OIB 123.

Shrimp, Fairy (Branchipus stagnalis) ALE 1:451; GAL 496 (bw).

Shrimp, Freshwater (Gammarus palex) GAL 494 (bw); OIB 135; (G. locusta) CSS pl IX.

Shrimp, Ghost or Skeleton (Caprella sp.) LAW 145; MOL 81; OIB 127.

Shrimp Plant (Beloperone guttata) BIP 33 (cp 32); DEW 2:182; DFP 54 (cp 431); EGG 91; EPF pl 860; FHP 107; HPW 250; LFW 230 (cp 513); LHP 44-45; MEP 144; MWF 56; PFW 14; RDP 20, 98; SPL cp 40; (B. comosa) LFW 230 (cp 515).

Shrimpfish —See Razorfish.

Siamang (Hylobates syndactylus) DAA
134; DSM 94; MLS 159 (bw); SMW 100 (bw);
(Symphalangus syndactylus) ALE 10:476.

Siberian Cardamon —See **Star-anise.**

Siberian Tea Leaves —See **Pig Squeak.**

Sibia, Black-headed (Heterophasia
melanoleuca) KBA 285 (pl 44); RBI 160.

Sibia, Langbian (Crocias langbianus) KBA
285 (pl 44); WBV 221.

Sicklebill (Eutoxeres aquila) ABW 171; CDB
122 (cp 473); SHU 76, 105; (E. condamini)
SSA cp 32; (Falculea palliata) ALE 10:283.

Sicklethorn (Asparagus falcatus) LHP 32;
RDP 88.

Sida (Sida spinosa) PFF·251; WUS 265; (S. sp.)
LFT cp 106; MEP 79.

Side Saddle Plant —See **Pitcher Plant.**

Sifaka, Verreaux's (Propithecus verreauxi)
ALE 10:251 (cp 5), 283; DSM 67; LVA 25;
LVS 57; MLS 118 (bw); PWC 233 (bw); SLS
181; WEA 337.

Sika (Cervus nippon) ALE 13:214; CAS 56
(bw); DAA 138; FBW 19; LEA 596 (bw); OBV
147; (Sika nippon) DAE 39; (C. n. taiouanus)
LVA 51; PWC 245 (bw).

Silk Tassel (Garrya elliptica) DFP 203 (cp
1619); EWF cp 155e; GDS cp 209; HSC 55,
60; MWF 139; OGF 1; PFW 120.

Silk Tree (Albizia julibrissin) EGT 36, 39, 97;
EJG 100; ELG 101; LFW 4 (cp 7); MEP 49;
SPL cp 65; SST cp 227; TGS 134; (A.
lophantha) DEW 1:304 (bw); MWF 35; (A.
lebbeck) MEP 49.

Silk-cotton Tree —See also **Kapok Tree.**

Silk-cotton Tree (Bombax sp.) DEW 1:218 (cp
129); EWF cp 57c, cp 115b, c; HFW 93; LFW
155 (cp 344); MEP 81; PFW 47; SST cp 231.

Silk-oak (Grevillea robusta) EPF 116; HPW
170; MWF 145; RDP 227; SST cp 123.

Silk-vine (Periploca graeca) EWF cp 36a;
HPW 225; HSC 86; PFE cp 97 (1009); TGS
70.

Silver Crown (Cotyledon undulata) BIP 63
(cps 85, 86); ECS 110; EPF pl 381; LCS cp
230; MEP 43; RDP 12, 157; SCS cp 75.

Silver Dollar Fish (Mylossoma argenteum)
FEF 74 (bw); WFW cp 81; (Metynnis sp.)
ALE 4:307; FEF 74, 75 (bw); LAW 314; SFW
80 (cp 11), 105 (pl 18); WFW cp 73.

Silver Dollar Plant (Crassula arborescens)
BIP 63 (cp 87); DFP 61 (cp 483); ECS 110;
LCS cp 234; RDP 158.

Silver Torch (Cleistocactus strausii) ECS
108; LCS cp 300; RDP 146; (C. baumannii)
SCS cp 11; (C. jujuyensis) LCS cp 14; (C.
wendlandiorum) DEW 1:201.

Silver Tree (Leucadendron argenteum) MWF
182; PFW 245.

Silvereye —See **White-eye.**

Silverfish (fish) (Monodactylus sebae) ALE
5:111; FEF 365 (bw); HFI 30.

Silverfish (insect) (Lepisma sp.) ALE 2:73;
BIN pl 3; GAL 474 (bw); KIW 33 (cp 1); LEA
127 (bw); OBI 1; PEI 30 (bw); (Machilis sp.)
ALE 2:73; GAL 476 (bw).

Silver-lace Vine (Polygonum aubertii) EGV
47, 131; EPG 138; MWF 239; PWF 86d; TGS
71.

Silversides (Bedotia geoyi) ALE 4:451, 455;
FEF 299, 330 (bw); WFW cp 214; (Pseudomugil signifer) ALE 4:451; FEF 327.

Silversword (Argyroxiphium sp.) CAU 181;
EWF cp 149A; MLP 237, 246.

Silverweed (Potentilla anserina) BWG 33;
DEW 1:297; EPF pl 405; KVP cp 81; MBF cp
27; MFF cp 42 (352); NHE 211; OWF 17;
PWF 48c; TEP 193; WFB 113 (7).

Sim-sim —See **Sesame** (Sesamum).

Singapore Holly (Malpighia coccigera) BIP
117 (cp 177); EGL 125; FHP pl 32; HPW 213;
MWF 194.

Siren, Broad-striped Dwarf (Pseudobranchus striatus) ALE 5:352; CAW 74 (cp
20).

Siren, Greater (Siren lacertina) ALE 5:352;
ART 42; ERA 78; HRA 120.

Sisal Plant (Agave sisalana) DEW 2:243;
MLP 251; WWT 98, 123; (A. rigida) PFF 385.

Siskin (Carduelis spinus) ABW 301; ALE
9:391; BBB 157; BBE 281; DPB 419; FBE
287; GAL 118 (cp 1); OBB 177; PBE 261 (cp
62); RBE 299.

Sitatunga (Tragelaphus spekei) ALE 13:302;
DDM 181 (cp 30); DSM 241; SMR 112 (cp 9).

Sitella (Neositta chrysoptera) ALE 9:319;
HAB 170; RBU 252; WAB 189; (N. pileata)
MWA 89 (bw).

Siva, Blue-winged (Minla cyanouroptera)
KBA 288 (cp 45); WBV 205.

Skate —See Ray (Raja sp.)

Skimmer, African (Rhynchops flavirostris)
ALE 8:212; GPB 169; LAW 449; PSA 176 (cp
19); SAO 135.

Skimmer, Black (Rynchops nigra) ABW 130;
CDB 95 (cp 342); SAO 139; SSA pl 23; WAB
71.

Skimmer, Indian (Rynchops albicollis) CAS
163; KBA 156 (pl 19).

Skimmia, Japanese (Skimmia japonica)
DFP 238 (cps 1902, 1903); EEG 111; EGE
147; EGW 141; EPF pl 654; ESG 142; GDS
cps 446, 447; HSC 113, 114; TGS 246.

Skink (Lygosoma sp.) ALE 6:260; HRA 17;
SIR 125 (bw); (Riopa sp.) ALE 6:244, 259;
HWR 121; WEA 339.

Skink, Blue-tongued (Tiliqua sp.) ALE
6:244, 259; 10:89; COG cp 31; ERA 196; HRA
15; MAR 145 (cp XI); SIR 148 (cps 65, 66);
WAW 75 (bw).

Skink, Bouton's Snake-eyed (Cryptoblepharus boutonii) ALE 6:267; COG cp 30.

Skink, Cylindrical or Eyed (Chalcides sp.)
ALE 6:267; DAE 145; HRA 17, 18; LAW 418;
SIR 131 (cp 64).

Skink, Mabuya (Mabuya sp.) ALE 6:244;
CAA 117 (bw); CAS 169 (bw); MAR 149 (pl
59).

Skink, Prehensile-tailed Giant (Corucia
zebrata) ALE 6:244; ERA 196.

Skink, Sand (Scincus scincus or S. philbyi)
ALE 6:244; HRA 16; SIR 132, 133 (bw); VRA
195.

Skink, Shingle-back (Trachydosaurus
rugosus) COG cp 27; HRA 16; MAR 142 (pl
58), 145 (cp XI); WAW 74 (bw).

Skink, Spiny-tailed (Egernia sp.) ALE 6:244;
COG cps 29, 33; HRA 15; LAW 418; MAR
149 (pl 59); SIR 134 (bw).

Skink, Water (Sphenomorphus sp.) COG cp
28; MWA 110.

Skipjack —See Bonito, Oceanic.

Skipper —See Saury.

Skua, Arctic (Stercorarius parasiticus) ALE
8:229; BBB 256; BBE 141; FBE 145; GPB
160; HAB 76; OBB 85; PBE 124 (pl 35); PSA
176 (cp 19); SAO 16, 97; WAB 101.

Skua, Great (Stercorarius skua or Catharacta skua) ABW 129; ALE 8:210; BBB

257; BBE 141; CDB 91 (cp 317); DAE 33;
FBE 145; HAB 77; HBG 109 (bw); LAW 456;
LEA 414 (bw); OBB 85; PBE 124 (pl 35); PSA
176 (cp 19); SAO 97; SAR 120; SSA pl 20;
WAB 223; WAN 127-131.

Skua, Long-tailed (Stercorarius longicaudatus) ALE 11:151; BBB 277; BBE 141; FBE
145; OBB 85; PBE 124 (pl 35); SAO 97.

Skua, Pomarine or Pomatorhine —See
Jaeger, Pomarine.

Skullcap (Scutellaria sp.) EPF pl 806; HPW
238; NHE 98; PFE 352; SNZ 82 (cp 234);
TGF 204.

Skullcap, Alpine (Scutellaria alpina)
EFC cp 49; NHE 281; PFE cp 108 (1107); (S.
orientalis) PFE cp 108 (1107).

Skullcap, Common (Scutellaria galericulata) BFB 17 (9); MBF cp 68; MFF cp 11
(78); NHE 98; OWF 177; PFE cp 108 (1108);
PWF 85g; WFB 197 (3).

Skullcap, Lesser (Scutellaria minor) MBF cp
68; MFF cp 28 (782); NHE 187; OWF 145;
WFB 197 (4).

Skunk, Hog-nosed (Conepatus mesoleucus)
ALE 12:84; DSM 174.

Sky Flower or Vine (Thunbergia grandiflora) BIP 175 (cp 271); EWF cp 122b; HPW
250; LFW 181 (cp 407); MEP 147; PFW 295.

Skylark (Alauda arvensis) ABW 215; ALE
9:169, 179, 10:223, 11:321; BBB 85; BBE
199; CDB 147 (cp 599); FBE 205; HAB 142;
LAW 459, 507; OBB 123; PBE 189 (cp 48);
RBE 216; SAB 10 (cp 4); WAB 113.

Skylark, Oriental or Small (Alauda
gulgula) DPB 231; KBA 400 (pl 59); WBV
147.

Slime-mold, Wood (Lycogala epidendrum)
NFP 29; ONP 173; PFF 66.

Slipper Flower —See Redbird Cactus.

Slipper Shell (Crepidula fornicata) CSS cp
17; GSS 41; MOL 69.

Slipperflower or Slipperwort —See
Pocketbook Flower.

Slit Shell (Pleurotomaria sp.) CTS 17 (cp 1);
GSS 21.

Sloe —See Blackthorn.

Sloth, Hoffman's (Choloepus hoffmanni)
BMA 216 (bw); CSA 19; JAW 66; MLS 170
(bw).

Sloth, Three-fingered or Three-toed
(Bradypus tridactylus) ALE 11:175, 190;
BMA 215; CSA 117; DSM 104; HMA 91;
SMW 108 (bw); (B. torquatus) LVS 67; (B.
cuculliger) LAW 579.

Sloth, Two-fingered or Two-toed
(Choloepus didactylus) ALE 8:359; 11:174,
190; BMA 215; DSM 104; HMA 91; HMW
181; LEA 516 (bw); SMW 107 (bw); TWA
141.

Slow-worm (Ophisaurus sp.) —See **Lizard,
Glass.**

Slow-worm (Pygopus sp.) ALE 6:179; COG cp
13; HRA 37; MEA 28.

Slow-worm, European (Anguis fragilis)
ALE 6:300, 311; DAE 147; ERA 197; HRA
25; LAW 421; LBR 76; MAR 125 (pl 51); OBV
123; SIR 165 (bw); VRA 189.

Slug, Field (Agriolimax agrestes) GAL 510
(bw); (A. reticulatus) OIB 63.

Slug, Garden (Arion sp.) ALE 3:89, 94; GAL
509 (bw); LAW 114; LEA 81; OIB 63; WEA
341.

Slug, Spotted (Limax sp.) ALE 3:89, 94; EPD
126; GAL 510 (bw); OIB 63.

Small-reed (Calamagrostis sp.) EGW 96;
MBF pl 96; MFF pl 134 (1228, 1229); MLP
239; NHE 196, 292; TEP 263.

Smartweed, Pale (Polygonum lapathi-
folium) BFB cp 12; MBF cp 73; MCW 26;
NHE 224; OWF 127; PFE cp 8 (82); PFF 154;
PWF 160c; WFB 41 (3a).

Smelt, European (Osmerus eperlanus) CSF
81; GAL 238 (bw); HFI 123; OBV 89.

Smew (Mergus albellus) ALE 7:61, 308; BBB
199; BBE 67; CDB 53 (cp 120); FBE 67; FBW
51; OBB 29; PBE 37 (cp 12), 40 (pl 13), 48 (pl
15); (Mergellus albellus) ABW 69.

Smoke Tree (Cotinus coggyria) DEW 2:66;
DFP 192 (cps 1535, 1536); EEG 94; EET 185;
EFS 34, 105; ELG 114; GDS cps 116-118;
HSC 35, 37; MWF 95; OGF 159; PFE cp 69
(706); PFF 234; PFW 27; SST cp 96; TGS 247;
(C. obovatus) MLP 128.

Smut, Corn (Ustilago maydis) EPF pl 77;
NFP 24, 39; PFF 75; (U. sp.) DEW 3:146;
EPF 76; PFF 75.

Snail (Helicella sp.) ALE 3:89, 100; GAL 505
(bw); OIB 53, 61.

Snail, Amber (Succinea putris) ALE 3:80, 94;
GAL 506 (bw); OIB 57.

Snail, Bladder (Physa fontinalis) ALE 3:80;
OIB 71.

Snail, Copse (Arianta arbustorum) ALE
3:89; GAL 505 (bw); OIB 55.

Snail, Edible or Roman (Helix pomatia)
ALE 3:99; 10:223; GAL 504 (bw); LEA 81, 83
(bw); OIB 55; WEA 343.

Snail, Grove (Cepaea nemoralis) ALE 3:89;
ALE 10:223; GAL 504 (bw); OIB 53; (C.
hortensis) ALE 3:94; OIB 53.

Snail, Land (Achatina sp.) ALE 3:89, 94;
LAW 123.

Snail, Moon (Natica sp.) CSS pl XV; GSS 50;
OIB 41.

Snail, Olive —See **Olive Shell.**

Snail, Pond (Lymnaea sp.) ALE 3:80; GAL
511 (bw); OIB 65.

Snail, Ram's Horn (Planorbis sp.) GAL 512
(bw); LEA 90 (bw); OIB 69.

Snail, River (Viviparus sp.) ALE 3:46; GAL
514 (bw); LAW 121; OIB 67.

Snail, Valve (Valvata sp.) GAL 514 (bw); OIB
67.

Snailfish (Liparis liparis) ALE 5:66; CSF 169;
OBV 81; (L. montagui) CSS cp 30.

Snake, Aesculapian (Elaphe longissima)
ALE 6:403, 406; HRA 60; MAR 164 (pl 66);
SIR 198 (cp 96); SSW 70.

Snake, Australian Black (Pseudechis
porphyriacus) COG cp 45; SSW 112.

Snake, Black-striped (Vermicella calonota)
MAR 179 (cp XIV); MEA 24.

Snake, Blind (Typhlops sp.) ALE 6:342, 383;
COG cp 34; ERA 209; FSA 69 (pl 2), 84 (cp 5);
HRA 72; LEA 321 (bw); SSW 40; (Rampho-
typhlops sp.) MEA 14.

Snake, Boomslang (Dispholidus typus) ALE
6:429; ART 160; CAF 183; ERA 212; FSA
109 (pl 12), 112 (cp 13); HWR 23; LEA 325
(bw); MAR 173 (pl 69); SIR 244 (cp 121); SSW
88.

Snake, Brazilian Coral (Micrurus frontalis)
ALE 6:430; MAR 169 (cp XIII); SIR 233 (bw).

Snake, Broad-headed Water (Homalopsis
buccata) ALE 6:406, 419; HRA 52.

Snake, Brown House (Boaedon fuliginosus)
ALE 6:429; FSA 85 (cp 6), 100 (pl 9); HRA 57.

Snakehead (Channa asiatica) ALE 5:45; HFW 245 (bw); SFW 649 (pl 178); (Ophicephalus sp.) FEF 331 (bw); HFI 61; SFW 624 (pl 171); WFW cps 237-239.

Snake-lizard, Burton's —See **Lizard, Legless.**

Snakeroot, White (Eupatorium rugosum) PFF 311; WUS 407.

Snake's Head —See **Fritillary, Snake's Head.**

Snakeweed —See **Bistort, Snakeroot.**

Snapdragon, Common (Antirrhinum majus) DFP 30 (cps 236, 237); EDP 135; EFC cp 59; EGA 95; EGG 89; EPF pl 839; LFW 12 (cps 24, 25); MFF cp 29 (682); MWF 43; PFF 296; PFW 275; SPL cp 471; TGF 220; TVG 33; WFB 211 (1).

Snapdragon, Large (Antirrhinum latifolium) KVP cp 169; PFE cp 121 (1198); PFM cp 171.

Snapdragon, Lesser (Misopates orontium or Antirrhinum orontium) MBF cp 62; MFF cp 29 (68); NHE 232; PFE cp 122 (1199); PFM 172; PWF 129e; WFB 211 (2).

Snapdragon Tree (Gmelina sp.) GFI 135 (cp 30); MEP 123.

Snapper (Lutjanus sp.) ALE 5:102; HFW 170 (cp 80); LMF 94; MOL 120; WFW cps 296,297.

Snapper, Red or Emperor (Lutjanus sebae) FEF 351 (bw); HFW 170 (cp 78); LAW 334.

Sneezeweed (Helenium autumnale) DFP 145 (cps 1155-1158); EWG 115; MWF 149; OGF 147; PFF 324; PFW 83; SPL cp 430; TGF 285; (H. sp.) TVG 95; WUS 417.

Sneezewort (Achillea ptarmica) BFB cp 19 (5); DFP 119 (cp 951); LFW 3 (cp 5); MBF cp 45; MFF pl 94 (889); NHE 189; OWF 99; PWF 99i; TEP 45; TGF 252; TVG 77; WFB 239 (3).

Snipe (Gallinago gallinago) ALE 7:255, 385; BBB 170; BBE 119; DPB 97; FBE 141; KBA 136, 137 (bw); LAW 504; LEA 411 (bw); OBB 67; PBE 84 (pl 27), 116 (cp 33); WBV 91; WFI 156 (bw); (G. stricklandii) SSA pl 23.

Snipe, Ethiopian (Gallinago nigripennis) CDB 86 (cp 284); PSA 160 (cp 17).

Snipe, Great (Gallinago media) BBB 271; BBE 119; FBE 141; OBB 67; PBE 84 (pl 27), 116 (cp 33).

Snipe, Jack (Lymnocryptes minimus) BBB 171; BBE 119; DPB 97; FBE 141; OBB 67; PBE 84 (pl 27), 116 (cp 33); WBV 91.

Snipe, New Zealand (Coenocorypha sp.) BNZ 141; FNZ 139 (bw).

Snipe, Painted (Rostratula bengalensis) ABW 117; ALE 8:170; CDB 79 (cp 252); DPB 81; FBE 141; GPB 138-139; HAB 60; KBA 129 (pl 14), 136; PSA 160 (cp 17); SNB 88 (cp 43); WAW 27 (bw); (Nycticryphes semicollaris) SSA pl 23.

Snipe, Pintail (Gallinago stenura) DPB 97; KBA 137 (bw); WBV 87; (G. megala) DPB 97; HAB 60.

Snipe, Seed —See **Seedsnipe.**

Snipe-eel (Nemichythys scolopaceus) ALE 4:176; HFI 84; (Cyema atrum) MOL 99.

Snipefish (Macroramphosus sp.) ALE 5:40; HFI 74; MOL 111; WFW cp 228.

Snoek (Thyrsites atun) LMF 57; MOL 95.

Snow Bush (Breynia nivosa) BIP 37 (cp 38); MEP 71; MWF 62.

Snowball, Japanese (Viburnum plicatum) EFS 143; ESG 147; GDS cps 496, 497; MWF 293; PFW 69; SPL cp 122; (V. tomentosum) DFP 244 (cps 1949, 1950); EEG 103; EFS 22, 31, 143; ESG 147; HSC 125; SPL 123; TGS 343.

Snowbell, Fragrant (Styrax obassia) DEW 1:254; SPL cp 118; (S. hemsleyana) DFP 240 (cp 1918).

Snowbell, Japanese (Styrax japonica) DFP 24 (cp 1919); EET 201; EGT 145; EJG 146; EPF pl 496; LFW 72 (cp 167); PFW 290; TGS 375.

Snowberry (Symphoricarpos albus) DEW 2:138; DFP 140 (cp 1920); EFS 29, 139; GDS cp 467; HSC 117, 118; MWF 277; PFW 68; PWF 81d; SPL cp 119; TGS 374.

Snowberry (Symphoricarpos rivularis) EPF pl 776; MFF cp 17 (823); NHE 25; PFE cp 135 (1301); PWF 165g.

Snowcock (Tetraogallus sp.) ALE 7:481; 12:103; FBE 101; RBI 7.

Snowdrop (Galanthus nivalis) DEW 2:200 (cp 135); DFP 93 (cp 744), 94 (cp 745); EFC cp 1; EGB 115; EGW 79, 113; EMB 101; EPF pl 924; MBF cp 83; MFF pl 81 (1044); MLP 258; NHE 192; OGF 7; PFE cp 173 (1663); PFF 384; PFW 23; PWF cp 12b; TEP 261; TGF 12; WFB 271 (6).

Snowdrop, Giant (Galanthus elwesii) OFP 93 (cp 742); EWF cp 42e; LFW 52 (cp 121); TVG 195.

Snowflake, Spring (Leucojum vernum) DFP 100 (cp 796); EFC cp 1; EPF pl 925; ESG 123; HPW 315; MBF cp 83; MFF pl 81 (1042); MWF 182; NHE 192; OGF 7; PFE cp 172 (1661); PFW 23; PWF cp 12e; RWF 20; TEP 263; TGF 12; TVG 199; WFB 271 (4).

Snowflake, Summer (Leucojum aestivum) BFB cp 23 (3); DEW 2:244; DFP 100 (cp 794); EGB 125; LFW 80 (cp 184); MBF cp 83; MFF pl 181 (1043); NHE 192; PFE cp 172 (1662); PWF cp 21n; WFB 271 (5).

Snowflake, Water —See **Water-lily, Fringed.**

Snow-in-summer (Arabis albida) DFP 2 (cp 13); EGC 121; ERW 97; MWF 44; SPL cp 352; TGF 92.

Snow-in-summer (Cerastium tomentosum) EGC 126; LFW 22 (cp 44); MWF 78; TGF 71; TVG 145.

Snow-on-the-mountain (Euphorbia marginata) EGA 118; EPF pl 629; PFF 232; TGF 109.

Soap-fish, Golden-lined —See **Grouper, Six-lined.**

Soapwort —See also **Bouncing Bett.**

Soapwort (Saponaria sp.) MCW 39; NHE 277; PFM cp 17; TVG 169.

Soapwort, Rock (Saponaria ocymoides) DFP 24 (cp 187); ERW 133; NHE 277; OGF 41; PFE cp 16 (180).

Soft-grass, Creeping (Holcus mollis) MBF pl 96; NHE 40; PFW 131.

Soldanella, Alpine (Soldanella alpina) DEW 1:229; DFP 27 (cp 210); HPW 135; NHE 278; WFB 174; (S. carpatica) TVG 175; (S. villosa) OGF 21.

Soldanella, Dwarf or Lesser (Soldanella pusilla) KVP cp 153; NHE 278; PFE cp 91 (956); RWF 17; (S. minima) NHE 278.

Soldanella, Mountain (Soldanella montana) DFP 27 (cp 211); EPF pl 264; NHE 278; TEP 249.

Soldierfish, Black-tipped (Myripristis murdjan) ALE 5:35; FEF 324 (bw); HFI 65.

Soldierfish, Deep-water (Ostichthys japonicus) HFI 65; HFW 85 (cp 40).

Soldiers and Sailors —See **Lungwort, Common.**

Sole (Solea solea) ALE 5:233; CSF 193; HFI 19; HFW 275 (bw); OBV 51; (S. fulvomarginata) ALE 5:234.

Sole, Lemon (Microstomus kitt) CSF 191; OBV 49.

Solenette (Buglossidium luteum) CSF 193; HFI 19; OBV 51; (Microchirus sp.) CSS pl XXIV; OBV 51.

Solomon's Seal (Polygonatum sp.) AMP 169; EWG 134; PFW 172.

Solomon's Seal, Common (Polygonatum multiflorum) MBF cp 84; MFF pl 81 (991); MWF 238; NHE 38; OGF 51; PFE cp 171 (1655); PWF 43e; SPL cp 489; TEP 261; WFB 267, 269 (1).

Solomon's Seal, False —See **Spikenard, False.**

Solomon's Seal, Great (Polygonatum canaliculatum) EWG 27; LFW 213 (cp 476).

Solomon's Seal, Sweet-scented (Polygonatum odoratum) KVP cp 54; MBF cp 84; NHE 38; PFE cp 171 (1654).

Solomon's Seal, Whorled (Polygonatum verticillatum) MBF cp 84; MFF pl 81 (990); NHE 290; WFB 268, 269 (2).

Son-before-the-father —See **Coltsfoot.**

Sonerila (Sonerila margaritacea) BIP 167 (cp 257); DFP 82 (cp 649); EPF pl 712; EWF cp 121a; HPW 164; PFW 187; RDP 366.

Song of India (Pleomele reflexa) DFP 79 (cp 628); EFP 19, 136; RDP 323.

Songlark, Brown (Cinclorhamphus cruralis) HAB 179; SAB 10 (cp 4).

Songlark, Rufous (Cinclorhamphus mathewsi) CDB 177 (cp 759); HAB 179; SAB 10 (cp 4).

Sore Eye Flower —See **Cape Poison.**

Sorrel, Common (Rumex acetosa) BWG 168; MWF cp 74; MFF pl 107 (540); NHE 183; OFP 191; OWF 59; PWF 37i; TEP 141; WFB 45 (1).

Sorrel, French —See **Dock, Shield.**

Sorrel, Mountain (Oxyria digyna) HPW 78; MBF cp 74; MFF pl 107 (538); NHE 276; OWF 59; PWF 117j; WFB 45 (8).

Sorrel, Sheep's (Rumex acetosella) EGC 73;
MBF cp 74; MCW 29; MFF pl 107 (539); NHE
183; OWF 59; PFF 152; PWF 37h; WFB 45
(2); WUS 129.

Sorrel, Upright Yellow (Oxalis europaea)
MBF cp 20; MFF cp 43 (254); NHE 31; PWF
61f.

Sour Fig —See **Fig, Hottentot.**

Soursop (Annona muricata or A. squamosa)
DEW 1:59 (cp 27), 82-83; HPW 30; OFP 97.

Souslik, Common European (Citellus
citellus) ALE 11:239; CAS 73; CEU 147
(bw); DAE 82; GAL 57 (bw); HMW 162; MLS
202 (bw); SMW 94 (cp 65); TWA 90.

Souslik, Spotted (Citellus suslica) ALE
11:239; CEU 123 (bw); DAE 82; HMW 162.

Southern Star —See **Star-of-the-
Argentine.**

Sowbane —See **Goosefoot.**

Sowbread —See **Cyclamen.**

Sowbug (Armadillidium sp.) ALE 1:496; GAL
479 (bw); GSP 152; LEA 170; OIB 165;
(Idotea sp.) CSS pl VII; OIB 125.

Sowbug (Porcellio sp.) ALE 2:51; EPD 127;
GAL 479 (bw); GSP 153; LAW 147; OIB 167;
(Oniscus sp.) ALE 1:496; GSP 153; OIB 153.

Sowbug, Water (Asellus sp.) GAL 495 (bw);
LAW 147; LEA 169 (bw); OIB 133.

Sow-thistle, Alpine or Blue (Cicerbita
alpina or A. macrophylla) EFC cp 39; EWF cp
22f; KVP cp 151; MBF cp 53; MFF cp 9 (945);
NHE 288; PFE cp 159 (1537); PWF 112b;
WFB 251 (6).

Sow-thistle, Common or Smooth (Sonchus
oleraceus) BFB cp 21 (10); BWG 55; MBF cp
53; MCW 128; MFF cp 53 (943); NHE 240;
OWF 35; PFF 333; PWF 87l; TEP 179; WFB
253i.

Sow-thistle, Field (Sonchus arvensis) BFB cp
21 (3); MBF cp 53; MCW 126; MFF cp 53
(942); NHE 240; OWF 35; PFE cp 159 (1538);
PWF 105i; WFB 253 (2); WUS 433.

Sow-thistle, Marsh (Sonchus palustris) MBF
cp 53; MFF cp 53 (941); NHE 101; PWF 146a;
WFB 253 (3).

Sow-thistle, Prickly (Sonchus asper) BWG
56; MBF cp 53; MCW 127; MFF cp 53 (944);
NHE 240; PFE cp 159 (1539); PWF 87k;
WFB 252; WUS 435.

Soybean or **Soya** (Glycine max or G. soya)
EPF pl 450; OFP 25; PFF 219.

Spadefoot, European (Pelobates fuscus)
ALE 5:375, 399; ART 75; CAW 76 (cp 24);
HRA 108.

Spanish Moss (Tillandsia usneoides) EWF cp
188a; PFF 368; RDP 383.

Spanish Shawl (Schizocentron elegans) FHP
143; RDP 351.

Sparrow, Cape (Passer melanurus) ALE
9:416; CDB 214 (cp 955); NBA 89; PSA 305
(cp 36).

Sparrow, Dead Sea (Passer moabiticus) ALE
9:416; FBE 299.

Sparrow, Desert (Passer simplex) ALE
9:416; FBE 299.

Sparrow, Hedge (Prunella modularis) ABW
262; ALE 9:221; BBB 56; BBE 221; CDB 161
(cp 674); FBE 221; FNZ 208 (cp 18); GAL 128
(cp 3); GPB 268; LAW 470, 472; OBB 145;
PBE 276 (cp 63); RBE 276; WEA 40.

Sparrow, House or English (Passer
domesticus) ABW 304; ALE 9:416; BBB 45;
BBE 277; CDB 214 (cp 954); DAE 133; FBE
299; GAL 118 (cp 1); GPB 324; HAB 256;
NBA 89; OBB 191; PBE 276 (cp 63); PSA 305
(cp 36); SAB 70 (cp 34).

Sparrow, Java (Padda oryzivora) ALE 9:439;
CDB 212 (cp 941); DPB 413; KBA 401 (pl 60);
WBV 221.

Sparrow, Orange-billed (Arremon aurantirostris) ALE 9:342; DTB 145 (cp 72); SSA cp
42; (A. schlegeli) CDB 195 (cp 851).

Sparrow, Plain-backed (Passer flaveolus)
KBA 433 (cp 64); WBV 221.

Sparrow, Rock (Petronia petronia) BBE 277;
CDB 215 (cp 957); DAE 78; FBE 301; PBE
276 (cp 63); (P. brachydactyla) FBE 301; (P.
xanthocollis) ALE 9:416; FBE 301.

Sparrow, Rufous-collared (Zonotrichia
capensis) SSA cp 18; WFI 202 (bw).

Sparrow, Spanish (Passer hispaniolensis)
BBE 277; FBE 299; PBE 276 (cp 63).

Sparrow, Tree (Passer montanus) ABW 304;
ALE 9:342; 10:223; BBB 87; BBE 277; CDB
214 (cp 956); FBE 299; DPB 413; GAL 118
(cp 1); HAB 256; KBA 433 (cp 64); OBB 191;
PBE 276 (cp 63); SAB 70 (cp 34); WBV 221.

Sparrow-hawk (Accipiter collaris) WAB 90;
(A. cirrocephalus) SNB 42 (cp 20), 50 (pl 24).

Sparrow-hawk, African Little (Accipiter minullus) ALE 7:355; PSA 112 (cp 11).

Sparrow-hawk, European or Northern (Accipiter nisus) ALE 7:332, 355; BBB 129; BBE 81; CDB 55 (cp 133); DAE 117; FBE 75; GAL 153 (bw); KBA 65 (pl 6); LEA 385; LVB 51; OBB 41; PBE 53 (cp 18), 73 (pl 24); RBE 11; WAB 106; WBV 61.

Sparrow-hawk, Levant (Accipiter brevipes) BBE 81; FBE 75.

Sparrow-hawk, Philippine (Accipiter virgatus) DPB 45; KBA 65 (pl 6).

Sparrow-weaver, White-browed (Ploce-passer mahali) CDB 215 (cp 959); PSA 305 (cp 36).

Spathularia (Spathularia flavida) CTM 71; DEW 3:124; LHM 45; RTM 239.

Spatterdock —See **Water-lily** (Nuphar sp.).

Spearmint (Mentha spicata) BWG 148; EGH 125; MBF cp 67; MFF cp 11 (748); OFP 141; OWF 143; PFF 291; PWF 142b; WFB 205 (3).

Spearwort, Great (Ranunculus lingua) BFB cp 2 (6); DFP 169 (cp 1345); MBF cp 3; MFF cp 39 (18); NHE 91; PFE cp 25 (244); PFW 252; PWF 107k; TEP 29; WFB 69 (6).

Spearwort, Lesser (Ranunculus flammula) MBF cp 3; MFF cp 39 (19); NHE 91; OWF 3; PWF 74b; WFB 69 (7).

Speedwell (Veronica sp.) DFP 177 (cps 1414, 1416); EMB 64; LFT cps 168-170; LFW 144 (cp 322), 145 (cp 325); MFF cps 2 (711, 712), 3 (715); NHE 233, 234, 281, 282; OGF 41, 161; TVG 125; WFB 215 (4c, 5); WUS 341.

Speedwell, Alpine (Veronica alpina) MBF cp 64; MFF cp 3 (709); NHE 281; WFB 215 (6b).

Speedwell, Buxbaum's (Veronica persica) BWG 1 (32); MBF cp 163; MFF cp 2 (718); NHE 234; OWF 175; PFE cp 124 (1223); PWF 29i; WFB 215 (4).

Speedwell, Common (Veronica officinalis) EGH 149; MBF cp 64; MFF cp 2 (704); NHE 233; OWF 175; PWF 68a; TEP 255; WFB 215 (7); WUS 339.

Speedwell, Creeping (Veronica filiformis) MBF cp 63; MFF cp 2 (721); PFE cp 124 (1224); PWF cp 15k; WFB 215 (4d); (V. repens) EGC 149.

Speedwell, Field (Veronica agrestis) MBF cp 63; MFF cp 2 (720); NHE 234; PWF 29g; WFB 215 (4a).

Speedwell, Fingered (Veronica triphyllos) MBF cp 63; MFF cp 2 (716); NHE 234; PFE 384.

Speedwell, Germander or Bird's-eye (Veronica chamaedrys) BWG 132; EPF pl 843; EWF cp 21e; MBF cp 64; MFF cp 2 (706); NHE 234; OWF 175; PWF 29j; TEP 121; WFB 215 (3).

Speedwell, Gray (Veronica polita) MBF cp 63; MFF cp 2 (719); NHE 234; WFB 215 (4b).

Speedwell, Ivy-leaved (Veronica hederi-folia) MBF 63; MFF cp 2 (717); NHE 234; PWF 29e.

Speedwell, Large (Veronica latifolia) PFE 384; TGF 220.

Speedwell, Long-leaved (Veronica longi-folia) DEW 2:155; DFP 178 (cp 1417); NHE 99.

Speedwell, Marsh (Veronica scutellata) MBF cp 64; MFF pl 105 (703); NHE 99; OWF 93; PWF 153i.

Speedwell, Pink Water (Veronica catenata) MBF cp 64; NHE 99; WFB 215 (10a).

Speedwell, Prostrate (Veronica prostrata) DFP 28 (cp 219); NHE 234; TVG 179.

Speedwell, Rock or Shrubby (Veronica fruticans) EPF pl 844; HPW 244; MBF cp 64; MFF cp 3 (708); NHE 281; PFE cp 124 (1220); WFB 215 (3b).

Speedwell, Saw-leaved (Veronica teucrium) DFP 178 (cps 1418, 1419); NHE 234; OGF 109; WFB 215 (2).

Speedwell, Spiked (Veronica spicata) EEG 139; MBF cp 64; MFF cp 9 (707); MWF 292; NHE 136; OGF 81; PWF 114e; TEP 123; TGF 220; WFB 215 (1).

Speedwell, Spring (Veronica verna) MBF cp 63; MFF cp 2 (714); NHE 234.

Speedwell, Thyme-leaved (Veronica serpyllifolia) MBF cp 64; MFF cp 2 (710); NHE 233; OWF 175; PFF 297; PWF 29h; WFB 215 (6).

Speedwell, Veronica (Veronica gentian-oides) DFP 177 (cp 1415); PFW 279; SPL cp 378.

Speedwell, Wall (Veronica arvensis) MBF cp 64; MFF cp 2 (713); NHE 233; OWF 175; PWF 29f; WFB 215 (8); WUS 337.

Speedwell, Water (Veronica anagallis-aquatica) MBF cp 64; MFF cp 11 (702); OWF 175; PWF 91i; WFB 215 (10).

Speedwell, Wood (Veronica montana) MBF cp 64; MFF cp 2 (705); NHE 36; PWF 28a; WFB 215 (3a).

Speedy Jenny —See **Wandering Jew** (Tradescantia sp.)

Spelerpes, Brown —See **Salamander, Brown Cave.**

Sphedamnocarpus (Sphedamnocarpus pruriens) HPW 213; LFT cp 91.

Spiceberry —See **Coral Berry.**

Spider (Achaearanea sp.) GSP 40; KBM 82 (pl 26); OIB 163; (Amaurobius sp.) GSP 111; OIB 151; (Gasteracantha sp.) ALE 1:400, 410; CAS 251 (bw); GSP 66; LEA 192; (Micrommata sp.) ALE 1:400; GSP 92; OIB 165.

Spider, Barrel (Galeodes sp.) LAW 140; PEI 14 (bw).

Spider, Black Widow (Latrodectus mactans tredecimguttatus) ALE 1:399; GSP 17, 42; LAW 134; PEI 12 (bw).

Spider, Cave Orb-weaver (Meta sp.) GSP 60; OIB 151.

Spider, Cellar (Segestria senoculata) ALE 1:399; GSP 27; OIB 151.

Spider, Crab (Misumena vatia) GSP 94; LAW 136; OIB 165 (Philodromus sp.) GSP 97; OIB 159; (Thomisus sp.) GSP 95; OIB 163; PEI 32 (cp 3); (Xysticus sp.) ALE 1:400; GAL 489 (bw); GSP 96; OIB 159.

Spider, European House (Tegenaria domestica) ALE 1:399; GAL 485 (bw); GSP 74; (T. saeva) OIB 149.

Spider Flower —See also **Glory Bush.**

Spider Flower (Cleome spinosa or C. hasslerana) DEW 1:171 (cp 9); DFP 34 (cp 270); EGA 107; EPF pl 239; EWF cp 174b; LFW 24d (cp 534); MEP 41; MWF 87; OGF 127; PFW 75; SPL cp 196; TGF 92; TVG 41; (C. sp.) EWF cp 52b; HPW 118; LFT cp 73.

Spider, Garden (Araneus sp.) ALE 1:399, 408-411; GAL 480 (bw); GSP 53, 56-59; LAW 132, 137; LEA 182 (bw); OIB 151; PEI 16 (cp 1b); WEA 159; (Singa sp.) OIB 163.

Spider, Golden Silk (Nephila sp.) GSP 65; PWI 10 (bw).

Spider, Harvest —See **Harvestman.**

Spider, Jumping (Hasarius adansoni) GSP 104; PEI 12 (bw); (Epiblemum scenicum or Salticus scenicus) ALE 1:399; GAL 490 (bw); GSP 99; (Plexippus paykulli) GSP 104.

Spider, Long-bodied Cellar (Pholcus phalangioides) ALE 1:415; GSP 33; OIB 149.

Spider, Pirate (Dolomedes fimbriatus) ALE 2:486; GSP 80; (Ero furcata) GSP 50; OIB 163.

Spider Plant (Chlorophytum comosum) BIP 53 (cp 63); DFP 56 (cp 447); EDP 132; EFP 99; EGL 94; LHP 56, 57; RDP 139.

Spider, Platform (Linyphia pusilla) GSP 47; (L. hortensis) OIB 159.

Spider, Sea —See **Sea Spider.**

Spider, Spitting (Scytodes sp.) GSP 28; OIB 149.

Spider, Wasp or Zebra (Argiope sp.) ALE 1:399, 408; CEU 214; GSP 53, 68, 69; LAW 135-137; MWA 107.

Spider, Water (Argyroneta aquatica) ALE 1:410; GAL 486 (bw); GSP 76; LAW 139; LEA 183 (bw); OIB 133; WEA 388.

Spider, Wolf (Lycosa amentata) GAL 487 (bw); OIB 151; (L. narboensis) PWI 10 (bw); (Pisaura mirabilis) ALE 1:399, 408; (Arctosa perita) OIB 163; (Geolycosa blackwalli) WEA 394; (Pardosa sp.) ALE 2:51; GSP 10.

Spiderfish (Bathypterois ventralis) ALE 4:266; HFI 116.

Spider-hunter, Lesser Yellow-eared (Arachnothera chrysogenys) ALE 9:320; KBA 401 (pl 60).

Spider-hunter, Little (Arachnothera longirostra) ABW 278; DPB 391; KBA 401 (pl 60); WBV 215.

Spider-hunter, Streaked (Arachnothera magna) CDB 188 (cp 819); KBA 401 (pl 60).

Spider-lily (Hymenocallis sp.) BIP 103 (cp 154); DEW 2:245; EGB 121; EPF pl 931; LFW 202 (cp 456); MWF 159; PFW 26.

Spider-monkey —See **Monkey, Spider.**

Spiderwort —See also **Wandering Jew** (Tradescantia sp.).

Spiderwort (Anthericum ramosum) NHE 38; WFB 260; (A. fasciculatum) LFT 30 (cp 14).

Spiderwort, Common (Tradescanthia virginiana) DEW 2:257; DFP 175 (cp 1400), 176 (cp 1401); EPF pl 1025; EWG 22, 142; MWF 286; OGF 109; SPL cp 349; TGF 36; (T. sp.) HPW 280; LFW 200 (cp 453); PFF 369.

Spiderwort, Mountain —See Lily, Snowdon.

Spiderwort, Queen's (Dichorisandra reginae) BIP 69 (cp 98); EWF cp 187b; RDP 175; (D. thrysiflora) MWF 111.

Spignel (Meum athamanticum) MBF cp 39; MFP pl 98 (500); NHE 182; PFE 287; PWF 83j; WFB 159 (7).

Spikenard, False (Smilacina racemosa) DFP 173 (cp 1380); EWG 139; OGF 51; PFF 381.

Spike-rush (Eleocharis sp.) MBF cp 91; NHE 105, 138; PFF 361.

Spike-rush, Common (Eleocharis palustris) MBF cp 91; MFF pl 116 (1112); NHE 105; PFE cp 185 (1846).

Spike-rush, Many-stemmed (Eleocharis multicaulis) MBF cp 91; MFF pl 116 (1111); NHE 77.

Spike-rush, Slender (Eleocharis acicularis) MBF cp 91; MFF pl 116 (1110); NHE 105.

Spindle Shell (Neptunea antiqua) CSS pl XVI; OIB 43; (Fusinus sp.) CTS 48 (cps 83, 84); GSS 86.

Spindle Tree, Common (Euonymus europaeus) AMP 73; BFB cp 7 (9); DFP 201 (cp 1604); EFC cp 7; EPF pls 661, 662; GDS cp 186; HSC 51; MBF cp 20; MFF pl 71 (264); NHE 22; OWF 181; PFE cp 71 (718); PFW 73; PWF 58a, 172c; TEP 281; TGS 177; TVG 225; WFB 141 (3).

Spindle Tree, Japanese (Euonymus japonica) EFP 110; EGL 107; GDS cp 190; RDP 196; SPL cp 87; TGS 176; TVG 225.

Spindle Tree, Winged (Euonymus alatus) DEW 2:76; DFP 201 (cp 1603); EFS 30, 111; ELG 115; EPG 120; HSC 51; TGS 182.

Spinebill, Eastern (Acanthorhynchus tenuirostris) CDB 193 (cp 835); HAB 233; SAB 56 (cp 27).

Spinebill, Western (Acanthorhynchus superciliosus) HAB 233; MBA 16, 18; MEA 61; MWA 34; RBU 275; SAB 56 (cp 27).

Spinefoot —See Rabbitfish (Siganus sp.).

Spinetail (Synallaxis sp.) ABW 198; ALE 9:125; SSA cp 35; WAB 90.

Spinifex-bird (Eremiornis carteri) HAB 195; SAB 26 (cp 12).

Spiral Flag (Costus sanguineus) EDP 144; EFP 103.

Spirea, Blue or Bluebeard (Caryopteris clandonensis) DFP 187 (cp 1496), 188 (cp 1497); EFS 100; GDS cp 53; HSC 24, 27; OGF 171; (C. incana) SPL cp 418; TGS 342.

Spirea, Bumalda (Anthony Waterer) (Spiraea x bumalda) DFP 240 (cp 1915); EEG 102; EFS 25, 78, 136; GDS cp 460; HSC 115; LFW 132 (cp 299); PFW 260; TGS 231; TVG 247.

Spirea, False —See also Astilbe.

Spirea, False (Sorbaria sorbifolia) EFS 24, 136; HSC 114; TGS 326.

Spirea, Japanese (Spiraea japonica) EPF pl 395; GDS 461; SPL cp 328.

Spirea, Willow (Spiraea salicifolia) MFF cp 17 (340); PFE cp 44 (419); PWF 109m.

Spleenwort, Black (Asplenium adiantum-nigrum) EPF pl 111; MFF pl 139 (1283); NHE 80; ONP 73.

Spleenwort, Ebony (Asplenium platy-neuron) CGF 91; EGF 96; EGW 91; NFP 136.

Spleenwort, Forked (Asplenium septentrio-nale) MFF pl 140 (1288); NHE 80; TEP 87.

Spleenwort, Green (Asplenium viride) CGF 99; NHE 293; ONP 73.

Spleenwort, Maidenhair (Asplenium trichomanes) CGF 95; EGF 15, 96; ESG 94; MFF pl 140 (1286); NFP 136; NHE 245; ONP 73; PFF 99; TEP 87.

Spleenwort, Narrow-leaved (Athyrium pycnocarpon) CGF 145; EGF 97.

Spleenwort, Scott's (Asplenium ebenoides) CGF 107; EMB 120.

Spleenwort, Sea (Asplenium marinum) MFF pl 140 (1285); ONP 1.

Spleenwort, Silvery (Athyrium thelypteri-oides) CGF 113; EGF 98; NFP 137.

Split Gill (Schizophyllum commune) DEW 3:157; DLM 158; KMF 59; LHM 105; NFP 71; ONP 125; PFF 86.

Split Rock —See Mimicry Plant.

Sponge (Calyx nicaeensis) ALE 1:149; LEA
17; (Axinella sp.) ALE 1:145; VWS 60 (bw);
(Hymeniacidon sp.) LAW 43, 44; OIB 5;
(Cliona sp.) LAW 40; OIB 5.

Sponge (Myxilla sp.) ALE 1:148; CSS cp 1;
(Tethya sp.) LAW 38; MOL 51.

Sponge, Bread-crumb (Halichondria sp.)
CSS cp 1; MOL 51; OIB 5.

Sponge, Calcareous (Clathrina sp.) ALE
1:149; LAW 36.

Sponge, Purse (Sycon sp.) CSS cp 1; OIB 5;
(Grantia sp.) CSS cp 1; OIB 5.

Spoonbill, African (Platalea alba) BBT 63;
CAA 31 (bw); CAF 112 (bw); LEA 374 (bw);
PSA 48 (cp 3).

Spoonbill, Black-faced or Lesser (Platalea
minor) DPB 25; KBA 33 (pl 2); WBV 41.

Spoonbill, European or White (Platelea
leucorodia) ALE 7:237; BBB 221; BBE 41;
CDB 48 (cp 84); CEU 141 (bw), 145; FBE 41;
FBW 61; GPB 75; LAW 482; NBB 93 (bw);
OBB 13; PBE 9 (cp 4); RBE 80; SBB 18 (pl
15); SLP 34; WAB 177.

Spoonbill, Roseate (Ajaia ajaja) ABW 60;
ALE 7:237; CDB 47 (cp 78); NBB 10; WAB
35, 66.

Spoonbill, Royal (Platalea regia) HAB 22;
SNB 28 (cp 13).

Spoonbill, Yellow-billed (Platalea flavipes)
HAB 22; SNB 28 (cp 13).

Spotted Dog —See **Lungwort, Common.**

Sprat (Sprattus sprattus) CSF 69; OBV 23;
(Clupea sprattus) HFI 119.

Spring Beauty (Montia perfoliata) BWG 70;
MBF cp 16; PFE cp 11 (124); PWF 41l; WFB
43 (7); (Claytonia perfoliata) HPW 75; MFF
pl 80 (199).

Spring Bell (Sisyrinchium striatum) DFP
173 (cp 1379); SPL cp 375.

Springbok or **Springbuck** (Antidorcas
marsupialis) ALE 13:391, 426; CAA 78-79
(bw); CAF 206 (bw); DDM 240 (cp 39); DSM
267; HMA 54; HMW 78; JAW 200; LEA 613
(bw); MLS 419 (bw); SMW 268 (bw); TWA 59
(bw).

Springhare (Pedetes sp.) ALE 11:284; DDM
49 (cp 4); DSM 115; HMW 164; LEA 532
(bw); MLS 207 (bw); SMR 144 (cp 11); SMW
121 (bw).

Springtail (Onchiurus sp.) ALE 2:73, 97; GAL
444 (bw); (Orchesella sp.) ALE 2:73, 97; PEI
24 (bw); (Sminthurus sp.) ALE 2:73; OBI 1;
PEI 25 (bw); (Arthropleona sp.) ZWI 99.

Sprosser —See **Nightingale, Thrush.**

Spruce, Likiang (Picea likiangensis) DFP
254 (cp 2028); MTB 129 (cp 10).

Spruce, Morinda or West Himalayan
(Picea smithiana) MTB 129 (cp 10); OBT 117.

Spruce, Norway (Picea abies) BKT cps 19,
29; DEW 1:32, 55 (cp 17); EET 108; EGW 6;
EPF pls 127, 128; GDS cp 316; MTB 144 (cp
11); NHE 18; OBT 100, 108; OWF 185; PFE
cp 1 (2); PFF 119; PWF 169l; SST cp 26; TEP
265; TGS 38; WFB 25 (2).

Spruce, Oriental (Picea orientalis) BKT cp
34; EET 111; MTB 144 (cp 11); OBT 117;
TGS 44.

Spruce, Sargent (Picea brachytyla) EET 111;
OBT 117.

Spruce, Serbian (Picea omorika) BKT cp 30;
DFP 254 (cp 2029); EET 110, 111; EGE 101;
EPF pl 129; MTB 144 (cp 11); OBT 109; SST
cp 27; TGS 43.

Spruce, Tiger-tail (Picea polita) EET 111;
MTB 129 (cp 10).

Spur Dog —See **Dogfish, Piked or Spiny.**

Spurge (Euphorbia wulfenii) MWF 130; OGF
13.

Spurge, Broad-leaved Glaucous
(Euphorbia myrsinites) DFP 9 (cp 69); PFM
cp 94; TVG 151.

Spurge, Caper (Euphorbia lathyris) BWG
166; MBF cp 75; MFF pl 109 (516); PFE cp 67
(676); PWF 43j; WFB 137 (4).

Spurge, Cushion (Euphorbia epithymoides)
DFP 142 (cp 1130); EDP 112; PFW 115; TVG
91.

Spurge, Cypress (Euphorbia cyparissias)
EGC 132; MBF cp 75; MCW 67; MFF pl 109
(525); NHE 223; PFF 232; PWF 56a; TEP 113.

Spurge, Dwarf (Euphorbia exigua) MBF cp
75; MFF pl 109 (521); NHE 224; PWF 111e;
WFB 137 (8).

Spurge, Greek Spiny (Euphorbia acanthothamnos) PFE cp 66 (670); PFM cp 89.

Spurge, Hungarian (Euphorbia esula) EPF
pl 627; MCW 68; MFF pl 109 (524); NHE
183; WUS 249.

Spurge, Irish (Euphorbia hyberna) MBF cp 75; MFF pl 109 (517); PWF 56b; WFB 137 (5).

Spurge, Japanese (Pachysandra terminalis) DFP 162 (cp 1294); EEG 145; EGC 36, 142; EJG 134; ELG 147; ESG 61, 134; HPW 184; HSC 82; LFW 213 (cp 477); PFW 57; TGF 122.

Spurge, Large Mediterranean or Hedge (Euphorbia characias) DFP 142 (cp 1129); GDS cp 191; KVP cp 179; PFE cp 66 (678); PFM cp 93.

Spurge, Leafy —See **Spurge, Hungarian.**

Spurge, Marsh (Euphorbia palustris) DFP 142 (cp 1132); NHE 96.

Spurge, Petty (Euphorbia peplus) BWG 167; NHE 224; PWF cp 15h; WFB 137 (7).

Spurge, Portland (Euphorbia portlandica) MBF cp 75; MFF pl 109 (522); PWF 70a; WFB 137 (3).

Spurge, Purple (Euphorbia peplis) MBF cp 75; MFF cp 27 (515); NHE 139; WFB 137 (9).

Spurge, Sea (Euphorbia paralias) KVP cp 107; MBF cp 75; MFF pl 109 (523); NHE 139; PFE cp 66 (684); PWF 124c; WFB 137 (2).

Spurge, Spiny or Thorny (Euphorbia spinosa) NHE 81; PFM cp 88.

Spurge, Sun (Euphorbia helioscopia) MBF cp 75; MFF pl 109 (519); NHE 223; OWF 63; PFE cp 67 (671); PWF cp 23k; TEP 177; WFB 137 (6).

Spurge, Tree (Euphorbia dendroides) NHE 81; PFE cp 66 (669); PFM cp 90.

Spurge, Wood (Euphorbia amygdaloides) BFB cp 14 (2); HPW 186; MBF cp 75; MFF pl 65 (526); NHE 33; OWF 63; PFE cp 66 (677); PWF cp 24a; WFB 137 (1).

Spurge-laurel (Daphne laureola) DEW 1:270; MBF cp 76; MFF pl 73 (422); NHE 267; OWF 191; PFE cp 74 (760); PFW 296; PWF cp 25l; TGS 263; WFB 139 (7); (D. striata) DFP 195 (cp 1557); NHE 267.

Spurrey (Spergula sp.) NHE 73, 185; PFE 81.

Spurrey, Cliff or Rock (Spergularia rupicola) MBF cp 16; PFE cp 13 (156); PWF 81b.

Spurrey, Corn (Spergula arvensis) MBF cp 16; MCW 42; MFF pl 87 (190); NHE 226; OWF 77; PWF 160b; WFB 57 (1); WUS 167.

Spurrey, Greater Sea (Spergularia media) KVP cp 109; MBF cp 16; NHE 135.

Spurrey, Lesser Sea (Spergularia marina) MBF cp 16; MFF cp 27 (192); NHE 135; PWF 106b; TEP 57; WFB 57 (2); (S. marginata) OWF 113.

Spurrey, Sand or Red (Spergularia rubra) MBF cp 16; MFF cp 27 (191); NHE 226; PFE cp 13 (157).

Squash (Cucurbita maxima) EGA 122; EMB 147; EPF pls 511, 512; EVF 107; MLP 234; OFP 123; PFF 308; (C. moschata) HPW 116.

Squashbug —See **Bug, Squash.**

Squid, Common (Loligo forbesi) CSF 205; CSS pl XXI; OIB 93; (L. vulgaris) ALE 3:188.

Squid, Jeweled (Lycoteuthis diadema) ALE 3:188; MOL 74.

Squid, Vampire (Vampyroteuthis infernalis) ALE 3:197; MOL 73.

Squill (crustacean) (Squilla sp.) ALE 1:451; LAW 147; LEA 157 (bw); VWS 53 (bw).

Squill (plant) (Scilla sp.) DFP 110 (cps 880, 882); ECS 142; EGB 138; LFT cp 14; PFM cp 244.

Squill, Alpine or Two-leaved (Scilla bifolia) DFP 110 (cp 878); EPF pl 906; EWF cp 24f; NHE 91; OGF 9; PFE cp 170 (1635); PFM cp 240; WFB 267 (3).

Squill, Autumn (Scilla autumnalis) MBF cp 85; MFF cp 6 (1005); NHE 191; PFE cp 168 (1636); PWF 168b.

Squill, Peruvian (Scilla peruviana) DFP 110 (cp 879); MWF 265; PFE cp 168 (1632); PFM cp 243; SPL cp 273.

Squill, Sea (Urginea maritima) AMP 79; PFE cp 169 (1630); PFF 375.

Squill, Spring (Scilla verna) MBF cp 85; MFF cp 6 (1004); OWF 169; PWF 26f; WFB 267 (2).

Squill, Striped (Puschkinia scilloides) DFP 110 (cp 873); EGB 136; EPF pl 913; ESG 138; EWF cp 151b; OGF 9; OGF 25; TVG 209.

Squinancywort (Asperula cynanchica) MBF cp 42; MFF pl 87 (810); NHE 75; OWF 93; PWF 93l; WFB 185 (7); (A. arcadiensis) OGF 43.

Squirrel, Bush (Paraxerus sp.) ALE 11:230; DDM 33 (cp 2); SMR 144 (cp 11).

Squirrel, Flying (Petaurista sp.) ALE
11:240; CAS 50, 197 (bw); DAA 104, 140;
HMW 159; MLS 204 (bw); (Pteromys volans)
ALE 11:240; (Idiurus zenkeri) DDM 48 (cp
3); (Sciuropterus russicus) DAE 112.

Squirrel, Flying or Scaly-tailed
(Anomalurus sp.) ALE 11:240; DDM 48 (cp
3); HMW 164; SMR 144 (cp 11).

Squirrel, Giant (Ratufa indica) CAS 193
(bw); DAA 106; DSM 113; HMW 159; MLS
200 (bw); (R. bicolor) ALE 11:230; DAA 106.

Squirrel, Giant Forest (Protoxerus stangeri)
ALE 11:230; DDM 33 (cp 2); LAW 582.

Squirrel, Ground (Xerus sp.) ALE 11:213,
230; CAA 77, 86 (bw); DDM 33 (cp 2).

Squirrel, Palm (Epixerus ebii) ALE 11:230;
DDM 33 (cp 2); (Funambulus palmarum)
CAS 194; DAA 106.

Squirrel, Pygmy (Myosciurus pumilio) ALE
11:230; DDM 33 (cp 2).

Squirrel, Red (Sciurus vulgaris) ALE 11:213,
229; 13:163; BMA 217; DAE 60; DSM 111;
GAL 57 (bw); HMW 158; LAW 583; OBV
137; TWA 91.

Squirrel, Striped (Funisciurus sp.) ALE
11:230; DDM 33 (cp 2); (Callosciurus
swinhoei) ALE 11:230.

Squirrel, Sun (Heliosciurus sp.) ALE 11:230;
DDM 33 (cp 2); SMR 144 (cp 11).

Squirrelfish or **Soldierfish** (Holocentrus sp.)
ALE 5:35; FEF 323, 324 (bw); HFI 65; HFW
84 (cp 38); WFW cps 219, 221.

Squirrelfish, Deep-water —See **Soldierfish, Deep-water.**

Squirrel-glider —See **Phalanger** (Petaurus
sp.).

Stachyurus (Stachyurus chinensis) DFP 240
(cp 1916); EWF cp 94b; (S. praecox) GDS cp
464; HSC 117; PFW 287.

Stag, European —See **Deer, Red.**

Stag, Hangul —See **Deer, Red.**

Stage-maker —See **Bowerbird, Tooth-
billed.**

Staggerwort or **Stammerwort** —See
Ragwort, Common.

Standard-wing, Wallace's (Semioptera
wallacei) ALE 9:480; RBU 307.

Star Flower —See **Carrion Flower.**

Star Flower, Spring (Ipheion uniflorum)
DFP 97 (cp 773); EGB 121; MWF 164; PFW
173.

Star of Bethlehem (Ornithogalum sp.) EPF
pls 908, 909; EWF cp 24b; FHP 136; KVP cp
161; LFT cp 15; MWF 218; PFE 504; PFM cp
237; NHE 191.

Star of Bethlehem, Common (Ornithogalum
umbellatum) DFP 109 (cp 871); EMB 104;
KVP cp 52; LFW 104 (cp 233); MBF cp 85;
MFF pl 95 (1001); NHE 191; PFE cp 170
(1639); PWF 64c; TGF 23; TVG 207; WFB
269 (6).

Star of Bethlehem, Nodding (Ornithogalum
nutans) DFP 109 (869); EFC cp 30; EGB 134;
MBF cp 85; MFF pl 81; NHE 191; PFM cp
241; WFB 269 (5).

Star of Bethlehem, Spiked (Ornithogalum
pyrenaicum) MBF cp 85; MFF pl 81 (1003);
NHE 191; PFE cp 169 (1640); WFB 269 (4).

Star of the Veldt —See **Marigold, Cape.**

Star-anise (Illicium vernum) AMP 155; DEW
1:57 (cp 19) Fig. 63.

Starfish (Astropecten irregularis) CSS cp 25;
MOL 86; OIB 183; VWS 40 (bw); (A.
aurantiacus) ALE 3:358, 386; LAW 248;
MOL 86.

Starfish (Acanthaster sp.) ALE 3:372, 380;
LAW 254; (Luidia sp.) ALE 3:359; LAW 255;
OIB 183; (Protoreaster linckii) ALE 3:379;
CAA 19 (bw); LAW 252; WEA 349.

Starfish (Palmipes membranaceus) CSS cp
25; MOL 86; (P. placenta) VWS 40 (bw);
(Ceramaster placenta) MOL 86; VWS 41.

Starfish, Basket —See **Basket-star.**

Starfish, Brittle —See **Brittle-star.**

Starfish, Common European (Asterias
rubens) ALE 3:385; CSS cp 24; LAW 266;
MOL 86; OIB 181; VWS 39 (bw).

Starfish, Cushion (Porania sp.) CSS cp 25;
LAW 254; OIB 183; VWS 41.

Starfish, Feather —See **Feather-star.**

Starfish Flower or **Plant** —See **Carrion
Flower.**

Starfish, Goose-foot (Anseropoda placenta)
ALE 3:385; OIB 183.

Starfish, Green or Spiny (Marthasterias
glacialis) ALE 3:380; CSS cp 24; OIB 181;
VWS 32, 39 (bw).

Starfish, Rose or Sun (Crossaster papposus) ALE 3:386; MOL 86.

Starfish, Scarlet (Henricia sp.) CSS cp 24; OIB 183.

Starfish, Sun (Solaster sp.) CSS cp 24; LAW 252; OIB 183; VWS 32, 33 (bw).

Starflower (Grewia occidentalis) LFT cp 104; MEP 76; (G. flava) LFT cp 104.

Star-fruit (Damasonium alisma) HPW 270; MBF cp 79; MFF 69 (961); WFB 259 (4).

Stargazer, Common or European (Uranoscopus scaber) ALE 5:161; HFI 43.

Star-grass (Hypoxis sp.) EWF cp 66a; EWG 119; LFT cp 34; MWF 160.

Starlet (Asterina gibbosa) LEA 190 (bw); OIB 181.

Starling, Amethyst (Cinnyricinclus leucogaster) ABW 273; PSA 288 (cp 33); WAB 139.

Starling, Common or European (Sturnus vulgaris) ABW 274; BBB 43; BBE 213; CDB 218 (cp 974); FBE 303; GAL 126 (cp 2); GPB 329; HAB 222; LAW 493; NBA 77; NBB 36 (bw); OBB 173; PBE 197 (cp 50); RBE 288; RBI 303; SAB 72; (cp 35).

Starling, Glossy (Lamprotornis sp.) ALE 9:460; BBT 72-73; CDB 217 (cp 970); NBA 79; PSA 288 (cp 33); (Lamprocolius chalybeus) CDB 217 (cp 969).

Starling, Golden-breasted (Cosmopsarus regius) ABW 274; CDB 216 (cp 966).

Starling, Gray-backed (Sturnus sinensis) DPB 367; KBA 416 (cp 61).

Starling, Pagoda (Temenuchus pagodarum) ALE 9:449; CDB 218 (cp 975).

Starling, Philippine Glossy (Aplonis panayensis) DPB 367; KBA 416 (cp 61).

Starling, Red-winged (Onychognathus morio) NBA 81; PSA 288 (cp 33).

Starling, Rose-colored or Rosy (Sturnus roseus) ABW 274; BBB 275; BBE 213; FBE 303; OBB 173; PBE 197 (cp 50); RBE 291; WAB 111; (Pastor roseus) ALE 9:449; FBW 77.

Starling, Shining (Aplonis metallica) HAB 221; RBU 320; SAB 72 (cp 35).

Starling, Spotless (Sturnus unicolor) BBE 213; FBE 303; PBE 197 (cp 50).

Starling, Superb (Spreo superbus) ABW 274; BBT 74; CDB 217 (cp 973); WEA 349.

Starling, Violet-backed (Sturnus philippensis) DPB 367; GPB 328.

Starling, Wattled (Creatophora cinerea) ALE :449; CDB 216 (cp 967); PSA 288 (cp 33).

Star-of-the-Argentine (Oxypetalum caeruleum) EGA 141; PFW 38.

Star-thistle, Red (Centaurea calcitrapa) BFB cp 20 (4); MBF cp 49; MFF cp 31 (921); PFE cp 156 (1500); PWF 113l; WFB 249 (6).

Star-thistle, Rough (Centaurea aspera) MBF cp 49; MFF cp 31 (920).

Starwort, Common Water (Callitriche stagnalis) MBF cp 34; NHE 96; OWF 53; PFE cp 105 (1088); PWF 47d; WFB 293 (4).

Starwort, Water (Callitriche sp.) HPW 240; MBF cp 34; MFF 66 (444); NHE 96; TEP 15.

Statice (Limonium suworowii) DFP 41 (cp 327); EGA 131; EWF cp 48b; TVG 57.

Steinbok (Raphicerus campestris) ALE 13:426; SMR 92 (cp 7); TWA 30; (R. melanotis) TWA 29; (R. sharpei) SMR 92 (cp 7).

Stephanandra, Dwarf Cut-leaved (Stephanandra incisa) EEG 102; EFS 138; EJG 145; TGS 231; (S. tanakae) HSC 115, 117.

Stereum (Stereum sp.) DEW 3:157; DLM 160; KMF 19; LHM 51; NFP 47; ONP 157; PFF 78.

Steriphoma (Steriphoma paradoxa) EWF cp 174c; PFW 66.

Sterlet (Acipenser ruthenus) ALE 4:145; FWF 17; HFI 131; OBV 89.

Stewartia (Stewartia sp.) EFS 138; EJG 146; EWF cp 101e; TGS 183.

Stick-insect (Clitarchus sp.) WEA 350; (Clitummus sp.) LAW 245; (Dixippus sp.) CTI 18 (cp 7); (Clonopsis sp.) PWI 63; (Micrarchus sp.) BIN cp 44; (Vetilia sp.) PEI 68.

Stick-insect, Mediterranean (Bacillus rossii) ALE 2:125; CIB 96 (cp 7).

Stickleback, Fifteen-spined (Spinachia spinachia) ALE 5:39; CSF 171; HFI 75; LAW 330; LEA 224 (bw); OBV 91; (S. vulgaris) CSS cp 31.

Stickleback, Nine- or Ten-spined (Pungitius pungitius or Pygosteus pungitius) ALE 5:39; CSF 171; FEF 266; HFI 75; OBV 91; SFW 077 (pl 100).

Stonecrop (Sedum kamtschaticum) SPL cps 373, 374; TVG 173; (S. middendorfianum) LWF 130 (cp 296); OGF 91.

Stonecrop, Alpine (Sedum alpestre) NHE 270; WFB 103 (5).

Stonecrop, Annual (Sedum annum) NHE 270; WFB 103.

Stonecrop, Blue (Sedum caeruleum) DFP 47 (cp 374); EWF cp 33a; PFE cp 40 (394).

Stonecrop, English (Sedum anglicum) MBF cp 33; MFF pl 88 (390); OWF 83; PWF 89f; WFB 103 (6).

Stonecrop, Great (Sedum maximum) DFP 172 (cp 1371); NHE 27.

Stonecrop, Hairy or Pink (Sedum villosum) MBF cp 33; MFF cp 23 (395); NHE 71; OWF 113; PFE cp 40 (394); PWF 150d; WFB 103 (7).

Stonecrop, Mossy (Crassula tillaea) MBF cp 33; MFF pl 105 (396); WFB 103 (9).

Stonecrop, Mossy or Wall-pepper (Sedum acre) BIP 163 (cp 248); BWG 37; DFP 26 (cp 202); MBF cp 33; MFF cp 46 (392); NHE 210; OWF 15; PFE cp 40 (389); PFF 191; PWF 84c; SPL cp 464; TEP 93; TGF 93; TVG 173; WFB 103 (2).

Stonecrop, Rock (Sedum forsteranum) MBF cp 33; MFF cp 46 (393).

Stonecrop, Rock or Reflexed (Sedum reflexum) MBF cp 33; MFF cp 46 (394); NHE 210; PFE cp 40 (388); PWF 100a; WFB 103 (4).

Stonecrop, Showy (Sedum spectabile) DFP 172 (cps 1372-1374); ECS 144; EEG 138; LFW cp 298; TGF 93.

Stonecrop, Thick-leaved (Sedum dasyphyllum) LCS cp 288; MBF cp 33; NHE 270.

Stonecrop, White or Worm-grass (Sedum album) BWG 73; EGC 146; ELG 148; KVP cps 79, 79a; MBF cp 33; MFF pl 88 (391); NHE 270; PFE cp 39 (391); PWF 90d; TEP 93; TGF 93; WFB 103 (3).

Stone-curlew —See **Curlew.**

Stonefish (Synanceja verrucosa) ALE 5:46, 60; WFW cp 253.

Stonefly —See **Fly, Stone.**

Stonewort (Chara sp.) NFP 23; ONP 93; PFF 65; WUS 5.

Storax (Styrax officinalis) EWF cp 31b; PFM cp 134.

Stork, African Openbill (Anastomus lamelligerus) CAA 30 (bw); PSA 48 (cp 3).

Stork, Asian Openbill (Anastomus oscitans) CDB 45 (cp 67); KBA 33 (pl 2).

Stork, Black (Ciconia nigra) ABW 57; ALE 7:222; 11:257; BBB 274; BBE 41; CDB 46 (cp 69); FBE 43; KBA 33 (pl 2); OBB 15; PBE 9 (cp 4); PSA 48 (cp 3); RBE 79; SBB 14 (pl 11); SLP 31.

Stork, Black-necked —See **Jabiru** (Xenorhynchus asiaticus).

Stork, Common or White (Ciconia ciconia) ABW 57; ALE 7:218-219, 222; BBB 269; BBE 41; DAE 131; FBE 43; GPB 71; LAW 455; NBB 17 (bw); OBB 15; PBE 9 (cp 4); PSA 48 (cp 3); RBE 76; SBB 13 (cp 9); WEA 353; (C. alba) CDB 45 (cp 68).

Stork, Marabou —See **Marabou.**

Stork, Painted (Ibis leucocephalus or Mycteria leucocephalus) CAS 163; CDB 47 (cp 76); KBA 33 (pl 2); LEA 368 (bw); RBI 72.

Stork, Saddle-billed (Ephippiorynchus senegalensis) ABW 57; ALE 7:222; CAF 187; CDB 46 (cp 71); LAW 443; PSA 161 (cp 18); SBB 13 (cp 10).

Stork, Shoebill or Whale-headed (Balaeniceps rex) ABW 56; ALE 7:216; CDB 47 (cp 77); GPB 69; SLP 225; WAB 155; WEA 335.

Stork, Wood or Yellow-billed —See **Ibis, Wood or Yellow-billed.**

Stork, Woolly-necked (Ciconia episcopus) DPB 25; KBA 33 (pl 2); (Dissoura episcopus) CDB 46 (cp 70).

Storksbill (Erodium sp.) DFP 9 (cp 67, 68); EMB 64; EPF pl 647; EWF cp 32d; HPW 208; MBF cp 20; MFF cp 22 (250); PFE cp 64 (652); TGF 109.

Storksbill, Common (Erodium cicutarium) EPF pl 645; MBF cp 20; MFF cp 22 (251); NHE 221; OWF 129; PFF 225; PWF 29i; TEP 177; WFB 133 (5).

Storksbill, Long-beaked (Erodium gruinum) EWF cp 32g; PFE 64 (652).

Storksbill, Sea (Erodium maritimum) MBF cp 20; MFF pl 89 (249); WFB 133 (6).

Storm-petrel —See **Petrel.**

Stranvaesia, Chinese (Stranvaesia davidiana) DFP 240 (cp 1917); EGW 142; GDS cp 466; HSC 115; TGS 279.

Strapwort (Corrigiola litoralis) MBF cp 71; MFF pl 104 (194); NHE 97; PWF 118a; WFB 57 (6).

Straw Flower —See **Everlasting** (Helichrysum sp.).

Strawberry, Barren (Potentilla sterilis) MBF cp 27; MFF cp 78 (350); NHE 27; OWF 81; PWF cp 22c; WFB 110, 111 (2).

Strawberry Geranium —See **Aaron's Beard.**

Strawberry, Hautbois (Fragaria moschata) NHE 27; WFB 110, 111 (1a).

Strawberry Herb —See **Sanicle, Wood.**

Strawberry Tomato —See **Cherry, Bladder** or **Chinese Lantern.**

Strawberry Tree (Arbutus unedo) AMP 65; BKT cps 175, 176; DEW 1:172 (cp 98), 211 (cp 113), 238; DFP 181 (cps 1447, 1448), 182 (cps 1450, 1451); EET 200, 201; MBF cp 55; MFF pl 77 (587); MTB 368 (cp 37); MWF 45; NHE 246; OBT 36; OFP 83; PFE cp 60 (924); PFM cp 119; PFW 108; SPL cp 381; SST cp 184; TGS 406; WFB 172, 173 (9).

Strawberry, Wild or Wood (Fragaria vesca) BWG 72; DEW 1:273 (cp 147), 296; EFC cp 56; EGA 119; EMB 148; EPF pls 407, 408; KVP cp 53; MBF cp 27; MFF pl 78 (359); NHE 27; OFP 75; OWF 81; PFF 199; PWF 33g, 93h; TEP 241; WFB 110, 111 (1); (F. indica) EGC 130; MWF 135.

Streber (Zingel streber) ALE 5:86; FWF 123; (Aspro asper) HFI 24; SFW 533 (pl 144).

String-of-buttons (Crassula perforata) ECS 45, 111; SCS cp 76.

Stromanthe (Stromanthe sanguinea) EWF cp 189c; HPW 300; (S. amabilis) RDP 375.

Strophanthus (Strophanthus sp.) DEW 2:101 (cp 51), 102 (cp 54); EWF cp 61a, e, cp 103a; PFW 31.

Strychnos (Strychnos sp.) DEW 2:101 (cp 50), 118; HPW 222; LFT cp 121; SST cp 295.

Stultitia (Stultitia sp.) LCS cp 294; LFT cp 131.

Sturgeon, Atlantic (Acipenser sturio) ALE 4:145; CSF 63; FEF 35 (bw); FWF 19; HFI 131; HFW 67 (bw); LAW 297; OBV 89; WEA 352.

Styphelia (Styphelia sp.) EWF cp 136c; HPW 128; MWF 277.

Succory, Blue or Cupidone (Catananche caerulea) DFP 127 (cp 1016); EGA 103; MWF 75; OGF 123; PFE cp 157 (1511); PFW 85; TGF 45.

Succory, Swine's or Lamb's (Arnoseris minima) MBF cp 50; MFF cp 58 (926); NHE 239; WFB 253 (8).

Succory, Wild —See **Chicory.**

Sucker (fish) (Catostomus sp.) ALE 4:256; LEA 213 (bw); (Lepadogaster sp.) CSS pl XXIII; HFI 12; OBV 79.

Sucker, Apple (insect) (Psylla mali) OBI 35; PEI 127 (bw); (Chermes mali) GAL 446 (bw).

Sugar Bush —See **Honeysuckle, Cape.**

Sugar Cane (Saccharum officinarum) DEW 2:270; MLP 298; OFP 15; PFF 346; WWT 103.

Sugar Glider —See **Phalanger.**

Sugarbird, Cape (Promerops cafer) ABW 277; ALE 9:320; CDB 195 (cp 849); PSA 304 (cp 35).

Sulphur Tuft (Hypholoma fasciculare) DLM 200; KMF 99; LHM 147; NFP 83; NHE 48; ONP 141; RTM 55; SMF 50.

Sumac, Venetian —See **Smoke Tree.**

Sun Moss or **Sun Plant** —See **Rose-moss.**

Sunbird (Anthreptes collaris) ALE 9:320; PSA 304 (cp 35); (A. platura) FBE 265; (Cyanomitra verticalis) CDB 189 (cp 824).

Sunbird, Black (Nectarinia amethystina) NBA 85; PSA 304 (cp 35).

Sunbird, Black-throated (Aethopyga saturata) KBA 417 (cp 62); WBV 215.

Sunbird, Brown-throated (Anthreptes malaccensis) CAS 211 (bw); DPB 377; KBA 417 (cp 62); LEA 449; WBV 215.

Sunbird, Copper-throated (Nectarinia calcostetha) DPB 377; KBA 417 (cp 62).

Sunbird, Double-collared (Nectarinia chalybea) NBA 83; PSA 304 (cp 35); (N. afra) NBA 83; (N. mediocris) WAB 143; (Cinnyris chalybeus) CDB 189 (cp 821).

Sunbird, False (Neodrepanis sp.) ABW 212; ALE 9:150; GPB 237; WAB 30, 159; (N. hypoxantha) WAB 30.

Sunbird, Golden-winged (Drepanorhynchus reichenowi) CDB 190 (cp 825); (Nectarinia reichenowi) ABW 278.

Sunbird, Gould's (Aethopyga gouldiae) KBA 417 (cp 62); RBI 264; WBV 215.

Swallow, Mosque (Hirundo striolata) DPB 231; (M. senegalensis) CDB 151 (cp 618).

Swallow, Pacific (Hirundo tahitica) DPB 231; KBA 245 (pl 34); RBI 271.

Swallow, Plain Bank (Riparia paludicola) CDB 151 (cp 621); DPB 231; FBE 207; KBA 245 (pl 34); PSA 208 (cp 23).

Swallow, Red-rumped (Hirundo daurica) ALE 9:180; BBE 201; CDB 150 (cp 613); FBE 207; GPB 252; KBA 245 (pl 34); PBE 196 (cp 49); WAB 178; (Cecropis daurica) RBI 268.

Swallow, Rough-winged (Psalidoprogne sp.) ALE 9:180; PSA 208 (cp 23).

Swallow, Striped (Hirundo abyssinica) CDB 150 (cp 610); (H. cucullata) CDB 150 (cp 612); PSA 208 (cp 23).

Swallow, Tree (Iridoprocne bicolor) ABW 216; ALE 9:180.

Swallow, Welcome (Hirundo tahitica neoxena) FNZ 208 (cp 18); HAB 143; RBU 279; SAB 8 (cp 8); SNB 130 (cp 64).

Swallow, White-backed (Cheramoeca leucosterna) HAB 143; SAB 8 (cp 8); SNB 130 (cp 64).

Swallow, White-throated (Hirundo albigularis) CDB 150 (cp 611); PSA 208 (cp 23).

Swallow, Wire-tailed (Hirundo smithii) CDB 51 (cp 619); KBA 245 (pl 34).

Swallow, Wood —See **Wood-swallow.**

Swallow-tanager (Tersina viridis) ABW 292; ALE 9:362; CDB 202 (cp 885); DTB 87 (cp 43); GPB 306; SSA cp 16.

Swallowwort (Vincetoxicum hirundinaria) NHE 34; PFE cp 99 (1010); WFB 179 (9); (V. officinale) EFP pl 690.

Swamp-hen (Porphyrio melanotus) BNZ 159; CAU 261; HAB 57; RBU 24; (P. poliocephalus bellus) RBU 27.

Swan, Black (Cygnus atratus) ABW 67; ALE 7:273; 10:89; BBB 191; CAU 61; CDB 52 (cp 114); HAB 33; LAW 503; LEA 377 (bw); MEA 68; MWA 108 (bw); RBU 60; SNB 34 (cp 16); WAB 181, 200; WAW 19 (bw); WEA 357.

Swan, Black-necked (Cygnus melanocoryphus) ABW 67; ALE 7:273; LEA 377 (bw); SBB 27 (pl 27); WEA 357.

Swan, Bewick's (Cygnus columbianus or C. bewickii) ALE 7:274, 385; BBB 190; BBE 45; FBE 45; NBB 48 (bw); OBB 35; PBE 16 (cp 5); SBB 26 (pl 26).

Swan, Coscoroba (Coscoroba coscoroba) ALE 7:273; CDB 52 (cp 113); SBB 29 (pl 29); (C. olor) LEA 344 (bw).

Swan Flower —See **Pea, Darling.**

Swan, Mute (Cygnus olor) ALE 7:244, 255, 274; 11:257; BBE 45; FBE 45; GPB 88; LEA 376 (bw); OBB 35; PBE 16 (cp 5), 20 (pl 7); RBE 112; SBB 25 (pl 24); WEA 357.

Swan, Whooper (Cygnus cygnus) ALE 7:274, 385; BBB 190; BBE 42, 45; FBE 45; LAW 436; OBB 35; PBE 16 (cp 5), 20 (pl 7); SBB 26 (pl 25).

Swede —See **Rape.**

Swedish Ivy (Plectranthus australis) EGL 137; EGV 84, 130; RDP 322; (P. coleoides marginatus) EDP 148; EFP 57, 136; RDP 322.

Swedish Turnip —See **Rutabaga.**

Sweep's Brush —See **Woodrush, Field.**

Sweet Bay (Laurus nobilis) AMP 27; DEW 1:92, 98 (cp 39); EET 162, 163; EFP 120; EGE 132; EGH 119; EPF pl 366; MTB 257 (cp 24); NHE 294; OBT 136; OFP 133; PFM cp 31; SPL cp 99; SST cp 133; TGS 247.

Sweet Briar (Rosa rubiginosa) EWF cp 9a; MBF cp 30; OWF 117; PWF 86a.

Sweet Cicely (Myrrhis odorata) EGH 126; EPF pl 760; EVF 142; MBF cp 38; MFF pl 99 (460); NHE 275; OFP 147; OWF 91; PFE cp 84 (863); PWF cp 23i.

Sweet Flag —See **Flag, Sweet.**

Sweet Gale (Myrica gale) HPW 58; MBF cp 76; NHE 69; OWF 191; PFE cp 4 (34); PWF 60a; WFB 31 (4).

Sweet Gum (Liquidambar styraciflua) BKT cps 102, 103; DFP 210 (cp 1677); EEG 118; EET 163; EGT 45, 69, 122; EGW 123; ELG 106; MTB 257 (cp 24); MWF 186; OBT 148; PFF 194; PFW 137; SST cp 286; TGS 182.

Sweet Olive —See **Tea Olive.**

Sweet Pea Shrub (Polygala myrtifolia) DEW 2:65; MWF 238; PFW 235.

Sweet Potato (Ipomaea batatas) EPF pl 785; EVF 18; OFP 183; PFF 284.

Sweet Sultan (Centaurea moschata) DFP 33 (cp 260); OGF 87; TVG 39.

Sweet William (Dianthus barbatus) DFP 36 (cp 283); EGA 114; EPF pl 307; LFW 44 (cps 100, 101); MWF 110; OGF 127; PFE 90; PFF 163; PFW 71; SPL cp 333; TGF 60; WFB 64.

Sweet-clover —See **Melilot.**

Sweet-fern (Comptonia peregrina or C. aspelenifolia) EFS 102; EGC 127; TGS 86.

Sweet-grass (Glyceria sp.) MBF pl 97; MFF pl 131 (1171), pl 134 (1172); NHE 106; PFE cp 180 (cps 1755, 1756); WFB 297 (4).

Sweetlips (Plectorhinchus sp.) FEF 318, 355, 356 (bw); HFW 168 (cp 74), 169 (cp 76); MOL 121; (Gaterin sp.) HFW 168 (cp 75); WFW cp 299.

Sweetshade (Hymenosporum flavum) MEP 45; MWF 159.

Sweetsop —See **Soursop.**

Sweetspire, Holly-leaf (Itea ilicifolia) DFP 208 (cp 1663); GDS cp 258.

Swift, Alpine (Apus melba) ALE 8:426; 13:463; BBB 273; BBE 181; FBE 187; OBB 119; PBE 196 (cp 49).

Swift, Common or Eurasian (Apus apus) ALE 8:426; BBB 46; BBE 181; CDB 120 (cp 464); FBE 187; GPB 200; LEA 337 (bw); OBB 119; PBE 196 (cp 49); RBE 196; WAB 137.

Swift, Fork-tailed (Apus pacificus) DPB 193; HAB 131; SNB 130 (cp 64).

Swift, House or Little (Apus affinis) DPB 193; FBE 187; LAW 488; PSA 208 (cp 23).

Swift, Pallid (Apus pallidus) ALE 8:426; BBE 181; FBE 187; LAW 488; PBE 196 (cp 49).

Swift, Palm (Cypsiurus parvus) ABW 166; ALE 8:426; DPB 193; GPB 199; PSA 208 (cp 23).

Swift, Spine-tailed (Chaetura sp.) ABW 165; DPB 193; PSA 208 (cp 23); SSA pl 24; (Hirundapus sp.) ALE 8:426; HAB 131; SNB 130 (cp 64).

Swift, Tree (Hemiprocne longipennis) ABW 165; ALE 8:426; WAB 166; (H. comata) DPB 193; KBA 245 (pl 34); (H. mystacea) RBI 147.

Swift, White-collared (Streptoprocne zonaris) SSA pl 24; WAB 91.

Swift, White-rumped (Apus caffer) ABW 165; BBE 181; FBE 187; PSA 208 (cp 23).

Swiftlet (Collocalia sp.) ABW 167; DPB 185; HAB 131; SNB 130 (cp 64); WAB 171.

Swine-cress, Common (Coronopus squamatus) MBF cp 10; MFF cp 89 (63); NHE 216; OWF 71; PWF 73l; WFB 99 (9).

Swine-cress, Lesser (Coronopus didymus) MBF cp 10; MFF cp 89 (64); PFE cp 36 (354); PWF 93e; WFB 99 (10).

Swiss Cheese Plant (Monstera deliciosa) EDP 126; EFP 124; EGV 78, 123; EPF pl 1006; LHP 126, 127; MEP 12; MWF 202; OFP 97; RDP 270; SPL cp 23; (M. pertusa) BIP 121 (cp 185); DFP 70 (cp 560).

Swordfish (Xiphias gladius) ALE 5:192; CSF 149; HFI 52; LMF 61.

Sycamore —See **Maple, Sycamore.**

Sylph (Aglaiocercus kingi) ABW 171; CDB 120 (cp 465); DTB 7 (cp 3); SHU 73, 89, 91; (A. coelestis) SSA cp 32.

T

Tadpole-fish —See **Forkbeard, Lesser.**

Tahr, Himalayan (Hemitragus jemlahicus) ALE 13:464; DAA 46; HMA 56; HMW 86; MLS 427 (bw); SMW 285 (cp 183).

Tahr, Nilgiri (Hemitragus hylocrius) DAA 46; LVS 131.

Tailorbird, Ashy (Orthotomus ruficeps) KBA 337 (cp 52); (O. sepium) DPB 323.

Tailorbird, Dark-necked (Orthotomus atrogularis) DPB 323; GPB 282; KBA 337 (cp 52).

Tailorbird, Long-tailed (Orthotomus sutorius) ABW 255; ALE 9:259; CAS 210 (bw); KBA 337 (cp 52); WAB 179; WBV 179; (O. longicaudus) RBI 252.

Tailorbird, Mountain (Orthotomus cuculatus) DPB 323; KBA 337 (cp 52).

Tailorbird, Rufous-tailed (Orthotomus sericeus) DPB 323; KBA 337 (cp 52).

Taipan (Oxyuranus scutellatus) HWR 146; SSW 112; WEA 359.

Takahe (Notornis mantelli) ALE 8:84, 97;
BNZ 163; CDB 75 (cp 230); FNZ 109 (bw);
GPB 30; LEA 405 (bw); LVB 43; LVS 218;
PWC 221 (bw); SLP 145; WAB 33, 211; (N.
hochstetteri) CAU 284 (bw).

Takin (Budorcas taxicolor) ALE 12:103;
13:469; DAA 23; DSM 270; HMW 81; LVS
137; MLS 425 (bw); SMW 270 (bw); WEA
358.

Talang Talang (Chorinemus sp.) LMF 77
(bw); MOL 120.

Talapoin —See **Guenon, Dwarf.**

Tallow-tree, Chinese (Sapium sp.) EGT 143;
GDS cp 441.

Tamandua —See **Anteater, Lesser or
Three-toed.**

Tamarack —See **Larch, American.**

Tamarau or **Tamaraw** (Anoa mindorensis)
PWC 246 (bw); SLS 157; VWA cp 29.

Tamarillo —See **Tree Tomato.**

Tamarin, Cotton-head —See **Marmoset,
Cotton-topped or Pinche.**

Tamarin, Emperor (Saguinus imperator)
ALE 10:349, 376; DSM 73; LAW 616; WEA
245.

Tamarin, Moustached (Saguinus mystax)
ALE 10:376; DSM 73.

Tamarind (Tamarindus indica) AMP 61;
DEW 1:308; SST cp 297.

Tamarisk (Tamarix sp.) DEW 1:170 (cp 95);
HPW 109; HSC 120; MWF 279; PFE cp 78
(807); PFW 291; TGS 270.

Tamarisk, Five-stamened (Tamarix
pentandra) DFP 241 (cp 1927); EFS 26, 141;
EPG 148; HSC 119; SST cp 174; TGS 278;
TVG 249.

Tamarisk, French or Salt Cedar (Tamarix
gallica) DEW 1:245; EPF pl 494; MBF cp 17;
MFF cp 17 (143); OWF 191; SST cp 173; TGF
278; WFB 141 (8).

Tamarisk, German (Myricaria germanica)
EPF pl 497; HPW 109; NHE 94; PFE 264.

Tambourissa (Tambourissa sp.) DEW 1:88;
HPW 34.

Tammar —See **Wallaby.**

Tana —See **Tree-shrew, Large or Tana.**

Tanager, Ant —See **Ant-tanager** and
Saltator.

Tanager, Bay-headed (Tangara gyrola)
CDB 202 (cp 882); DTB 105 (cp 52).

Tanager, Black-eared (Tangara parzudakii)
ABW 293; ALE 9:361; BBT 104.

Tanager, Blue-gray (Thraupis episcopus or
T. virens) ABW 293; ALE 9:361; CDB 202
(cp 884); CSA 64.

Tanager, Brazilian (Ramphocelus bresilius)
ALE 9:335; BBT 105; WAB 91; (R. carbo)
ALE 9:361.

Tanager, Diademed (Stephanophorus
diadematus) ALE 9:361; SSA cp 16.

Tanager, Flame-crowned or -faced —See
Tanager, Black-eared.

Tanager, Glistening-green (Chlorochrysa
phoenicotis) CDB 201 (cp 875); DTB 91 (cp
45); SSA cp 41; (Chlorospingus phoenicotis)
ALE 9:335.

Tanager, Golden-crowned (Iridosornis
rufivertex) CDB 201 (cp 876); DTB 111 (cp
55); SSA cp 40; (I. analis) DTB 109 (cp 54).

Tanager, Grass-green (Chlorornis riefferii)
DTB 133 (cp 66); SSA cp 40.

Tanager, Magpie (Cissopis leveriana) ALE
9:361; BBT 108; DTB 135 (cp 67); SSA pl 30.

Tanager, Masked (Tangara nigrocincta)
ABW 293; ALE 9:361.

Tanager, Masked Crimson (Ramphocelus
nigrogularis) CDB 21 (cp 879); DTB 123 (cp
61).

Tanager, Moss-backed (Bangsia edwardsi)
DTB 119 (cp 59); SSA cp 40.

Tanager, Mountain —See **Mountain-tanager.**

Tanager, Multi-colored (Chlorochrysa
nitidissima) DTB 93 (cp 46); SSA cp 41.

Tanager, Paradise (Tangara chilensis) ABW
293; ALE 9:361; DTB 95 (cp 47); SSA cp 41.

Tanager, Scarlet-and-white (Erythrothlypis salmoni) DTB 129 (cp 64); SSA cp 40.

Tanager, Scarlet-rumped (Ramphocelus
passerinii) ABW 293; ALE 9:361; WAB 97;
(R. dimidiatus) DTB 121 (cp 60).

Tanager, Seven-colored (Tangara fastuosa)
ALE 9:335; BBT 109; GPB 305.

Tanager, Silver-throated (Tangara chrysophrys) BBT 103; (T. icterocephala) WAB 90.

Tanager, Swallow —See **Swallow-tanager.**

Tanager, Thrush —See **Thrush-tanager.**

Tanager, Yellow-rumped (Ramphocelus
icteronotus) CDB 201 (cp 878); DTB 125 (cp
62).

Tang, Blue (Acanthurus coeruleus) ALE 5:201; HFI 48; HFW 215 (cp 99); WFW cp 437.

Tangalunga —See **Civet** (Viverra sp.)

Tangerine (Citrus reticulata or C. nobilis) EPF pl 657; MWF 86; OFP 87; PFF 229; SST cp 194.

Tansy (Chrysanthemum vulgare) CTH 56 (cp 88); EPF pl 558; MBF cp 46; MFF cp 52 (894); NHE 237; PFE cp 147 (1426); TEP 201.

Tansy (Tanacetum vulgare) AMP 147; BFB cp 19 (8); EFC cp 36; EGH 144; MWF 280; OFP 145; OWF 35; PFF 328; PWF 99k; SPL cp 407; TGF 237; WFB 241 (3).

Tapaculo (Acropternis orthonyx) ABW 202; (Scytalopus unicolor) SSA pl 47; WAB 92.

Tape-grass (Vallisneria spiralis) DEW 2:229, 230; HPW 271; MFF pl 68 (968); PFE 480; WFB 293 (6).

Tapeworm (Taenia sp.) GAL 527 (bw); LAW 76; OIB 27; (Echinococcus sp.) ALE 1:290; OIB 27.

Tapioca Plant —See **Cassava.**

Tapir, Brazilian or South American (Tapirus terrestris) ALE 8:359; 13:21, 22; DSM 215; HMA 94; HMW 36; JAW 159; MLS 355 (bw); SMW 232 (cp 141); WEA 360.

Tapir, Malayan or Saddle-backed (Tapirus indicus) ALE 13:22; DAA 124; DSM 214; HMA 94; HMW 36; JAW 160; LAW 556; LVS 115; MLS 354 (bw); SLP 195; SMW 224 (bw); TWA 169.

Tapir, Mountain (Tapirus pinchaque) ALE 13:22; DSM 215; SLS 47; (T. roulini) SMW 224 (bw).

Tarantula (Lycosa tarentula) ALE 1:425; CSA 135; GSP 83; PEI 13 (bw); (Tarentula inquilina) GAL 487 (bw).

Tarpan —See **Horse, Wild.**

Tarpon (Tarpon atlanticus) HFW 73 (bw); LAW 300; LMF 27.

Tarragon (Artemisia dracunculus) EGH 101; EVF 147; OFP 145.

Tarsier (Tarsius sp.) ALE 10:253 (cp 7), 254, 265; BMA 219; CAS 253; DSM 68; HMA 94; HMW 29; JAW 41; LEA 503 (bw); MLS 126 (bw); SMW 46 (cp 32); TWA 157 (bw).

Tassel-flower, Mountain —See **Soldanella, Mountain.**

Tasselweed —See **Pondweed, Tassel.**

Tasmanian Devil (Sarcophilus harrisii) ALE 10:102; BMA 220, 221; CAU 61; DSM 27; JAW 22; LEA 483 (bw); MLS 53 (bw); SMW 37 (cp 8); WAW 53 (bw); WEA 361; (S. ursinus) TWA 197.

Tasmanian Wolf (Thylacinus cynocephalus) ALE 10:71; HMW 199; JAW 23; LAW 525; LVA 57; LVS 13 (bw); MLS 54 (bw); SLS 227; TWA 196; VWA cp 31.

Tattler, Gray-tailed or Siberian (Tringa brevipes) CDB 87 (cp 294); FNZ 129 (pl 12); HAB 68; SNB 70 (cp 34), 72 (pl 35).

Tattler, Wandering (Tringa incana) DPB 91; HAB 68; SNB 70 (cp 34), 72 (pl 35).

Taurepo or **Tauropo** (Rhabdothamus solandri) EWF cp 139g; SNZ 32 (cps 53, 54).

Tayra (Eira barbara or Tayra barbara) ALE 12:66; DSM 169; JAW 115; MLS 286 (bw); SMW 204 (bw); WEA 363.

Tea Olive (Osmanthus fragrans) EGH 130; EJG 133; FHP 136; MWF 219; SPL cp 107; TGS 327.

Tea Tree (Camellia sinensis or Thea sinensis) AMP 127; EFP pl 472; MLP 151; OFP 113; PFF 252; SST cp 270.

Tea Tree, Manuka (Leptospermum scoparium) DFP 13 (cp 99), 210 (cp 1674); EGE 133; EPG 128; HSC 74, 77; MEP 98; MWF 181; OGF 103; PFW 191; SNZ 38 (cps 72-75).

Tea Tree, White (Leptospermum ericoides) EET 245; SNZ 39 (cps 76, 77).

Teak (Tectona grandis) EET 211; SST cp 175.

Teal, Baikal or Formosa (Anas formosa) ABW 68; ALE 7:306; BBE 55; FBE 55; SBB 40 (cp 45).

Teal, Black —See **Scaup, New Zealand.**

Teal, Blue-winged (Anas discors) ALE 7:304; BBE 55; FBE 55; HBG 80 (pl 3); WAB 61.

Teal, Brazilian (Amazonetta brasiliensis) ALE 7:319; SSA cp 2.

Teal, Chestnut-breasted (Anas castanea) ALE 7:307; HAB 28; SNB 39 (cp 18).

Teal, Common or European (Anas crecca) ALE 7:305; 11:257; BBB 195; BBE 55; DPB 29; FBE 55; KBA 80 (pl 7); OBB 19; PBE 28 (cp 9), 40 (pl 13), 48 (pl 15); RBE 84; SBB 39 (pl 44); WAB 61, 101; WBV 49.

Teal, Cotton —See **Goose, Indian Pygmy.**

Teal, Falcated (Anas falcata) ALE 7:304; BBE 55; FBE 55.

Teal, Garganey —See **Garganey.**

Teal, Gray (Anas gibberifrons) ALE 7:307; FNZ 161 (cp 15); HAB 32; SNB 39 (cp 18).

Teal, Hottentot (Anas punctata) ALE 7:307; (A. hottentotta) PSA 49 (cp 4), 64 (pl 5).

Teal, Madagascar (Anas bernieri) ALE 7:307; LVB 13.

Teal, Marbled (Anas angustirostris) ALE 7:304; BBE 57; CDB 52 (cp 118); FBE 57; PBE 28 (cp 9).

Teal, Red-bill (Anas erythrorhyncha) ALE 7:307; PSA 49 (cp 4), 64 (pl 5).

Teal, Ringed (Calonetta leucophrys) ALE 7:307; CDB 51 (cp 108); (Anas leucophrys) SSA cp 3.

Teal, Silver (Anas versicolor) ALE 7:303; WFI 120.

Teal, South American Green-winged (Anas flavirostris) ALE 7:306; WFI 124 (bw).

Teasel, Common (Dipsacus fullonum) BFB cp 11 (6); BWF 150; CTH 52 (cp 79); EPF pls 780, 781; EWF cp 17b; HPW 262; MBF cp 43; MFF cp 9 (833); NHE 235; PFW 101; PWF 131h; TEP 199; WFB 225 (5).

Teasel, Small (Dipsacus pilosus) MBF cp 43; MFF pl 107 (834); NHE 235; PWF 144d; WFB 225 (4).

Teasel, Wild (Dispsacus sylvestris) DEW 2:110 (cp 73); EGH 113; KVP cp 82; PFF 306; WUS 361; (D. laciniatus) PFE cp 137 (1318).

Teddy-bear Plant (Cyanotis kewensis) BIP 65 (cp 90); RDP 165.

Tegu, Common or Great (Tupinambis sp.) ALE 6:262, 277; ART 235; HRA 19; LAW 421; LEA 309 (bw); SIR 110 (cp 55).

Tellin (Tellina sp.) CSS pl XVII; GSS 19, 149; MOL 71; OIB 79; (Macoma sp.) ALE 3:168; GSS 150; OIB 79.

Tembusu (Fagraea sp.) EWF cp 121e; HPW 222; MEP 106.

Temple Bells (Smithiantha hybrid) BIP 165 (cp 253); EGB 139; EGG 142; EGL 145; FHP 145; PFW 129; RDP 364.

Temple Tree —See **Frangipani.**

Tench (Tinca tinca) ALE 4:327, 336, 341; FEF 148 (bw); FWF 73; GAL 233 (bw); HFI 100; HFW 41 (cp 19); OBV 95; SFW 221 (pl 52); WFW cp 143.

Ten-pounder —See **Ladyfish.**

Tenrec (Centetes ecaudatus) HMW 190; LEA 491 (bw); (Geogale aurita) ALE 10:178; (Ericulus setosus) HMW 190.

Tenrec, Hedgehog (Echinops telfairi) ALE 10:168, 177; DSM 44; (Setifer setosus) ALE 10:177; MLS 79 (bw); SMW 42 (cp 19).

Tenrec, Long-tailed (Microgale longicauda) ALE 10:178; DSM 45.

Tenrec, Striped (Hemicentetes semispinosus) ALE 10:177, 283; DSM 45.

Tenrec, Tailless (Tenrec ecaudatus) ALE 10:177, 283; HMA 59.

Terebinth —See **Turpentine Tree.**

Termite (Bellicositermes natalensis) ALE 2:145; ZWI 212; (Neotermes sp.) WEA 364; (Termes bellicosus) PEI 82-84 (bw).

Termite, Yellow-necked (Callotermes flavicollis) LAW 191; PEI 81 (bw).

Tern, Antarctic (Sterna vittata) FNZ 145 (pl 14); SAO 133; SAR 122.

Tern, Arctic (Sterna paradisaea) ABW 134; ALE 8:187, 212; BBE 159; CDB 94 (cp 338); FBE 161; OBB 97; PBE 148 (pl 39), 156 (cp 41); SAO 133; WAB 213; (S. macrura) FNZ 145 (pl 14); NBB 59 (bw).

Tern, Black (Chlidonias niger) ALE 8:211; BBB 214; BBE 155; CDB 91 (cp 320); FBE 163; FBW 59; GPB 167; OBB 95; PBE 149 (pl 40), 156 (cp 41); (Sterna nigra) SAO 129.

Tern, Black-fronted (Chlidonias albostriatus) BNZ 153; FNZ 145 (pl 14).

Tern, Black-naped (Sterna sumatrana) CDB 95 (cp 339); DPB 109; HAB 82; KBA 157 (pl 20).

Tern, Bridled (Sterna anaetheta) DPB 113; FBE 163; HAB 82; KBA 156 (pl 19).

Tern, Caspian (Hydroprogne caspia) ALE 7:61; CDB 92 (cp 323); DPB 109; HAB 82; KBA 156 (pl 19); PSA 176 (cp 19); SSA pl 21; WAB 71; (H. tschegrava) BBB 271; BBE 157; FBE 159; PBE 148 (pl 39), 156 (cp 41); (Sterna caspia) OBB 99;

Tern, Common (Sterna hirundo) ABW 130; ALE 7:255, 385; 8:205; BBB 242; BBE 159; CDB 44 (cp 337); DAE 32; DPB 113; FBE 161; HBG 140 (pl 11); KBA 157 (pl 20); LEA 415 (bw); OBB 97; PBE 148 (pl 39), 156 (cp 41); PSA 176 (cp 19); RBE 176; SAO 133; WEA 365.

Tern, Crested or Swift (Sterna bergii) DPB 113; FBE 159; HAB 81; KBA 156 (pl 19); PSA 176 (cp 19); WAW 25 (bw); (Thalasseus bergii) RBU 47.

Tern, Damara (Sterna balaenarum) PSA 176 (cp 19); SAO 129.

Tern, Fairy or White (Gygis alba) ABW 130; ALE 8:211; BBT 126; CDB 92 (cp 322); FNZ 145 (pl 14); GPB 166; (Sterna nereis) FNZ 145 (pl 14); HAB 81; SAO 135; WAN 126.

Tern, Gull-billed (Gelochelidon nilotica) ALE 8:211; BBB 269; BBE 157; DPB 109; FBE 159; FBW 59; HAB 81; KBA 156 (pl 19); OBB 99; PBE 148 (pl 39), 156 (cp 41); SAO 135; WAB 71.

Tern, Inca (Larosterna inca) ALE 8:211; CDB 92 (cp 324); SSA pl 21.

Tern, Large-billed (Phaetusa simplex) ALE 8:211; SSA pl 21.

Tern, Lesser Crested (Sterna bengalensis) FBE 159; HAB 81; SAO 135; WAB 161.

Tern, Little (Sterna albifrons) ALE 8:187; BBB 244; BBE 159; CDB 94 (cp 334); DPB 109; FBE 161; FBW 59; FNZ 145 (pl 14); HAB 81; KBA 157 (pl 20); LAW 462; LEA 343 (bw), 416; OBB 99; PBE 149 (pl 40), 156 (cp 41); SAO 125; WAB 133; WBV 97.

Tern, Noddy (Anous stolidus) ABW 130; ALE 8:211; CDB 91 (cp 318); DPB 113; GPB 166; HAB 82; HBG 140 (pl 11); KBA 156 (pl 19); RBU 48; SAO 135.

Tern, Roseate (Sterna dougallii) ABW 134; BBB 243; BBE 159; CDB 94 (cp 335); DPB 113; FBE 161; FBW 59; HAB 81; KBA 157 (pl 20); OBB 97; PBE 148 (pl 39), 156 (cp 41); SAO 131.

Tern, Royal (Sterna maxima) FBE 159; GPB 168; HBG 116 (bw), 140 (pl 11); (Thalasseus maximus) ABW 130; WAB 71.

Tern, Sandwich (Sterna sandvicensis) ALE 8:187; BBB 245; BBE 157; FBE 159; GPB 168; OBB 99; PBE 148 (pl 39), 156 (cp 41); PSA 176 (cp 19); SAO 133; SLP 33.

Tern, Sooty (Sterna fuscata) BBE 155; CAU 132, 227 (bw); CDB 94 (cp 336); DPB 113; FBE 163; GPB 165; HAB 82; HBG 118 (bw) 140 (pl 11); KBA 156 (pl 19); SAO 129; WBV 97.

Tern, South American (Sterna hirundinacea) SAO 133; WFI 174 (bw).

Tern, Whiskered (Chlidonias hybrida) BBB 273; BBE 155; CDB 91 (cp 319); DPB 109; FBE 163; KBA 157 (pl 20); LAW 505; OBB 95; PBE 149 (pl 40), 156 (cp 41); WBV 97; (Sterna hybrida) SAO 129.

Tern, White-capped Noddy (Anous tenuirostris) DPB 113; HAB 82; PWC 27 (bw); SAO 135; (A. minutus) CAU 132.

Tern, White-fronted (Sterna striata) BNZ 153; FNZ 145 (pl 14); HAB 82; WAN 125.

Tern, White-winged Black (Chlidonias leucoptera) BBB 275; BBE 155; DPB 109; FBE 163; FNZ 145 (pl 14); HAB 81; KBA 157 (pl 20); OBB 95; PBE 149 (pl 40), 156 (cp 41); PSA 176 (cp 19); WBV 97; (Sterna leucoptera) SAO 129.

Terrapin —See also **Tortoise** and **Turtle**.

Terrapin, Chinese (Chinemys reevesii) ALE 6:106; MAR 73 (pl 27).

Terrapin, Cope's (Hydromedusa tectifera) ALE 6:125; HRA 92.

Terrapin, Dura or Roofed (Kachuga tecta tecta) ALE 6:91, 106; HRA 86; MAR 69 (cp V); VRA 251.

Terrapin, Large-headed (Platysternon megacephalum) ALE 6:74; ERA 121; HRA 87; MAR 65 (pl 25).

Terrapin, Matamata (Chelys fimbriatus) ALE 6:125; ART 134; HRA 92; LAW 406; MAR 79 (pl 31); SIR 39 (bw).

Terrapin, Spiny (Geoemyda spinosa) ALE 6:91, 105; HRA 86.

Tetra (Hyphessobrycon rubrostigma) ALE 4:286; FEF 94 (bw), 115; WFW cp 69.

Tetra (Moenkhausia pittieri) FEF 76 (bw); SFW 113 (pl 20); (M. sanctaefilomenae) CTF 21 (cp 8); FEF 80.

Tetra, Black Phantom (Megalamphodus megalopterus) FEF 96 (bw); WFW cp 71.

Tetra, Buenos Aires (Hemigrammus caudovitatus) FEF 82 (bw); SFW 124 (cp 23); WFW cp 55.

Tetra, Cardinal (Cheirodon axelrodi) ALE 4:300; CTF 21 (cp 9); FEF 127; LAW 313; SFW 125 (cp 24); WFW cp 53.

Tetra, Congo (Micralestes interruptus) FEF 114 (bw); SFW 125 (cp 24); WFW cp 75.

Tetra, Emperor (Nematobrycon palmeri) ALE 4:286; FEF 153; WFW cp 82.

Tetra, Flag (Hyphessobrycon heterorhabdus) ALE 4:286; SFW 124 (cp 23); WFW cp 65.

Tetra, Flame or Red (Hyphessobrycon flammeus) ALE 4:286; FEF 88-90 (bw); HFI 87; SFW 81 (cp 12); WFW cp 63.

Tetra, Glow-light (Hemigrammus erythrozonus) FEF 82 (bw); SFW 117 (cp 22); WFW cp 56.

Tetra, Jewel (Hyphessobrycon callistus) ALE 4:286; CTF 19 (cp 4); HFI 87; HFW 37 (cp 11); SFW 81 (cp 12); WFW cp 61.

Tetra, Lemon (Hyphessobrycon pulchripinnis) CTF 19 (cp 5); SFW 116 (cp 21); WFW cp 67.

Tetra, Neon (Hyphessobrycon herbertaxelrodi) FEF 93 (bw); WFW cp 64; (H. innesi) CTF 20 (cp 6); SFW 117 (cp 22).

Tetra, Neon (Paracheirodon innesi) FEF 126; LAW 313.

Tetra, Ornate (Hyphessobrycon ornatus) HFI 87; SFW 81 (cp 12); WFW cp 66.

Tetra, Pretty (Hemigrammus pulcher) ALE 4:286; CTF 21 (cp 7); FEF 85 (bw); HFI 88; SFW 133 (pl 26); WFW cp 58.

Tetra, Red- or Rummy-nosed (Hemigrammus rhodostomus) ALE 4:286; SFW 137 (pl 28); WFW cp 59.

Tetra, Red Phantom (Megalamphodus sweglesi) FEF 96 (bw), 128; WFW cp 72.

Tetra, Serpae (Hyphessobrycon serpae) FEF 97; LAW 313; SFW 116 (cp 21); WFW cp 70.

Thatcheria, Miraculous (Thatcheria mirabilis) CTS 44 (cp 72); GSS 21.

Thick Plant —See **Moonstones.**

Thick-knee, Double-striped (Burhinus bistriatus) CDB 89 (cp 306); SSA pl 23.

Thick-knee, Great (Esacus magnirostris recurvirostris) DPB 107; KBA 149 (pl 18).

Thick-knee, Spotted (Burhinus capensis) CDB 89 (cp 307); LEA 414 (bw); PSA 145 (pl 16).

Thimble Flower, Blue —See **Gilia, Blue.**

Thistle, Blessed (Cnicus benedictus) AMP 17; EGH 109; PFE cp 160 (1509).

Thistle, Cabbage (Circium oleraceum) NHE 101; PFE cp 154 (1482); PWF 131e; TEP 47; WFB 245 (8).

Thistle, Carline (Carlina vulgaris) BFB cp 20 (1); EFC cp 32; EWF cp 22b; KVP cp 90; MBF cp 47; MFF cp 52 (901); NHE 189; OWF 39; PWF 1240; TEP 125; WFB 245 (2).

Thistle, Creeping or Canada (Circium arvense) BWG 152; EGC 73; EPF pl 584; MBF cp 48; MCW 107; MFF cp 31 (909); NHE 238; OWF 153; PFF 331; PWF 87j; TEP 181; WFB 247; WUS 397.

Thistle, Great Marsh or Mountain (Carduus personata) NHE 287; PFE cp 153 (1479); WFB 247 (8).

Thistle, Holy —See **Milk-thistle.**

Thistle, Marsh (Circium palustre) BFB cp 18 (4); MBF cp 48; MFF cp 31 (908); NHE 101; OWF 153; PWF 99j.

Thistle, Meadow (Circium dissectum) BFB cp 20 (2); MBF cp 48; MFF cp 31 (912); OWF 153; PWF 107f; WFB 247 (4).

Thistle, Melancholy (Circium heterophyllum) BFB cp 20 (3); MBF cp 48; MFF cp 30 (911); NHE 288; OWF 153; WFB 246; (C. helenoides) PWF 79j.

Thistle, Milk —See **Milk-thistle.**

Thistle, Musk or Nodding (Carduus nutans) BFB cp 18 (5); KVP cp 92; MBF cp 48; MFF cp 30 (904); NHE 238; OWF 151; PFE cp 152 (1478); PWF 95m; TEP 199; WFB 247 (8).

Thistle, Russian —See **Saltwort.**

Thistle, St. Barnaby's (Centaurea solstitialis) MBF cp 49; MFF cp 56 (922); PFE cp 155 (1499); PFM 46; WFB 249 (5); WUS 387.

Thistle, Scotch or Cotton (Onopordum acanthium) EPF pl 583; MBF cp 49; MFF cp 30 (915); NHE 239; OWF 153; PFE cp 155 (1494); PFW 86; PWF 113k; WFB 245 (6); (O. illyricum) PFE cp 155 (1495).

Thistle, Slender or Seaside (Carduus tenuiflorus) BFB cp 18 (3); MBF cp 48; MFF cp 30 (903); OWF 151; PWF 110e; WFB 247 (7).

Thistle, Spear or Bull (Circium vulgare) BFB cp 1 (4); BWG 152; EPF pl 585; MCW 108; MFF cp 30 (907); MLP 238; NHE 238; OWF 151; PWF 112c; WFB 247 (2); WUS 399.

Thistle, Spiniest (Circium spinosissimum)
KVP cp 154; NHE 288; PFE cp 154 (1483).

Thistle, Stemless Carline (Carlina acaulis)
CTH 61 (cp 99); EPF pl 578; NHE 189; PFE
cp 151 (1467); RWF 23; SPL cp 475; TEP 123;
WFB 245 (3).

Thistle, Stemless or Dwarf (Circium acaule
or acaulon) BFB cp 20 (6); MBF cp 48; MFF
cp 30 (910); NHE 238; OWF 153; PFE cp 154
(1490); PWF 106d; WFB 247 (5).

Thistle, Sticky (Cirsium erisithales) NHE
288; WFB 244.

Thistle, Syrian (Notobasis syriaca) PFE cp
153 (1481); PMF cp 206.

Thistle, Tuberous (Cirsium tuberosum) MFI
cp 30 (913); PFE cp 153 (1489); PWF 113e;
WFB 246.

Thistle, Welted (Carduus acanthoides) MBF
cp 48; MCW 101; NHE 238; PWF 77j; TEP
99; WFB 247 (9); (C. crispus) MFF cp 31
(905); NHE 238; OWF 151.

Thistle, Woolly (Cirsium eriophorum) DEW
2:204; EFC cp 33; EPF pls 579, 581; KVP cp
89; MBF cp 48; MFF cp 30 (906); NHE 37;
OWF 151; PFE cp 154 (1485); PWF 131j;
WFB 247 (3).

Thorn, Tambookie —See Coral Tree.

Thorn, Whistling (Acacia drepanolobium)
CAF 71; EET 229; MLP 108.

Thorn-apple (Datura stramonium) AMP 121;
BWG 84; CTH 49 (cp 71); DEW 2:151; EFC
cp 24; EHP 143; EPF pls 828, 829; HPW 228;
KVP cp 44; MFF pl 102 (675); MLP 214;
NHE 232; PFE cp 118 (1186); PFF 295; PWF
145i, 164h; TEP 197; WFB 207 (7); WUS
319.

Thornbill, Buff-tailed (Acanthiza regu-
loides) HAB 184; SAB 30 (cp 14).

Thornbill, Chestnut-tailed (Acanthiza
uropygialis) CDB 176 (cp 756); SAB 30 (cp14).

Thornbill, Little (Acanthiza nana) HAB 185;
RBU 215; SAB 30 (cp 14).

Thornbill, Rufous-fronted (Phacellodomus
rufifrons) ABW 198; ALE 9:125.

Thornbill, Western (Acanthiza inornata)
HAB 185; SAB 30 (cp 14).

Thornbill, Yellow-tailed (Acanthiza
chrysorrhoa) ALE 9:259; CDB 176 (cp 757);
HAB 184; RBU 216; SAB 30 (cp 14); WAB 203.

Thorn-tree (Acacia sp.) CAU 180; EET 229;
EGG 85; LFT 155 (cp 77); LFW 180 (cp 404);
PFM cp 50; WAW 114, 121.

Thorow-wax (Bupleurum rotundifolium)
MBF cp 36; MFF cp 58 (469); NHE 222; WFB
167 (7).

Thousand-jacket —See Lacebark.

Threadfin (Polynemus sp.) ALE 5:138; LMF
37 (bw).

Threadfish (Alectis ciliaris) FEF 350 (bw);
HFW 185 (bw).

Three-faces-under-a-hood —See Pansy,
Wild.

Thresher —See Shark, Thresher.

Thrift (Armeria caespitosa) DFP 2 (cp 14);
SPL cp 353; TVG 141; (A. fasciculata) PFE
cp 94 (976).

Thrift, Alpine (Armeria alpina) DEW 1:228;
NHE 279.

Thrift, Estorial (Armeria pseudarmeria)
HPW 75; NHE 186.

Thrift, Prickly (Acantholimon venustum)
EPF pl 310; EWF cp 48d; PFW 232; (A.
glumaceum) TGF 62 (bw).

Thrips, Onion (Thrips tabaci) EPD 128; OBI
19.

Throatwort (Trachelium caeruleum) BIP 177
(cp 275); EGA 158; HPW 255.

Thrumwort —See Star-fruit.

Thrush, Ant —See Ant-thrush.

Thrush, Babbling —See Babbler.

Thrush, Black-throated (Turdus ruficollis)
BBE 263; FBE 257; PBE 220 (cp 53).

Thrush, Cape —See Thrush, Olive.

Thrush, Chestnut-capped Ground (Zoothera interpres) DPB 299; KBA 336 (cp 51).

Thrush, Eye-browed (Turdus obscurus)
BBE 261; DPB 305; FBE 260; KBA 336 (cp
51); PBE 220 (cp 53).

Thrush, Golden or Scaly Ground (Zoothera
dauma) BBE 265; DPB 299; FBE 259; KBA
385 (pl 58); PBE 220 (cp 53); SAB 14 (cp 6).

Thrush, Kurrichane (Turdus libonyanus)
CDB 167 (cp 705); NBA 57.

Thrush, Laughing (Garrulax sp.) ABW 240;
ALE 9:221, 12:103; CDB 168 (cp 710); KBA
285 (pl 44), 288 (cp 45); RBI 167, 168, 171,
172, 175, 176; WBV 191, 205.

Thrush, Mistle (Turdus viscivorus) ALE 9:290; BBB 53; BBE 265; CDB 168 (cp 709); FBE 259; GAL 126 (cp 2); OBB 137; PBE 220 (cp 53); RBE 252.

Thrush, Naumann's (Turdus naumanni) BBE 261; FBE 260; PBE 220 (cp 53).

Thrush, New Zealand (Turnagra capensis) BNZ 19; FNZ 176 (cp 16).

Thrush, Nightingale (Catharus aurantiirostris) ABW 251; ALE 9:221; CSA 55; (C. fuscater) CDB 161 (cp 675); (C. dryas) SSA pl 29.

Thrush, Olive (Turdus olivaceus) CDB 167 (cp 708); NBA 57; PSA 256 (cp 29).

Thrush, Orange-headed Ground (Zoothera citrina) ALE 9:221, 289; KBA 336 (cp 51).

Thrush, Quail —See **Quail-thrush.**

Thrush, Ring —See **Ouzel, Ring.**

Thrush, Rock (Monticola sp.) ABW 253; ALE 9:289; BBE 255; CDB 165 (cps 690-692); DAE 78; DPB 299; FBE 243; KBA 336 (cp 51); PBE 221 (cp 54); PSA 256 (cp 29); RBI 200; WAB 116.

Thrush, Shrike —See **Shrike-thrush.**

Thrush, Siberian (Turdus sibiricus) BBE 263; FBE 261; PBE 220 (cp 53); (Zoothera sibirica) KBA 336 (cp 51).

Thrush, Song (Turdus philomelos) ALE 9:290; 13:163; BBB 52; BBE 265; DAE 134; FBE 259; GAL 126 (cp 2); HAB 174; OBB 137; PBE 220 (cp 53); RBE 255; SAB 14 (cp 6); (T. ericetorum) NBB 70 (bw), 95 (bw).

Thrush, Whistling (Myiophoneus caeruleus) ALE 9:280; KBA 385 (pl 58); WAB 161; (M. blighi) RBI 207.

Thrush, White's —See **Thrush, Golden or Scaly Ground.**

Thrush-nightingale —See **Nightingale, Thrush.**

Thrush-tanager, Rose-breasted (Rhodinocichla rosea) ABW 292; ALE 9:362; DTB 127 (cp 63); SSA cp 16.

Thylacine —See **Tasmanian Wolf.**

Thyme (Thymus sp.) EMB 142; EPF pl 816; EWF cp 5b, cp 20g; GDS cp 481; PWF 44c.

Thyme, Common (Thymus vulgaris) AMP 23; EGH 146; EVF 147; OFP 141; PFE cp 115 (1163); TGF 201.

Thyme, Large Wild (Thymus pulegioides)

MBF cp 68; PWF 121e; (T. drucei) MFF cp 28 (753); OWF 145.

Thyme, Lemon (Thymus citriodorus) EGH 145; EGL 147; OFP 141.

Thyme, Wild (Thymus serpyllum) AMP 105; DFP 27 (cp 214); EGC 148; EGH 145; MBF cp 68; NHE 74; OGF 89; PFE cp 115 (1164); PWF 125k; TEP 117; TVG 177; WFB 205 (10).

Thymelaea (Thymelaea hirsuta) PFE cp 73 (753); PFM cp 112; (T. passerina) NHE 220; WFB 139 (8); (T. tartonraira) PFE cp 73 (754); PFM cp 114.

Ti Plant (Cordyline terminalis) BIP 61 (cp 83); DFP 61 (cp 481); EFP 102; LHP 70, 71; MEP 16; MWF 93; RDP 156; SPL cp 12.

Tick, Bird (Argas sp.) ALE 1:437; GAL 464 (bw); GSP 138; OIB 143; PEI 15 (bw).

Tick Bush (Kunzea sp.) MLP 175; MWF 175.

Tick, Castor Bean or Wood (Ixodes sp.) GAL 463 (bw); GSP 139; LAW 139; OIB 143; PEI 17 (bw).

Tick, Dog (Rhipicephalus sp.) GSP 139; PEI 33 (cp 4).

Tick, Sheep (Melophagus ovinus) CIB 240 (cp 35); GAL 459 (bw); LEA 145 (bw); OBI 141.

Tick, Shield or Wood (Dermacentor sp.) GSP 138; OIB 143; PEI 18, 19 (bw).

Tickseed or **Tickweed** (Coreopsis sp.) DFP 35 (cp 274), 132 (cp 1056); EDP 114; EEG 133; EGA 110; EWG 100; LFW 186 (cp 416), 193 (cp 436); MWF 93; OGF 147; SPL cp 197; TGF 284; TVG 43, 85.

Tidy Tips (Layia sp.) DFP 41 (cp 323); EGA 131; EWG 121.

Tiger, Bengal or Indian (Panthera tigris or Felis tigris) ALE 12:345, 346, 356; 13:285; BMA 222-224; CAS 166, 178 (bw); DAA 62; DSM 196-197; HMA 94; HMW 119; JAW 136-137; LAW 544; LEA 571 (bw); LVS 92; MLS 319 (bw); PWC 138; SLS 115-116; SMW 179 (cps 101, 102); TWA 171; WEA 366.

Tiger Flower (Tigridia pavonia) DEW 2:250; DFP 111 (cp 886); EDP 146; EGB 141; EPF pl 948; LFW 269 (cp 606); MEP 23; MWF 284; OGF 117; PFW 146; TGF 29; TVG 209.

Tiger, Siberian or Manchurian (Panthera tigris altaica or Felis tigris altaica) ALE 12:346, 356; CAS 54 (bw); DAA 30; JAW 135; LVA 37; VWA cp 16.

Tiger, Tasmanian —See **Wolf, Tasmanian**.

Tiger-cat (Dasyurus maculatus or Dasyurops maculatus) ALE 10:71; CAU 59 (bw); DSM 26; LEA 483 (bw); MEA 90.

Tiger-moth —See **Moth**.

Tiger's Jaws (Faucaria tigrina) ECS 118; EGL 109; EWF cp 73e; HPW 66; RDP 203; (F. lupina) SCS cp 82; (F. tuberculosa) RDP 203.

Tilapia —See **Mouthbrooder**.

Tillandsia (Tillandsia cyanea) BIP 175 (cp 273); FHP 148; MEP 14; MWF 285; PFW 52; RDP 383, 384; (T. sp.) EGL 147; EPF pl 1034; EWF cp 188c; PFW 52.

Tinamou (Tinamus sp.) ALE 7:87; CDB 34 (cp 7); SSA cp 1; WAB 82; (Nothoprocta perdicaria) CDB 34 (cp 6).

Tinamou, Crested (Eudromia elegans) ABW 21; ALE 7:87.

Tinamou, Puna (Tinamotis pentlandii) ALE 7:87; SSA cp 1.

Tinamou, Red-winged (Rhynchotus rufescens) ALE 7:87; SSA cp 1.

Tinamou, Variegated (Crypturellus variegatus) ABW 21; SSA cp 1.

Tinkerbird, Golden-rumped (Pogoniulus bilineatus) ALE 9:97; CDB 133 (cp 528).

Tinkerbird, Yellow-fronted (Pogoniulus chrysoconus) CDB 134 (cp 529); GPB 223; PSA 225 (cp 26).

Tinnea (Tinnea aethiopica) EWF cp 63d; (T. galpinii) LFT cp 142.

Tit, Azure (Parus cyanus) ALE 9:307; BBE 269; FBE 267; PBE 213 (cp 52).

Tit, Bearded (Panurus biarmicus) ALE 9:221; BBB 181; BBE 271; CDB 170 (cp 721); FBE 265; FBW 63; GPB 277; OBB 131; PBE 213 (cp 52); RBE 243; WAB 124; WEA 369.

Tit, Blue (Parus caeruleus) ABW 235; ALE 9:307; BBB 59; BBE 269; CDB 185 (cp 803); FBE 267; LAW 509; OBB 131; PBE 213 (cp 52); WAB 136.

Tit, Coal (Parus ater) ALE 9:307; 13:163; BBB 159; BBE 269; FBE 267; PBE 213 (cp 52).

Tit, Crested (Parus cristatus) ALE 9:307; 13:163; BBB 158; BBE 269; CDB 185 (cp 804); FBE 269; GPB 290; NBB 2 (bw); OBB 131; PBE 213 (cp 52); (P. dichrous) RBI 240.

Tit, Gray (Parus afer) ALE 9:308; PSA 241 (cp 28).

Tit, Great (Parus major) ABW 235; ALE 9:307; BBB 58; BBE 269; CDB 185 (cp 805); DAE 133; FBE 267; KBA 320 (cp 49); PBE 213 (cp 52); RBE 239; RBI 239.

Tit, Long-tailed (Aegithalos caudatus) ABW 236; ALE 9:308; BBB 144; BBE 271; CDB 185 (cp 801); FBE 265; GPB 292; OBB 131; PBE 213 (cp 52); RBE 240.

Tit, Marsh (Parus palustris) ALE 9:302, 307; BBB 145; BBE 267; CDB 186 (cp 808); FBE 269; PBE 213 (cp 52).

Tit, Penduline (Remiz pendulinus) ABW 236; ALE 9:308; BBE 271; CDB 186 (cp 809); FBE 265; LEA 348 (bw); PBE 213 (cp 52); (Anthoscopus sp.) ALE 9:308; PSA 241 (cp 28).

Tit, Red-headed (Aegithalos concinnus) ABW 236; ALE 9:308; KBA 320 (cp 49); RBI 247; WBV 215.

Tit, Siberian (Parus cinctus) ALE 9:307; BBE 267; FBE 269; PBE 213 (cp 52).

Tit, Somber (Parus lugubris) BBE 267; FBE 269; PBE 213 (cp 52).

Tit, Southern Black (Parus niger) ALE 9:307; CDB 186 (cp 807); PSA 241 (cp 28).

Tit, Sultan (Melanochlora sultanea) DAA 145; KBA 320 (cp 49); RBI 243; WAB 164; WBV 215.

Tit, Varied (Parus varius) ABW 235; ALE 9:308; BBE 267; RBI 236.

Tit, Willow (Parus montanus) ALE 9:307; BBB 145; CDB 186 (cp 806); FBE 269; PBE 213 (cp 52).

Tit-babbler —See **Babbler, Tit**.

Titi, Dusky (Callicebus moloch) ALE 10:337; BMA 225; DSM 74; SMW 72 (bw).

Titi, Red (Callicebus cupreus) ALE 10:337; MLS 128 (bw).

Titmouse —See **Tit**.

Titoki (Alectryon sp.) DEW 2:58 (cps 25, 29); SNZ 87 (cp 255).

Tittle-my-fancy —See **Pansy, Wild**.

Tityra, Black-tailed (Tityra cayana) ALE 9:144; CDB 147 (cp 596); DTB 59 (cp 29).

Tityra, Masked (Tityra semifasciata) ABW 204; SSA pl 45.

Toad, Blomberg's (Bufo blombergi) CAW 80 (cp 32); MAR 26 (pl 8).

Toad, British or Natterjack (Bufo calamita) ALE 5:439, 455; CAW 189 (bw); ERA 104, 105; HRA 107; HWR 148; LAW 361; OBV 119.

Toad, Clawed (Xenopus laevis) ALE 5:375; ART 43; CAW 50-53 (bw); ERA 84, 85; HRA 110; HWR 160; LAW 354; LEA 271; (X. fraseri) MAR 17 (pl 3).

Toad, Cricket-voiced (Ansonia grillivoca) ALE 5:439; CAW 82 (cp 35).

Toad, Darwin's —See **Frog, Darwin's Dwarf.**

Toad, European (Bufo bufo) ALE 5:439, 456; 10:223; ART 37, 67; CAW 101, 190 (bw); ERA 86-89; HRA 107; HRW 145, 149; LAW 365; OBV 119.

Toad, False (Pseudobufo subasper) ALE 5:439; CAW 105 (bw).

Toad, Fire-bellied (Bombina bombina) ALE 5:361; CAW 54 (bw); ERA 92; HRA 109.

Toad Flower or Plant —See **Carrion Flower.**

Toad, Giant or Marine (Bufo marinus) ALE 5:439; ART 19; CAW 79 (cp 30); ERA 97, 98; HRA 108; LAW 362; MAR 26 (pl 8); VRA 139.

Toad, Green (Bufo viridis) ALE 5:439; CAA 122 (bw); CAW 78 (cp 29), 100 (bw); HRA 107; VRA 137.

Toad, Midwife (Alytes obstetricans) ALE 5:361, 376; ART 63; CAW 55 (bw); ERA 102; HRA 109; HWR 103; LEA 278 (bw); MAR 18 (pl 4).

Toad, Oriental Fire-bellied (Bombina orientalis) ALE 5:361, 376; CAW 76 (cp 23).

Toad, Spadefoot —See **Spadefoot, European.**

Toad, Surinam (Pipa pipa) ALE 5:361, 375; CAW 48-49 (bw), 75 (cp 22); ERA 113, 114; HRA 110; MAR 17 (pl 3).

Toad, Yellow-bellied (Bombina variegata) ALE 5:361, 376; 13:463; CAW 56 (bw); ERA 91; LAW 381; MAR 18 (pl 4).

Toadfish, Spotted (Thalassophryne maculosa) ALE 4:418; HFI 12.

Toadflax (Linaria sp.) EWF cp 38b; MFF cp 61 (686); NHE 136, 232; PFE cp 122 (1204); PFM cp 174.

Toadflax, Alpine (Linaria alpina) NHE 281; PFE cp 122 (1204); WFB 211 (5b).

Toadflax, Alpine Bastard (Thesium alpinum) NHE 183, 276; PFE cp 6 (71).

Toadflax, Annual (Linaria maroccana) DFP 41 (cp 328); EGA 132; MWF 186; OGF 135; TGF 220.

Toadflax, Bastard (Thesium humifusum) MBF cp 75; MFF cp 50 (447); OWF 65; PWF 110b.

Toadflax, Broad-leaved —See **Snapdragon, Large.**

Toadflax, Common (Linaria vulgaris) AMP 160; BFB cp 16; EPF pl 838; HPW 244; KVP cp 85; MBF cp 62; MCW 88; MFF cp 61 (685); NHE 232; OWF 25; PFF 296; PWF 87i; TEP 121; WFB 211 (3); WUS 329.

Toadflax, Daisy-leaved (Anarrhinum bellidifolium) NHE 232; PFE cp 123 (1211); WFB 211 (10).

Toadflax, Dalmatian (Linaria dalmatica) PFE cp 121 (1205); WUS 327.

Toadflax, Ivy-leaved (Cymbalaria muralis) BWG 146; EGL 101; MBF cp 62; MFF cp 7 (691); NHE 232; OWF 141; PFE cp 122 (1210); PFF 296; PWF 42b; TGF 221; WFB 211 (7).

Toadflax, Pale (Linaria repens) MBF cp 62; MFF cp 7 (684); PFE cp 122 (1202); PWF 129f; WFB 211 (5).

Toadflax, Purple (Linaria purpurea) BWG 145; DFP 156 (cp 1241); MBF cp 62; MFF cp 7 (683); PWF 93j; WFB 211 (6).

Toadflax, Sand or French (Linaria arenaria) MBF cp 62; MFF cp 61 (687); WFB 211 (4).

Toadflax, Small (Chaenorhinum minus) MBF cp 62; MFF cp 7 (688); NHE 232; OWF 141; PWF 138b; WFB 211 (9).

Toadflax, Three-leaved (Linaria triphylla) EWF cp 38e; PFE cp 122 (1208); PFM cp 173.

Toadlet, Corroboree (Pseudophryne corroboree) ALE 5:449, 455; CAW 94 (bw); WAW 100.

Toadstool, Soap-scented (Tricholoma saponaceum) DLM 262; LHM 85; NFP 60; ONP 135; PMU cp 73; PUM cp 67; REM 116 (cp 84); RTM 103; SMF 45.

Toadstool, Verdigris (Stropharia aeruginosa) DEW 3:cp 67; DLM 69, 254; KMF 103; LHM 145; NFP 83; ONP 3; PUM cp 133; RTM 55, 301; SMF 49; SMG 218.

Toatoa (Haloragis sp.) HPW 155; SNZ 136 (cp 422).

Tobacco, Large (Nicotiana tabacum) CTH 49 (cp 72); EPF pl 831; HPW 228; MWF 213; PFE cp 119 (1188); PFF 294.

Tobacco, Small (Nicotiana rustica) DEW 2:153; EPF pl 832; PFE cp 118 (1188); WFB 207 (8); (N. sp.) MEP 129; PFE cp 119 (1187); TVG 61.

Tobira —See Pittosporum, Japanese.

Toby (Arothron) —See Blowfish.

Tokay —See Gecko, Great House.

Tolpis (Tolpis barbata) PFE cp 157 (1515); PFM cp 214.

Tomato (Lycopersicum esculentum) EMB 149; EPF pl 818; EVF 39-41, 109; OFP 125; PFF 293; WFB 206.

Tomtit (Petroica macrocephala) BNZ 25; FNZ 201 (bw); WAN 134.

Tongue-leaf (Glottiphyllum sp.) EWF cp 73d; LCS cps 258, 260.

Tooth Shell —See Tusk Shell.

Toothcarp (Aphanius sp.) ALE 4:450; FEF 272 (bw); SFW 409 (pl 108), 420 (pl 109).

Toothwort (Lathraea squamaria) BFB 234; MBF cp 66; MFF cp 36 (732); NHE 36; OWF 125; PFE cp 130 (1269); PWF 33i; TEP 255; WFB 219 (1), 286.

Toothwort, Purple (Lathraea cladestina) PFE cp 130 (1269); WFB 219 (1a).

Top Shell (Gibbula sp.) CSS pl XIV; CTS 18 (cp 6); OIB 37.

Top Shell, Grooved (Cantharides sp.) CSS pl XIV; OIB 37.

Top Shell, Painted (Calliostoma sp.) ALE 3:45, 74; CSS pl XIV; CTS 18 (cp 5); GSS 31; OIB 37.

Top Shell, Pheasant (Tricolia pullus) CSS pl XIV; OIB 37.

Top Shell, Strawberry (Clanculus pharaonis) CTS 19 (cp 7); GSS 28.

Tope (Galeorhinus galeus) ALE 4:94; CSF 41; HFI 147; OBV 9.

Topi (Damaliscus korrigum) DDM 209 (cp 36); DSM 261; HMW 74; LAW 513; (D. lunatus topi) ALE 13:302, 414-415.

Topknot, Common (Zeugopterus punctatus) CSF 177; CSS pl XXIV; OBV 53; WFW cp 465.

Topknot, Norwegian (Phrynorhombus norvegicus) CSF 177; OBV 53.

Tor-grass —See False-brome.

Tormentil (Potentilla erecta) MBF cp 27; MFF cp 42 (356); NHE 179; OWF 17; PWF 57i; WFB 113 (2).

Torsk —See Cusk.

Tortoise —See also Terrapin and Turtle.

Tortoise, Australian Snake-necked (Emydura sp.) ALE 6:125; COG cp 3; SIR 39 (bw).

Tortoise, European Pond (Emys orbicularis) ALE 6:105; 11:257; CEU 33, 89; HRA 85; LAW 404; MAR 66 (pl 26); SIR 50 (cp 4); VRA 249.

Tortoise, Galapagos Giant (Testudo sp.) ALE 6:92; CSA 251; HRA 84; LBR 14, 15; LVS 160-161; MAR 74 (pl 28); SIR 29 (bw), 55 (cp 15); SLP 106; VRA 253; (Geochelone elephantopus) ART 100; ERA 142-143.

Tortoise, Greaved (Podocnemis sp.) ALE 6:116; ERA 124, 125; HRA 91.

Tortoise, Hinged (Kinixys sp.) ALE 6:91, 115; CAA 118 (bw); LEA 293 (bw); SIR 53 (cp 11).

Tortoise, Leopard (Testudo pardalis) ALE 6:115; CAA 118 (bw); ERA 141; HRA 83; MAR 77 (pl 29).

Tortoise, Long- or Snake-necked (Chelodina longicollis) ALE 6:91, 125; CAU 148 (bw); COG cp 2; ERA 131; HRA 92; LEA 295 (bw); MAR 78 (pl 30).

Tortoise, Mediterranean (Testudo hermanni) ALE 6:115; LAW 405; LBR 67; MAR 73 (pl 27); SIR 54 (cp 13).

Tortoise, Mediterranean or Greek (Testudo graeca) ALE 6:91; HRA 82; HWR 149; LAW 406; LBR cps 50, 66; LEA 292 (bw); SIR 54 (cp 12).

Tortoise, Pancake or Soft-shelled (Malacochersus tornieri) ALE 6:115; LEA 293 (bw).

Tortoise, Starred (Testudo elegans) ALE 6:91; HRA 82; LBR 56, 57; LEA 291 (bw); SIR 52 (cp 9); VRA 255.

Toucan, Ariel (Ramphastos ariel) ALE 8:359; 9:98; SBB 62 (cp 78).

Toucan, Cuvier's (Ramphastos cuvieri) ABW 188; ALE 9:98.

Treefrog, White's Green (Hyla caerulea) CAW 127 (cp 59), 139 (bw); VRA 147; WAW 99, 101 (bw).

Treehopper (Cyphonia sp.) CTI 22 (cp 16), 24 (cp 20); (Heteronotus sp.) CTI 24 (cp 21), 25 (cp 23); KIW 48 (cp 28); (Umbonia sp.) CTI 24 (cp 22); KIW 78 (bw); PEI 119 (bw).

Treehopper, Horned (Centrotus cornutus) GAL 446 (bw); OBI 33.

Treehunter (Thripadectes virgaticeps) CDB 141 (cp 564); (T. flammulatus) SSA cp 10.

Tree-kangaroo (Dendrolagus sp.) ALE 10:135 (cp 5); CAU 246 (bw); DSM 41; HMW 206; JAW 31; LEA 487 (bw); MLS 65 (bw); TWA 190; WAW 44 (bw).

Treepie, Racket-tailed (Crypsirina temia) KBA 276 (pl 41); WBV 155; (C. cucullata) RBI 276.

Treepie, Rufous (Dendrocitta vagabunda) CAS 195; KBA 276 (pl 41).

Tree-runner, Pearled (Margarornis squamiger) CDB 141 (cp 563); DTB 37 (cp 18); SSA cp 35.

Tree-shrew (Tupaia glis) ALE 10:253 (cp 7); DSM 54; MLS 113 (bw); SMW 46 (cp 33); (T. tana) HMW 193; (Ptilocercus lowii) ALE 10:253 (cp 7); DAA 142; (Urogale everetti) DSM 54.

Tree-snake —See **Snake.**

Tree-swift —See **Swift, Tree.**

Trefoil —See also **Clover** and **Hop-trefoil.**

Trefoil, Alpine (Trifolium alpinum) KVP cp 149; NHE 272; PFE cp 60 (603).

Trefoil, Bird's-foot (Lotus sp.) MBF cp 23; OFP 43.

Trefoil, Brown (Trifolium badium) NHE 272; PFE cp 60 (593).

Trefoil, Common Bird's-foot (Lotus corniculatus) BFB cp 8; BWG 32; DFP 156 (cp 1244); EGC 138; MBF cp 23; MFF cp 49 (307); NHE 180; OWF 21; PFF 213; PWF 38a; TEP 145; WFB 129 (1).

Trefoil, Greater Bird's-foot (Lotus uliginosus) MFF cp 49 (308); NHE 92; PFE cp 61 (614).

Trefoil, Hairy Bird's-foot (Lotus hispidus) MBF cp 23; MFF cp 49 (309); (L. subbiflorus) WFB 129 (2).

Trefoil, Least or Slender Bird's-foot (Lotus

angustissimus) MBF cp 23; MFF cp 49 (310); OWF 21.

Trefoil, Marsh Bird's-foot (Lotus pedunculatus) MBF cp 23; PWF 101f, 135m.

Trefoil, Pitch (Psoralea bituminosa) NHE 199; PFE cp 54 (537); PFM cp 73; (P. pinnata) MWF 246.

Trefoil, Silvery or Southern Bird's-foot (Lotus creticus) NHE 199; PFE 198.

Trefoil, Slender Yellow (Trifolium micranthum) MBF cp 23; MFF cp 48 (305); NHE 180.

Trevally (Caranx speciosus) LMF 71; WFW cps 294, 295.

Tricholoma, Leopard (Tricholoma pardinum) DLM 261; NFP 60; PUM cp 68; REM 118 (cp 86); RTM 98; SMG 150.

Triggerfish (Oxymonacanthus longirostris) ALE 5:256; MOL 115.

Triggerfish, Common (Balistes capriscus or B. carolinensis) HFI 13; LAW 352; OBV 43; (B. aculeatus) VWS 28.

Triggerfish, Pink-tail (Xanichthys ringens) HFW 258 (cp 121); WFW cp 480; (Balistes vidua) ALE 5:256.

Triggerfish, Queen (Balistes vetula) FEF 518 (bw); HFI 13; HFW 259 (cp 122), 276 (bw); WFW cp 470.

Triggerfish, Spotted (Balistoides conspicillum) FEF 527; HFI 13; HFW 258 (cp 120); WFW cp 471.

Triggerfish, Undulate (Balistapus undulatus) CTF 62 (cp 83); HFI 13; HFW 257 (cp 119); MOL 117; WFW cp 469.

Triller, Pied (Lalage nigra) DPB 243; KBA 252 (pl 35).

Triller, Varied (Lalage leucomela) HAB 147; SAB 12 (cp 5).

Triller, White-winged (Lalage sueurii) ABW 219; ALE 9:450; HAB 147; SAB 12 (cp 5); WAB 191.

Triton, Hairy (Cymatium sp.) CTS 36 (cps 51, 55, 56); GSS 69.

Triton's-trumpet (Charonia sp.) ALE 3:55; CTS 36 (cp 54), 38 (cp 57); GSS 68; VWS 88.

Triumfetta (Triumfetta sp.) HPW 91; LFT cp 104.

Trogon, Black-throated (Trogon rufus) CDB 135 (cp 483); DTB 11 (cp 5), WAB 91.

Trogon, Indian or Malabar (Harpactes fasciatus) RBI 111; WAB 167.

Trogon, Orange-breasted (Harpactes oreskios) ABW 172; KBA 176 (cp 23); RBI 115.

Trogon, Philippine (Harpactes ardens) DPB 195; RBI 112.

Tropic Bird, Red-billed (Phaethon aethereus) ABW 39; ALE 7:157; FBE 31; GPB 50; HBG 65 (pl 2); SAO 67; SSA pl 20; WAB 215.

Tropic Bird, Red-tailed (Phaethon rubricauda) ABW 39; ALE 7:157; BBT 113; CAS 222; HAB 19; MBA 56, 57; MEA 68; RBU 92; SNB 22 (pl 10).

Tropic Bird, White-tailed or Yellow-billed (Phaethon lepturus) ABW 39; ALE 7:157; BBT 113; CDB 38 (cp 38); HAB 19; SAO 65; SNB 22 (pl 10).

Trough Shell (Mactra sp.) ALE 3:168; CTS 62 (cp 120); GSS 151; OIB 77; (Spisula sp.) CSS pl XVIII; OIB 77.

Troupial (Icterus icterus) ABW 288; ALE 9:381; LEA 460 (bw); (I. chrysater) SSA cp 40.

Trout, Brook (Salvelinus fontinalis) ALE 4:224; GAL 246 (bw); FEF 26, 58 (bw); FWF 45; HFI 120; HFW 34 (cp 4); OBV 103; SFW 73 (cp 10); WFW cp 38.

Trout, Brown (Salmo trutta) ALE 4:224; 11:105; CSF 78, 79; FEF 26; FWF 25, 27, 29, 31, 33, 35; GAL 245 (bw); HFI 120; LEA 240 (bw); OBV 107, 109.

Trout, Rainbow (Salmo gairdneri) ALE 4:221-224, 237; CSF 79; FEF 53; FWF 37, 39; LEA 240 (bw); OBV 103; WFW cp 37; (S. irideus) HFI 121.

Truffle (Choiromyces meandriformis) EPF pl 47; PMU cp 5a.

Truffle, Black or Perigord (Tuber melanosporum) CTM 80; DEW 3:123; cp 48; DLM 98, 99, 128; REM 185 (cp 153); RTM 240.

Truffle, Deer (Elaphomyces granulatus) DLM 122; LHM 49; RTM 240.

Truffle, False (Rhizopogon luteolus) DLM 138; KMF 79; LHM 215.

Truffle, Summer (Tuber aestivum) DLM 127; EPF pl 46; LHM 43; NFP 37; OFP 189; PFF 71; RTM 240.

Truffle, White (Tuber magnatum) CTM 80; DLM 128; RTM 240.

Truffle, Winter (Tuber brumale) REM 185 (cp 153); RTM 240.

Trumpet Creeper (Campsis tagliabuana) DFP 245 (cp 1958); ELG 149; EPG 111; SPL cp 158.

Trumpet Creeper or **Trumpet Vine** (Campsis radicans) EGV 98; EPF pl 852; EWF cp 162b; HSC 23, 24; LFW 24 (cp 48); PFW 44; TGS 39; WUS 343.

Trumpet Creeper, Chinese (Campsis grandiflora) EWF cp 103e; MEP 134; MWF 72.

Trumpet Flower, China (Incarvillea delavayi) DEW 2:178; DFP 148 (cp 1183); EWF cp 103d; MWF 163; PFW 46; TGF 236; TVG 101.

Trumpet Flower (Incarvillea grandiflora) DFP 148 (cp 1184); EPF pl 856; SPL cp 214; (I. younghusbandii) EWF cp 103c.

Trumpet Tree (Tabebui sp.) LFW 266 (cp 597); MEP 139; SST cp 266.

Trumpet Vine, Yellow —See **Cat's Claw.**

Trumpetbird —See **Manucode.**

Trumpeter (Psophia leucoptera) ABW 106; ALE 8:100, 359; CDB 76 (cp 238); GPB 129; SSA cp 1; (P. crepitans) WAB 91.

Trumpetfish (Aulostomus sp.) ALE 5:26, 40; HFI 72; HFW 82 (cp 35), 153 (bw).

Trumpets —See **Pitcher Plant.**

Trunkfish —See **Boxfish.**

Tsessebe —See **Sassaby.**

Tuatara (Sphenodon punctatus) ALE 6:140; ERA 117-120; HRA 9; HWR 153; LBR cp 27; LEA 291 (bw); LVS 171; MAR 92 (pl 36), 129 (cp IX); PWC 183 (bw); SIR 57 (cp 18); SLP 154.

Tubfish —See **Gurnard, Tub or Yellow.**

Tubifex —See **Worm, Tubifex.**

Tucotuco (Ctenomys mendocinus) DSM 134; LEA 522 (bw); MLS 234 (bw); SMW 152 (bw); (C. knighti) ALE 11:435.

Tui (Prosthemadera novaeseelandiae) ABW 276; ALE 9:320; BNZ 53; CAU 264 (bw); CDB 195 (cp 850); FNZ 176 (cp 16); LEA 457 (bw); WAB 210; WAN 133.

Tulbaghia, Fragrant (Tulbaghia fragrans) EGB 142; FHP 149; LFT cp 11; (T. acutiloba) LFT cp 10.

Tulip, Southern (Tulipa australis) KVP cp 162; PFE cp 168 (1625).

Tulip Tree (Liriodendron tulipifera) BKT cps 113, 114; DEW 1:77, 78; DFP 210 (cp 1678); EEG 118; EET 160; EGT 122; EPF cp III, pl 180; EWF cp 150e; HPW 27; MTB 257 (cp 24); MWF 187; OBT 145; PFF 176; PFW 181; SST cp 250; TGS 183.

Tulip Tree, African (Spathodea campanulata) DEW 2:168 (cp 91); EET 30; EWF cp 62b; GFI 143 (cp 32); LFW 176 (cp 398); MEP 138; MLP 210; MWF 271; PFW 45; SST cp 264.

Tulip, Wild (Tulipa sylvestris) MBF cp 85; MFF pl 63 (998); NHE 192; PFM cp 226; WFB 263 (8).

Tulp, Yellow (Moraea sp.) LFT cp 36; MWF 203.

Tumblebug —See **Beetle, Dung** (Phanaeus sp.).

Tun Shell, Giant (Tonna galea) ALE 3:55; GSS 70.

Tuna, Bluefin (Thunnus thynnus) ALE 5:191; CSF 143; HFI 20; LMF 39 (bw); OBV 27.

Tuna, Long-fin —See **Albacore.**

Tuna, Yellowfin (Thunnus albacares) HFI 20; LMF 41 (bw); MOL 96.

Tunic Flower (Petrorhagia saxifraga or Tunica saxifraga) EWF cp 13a; NHE 276; TGF 71; TVG 177; WFB 63 (7); (P. prolifera) NHE 225; WFB 63 (8).

Tunicate —See **Sea Squirt.**

Tunny —See **Tuna, Bluefin.**

Tupai —See **Tree-shrew.**

Tur, Caucasian (Capra caucasica) DAA 20-21; HMW 85.

Turaco (Tauraco leucotis) CAA 33; LEA 417; (T. persa) LAW 504; (T. ruspollii) LVB 9; (T. schuttii) WAB 143.

Turaco, Gray —See **Go-away Bird, Common.**

Turaco, Great Blue (Corytheola cristata) ALE 8:345; CDB 109 (cp 410); GPB 181.

Turaco, Knysna (Tauraco corythaix) ABW 150; ALE 8:345; PSA 192 (cp 21).

Turaco, Livingstone's (Tauraco livingstonii) ALE 8:345; CDB 111 (cp 416); WEA 375.

Turaco, Ross' (Musophaga rossae) BBT 81; CDB 111 (cp 415).

Turaco, Violet-crested (Tauraco porphyreolophus) ALE 8:345; PSA 192 (cp 21).

Turban Shell (Turbo sp.) CTS 18 (cp 4); GSS 32, 34, 35; MOL 69.

Turbot (Scophthalmus maximus) ALE 5:233; CSF 173; FEF 510 (bw); HFI 19; MOL 107; OBV 55.

Turkey, Brush (Alectura lathami) ABW 85; ALE 7:435; CAU 116; HAB 51; SNB 58 (pl 28).

Turkey Fish —See **Lionfish.**

Turk's Cap (Melocactus sp.) ECS 31, 77, 131; LCS cp 308.

Turnip (Brassica rapa) BFB cp 3 (11); EVF 110; MBF cp 9; NHE 215; OFP 173; PFF 186; PWF 51d; WFB 89 (2).

Turnstone, Ruddy (Arenaria interpres) ABW 120; ALE 7:61; BBB 237; BBE 117; CDB 84 (cp 275); DPB 91; FBE 123; FNZ 129 (pl 12); GPB 148; HAB 70; HBG 93 (bw), 133 (pl 10); KBA 129 (pl 14), 149 (pl 18); OBB 79; PBE 93 (pl 30), 100 (cp 31); RBE 135; SNB 76 (cp 37), 78 (pl 38); WBV 91.

Turpentine Tree (Pistacia terebinthus) AMP 171; PFE cp 69 (704); PFM cps 95, 96.

Turtle —See also **Terrapin** and **Tortoise.**

Turtle, Green (Chelonia mydas) ALE 6:89; ART 240-241; CAA 119 (bw); CAU 141 (bw); COG cp 4; HRA 89; LBR 94 (bw), 106; LEA 295 (bw); LVS 156; MAR 62 (pl 24); SIR 32 (bw).

Turtle, Hawksbill (Eretmochelys imbricata) ALE 6:116; ART 239; ERA 122, 123; HRA 89; HWR 155; LAW 406; LBR 102, 103, 119; LVS 156; VRA 257; WEA 376.

Turtle, Leatherback (Dermochelys coriacea) ALE 6:90, 116; ART 166; CAS 282; ERA 126, 127; HRA 91; HWR 156; LAW 407; LBR 90-91 (bw); LEA 296 (bw); LVS 158; MAR 62 (pl 24); OBV 121; SIR 30 (bw); SLP 199.

Turtle, Loggerhead (Caretta caretta) ALE 6:90; HRA 89; LAW 386; LEA 296 (bw); LVS 157; OBV 121; WAW 93 (bw).

Turtle, Soft-shelled (Trionyx sp.) ALE 6:116; ERA 136, 137; HRA 93; HWR 152; MAR 80 (pl 32).

Turtle, Swamp —See **Tortoise, European Pond.**

Turtle-dove —See **Dove.**

Turutu (Blueberry) (Dianella intermedia)
SNZ 88 (cp 258), 89 (cp 259); (D. tasmanica)
EWF cp 142b.

Tusk Shell (Dentalium sp.) ALE 3:139; OIB
33; (Fissidentalium sp.) CTS 55 (cp 102).

Tutsan (Hypericum androsaemum) BFB cp 5
(8); MBF cp 17; MFF cp 41 (130); PFE cp 74
(763); PWF 76c, 163h; WFB 145 (1).

Twayblade, Common (Listera ovata) BFB cp
24 (6); EWF cp 26b; MBF cp 80; MFF pl 65
(1067); NHE 40; OWF 45; PWF 92c; WFB
281 (8).

Twayblade, Lesser (Listera cordata) EPF pl
974; MBF cp 80; MFF pl 83 (1068); NHE 79;
OWF 45; PFE cp 192 (1923); WFB 281 (7).

Twinflower (Linnaea borealis) DEW 2:140;
EWG 29; KVP cp 157; MBF cp 41; MFF cp 23
(824); NHE 25; PFE cp 135 (1307); PFF 304;
PWF 88a; TGF 188; WFB 223 (8).

Twin-leaf (Jeffersonia dubia) DFP 13 (cp 97);
HPW 45; PFW 233; (J. diphylla) ERW 124;
EWG 120; TGF 86.

Twinspot, Green-backed (Mandingoa
nitidula) ALE 9:440; PSA 316 (cp 39);
(Estrilda nitidula) ABW 303.

Twisted Stalk (Streptopus amplexifolius)
NHE 38; PFE 504; (S. roseus) EWG 140.

Twitch —See **Couch** and **Quackgrass.**

Twite (Acanthis flavirostris) ALE 9:391;
12:103; BBB 125; BBE 283; FBE 291; PBE
260 (cp 61); (Carduelis flavirostris) OBB
179.

Tyrant, Ground (Muscisaxicola sp.) SSA pl
48; WFI 183 (bw).

Tyrant, Pygmy (Myiornis auricularis) ALE
9:143; (Pseudotriccus sp.) SSA cps 12, 39.

Tyrant, Water (Flavicola pica) ALE 9:143;
SSA pl 48; (Ochthornis littoralis) SSA pl 28.

U

Uakari, Bald or Red-faced (Cacajao calvus
or C. rubicundus) ALE 8:359; 10:327; BMA
226-228; DSM 74; HMA 71; LAW 612; LEA
504 (bw); LVS 50, 51; MLS 129 (bw); PWC
170; SMW 74 (bw).

Ulluco (Ullucus tuberosus) HPW 76; OFP
179.

Ulua —See **Trevally.**

Umbrella Plant (Cyperus alternifolius) EFP
105; EPF pl 955; ERW 140; LHP 78; RDP 8,
169, 170; SPL cp 499.

Umbrella Plant, Dwarf (Cyperus diffusus)
BIP 67 (cp 92); EGL 102; LHP 79.

Umbrella Tree, Queensland (Brassaia
actinophylla or Schefflera actinophylla)
EDP 126; EFP 18, 94; LHP 164, 165; MEP
102; MWF 61; RDP 103; WAW 115.

Umbrella-bird, Amazonian (Cephalopterus
ornatus) ABW 205; ALE 9:144; CDB 146 (cp
590); WAB 91; (C. penduliger) SSA pl 27.

Unau —See **Sloth, Two-fingered or
Two-toed.**

Uncarina (Uncarina sp.) DEW 2:180; EWF cp
71c.

Unicorn Fish, Striped-face (Naso lituratus)
ALE 5:201; WFW cp 445.

Upside-down Tree —See **Baobob.**

Urial —See **Sheep, Wild.**

Urn Plant (Aechmea fasciata or Billbergia
rhodocyanea) BIP 19 (cp 6); EDP 136; EGL
81; EPF pl 1030; FHP 103; LHP 17-18; PFW
53; RDP 20, 71; SPL cp 36.

Urn Plant (Aechmea fulgens) BIP 19 (cp 5);
DFP 51 (cp 402); LHP 18; PFW 53.

Ursinia (Ursinia sp.) DFP 49 (cps 389, 390);
EGA 159; HPW 266; OGF 139; RWF 64; TGF
268; TVG 73.

Urutu (Bothrops alternatus) HRW 105; MAR
pl 77; SIR cp 137.

V

Vetch, Hairy (Vicia hirsuta) MBF cp 24; MFF cp 5 (323); NHE 213; OWF 135; PWF 37f; WFB 123 (6).

Vetch, Horsehoe (Hippocrepis comosa) MBF cp 24; MFF cp 49 (321); NHE 72; OWF 23; PWF 38d; WFB 129 (4).

Vetch, Narrow-leaved (Vicia angustifolia) BFB cp 8 (5); MBF cp 25; WUS 239.

Vetch, Sand or Fodder (Vicia villosa) EWF cp 10a; NHE 213; PFE cp 55 (545); PFF 217; WFB 123 (1c).

Vetch, Tufted or Cow (Vicia cracca) BFB cp 8 (9); BWG 140; MBF cp 24; MCW 66; MFF cp 5 (324); NHE 213; OWF 135; PFE cp 55 (545); PWF 85i; TEP 143; WFB 123 (1).

Vetch, Wood (Vicia sylvatica) MBF cp 24; MFF cp 5 (326); NHE 28; PWF 116c; WFB 123 (2).

Vetch, Yellow (Vicia lutea) EWF cp 29d; KVP cp 183; MBF cp 25; MFF cp 49 (328); NHE 213; WFB 123 (7).

Vetchling, Grass (Lathyrus nissolia) MBF cp 25; MFF cp 48 (332); NHE 213; OWF 133; PFE cp 55 (556); PWF 103e; WFB 125 (9).

Vetchling, Hairy (Lathyrus hirsutus) MFF cp 18 (333); NHE 213; PFE 198; WFB 125 (8).

Vetchling, Meadow (Lathyrus pratensis) KVP cp 29; MBF cp 25; MFF cp 18 (334); NHE 181; OWF 23; PWF 84b; TEP 143; WFB 125 (2).

Vetchling, Yellow (Lathyrus aphaca) MBF cp 25; MFF cp 48 (331); NHE 213; OWF 21; PFE cp 55 (553); PFM cp 77; PWF 48b; WFB 125 (10).

Viburnum, Leatherleaf (Viburnum rhytidophyllum) DFP 244 (cp 1947); EGE 148; EGW 146; EPF pl 775; TGS 343.

Victorian Christmas Bush —See **Mintbush.**

Vicuna (Lama vicuna or Vicugna vicugna) ALE 13:133; DSM 225; JAW 172; LVA 31; LVS 133; MLS 379 (bw); SLP 108; SLS 51; SMW 248 (bw).

Violet —See also **Pansy** (Viola sp.).

Violet (Viola sp.) DFP 28 (cp 220), 50 (cps 393-394); 178 (cp 1424); EDP 148; EGA 161; EGG 147; EPF pl 508; EWF 159a, b, c; EWG 145; LFW 148 (cps 329-333); MBF 12; MLP 151; NHE 30, 181, 220; OGF 55; PFW 306; SNZ 111 (cp 340); SPL cp 278; TGF 158; TVG 75; WFB 147 (4, 5).

Violet, African (Saintpaulia hybrids) EGL 56, 141; EMB 27, 42, 129; EPF pl 864; RDP 26, 344, 345.

Violet, African (Saintpaulia ionantha) BIP 157 (cp 241); DEW 2:169 (cp 93), 179; DFP 80 (cps 637-640); EDP 120; EGG 139; EPF pl 865; FHP 142; LFW 174 (cp 391), 175 (cps 392, 393); LHP 158, 159; MEP 41; MWF 260; PFW 127; SPL cp 55.

Violet, Arabian or German (Exacum affine) BIP 81 (cp 120); EGL 108; FHP 120; PFW 122; RDP 200.

Violet, Australian (Viola hederacea) EWF cp 135b; HPW 100; PFW 306.

Violet, Common Dog (Viola riviniana) EWF cp 18e; MBF cp 12; MFF cp 1 (117); NHE 30; OWF 165; PFE cp 76 (777); PWF cp 22b; TEP 241; WFB 147 (2).

Violet, Fen (Viola stagnina) MBF cp 12; MFF cp 1 (121); NHE 181.

Violet, Hairy (Viola hirta) MBF cp 12; MFF cp 1 (116); NHE 220; OWF 165.

Violet, Heath Dog (Viola canina) MBF cp 12; MFF cp 1 (119); NHE 181; OWF 165; WFB 147 (3).

Violet, Marsh or Bog (Viola palustris) EWF cp 4e; MBF cp 12; MFF cp 1 (122); NHE 181; OWF 165; PFE cp 76 (773); PWF 26; WFB 147 (6).

Violet, Pale Heath (Viola lactea) MBF cp 12; MFF cp 1 (120); WFB 147 (3a).

Violet, Pale Wood (Viola reichenbachiana) MBF cp 12; MFF cp 1 (118); NHE 30; OWF 165; PWF cp 19j; WFB 147 (2b).

Violet, Philippine (Barleria cristata) MEP 143; MWF 53; (B. sp.) LFT cps 158, 159.

Violet, Sweet (Viola odorata) AMP 89; DEW 1:243; EGH 149; ESG 148; MBF cp 12; MFF cp 1 (115); MWF 294; NHE 30; OWF 165; PFE cp 76 (774); PFW 305; PWF cp 17g; SPL cp 277; TEP 239; TGF 140; WFB 147 (1).

Violet, Teesdale (Viola rupestris) MBF cp 12; NHE 72.

Violet, Upright (Viola elatior) NHE 30; PFE 257.

Violet, Yellow Wood (Viola biflora) EPF pl
505; KVP cp 140; NHE 275; PFE cp 76 (782);
WFB 147 (7).

Violet Tree (Securidaca longipedunculata)
LFT cp 92; MEP 68.

Violet-ear, Sparkling (Colibri coruscans)
ABW 170; ALE 8:453; BBT 87; CDB 122 (cp
470); GPB 201; SHU 59, 60; SSA cp 32.

Viper —See also **Adder.**

Viper, Asp (Vipera aspis) ALE 6:467; DAE
27; ERA 207; HRA 43; LAW 425, 431; MAR
184 (pl 74); SIR 258 (bw); SSW 123.

Viper, Bush or Tree (Atheris squamiger)
ALE 6:467; HWR 159; MAR 189 (cp XV).

Viper, Horned (Cerastes cerastes) ALE
6:478; ART 117; HRA 45; SSW 125; (C.
cornutus) LEA 327 (bw); SIR 263 (bw).

Viper, Long-nosed or Sand (Vipera
ammodytes) HRA 43; MAR 183 (pl 73); SIR
259 (bw); SSW 122; VRA 241.

Viper, Northern (Vipera berus) ALE 6:467;
13:163; ART 190; DAE 148; ERA 203, 204;
HRA 42; HWR 158; LAW 431; LBR 47; MAR
175 (pl 71), 176 (pl 72); OBV 125; SIR 148 (cp
128); SSW 121.

Viper, Orsini's (Vipera ursinii) ALE 6:467;
DAE 27; SIR 257 (bw).

Viper, Pit (Bothrops schlegeli) ERA 230, 231;
MAR 193 (cp XVI); SIR 269 (bw).

Viper, Pit or Tree (Trimeresurus sp.) ALE
6:468, 477, 478; CAS 200, 216 (bw); ERA 229;

Viper, Rhinoceros (Bitis nasicornis) ALE
6:467; ART 163; CAA 109; ERA 232, 233;
HRA 44; SIR 250 (cp 133).

Viper, Russell's (Vipera russelli) ART 124;
DAA 153; HRA 44; SIR 261 (bw); SSW 124;
VRA 243.

Viper, Saw-scaled (Echis carinatus) CAS
103 (bw), 104; SSW 126.

Vipergrass, Common (Scorzonera humilis)
MFF cp 58 (937); NHE 190; WFB 251 (4); (S.
hispanica) OFP 173; (S. austriaca) PFE 469.

Vipergrass, Purple (Scorzonera purpurea)
NHE 190; WFB 251 (3).

Vireo, Shrike —See **Shrike-vireo.**

Virginia Creeper (Parthenocissus quinquefolia) EGC 143; EGV 125; EPF pl 725; ESG
134; PFF 246; SPL cp 177; TGS 75; TVG 235.

Viscacha, Plains (Lagostomus maximus)
ALE 11:438; DSM 134; HMW 176; MLS 231
(bw); SMW 154 (bw).

Vole, Alpine or Snow (Microtus nivalis)
ALE 11:300; GAL 53 (bw); LAW 588.

Vole, Bank (Clethrionomys glareolus) ALE
11:300; GAL 52 (bw); HMW 170; LAW 588;
LEA 561; OBV 131; (C. rufocanus) CAS 31
(bw).

Vole, Common (Microtus arvalis) ALE
10:223; 11:300, 321; DAE 64; GAL 51 (bw);
OBV 131; (M. orcadensis) MLS 211 (bw).

Vole, Field or Short-tailed (Microtus
agrestis) GAL 50 (bw); HMW 170; LAW 588;
OBV 131; SMW 131 (cp 74).

Vole, Water (Arvicola terrestris or A.
amphibius) ALE 11:105, 257; CEU 200;
DAE 64; DSM 127; GAL 54 (bw); HMW 169;
LEA 537 (bw); OBV 131; TWA 97 (bw).

Volute (Volute sp.) GSS 97; VWS 88.

Volute, Australian (Amoria sp.) CTS 46 (cp
76), 47 (cp 77); GSS 104.

Volute, Bat (Aulica vespertilio) GSS 98; MOL
67.

Volute, Imperial (Aulica imperialis) GSS 17,
98; MOL 67.

Volute, West African (Cymbium sp.) CTS 47
(cps 79, 80); GSS 103.

Volvaria (Volvaria sp.) DLM 264; NFP 77;
PUM cp 144b; REM 67 (cp 35); RTM 40-42,
303.

Vorticella (Vorticella sp.) ALE 2:305; OIB 1.

Vulture, African White-backed (Gyps
africanus or Pseudogyps africanus) ALE
7:329, 333; CDB 58 (cp 148); GPB 93; PSA 80
(pl 7); WAB 139.

Vulture, Bearded (Gypaetus barbatus) ALE
7:397; 12:103; 13:463; BBE 71; CDB 57 (cp
146); DAE 14-15; FBE 83; GAL 203 (bw);
GBP 345 (bw), 346 (bw); GPB 93; LAW 498;
LVB 47; LVS 239; RBE 40; WAB 101, 163;
WEA 379.

Vulture, Black (Aegypius monachus) BBE
71; FBE 85; GBP 334 (bw); LEA 382 (bw);
RBE 39.

Vulture, Cape (Gyps coprotheres) CAF 227
(bw); PSA 80 (pl 7), 96 (cp 9).

Vulture, Eared —See **Vulture, Lappet-
faced.**

W

Wallaby, Parma or White- throated (Macropus parma or Wallabia parma) ALE 10:136 (cp 6); LVS 31; SLS 231.

Wallaby, Pretty-faced (Wallabia parryi) DSM 38; WAW 42 (bw); (W. canguru) ALE 10:137 (cp 7); (W. elegans) LEA 487 (bw); (Protemnodon elegans) SMW 40 (cp 14).

Wallaby, Rock —See **Rock-wallaby.**

Wallaroo, Gray (Macropus robustus) ALE 10:138 (cp 8); WAW 41 (bw).

Wall-creeper (Tichodroma muraria) ABW 238; ALE 9:319; 13:463; BBE 275; DAE 47; FBE 273; OBB 133; PBE 212 (cp 51); RBE 247.

Wallflower (Cheiranthus cheiri) AMP 151; BFB cp 3 (6); DFP 5 (cp 34), 33 (cp 262), 128 (cps 1021-1023); EGA 105; LFW 24 (cp 47); MFF cp 41 (105); MWF 82; NHE 219; OGF 33; PFE cp 32 (299); PFW 97; PWF 49h; SPL cp 332; TGF 77; TVG 39; WFB 87 (1).

Wallflower, Siberian (Cheiranthus allionii) DFP 33 (261); EGA 105; OGF 33.

Wall-rue (Asplenium ruta-muraria) CGF 101; DEW 3: Fig. 375; MFF pl 139 (1287); NFP 136; NHE 245; ONP 73; TEP 89.

Walnut, Black (Juglans nigra) BKT cp 86, 88; EET 137; EPG 126; EVF 136; MTB 192 (cp 17); OFP 29; PFF 133; SST cp 205; TGS 81.

Walnut, English or Persian (Juglans regia) BKT cp 90; DEW 1:190; EET 136, 137; EGT 119; EFP pls 327, 328; EVF 137; HPW 205; MLP 75; MTB 192 (cp 17); NHE 294; OBT 193; OFP 29; PFE cp 4 (35); PFF 133; PWF 25j, 175k; SST cp 206; TGS 86; WFB 31 (5).

Walrus (Odobenus rosmarus) ALE 12:374, 399; CAS 19 (bw); DSM 202; HMW 116; JAW 144; LEA 578 (bw); OBV 181; SLP 80, 81; TWA 147 (bw); VWA cp 18; WEA 383; WID 90.

Wand Flower, South African (Dierama pulcherrima) DFP 91 (cp 725); MWF 113; PFW 150; (D. medium) LFT 79 (cp 39).

Wandering Jew (Tradescantia fluminensis) BIP 177 (cp 277); EFP 146; EGV 140; LHP 182; MWF 286; PFW 77; RDP 386, 387; SPL cp 348.

Wandering Jew (Zebrina pendula) BIP 183 (cp 284); DEW 2:256; EDP 144; EFP 146; EGL 149; EGV 145; EPF pl 1026; EWF cp 187d; HPW 280; LHP 182; PFF 369; PFW 77; RDP 398.

Wandering Jew, White Velvet (Tradescantia sillamontana) EGL 148; EWF cp 187e; HPW 280; RDP 386.

Waratah (Telopea speciosissima) MEP 34; MWF 280; PFW 247; WAW 118.

Warbler (Seicercus sp.) DPB 315; KBA 337 (cp 52).

Warbler, Aquatic (Acrocephalus paludicola) ALE 9:231; BBB 269; BBE 229; FBE 227; OBB 147; PBE 228 (cp 55).

Warbler, Arctic (Phylloscopus borealis) BBB 270; BBE 245; CEU 247; DPB 315; FBE 239; KBA 352 (cp 53); OBB 157; PBE 244 (cp 57).

Warbler, Australian Reed (Acrocephalus australis) HAB 182; RBU 220; SAB 26 (cp 12).

Warbler, Barred (Sylvia nisoria) ALE 9:232; BBB 275; BBE 235; CDB 176 (cp 754); FBE 231; GAL 128 (cp 3); OBB 151; PBE 229 (cp 56); (Camaroptera fasciolata) PSA 272 (cp 31).

Warbler, Black-throated (Gerygone palpebrosa) ALE 9:259; HAB 178; SAB 28 (cp 13).

Warbler, Blyth's Reed (Acrocephalus dumetorum) BBE 231; FBE 225.

Warbler, Bonelli's (Phylloscopus bonelli) ALE 9:259; BBE 245; FBE 237; OBB 157; PBE 244 (cp 57).

Warbler, Booted (Hippolais caligata) BBE 233; FBE 229.

Warbler, Bush (Cettia diphone) ABW 257; DPB 307; (C. squameiceps) CDB 172 (cp 733); (Bebrornis sechellensis) LVB 63.

Warbler, Canegrass —See **Grassbird.**

Warbler, Cetti's (Cettia cetti) ALE 9:231; BBE 225; CDB 172 (cp 732); FBE 227; PBE 228 (cp 55).

Warbler, Clamorous Reed (Acrocephalus stentoreus) CDB 171 (cp 726); FBE 225.

Warbler, Dartford (Sylvia undata) BBB 104; BBE 241; DAE 79; FBE 235; FBW 85; OBB 153; PBE 229 (cp 56); WAB 116.

Warbler, Desert (Sylvia nana) BBE 239; FBE 235.

Warbler, Dusky (Phylloscopus fuscatus)
BBE 243; FBE 237; KBA 352 (cp 53); (Gerygone tenebrosa) HAB 183; SAB 28 (cp 13).

Warbler, Fan-tailed (Cisticola juncidis)
ABW 256; ALE 9:231; BBE 229; DPB 315;
FBE 227; GPB 280; KBA 353 (cp 54); PBE
228 (cp 55); PSA 272 (cp 31); SAB 26 (cp 12);
WBV 179; (C. aridula) CDB 173 (cp 734).

Warbler, Garden (Sylvia borin) ALE 9:232;
BBB 141; BBE 235; CDB 175 (cp 750); FBE
231; GAL 128 (cp 3); OBB 151; PBE 244 (cp
57); PSA 272 (cp 31).

Warbler, Grasshopper (Locustella naevia)
ALE 9:231; BBB 105; BBE 227; CDB 174 (cp
742); FBE 223; GAL 128 (cp 3); OBB 147;
PBE 228 (cp 55).

Warbler, Gray (Gerygone igata) BNZ 71;
CDB 177 (cp 761); FNZ 193 (cp 17).

Warbler, Gray's Grasshopper (Locustella
fasciolata) CDB 174 (cp 741); DPB 313.

Warbler, Great Reed (Acrocephalus arundinaceus) ABW 256; ALE 9:231, 242; 11:257;
BBB 269; BBE 231; DPB 313; FBE 225; GAL
128 (cp 3); KBA 353 (cp 54); OBB 149; PBE
228 (cp 55); RBE 264; SAB 26 (cp 12).

Warbler, Green (Phylloscopus nitidus) BBE
243; FBE 239.

Warbler, Green-backed (Gerygone
chloronota) HAB 183; SAB 28 (cp 13).

Warbler, Greenish (Phylloscopus trochiloides) ALE 9:259; BBB 275; BBE 245; FBE
239; OBB 157; PBE 244 (cp 57); WBV 179.

Warbler, Icterine (Hippolais icterina) ALE
9:232; BBB 269; BBE 233; FBE 229; GAL 128
(cp 3); OBB 157; PBE 244 (cp 57); RBE 267.

Warbler, Lanceolated Grasshopper
(Locustella lanceolata) BBE 227; DPB 313;
FBE 223; PBE 284 (cp 65).

Warbler, Large-billed (Gerygone magnirostris) HAB 183; SAB 28 (cp 13).

Warbler, Leaf (Phylloscopus olivaceus) DPB
315; (P. tenellipes) KBA 352 (cp 53).

Warbler, Marmora's (Sylvia sarda) BBE
241; FBE 235; PBE 229 (cp 56).

Warbler, Marsh (Acrocephalus palustris)
ALE 9:232; BBB 185; BBE 231; CDB 171 (cp
724); FBE 225; GAL 128 (cp 3); GPB 281;
OBB 149; PBE 228 (cp 55); (A. baeticatus)
PSA 272 (cp 31).

Warbler, Melodious (Hippolais polyglotta)
BBB 269; BBE 233; CDB 174 (cp 739); FBE
229; GPB 282; OBB 157; PBE 244 (cp 57).

Warbler, Moustached (Acrocephalus melanopogon) ALE 9:231; FBE 227; (Lusciniola
melanopogon) BBE 229; OBB 147; PBE 228
(cp 55).

Warbler, Olivaceous (Hippolais pallida)
BBE 233; FBE 229; PBE 244 (cp 57).

Warbler, Olive-tree (Hippolais olivetorum)
BBE 233; FBE 229; PBE 244 (cp 57).

Warbler, Orphean (Sylvia hortensis) BBE
235; CDB 176 (cp 752); FBE 233; PBE 229 (cp
56).

Warbler, Paddy-field (Acrocephalus
agricola) BBE 225; FBE 225.

Warbler, Pallas' Grasshopper (Locustella
certhiola) ABW 257; ALE 9:231; BBE 225;
DPB 313; FBE 223; PBE 284 (cp 65).

Warbler, Pallas' Leaf (Phylloscopus proregulus) BBE 247; FBE 239; KBA 352 (cp 53);
PBE 284 (cp 65).

Warbler, Radde's Willow (Phylloscopus
schwarzi) BBE 243; FBE 237; KBA 352 (cp
53).

Warbler, Reed (Acrocephalus scirpaceus)
ALE 9:232; BBB 185; BBE 231; CEU 144;
FBE 225; GAL 128 (cp 3); NBB 71, 90 (bw);
OBB 149; PBE 228 (cp 55); (A. sorghophilus)
DPB 313.

Warbler, River (Locustella fluviatilis) BBE
225; FBE 223; PBE 228 (cp 55).

Warbler, Rock (Origma solitaria) HAB 178;
RBU 219; SAB 32 (cp 15).

Warbler, Rufous (Cercotrichas galactotes)
BBE 241; CDB 162 (cp 677); FBE 255; PBE
229 (cp 56); (Agrobates galactotes) DAE 79.

Warbler, Rufous-capped (Basileuterus
rufifrons) CDB 203 (cp 890); (B. coronatus)
SSA cp 39.

Warbler, Ruppell's (Sylvia ruppelli) BBE
237; FBE 233; PBE 229 (cp 56).

Warbler, Sardinian (Sylvia melanocephala)
BBE 237; CDB 176 (cp 753); FBE 233; GPB
279; PBE 229 (cp 56).

Warbler, Savi's (Locustella luscinioides) ALE
9:231; BBB 267; BBE 225; FBE 223; OBB
147; PBE 228 (cp 55).

Warbler, Scrub (Bradypterus sp.) CDB 172 (cps 728, 729); DPB 307; (Scotocerca inquicta) FBE 227.

Warbler, Sedge (Acrocephalus schoenobaenus) ALE 9:231; BBB 184; BBE 229; CDB 171 (cp 725); FBE 227; GAL 128 (cp 3); OBB 147; PBE 228 (cp 55).

Warbler, Speckled (Chthonicola sagittata) CDB 177 (cp 758); HAB 182; SAB 32 (cp 15).

Warbler, Spectacled (Sylvia conspicillata) BBE 239; FBE 235; PBE 229 (cp 56).

Warbler, Subalpine (Sylvia cantillans) BBB 272; BBE 239; CDB 175 (cp 751); FBE 235; GPB 280; OBB 153; PBE 229 (cp 56).

Warbler, Western or White-tailed (Gerygone fusca) HAB 178; MBA 62; SAB 28 (cp 13).

Warbler, White-throat —See **Whitethroat.**

Warbler, White-throated (Gerygone olivacea) CDB 177 (cp 762); HAB 178; SAB 28 (cp 13).

Warbler, Willow (Phylloscopus trochilus) ABW 256; ALE 9:259; BBE 142; BBE 243; CDB 174 (cp 745); FBE 237; GAL 128 (cp 3); OBB 155; PBE 244 (cp 57); PSA 272 (cp 31); RBE 271; WAB 106.

Warbler, Wood (Phylloscopus sibilatrix) ALE 9:259; BBB 143; BBE 243; CDB 174 (cp 744); FBE 237; GAL 128 (cp 3); GPB 283; OBB 155; PBE 244 (cp 57); WEA 382.

Warbler, Wren (Prinia gracilis) FBE 227; (P. inornata) WBV 179.

Warbler, Yellow-bellied (Eremomela icteropygialis) CDB 173 (cp 738); PSA 272 (cp 31).

Warbler, Yellow-breasted Wren (Gerygone sulphurea) DPB 307; KBA 337 (cp 52).

Warbler, Yellow-browed (Phylloscopus inornatus) ALE 9:259; BBB 275; BBE 245; FBE 239; FBW 83; KBA 352 (cp 53); OBB 159; PBE 244 (cp 57).

Warbonnet, Atlantic —See **Blenny, Yarrell's.**

Wart Plant —See **Haworthia.**

Wart-cress—See **Swine-cress.**

Warthog (Phacochoerus aethiopicus) ALE 13:92; CAA 68, 82 (bw); CAF 96 (bw); DDM 161 (cp 26); DSM 221; HMA 61; HMW 51; JAW 164; LAW 560; LEA 590 (bw); MLS 370 (bw); SMR 85 (cp 6); SMW 234 (cp 146), 245 (bw); TWA 38.

Wasp (Blastophaga psenes) ZWI 178; (Ceropales maculatus) ALE 2:475; (Gravenhorstia sp.) BIN cp 109; (Hedychrum nobile) PEI 160 (cp 15); (Sphecius speciosus) KIW 268 (cp 144).

Wasp, Bee-killer (Philanthus triangulum) GAL 380 (cp 14); PEI 173 (bw).

Wasp, Common (Vespula sp.) CIB 269 (cp 42); CTI 64 (cp 89); GAL 380 (cp 14); KIW 250 (bw); LEA 144; OBI 155; PWI 169.

Wasp, Cuckoo (Chrysis sp.) ALE 2:475; CIB 260 (cp 39); LAW 226; OBI 149; (Nysson spinosus) OBI 157; (Stilbum sp.) CTI 58 (cp 80); LEA 145; (Terachrysis semicincta) CTI 58 (cp 79).

Wasp, Digger (Mellinus arvensis) CIB 261 (cp 40); OBI 157; (Scolia flavifrons) ALE 2:475; CIB 260 (cp 39); (S. flavescens) PEI 166 (bw).

Wasp, Ichneumon (Amblyteles armatorius) CIB 257 (cp 38); OBI 147; (Heteropelma sp.) KBM 89 (pl 29); (Ichneumon pisorius) GAL 380 (cp 14); (I. suspiciosus) CIB 257 (cp 38); (Paniscus flavidella) BIN pl 107; (Ophion sp.) CIB 257 (cp 38); KIW 241 (bw); (Therion circumflexum) ALE 2:447; GAL 380 (cp 14).

Wasp, Ichneumon (Rhyssa persuasoria) ALE 2:447, 465; CIB 257 (cp 38); OBI 147; PEI 155 (bw); PWI 106; WEA 205.

Wasp, Marble Gall (Andricus kollari) CIB 256 (cp 37); OBI 149.

Wasp, Mossy Rose Gall (Diplolepis rosae) ALE 2:438; CIB 256 (cp 37); OBI 149; PEI 160 (bw).

Wasp, Mud-dauber (Sceliphron sp.) ALE 2:475; CIB 268 (cp 41); KIW 252 (bw).

Wasp, Paper (Polistes sp.) BIN pls 110, 111; CIB 268 (cp 41); KIW 271 (cp 150); MWA 82; PWI 97 (bw); ZWI 198-199.

Wasp, Potter (Eumenes sp.) ALE 2:476; CIB 268 (cp 41); GAL 380 (cp 14); KIW 246 (bw), 268 (cp 145); LAW 226; OBI 155; PEI 171.

Wasp, Sand (Ammophila or Sphex sp.) ALE 2:475, 476; CIB 261 (cp 40), 268 (cp 41); GAL 380 (cp 14); KIW 253 (bw), 267 (cp 143); (Bembix sp.) CIB 268 (cp 41); GAL 380 (cp 14); OBI 155; PEI 171-173 (bw); ZWI 81.

Wasp, Sand-tailed Digger (Cerceris arenaria) ALE 2:475; CIB 261 (cp 40); GAL 380 (cp 14); OBI 157.

Wasp, Slender-bodied Digger (Crabro cribarius) CIB 261 (cp 40); OBI 157.

Wasp, Spangle Gall (Neuroterus quercusbaccarum) CIB 256 (cp 37); OBI 149.

Wasp, Spider-hunting (Anoplius sp.) ALE 2:475; CIB 261 (cp 40); OBI 155; (Cryptocheilus sp.) ALE 2:475; CIB 261 (cp 40); LAW 226.

Wasp, Tree (Vespa silvestris) GAL 380 (cp 14); OBI 155.

Wasp, Velvet-ant (Mutilla sp.) CIB 260 (pl 39); PEI 167 (bw); (Dasymutilla occidentalis) KIW 269 (cp 146).

Wasp, Wood (Urocerus sp.) ALE 2:437; CIB 241 (cp 36); GAL 380 (cp 14); LEA 114 (bw); OBI 143; (Tremex sp.) PEI 153 (bw); PWI 169.

Watches —See **Pitcher Plant.**

Water Bear (Echiniscus blumi) ALE 1:382; (Macrobiotus sp.) LEA 192; OIB 137.

Water Blinks (Montia fontana) HPW 75; MBF cp 16; MFF pl 105 (198); NHE 96; PWF 156c; WFB 43 (6).

Water Boatman —See **Backswimmer.**

Water Celery —See **Tape-grass.**

Water Poppy (Hydrocleys nymphaeoides) DEW 2:212 (cp 121); EPF pl 874; ERW 142.

Water Rat —See **Vole, Water.**

Water Soldier (Stratiotes aloides) DEW 2:228; EPF pl 876; KVP cp 9; MBF cp 79; MFF pl 69 (965); NHE 102; OWF 103; PFE cp 161 (1565); PWF 71e; TEP 19; WFB 259 (9).

Water Strider (Gerris sp.) ALE 2:188; CIB 128 (cp 11); GAL 302 (bw); KIW 71 (bw); LAW 230; LEA 113; OBI 31; PEI 108, 109 (bw); ZWI 97 (Halobates sp.) MOL 59.

Water Violet (Hottonia palustris) MBF cp 57; MFF pl 94 (616); NHE 97; OWF 109; PFE cp 92 (957); PWF 74d; TEP 17; WFB 177 (1).

Waterbuck, Common (Kobus ellipsiprymnus) ALE 13:402; DDM 189 (cp 32); DSM 254; HMA 97; SMR 92 (cp 7); SMW 282 (cp 175); TWA 35.

Waterbuck, Defassa or Sing-sing (Kobus defassa) ALE 13:402, 413; DDM 189 (cp 32); SMR 92 (cp 7); WEA 387.

Waterbuck, Lechwe (Kobus leche or Hydrotragus leche) ALE 13:402; DDM 192 (cp 33); DSM 254; HMW 71; LVS 124; SMR 97 (cp 8).

Water-buffalo —See **Buffalo.**

Waterbug, Giant (Belostoma sp.) ALE 2:188; KIW 44 (cp 22); LAW 241; PEI 111 (bw).

Water-chestnut (Trapa natans) DEW 2:26, 53 (cp 14); EPF pl 720; NHE 93; PFE cp 79 (823); OFP 33; TEP 17; WFB 258.

Water-clover (Marsilea quadrifolia) CGF 179; DEW 3: cp 155; EGF 121; EPF pl 118; NFP 146; NHE 108; PFF 106; (M. vestita) EGF 13.

Watercock (Gallicrex cinerea) ALE 8:84; DPB 73; KBA 80 (pl 7).

Watercress (Rorippa nasturtium) MBF cp 7; MFF pl 93 (98).

Watercress, Common (Nasturium officinale) EGH 127; EWG 78; NHE 94; OFP 153; OWF 69; PFF 188; PWF 91l; WFB 91 (8); WUS 211.

Watercress, Fool's (Apium nodiflorum) MBF cp 37; MFF pl 97 (475); OWF 89; PFE cp 85 (886); PWF 91j; WFB 163 (2).

Water-cricket (Velia caprai) CIB 128 (cp 11); LEA 135 (bw); OBI 31; (V. saulii) PEI 110 (bw).

Water-fern (Salvinia natans) DEW 3: cp 156; 294; EGF 140; EPF pl 119; NHE 108; (S. sp.) EGF 13; NFP 146.

Water-flea (Daphnia sp.) ALE 1:444, 451; GAL 498 (bw); LAW 153; OIB 137; WEA 386.

Water-hemlock (Cicuta maculata) MCW 76; PFF 264; WUS 279; (C. bolanderi) MLP 181.

Waterhen, White-breasted (Amaurornis phoenicurus) ALE 8:84; KBA 80 (pl 7); RBI 56; WBV 75.

Water-hyacinth (Eichhornia crassipes) DEW 2:218 (cp 132), 219 (cp 133); EPF pl 1029; ERW 141; MLP 293; MWF 121; PFF 368; PFW 237; SPL cp 501; TVG 131; WUS 103; WWT 52; (E. paniculata) HPW 310.

Water-lettuce (Pistia stratiotes) DEW 2:315; ERW 147; HPW 308; SPL cp 509.

Water-lily (Nuphar sp.) MBF cp 5; MLP 53; NHE 91; PFF 167.

Water-lily (Nymphaea sp.) DFP 160, 161 (cps 1279-1286); EJG 132; EPF cp IVb; ERW 145, 146; LFW 152 (cp 338), 172 (cp 388); MWF 215; NHE 91; PFW 196, 197; SPL cp 506; TVG 133.

Water-lily, Cape Blue (Nymphaea capensis) DFP 71 (cp 568); MEP 38; MWF 215.

Water-lily, Fringed (Nymphoides peltata) BFB cp 15 (9); DEW 2:122; EPF pl 679; HPW 231; KVP cp 4; MBF cp 59; MFF cp 59; NHE 97; PFE cp 97 (1004); PFW 188; PWF 140c; TEP 18; WFB 179 (7); (N. sp.) LFT cp 122; MLP 120.

Water-lily, Red (Nymphaea rubra) DFP 72 (cps 570-573); MEP 38; SPL cp 508.

Water-lily, Royal (Victoria amazonica) DEW 1:127; EPF pl 189; EWF cp 169a; HPW 44; MEP 38; MLP 56; RWF 43; SPL cp 513; WWT 53; (V. cruziana) DEW 1:129; EPF cp Vb, pls 189, 192-193; (V. regia) MLP 50.

Water-lily, White (Nymphaea alba) AMP 119; BFB cp 2 (7); EPF pls 186, 187; KVP cp 1; MBF cp 5; MFF pl 69 (32); MWF 215; NHE 91; OWF 67; PFE cp 18 (196); PWF 133h; TEP 13; WFB 67 (6).

Water-lily, Yellow (Nuphar lutea) BFB cp 2 (9); DEW 1:62 (cp 31); EPF pl 184, 185; EWG 57; KVP cp 8; MBF cp 5; MFF cp 59 (33); NHE 91; OWF 7; PFE cp 18 (197); PWF 133i; SPL cp 505; TEP 13; WFB 67 (5); WUS 175.

Water-measurer (Hydrometra stagnorum) CIB 128 (cp 11); OBI 31; PEI 110 (bw).

Watermelon (Citrullus vulgaris) EVF 99; OFP 121; PFF 308; (C. lanatus) EMB 145.

Water-milfoil (Myriophyllum sp.) HPW 155; MBF cp 34; MWF 206; NHE 93; PFF 263; WUS 273, 275.

Water-milfoil, Spiked (Myriophyllum spicatum) DEW 2:26; HPW 155; MBF cp 34; MFF pl 68 (442); NHE 93; OWF 53; PFE cp 82 (843); WFB 293 (2); WUS 277.

Water-milfoil, Whorled (Myriophyllum verticillatum) MBF cp 34; MFF pl 68 (441); NHE 93; TEP 17; WFB 293 (2a).

Water-opossum —See **Opossum, Water.**

Water-parsnip, Lesser (Berula erecta) MBF cp 38; MFF pl 97 (490); NHE 95; OWF 89; PWF 91k; WFB 163 (1).

Water-pepper (Polygonum hydropiper) AMP 167; MBF cp 73; MFF pl 103 (532); NHE 183; OWF 59; PFE cp 8 (81); PWF 170b; WFB 41 (5); WUS 123.

Water-pepper, Least (Polygonum minus) MBF cp 73; NHE 97; PFE 64.

Water-plantain (Alisma sp.) MBF cp 79; MFF pl 69 (960); NHE 102.

Water-plantain, Common (Alisma plantago-aquatica) BFB cp 22 (10); HPW 270; MBF cp 79; MFF pl 69 (959); OWF 109; PWF 102d; TEP 33; WFB 259 (1).

Water-plantain, Floating (Luronium natans) MBF cp 79; MFF pl 69 (958); NHE 102; WFB 259 (3).

Water-plantain, Lesser (Baldellia ranunculoides) MBF cp 79; MFF pl 69 (957); NHE 102; PWF 115d; WFB 259 (2).

Water-plantain, Parnassus-leaved (Caldesia parnassifolia) NHE 102; WFB 259 (5).

Water-scorpion (Nepa sp.) ALE 2:188; CIB 128 (cp 11); GAL 299 (bw); LAW 172; LEA 134 (bw); OBI 31; PEI 112 (bw); ZWI 195.

Water-scorpion or **Water-stick Insect** (Ranatra sp.) ALE 2:188; CIB 128 (cp 11); GAL 300 (bw); OBI 31; PEI 112, 113 (bw); PWI 145.

Water-snake —See **Snake.**

Waterweed, Canadian —See **Pondweed, Canadian.**

Waterweed, Esthwaite (Elodea nuttallii) DEW 2:229; MFF pl 68 (967); WFB 293 (5a).

Waterwort (Elatine sp.) HPW 87; MBF cp 17; MFF cp 27 (145); NHE 93; WFB 291 (2).

Wattle (Acacia cultiformis) BKT cp 137; EWF cp 129e.

Wattle, Cootamundra (Acacia baileyana) EGE 111; ELG 125; EPG 103; MEP 49; MWF 27; SST cp 223.

Wattle, Queensland (Acacia podalyriifolia) EWF cp 130a; HPW 151 (12); MWF 27.

Wattle, Silver (Acacia decurrens or A. dealbata) BIP 17 (cp 2); BKT cp 136; DFP 179 (cp 1429); EET 242; OBT 184; SPL cp 61; SST cp 224.

Wattle, Sydney (Acacia longifolia) LFW 180 (cp 405); MWF 27; PFE cp 49 (488); SPL cp 62.

Wattlebird, Little (Anthochaera chrysoptera) CDB 193 (cp 836); HAB 245; SAB 66 (cp
32).

Wattlebird, Red (Anthochaera carunculata)
HAB 244; SAB 66 (cp 32).

Wattlebird, Yellow (Anthochaera paradoxa)
HAB 245; RBU 264; SAB 66 (cp 32).

Wax Flower (Stephanotis floribunda) BIP
171 (cp 263); DFP 82 (cp 651); EGV 137; FHP
146; LHP 176, 177; MEP 114; MWF 273;
PFW 37; RDP 371.

Wax Pink —See **Rose-moss.**

Wax Plant or **Wax Flower** (Hoya carnosa)
BIP 101 (cp 151); EDP 143; EGG 115; EGV
114; FHP 125; LCS cp 270; LFW 168 (cps
375, 376); MWF 157; PFW 38; RDP 242.

Wax Plant, Miniature (Hoya bella) DEW
2:107 (cp 66); EGL 119; EPF pl 691; LHP
110; MWF 157; (H. sp.) EWF cp 120a, b, c;
RDP 242.

Waxbell, Yellow (Kirengeshoma palmata)
DFP 153 (cp 1223); OGF 111; PFW 141.

Waxbill, Common or St. Helena (Estrilda
astrild) ALE 9:440; CDB 211 (cp 937); FBE
301; PSA 316 (cp 39).

Waxbill, Cordon Bleu or Red-cheeked
—See **Cordon Bleu, Red-cheeked.**

Waxbill, Golden- or Orange-breasted
(Amandava subflava) ALE 9:440; BBT 69;
PSA 316 (cp 39).

Waxbill, Swee or Yellow-bellied (Estrilda
melanotis) ABW 303; ALE 9:440; BBT 68;
PSA 316 (cp 39).

Waxbill, Violet-eared (Uraeginthus granatinus) ALE 9:440, 460; PSA 316 (cp 39).

Waxflower (Eriostemon sp.) MLP 114; MWF 125.

Waxwing, Bohemian (Bombycilla garrulus)
ABW 265; ALE 9:196; BBB 65; BBE 213;
CDB 159 (cp 660); FBE 217; FBW 79; GPB
262; OBB 175; PBE 253 (cp 60); RBE 284.

Wayfaring Tree (Viburnum lantana) DEW
2:139; EPF pl 773, 774; MBF cp 41; MFF pl
74 (821); NHE 25; OBT 33; OWF 193; PFE cp
134 (1299); PWF 34b, 163j; WFB 223 (7).

Weasel, European (Mustela nivalis) ALE
10:223, 11:321; 12:40, 56; CEU 192 (bw);
DAE 16; GAL 27 (bw); HMW 135; LEA 561;
MLS 283 (bw); OBV 171; SMW 225 (cp 127);
TWA 87; WEA 389.

Weasel, Libyan Striped or Banded
(Poecilictis libyca) ALE 12:66; DDM 112 (cp
19); (Ictonyx libyca) HMW 137.

Weasel, Long-tailed (Mustela frenata) ALE
11:55; HMA 97.

Weasel, Short-tailed (Mustela erminea)
ALE 11:105; 12:40, 55; BMA 87-89; CAS 55
(bw); CEU 271 (bw); DAE 49; DSM 166; FBW
33; GAL 27 (bw); HMA 97; HMW 135; JAW
111; LAW 516; MLS 281 (bw); OBV 171;
TWA 86.

Weasel, Siberian (Mustela sibirica) ALE
12:56; CAS 41 (bw); DAA 78.

Weasel Snout —See **Archangel, Yellow.**

Weasel, Striped (Poecilogale albinucha)
ALE 12:66; DDM 112 (cp 19); SMR 33 (cp 2).

Weasel's Snout —See **Snapdragon, Lesser.**

Weatherfish —See **Loach, Pond.**

Weather-thistle, Mediterranean (Carlina
acanthifolia) DEW 2:206; EPF pl 577.

Weaver, Baya (Ploceus philippinus) ABW
304; CAS 195; KBA 433 (cp 64); LAW 463.

Weaver, Buffalo (Bubalornis albirostris)
ABW 304; PSA 305 (cp 36).

Weaver, Cape (Textor capensis) ALE 9:430;
(Ploceus capensis) PSA 308 (cp 37).

Weaver, Grenadier (Euplectes orix) ALE
9:430; CDB 214 (cp 950); PSA 309 (cp 38);
SAB 70 (cp 34); (Pyromelana orix) ABW 305.

Weaver, Manyar or Streaked (Ploceus
manyar) ALE 9:430; CDB 215 (cp 961); KBA
433 (cp 64); WAB 175.

Weaver, Masked (Ploceus velatus) NBA 91;
PSA 308 (cp 37); (P. intermedius) CAF 177.

Weaver, Napoleon (Euplectes afra) CAA 33;
PSA 309 (cp 38).

Weaver, Red-headed (Anaplectes rubriceps)
ALE 9:430; PSA 308 (cp 37).

Weaver, Scaly (Sporopipes squamifrons)
ALE 9:430; PSA 305 (cp 36).

Weaver, Sociable (Philetarius socius) ABW
305; CAA 51 (bw); CDB 215 (cp 958); PSA
305 (cp 36).

Weaver, Sparrow —See **Sparrow-weaver.**

Weaver, Spectacled (Ploceus ocularius)
NBA 93; PSA 308 (cp 37); (Hyphanturgus
ocularis) ABW 305.

Weaver, Spotted-backed (Ploceus cucullatus) ABW 305; CDB 215 (cp 960); NBA 93; PSA 308 (cp 37); (Textor cucullatus) ALE 9:430.

Weaver, Thick-billed (Amblyospiza albifrons) ALE 9:430; PSA 308 (cp 37).

Weaver, White-headed Buffalo (Dinemellia dinemelli) ALE 9:430; CDB 213 (cp 948); GPB 326.

Web-spinner (Embia sp.) CIB 93 (bw); LAW 197.

Wedgebill (Sphenostoma cristatum) HAB 250; RBU 232; (Psophodes cristatus) SAB 18 (cp 8).

Wedgebill, Geoffroy's (Schistes geoffroyi) ABW 170; SSA cp 5.

Weebill, Brown (Smicrornis brevirostris) CDB 179 (cp 772); HAB 180; SAB 30 (cp 14).

Weeping Widow (Lacrymaria velutina) PUM cp 137; RTM 53.

Weever, Greater (Trachinus draco) ALE 5:161; CSF 135; FEF 469 (bw); HFI 42; OBV 59.

Weever, Lesser (Trachinus vipera) CSF 135; CSS pl XXIII; HFI 42; OBV 59.

Weevil —See also **Beetle**.

Weevil (Byctiscus sp.) CIB 317 (cp 56); GAL 276 (cp 6); ZWI 132; (Cyphus augustus) CTI 55 (cp 74); LAW 240; (Entimus imperialis) PEI 240 (cp 25a); RBT 173; (Apoderus coryli) GAL 276 (cp 6); (Bryochaeta quadrimaculata) LAW 209; (Pachyrrhinchus sp.) ALE 2:256; RBT 175.

Weevil, Apple Blossom (Anthonomus sp.) GAL 276 (cp 6); OBI 189.

Weevil, Brazilian (Rhina barbirostris) ALE 2:256; PEI 249 (bw).

Weevil, Coal-black (Otiorrhynchus clavipes) CIB 317 (cp 56); OBI 189; (O. laevigatus) PEI 246 (bw).

Weevil, Figwort (Cionus hortulanus) CIB 317 (cp 56); OBI 191.

Weevil, Giraffe (Lasiorrhynchus barbicornis) ALE 2:256; BIN pls 130, 131.

Weevil, Granary (Calandra granaria or Sitophilus granarius) CIB 317 (cp 56); GAL 472 (bw); OBI 191; PWI 149 (bw).

Weevil, Large Pine (Hylobius abietis) OBI 191; PEI 251 (bw).

Weevil, Leaf-roller (Rhynchites sp.) OBI 189; PWI 142 (bw).

Weevil, Nut (Curculio sp.) ALE 2:277; CIB 317 (cp 56); CTI 54 (cp 73); KIW 129 (bw); OBI 189; PEI 250 (bw); (Balaninus sp.) GAL 276 (cp 6); LAW 207, 208.

Weevil, Oak-leaf Roller (Attelabus nitens) GAL 276 (cp 6); OBI 189; ZWI 133.

Weevil, Pea (Bruchus sp.) CIB 316 (cp 55); GAL 470 (bw); OBI 189.

Weevil, Tree (Phyllobius sp.) CIB 317 (cp 56); GAL 276 (cp 6); OBI 189.

Weigela (Weigela florida) DEW 2:137; DFP 244 (cp 1952-1954); EEG 103; EPF pl 778; GDS cp 501; HSC 122, 123; MWF 297; OGF 59; SPL cp 124; TGS 342; (W. sp.) EFS 24, 145; EWG cp 94g; GDS cps 500, 502; HPW 259; LFW 151 (cp 335); PFW 68; TVG 249.

Weinmannia (Weinmannia sp.) HPW 137; SNZ 61 (cps 154, 155), 62 (cps 159, 160).

Weka —See **Rail, Weka**.

Welcome-home-husband-though-never-so-drunk —See **Houseleek, Common**.

Weld —See **Greenweed, Dyer's** (Reseda sp.)

Wels, European (Silurus glanis) ALE 4:394; FEF 222, 223 (bw); FWF 107; GAL 229 (bw); HFI 107; HFW 126 (bw); OBV 93; SFW 325 (pl 82); WFW cp 155.

Wels, Indian Glass —See **Catfish, Glass**.

Welwitschia (Welwitschia mirabilis) DEW 1:67, 68; EPF pl 175; MLP 44; RWF 65.

Wendletrap, Common (Epitonium sp.) ALE 3:46; CTS 19 (cp 8), 20 (cp 9); GSS 40; MOL 69; (Clathrus clathrus) CSS pl XIV; OIB 41.

Whale, Blue (Balaenoptera musculus or Sibbaldus musculus) ALE 11:465; DSM 137; HMA 98-99; HMW 109; JAW 91; LEA 540 (bw); OBV 191; SAR 130; SLS 250; VWA (cp 42).

Whale, Bottle-nosed (Hyperoodon ampulatus or H. rostratus) ALE 11:480; HMA 99; HMW 108; LEA 540 (bw); OBV 189.

Whale, Finback (Balaenoptera physalis) ALE 11:464; HMA 99; LEA 543 (bw); OBV 191; SAR 130.

Whale, Humpback (Megaptera novae-angliae) ALE 11:465; DSM 137; HMA 99; JAW 90; LEA 540 (bw), 545; LVS 102-103; SAR 131.

Whale, Killer (Grampus orca or Orcinus orca) ALE 8:229; 11:480; BMA 125-127; DSM 138; HMA 100; JAW 89; LEA 545 (bw); OBV 187; SAR 131, 135.

Whale, Pilot (Globicephala melaena) ALE 11:480, 498; DSM 139; HMW 108; OBV 187.

Whale, Right (Eubalaena glacialis) ALE 11:463; HMA 99; HMW 110; (Neobalaena marginata) ALE 11:463; LEA 540 (bw).

Whale, Sei —See **Rorqual.**

Whale, Sperm (Physeter catodon) ALE 11:474; DSM 136; HMA 100; HMW 108; JAW 87; LEA 540 (bw); OBV 189; SAR 131; (P. macrocephalus) TWA 150 (bw).

Whale, Sulphur-bottom —See **Whale, Blue.**

Whalefish (Cetomimus regani) HFI 117; MOL 98.

Whalehead —See **Stork, Shoebill.**

Wheat (Triticum aestivum) EPF pls 1041, 1053, 1054; OFP 3; PFF 355; TEP 168; (T. sp.) DEW 2:267; EPF pl 1055; MLP 315; OFP 3; PFE cp 178 (1732).

Wheatear (Oenanthe oenanthe) ABW 254; ALE 9:270, 280; BBB 99; BBE 251; CDB 165 (cp 697); DPB 299; FBE 245; GAL 126 (cp 2); LEA 454 (bw); NBB 54 (pl a); OBB 141; PBE 221 (cp 54).

Wheatear, Black (Oenanthe leucura) BBE 253; FBE 249; PBE 221 (cp 54).

Wheatear, Black-eared (Oenanthe hispanica) BBB 272; BBE 253; CDB 165 (cp 694); CEU 44 (bw); FBE 245; PBE 221 (cp 54).

Wheatear, Capped —See **Chat, Pied.**

Wheatear, Desert (Oenanthe deserti) BBE 253; FBE 245; PBE 284 (cp 65); RBI 212; WAB 120.

Wheatear, Hooded (Oenanthe monacha) CDB 165 (cp 695); FBE 247.

Wheatear, Isabelline (Oenanthe isabellina) BBE 251; FBE 245.

Wheatear, Pied (Oenanthe pleschanka) BBE 253; FBE 247; PBE 221 (cp 54).

Wheatear, Red-tailed (Oenanthe xanthoprymna) FBE 245; WAB 160.

Wheatear, White-crowned Black (Oenanthe leucopyga) CAS 92 (bw); FBE 249.

Wheel of Fire —See **Firewheel Tree.**

Whelk, Dog —See **Basket Shell** (Nassarius sp.) and **Dogwinkle** (Nucella sp.).

Whelk, Needle (Bittium reticulatum) ALE 3:46; CSS pl XIV; OIB 41.

Whelk, Northern or Waved (Buccinum undatum) ALE 3:56, 71; GSS 81.

Whiff —See **Megrim.**

Whimbrel (Numenius phaeopus) ALE 8:170; BBB 223; BBE 121; CDB 87 (cp 290); DPB 87; FBE 139; GPB 146; HAB 67; HBG 105 (bw), 132 (pl 9); KBA 128 (pl 13), 149 (pl 18); OBB 69; PBE 92 (cp 29), 101 (cp 32); SNB 84 (cp 41), 86 (pl 42); WBV 75; (N. variegatus) FNZ 129 (pl 12).

Whinchat (Saxicola rubetra) ALE 10:223; BBB 101; BBE 255; FBE 243; GAL 126 (cp 2); NBB 53 (bw), 60 (bw); OBB 141; PBE 221 (cp 54).

Whipbird, Eastern (Psophodes olivaceous) CDB 170 (cp 720); HAB 250; MBA 6; RBU 231; SAB 18 (cp 8); WAB 181.

Whipbird, Western (Psophodes nigrogularis) HAB 250; SAB 18 (cp 8).

Whip-scorpion (Damon medius johnstoni) ALE 1:416; (Hypoctonus rangunensis) PEI 11 (bw).

Whipsnake (Coluber viridiflavus) ALE 6:420; ERA 249; LAW 387, 426; LBR 79; SIR 219 (bw); (C. sp.) CEU 93; DAE 150, 151; HRA 58; SIR 194 (cp 89); SSW 73.

Whipsnake, Little (Denisonia gouldii) MEA 25; MWA 58; (D. suta) COG cp 48.

Whistler, Brown (Pachycephala simplex) HAB 217; SAB 44 (cp 21).

Whistler, Golden (Pachycephala pectoralis) ABW 259; ALE 9:260; CDB 185 (cp 800); HAB 212; MBA 63; MWA 78; SAB 44 (cp 21); WAB 185.

Whistler, Mangrove or White-bellied (Pachycephala cinerea) DPB 353; KBA 396 (bw).

Whistler, Red-throated (Pachycephala rufogularis) HAB 216; RBU 235; SAB 44 (cp 21).

White Sails (Spathiphyllum wallisii) DFP 82 (cp 650); LHP 174, 175; PFW 35; SPL cp 58; (S. sp.) EPF pl 1007; FHP 145; LHP 174; MWF 271; RDP 368.

Whitebeam (Sorbus aria) BFB cp 10 (4); BKT cp 6; EET 168, 169; EPF pls 438, 439; HPW 143; MBF cp 31; MFF pl 74 (384); MTB 288 (cp 25); NHE 19; OBT 20, 24; OWF 195; PFE cp 47 (467); PWF 58d, 175e; TEP 96; WFB 115 (3).

Whitebeam, Broad-leaved (Sorbus latifolia) MBF cp 31; OBT 173.

Whitebeam, Cut-leaved (Sorbus intermedia) MTB 288 (cp 25); NHE 294.

Whitebeam, Dwarf (Sorbus chamaemespilus) NHE 19; PFE 171.

White-eye (Aythya sp.) —See **Duck.**

White-eye, Gray-breasted (Zosterops lateralis) BNZ 45; CAU 257 (bw); CDB 191 (cp 831); FNZ 193 (cp 17); HAB 255; LEA 456 (bw); RBU 276; SAB 54 (cp 26); WAN 134.

White-eye, Indian or Oriental (Zosterops palpebrosa) CDB 191 (cp 833); KBA 432 (cp 63); LAW 509.

White-eye, Mountain or Hill (Zosterops montana) ALE 9:372; DPB 403; (Z. ceylonensis) ABW 281.

White-eye, Pale (Zosterops pallidus) CDB 191 (cp 832); GPB 298; (Z. albiventris) ALE 10:89; HAB 255; PSA 304 (cp 35); (Z. chloris) SAB 54 (cp 26).

White-eye, Western (Zosterops gouldi) HAB 255; MEA 67; SAB 54 (cp 26).

White-eye, Yellow (Zosterops lutea) HAB 255; SAB 54 (cp 26); (Z. nigrorum) DPB 403.

Whiteface, Eastern or Southern (Aphelocephala leucopsis) HAB 179; SAB 32 (cp 15).

Whitefish (Coregonus sp.) ALE 4:255; FWF 49.

Whitefly —See **Fly, White.**

Whitehead (Mohoua albicilla) BNZ 31; FNZ 193 (cp 17).

White-helleborine —See **False-helleborine.**

White-rot —See **Pennywort, Marsh.**

Whitethroat (Sylvia communis) ALE 9:232; 11:321; BBB 102; BBE 237; FBE 231; GAL 128 (cp 3); NBB 58 (pl a); OBB 153; PBE 229 (cp 56); WAB 150.

Whitethroat, Lesser (Sylvia curruca) ALE 9:232; BBB 103; BBE 237; FBE 231; GAL 128 (cp 3); OBB 153; PBE 229 (cp 56).

Whiting (Merlangius merlangus) CSF 107; OBV 35; WEA 392; WFW cp 188; (Gadus merlangus) LEA 251 (bw).

Whiting, Blue (Micromesistius sp.) CSF 109; HFI 77.

Whitlow-grass (Draba sp.) DFP 8 (cps 58, 59); EMB 64, 65, 137; ERW 111; EWF cp 1g; MBF cp 8; NHE 274; WFB 97 (6d).

Whitlow-grass, Common (Erophila verna) MBF cp 8; MFF pl 90 (84); NHE 218; OWF 71; PWF 41h; WFB 97 (7).

Whitlow-grass, Hoary (Draba incana) MBF cp 8; MFF pl 90 (82); PFE cp 35 (333); PWF 711c.

Whitlow-grass, Wall (Draba muralis) MBF cp 8; MFF pl 90 (83); NHE 218; WFB 97 (6).

Whitlow-grass, Yellow (Draba aizoides) EPF pl 248; EWF cp 11e; MBF cp 8; MFF cp 43 (81); NHE 274; OGF 21; PFE 34 (331); PWF cp 20b; TGS 92; TVG 149; WFB 85 (1).

Whorl-grass, Water (Catabrosa aquatica) MBF pl 97; NHE 107.

Whortleberry, Bog (Vaccinium uliginosum) MBF cp 55; MFF pl 77 (598); NHE 70; TEP 79.

Whydah (Coliuspasser sp.) ALE 9:430; SAB 70 (cp 34); (Tetraenura fischeri) ALE 9:429; (Vidua paradisea) ABW 305.

Whydah, Pin-tailed (Vidua macroura) CDB 213 (cp 947); LAW 510; PSA 309 (cp 38); WAB 139.

Whydah, Sharp-tailed Paradise (Steganura paradisea) ALE 9:429; CDB 213 (cp 944); PSA 309 (cp 38).

Widgeon, Cape (Anas capensis) ALE 7:307; CAF 262 (bw); PSA 49 (cp 4), 64 (pl 5).

Widgeon, Chiloe (Anas sibilatrix) ALE 7:303; WFI 120.

Widgeon, European (Anas penelope) ABW 68; ALE 7:303, 385; BBB 203; BBE 53; DPB 33; FBE 53; KBA 80 (pl 7); OBB 19; PBE 28 (cp 9), 40 (pl 13), 48 (pl 15); SBB 41 (cp 46); WBV 49.

Widow, Long-tailed (Euplectes progne) CDB 214 (cp 951); PSA 305 (cp 36); WAB 147.

Widow-finch (Hypochera sp.) ALE 9:429; PSA 305 (cp 36).

Widow's Tear —See **Magic Flower.**

Wig Tree —See **Smoke Tree.**

Wild Calla —See **Arum, Bog or Water.**

Wild Dagga —See **Lion's Ear.**

Wild Pear or **Wild Plum** (Dombeya sp.) EPF pl 605; EWF cp 57a; HPW 92; LFT cp 108; MEP 83; MWF 116.

Wildcat, African —See **Cat, Caffer Wild.**

Wildebeest (Connochaetes taurinus or Gorgon taurinus) ALE 13:416; CAA 72, 75 (bw); CAF 149 (bw); DDM 224 (cp 37); DSM 261; HMA 55; HMW 75; JAW 194; LAW 567; LEA 612 (bw); MLS 412 (bw); SMR 112 (cp 9); TWA 28.

Willow, Almond (Salix triandra) MBF cp 77; MFF pl 72 (570); OBT 89; PFE 45; WFB 27 (2c).

Willow, Basket —See **Osier, Common.**

Willow, Bay or **Laurel** (Salix pentandra) MBF cp 77; MFF pl 72 (568); OBT 88; PFE 45; WFB 27 (1).

Willow, Black (Salix nigra) PFF 128; SST cp 164; TGS 95.

Willow, Blunt-leaved (Salix retusa) NHE 267; PFE cp 3 (20).

Willow, Corkscrew (Salix matsudana) DFP 237 (cp 1895); EGW 139; SST cp 163; TGS 96.

Willow, Crack or **Yellow** (Salix fragilis) EET 132-135; MBF cp 77; MFF pl 72 (569); MTB 177 (cp 16); NHE 24; OBT 85, 88; OWF 199; PFF 129; PWF 167e; TEP 49; WFB 27 (2).

Willow, Creeping (Salix repens) MBF cp 78; MFF pl 73 (577); NHE 134; OWF 199; PFE 45; PWF 167m; WFB 29 (5).

Willow, Dark-leaved (Salix nigricans) MBF cp 78; WFB 29 (1).

Willow, Downy (Salix lapponum) MBF cp 77; MFF pl 73 (578); WFB 29 (2).

Willow Fomes (Phellinus igniarius) DLM 155; LHM 63; (P. sp.) DLM 155; LHM 63; ONP 157; RTM 212.

Willow, Goat —See **Sallow, Great.**

Willow, Least (Salix herbacea) MBF cp 78; MFF pl 73 (580); NHE 267; PFE 45; WFB 29 (6).

Willow, Mountain or **Little Tree** (Salix arbuscula) MBF cp 78; WFB 29 (4).

Willow, Myrtle-leaved (Salix myrsinites) MBF cp 78; MFF pl 73 (579); PFE 45; WFB 29 (4b).

Willow, Purple (Salix purpurea) MBF cp 77; MFF pl 72 (571); OBT 89; OWF 187; PFE 45; PFF 129; PWF 167i; TEP 50; TGS 95; WFB 27 (4).

Willow, Pussy (Salix discolor) PFF 128; TGS 96.

Willow, Reticulate or **Netted** (Salix reticulata) DFP 24 (cp 185); EMB 123; EPF pl 504; MBF cp 78; MFF pl 73 (581); MLP 59; NHE 267; PFE cp 3 (19); SPL cp 463; WFB 29 (6b).

Willow, Tea-leaved (Salix phylicifolia) MBF cp 78; MFF pl 72 (576); WFB 29 (1a).

Willow, Violet (Salix daphnoides) DFP 237 (cp 1894); OBT 164; WFB 27 (4a).

Willow, Weeping or **Poet's** (Salix babylonica) EET 135; EGE 36; EGT 142; EJG 143; EPG 145; MWF 261; OBT 165; PFF 130; PWF 167f; SST cp 162; TGS 86; WFB 27 (2b).

Willow, White (Salix alba) AMP 107; BKT cp 89; DFP 237 (cp 1892); EET 133-135; ELG 111; EPF pl 501; MBF cp 77; MTB 177 (cp 16); NHE 24; OBT 84, 88, 164; PFE cp 3 (25); PFF 127; PWF 167h; SST cp 161; TEP 48; WFB 27 (2a).

Willow, Woolly (Salix lanata) GDS cp 439; MBF cp 78; WFB 29 (3).

Willow-herb —See also **Fireweed.**

Willow-herb (Epilobium sp.) DEW 2:22; EWF cp 4; MBF cp 35; MFF cps 20, 23, 105; NHE 29, 72, 92; OWF 111; SNZ 121, 125, 135; WFB 151.

Willow-herb, Broad-leaved (Epilobium montanum) BWG 113; MBF cp 35; MFF cp 20 (427); NHE 29; OWF 111; PFE cp 81 (839); PWF 79g; WFB 151 (3).

Willow-herb, Hairy (Epilobium hirsutum) BFB cp 10 (6); EFC cp 18; HPW 163; MBF cp 35; MFF cp 20 (425); NHE 92; OWF 111; PFE cp 81 (837); PWF 105h; WFB 151 (2).

Willow-herb, Marsh (Epilobium palustre) MBF cp 35; MFF cp 20 (431); NHE 92; OWF 111; WFB 151 (4).

Willow-herb, Rosebay (Chamaenerion angustifolium) BFB cp 10 (7); EPF pls 715, 716; MBF cp 35; MFF cp 20 (435); NHE 29, 273; OWF 111.

Wind Ball —See **Cape Poison.**

Windflower —See also **Anemone.**

Windflower, Snowdrop (Anemone sylvestris) DEW 1:62 (cp 33); EPF pl 212; NHE 71; TEP 233; TGF 61; TGV 79; WFB 77 (7).

Wine Cup —See **Mallow, Poppy.**

Wineberry (Aristotelia sp.) HPW 90; SNZ 66 (cp 174-176), 140 (cp 438).

Winter Creeper (Euonymus fortunei) EDP 149; EEG 106, 142; EGC 131; EGE 124; EGV 107; EGW 111; ESG 111; GDS cps 187-189; HSC 51; PFF 239; TGS 60.

Winter Hazel, Buttercup (Corylopsis pauciflora) DFP 192 (cp 1530); GDS cp 112; OGF 27; TGS 232.

Winter Hazel, Chinese (Corylopsis sinensis) DFP 192 (cp 1531); EWF cp 94f.

Winter Hazel, Spike (Corylopsis spicata) DFP 192 (cp 1532); EFS 104; EJG 111; ESG 106; HSC 35; TGS 230; (C. veitchiana) GDS cp 113.

Winterberry (Ilex verticillata) EEG 98; EFS 28, 118; EGW 118.

Wintergreen (Chimaphila sp.) EWG 96; HPW 130; NHE 34; PFF 271; WFB 169 (6).

Wintergreen, Chickweed (Trientalis europaea) KVP cp 158; MBF cp 57; MFF pl 80 (622); NHE 34; PFE cp 93 (965); PWF 41j; WFB 177 (3).

Wintergreen, Common or Lesser (Pyrola minor) BFB cp 13 (9); EWF cp 161g; MBF cp 56; MFF pl 80 (600); NHE 34; OWF 85; PFE cp 86 (912); PWF 88e; WFB 169 (3).

Wintergreen, Green-flowered or Yellow (Pyrola chlorantha) NHE 34; WFB 169 (5).

Wintergreen, Intermediate (Pyrola media) MBF cp 55; NHE 34; PFE cp 87 (912); PWF 154b.

Wintergreen, Large or Round-leaved (Pyrola rotundifolia) EPF pl 474; HPW 130; KVP cp 51; MBF cp 56; NHE 34; PWF 154a; TEP 251; WFB 169 (4).

Wintergreen, Nodding or Serrated or Toothed (Orthilia secunda) MBF cp 56; MFF pl 80 (601); NHE 34; PFE cp 86 (914); PWF 136e; WFB 169 (2).

Wintergreen, One-flowered (Moneses uniflora) MBF cp 56; MFF pl 84 (602); NHE 34; PFE cp 87 (915); PWF 88f; TEP 251; WFB 169 (1).

Wintergreen, One-flowered (Pyrola uniflora) DEW 1:240; KVP cp 123; (P. grandiflora) EWF cp 6c.

Winter's Bark (Drimys winteri) DEW 1:79; GDS cp 156; HPW 29; HSC 45; PFW 308.

Winter-sweet —See **Marjoram, Pot or Wild.**

Wintersweet (Chimonanthus praecox) DFP 189 (cp 1511); EGW 101; EWF cp 92c; GDS cp 75; HPW 35; HSC 28; OGF 1; PFW 60; SPL cp 74; SST cp 237; TGS 150.

Wintersweet, African (Acokanthera spectabilis) EPF pl 688; MEP 107; MWF 30.

Wiregrass —See **Goosegrass** (Eleusine sp.).

Wiretail, Des Mur's (Sylviorthorhynchus desmursii) ALE 9:125; SSA pl 25.

Wire-vine —See **Pohuehue.**

Wisent —See **Bison, European.**.

Wishbone Flower (Torenia fournieri) DFP 49 (cp 386); EGA 157; ESG 144; MWF 286; PFW 276; TGF 220.

Wisteria, Chinese (Wisteria sinensis) DEW 1:315; DFP 250 (cp 1999); EGV 145; EHP 20; EJG 149; ELG 152; EPF pl 452; EPG 153; GDS cp 503; HSC 123, 126; MWF 297; OGF 63; PFW 162; SPL 184; TGS 71; TVG 249.

Wisteria, Japanese (Wisteria floribunda) DFP 250 (cp 2000); EGV 40, 41; LFW 151 (cp 337); MWF 297; TGS 71.

Wisteria Tree —See **Epaulette Tree.**

Witch (Glyptocephalus cynoglossus) ALE 5:235; CSF 191; OBV 47; (G. cynocephalus) HFI 18.

Witch Hazel (Hamamelis virginiana) AMP 81; EPF pl 346; PFF 194; TGS 224; (H. intermedia) GDS cp 216; OBT 148; PFW 136.

Witch Hazel, Chinese (Hamamelis mollis) DEW 1:152; DFP 204 (cp 1632), 205 (cps 1633, 1634); EEG 96; EFS 20, 114; EGW 114; EJP 118; EPG 123; EWF cp 94a; GDS cps 217, 218; HPW 56; HSC 59; MWF 147; OBT 148; OGF 3; PFW 136; TVG 225.

Witch Hazel, Japanese (Hamamelis japonica) DEW 1:100 (cp 46); DFP 204 (cps 1630, 1631); EPF pl 345; HSC 59, 64; MLP 107; OGF 3; TGS 224.

Witch-alder (Fothergilla sp.) DEW 1:153; DFP 202 (cps 1613, 1614); EEG 96; EFS 22, 112; ESG 112; GDS cps 200, 201; HPW 56; HSC 54, 57.

Witches' Butter (Exidia sp.) DLM 141; LHM 225; ONP 149; RTM 245.

Witchweed (Striga elegans) EWF cp 81f; LFT cp 149; (S. lutea) WUS 331.

Woad (Isatis tinctoria) BFB cp 4 (6); EGH 118; MBF cp 10; MFF 50 (66); NHE 216; OWF 11; PFE cp 32 (292); PWF 42k; WFB 87 (4).

Woadwaxen, Ashy (Genista cinera) DFP 203
(cp 1621); GDS cp 211; PFM cp 57.

Wobbegong (Orectolobus sp.) ALE 4:103,
114; HFI 142.

Wolf, Gray or Timber (Canis lupus) ALE
12:205, 229; 13:163; BMA 237; CEU 255
(bw); DAA 68; DAE 90-91; GAL 22 (bw);
HMA 101; HMW 141; JAW 94; LVS 95; MLS
255 (bw); SMW 184 (cp 112); TWA 82.

Wolf, Maned (Chrysocyon brachyurus or C.
jubatus) ALE 12:257; BMA 147; CSA 152
(bw); DSM 152; HMW 144; LEA 552 (bw);
LVA 28; LVS 94; MLS 262 (bw); PWC 237
(bw); SMW 198 (bw).

Wolf, Tasmanian —See **Tasmanian Wolf.**

Wolf-fish —See **Catfish, Sea.**

Wolf's Milk —See **Spurge, Hungarian.**

Wolfsbane —See **Monkshood.**

Wolverine (Gulo gulo) ALE 12:41, 83; BMA
239, 240; CEU 254 (bw); DAE 111; DSM 170;
HMW 137; LAW 537 (bw); MLS 289 (bw);
SLP 77; SMW 205 (bw); TWA 70 (bw).

Wombat (Phascolomys ursinus) BMW 242;
JAW 28; SMW 38 (cp 10); TWA 195; (P.
mitchelli) HMW 202; WAW 55.

Wombat, Hairy-nosed or Soft-furred
(Lasiorhinus latifrons) ALE 10:101 (cp 3);
BMA 241; DSM 35; LVA 57; LVS 29.

Wombat, Tasmanian (Wombatus ursinus)
ALE 10:101 (cp 3), 103; CAU 59 (bw); DSM
35; LAW 529; MLS 61 (bw); (W. hirsutus)
LEA 488 (bw).

Wonga Wonga Vine —See **Bower-plant.**

Wood-betony —See **Lousewort.**

Woodbine —See **Honeysuckle.**

Woodcock, Eurasian (Scolopax rusticola)
ABW 122; ALE 8:170; 11:105; BBB 128;
BBE 119; CDB 87 (cp 293); DAE 124; DPB
97; FBE 141; GPB 147; KBA 136 (bw), 137
(bw); OBB 67; PBE 84 (pl 27), 116 (cp 33);
RBE 136; WAB 109 (bw); WBV 87.

Woodcreeper (Lepidocolaptes sp.) ALE
9:120; SSA pl 25; WAB 80; (Xiphorhynchus
sp.) CDB 140 (cp 560); SSA cp 35.

Woodcreeper, Barred (Dendrocolaptes
certhia) ABW 197; SSA cp 35.

Woodcreeper, Olivaceous (Sittasomus
griseicapillus) ABW 197; ALE 9:120; SSA cp 35.

Woodcreeper, Plain Brown (Dendrocincla
fulginosa) ALE 9:120; GPB 230.

Woodcreeper, Strong-billed (Xiphocolaptes
promeropirhynchus) ALE 9:120; SSA cp 35.

Woodcreeper, Wedge-billed (Glyphorhynchus spirurus) ALE 9:120; CDB 140 (cp 559).

Wood-hoopoe —See **Hoopoe.**

Woodlark (Lullula arborea) ALE 9:179; BBB
107; BBE 199; CDB 149 (cp 607); FBE 205;
GPB 248; OBB 123; PBE 189 (cp 48).

Woodlouse —See **Sowbug.**

Wood-nymph (Thalurania furcata) ALE 90.
8:453; CDB 123 (cp 477); SSA cp 32; WAB

Woodpecker (Celeus sp.) SSA cp 9; WAB 90;
(Blythipicus sp.) KBA 240 (cp 31); (Meig-
lyptes sp.) KBA 244 (pl 33); (Brachyternus benghalensis) ALE 13:285; (Dendropicos fuscescens) PSA 225 (cp 26).

Woodpecker, Bearded (Thripias namaquus)
CDB 139 (cp 555); PSA 225 (cp 26).

Woodpecker, Black (Dryocopus martius)
ALE 9:108; 13:163; BBB 269; BBE 191; DAE
130; FBE 193; GAL 148 (cp 4); PBE 188 (cp
47); RBE 212.

Woodpecker, Crimson (Picus puniceus)
ALE 9:119; KBA 241 (cp 32); (P. rivolii) SSA
cp 34.

Woodpecker, Crimson-backed (Chrysocolaptes lucidus) ABW 193; ALE 9:119; DPB
223; KBA 241 (cp 32); RBI 136.

Woodpecker, Gray-headed (Picus canus)
ABW 194; ALE 9:84, 119; BBE 187; CDB 138
(cp 553); FBE 193; GAL 148 (cp 4); KBA 240
(cp 31); PBE 188 (cp 47); WBV 131.

Woodpecker, Great Slaty (Mulleripicus
pulverulentus) DAA 150; DPB 223; KBA 232
(bw).

Woodpecker, Great Spotted (Dendrocopus
major) ALE 9:83, 84, 119; BBB 135; BBE
189; CDB 137 (cp 544); DAE 130; FBE 195;
GAL 148 (cp 4); GPB 229; LAW 506; OBB
115; PBE 188 (cp 47); RBE 208; WAB 108;
WEA 396.

Woodpecker, Green (Picus viridus) ALE
9:82, 119; 13:163; BBB 132; BBE 187; DAE
130; FBE 193; GAL 148 (cp 4); GPB 227;
LEA 335 (bw); OBB 115; PBE 188 (cp 47);
RBE 207; WEA 397.

Woodpecker, Ground (Geocolaptes olivaceus) ALE 9:108; CDB 138 (cp 548); PSA 225 (cp 26).

Woodpecker, Lesser Spotted (Dendrocopus minor) ALE 9:119; BBB 134; BBE 189; CDB 137 (cp 545); DAE 130; FBE 197; GAL 148 (cp 4); OBB 115; PBE 188 (cp 47).

Woodpecker, Middle Spotted (Dendrocopus medius) ALE 9:119; BBE 189; FBE 195; GAL 148 (cp 4); LAW 443, 506; PBE 188 (cp 47).

Woodpecker, Pygmy (Dendrocopus canicapillus) RBI 132; (D. maculatus) DPB 223.

Woodpecker, Spotted (Hemicircus sp.) KBA 244 (pl 33); RBI 135.

Woodpecker, Syrian (Dendrocopus syriacus) ALE 9:119; BBE 189; CDB 137 (cp 546); FBE 195; PBE 188 (cp 47).

Woodpecker, Three-toed (Dinopium javanense) DPB 223; KBA 241; WBV 131.

Woodpecker, Three-toed (Northern) (Picoides tridactylus) ALE 9:119; BBE 191; FBE 197; PBE 188 (cp 47); RBE 211.

Woodpecker, Tristam's (Dryocopus javensis) DPB 223; WAB 29.

Woodpecker, White-backed (Dendrocopus leucotus) BBE 191; FBE 197; PBE 188 (cp 47); RBI 131.

Woodpecker, Yellow-naped (Picus flavinucha) ABW 193; KBA 241 (cp 32).

Wood-pigeon —See Pigeon, Wood.

Woodruff (Asperula orientalis) EGA 96; OGF 135; (A. arvensis) WFB 185 (9); (A. suberosa) DFP 2 (cp 16); HPW 258; (A. tinctoria) NHE 75.

Woodruff, Sweet (Asperula odorata or Galium odoratum) AMP 129; EEG 140; EGC 123; EGH 102; ERW 100; MBF cp 42; MFF pl 103 (809); NHE 36; OWF 93; PFE 99 (1027); PFF 301; PWF 32d; TEP 255; WFB 185 (8).

Woodrush (Luzula sp.) HPW 291; MBF cp 87; MFF pl 113 (1041); NHE 39, 290; PFE cp 177 (1712), 519.

Woodrush, Curved or Arctic (Luzula arcuata) MBF cp 87; MFF pl 113 (1039).

Woodrush, Field (Luzula campestris) MBF cp 87; MFF pl 113 (1040); NHE 39; PFE cp 178 (1714); TEP 125.

Woodrush, Greater (Luzula sylvatica) EWF cp 25a; MBF cp 87; MFF pl 114 (1037); NHE 39.

Woodrush, Hairy (Luzula pilosa) EPF pl 952; MBF cp 87; MFF pl 113 (1036); NHE 39.

Woodrush, Spiked (Luzula spicata) MBF cp 87; MFF pl 113 (1038).

Wood-shrike (Tephrodornis gularis) GPB 255; (T. pondicerianus) KBA 252 (pl 35).

Wood-shrike, Bald-headed (Pityriasis gymnocephala) ALE 9:449; WAB 170.

Woodsia, Oblong (Woodsia ilvensis) CGF 147; EGC 132; EGF 144; MFF 139 (1293); NHE 293.

Wood-sorrel (Oxalis adenophylla) BIP 137 (cp 208); DFP 16 (cp 123); EGB 134; EPF pl 640; HPW 209; TGF 109.

Wood-sorrel, Common (Oxalis acetosella) DEW 2:32; EPF pl 638; EWF cp 14a; EWG 130; MBF cp 20; MFF pl 84 (252); NHE 31; PFE cp 62 (638); PFF 223; PWF cp 19m; TEP 245; WFB 133 (1).

Wood-sorrel, Procumbent Yellow (Oxalis corniculata) BWG 25; MBF cp 20; MFF cp 43 (253); NHE 31; PFE 220; PFF 223; PWF 61e; WFB 133 (2).

Woodstar, White-bellied (Acestrura mulsanti) SHU 90, 124; SSA cp 7.

Wood-swallow, Black-faced (Artamus cinereus) HAB 219; MBA 64; SAB 74 (cp 36).

Wood-swallow, Dusky (Artamus cyanopterus) HAB 219; SAB 74 (cp 36).

Wood-swallow, Masked (Artamus personatus) HAB 219; RBU 288; SAB 74 (cp 36).

Wood-swallow, White-breasted (Artamus leucorhynchus) CDB 220 (cp 982); DPB 363; GPB 333; HAB 219; RBU 291; SAB 74 (cp 36).

Wood-swallow, White-browed (Artamus superciliosus) ABW 267; ALE 9:196; HAB 219; SAB 74 (cp 36); WAB 195; WAW 29 (bw).

Woolly Bears —See Mulla Mulla (Trichinium).

Woollyhead (Craspedia uniflora) EWF cp 140a; SNZ 17 (cp 2), 161 (cp 513).

Worm (Amphitrite johnstoni) CSS cp 12; OIB 103; (Cephalothrix sp.) CSS cp 8; OIB 29; (Flabelligera sp.) CSS cp 11; OIB 107; (Lumbriculus sp.) CSS cp 10; OIB 115.

Worm (Potamilla reniformis) CSS cp 13; OIB 109; (Eulalia viridis) CSS cp 9; OIB 97; (Thalassema sp.) MOL 61; OIB 121.

303 Worm, Acorn / Woundwort, Wood

Worm, Acorn (Balanoglossus sp.) ALE 3:427;
WEA 41; (Glossobalanus sp.) ALE 3:427;
OIB 169; (Saccoglossus sp.) ALE 3:427; CSS
cp 28.

Worm, Arrow (Sagitta sp.) ALE 3:271; MOL
133 (bw).

Worm, Bootlace (Lineus geniculatus) ALE
1:301; LAW 93; OIB 29; (L. sp.) CSS cp 8;
MOL 62.

Worm, Burrowing (Megalomma vesiculosum) ALE 1:374; OIB 105.

Worm, Chalk-tube (Serpula vermicularis)
ALE 1:376; LAW 97; OIB 111; (Protula
tubularia) ALE 1:376; LAW 97.

Worm, Clam (Nereis or Eunereis sp.) ALE
1:373; CSS cp 9; LAW 94; LEA 100 (bw);
MOL 63; OIB 101.

Worm, Earth —See **Earthworm.**

Worm, Echiurid (Bonellia sp.) ALE 1:302,
351; OIB 121.

Worm, Fan —See **Fanworm.**

Worm, Fish (Echinorhynchus clavula) LEA
75 (bw); OIB 31.

Worm, Horsehair (Gordius sp.) LAW 92; OIB
31.

Worm, Mud or River —See **Worm, Tubifex.**

Worm, Paddle (Phyllodoce lamelligera)
LAW 101; OIB 97; (P. maculata) CSS cp 9.

Worm, Peacock (Sabella pavonina) ALE
1:376; CSS cp 13, pl V; OIB 105; VWS 34
(bw); (S. penicillus) LEA 104 (bw).

Worm, Pink Ribbon (Amphiporus sp.) CSS
cp 8; OIB 29.

Worm, Ribbon (Tubulanus sp.) ALE 1:301;
MOL 62; OIB 29; (Cerebratulus sp.) ALE
1:302; (Paranemertes plana) MOL 62.

Worm, Rock (Marphysa sanguinea) CSS cp
10; LAW 101; OIB 99.

Worm, Sabellid (Spirographis spallanzanii)
ALE 1:374, 375; LAW 104.

Worm, Sand-mason (Lanice conchilega) CSS
cp 12; OIB 105.

Worm, Slow —See **Slow-worm.**

Worm, Tape —See **Tapeworm.**

Worm, True (Myzostoma sp.) MOL 61; OIB
121.

Worm, Tubifex (Tubifex sp.) GAL 519 (bw);
LAW 103; OIB 115.

Worm-grass —See **Stonecrop, White.**

Worm-lizard (Amphisbaena alba) ALE 6:341;
SIR 111 (cp 57); (A. angustifrons) ALE
6:336.

Worm-lizard, Arabian (Diplometopon
zarudnyi) MAR 149 (pl 59); SIR 162 (bw).

Worm-lizard, Two-handed or Two-legged
(Bipes biporus or B. canaliculatus) ALE
6:336, 341; HRA 18.

Worm-lizard, Wiegmann's (Trogonophis
wiegmanni) ALE 6:336, 341; LAW 418.

Wormwood, Absinthe (Artemisia absinthium) AMP 55; CTH 57 (cps 89, 90); DFP 122
(cp 776); EGH 85, 101; EPF pl 560; MBF cp
46; MCW 100; MFF pl 101 (897); NHE 238;
OFP 137; OWF 33; PFE cp 149 (1436); PFF
328; PWF 143k; SPL cp 384.

Wormwood, Alpine (Artemisia mutellina)
CTH 58 (cp 91); NHE 289; (A. borealis) EWF
cp 6f.

Wormwood, Sea (Artemisia maritima) DFP
123 (cp 978); MBF cp 46; MFF pl 101 (898);
NHE 136; OWF 61; PWF 146d; WFB 239 (5).

Wormwood, Shining (Artemisia nitida)
NHE 289; TVG 143.

Wormwood, Shrubby (Artemisia arborescens) DFP 182 (cp 1452); GDS cp 11; HSC
14.

Wortleberry —See **Whortleberry.**

Woundwort, Alpine or Limestone (Stachys
alpina) MBF cp 69; PWF 97j.

Woundwort, Downy (Stachys germanica)
MBF cp 69; MFF cp 29 (765); NHE 231; PFE
cp 112 (1140); PWF 129j; WFB 203 (6).

Woundwort, Field (Stachys arvensis) MBF
cp 69; MFF cp 10 (764); NHE 231; OWF 147;
PWF 50c; WFB 203 (8).

Woundwort, Marsh (Stachys palustris) BFB
cp 17 (4); KVP cp 22; MBF cp 69; MFF cp 29
(766); NHE 98; OWF 147; PFE cp 112 (1138);
PWF 128c.

Woundwort, Perennial Yellow (Stachys
recta) EPF pl 813; NHE 231; PFE cp 113
(1136); WFB 203 (7).

Woundwort, Saracen's —See **Ragwort,
Broad-leaved.**

Woundwort, Wood (Stachys sylvatica) BFB
cp 17 (6); BWG 121; HPW 238; MBF cp 69;
MFF cp 29 (767); NHE 35; OWF 147; PWF
59l; TEP 253; WFB 203 (5).

Wow-wow —See **Gibbon, Gray or Silvery.**

Wrack, Channelled (Pelvetia canaliculata) EPF pl 37; ONP 3, 27.

Wrack, Knotted or Bladder (Ascophyllum nodosum) EPF pl 38; NHE 140; OFP 187; ONP 5; PFF 57.

Wrack, Serrated (Fucus serratus) DEW 3: cp 22; NHE 140; ONP 5; (F. sp.) DEW 3: cps 20, 21; NFP 16; NHE 140; ONP 3, 5, 27; PFF 58.

Wrasse, Ballan (Labrus bergylta) CSF 129; CSS cp 30; FEF 467 (bw); OBV 63.

Wrasse, Cleaner (Labroides dimidiatus) ALE 5:115, 143; HFI 41; MOL 124; WFW cp 247; (L. phthirophagus) HFW 45 (cp 27).

Wrasse, Corkwing (Crenilabrus sp.) CSF 131; CSS cp 30; HFI 40; OBV 63.

Wrasse, Cuckoo (Labrus mixtus or L. ossifagus) ALE 5:143; CSF 129; HFI 41; HFW 164 (cp 67); LEA 255 (bw); OBV 63; VWS 25; WFW cps 406, 407.

Wrasse, Goldsinny —See **Goldsinny.**

Wrasse, Moon (Thalassoma lunare) CTF 64 (cp 88); FEF 468 (bw), 482; WFW cp 412.

Wrasse, Olive Club-nosed (Gomphosus varius) FEF 465 (bw); HFI 41.

Wrasse, Rainbow (Coris julis) ALE 5:143; CSF 133; FEF 461 (bw); HFI 40; LAW 343; OBV 63; WFW cp 402.

Wrasse, Twin-spot (Coris angulata) FEF 462 (bw), 464; HFW 166 (cp 70); LAW 339; WFW cp 400; (C. gaimardi) ALE 5:143; WFW cp 401.

Wren (Campylorhynchus sp.) CDB 159 (cp 665); SSA pl 29; (Cyphorinus thoracicus) CDB 160 (cp 666).

Wren, Banded or Red-winged (Malurus elegans) ALE 10:89; MBA 49; MEA 68; MWA 18; SAB 22 (cp 10).

Wren, Black-and-white (Malurus leucopterus) HAB 200; SAB 22 (cp 10).

Wren, Black-backed (Malurus melanotus) CDB 178 (cp 769); HAB 200; SAB 20 (cp 9).

Wren, Blue (Malurus cyaneus) CDB 178 (cp 765); HAB 200; RBU 203; SAB 20 (cp 9); WAB 202.

Wren, Blue-and-white (Malurus cyanotus) CDB 178 (cp 766); HAB 200; MBA 48; (M. leuconotus) MWA 19.

Wren, Blue-breasted (Malurus pulcherrimus) HAB 202; RBU 204; SAB 22 (cp 10).

Wren, Bush (Xenicus longipes) ABW 212; ALE 9:150; BNZ 61; FNZ 193 (cp 17).

Wren, Dusky Grass (Amytornis purnelli) HAB 193; SAB 24 (cp 11).

Wren, Emu —See **Emu-wren.**

Wren, Eyrean Grass (Amytornis goyderi) SAB 24 (cp 11); WAB 33.

Wren, Fern (Oreoscopus gutteralis) HAB 180; SAB 34 (cp 16); WAB 183.

Wren, Heath —See **Heath-wren.**

Wren, Lilac-crowned (Malurus coronatus) HAB 202; SAB 20 (cp 9).

Wren, Lovely (Malurus amabilis) HAB 202; SAB 22 (cp 10).

Wren, Purple-backed (Malurus assimilis) CDB 177 (cp 763); HAB 200; SAB 22 (cp 10).

Wren, Red-backed (Malurus melanocephalus) CDB 178 (cp 768); HAB 202; MBA 46, 47; SAB 22 (cp 10).

Wren, Rock (Xenicus gilviventris) BNZ 61; CDB 143 (cp 576); FNZ 193 (cp 17); GPB 238.

Wren, Rufous Field (Calamanthus campestris or C. fuliginosus) HAB 186, 187; SAB 32 (cp 15).

Wren, St. Kilda (Troglodytes troglodytes hirtensis) LVS 235; OBB 135.

Wren, Scrub —See **Scrub-wren.**

Wren, Splendid (Malurus splendens) CAU 42; HAB 200; MBA 48; MWA 15, 17 (bw); SAB 20 (cp 9).

Wren, Striated Grass (Amytornis striatus) SAB 24 (cp 11); WAB 196.

Wren, Turquoise (Malurus callainus) CDB 178 (cp 764); HAB 202; RBU 207; SAB 20 (cp 9).

Wren, Variegated (Malurus lamberti) ABW 258; ALE 9:259; CDB 178 (cp 767); GPB 284; HAB 200; MBA 48; MWA 22; SAB 22 (cp 10).

Wren, Western Grass (Amytornis textilis) HAB 192; RBU 208; SAB 24 (cp 11).

Wren, Winter (Troglodytes troglodytes) ABW 247; ALE 9:190; 11:105; BBB 57; BBE 221; CDB 160 (cp 668); FBE 273; LAW 507; OBB 135; PBE 212 (cp 51); RBE 248; (T. t. nipalensis) RBI 227.

Wrinklenut (Nonea pulla) EPF pl 799; NHE 229; PFE 341; TEP 117; WFB 189 (5).

Wrybill —See **Plover, Wrybill.**

Wryneck, Eurasian (Jynx torquilla) ABW 194; ALE 9:82, 108; BBB 133; BBE 191; CAA 53 (bw); CDB 138 (cp 549); DAE 128; FBE 197; GAL 148 (cp 4); GPB 226; KBA 240 (cp 31); LEA 439 (bw); OBB 117; PBE 188 (cp 47); WBV 131.

Wutu —See Urutu.

X-Y

Xenops (Xenops sp.) ABW 198; ALE 9:125; CDB 141 (cp 565).

X-ray Fish (Pristella riddlei) ALE 4:286; FEF 81; SFW 132 (pl 25); WFW cp 84.

Xylosma (Xylosma senticosa) EGE 148; ELG 145.

Yak, Wild (Bos grunniens) CAS 119, 122 (bw); DSM 251; HMA 102; HMW 93; JAW 186; LVA 53; MLS 403 (bw); SMW 278 (cp 167); TWA 165; (B. mutus) ALE 13:341; LVS 134; (Poephagus grunniens) DAA 48.

Yam (Dioscorea sp.) EGV 80, 105; OFP 183; PFF 385; RDP 177.

Yapok —See Opossum, Water.

Yarrow —See also Milfoil.

Yarrow (Achillea millefolium) AMP 167; BFB cp 19 (6); BWG 95 (bw); CTH 59 (cp 95); DFP 119 (cp 949); EGC 73; EGH 92; EGP 92; EPF pl 949; EWG 77; MBF cp 45; MCW 94; MFF pl 95 (888); NHE 189; OWF 99; PFF 326; PWF 67e; TEP 125; WFB 239 (2); WUS 363 (bw).

Yarrow, Common (Achillea filipendula) LFW 2 (cp 4); OGF 151; PFW 85.

Yarrow, Dark (Achillea atrata) KVP cp 134; NHE 286.

Yarrow, Fernleaf (Achillea filipendulina) DFP 119 (cps 947-948); EDP 114; EGP 92; LFW 2 (cp 4); MWF 29; SPL cp 186; TVG 77.

Yarrow, Woolly (Achillea tomentosa) EEG 130; EGC 120; ERW 92; PFE cp 147 (1421).

Yaupon (Ilex vomitoria) EGC 34, 35; EGE 131.

Yavering Bells —See Wintergreen, Nodding.

Yellow Elder (Stenolobium stans) LFW 240 (cp 533); MEP 138; MWF 273.

Yellow Horn (Xanthoceras sorbifolium) EPF pl 733, 734; EWF cp 95b; MWF 298; TGS 326.

Yellow Sage —See Lantana, Common.

Yellow Stainer (Agaricus xanthodermus) DLM 70, 164; LHM 135; PUM cp 143.

Yellow-cress, Creeping (Rorippa sylvestris) MBF cp 7; MFF cp 44 (99); NHE 94; WFB 87.

Yellow-cress, Great (Rorippa amphibia) BFB cp 3 (9); MBF cp 7; MFF pl 44 (101); NHE 95; PFE cp 33 (303); PWF 75j; WFB 87.

Yellow-cress, Marsh (Rorippa islandica) MBF cp 7; MFF cp 44 (100); NHE 94; OWF 11.

Yellowhammer (Emberiza citrinella) ALE 9:341; 10:223; BBB 86; BBE 295; CDB 196 (cp 855); FBE 277; FNZ 208 (cp 18); GAL 118 (cp 1); OBB 187; PBE 277 (cp 64).

Yellowhead (Mohoua ochrocephala) BNZ 31; FNZ 193 (cp 17).

Yellow-palm —See Palm, Butterfly.

Yellow-rattle (Rhinanthus minor) BFB cp 16 (2); HPW 244; MBF cp 65; NHE 187; OWF 25; PFE cp 127 (1247); PWF 90b; TEP 145; WFB 217 (2).

Yellow-rattle (Rhinanthus sp.) EPF pl 849; MBF cp 65; MFF cp 61 (723); NHE 187, 282.

Yellow-rattle, Greater (Rhinanthus serotinus) KVP cp 31; MBF cp 65; NHE 187.

Yellow-star-of-Bethlehem (Gagea lutea) MBF cp 86; MFF pl 63 (1000); PWF cp 191; WFB 263 (4).

Yellow-star-of-Bethlehem (Gagea sp.) NHE 38, 192, 241, 290; PFE cp 164 (1601); PFM cp 236; WFB 263 (5-7).

Yellowvein Bush (Pseuderanthemum reticulatum) MEP 142; MWF 245.

Yellow-wort (Blackstonia perfoliata) BFB cp 15 (1); EWF 15e; MBF cp 58; MFF cp 40 (638); NHE 74; OWF 31; PFE cp 94 (988); PWF 93f; WFB 181 (4).

Yesterday, Today and Tomorrow (Brunfelsia sp.) BIP 37 (cp 40); DEW 2:162 (cp 80); DFP 55 (cp 436); EGE 114; EHP 140, 141; EPF pl 835; EWF cp 183a; FHP 108; MEP 127; MWF 63; PFW 286; RDP 109.

Yew (Taxus sp.) EEG 112; EGC 147; EGE 35, 107, 108; EGW 142, 143; EJG 147; ELG 133; EPG 149; ESG 143; NFP 152; OBT 132; PFF 114; TGS 9.

Yew, English (Taxus baccata) AMP 149; BFB cp 14 (7); BKT cps 5, 6; CTH 21 (cp 8); DFP 256 (cps 2042, 2043); EET 1, 86, 87; EGE 41; EPF pls 163, 165, cp IVa; MBF pl 100; MFF cp 64 (1259); MTB 48 (cp 1); NHE 18; OBT 21, 25, 105; OWF 185; PFE cp 2 (17); PFF 114; PWF 18d, 161k; SST cp 41; TGS 22; TVG 255; WFB 25 (7).

Yew, Japanese (Podocarpus macrophyllus) EDP 141; EFP 137; EGE 105; EJG 139; NFP 152; RDP 325; TGS 6.

Yorkshire Fog (Holcus lanatus) DEW 2:262; MBF pl 96; MFF pl 131 (1221); NHE 195; PFE cp 181 (1770); TEP 153.

Youth-and-Old-Age —See **Aichryson.**

Youth-on-Age (Tolmiea menziesii) BIP 177 (cp 274); EPF 20, 49, 145; RDP 385; WFB 300.

Ysano —See **Nasturium, Tuberous.**

Yuhina (Yuhina sp.) ALE 12:103; KBA 289 (cp 46), 352 (cp 53); RBI 179, 180.

Yulan Tree —See **Lily Tree.**

Z

Zaluzianskya (Zaluzianskya sp.) EWF cp 81d; LFT cp 147.

Zander —See **Pike-perch.**

Zebra, Cape Mountain (Equus zebra) ALE 12:538, 13:61; BMA 249; DDM 141 (cp 24); HMW 39; LVA 19; LVS 119; MLS 350 (bw); PWC 242 (bw); SLP 232; TWA 19; VWA cp 5.

Zebra, Common (Equus burchelli) BMA 246-248; CAA 70-71 (bw), 72, 76-77; CAF 152-153, 178-179, 198-199; DDM 141 (cp 24); DSM 212; HMA 102; HMW 39; JAW 156; LAW 512; LEA 585 (bw); MLS 351 (bw); SLP 236, 237; SMR 85 (cp 6); SMW 231 (cps 138, 140).

Zebra, Grevy's (Equus grevyi) ALE 12:538; BMA 248-249; CAF 84-85 (bw); DDM 141 (cp 24); DSM 213; HMW 39; LEA 585 (bw); MLS 353 (bw); SMW 231 (cp 139).

Zebra Plant (Aphelandra squarrosa) BIP 25 (cp 17); EDP 144; EGL 87; FHP 105; LHP 26-27; MWF 43; PFW 13; RDP 84; SPL cp 39.

Zebra Plant (Calathea zebrina) BIP 29 (cp 22); PFF 389; RDP 124.

Zebrafish —See **Scorpionfish** (Dendrochirus sp.).

Zebu —See **Cattle, Brahman.**

Zelkova, Caucasian (Zelkova sp.) EET 158; OBT 160, 161; SST cp 182.

Zelkova, Japanese (Zelkova serrata) EET 158; EGT 147; EPF pl 358; TGS 134.

Ziege (Pelecus cultratus) FWF 91; HFI 96.

Zigzag Plant —See **Redbird Cactus.**

Zingel (Zingel zingel) ALE 5:86; FEF 350; FWF 123.

Zobel or Zope (Abramis sp.) FEF 160; HFI 98; SFW 205 (pl 48), 221 (pl 52).

Zonure, Lord Derby's —See **Sungazer.**

Zorilla (Ictonyx striatus) ALE 12:66; DDM 112 (cp 19); MLS 288 (bw); SMR 33 (cp 2); SMW 205 (bw).

Zoysia Grass (Zoysia matrella) EGC 76, 119; EJG 150; PFF 360.

Scientific Name Index

Acanthobrahmaea europaea — Moth

Acanthochitona sp. — Chiton

Acanthocinus aedilis — Beetle, Timberman

Acanthocybium solandri — Wahoo

Acanthodactylus sp. — Lizard, Fringe-toed

Acanthodoris pilosa — Nudibranch

Acantholimon glumaceum or A. venustum — Thrift, Prickly

Acanthopanax sieboldianus — Aralia, Five-leaved

Acanthophis antarcticus — Adder, Death

Acanthophthalmus kuhlii — Loach, Coolie

Acanthorhynchus superciliosus — Spinebill, Western

Acanthorhynchus tenuirostris — Spinebill, Eastern

Acanthurus coeruleus — Tang, Blue

Acanthurus leucosternon — Surgeonfish, White-breasted or White-throated

Acanthurus lineatus — Surgeonfish, Lined or Striped

Acanthus longifolius — Bear's Breeches

Acanthus mollis — Bear's Breeches

Acanthus montanus — Bear's Breeches

Acanthus spinosus — Bear's Breeches, Spiny

Acantostracion tricornis — Boxfish

Accipiter brevipes — Sparrow-hawk, Levant

Accipiter cirrocephalus — Sparrow-hawk

Accipiter collaris — Sparrow-hawk

Accipiter fasciatus — Goshawk, Brown

Accipiter gentilis — Goshawk

Accipiter melanochlamys — Goshawk, Black

Accipter melanoleucus — Goshawk, Black

Accipiter minullus — Sparrow-hawk, African Little

Accipiter nisus — Sparrow-hawk, European or Northern

Accipiter novaehollandiae — Goshawk, Gray or White

Accipiter soloensis — Goshawk, Chinese

Accipiter trivirgatus — Goshawk, Crested

Accipiter virgatus — Sparrow-hawk, Philippine

Acer campestre — Maple, Common or Field

Acer cappadocicium — Maple, Cappadocian

Acer ginala — Maple, Amur

Acer griseum — Maple, Paperbark

Acer japonicum — Maple, Japanese

Acer macrophyllum — Maple, Big-leaf or Oregon

Acer monspessulanum — Maple, Montpelier or French

Acer negundo — Box-elder

Acer opalus — Maple, Italian

Acer palmatum — Maple, Japanese

Acer pensylvanicum — Maple, Striped

Acer platanoides — Maple, Norway

Acer pseudoplatnus — Maple, Sycamore

Acer rubrum — Maple, Red or Scarlet or Swamp

Acer saccharinum — Maple, Silver or Soft

Acer saccharum — Maple, Sugar or Rock

Acer tataricum — Maple, Tartar

Aceras anthropophorum — Orchid, Man

Acerentomon sp. — Half-insect

Acerina cernua — Ruffe

Acerina schraetzer — Schraetzer

Aceros sp. — Hornbill

Acestrura mulsanti — Woodstar, White-bellied

Achaearanea sp. — Spider

Achatina sp. — Snail, Land

Acherontia atropos — Hawkmoth, Death's-head

Acheta domesticus — Cricket, House

Achillea atrata — Yarrow, Dark

Achillea filipendula — Yarrow, Common

Achillea filipendulina — Yarrow, Fernleaf

Achillea millefolium — Yarrow

Achillea nana — Milfoil, Dwarf

Achillea ptarmica — Sneezewort

Achillea tomentosa — Yarrow, Woolly

Achimenes erecta — Magic Flower

Achimenes hybrids — Magic Flower

Achimenes longiflora — Achimenes, Trumpet

Achras sapota — Sapodilla

Acidanthera bicolor murielae — Acidanthera

Acilius sulcatus — Beetle, Predaceous Diving

Acinonyx jubatus — Cheetah

Acinos arvensis — Basil-thyme

Acipenser ruthenus — Sterlet

Acipenser sturio — Sturgeon, Atlantic

Acmaea sp. — Limpet, Tortoise-shell

Acmaeops collaris — Beetle

Acokanthera spectabilis — Wintersweet, African

Acomys sp. — Mouse, Spiny

Aconitum anglicum — Monkshood

Aconitum cammarum or A. henryi — Monkshood, Spark's Variety

Aconitum napellus — Monkshood, Common

Aconitum paniculatum — Monkshood, Panicled
Aconitum variegatum — Monkshood, Lesser
Aconitum vulparia — Monkshood, Yellow
Acorus calamus — Flag, Sweet
Acorus gramineus — Flag, White-striped Sweet
Acraea sp. — Butterfly, Acraea
Acrantophis madagascariensis — Boa, Madagascar
Acridotheres cristatellus — Mynah, Crested
Acridotheres tristis — Mynah, Common or Indian
Acrobates pygamaeus — Phalanger, Pygmy Flying
Acrocephalus agricola — Warbler, Paddy-field
Acrocephalus arundinaceus — Warbler, Great Reed
Acrocephalus australis — Warbler, Australian Reed
Acrocephalus baeticatus — Warbler, Marsh
Acrocephalus dumetorum — Warbler, Blyth's Reed
Acrocephalus melanopogon — Warbler, Moustached
Acrocephalus paludicola — Warbler, Aquatic
Acrocephalus palustris — Warbler, Marsh
Acrocephalus schoenobaenus — Warbler, Sedge
Acrocephalus scirpaceus — Warbler, Reed
Acrocephalus sorghophilus — Warbler, Reed
Acrocephalus stentoreus — Warbler, Clamorous Reed
Acrochordus javanicus — Snake, Elephant's-trunk
Acrocinus longimanus — Beetle, Harlequin Longhorn
Acroclinium roseum — Daisy, Everlasting
Acronycta sp. — Moth, Dagger
Acropora sp. — Coral, Stag's Horn
Acropternis orthonyx — Tapaculo
Acrostichum aureum — Fern, Leather
Acryllium vulturinum — Guinea-fowl, Vulturine
Actaea spicata — Baneberry, Common
Actaeon tornatalis — Bubble Shell
Actias isis — Moth, Indian or Luna
Actias selene — Moth, Indian or Luna
Actinia sp. — Sea Anemone, Beadlet
Actinidia arguta — Gooseberry, Chinese
Actinidia chinensis — Gooseberry, Chinese
Actinidia kolomikta — Kolomikta

Actinotus helianthi — Flannel Flower
Actitis hypoleucos — Sandpiper, Common
Actophilornis africanus — Jacana, African
Adalia sp. — Beetle, Ladybird
Adansonia digitata — Baobob
Addax nasomaculatus — Addax
Adelges abietis — Aphid, Spruce Pineapple Gall
Adenium obesum — Impala Lily
Adenostyles alliariae — Adenostyles, Gray
Adenostyles glabra — Adenostyles, Glabrous
Adenota kob — Kob
Adiantum capillis-veneris — Fern, Southern Maidenhair
Adiantum hispidulum — Fern, Rosy Maidenhair
Adiantum pedatum — Fern, Northern Maidenhair
Adiantum raddianum — Fern, Delta Maidenhair
Adiantum reniforme — Fern, Maidenhair
Adiantum tenerum wrightii — Fern, Fan Maidenhair
Adonis aestivalis — Pheasant's Eye, Summer
Adonis amurensis — Golden Cup
Adonis annua — Pheasant's Eye, Winter
Adonis chrysocyanthus — Golden Cup
Adonis flammea — Pheasant's Eye, Scarlet or Large
Adonis vernalis — Pheasant's Eye, Spring or Yellow
Adoxa moschatellina — Moschatel
Adromischus sp. — Leopard's Spots
Aechmea caudata — Aechmea
Aechmea chantinii — Aechmea
Aechmea fasciata — Urn Plant
Aechmea fulgens — Urn Plant
Aechmea mariae-reginae — Aechmea
Aechmea miniata — Aechmea
Aechmea nudicaulis — Aechmea
Aechmea tillandsioides — Aechmea
Aedes sp. — Mosquito
Aegeria sp. — Clearwing
Aegintha temporalis — Finch, Red-browed
Aegithalos caudatus — Tit, Long-tailed
Aegithalos concinnus — Tit, Red-headed
Aegithina tiphia — Iora, Common
Aegolius funereus — Owl, Boreal
Aegolius harrisii — Owl, Boreal
Aegopodium podagraria — Goutweed
Aegotheles cristatus — Nightjar, Owlet
Aegypius monachus — Vulture, Black

Aegypius occipitalis — Vulture, White-headed

Aegypius tracheliotus — Vulture, Lappet-faced

Aeolidia papillosa — Nudibranch, Gray

Aeoliscus punctualatus — Razorfish

Aeoliscus strigatus — Razorfish

Aepyceros melampus — Impala

Aequidens sp. — Cichlid

Aeschna sp. — Dragonfly, Aeschna

Aeschynanthus sp. — Lipstick Plant

Aesculus carnea — Horse-chestnut, Red

Aesculus flava — Buckeye, Yellow

Aesculus hippocastanum — Horse-chestnut, Common

Aesculus indica — Horse-chestnut, Indian

Aesculus octandra — Buckeye, Sweet

Aesculus parviflora — Horse-chestnut, Dwarf

Aethechinus algirus — Hedgehog, Algerian

Aethopyga gouldiae — Sunbird, Gould's

Aethopyga saturata — Sunbird, Black-throated

Aethopyga shelleyi — Sunbird, Lovely

Aethopyga siparaja — Sunbird, Yellow-backed

Aethusa cynapium — Parsley, Fool's

Afribyx senegallus — Plover, Wattled

Afropavo congensis — Peafowl, African or Congo

Afrotis afra — Korhaan, Black

Agabus sp. — Beetle, Water

Agama agama — Lizard, Agama

Agama atricollis — Lizard, Agama

Agama stellio — Lizard, Starred Agama

Agamia agami — Heron, Chestnut-bellied

Agapanthus africanus — Lily, Blue African

Agapanthus sp. — Lily, Blue African

Agapornis fischeri — Lovebird, Fischer's

Agapornis nigrigensis — Lovebird, Black-cheeked

Agapornis personata — Lovebird, Yellow-collared

Agapornis roseicollis — Lovebird, Rosy-faced

Agapornis swinderniana — Lovebird, Black-collared

Agaricus arvensis — Mushroom, Horse

Agaricus augustus — Mushroom, Big or Shaggy

Agaricus bisporus — Mushroom

Agaricus bitorquis — Mushroom, Pavement

Agaricus campestris — Mushroom, Field

Agaricus hortensis — Mushroom, Garden

Agaricus silvaticus — Mushroom, Wood

Agaricus sivicola — Mushroom, Wood

Agaricus xanthodermus — Yellow Stainer

Agarista agricola — Moth, Owlet

Agathis australis — Pine, Kauri

Agathis palmerstonii — Pine, Kauri

Agave americana — Century Plant, American

Agave attenuata — Agave

Agave filifera — Century Plant, Thread-bearing

Agave rigida — Sisal Plant

Agave sisalana — Sisal Plant

Agave univittata — Agave

Agave victoria reginae — Century Plant, Queen Victoria

Agave victoriae americanae — Agave, Victoria-American

Ageratum houstonianum — Floss Flower

Aglaiocercus coelestis — Sylph

Aglaiocercus kingi — Sylph

Aglais urticae — Butterfly, Small Tortoise-shell

Aglaonema commutatum — Chinese Evergreen, Spotted

Aglaonema crispum or A. robelinii — Painted Drop-tongue

Aglaonema pseudo-bracteata — Chinese Evergreen

Aglia tau — Moth, Tau Emperor

Agonus cataphractus — Poacher or Pogge

Agrias sp. — Butterfly

Agrimonia eupatoria — Agrimony, Common

Agrimonia odorata — Agrimony

Agrimonia pilosa — Agrimony

Agriolimax agrestes or A. reticulatus — Slug, Field

Agrion sp. — Damselfly, Agrion

Agriotes sp. — Beetle, Wireworm

Agrius convolvuli — Hawkmoth, Convolvulus

Agrobates galactotes — Warbler, Rufous

Agrocybe praecox — Agaric, Spring

Agropyron caninum — Couch, Bearded

Agropyron junceiforme — Couch, Sand

Agropyron pungens — Couch, Sea

Agropyron repens — Quackgrass

Agrostemma githago — Corn Cockle

Agrostis alba — Redtop

Agrostis canina — Bent-grass, Velvet

Agrostis nebulosa — Grass, Cloud

Agrostis stolonifera — Bent-grass, Creeping or White

Agrostis tenuis — Bent-grass, Common

Agrotis exclamationis — Moth, Heart and Dart

Ahaetulla sp. — Snake, Indian Tree

Aichryson sp. — Aichryson
Aidemosyne modesta — Finch, Plum-headed
Ailanthus altissima — Tree of Heaven
Ailanthus excelsa — Tree of Heaven
Ailuroedus crassirostris — Catbird, Green
Ailuroedus melanotis — Catbird, Spotted
Ailuropoda melanoleuca — Panda, Giant
Ailurus fulgens — Panda, Lesser or Red
Aipyanthes echioides — Prophet Flower
Aipysurus laevis — Sea-snake
Aira carophyllea — Hair-grass, Silver
Aira praecox — Hair-grass, Early
Aix galericulata — Duck, Mandarin
Ajaia ajaja — Spoonbill, Roseate
Ajuga chamaepitys — Ground-pine
Ajuga genevensis — Bugle, Blue or Geneva
Ajuga pyramidalis — Bugle, Pyramidal
Ajuga reptans — Bugle, Common or Carpet
Akebia quinata — Akebia, Five-leaf
Akera bullata — Bubble Shell, Paper
Akis italica — Beetle
Alabonia geoffrella — Moth, Geoffroy's Tubic
Alaemon alaudipes — Lark, Bifasciated or
    Hoopoe
Alauda arvensis — Skylark
Alauda gulgula — Skylark, Oriental or Small
Albizia julibrissin — Silk Tree
Albizia lebbeck — Silk Tree
Albizia lophantha — Silk Tree
Albugo sp. — Rust, White
Alburnoides bipunctatus — Schneider
Alburnus alburnus — Bleak
Alburnus bipunctatus — Schneider
Alca torda — Auk, Razorbill
Alcea sp. — Hollyhock
Alcedo atthis — Kingfisher, Eurasian
Alcedo cristata — Kingfisher, Malachite
Alcedo meninting — Kingfisher, Blue-eared or
    Malaysian
Alcelaphus buselaphus — Hartebeest, Bubal or
    Red
Alcelaphus cokei — Hartebeest, Coke's
Alcelaphus lichtensteini — Hartebeest,
    Lichtenstein's
Alces alces — Elk, European
Alchemilla alpina — Lady's Mantle, Alpine
Alchemilla sp. — Lady's Mantle
Alchemilla vulgaris — Lady's Mantle, Common
Alcidis agathyrsus — Moth, Australian
    Day-flying

Alcidis metaurus — Moth, Australian
    Day-flying
Alcippe sp. — Babbler, Nun
Alcyone azurea — Kingfisher, Azure
Alcyone pusilla — Kingfisher, Little
Alcyonidium sp. — Bryozoan, Marine
Alcyonium sp. — Dead Men's Fingers (coral)
Alectis ciliaris — Threadfish
Alectoria sp. — Lichen, Horsehair
Alectoris barbara — Partridge, Barbary
Alectoris chukar — Chukar
Alectoris graeca — Chukar
Alectoris rufa — Partridge, Red-legged
Alectryon sp. — Titoki
Alectura lathami — Turkey, Brush
Alestes longipinnis — Characin, Long-finned
Alestopetersius caudalis — Characin, Yellow
    Congo
Aleuria aurantia — Orange Peel or Orange Cup
Aleurochiton complanatus — Fly, White
Algyroides nigropunctatus — Lizard, Corfu
Alisma plantago-aquatica — Water-plantain,
    Common
Alisma sp. — Water-plantain
Alisterus scapularis — Parrot, King
Alkanna sp. — Alkanet
Allactaga sp. — Jerboa
Allamanda cathartica — Golden Trumpet
Allamanda nerifolia — Allamanda, Yellow
Allamanda violacea — Allamanda, Purple
Alliaria petiolata or A. officinalis— Mustard,
    Garlic
Alligator sinensis — Alligator, Chinese
Allium albopilosum — Allium, Giant
Allium ampeloprasum — Leek, Wild
Allium ascalonicum — Shallot
Allium carinatum — Garlic, Keeled
Allium cepa — Onion, Common
Allum cernuum — Onion, Nodding or Flowering
Allium fistulosum — Onion, Welsh
Allium flavum — Onion, Yellow
Allium giganteum — Onion, Ornamental
Allium karataviense — Onion, Turkestan
Allium moly — Garlic, Golden
Allium montanum — Garlic, Mountain
Allium neapolitanum — Garlic, Naples
Allium oleraceum — Garlic, Field
Allium paradoxum — Leek, Few-flowered
Allium porrum — Leek
Allium roseum — Garlic, Rose

Allium sativum — Garlic
Allium schoenoprasum — Chives
Allium scorodoprasum — Leek, Sand
Allium sphaerocephalon — Leek, Round-headed
Allium triquetrum — Leek, Tree-cornered
Allium ursinum — Ransons
Allium victorialis — Garlic, Long-rooted
Allium vineale — Garlic, Field or Crow
Allochernes panzeri — Scorpion, Book or False
Alluaudia sp. — Alluaudia
Alnus cordata — Alder, Italian
Alnus glutinosa — Alder, European or Common
Alnus incana — Alder, Gray Speckled
Alnus rhombofolia — Alder, White
Alnus viridis — Alder, Green
Alocasia macrorhiza — Lily, Cunjevoi or Spoon
Aloe arborescens — Aloe, Candelabra or Tree
Aloe aristata — Aloe, Lace
Aloe dichotoma — Quiver or Dragon Tree
Aloe humilis — Crocodile Jaws
Aloe speciosa — Aloe
Aloe variegata — Aloe, Partridge-breasted
Aloe vera — Aloe, True
Alonsoa warscewiczii — Mask Flower
Alopecurus sp. — Foxtail
Alopex corsac — Fox, Corsac
Alopex lagopus — Fox, Arctic or White
Alopias vulpinus — Shark, Thresher
Alopochen aegyptiacus — Goose, Egyptian
Alosa alosa — Shad, Allis
Alosa fallax — Shad, Twaite
Alouatta seniculus — Monkey, Howler
Alouatta villosa — Monkey, Howler
Alpinia purpurata — Ginger Blossom, Red
Alpinia speciosa — Shell Flower
Alstroemeria pulchella — Christmas Bell, New
    Zealand
Alstroemeria sp. — Lily, Peruvian
Alternanthera amoena — Alternanthera
Alternanthera philoxeroides — Alternanthera
Althaea hirsuta — Mallow, Rough
Althaea officinalis — Mallow, Marsh
Althaea rosea — Hollyhock
Alucita hexadactyla — Moth, Many-plumed
Alucita pentadactyla — Moth, Many-plumed
Alytes obstetricans — Toad, Midwife
Amadina erythrocephala — Finch, Red-headed
Amadina fasciata — Finch, Cut-throat
Amandava amandava — Avadavat, Red

Amandava subflava — Waxbill, Golden- or
    Orange-breasted
Amanita caesarea — Mushroom, Caesar's
Amanita citrina — Death Cap, False
Amanita fulva — Grisette
Amanita muscaria — Mushroom, Fly
Amanita pantherina — Panther Cap
Amanita phalloides — Death Cap
Amanita rubescens — The Blusher
Amanita spissa — Agaric, Stout
Amanita vaginata — Grisette
Amanite verna — Destroying Angel
Amanita virosa — Destroying Angel
Amanitopsis sp. — Mushroom, Amanitopsis
Amaranthus albus — Pigweed, White
Amaranthus caudatus — Love-lies-bleeding
Amaranthus retroflexus — Pigweed, Redroot or
    Common
Amaranthus tricolor — Joseph's Coat
Amaryllis belladonna — Lily, Belladonna
Amathes c-nigrum — Moth, Square-spot Rustic
Amathes xanthographa — Moth, Square-spot
    Rustic
Amaurobius sp. — Spider
Amaurornis akool — Bush-hen
Amaurornis olivaceus — Bush-hen
Amaurornis phoenicurus — Waterhen,
    White-breasted
Amaurornis ruficrissus — Bush-hen
Amazilia franciae — Emerald
Amazilia iodura — Emerald
Amazilia leucogaster — Emerald
Amazilia tzacatl — Emerald
Amazon aestiva — Parrot, Amazon
Amazon amazonica — Parrot, Amazon
Amazon ochrocephala — Parrot, Amazon
Amazonetta brasiliensis — Teal, Brazilian
Amblonyx cinerea — Otter, Asian Short-clawed
Amblydoras hancocki — Amblydoras, Hancock's
Amblyornis inornatus — Bowerbird, Crestless
Amblyornis macgregoriae — Bowerbird,
    MacGregor's
Amblyornis subalaris — Bowerbird,
    Orange-crested
Amblyospiza albifrons — Weaver, Thick-billed
Amblyrhynchus cristatus — Iguana, Marine
Amblysomus sp. — Mole, Golden
Amblyteles armatorius — Wasp, Ichneumon
Amelanchier sp. — Serviceberry.

313

Ameles sp. — Mantis
Amherstia nobilis — Pride of Burma
Ammobium alatum — Everlasting, Winged
Ammodorcas clarkei — Dibatag
Ammodytes lancea or A. tobianus — Sand-eel,
    Lesser
Ammodytes lanceolatus — Sand-eel, Greater
Ammomanes cincturus — Lark, Desert
Ammomanes deserti — Lark, Desert
Ammoperdix griseogularis — Partridge, See-see
Ammophila arenaria — Marram-grass
Ammophila sp. — Wasp, Sand
Ammotragus lervia — Sheep, Barbary or Maned
Amoeba sp. — Amoeba
Amoria sp. — Volute, Australian
Amorphophallus sp. — Amorphophallus
Ampedus cinnabarinus — Beetle, Click
Ampeliceps coronatus — Mynah, Golden-crested
Ampelopsis brevipedunculata — Ampelopsis
Amphibolurus barbatus — Lizard, Bearded
Amphipholis sp. — Brittle-star
Amphiphorus sp. — Worm, Pink Ribbon
Amphiprion sp. — Anemone-fish
Amphisbaena alba — Worm-lizard
Amphisbaena angustifrons — Worm-lizard
Amphitrite johnstoni — Worm
Amusium japonicum or A. pleuronectes —
    Scallop, Moon
Amytornis goyderi — Wren, Eyrean Grass
Amytornis purnelli — Wren, Dusky Grass
Amytornis striatus — Wren, Striated Grass
Amytornis textilis — Wren, Western Grass
Anabas testudineus — Perch, Climbing
Anacampseros sp. — Anacampseros
Anacamptis pyramidalis — Orchid, Pyramidal
Anacardium occidentale — Cashew
Anacyclus depressus — Daisy, Atlas
Anaea sp. — Butterfly, Leaf-wing
Anagallis arvensis — Pimpernel, Scarlet
Anagallis foemina — Pimpernel, Blue
Anagallis linifolia — Pimpernel, Blue
Anagallis minima — Chaffweed
Anagallis monelli — Pimpernel, Blue
Anagallis tenella — Pimpernel, Bog
Anagyris foetida — Stinking Wood
Ananas comosus — Pineapple
Ananas sp. — Pineapple
Anaphalis sp. — Pearly Everlasting
Anaplectes rubriceps — Weaver, Red-headed

Anaptychia sp. — Anaptychia
Anarhichas lupus — Catfish, Sea
Anarhichas minor — Catfish, Sea
Anarhynchus frontalis — Plover, Wrybill
Anarrhinum bellidifolium — Toadflax,
    Daisy-leaved
Anartia amathea — Butterfly, Peacock
Anas acuta — Pintail, Common
Anas angustirostris — Teal, Marbled
Anas aucklandica — Duck, Auckland Island
Anas bahamensis — Duck, Bahama
Anas bernieri — Teal, Madagascar
Anas capensis — Widgeon, Cape
Anas castanea — Teal, Chestnut-breasted
Anas clypeata — Shoveler
Anas crecca — Teal, Common or European
Anas discors — Teal, Blue-winged
Anas erythrorhyncha — Teal, Red-bill
Anas falcata — Teal, Falcated
Anas flavirostris — Teal, South American
    Green-winged
Anas formosa — Teal, Baikal or Formosa
Anas gibberifrons — Teal, Gray
Anas hottentotta — Teal, Hottentot
Anas leucophrys — Teal, Ringed
Anas penelope — Widgeon, European
Anas platyrhynchos — Mallard
Anas poecilorhyncha — Duck, Spotbill
Anas punctata — Teal, Hottentot
Anas querquedula — Garganey
Anas rhynchotis — Shoveler, Blue-winged
Anas rhynchotis variegata — Shoveler, New
    Zealand
Anas sibilatrix — Widgeon, Chiloe
Anas sparsa — Duck, African Black
Anas strepera — Gadwall
Anas superciliosa — Duck, Australian Black
Anas undulata — Duck, Yellow-billed
Anas versicolor — Teal, Silver
Anastatica hierochuntica — Rose of Jericho
Anastomus lamelligerus — Stork, African
    Openbill
Anastomus oscitans — Stork, Asian Openbill
Anatis ocellata — Beetle, Eyed Ladybird
Anax imperator — Dragonfly, Emperor
Anchusa azurea — Alkanet, Large Blue
Anchusa capensis — Forget-me-not, Summer
Anchusa officinalis — Alkanet, True
Ancylastrum fluviatile — Limpet, River

Ancyluris sp. — Butterfly
Ancylus fluviatilis — Limpet, River
Andigena sp. — Toucan, Mountain
Andrena sp. — Bee, Mining
Andrias japonicus — Salamander, Giant
Andricus kollari — Wasp, Marble Gall
Andromeda glaucophylla — Rosemary, Bog
Andromeda polifolia — Rosemary, Bog
Andropadus importunus — Bulbul, Somber
Andropogon sp. — Broomsedge
Androsace alpina — Rock-jasmine, Alpine
Androsace carnea — Rock-jasmine, Flesh-red
Androsace elongata — Rock-jasmine,
    Long-stalked
Androsace helvetica — Rock-jasmine, Swiss
Androsace lactea — Rock-jasmine, Milk
Androsace maxima — Rock-jasmine, Great
Androsace septentrionalis — Rock-jasmine,
    Northern
Androsace sp. — Rock-jasmine
Androsace villosa — Rock-jasmine, Woolly
Andryala integrifolia — Andryala
Anemone appennina — Anemone, Blue Wood
Anemone blanda — Anemone, Greek
Anemone coronaria — Anemone, Poppy or
    Crown
Anemone hortensis — Anemone, Star
Anemone japonica or hupehensis — Anemone,
    Japanese
Anemone narcissiflora — Anemone, Daffodil
Anemone nemorosa — Anemone, Wood
Anemone pavonina — Anemone, Great Peacock
    of Greece
Anemone ranunculoides — Anemone, Yellow
Anemone sylvestris — Windflower, Snowdrop
Anemonella thalictroides — Rue-anemone
Anemonia sp. — Sea Anemone, Snakelocks
Anergates atratulus — Ant
Anethum graveolens — Dill
Angelica archangelica — Angelica, Common or
    Eurasian
Angelica sylvestris — Angelica, Wild
Angraecum sesquipedale — Orchid,
    Star-of-Bethlehem
Angraecum sp. — Orchid
Anguilla anguilla — Eel, Common or European
Anguis fragilis — Slowworm, European
Anhima cornuta — Screamer, Horned
Anhinga melanogaster — Darter, Indian or
    Oriental

Anhinga novaehollandiae — Darter, Australian
Anhinga rufa — Anhinga
Anigozanthos sp. — Kangaroo-paw or
    Kangaroo-flower
Anisognathus sp. — Mountain-tanager
Annona cherimola — Custard Apple
Annona muricata — Soursop
Annona squamosa — Soursop
Anoa depressicornis — Anoa, Lowland
Anoa mindorensis — Tamarau or Tamaraw
Anobium punctatum — Beetle, Furniture
Anodonta cygnea — Mussel, Swan
Anodorhynchus hyacinthinus — Macaw, Blue or
    Hyacinthine
Anodorhynchus leari — Macaw, Blue or
    Hyacinthine
Anoectochilus sp. — Orchid, Jewel
Anogramma sp. — Fern, Jersey
Anole sp. — Anole
Anomalurus sp. — Squirrel, Flying or
    Scaly-tailed
Anomia ephippium — Oyster, Saddle
Anopheles sp. — Mosquito, Malaria
Anoplius sp. — Wasp, Spider-hunting
Anostomus anostomus — Anostomus, Striped
Anostomus trimaculatus — Anostomus,
    Three-spot
Anous minutus — Tern, White-capped Noddy
Anous stolidus — Tern, Noddy
Anous tenuirostris — Tern, White-capped Noddy
Ansellia sp. — Orchid, Leopard
Anser albifrons — Goose, White-fronted
Anser anser — Goose, Graylag
Anser arvensis — Goose, Bean
Anser brachyrhynchus — Goose, Pink-footed
Anser caerulescens — Goose, Snow
Anser cygnoides — Goose, Swan
Anser erythropus — Goose, Lesser White-fronted
Anser fabalis — Goose, Bean
Anser hyperboreus — Goose, Snow
Anser indicus — Goose, Bar-headed
Anser leucopsis — Goose, Barnacle
Anseranas semipalmata — Goose, Magpie or
    Pied
Anseropoda placenta — Starfish, Goose-foot
Ansonia grillivoca — Toad, Cricket-voiced
Antechinomys laniger — Mouse, Jerboa-like
    Pouched
Antechinomys spenceri — Mouse, Jerboa-like
    Pouched

Antechinus sp. — Mouse, Marsupial
Antedon sp. — Feather-star
Antennaria dioica — Cat's-foot
Antennarius sp. — Frogfish
Antennularia antennina — Hydroid
Anthemis arvensis — Chamomile, Corn or Field
Anthemis cotula — Chamomile, Stinking or
    Mayweed
Anthemis nobilis — Chamomile, Roman or
    English
Anthemis sp. — Chamomile
Anthemis tinctoria — Chamomile, Yellow
Antherea sp. — Moth, Oak Silk
Anthericum fasciculatum — Spiderwort
Anthericum liliago — Lily, St. Bernard's
Anthericum ramosum — Spiderwort
Anthia thoracica — Beetle, Desert Ground
Anthia collaris — Beetle, Desert Ground
Anthias anthias — Sea-perch
Anthidium sp. — Bee
Anthochaera carunculata — Wattlebird, Red
Anthochaera chrysoptera — Wattlebird, Little
Anthochaera paradoxa — Wattlebird, Yellow
Anthochaera rufogularis — Honey-eater,
    Spiny-cheeked
Anthocharis cardamines — Butterfly,
    Orange-tip
Anthocharis genutia — Butterfly, Orange-tip
Anthocoris nemorum — Bug, Flower
Antholyza sp. — Antholyza
Anthonomus sp. — Weevil, Apple Blossom
Anthophora sp. — Bee, Flower or Potter
Anthornis melanura — Bellbird
Anthoscopus sp. — Tit, Penduline
Anthoxanthum odoratum — Vernal-grass,
    Sweet
Anthracoceros albirostris — Hornbill, Indian
    Pied
Anthracoceros malabaricus — Hornbill, Indian
    Pied
Anthracoceros marchei — Hornbill, Palawan
Anthracoceros undulatus — Hornbill, Palawan
Anthracothorax nigricollis — Mango,
    Black-throated
Anthrenus sp. — Beetle, Carpet or Museum
Anthreptes collaris — Sunbird
Anthreptes malaccensis — Sunbird,
    Brown-throated
Anthreptes platura — Sunbird

Anthricus caucalis — Chervil, Bur
Anthricus cerefolium — Chervil
Anthricus sylbestris — Cow Parsley
Anthropoides paradisea — Crane, Blue
Anthropoides virgo — Crane, Demoiselle
Anthurium andreanum — Flamingo Flower
Anthurium crystallinum — Anthurium, Crystal
Anthurium scherzeranum — Flamingo Flower
Anthurus sp. — Anthurus
Anthus australis — Pipit, New Zealand or
    Richard's
Anthus caffer — Pipit, Meadow
Anthus campestris — Pipit, Tawny
Anthus cervinus — Pipit, Red-throated
Anthus gustavi — Pipit, Pechora
Anthus hodgsoni — Pipit, Indian or Olive Tree
Anthus novaeseelandiae — Pipit, New Zealand
    or Richard's
Anthus pratensis — Pipit, Meadow
Anthus richardi — Pipit, New Zealand or
    Richard's
Anthus spinoletta — Pipit, Rock or Water
Anthus trivialis — Pipit, Tree
Anthyllis vulneraria — Kidney-vetch
Antidorcas marsupialis — Springbok or
    Springbuck
Antigonon leptopus — Coral Vine
Antilope cervicapra — Blackbuck
Antirrhinum latifolium — Snapdragon, Large
Antirrhinum majus — Snapdragon, Common
Antirrhinum orontium — Snapdragon, Lesser
Anumbius annumbi — Firewood-gatherer
Aonyx sp. — Otter, Clawless
Aotus trivirgatus — Ape, Night
Apatele aceris — Moth, Sycamore
Apatele alni — Moth, Alder
Apatele psi — Moth, Alder
Apatura ilia — Butterfly, Purple Emperor
Apatura iris — Butterfly, Purple Emperor
Apera interrupta — Bent
Apera spica-venti — Bent, Silky
Aphanes arvensis — Parsley Piert
Aphanis sp. — Toothcarp
Aphantopus hyperanthus — Butterfly, Ringlet
Aphelandra sp. — Aphelandra
Aphelandra squarrosa — Zebra Plant
Aphelocephala leucopsis — Whiteface, Eastern
    or Southern
Aphis fabae — Aphid, Bean

316

Aphodius sp. — Beetle, Dung
Aphomia sociella — Moth, Bee
Aphrodite aculeata — Sea Mouse
Aphrophora alni — Froghopper, Alder
Aphya minuta — Goby, Transparent
Aphyllanthes monspeliensis — Aphyllanthes
Aphyocharax rubripinnis — Bloodfin
Aphyosemion australe — Lyretail, Cape Lopez
Aphyosemion calabaricus — Lyretail, Calabar
Aphyosemion calliurum — Lyretail, Banner
Aphyosemion christyi — Lyretail
Aphyosemion gardneri — Lyretail, Steel-blue
Aphyosemion gulare caeruleum — Gularis, Blue
Aphyosemion lujae — Lyretail, Red-striped
Aphyosemion walkeri — Lyretail
Apis sp. — Bee, Honey
Apistogramma sp. — Cichlid, Dwarf
Apium graveolens — Celery, Wild
Apium graveolens var. dulce — Celery
   (Cultivated)
Apium graveolens var. rapaceum — Celeriac
Apium inundatum — Marshwort
Apium nodiflorum — Watercress, Fool's
Aplocheilus sp. — Panchax
Aplonis metallica — Starling, Shining
Aplonis panayensis — Starling, Philippine
   Glossy
Aplysia sp. — Sea Horse
Apodemus agrarius — Mouse, Striped Field
Apodemus flavicollis — Mouse, Yellow-necked
Apodemus sylvaticus — Mouse, Long-tailed
   Field
Apoderus coryli — Weevil
Aponogeton distachys — Pondweed,Cape
Aponogeton fenestralis or A. madagascariensis
   — Lace Plant, Madagascar
Aponogeton natans — Pondweed, Cape
Aporia crataegi — Butterfly, Veined White
Aporocactus flagelliformis — Cactus, Rattail
Aporrhais pespelicani — Pelican's-foot Shell
Aprosmictus erythropterus — Parrot,
   Red-winged
Aprosmictus scapularis — Parrot, King
Aptenodytes forsteri — Penguin, Emperor
Aptenodytes patagonica — Penguin, King
Aptenoides forsteri — Penguin, Emperor
Aptenoides patagonicus—Penguin, King
Apteryx australis — Kiwi, Brown or Common
Apteryx oweni — Kiwi, Little Spotted or Owen's

Apus affinis — Swift, House or Little
Apus apus — Swift, Common or Eurasian
Apus caffer — Swift, White-rumped
Apus melba — Swift, Alpine
Apus pacificus — Swift, Fork-tailed
Apus pallidus — Swift, Pallid
Aquila audax — Eagle, Wedge-tailed
Aquila chrysaetos — Eagle, Golden
Aquila clanga — Eagle, Greater Spotted
Aquila heliaca — Eagle, Imperial
Aquila heliaca adelberti — Eagle, Spanish
   Imperial
Aquila nipalensis — Eagle, Steppe
Aquila pomarina — Eagle, Lesser Spotted
Aquila rapax — Eagle, Tawny
Aquila verreauxi — Eagle, Black or Verreaux's
Aquila wahlbergi — Eagle, Wahlberg's
Aquilega canadensis — Columbine, American
Aquilega chrysantha — Columbine, Golden
Aquilega flabellata — Columbine, Fan
Aquilega hybrid — Columbine
Aquilega vulgaris — Columbine, Common
Ara ararauna — Macaw, Blue-and-yellow
Ara chloroptera — Macaw, Red-and-blue
Ara macao — Macaw, Scarlet
Ara militaris — Macaw, Military Green
Ara rubrogenys — Macaw, Scarlet
Arabidopsis thaliana — Cress, Thale
Arabis albida — Snow-in-summer
Arabis alpina — Rock-cress, Alpine
Arabis caucasica — Arabis, Garden
Arabis glabra — Mustard, Tower
Arabis hirsuta — Rock-cress, Hairy
Arabis recta — Rock-cress, Upright
Arabis stricta — Rock-cress, Bristol
Arachis hypogaea — Peanut
Arachnis flos-aeris — Orchid, Scorpion
Arachnothera chrysogenys — Spider-hunter,
   Lesser Yellow-eared
Arachnothera longirostra — Spider-hunter,
   Little
Arachnothera magna — Spider-hunter,
   Streaked
Aradus sp. — Flatbug
Aralia elata — Angelica Tree, Japanese
Aralia japonica — Fatsia, Japanese
Aramides cajanea — Rail, Wood
Aramides ypecaha — Rail, Wood
Aramus guarauna — Limpkin

Araneus sp. — Spider, Garden
Arapaima gigas — Arapaima
Araschnia levana — Butterfly, Map
Aratinga guarouba — Parakeet, Golden
Aratinga mitrata — Parakeet, Golden
Araucaria araucana — Monkey Puzzle Tree
Araucaria bidwillii — Bunya-bunya
Araucaria excelsa — Norfolk Island Pine
Auraucaria heterophylla — Norfolk Island Pine
Araucaria imbricata — Monkey Puzzle Tree
Arborophila sp. — Partridge, Tree
Arbutus menziesii — Madrona Tree
Arbutus unedo — Strawberry Tree
Arca sp. — Noah's Ark
Arcella sp. — Amoeba
Arctia caja — Moth, Garden Tiger
Arctia villica — Moth, Garden Tiger
Arctictis binturong — Binturong
Arctium lappa — Burdock, Great
Arctium minus — Burdock, Lesser or Common
Arctium tomentosum — Burdock, Woolly or
  Downy.
Arctocebus calabarensis — Potto, Golden
Arctocephalus sp. — Seal, Fur
Arctogalidia trivirgata — Palm-civet,
  Small-toothed
Arctonetta fischeri — Eider, Spectacled
Arctonyx collaris — Badger, Hog or Sand
Arctostaphylos uva-ursi — Bearberry
Arctotis sp. — Daisy, African
Arctous alpinus — Bearberry, Alpine or Black
Ardea cinerea — Heron, Common or Gray
Ardea goliath — Heron, Goliath
Ardea melanocephala — Heron, Black-headed
Ardea novaehollandiae — Heron, White-faced
Ardea pacifica — Heron, White-necked Pacific
Ardea picata — Heron, Pied
Ardea purpurea — Heron, Purple
Ardea sumatrana — Heron, Great-billed or
  Giant
Ardeola bacchus — Heron, Chinese Pond
Ardeola grayii — Heron, Indian Pond
Ardeola ibis — Egret, Cattle
Ardeola ralloides — Heron, Squacco
Ardeotis kori — Bustard, Kori
Ardisia crenata — Coral Berry or Spiceberry
Ardisia sp. — Ardisia
Areca catechu — Palm, Betel
Arenaria ciliata — Sandwort, Irish

Arenaria interpres — Turnstone, Ruddy
Arenaria leptoclados — Sandwort, Slender
Arenaria montana — Sandwort, Mountain
Arenaria norvegica — Sandwort, Arctic
Arenaria serpyllifolia — Sandwort,
  Thyme-leaved
Arenga sp. — Palm, Sugar
Arenicola marina — Lugworm
Argas sp. — Tick, Bird
Arge sp. — Sawfly, Rose
Argema mimosae — Moth, Malagasy Silk
Argema mittrei — Moth, Malagasy Silk
Argemone grandiflora — Poppy, Prickly
Argemone hispida — Poppy, Prickly
Argemone mexicana — Poppy, Prickly
Argentina sp. — Argentine
Argiope sp. — Spider, Wasp or Zebra
Argonauta argo or A. hians — Paper Nautilus
Argulus foliaceus — Louse, Fish
Argusianus argus — Pheasant, Great Argus
Argynnis paphia — Butterfly, Silver-washed
  Fritillary
Argynnis sp. — Butterfly, Fritillary
Argyreia nervosa — Morning-glory, Woolly
Argyroneta aquatica — Spider, Water
Argyropelecus sp. — Hatchetfish, Deep-sea
Argyroxiphium sp. — Silversword
Arianta arbustorum — Snail, Copse
Aricia agestis — Butterfly, Brown Argus
Ariocarpus fissuratus — Cactus, Star
Ariocarpus retusus — Cactus, Star
Arion sp. — Slug, Garden
Arisaema candidissimum — Jack-in-the-pulpit
Arisaema dracontium — Dragonroot or Green
  Dragon
Arisaema fimbriatum — Jack-in-the-pulpit
Arisaema ringens — Jack-in-the-pulpit
Arisaema sikokiamum — Jack-in-the-pulpit
Arisaema triphyllum — Jack-in-the-pulpit
Arisarum vulgare — Friar's Cowl
Aristolochia brasiliensis — Dutchman's Pipe
Aristolochia clematitis — Birthwort
Aristolochia durior — Dutchman's Pipe
Aristolochia elegans — Calico Flower
Aristolochia gigantea — Pelican Flower
Aristolochia grandiflora — Pelican Flower
Aristolochia macrophyllum — Dutchman's Pipe
Aristolochia rotunda — Birthwort,
  Round-leaved

Aristolochia saccata — Dutchman's Pipe
Aristolochia salpinx — Dutchman's Pipe
Aristotela sp. — Wineberry
Armadillidium sp. — Sowbug
Armandia lidderdalli — Butterfly, Bhutan Glory
Armeria alpina — Thrift, Alpine
Armeria caespitosa — Thrift
Armeria fasciculata — Thrift
Armeria maritima — Sea Pink or Thrift
Armeria pseudarmeria — Thrift, Estorial
Armillaria mellea — Fungus, Honey
Armoracia rusticana — Horseradish
Arnebia echioides — Prophet Flower
Arnica montana — Arnica, Mountain
Arnoglossus laterna — Scaldfish
Arnoldichthys spilopterus — Characin, Red-eyed
Arnoseris minima — Succory, Swine's or Lamb's
Aromia moschata — Beetle, Musk
Aronia arbutifolia — Chokeberry, Red
Arothron sp. — Blowfish
Arremon aurantirostris — Sparrow,
   Orange-billed
Arremon schlegeli — Sparrow, Orange-billed
Arrhenatherum elatius — Oat Grass
Arrhenurus superior or A. caudatus — Mite,
   Water
Arses kaupi — Flycatcher, Pied
Arses lorealis — Flycatcher, Frill-necked
Artamus cinereus — Wood-swallow, Black-faced
Artamus cyanopterus — Wood-swallow, Dusky
Artamus leucorhynchus — Wood-swallow,
   White-breasted
Artamus personatus — Wood-swallow, Masked
Artamus superciliosus — Wood-swallow,
   White-browed
Artemisia absinthium — Wormwood, Absinthe
Artemisia arborescens — Wormwood, Shrubby
Artemisia borealis — Wormwood, Alpine
Artemisia campestris — Mugwort, Breckland
Artemisia dracunculus — Tarragon
Artemisia maritima — Wormwood, Sea
Artemisia mutellina — Wormwood, Alpine
Artemisia nitida — Wormwood, Shining
Artemisia vulgaris — Mugwort
Arthropleona sp. — Springtail
Artocarpus altilis — Breadfruit Tree
Artocarpus communis — Breadfruit Tree
Artocarpus heterophyllus — Breadfruit Tree
Arum italicum — Arum, Italian
Arum maculatum — Cuckoo-pint

Aruncus dioicus — Goatsbeard
Aruncus sylvester — Goatsbeard
Arundinaria sp. — Bamboo
Arundo sp. — Reed
Arvicola amphibius — Vole, Water
Arvicola terrestris — Vole, Water
Asarum canadense — Ginger, Western Wild
Asarum caudatum — Ginger, Western Wild
Asarum europaeum — Ginger, European Wild
Ascalaphus sp. — Ant-lion
Ascidia mamillata — Sea Squirt
Ascidia mentula — Sea Squirt
Asclepias currassavia — Blood Flower
Asclepias fruticosa — Milkweed
Asclepias sp. — Milkweed
Asclepias syriaca — Milkweed, Common
Asclepias tuberosa — Butterfly Weed
Ascocenda hybrid — Orchid
Ascophyllum nodosum — Wrack, Knotted or
   Bladder
Asellus sp. — Sowbug, Water
Ashbyia lovensis — Chat, Gibber
Asilus sp. — Fly, Robber
Asimina triloba — Papaw or Pawpaw
Asio capensis — Owl, African Marsh
Asio flammeus — Owl, Short-eared
Asio otus — Owl, Long-eared
Asparagus acutifolius — Asparagus,
   Sharp-leaved
Asparagus densiflora — Asparagus-fern
Asparagus falcatus — Sicklethorn
Asparagus myersii — Asparagus-fern, Foxtail
Asparagus officinalis — Asparagus
Asparagus plumosus — Asparagus-fern
Asparagus setaceus — Asparagus-fern
Asparagus sprengeri — Asparagus-fern
Asperugo procumbens — Madwort
Asperula arcadiensis — Squinancywort
Asperula arvensis — Woodruff
Asperula cyanchica — Squinancywort
Asperula odorata — Woodruff, Sweet
Asperula orientalis — Woodruff
Asperula suberosa — Woodruff
Asperula tinctoria — Woodruff
Asphodeline lutea — King's Spear
Asphodelus aestivus — Asphodel, White
Asphodelus albus — Asphodel, White
Asphodelus fistulosus — Asphodel,
   Hollow-stemmed
Asphodelus sp. — Asphodel

Aspidelaps sp. — Snake, Shield-nose
Aspidistra elatior — Cast-iron Plant
Aspidistra lurida — Cast-iron Plant
Aspidomorpha sp. — Beetle, Tortoise
Aspitrigla cuculus — Gurnard, Red
Aspius aspius — Asp (fish)
Aspius repax — Asp (fish)
Asplenium adiantum-nigrum — Spleenwort,
    Black
Asplenium bulbiferum — Fern, Mother or
    Parsley
Asplenium ebenoides — Spleenwort, Scott's
Asplenium marinum — Spleenwort, Sea
Asplenium nidus — Fern, Bird's-nest
Asplenium nidus-avis — Fern, Bird's nest
Asplenium platyneuron — Spleenwort, Ebony
Asplenium ruta-muraria — Wall-rue
Asplenium scolopendrium — Fern,
    Hart's-tongue
Asplenium septentrionale — Spleenwort, Forked
Asplenium trichomanes — Spleenwort,
    Maidenhair
Asplenium viride — Spleenwort, Green
Aspro asper — Streber
Astacus astacus — Crab, Fresh-water
Astacus fluviatilis—Crab, Fresh-water
Astarte borealis — Astarte
Astarte sulcata — Astarte
Astelia solandri — Lily, Perching
Astelia sp. — Astelia
Aster alpinus — Aster, Alpine
Aster amellus — Daisy, European Michaelmas
Aster bellidiastrum — Aster, Daisy-star
Aster ericoides — Aster, heath
Aster nova-anglieae — Aster, New England
Aster novi-belgii — Daisy, Michaelmas
Aster salignus — Aster, Willow-leaved
Aster tripolium — Aster, Sea
Asterias rubens — Starfish, Common European
Asterina gibbosa — Starlet
Asterophora lycoperdioides — Asterophora
Asterophora parasitica — Asterophora
Astilbe sp. — Astilbe
Astragalus alpinus — Milk-vetch, Alpine
Astragalus cicer — Milk-vetch, Yellow
Astragalus danicus — Milk-vetch, Purple
Astragalus frigidus — Milk-vetch, Alpine
Astragalus glycphyllos — Liquorice, Wild
Astragalus penduliflorus — Milk-vetch,
    Drooping

Astragalus sp. — Milk-vetch
Astrantia major or maxima — Masterwort,
    Great
Astrapia stephaniae — Bird of Paradise,
    Princess Stephanie's
Astronotus ocellatus — Cichlid, Oscar's or Velvet
Astropecten aurantiacus or A. irregularis —
    Starfish
Astrophytum asterias — Sand Dollar
Astrophytum capricorne — Cactus, Star
Astrophytum myriostigma — Bishop's Cap
Astrophytum ornatum — Cactus, Star
Asystasia gangetica — Primrose, Ganges
Atelerix sp. — Hedgehog, Tropical
Ateles ater — Monkey, Spider
Ateles geoffroyi — Monkey, Spider
Ateles paniscus — Monkey, Black Spider
Atelopus sp. — Frog
Atelornis sp. — Roller, Ground
Athene brama — Owl, Spotted Little
Athene noctua — Owl, Little
Atherina sp. — Sand-smelt
Atheris squamiger — Viper, Bush or Tree
Atherurus africanus — Porcupine, Brush-tailed
Athous haemorrhoidalis — Beetle, Red-brown
    Skipjack
Athyrium filix-femina — Fern, Lady
Athyrium georingianum pictum — Fern,
    Japanese Painted
Athyrium pycnocarpon — Spleenwort,
    Narrow-leaved
Athyrium thelypterioides — Spleenwort, Silvery
Atilax paludinosus — Mongoose, Marsh or Water
Atlapetes semirufous — Finch, Ochre-breasted
    Brush
Atrichornis clamosus — Scrub-bird, Noisy
Atrichornis rufescens — Scrub-bird, Rufous
Atrichum undulatum — Moss, Catherine's or
    Spineleaf
Atriplex halimus — Orache, Shrubby
Atriplex hastata — Orache, Halberd-leaved
Atriplex hortensis — Orache
Atriplex laciniata — Orache, Frosted
Atriplex littoralis — Orache, Grass-leaved or
    Shore
Atriplex nitens — Orache, Shining
Atriplex patula — Orache, Common
Atropa belladonna — Nightshade, Deadly
Atta sp. — Ant, Leaf-cutter
Attacus atlas — Moth, Atlas

Attacus edwardsii — Moth, Atlas
Attelabus nitens — Weevil, Oak-leaf Roller
Aubrieta deltoidea — Rock Cress, Purple
Auchenoglanis occidentalis — Catfish
Aucuba japonica — Laurel, Spotted or Japanese
Aulacomnium androgynum — Aulacomnium
Aulacorhynchus haematopygius — Toucanet,
    Red-rumped Green
Aulica imperialis — Volute, Imperial
Aulica vespertilio — Volute, Bat
Aulostomus sp. — Trumpetfish
Aurelia sp. — Jellyfish, Moon
Auricularia auricula — Jew's Ear
Auriscalpium vulgare — Fungus, Earspoon
Autocrates aeneus — Beetle
Auxis thazard — Mackerel, Frigate
Avahi laniger — Lemur, Avahi
Avena fatua — Oat, Wild
Avena sativa — Oat
Avena sempervirens — Oat, Wild
Avena sterilis — Oat, Animated
Averrhoa carambola — Carambola Tree
Aviceda cuculoides — Hawk, Crested
Aviceda jerdoni — Hawk, Crested
Aviceda leuphotes — Hawk, Crested
Aviceda subcristata — Hawk, Crested
Avicennia sp. — Mangrove
Axinella sp. — Sponge
Axis axis — Deer, Axis
Axis porcinus — Deer, Hog
Aythya australis — Duck, Australian
    White-eyed
Aythya ferina — Pochard, Common
Aythya fuligula — Duck, Tufted
Aythya marila — Scaup, Greater
Aythya novaeseelandiae — Scaup, New Zealand
Aythya nyroca — Duck, Ferruginous
Azara sp. — Azara
Azolla caroliniana — Fern, Mosquito
Azolla filiculoides — Fern, Water

B

Babiana stricta — Baboon Flower
Babirussa babyrussa — Babirusa
Bacillus rossii — Stick-insect, Mediterranean
Badis badis — Badis
Baeomyces roseus — Lichen, Pink Earth
Balaeniceps rex — Stork, Shoobill

Balaenoptera acutorostrata — Rorqual
Balaenoptera borealis— Rorqual
Balaenoptera musculus — Whale, Blue
Balaenoptera physalis — Whale, Finback
Balaninus sp. — Weevil, Nut
Balanoglossus sp. — Worm, Acorn
Balanophora sp. — Balanophora
Balanus sp. — Barnacle, Acorn
Baldellia ranunculoides — Water-plantain,
    Lesser
Balearica pavonina — Crane, Crowned
Balistapus undulatus — Triggerfish, Undulate
Balistes aculeatus — Triggerfish
Balistes capriscus or B. carolinensis —
    Triggerfish, Common
Balistes vetula — Triggerfish, Queen
Balistes vidua — Triggerfish, Pink-tail
Balistoides conspicillum — Triggerfish, Spotted
Ballotta nigra — Horehound, Black
Bambusa sp. — Bamboo
Bangsia edwardsi — Tanager, Moss-backed
Banksia coccinea — Banksia, Scarlet
Banksia ericifolia — Banksia, Heath
Banksia integrifolia — Banksia, Coast
Banksia serrata — Banksia, Saw
Banksia sp. — Honeysuckle, Australian or
    Honeysuckle Tree
Baptisia australis or B. tinctoria — Indigo
Barbarea vulgaris — Cress, Winter
Barbastella barbastellus — Bat, Barbastelle
Barbus arulius — Barb
Barbus barbus — Barb
Barbus conchonius — Barb, Red or Rosy
Barbus everetti — Barb, Clown
Barbus filamentosus — Barb, Black-spot
Barbus nigrofasciatus — Black Ruby
Barbus pentazona — Barb, Five-banded
Barbus schwanenfeldi — Barb, Schwanenfeld's
    or Tinfoil
Barbus tetrazona — Barb, Tiger
Barbus ticto — Barb, Two-spot
Barbus titteya — Barb, Cherry
Barclaya sp. — Barclaya
Barilius christyi — Barilius, Goldlip
Barleria cristata — Violet, Philippine
Barnardius barnardi — Parrot, Ringnecked
Barnardius macgillivrayi — Parrot, Cloncurry
Barnardius zonarius semitorquatus — Parrot,
    Twenty-eight

321

Barnea candida — Piddock, White
Barringtonia sp. — Barringtonia
Bartsia alpina — Bartsia, Alpine
Baryphthengus ruficapillus — Motmot, Rufous
Basileuterus coronatus — Warbler, Rufous-capped
Basileuterus rufifrons — Warbler, Rufous-capped
Basiliscus basiliscus — Basilisk
Basiliscus plumbifrons — Basilisk, Double-crested or Plumed
Bassaricyon sp. — Olingo
Bassia hirsuta — Seablite, Hairy
Bathilda ruficauda — Finch, Star
Bathypterois ventralis — Spiderfish
Batis capensis — Flycatcher, Cape or Puff-back
Batis molitor — Flycatcher, Chinspot
Batrachostomus javensis — Frogmouth, Javanese
Bauera rubioides — River Rose
Bauhinia galpinii — Pride of De Kaap
Bauhinia sp. — Orchid Tree
Bauhinia tomentosa — St. Thomas Tree
Bdeogale nigripes — Mongoose, Black-legged
Beaucarnea recurvata — Elephant-foot Tree
Beaumontia grandiflora — Herald's Trumpet
Bebrornis sechellensis — Warbler, Bush
Bedotia geayi — Silversides
Begonia masoniana — Begonia, Iron Cross
Begonia rex — Begonia, Rex
Begonia semperflorens — Begonia, Wax or Bedding
Begonia sp. — Begonia
Begonia tuberhybrida — Begonia, Tuberous
Belamcanda chinensis — Lily, Blackberry
Bellardia trixago — Bellardia
Bellicositermes natalensis — Termite
Bellidastrum michelii — Daisy, False or Alpine
Bellis perennis — Daisy, English or Lawn
Belone bellone — Garfish
Beloperone guttata — Shrimp Plant
Belostoma sp. — Waterbug, Giant
Bembix sp. — Wasp, Sand
Bena fagana — Moth, Green-silver Lines
Bena prasinana — Moth, Green-silver Lines
Berberis darwinii — Barberry, Darwin's
Berberis julianae — Barberry Wintergreen
Berberis sp. — Barberry
Berberis stenophylla — Barberry, Rosemary

Berberis thunbergii — Barberry, Japanese
Berberis vulgaris — Barberry, Common or European
Berenicornis comatus — Hornbill, Asian White-crested
Bergenia cordifolia — Saxifrage, Elephant-leaved
Bergenia crassifolia — Pig Squeak
Bergenia purpurascens — Bergenia
Beroe sp. — Comb-jelly
Berteroa incana - Alison, Hoary
Bertholletia excelsa — Brazil-nut
Bertolonia sp. — Bertolonia
Berula erecta — Water-parsnip, Lesser
Beta vulgaris — Beet, Garden
Beta vulgaris cicla — Chard, Swiss
Beta vulgaris subsp. maritima — Beet, Sea
Betonica officinalis — Betony, Wood
Betta splendens — Fighting-fish, Siamese
Bettongia sp. — Rat-kangaroo
Betula albosinensis — Birch, Northern Chinese Red-barked
Betula costata — Birch, Manchurian
Betula lutea — Birch, Yellow
Betula nana — Birch, Dwarf
Betula nigra — Birch, River
Betula papyrifera — Birch, Paper or Canoe
Betula pendula — Birch, Silver or White
Betula pendula gracilis — Birch, Cut-leaved White
Betula pubescens — Birch, Downy or Hairy
Betula verrucosa — Birch, Silver
Bhutanitis lidderdalii — Butterfly, Bhutan Glory
Biarum tenuifolium — Biarum
Bibio sp. — Hairfly
Bibos frontalis — Cayal
Bibos sondaicus — Banteng
Bidens cernua — Bur-marigold, Nodding
Bidens frondosa — Beggar-ticks or Sticktight
Bidens tripartita — Bur-marigold, Trifid
Bignonia capreolata — Cross-vine
Bilderdykia convolvulus — Bindweed, Black
Billbergia nutans — Queen's Tears
Billbergia rhodocyanea — Urn Plant
Billbergia sp. — Billbergia
Bipes biporus — Worm-lizard, Two-handed or Two-legged

Bipes caniculatus — Worm-lizard, Two-handed or Two-legged

Biplex perca — Frog Shell

Birgus latro — Crab, Robber

Biscutella laevigata — Mustard, Buckler

Bison bonasus — Bison, European

Bispira volutacornis — Fanworm, Twin

Biston betularia — Moth, Peppered

Biston stratarius — Moth, Oak or Poplar Beauty

Bitis arietans — Adder, Puff

Bitis caudalis or B. cornuta — Adder, Horned

Bitis gabonica — Adder, Gaboon or West African

Bitis nasicornis — Viper, Rhinoceros

Bitis peringueyi — Adder, Desert

Bittium reticulatum — Whelk, Needle

Bixa orellana — Annatto Tree

Biziura lobata — Duck, Musk

Blackstonia perfoliata — Yellow-wort

Blandfordia sp. — Christmas Bells

Blaniulus guttulatus — Millipede, Spotted

Blaps sp. — Beetle, Cellar or Churchyard

Blastocerus dichotomus — Deer, Marsh

Blastophaga psenes — Wasp

Blatella germanica — Cockroach, German

Blatta orientalis — Cockroach, Common or Oriental

Blechnum spicant — Fern, Hard or Deer

Blennius gattorugine — Blenny, Tompot

Blennius montagui — Blenny, Montagu's

Blennius ocellaris — Blenny, Butterfly

Blennis pavo — Blenny, Peacock

Blennius pholis — Blenny, Common

Blennius rouxi — Blenny, Long-striped or Roux's

Blennius tentacularius — Blenny, Tentacled

Bletilla hyacinthina — Orchid, Chinese Ground

Bletilla striata — Orchid, Chinese Ground

Blicca bjoerkna — Bream, Silver

Blysmus sp. — Club-rush

Blythipicus sp. — Woodpecker

Boa canina — Boa, Emerald Tree

Boa constrictor constrictor — Boa Constrictor

Boa cooki — Boa, Cook's Tree

Boaedon fuliginosus — Snake, Brown House

Boiga sp. — Snake, Mangrove

Boissonneaua flavescens — Hummingbird, Coronet

Boissonneaua jardini — Hummingbird, Coronet

Bolbitius vitellinus — Mushroom, Yellow Cow-pat

Boletinus sp. — Boletinus

Boletus badius — Boletus, Sweet-chestnut

Boletus bovinus — Boletus, Cow

Boletus chrysenteron — Boletus, Red-cracked

Boletus edulis — Boletus, Edible or Cep

Boletus edulis var. reticulatus — Boletus, Edible

Boletus elegans — Boletus, Elegant

Boletus felleus — Boletus, Bitter

Boletus granulatus — Boletus, Granulated

Boletus luridus — Boletus, Lurid

Boletus luteus — Boletus, Brown-yellow

Boletus parasiticus — Boletus, Parasitic

Boletus purpureus — Boletus, Purple

Boletus satanas — Boletus, Satan's

Boletus scaber — Boletus, Rough-stemmed

Boletus subomentosus — Boletus, Yellow-cracked

Boloria sp. — Butterfly, Fritillary

Bomarea sp. — Bomarea

Bombax sp. — Silk-Cotton Tree

Bombina bombina — Toad, Fire-bellied

Bombina orientalis — Toad, Oriental Fire-bellied

Bombina variegata — Toad, Yellow-bellied

Bombus sp. — Bee, Bumble

Bombycilla garrulus — Waxwing, Bohemian

Bombylius sp. — Fly, Bee

Bombyx mori — Moth, True Silk

Bonellia sp. — Worm, Echiurid

Boocercus euryceros — Bongo

Boophone disticha — Cape Poison

Borago laxiflora — Borage

Borago officinalis — Borage

Borassus sp. — Palm, Palmyra

Boreus hyemalis — Flea, Snow

Boronia sp. — Boronia

Bos banteng — Banteng

Bos bonasus — Bison, European

Bos bubalis — Buffalo, Indian Water

Bos grunniens — Yak, Wild

Bos frontalis — Gayal

Bos gaurus — Gaur

Bos indicus — Cattle, Brahman

Bos javanicus lowi — Banteng

Bos mutus — Yak, Wild

Bos sauveli — Kouprey

Bos taurus — Cattle

Boselaphus tragocamelus — Nilghai

Bostrychia hagedash — Ibis, Hadeda

Boswellia sp. — Boswellia

Botaurus poiciloptilus — Bittern, Australian or Brown

Botaurus stellaris — Bittern, Common or Eurasian

Bothriochloa ischaemum — Beard-grass

Bothrops alternatus — Urutu

Bothrops atrox — Fer-de-lance

Bothrops schlegeli — Viper, Pit

Bothus lunatus — Flounder, Peacock

Botia horae — Loach, Hora's Clown

Botia hymenophysa — Loach, Tiger

Botia macracantha — Loach, Clown

Botia modesta — Loach, Blue

Botrychium lunaria — Moonwort

Botrychium sp. — Fern, Grape

Botrydium granulatum — Alga, Yellow

Botryllus schlosseri — Sea Squirt, Star

Botryllus violaceus — Sea Squirt

Bougainvillea glabra — Bougainvillea

Bougainvillea species — Bougainvillea

Bougainvillea spetabilis — Bougainvillea

Boulengerella sp. — Characin, Pike

Bouvardia sp. — Bouvardia

Bovista sp. — Bovista

Bowdleria punctata — Fernbird

Brachinus crepitans — Beetle, Bombadier

Brachychiton sp. — Flame Tree of Australia

Brachycome iberidifolia — Daisy, Swan River

Brachydanio albolineatus — Danio, Pearl

Brachydanio kerri — Danio

Brachydanio nigrofasciatus — Danio, Spotted

Brachydanio rerio — Danio, Zebra

Brachygobius xanthozona — Goby, Golden-banded or Wasp

Brachylophus fasciatus — Iguana, Fiji Banded

Brachypodium pinnatum — False-brome, Chalk

Brachypodium sylvaticum — False-brome, Wood

Brachypteracias sp. — Roller, Ground

Brachypteryx sp. — Shortwing

Brachystelma barberiae — Brachystelma

Brachyteles arachnoides — Monkey, Spider

Brachyternus benghalensis — Woodpecker

Brachythecium sp. — Brachythecium

Bradypterus sp. — Warbler, Scrub

Bradypus cuculliger — Sloth, Three-fingered or Three-toed

Bradypus torquatus — Sloth, Three-fingered or Three-toed

Bradypus tridactylus — Sloth, Three-fingered or Three-toed

Brahmaea wallichii — Moth

Brama brama — Bream, Ray's

Branchiocerianthus imperator — Hydroid

Branchiostoma sp. — Amphioxus

Branchipus stagnalis — Shrimp, Fairy

Branta bernicla — Brant

Branta canadensis — Goose, Canada

Branta ruficollis — Goose, Red-breasted

Brasilocactus sp. — Brasilocactus

Brassaia actinophylla — Umbrella Tree, Queensland

Brassavola nodosa — Orchid, Lady-of-the-night

Brassia sp. — Orchid, Spider

Brassica alba — Mustard, White

Brassica campestris — Cabbage, Field

Brassica chinensis — Pak-choi

Brassica kaber — Mustard, Field or Wild

Brassica napobrassica — Rutabaga

Brassica napus — Rape

Brassica nigra — Mustard, Black

Brassica oleracea — Cabbage, Wild

Brassica oleracea botrytis — Cauliflower

Brassica oleracea caulorapa — Kohlrabi

Brassica oleracea gemnifera — Brussel Sprouts

Brassica rapa — Turnip

Brassocattleya sp. — Orchid

Brassolaeliocattleya sp. — Orchid

Braula sp. — Louse, Bee

Breviceps sp. — Frog, Short-headed

Breynia nivosa — Snow Bush

Brintesia circe — Butterfly, Great Banded Grayling

Briza maxima — Quaking-grass, Large

Briza media — Quaking-grass, Common

Brochis coeruleus — Catfish

Brodiaea laxa — Grass Nut

Bromus commutatus — Brome, Meadow

Bromus erectus — Brome, Upright

Bromus inermis — Brome, Awnless or Smooth

Bromus madritensis — Brome, Madrid

Bromus mollis — Brome, Soft

Bromus ramosus — Brome, Hairy

Bromus secalinus — Brome, Rye

Bromus sterilis — Brome, Barren

Bromus tectorum — Brome, Drooping or Downy

Brookesia spectrum — Chameleon, Leaf

Brosme brosme — Cusk

Brotogeris jugularis — Parakeet,
    Orange-chinned
Broussonetia papyrifera — Mulberry, Paper
Browallia speciosa — Browallia
Browningia candelaris — Cactus, Hedge
Bruchus sp. — Weevil, Pea
Bruckenthalia spiculifolia — Heath, Spike
Bruguiera gymnorhiza — Bruguiera
Brunfelsia sp. — Yesterday, Today and
    Tomorrow
Brunnera macrophylla — Brunnera
Bryochaeta quadrimaculata — Weevil
Bryonia sp. — Bryony
Bryophyllum sp. — Lifeplant
Bryophyllum tubiflorum — Friendly Neighbor
Bryum sp. — Bryum
Bubalornis albirostris — Weaver, Buffalo
Bubalus arnee — Buffalo, Indian Water
Bubalus bubalis — Buffalo, Indian Water
Bubalus bulaus — Buffalo, Indian Water
Bubalus caffer—Buffalo, African or Cape
Bubalus depressicornis — Anoa, Lowland
Bubalus lichtensteini — Hartebeest,
    Lichtenstein's
Bubo africanus — Owl, Spotted Eagle
Bubo bubo — Owl, Eagle
Bubo capensis — Owl, Eagle
Bubo coromandus — Owl, Dusky Eagle
Bubo ketupa — Owl, Malay Fish
Bubo lacteus — Owl, Milky
Bubo nipalensis — Owl, Forest Eagle
Bubo philippensis — Owl, Philippine Eagle
Bubo sumatranus — Owl, Barred Eagle
Bubulcus egret — Egret, Cattle
Bubulcus ibis — Egret, Cattle
Buccinum undatum — Whelk, Northern or
    Waved
Bucco capensis — Puffbird, Collared
Bucco macrodactylus — Puffbird,
    Chestnut-capped
Bucephala clangula — Goldeneye
Buceros bicornis — Hornbill, Great Indian
Buceros hydrocorax — Hornbill, Great Indian
Bucorvus abyssinicus — Hornbill, Ground
Bucorvus leadbeateri — Hornbill, Ground
Buddleia alternifolia — Butterfly Bush,
    Fountain
Buddleia davidii — Butterfly Bush
Buddleia globosa — Butterfly Bush, Globe

Buddleia sp. — Buddleia
Budorcas taxicolor — Takin
Buellia sp. — Buellia
Bufo blombergi — Toad, Blomberg's
Bufo bufo — Toad, European
Bufo calamita — Toad, British or Natterjack
Bufo marinus — Toad, Giant or Marine
Bufo viridis — Toad, Green
Bugeranus carunculatus — Crane, Wattled
Buglossidium luteum — Solenette
Bulbocodium vernum — Saffron, Spring Meadow
Bulbophyllum medusae — Orchid,
    Medusa's-head
Bulbophyllum sp. — Orchid
Bulgaria inquinans — Black Bulgar or
    Black-stud Fungus
Bulweria bulwerii — Petrel, Bulwer's
Bungarus sp. — Krait
Bunias erucago — Cabbage, Crested
Bunias orientalis — Cabbage, Warty
Bunium bulbocastanum — Earth-nut, Great
Bunodactis sp. — Sea Anemone, Gem
Bupalus piniaria — Moth, Bordered White
    Beauty
Buphagus africanus — Oxpecker, Yellow-billed
Buphagus erythrorhynchus — Oxpecker,
    Red-billed
Buphthalmum salicifolium — Ox-eye, Yellow
Bupleurum baldense — Hare's-ear, Small
Bupleurum falcatum — Hare's-ear, Sickle
Bupleurum fruticosum — Hare's-ear, Shrubby
Bupleurum rotundifolium — Thorow-wax
Bupleurum tenuissimum — Hare's-ear, Slender
Buprestris sp. — Beetle, Wood-borer
Burchella bubalina — Buffledoorn
Burhinus bistriatus — Thick-knee,
    Double-striped
Burhinus capensis — Thick-knee, Spotted
Burhinus magnirostris — Curlew, Bush or
    Scrub
Burhinus oedicnemus — Curlew, Stone
Burramys parvus — Possum, Pygmy
Butastus indicus — Buzzard, Gray-faced
Butastus teesa — Buzzard, Gray-faced
Butea frondosa — Flame-of-the-forest
Buteo auguralis — Buzzard, Red-tailed
Buteo buteo — Buzzard, Common
Buteo galapagoensis — Hawk, Galapagos
Buteo lagopus — Hawk, Rough-legged

Buteo magnirostris — Hawk, Roadside
Buteo rufinus — Buzzard, Long-legged
Buteo rufofuscus — Buzzard, Jackal
Buthraupis montana — Mountain-tanager
Buthus sp. — Scorpion
Butomus umbellatus — Rush, Flowering
Butorides striatus — Heron, Mangrove or
    Striated
Butorides sundevalli — Heron, Green or Lava
Buxus microphylla — Box, Small or Littleleaf
Buxus sempervirens — Box, Common
Byctiscus sp. — Weevil
Byturus sp. — Beetle, Raspberry

C

Cacajao calvus — Uakari, Bald or
    Red-faced
Cacajao rubicundus — Uakari, Bald or
    Red-faced
Cacatua galerita — Cockatoo, Sulphur-
    crested
Cacatua leadbeateri — Cockatoo, Lead-
    beater's or Pink
Cacatua roseicapilla — Cockatoo, Roseate
Cacatua sanguinea — Cockatoo, Little
Cacatua tenuirostris — Cockatoo, Slender-
    billed
Cacicus cela — Cacique
Cacomantis castaneiventris — Cuckoo,
    Chestnut-breasted
Cacomantis merulinus — Cuckoo, Plaintive
Cacomantis pyrrhophanus — Cuckoo,
    Fan-tailed
Cacomantis sonneratii — Cuckoo, Banded Bay
Cacomantis variolosus — Cuckoo, Brush
Cactospiza pallida — Finch, Woodpecker
Caecobarbus geertsi — Barb, Blind Cave
Caesalpinia gilliesii — Poinciana
Caesalpinia japonica — Poinciana
Caesalpinia pulcherrima — Pride of Barbados
Caesalpinia sepiaria — Mysore-thorn
Caiman crocodylus — Caiman, Spectacled
Caiman sclerops — Caiman, Spectacled
Cairina moschata — Duck, Muscovy
Cakile maritima — Sea Rocket
Caladenia sp. — Orchid
Caladium sp. — Caladium
Calamagrostis sp. — Small-reed

Calamanthus campestris — Wren, Rufous
    Field
Calamanthus fuliginosus — Wren, Rufous
    Field
Calamia tridens — Moth, Burren Green
Calamintha ascendens — Calamint, Common
Calamintha nepeta or C. nepetoides —
    Calamint, Lesser
Calamintha sylvatica — Calamint, Wood
Calamoichthys calabaricus — Reedfish
Calandra granaria — Weevil, Granary
Calandrella brachydactyla — Lark, Short-
    toed
Calandrella cinerea — Lark, Short-toed
Calandrella rufescens — Lark, Lesser
    Short-toed
Calandrinia sp. — Purslane
Calanthe vestita — Orchid, Calanthe
Calathea makoyana — Peacock Plant
Calathea sp. — Calathea
Calathea zebrina — Zebra Plant
Calcarius lapponicus — Bunting, Lapland
Calceolaria sp. — Pocketbrook Flower
Caldesia parnassifolia — Water-plantain,
    Parnassus-leaved
Calendula arvensis — Marigold, Field
Calendula officinalis — Marigold, Pot
Calicotome sp. — Broom, Spiny or Thorny
Calidris acuminata — Sandpiper, Sharp-
    tailed
Calidris alba — Sanderling
Calidris alpina — Dunlin
Calidris canutus — Knot, European or
    Gray-crowned
Calidris ferruginea — Sandpiper, Curlew
Calidris fuscicollis — Sandpiper, White-
    rumped
Calidris maritima — Sandpiper, Purple
Calidris minuta — Stint, Little
Calidris ruficollis — Stint, Rufous-necked
Calidris subminuta — Stint, Long-toed
Calidris temminckii — Stint, Temminck's
Calidris tenuirostris — Knot, Great or
    Stripe-crowned
Caligo sp. — Butterfly, Owl
Calla palustris — Arum, Bog or Water
Callaeas cinerea — Kokako
Calliactis parasitica — Sea Anemone, Hermit
    Crab

Calliandra haematocephala — Powder Puff,
Red
Calliandra inaequilatera — Powder Puff,
Pink
Calliandra surinamensis — Powder Puff, Pink
Callicarpa sp. — Beauty Berry
Callicebus cupreus — Titi, Red
Callicebus moloch — Titi, Dusky
Callichthys callichthys — Catfish, Armored
Callimorpha sp. — Moth, Tiger
Callionymus lyra — Dragonet, Common
Callionymus maculatus — Dragonet, Spotted
Calliphora sp. — Fly, Blow or Bluebottle
Callirhoe involucrata — Mallow, Poppy
Callisia elegans — Inch Plant
Callistemon citrinus — Bottlebrush, Lemon
Callistemon coccineus — Bottlebrush
Callistemon sp. — Bottlebrush
Callistemon specious — Bottlebrush, Showy
Callistephus chinensis — Aster, China
Callistoma sp. — Top Shell, Painted
Callithrix argentata — Marmoset, Silvery
Callithrix aurita — Marmoset, White-plumed
Callithrix penicillata — Marmoset, Black-
plumed
Callithrix pygmaea — Marmoset, Pygmy
Callitriche sp. — Starwort, Water
Callitriche stagnalis — Starwort, Common
Water
Callocephalon fimbriatus — Cockatoo,
Gang-gang
Callophrys rubi — Butterfly, Green
Hairstreak
Callosciurus swinhoei — Squirrel, Striped
Callotermes flavicollis — Termite, Yellow-
necked
Calluna vulgaris — Heather
Calocera viscosa — Fungus, Sticky Coral
Calochortus sp. — Globe Tulip
Calodema sp. — Beetle, Jewel
Calodendron capense — Chestnut, Cape
Caloenas nicobarica — Pigeon, Nicobar
Calonetta leucophrys — Teal, Ringed
Caloplaca sp. — Lichen, Orange
Calopteryx sp. — Dragonfly
Calosoma sp. — Beetle, Fiery Searcher or
Hunter
Calotes sp. — Lizard, Changeable
Calothamnus sp. — Bottlebrush, One-sided

Calotropis procera — Milkweed, Giant
Calpurnia sp. — Laburnum, East Africa
Caltha palustris — Marsh-marigold, Common
Caluromys sp. — Opossum, Woolly
Calvatia uteriformis — Puffball, Mosaic
Calycanthus floridus — Carolina Allspice
Calycanthus occidentalis — Carolina Allspice
Calycella citrina — Calycella
Calyptocephalella gayi — Frog, Chilean
Water
Calyptomena viridis — Broadbill, Green
Calyptorhynchus banksii — Cockatoo,
Red-tailed Black
Calyptorhynchus baudinii — Cockatoo,
White-tailed Black
Calyptorhynchus funereus — Cockatoo,
Yellow-tailed
Calyptorhynchus lathami — Cockatoo,
Glossy Black
Calyptorhynchus magnificus — Cockatoo,
Red-tailed Black
Calystegia sepium — Bindweed, Great
Calystegia silvatica — Bindweed, Great
Calystegia soldanella — Bindweed, Sea
Calyx nicaeensis — Sponge
Camarhynchus pallida — Finch, Woodpecker
Camarhynchus parvulus — Finch, Small
Insectivorus Tree
Camaroptera fasciolata — Warbler, Barred
Camelina microcarpa — Falseflax, Smallseed
Camelina sativa — Gold of Pleasure
Camellia hybrids — Camellia
Camellia japonica — Camellia
Camellia reticulata — Camellia, Netvein
Camellia sinensis — Tea Tree
Camellia williamsii — Camellia
Camelus bactrianus ferus — Camel, Bactrian
Camelus dromedarius — Camel, Arabian
Campaea margaritata — Moth, Light
Emerald
Campanula alpina — Bellflower, Alpine
Campanula barbata — Bellflower, Bearded
Campanula bononiensis — Bellflower, Downy
Campanula carpatica — Bellflower, Tussock
Campanula cochlearifolia — Fairy's Thimble
Campanula glomerata — Bellflower,
Clustered
Campanula isophylla — Bellflower, Italian
Campanula lactiflora — Bellflower, Milky

327

Campanula latifolia — Bellflower, Large or Giant
Campanula medium — Canterbury Bells
Campanula patula — Bellflower, Spreading
Campanula persicifolia — Bellflower, Narrow-leaved or Peach-leaved
Campanula portenschlagiana — Bellflower
Campanula poscharskyana — Bellflower, Serbian
Campanula rapunculoides — Bellflower, Creeping
Campanula rapunculus — Rampion
Campanula rhomboidalis — Bellflower, Rhomboidal
Campanula rotundifolia — Harebell or Bluebell, Scottish
Campanula scheuchzeri — Bellflower, Scheuchzer's
Campanula thyrsoides — Bellflower, Yellow
Campanula trachelium — Bats-in-the-belfry
Campanula zoysii — Bellflower, Dwarf
Campephaga phoenicia — Cuckoo-shrike, Black
Campephaga sulphurata — Cuckoo-shrike, Black
Campodea sp. — Bristletail
Camponotus sp. — Ant, Carpenter
Campsis grandiflora — Trumpet Creeper, Chinese
Campsis radicans — Trumpet Creeper or Vine
Campsis tagliabuana — Trumpet Creeper
Campylopterus sp. — Sabre-wing
Campylorhamphus falcularius — Scythebill
Campylorhamphus trochilirostris — Scythebill
Campylorhynchus sp. — Wren
Canarina sp. — Canarina
Cancer pagurus — Crab, Edible or Rock
Canis adustus — Jackal, Side-striped
Canis aureus — Jackal, Common or Golden
Canis dingo — Dingo
Canis lupus — Wolf, Gray or Timber
Canis mesomelas — Jackal, Black-backed
Canis simensis — Fox, Semien
Canna sp. — Canna
Cannabis indica — Hemp, Indian
Cannabis sativa — Hemp or Marijuana
Cantharellus cibarius — Chanterelle, Yellow
Cantharellus cinereus — Chanterelle

Cantharellus floccosus — Chanterelle, Scaly
Cantharellus infundibuliformis — Chanterelle
Cantharidus sp. — Top Shell, Grooved
Cantharis sp. — Beetle, Soldier
Canthigaster sp. — Pufferfish
Capito sp. — Barbet
Capparis spinosa — Caper Bush
Capra caucasica — Tur, Caucasian
Capra falconeri — Markhor
Capra hircus — Goat
Capra ibex — Ibex, Alpine
Capra nubiana — Ibex, Nubian
Capra sibirica — Ibex, Siberian
Capra walia — Ibex, Walia
Caprella sp. — Shrimp, Ghost or Skeleton
Capreolus capreolus — Deer, Roe
Capreolus duvauceli — Deer, Swamp
Capreolus pygargus — Deer, Roe
Capricornis crispus — Serow
Capricornis sumatraensis — Serow
Caprimulgus aegyptius — Nightjar, Egyptian
Caprimulgus affinis — Nightjar, Savanna
Caprimulgus europaeus — Nightjar, European
Caprimulgus indicus — Nightjar, Gray or Jungle
Caprimulgus macrurus — Nightjar, Large-tailed
Caprimulgus pectoralis — Nightjar, Dusky or South African
Caprimulgus ruficollis — Nightjar, Red-necked
Capromys prehensilis — Hutia
Capros aper — Boar-fish
Capsella bursa-pastoris — Shepherd's Purse
Capsicum annum — Pepper, Red
Carabus sp. — Beetle, Ground
Caracal caracal — Caracal
Caragana aborescens — Pea-tree
Caralluma sp. — Caralluma
Caranx sp. — Trevally
Carapus acus — Pearlfish
Carassius auratus — Goldfish
Carassius carassius — Carp, Crucian
Carcharinus sp. — Shark
Carcharodon carcharias — Shark, White
Carcinus maenas — Crab, Common Shore
Cardamine amara — Bitter-cress, Large

Cardamine bulbifera — Coral-wort, Common
Cardamine enneaphyllos — Coral-wort, Pale
Cardamine flexuosa — Bitter-cress, Wavy
Cardamine hirsuta — Bitter-cress, Hairy
Cardamine impatiens — Bitter-cress, Narrow-leaved
Cardamine pentaphyllos — Coral-wort, Five-leaved
Cardamine pratensis — Cuckoo-flower
Cardaminopsis arenosa — Rock-cress, Tall or Sand
Cardaminopsis petraea — Rock-cress, Mountain or Northern
Cardaria draba — Cress, Hoary
Cardium edule — Cockle, Edible
Cardium exiguum — Cockle, Little
Cardium sp. — Cockle
Carduelis ambigua — Greenfinch, Black-headed
Carduelis cannabina — Linnet
Carduelis carduelis — Goldfinch
Carduelis chloris — Greenfinch, European
Carduelis flammea — Redpoll, Common or Lesser
Carduelis flavirostris — Twite
Carduelis hornemanni — Redpoll, Arctic
Carduelis monguilloti — Greenfinch, Black-headed
Carduclis spinus — Siskin
Carduus acanthoides — Thistle, Welted
Carduus crispus — Thistle, Welted
Carduus nutans — Thistle, Musk or Nodding
Carduus personata — Thistle, Great Marsh or Mountain
Carduus tenuiflorus — Thistle, Slender or Seaside
Caretta caretta — Turtle, Loggerhead
Carex acutiformis — Sedge, Lesser Pond
Carex arenaria — Sedge, Sand
Carex atrata — Sedge, Dark or Black
Carex flava — Sedge, Yellow
Carex riparia — Sedge, Great Pond
Carex rostrata — Sedge, Bottle or Beaked
Carex rupestris — Sedge, Rock
Cariama cristata — Seriema, Crested or Red-legged
Carica papaya — Papaya
Carissa grandiflora or C. macrocarpa — Palm, Natal

Carlina acanthifolia — Weather-thistle, Mediterranean
Carlina acaulis — Thistle, Stemless Carline
Carlina vulgaris — Thistle, Carline
Carludovica sp. — Carludovica
Carnegiella marthae — Hatchetfish, Black-winged or Silver
Carnegiella strigata — Hatchetfish, Marbled
Carophyllia sp. — Coral, Cup
Carpinus betulus — Hornbeam, European
Carpobrotus acinaciformis — Fig, Red Hottentot
Carpobrotus edulis — Fig, Yellow Hottentot
Carpococcyx radiceus — Cuckoo, Ground
Carpococcys renauldi — Cuckoo, Ground
Carpodacus erythrinus — Grosbeak, Scarlet
Carpodectes hopkei — Cotinga
Carterocephalus palaemon — Butterfly, Checkered Skipper
Carthamus arborescens — Safflower
Carthamus lanatus — Safflower
Carthamus tinctoris — Safflower or False Saffron
Cartioderma cor — Bat, False Vampire
Carum carvi — Caraway, Common
Carum verticillatum — Caraway, Whorled
Carya cordiformis — Hickory, Bitternut
Carya illinoiensis — Pecan
Carya ovata — Hickory, Shagbark or Shellbark
Carya tomentosa — Hickory, Mockernut
Caryopteris clandonensis — Spirea, Blue or Bluebeard
Caryopteris incana — Spirea, Blue or Bluebeard
Caryota sp. — Palm, Fishtail
Casarca ferruginea — Shelduck, Ruddy
Casarca tadornoides — Shelduck, Australian
Casmaria vibex — Bonnet Shell
Casmerodius albus — Egret, Common or Great White
Cassia artemisioides — Cassia, Feathery or Silvery
Cassia corymbosa — Buttercup Tree
Cassia javanica — Shower, Pink-and-white
Cassia marilandica — Senna, Wild
Cassida sp. — Beetle, Tortoise
Cassidaria echinophora — Helmet Shell
Cassiope sp. — Heather

Cassis sp. — Helmet Shell
Castanea mollissima — Chestnut, Chinese
Castanea sativa — Chestnut, Spanish or
　Sweet
Castanospermum australe — Chestnut,
　Moreton Bay
Castor fiber — Beaver
Casuarina sp. — Beefwood
Casuarius bennetti — Cassowary, Bennett's
Casuarius casuarius — Cassowary,
　Double-wattled
Casuarius unappendiculatus — Cassowary,
　One-wattled
Catabrosa aquatica — Whorl-grass, Water
Catajapyx confusus — Double-tail
Catamblyrhynchus diadema — Finch,
　Plush-capped
Catamenia analis — Seedeater, Band-tailed
Catananche caerulea — Succory, Blue
Catapodium marinum — Fescue, Darael
Catapodium rigidum — Grass, Fern
Catathelasma imperialis — Catathelasma
Catharacta skua — Skua, Great
Catharus aurantiirostris — Thrush,
　Nightingale
Catharus dryas — Thrush, Nightingale
Catharus fuscater — Thrush, Nightingale
Catocala fraxini — Moth, Clifden Nonpareil
Catocala ilia — Moth, Ilia Underwing
Catocala nupta — Moth, Red Underwing
Catostomus sp. — Sucker
Cattleya hybrids — Orchid, Cattleya
Cattleya sp. — Orchid, Cattleya
Catreus wallichi — Pheasant, Cheer
Caucalis sp. — Bur-parsley
Caulophyllum thalictroides — Cohosh, Blue
Causus rhombeatus — Adder, Night
Cavia aperea tschudii — Guinea-pig
Cavia porcellus — Guinea-pig
Cebuella pygmaea — Marmoset, Pygmy
Cebus albifrons — Capuchin, White-fronted
Cebus apella — Capuchin, Black-capped
Cebus capuchinus — Capuchin,
　White-throated
Cebus nigrivittatus — Capuchin,
　Black-capped or Weeper
Cecropis daurica — Swallow, Red-rumped
Cedrus atlantica — Cedar, Atlantic or Atlas
Cedrus deodara — Cedar, Deodar

Cedrus libani — Cedar of Lebanon
Cedrus libanotica — Cedar of Lebanon
Ceiba pentandra — Kapok Tree
Celastrina argiolus — Butterfly, Holly Blue
Celastrus sp. — Bittersweet
Celerio euphorbiae — Hawkmoth, Spurge
Celerio galii — Hawkmoth, Bedstraw
Celerio livornica — Hawkmoth, Striped
Celeus sp. — Woodpecker
Celmisia sp. — Daisy, New Zealand Mountain
Celosia sp. — Cockscomb
Celtis sp. — Hackberry
Centaurea aspera — Star-thistle, Rough
Centaurea calcitrapa — Star-thistle, Red
Centaurea cyanus — Cornflower
Centaurea diffusa — Knapweed, Diffuse
Centaurea jacea — Knapweed, Brown
Centaurea maculosa — Knapweed, Spotted
Centaurea montana — Bluet, Mountain
Centaurea moschata — Sweet Sultan
Centaurea nigra — Knapweed, Black or
　Lesser
Centaurea salonitana — Knapweed, Yellow
Centaurea scabiosa — Knapweed, Greater
Centaurea solstitialis — Thistle, St.
　Barnaby's
Centaurium erythraea — Centaury, Common
Centaurium littorale — Centaury, Seaside
Centaurium maritimum — Centaury, Seaside
Centaurium minus — Centaury, Common
Centaurium pulchellum — Centaury, Lesser
Centaurium vulgare — Centaury, Seaside
Centetes ecaudatus — Tenrec
Centradenia sp. — Centradenia
Centranthus ruber — Valerian, Red
Centrolabrus exoletus — Rock Cook
Centronotus gunnellus — Butterfish
Centropus bengalensis — Coucal, Lesser
Centropus grilli — Coucal, Black
Centropus melanops — Coucal, Black
Centropus phasianinus — Coucal, Pheasant
Centropus senegalensis — Coucal, Senegal
Centropus sinensis — Coucal, Greater
Centropus superciliosus — Coucal,
　White-browed or Burchell's
Centrotus cornutus — Treehopper, Horned
Centunculus minimus — Chaffweed
Cepaea nemoralis or C. hortensis — Snail,
　Grove

Cephalanthera damasonium — Helleborine, White

Cephalanthera longifolia — Helleborine, Long-leaved

Cephalanthera rubra — Helleborine, Red

Cephalocereus sp. — Cactus, Old Man

Cephalopholis argus — Argus-fish

Cephalophus dorsalis — Duiker, Bay

Cephalophus jentincki — Duiker, Jentink's

Cephalophus monticola — Duiker, Blue

Cephalophus natalensis — Duiker, Red

Cephalophus niger — Duiker, Black

Cephalophus nigrifrons — Duiker, Black-fronted

Cephalophus rufilatus — Duiker, Red-flanked

Cephalophus sylvicultor — Duiker, Yellow-backed or Giant

Cephalophus zebra — Duiker, Banded

Cephalopterus ornatus — Umbrella-bird, Amazonian

Cephalopterus penduliger — Umbrella-bird, Amazonian

Cephalorhynchus commersonii — Dolphin

Cephalotaxus sp. — Pine, Cow's-tail

Cephalothrix sp. — Worm

Cephalotus follicularis — Fly-catcher Plant

Cephus pygmaeus — Sawfly, Wheat-stem

Cepphus grylle — Guillemot, Black

Ceramaster placenta — Starfish

Cerambyx cerdo — Beetle, Great Oak Longhorn

Cerastes cerastes — Viper, Horned

Cerastes cornutus — Viper, Horned

Cerastium alpinum — Chickweed, Alpine Mouse-ear

Cerastium arcticum — Chickweed, Arctic Mouse-ear

Cerastium arvense — Chickweed, Field Mouse-ear

Cerastium fontanum — Chickweed, Common Mouse-ear

Cerastium glomeratum — Chickweed, Sticky

Cerastium holosteoides — Mouse-ear, Common

Cerastium tomentosum — Snow-in-Summer

Cerastium vulgatum —Chickweed, Mouse-ear

Cerastoderma edule — Cockle, Edible

Ceratias holboelli — Angler, Ceratiid or Holboell's

Ceratium sp. — Ceratium

Ceratodon purpureus — Moss, Purple Horn-tooth

Ceratogymna atrata — Hornbill

Ceratonia siliqua — Carob

Ceratopetalum gummiferum — Christmas Bush or Red Bush

Ceratophora sp. — Lizard, Horned Agama

Ceratophrys ornata — Escuerzo, Painted

Ceratophrys sp. — Frog, Horned

Ceratophyllum demersum — Hornwort, Common

Ceratostigma plumbaginoides — Leadwort, Blue

Ceratostigma willmottianum — Leadwort, Willmott Blue

Ceratotherium simus — Rhinoceros, Square-lipped or White

Ceratozamia mexicana — Ceratozamia

Cercartetus concinnus — Possum, Pygmy

Cerceris arenaria — Wasp, Sand-tailed Digger

Cercidiphyllum japonicum — Katsura Tree

Cercis chinenis — Judas Tree, Chinese

Cercis siliquastrum — Judas Tree

Cercocebus albigena — Mangabey, Gray-cheeked

Cercocebus aterrimus — Mangabey, Black

Cercocebus galeritus — Mangabey, Crested or Agile

Cercocebus torquatus — Mangabey, Sooty or White-collared

Cercomela familiaris — Chat, Familiar

Cercopis sp. — Froghopper

Cercopithecus aethiops — Monkey, Vervet

Cercopithecus ascanius — Monkey, Black-cheeked White-nosed

Cercopithecus brazzae — Monkey, Brazza's

Cercopithecus cephus —Guenon, Moustached

Cercopithecus diana — Monkey, Diana

Cercopithecus erythrogaster — Guenon, Red-bellied

Cercopithecus erythrotis — Monkey, Red-eared Nose-spotted

Cercopithecus hamlyni — Guenon, Owl-faced

Cercopithecus l'hoesti — Monkey, L'Hoest's

Cercopithecus mitis — Guenon, Diademed

Cercopithecus mona — Monkey, Mona

Cercopithecus neglectus — Monkey, Brazza's

Cercopithecus nicticans — Monkey, Greater White-nosed

Cercopithecus petaurista — Monkey, Lesser White-nosed

Cercopithecus pygerythrus — Monkey, Vervet

Cercopithecus sabaeus — Monkey, Green

Cercopithecus talapoin — Guenon, Dwarf

Cercotrichas galactotes — Warbler, Rufous

Cerebratulus sp. — Worm, Ribbon

Cereopsis novaehollandiae — Goose, Cape Barren

Cerianthus sp. — Sea Anemone

Cerinthe sp. — Honeywort

Ceropales maculatus — Wasp

Ceropegia sp. — Ceropegia

Ceropegia stapeliiformis — Candelabra Flower

Ceropegia woodii — Rosary Vine

Certhia brachydactyla — Tree-creeper, Short-toed

Certhia discolor — Tree-creeper, Brown-throated

Certhia familiaris — Tree-creeper, Brown or Common

Certhidea olivacea — Finch, Warbler

Certhionyx variegatus — Honey-eater, Pied

Cerura sp. — Moth, Puss

Cervicapra cervicapra — Blackbuck

Cervus axis — Deer, Axis

Cervus eldi — Deer, Brow-antlered or Eld's

Cervus duvauceli — Deer, Swamp

Cervus elaphus maral — Maral

Cervus elaphus var. — Deer, Red

Cervus mariannus — Sambar

Cervus nippon — Sika

Cervus nippon taiouanus — Sika

Cervus unicolor — Sambar

Ceryle rudis — Kingfisher, Pied

Cestrum aurantiacum — Cestrum, Orange

Cestrum nocturnum — Jessamine, Night-flowering

Cestrum purpureum — Jessamine, Purple

Cestrum sp. — Cestrum

Cestus sp. — Venus' Girdle

Ceterach officinarum — Fern, Rusty-back

Cetomimus sp. — Whalefish

Cetonia aurata — Beetle, Rose Chafer

Cetonia cupraea — Beetle, Rose Chafer

Cetorhinus maximus — Shark, Basking

Cetraria sp. — Moss, Iceland

Cettia cetti — Warbler, Cetti's

Cettia diphone — Warbler, Bush

Cettia squameiceps — Warbler, Bush

Ceyx azureus — Kingfisher, Azure

Ceyx erithacus — Kingfisher, Indian Forest or Three-toed

Chaca chaca — Chaca

Chaenomeles japonica — Quince, Japanese Flowering

Chaenomeles lagenaria — Quince, Japanese Flowering

Chaenomeles speciosa — Quince, Flowering

Chaenomeles superba — Quince, Flowering

Chaenorhinum minus — Toadflax, Small

Chaerophyllum temulentum — Chervil

Chaetodon sp. — Butterflyfish

Chaetodontoplus mesoleucus — Butterflyfish, Emperor

Chaetophractus villosus — Armadillo, Hairy

Chaetura sp. — Swift, Spine-tailed

Chalamydoselachus anguineus — Shark, Frilled

Chalcides sp. — Skink, Cylindrical or Eyed

Chalcites basalis — Cuckoo, Bronze

Chalcites lucidus — Cuckoo, Shining Bronze

Chalcites minutillus — Cuckoo, Little Bronze

Chalcites plagosus — Cuckoo, Golden Bronze

Chalcomitra senegalensis — Sunbird, Scarlet-chested

Chalcophaps chrysochlora — Dove, Emerald or Green-winged

Chalcophaps indica — Dove, Emerald or Green-winged

Chalcopsitta sp. — Cockatoo-parrot

Chalcosoma atlas — Beetle, Atlas

Chalcostigma herrani — Hummingbird, Herran's or Rainbow-bearded Thornbill

Chalicodoma muraria — Bee, Mason

Chamaecereus silvestrii — Cactus, Peanut

Chamaecyparis lawsoniana — Cypress, Lawson

Chamaecyparis nootkatensis — Cypress, Nootka

Chamaecyparis obtusa — Cypress, Hinoki

Chamaecyparis pisifera — Cypress, Sawara

Chamaecyparis thyoides — Cedar, White

Chamaedaphne calyculata — Leatherleaf

Chamaedorea elegans — Palm, Parlor

Chamaedorea sp. — Palm, Feather

Chamaeleo bitaeniatus ellioti—Chameleon, Elliot's Dwarf

Chamaeleo chamaeleo — Chameleon, Common

Chamaeleo dilepis — Chameleon, Flap-necked

Chamaeleo fischeri — Chameleon, Fischer's

Chamaeleo jacksoni — Chameleon, Jackson's or Three-horned

Chamaeleo melleri — Chameleon, Giant One-horned or Meller's

Chamaeleo pumila — Chameleon, Dwarf

Chamaeleo sp. — Chameleon

Chamaemelum nobile — Chamomile, Sweet

Chamaenerion angustifolium — Willow-herb, Rosebay

Chamaepericlymenum suecicim — Cornel, Dwarf

Chamaepetes goudotii — Guan

Chamaerops humilis — Palm, Dwarf Fan

Chamaesaura anguina — Lizard, Cape Snake

Chamaespartium sagittale — Broom, Winged

Chanda ranga — Glassfish, Indian

Chanda wolfii — Glassfish, Wolff's

Channa asiatica — Snakehead

Channallabes apus — Catfish, Eel

Chanos chanos — Milkfish

Chaparrudo flavescens — Goby, Spotted

Chara sp. — Stonewort

Characidium fasciatum — Characin, Darter

Charadrius alexandrinus — Plover, Snowy

Charadrius asiaticus — Plover, Oriental

Charadrius bicinctus — Dotterel, Double-banded

Charadrius cinctus — Dotterel, Red-kneed

Charadrius cucullatus — Dotterel, Hooded

Charadrius dominica — Plover, Golden

Charadrius dubius — Plover, Little Ringed

Charadrius hiaticula — Plover, Ringed

Charadrius leschenaultii — Sandplover

Charadrius melanops — Dotterel, Black-fronted

Charadrius mongolus — Plover, Mongolian

Charadrius morinellus — Dotterel

Charadrius obscurus — Dotterel

Charadrius pecuarius — Plover, Kittlitz's

Charadrius peronii — Plover, Malaysian

Charadrius ruficapillus — Plover, Snowy

Charadrius squatarola — Plover, Gray

Charadrius tricollaris — Plover, Three-banded

Charadrius veredus — Plover, Oriental

Charagia sp. — Moth

Charaxes sp. — Butterfly

Charonia sp. — Triton's-trumpet

Chauna chavaria — Screamer, Black-necked

Chauna torquata — Screamer, Crested or Southern

Cheilanthes sp. — Lipfern

Cheiranthus allonii — Wallflower, Siberian

Cheiranthus cheiri — Wallflower

Cheirodon axelrodi — Tetra, Cardinal

Cheirogaleus major — Lemur, Greater Dwarf

Cheirogaleus medius — Lemur, Fat-tailed Dwarf

Cheiromeles torquatus — Bat, Naked

Chelictinia riocourii — Kite, African Swallow-tailed

Chelidonium majus — Celandine

Chelifer cancroides — Scorpion, Book or False

Chelmon rostratus — Butterflyfish, Long-nosed

Chelodina longicollis — Tortoise, Long-necked

Chelonia mydas — Turtle, Green

Chelorrhina polyphemus — Beetle, Flower

Chelorrhina savagei — Beetle, Flower

Chelys fimbriatus — Terrapin, Matamata

Chenonetta jubata — Goose, Maned

Chenopodium album — Fat Hen

Chenopodium bonus-henricus — Good King Henry

Chenopodium botrys — Goosefoot, Sticky

Chenopodium ficifolium — Goosefoot, Fig-leaved

Chenopodium foliosum — Goosefoot, Strawberry

Chenopodium glaucum — Goosefoot, Oak-leaved

Chenopodium hybridum — Goosefoot, Maple-leaved

Chenopodium murale — Goosefoot, Nettle-leaved

Chenopodium polyspermum — Goosefoot, Many-seeded

Chenopodium rubrum — Goosefoot, Red

Chenopodium sp. — Chenopodium

Chenopodium vulvaria — Goosefoot, Stinking

Cheramoeca leucosterna — Swallow, White-backed

Cherleria sedoides — Cyphal

Chermes mali — Sucker, Apple

Chersophilus duponti — Lark, Du Pont's

Chettusia gregaria — Plover, Sociable

Chiasognathus granti — Beetle, Stag

Chibea bracteata — Drongo, Spangled

Chilodus punctatus — Headstander, Spotted

Chilomenes lunata — Beetle, Ladybird

Chimaera monstrosa — Rabbitfish

Chimaphila sp. — Wintergreen

Chimonanthus praecox — Wintersweet

Chinchilla brevicaudata — Chinchilla

Chinchilla chinchilla — Chinchilla

Chinchilla laniger — Chinchilla

Chinemys reevesii — Terrapin, Chinese

Chioglossa lusitanica — Salamander, Spanish

Chionanthus sp. — Fringe Tree

Chionis alba — Sheathbill, Snowy or Yellow-billed

Chionodoxa gigantea rosea — Glory of the Snow

Chionodoxa luciliae — Glory of the Snow

Chionodoxa sardensis — Glory of the Snow

Chirita sp. — Chirita

Chirolophis ascanii — Blenny, Yarrell's

Chirolophis galerita — Blenny, Yarrell's

Chiromantis sp. — Frog

Chironectes minimus — Opossum, Water

Chironia sp. — Chironia

Chironius sp. — Snake, Harlequin

Chironomus annularis — Midge, Feather

Chironomus plumosus — Midge, Feather

Chiropotes sp. — Saki

Chiroxiphia sp. — Manakin

Chlamydera cerviniventris — Bowerbird, Fawn-breasted

Chlamydera guttata — Bowerbird, Western

Chlamydera maculata — Bowerbird, Spotted

Chlamydera nuchalis — Bowerbird, Great

Chlamydosaurus kingi — Lizard, Frilled

Chlamydotis undulata — Bustard, Houbara

Chlamys distorta — Scallop, Hunchback

Chlamys nobilis — Scallop, Noble

Chlamys opercularis — Scallop, Queen

Chlamys swifti — Scallop, Swift's

Chlamys varia — Scallop, Variegated

Chlidonias albostriatus — Tern, Black-fronted

Chlidonias hybrida — Tern, Whiskered

Chlidonias leucoptera — Tern, White-winged Black.

Chlidonias niger — Tern, Black

Chloebia gouldiae — Finch, Gouldian

Chloephaga hybrida — Goose, Kelp

Chloephaga leucoptera — Goose, Magellan or Upland

Chloephaga melanoptera — Goose, Andean

Chloephaga picta — Goose, Magellan or Upland

Chloephaga poliocephala — Goose, Ashy-headed

Chlorippe sp. — Butterfly

Chloris chloris — Greenfinch, European

Chloroceryle amazona — Kingfisher, Amazon

Chlorochrysa nitidissima — Tanager, Multi-colored

Chlorochrysa phoenicotis — Tanager, Glistening-green

Chloroperla torrentium — Fly, Stone

Chlorophanes spiza — Honeycreeper, Green

Chlorophonia sp. — Chlorophonia

Chlorophytum comosum — Spider Plant

Chloropsis aurifrons — Leafbird, Gold-fronted

Chloropsis hardwickii — Leafbird, Hardwicke's or Orange-bellied

Chloropsis jerdonii — Bulbul

Chlorornis riefferii — Tanager, Grass-green

Chlorosplenium sp. — Chlorosplenium

Chlorostilbon sp. — Emerald

Choeropsis liberiensis — Hippopotamus, Pygmy

Choiromyces meandriformis — Truffle

Choloepus didactylus — Sloth, Two-fingered or Two-toed

Choloepus hoffmanni — Sloth, Hoffman's

Chondrilla sp. — Gum-succory

Chondrohierax uncinatus — Kite, Hook-billed

Chondropython viridis — Python, Green Tree

Chondrostomus nasus — Nase or Nasling

Chorda sp. — Sea Lace

Chorinemus sp. — Talang Talang

Choriotis australis — Bustard, Australian

Choriotis kori — Bustard, Kori

Choriotis nigriceps — Bustard, Great Indian

Chorisia speciosa — Floss-silk Tree
Chorisia ventricosa — Floss-silk Tree
Chorizema cordatum — Flame Pea
Chorizema ilicifolium — Flame Pea
Chorthippus sp. — Grasshopper, Field or
   Meadow
Chromis chromis — Damselfish, Blue-green
Chrysalidocarpus lutescens — Palm,
   Butterfly
Chrysanthemum alpinum — Chrysanthe-
   mum, Alpine
Chrysanthemum carinatum — Chrysanthe-
   mum, Annual
Chrysanthemum coronarium — Daisy, Crown
Chrysanthemum frutescens — Marguerite
Chrysanthemum leucanthemum — Marguer-
   ite or Daisy, Ox-eye
Chrysanthemum maximum — Daisy, Shasta
Chrysanthemum morifolium — Chrysanthe-
   mum, Florist's
Chrysanthemum parthenium — Feverfew
Chrysanthemum segetum — Marigold, Corn
Chrysanthemum sp. — Chrysanthemum
Chrysanthemum vulgare — Tansy
Chrysaora sp. — Jellyfish
Chrysis sp. — Wasp, Cuckoo
Chrysochloris sp. — Mole, Golden
Chrysochroa sp. — Beetle, Jewel
Chrysococcyx basalis — Cuckoo, Bronze
Chrysococcyx caprius — Cuckoo, Didric
Chrysococcyx cupreus — Cuckoo, Emerald
Chrysococcyx lucidus —Cuckoo, Shining Bronze
Chrysococcyx maculatus — Cuckoo, Emerald
Chrysococcyx malayanus — Cuckoo, Malayan
   Bronze
Chrysococcyx minutillus — Cuckoo, Little
   Bronze
Chrysococcyx osculans — Cuckoo,
   Black-eared
Chrysococcyx plagosus — Cuckoo, Golden
   Bronze
Chrysococcyx xanthorhynchus — Cuckoo,
   Violet
Chrysocolaptes lucidus — Woodpecker,
   Crimson-backed
Chrysocyon brachyurus — Wolf, Maned
Chrysocyon jubatus — Wolf, Maned
Chrysolampis mosquitus — Hummingbird,
   Ruby-topaz

Chrysolophus amherstiae — Pheasant, Lady
   Amherst's
Chrysolophus pictus — Pheasant, Golden
Chrysomela sp. — Beetle, Leaf
Chrysopa sp. — Lacewing, Green
Chrysopelea ornata — Snake, Golden Tree
Chrysopelea paradisi — Snake, Paradise Tree
Chrysophora chrysochloa — Beetle, Garden
   Chafer
Chrysops sp. — Fly, Horse
Chrysosplenium alternifolium — Saxifrage,
   Alternate-leaved Golden
Chrysosplenium oppositifolium — Saxifrage,
   Opposite-leaved Golden
Chthonicola sagittata — Warbler, Speckled
Chthonius sp. — Scorpion, Book or False
Cibotium schiedei — Fern, Mexican Tree
Cicada sp. — Cicada
Cicadatra atra — Cicada
Cicadella viridus — Leafhopper
Cicadetta montana — Cicada, Forest or
   Mountain
Cicadula sexnotata — Leafhopper
Ciccaba huhula — Owl, Black-banded
Ciccaba virgata — Owl, Mottled
Ciccaba woodfordi — Owl, Wood
Cicendia filiformis — Gentianella, Yellow
Cicerbita alpina or C. macrophylla—
   Sow-thistle, Alpine or Blue
Cichla ocellaris — Cichlid, Perch
Cichlasoma biocellatus — Jack Dempsey
Cichlasoma facetum — Cichlid, Chameleon
Cichlasoma festivum — Cichlid, Barred or Flag
Cichlasoma hellabrunni — Cichlid,
   Hellabrium
Cichlasoma meeki — Cichlid, Firemouth
Cichlasoma severum — Cichlid, Banded
Cichorium intybus — Chicory
Cicindela sp. — Beetle, Tiger
Cicinnurus regius — Bird of Paradise, King
Ciconia ciconia — Stork, Common or White
Ciconia episcopus — Stork, Woolly-necked
Ciconia nigra — Stork, Black
Cicuta bolanderi — Water-hemlock
Cicuta maculata — Water-hemlock
Cicuta virosa — Cowbane
Cidaris sp. — Sea Urchin, Pencil
Ciliata mustela — Rockling, Five-bearded
Cilix glaucata — Moth, Chinese Character

Cimbex femorata — Sawfly, Birch-leaf
Cimex lectularius — Bedbug
Cinchona calisaya — Quinine Tree
Cinclodes sp. — Shaketail
Cinclorhamphus cruralis — Songlark, Brown
Cinclorhamphus mathewsi — Songlark Rufous
Cinclosoma castaneothorax — Quail-thrush, Chestnut-breasted
Cinclosoma castanotum — Quail-thrush, Chestnut
Cinclosoma cinnamomeum — Quail-thrush, Cinnamon
Cinclosoma punctatum — Quail-thrush, Spotted
Cinclus cinclus — Dipper, Eurasian
Cinclus leucocephalus — Dipper
Cinclus pallasii — Dipper
Cinnamomum burmani — Cinnamon
Cinnamomum camphora — Camphor
Cinnamomum litseifolium — Cinnamon
Cinnamomum zeylanicum — Cinnamon
Cinnyricinclus leucogaster — Starling, Amethyst
Cinnyris chalybeus — Sunbird, Double-collared
Cinnyris regius — Sunbird, Royal
Cinnyris talatala — Sunbird, White-bellied
Cinnyris venustus — Sunbird, Variable
Ciona intestinalis — Sea Squirt
Cionus hortulanus — Weevil, Figwort
Circaea alpina — Nightshade, Alpine Enchanter's
Circaea intermedia — Nightshade, Upland Enchanter's
Circaea lutetiana — Nightshade, Common Enchanter's
Circaetus cinereus — Eagle, Snake
Circaetus gallicus — Eagle, Short-toed
Circus aeruginosus — Harrier, Marsh
Circus approximans — Harrier, Swamp
Circus assimilis — Harrier, Spotted
Circus cyaneus — Hen-harrier
Circus macrourus — Harrier, Pallid
Circus melanoleucus — Harrier, Pied
Circus pygargus — Harrier, Montagu's
Circus ranivorus — Harrier, African Marsh
Cirrhopetalum sp. — Orchid
Cirsium acaule or acaulon — Thistle, Stemless or Dwarf

Cirsium arvense — Thistle, Creeping or Canada
Cirsium dissectum — Thistle, Meadow
Cirsium eriophorum — Thistle, Woolly
Cirsium erisithales — Thistle, Sticky
Cirsium helenoides — Thistle, Melancholy
Cirsium heterophyllum — Thistle, Melancholy
Cirsium oleraceum — Thistle, Cabbage
Cirsium palustre — Thistle, Marsh
Cirsium rivulare atropurpureum — Cirsium
Cirsium spinossissimum — Thistle, Spiniest
Cirsium tuberosum — Thistle, Tuberous
Cirsium vulgare — Thistle, Spear or Bull
Cissa chinensis — Magpie, Green
Cissa erythrorhyncha — Magpie, Red-billed Blue
Cissa ornata — Magpie, Ceylon Blue
Cissa thalassina — Magpie, Short-tailed Green
Cissa whiteheadi — Magpie, Whitehead's Blue
Cissopis leveriana — Tanager, Magpie
Cissus antarctica — Kangeroo Ivy or Vine
Cissus discolor — Begonia Treebine
Cissus quadrangularis— Veldt Ivy
Cissus rhombifolia — Grape Ivy
Cissus sp. — Cissus
Cisticola aridula — Warbler, Fan-tailed
Cisticola exilis — Cisticola, Bright-capped or Golden-headed
Cisticola juncidis — Warbler, Fan-tailed
Cistus albidus — Cistus, Gray-leaved
Cistus incanus — Rockrose, Pink
Cistus ladaniferus — Cistus, Gum
Cistus laurifolius — Cistus, Laurel-leaved
Cistus monspeliensis — Cistus, Narrow-leaved
Cistus purpureus — Rockrose, Purple
Cistus salvifolius — Cistus, Sage-leaved
Cistus sp. — Cistus
Citellus citellus — Souslik, Common European
Citellus suslica — Souslik, Spotted
Citrullus sp. — Cucumber, Bitter
Citrullus vulgaris — Watermelon
Citrus aurantifolia — Lime
Citrus aurantium — Orange, Seville or Sour
Citrus limon — Lemon

Citrus medica — Citron
Citrus mitis — Orange, Calamondium
Citrus paradisi — Grapefruit
Citrus reticulata — Tangerine
Citrus sinensis — Orange, Sweet
Civettictis civetta — Civet, African
Cladanthus arabicus — Cladanthus
Cladium mariscus — Sedge, Fen
Cladium tetraquetrum — Sedge, Fen
Cladognathus giraffa — Beetle, Giraffe Stag
Cladonia pyxidata — Pixie Cup
Cladonia rangiforina — Moss, Reindeer
Cladonia sp. — Cladonia
Cladorhynchus leucocephalus — Stilt,
    Banded
Clamator coromandus — Cuckoo, Red-winged
    Indian or Crested
Clamator glandarius — Cuckoo, Great
    Spotted
Clamator jacobinus — Cuckoo, Jacobin or Pied
Clanculus pharaonis — Top Shell, Strawberry
Clangula hyemalis — Old-squaw
Clathrina sp. — Sponge, Calcareous
Clava sp. — Hydroid
Clavaria argillacea — Fungus, Field or Moor
    Club
Clavaria aurea — Clavaria, Golden
Clavaria formosa — Clavaria, Beautiful
Clavaria sp. — Clavaria
Clavellina lepadiformis — Sea Squirt
Clavulina sp. — Clavulina
Clavulinopsis sp. — Clavulinopsis
Claytonia alsinoides — Purslane, Pink
Cleistocactus sp. — Silver Torch
Clematis alpina — Clematis, Alpine
Clematis flammula — Clematis, Fragrant
Clematis hybrids — Clematis
Clematis montana — Clematis, Mountain
Clematis orientalis — Clematis, Orange or
    Lemon Peel
Clematis recta — Clematis, Upright
Clematis sp. — Clematis
Clematis tangutica — Clematis, Golden
Clematis texensis — Clematis, Scarlet
Clematis vitalba — Traveler's Joy or Old
    Man's Beard
Clematis viticella — Clematis, Purple
Clematopsis sp. — Clematopsis
Cleome hasslerana — Spider Flower

Cleome spinosa — Spider Flower
Clerodendrum bungei — Clerodendrum
Clerodendrum capitatum — Glory Bower
Clerodendrum speciosissimum — Glory
    Bower
Clerodendrum sp. — Clerodendrum
Clerodendrum splendens—Glory Bower
Clerodendrum thomsoniae — Glory Bower
Clerodendrum trichotomum — Glory Bower,
    Harlequin
Clethra sp. — Pepperbush
Clethrionomys glareolus — Vole, Bank
Clethrionomys rufocanus — Vole, Bank
Clianthus dampieri — Glory Pea
Clianthus formosus — Desert Pea, Sturt's
Clianthus puniceus — Parrot's Bill
Climacium dendroides — Moss, Tree
Climacteris affinis — Tree-creeper,
    White-browed
Climacteris erythrops — Tree-creeper,
    Red-browed
Climacteris picumnus — Tree-creeper
Climacteris rufa — Tree-creeper, Rufous
Clinopodium vulgare — Basil, Wild
Cliona sp. — Sponge
Clitarchus sp. — Stick-insect
Clitocybe sp. — Clitocybe
Clitopilus sp. — Mushroom, Sweetbread
Clitoria ternatea — Pea, Butterfly
Clitummus sp. — Stick-insect
Clivia caulsescens — Lily, Kaffir
Clivia miniata — Lily, Kaffir
Cloeon dipterum — Fly, May
Clonopsis sp. — Stick-insect
Clossiana sp. — Butterfly, Fritillary
Clostera sp. — Moth, Chocolate-tip
Clupea harengus — Herring, Atlantic
Clupea sprattus — Sprat
Clusia sp. — Clusia
Clytra laeviscula — Beetle, Sack
Clytra quadripunctata — Beetle, Sack
Clytus arietis — Beetle, Wasp
Cnestis sp. — Cnestis
Cnicus benedictus — Thistle, Blessed
Cnidium dubium — Cnidium
Cobaea scandens — Cup and Saucer Vine
Cobitis taenia — Loach, Spined
Coccinella sp. — Beetle, Ladybird

Coccinia sp. — Coccinia
Coccoloba sp. — Grape, Seaside
Coccothraustes coccothraustes — Hawfinch
Coccyzus sp. — Cuckoo
Cochlearia anglica — Scurvy-grass, English or Long-leaved
Cochlearia danica — Scurvy-grass, Danish or Early
Cochlearia officinalis — Scurvy-grass, Common
Cochlearius cochlearius — Heron, Boat-billed
Cochliostema jacobianum — Cochliostema
Cochlospermum sp. — Cochlospermum
Cochoa purpurea — Cochoa, Purple
Cochoa viridis — Cochoa, Green
Cocos nucifera — Palm, Coconut
Codakia sp. — Lucine
Codiaeum variegatum — Croton
Codonopsis ovata — Codonopsis
Coeligena sp. — Hummingbird, Inca
Coelogenys paca — Paca, Spotted
Coeloglossum viride — Orchid, Frog
Coelogyne cristata — Orchid, Angel
Coelogyne pandurata — Orchid, Black
Coelogyne sp. — Orchid
Coelopa frigida — Fly, Kelp or Seaweed
Coenagrion puella — Damselfly, Coenagrion
Coendou prehensilis — Porcupine, Brazilian Tree
Coendou villosus — Porcupine, Brazilian Tree
Coenocorypha sp. — Snipe, New Zeland
Coenonympha sp. — Butterfly, Heath or Pearl-grass
Coereba flaveola — Bananaquit
Coffea arabica — Coffee Tree
Cola sp. — Cola Nut or Kolanut
Colaptes campestris — Flicker, South American
Colaptes rupicola — Flicker, South American
Colchium autumnale — Saffron, Meadow
Colchium sp. — Crocus, Autumn
Coleus sp. — Coleus
Colias australis — Butterfly, Berger's Clouded Yellow
Colias croceus — Butterfly, Clouded Yellow.
Colias hyale — Butterfly, Pale Clouded Yellow.
Colibri coruscans — Violet-ear, Sparkling
Colisa fasciata — Gourami, Giant or Striped

Colisa lalia — Gourami, Dwarf
Colius colius — Mousebird, White-backed
Colius indicus — Mousebird, Red-faced
Colius striatus — Mousebird, Speckled
Coliuspasser sp. — Whydah
Colletia sp. — Colletia
Collocalia sp. — Swiftlet
Colluricincla harmonica — Shrike-thrush
Colluricincla rufiventris — Shrike-thrush
Colybia dryophila — Agaric, Wood
Collybia fusipes — Agaric, Spindle-stem
Collybia maculata — Fungus, Rust Spot
Collybia platyphylla — Agaric, Broad-gilled
Collybia sp. — Collybia
Colobus abyssinicus — Colobus, Northern Black-and-White
Colobus angolensis — Colobus, Black-and-White
Colobus badius — Colobus, Red
Colobus kirkii — Colobus, Red
Colobus pennanti — Colobus, Red
Colobus polykomos — Colobus, Southern Black-and-White
Colobus verus — Colobus, Olive
Colocasia esculenta var. antiquorum — Elephant's Ears
Cololabis saira — Saury
Colossoma nigripinnis — Pacu
Colostygia pectinataria — Moth, Green Carpet
Colpothrinax wrightii — Palm, Pot-bellied or "Barrel"
Coltrichia perennis — Coltrichia
Coluber sp. — Whipsnake
Coluber viridiflavus — Whipsnake
Columba guinea — Pigeon, Speckled or Rock
Columba livia — Pigeon, Common
Columba livia intermedia — Pigeon, Common
Columba norfolciensis — Pigeon, White-headed
Columba oenas — Dove, Stock
Columba palumbus — Pigeon, Wood
Columba punicea — Pigeon, Pale-capped or Purple Wood
Columba vitiensis — Pigeon, Wood
Columbicola columbae — Louse, Pigeon
Columnea sp. — Columnea
Colutea arborescens — Bladder-senna
Combretum sp. — Combretum

Commelina sp. — Day-flower

Comparettia sp. — Orchid

Comptonia peregrina or C. asplenifolia — Sweet-fern

Conepatus mesoleucus — Skunk, Hog-nosed

Conger conger — Eel, Conger

Conirostrum sitticolor — Conebill, Blue-backed

Conium maculatum — Hemlock (Poison)

Connochaetes gnu — Gnu, White-tailed

Connochaetes talbojubatus — Gnu, White-bearded

Connochaetes taurinus — Wildebeest

Conocephalum conicum — Liverwort, Great Scented

Conocephalus sp. — Bush-cricket, Conehead

Conolophus subcristatus — Iguana, Galapagos Land

Conopophaga sp. — Gnat-eater

Conopophila picta — Honey-eater, Painted

Conopophila rufogularis — Honey-eater, Rufous-throated

Conophytum sp. — Living Stones

Conopodium majus — Pignut

Conringia orientalis — Mustard, Hare's-ear

Conus sp. — Cone Shell

Convallaria majalis — Lily-of-the-valley

Convolvulus althaeoides — Bindweed, Mallow-leaved

Convolvulus arvensis — Bindweed, Field or Lesser

Convolvulus cantabricus — Convolvulus, Pink

Convolvulus cneorum — Convolvulus, Shrubby

Convolvulus elegantissimus — Bindweed, Elegant

Convolvulus mauritanicus — Morning Glory, Ground

Convolvulus sepium — Bindweed, Greater or Hedge

Convolvulus sp. — Convolvulus

Convolvulus tricolor — Convolvulus, Dwarf

Conyza canadensis — Fleabane, Canadian

Copeina arnoldi — Characin, Spraying

Copeina guttata — Copeina, Red-spotted

Copella arnoldi — Characin, Spraying

Copiapoa sp. — Copiapoa

Coprinus atramentarius — Ink-cap, Common

Coprinus comatus — Ink-cap, Shaggy

Coprinus disseminatus — Crumble-cups, Trooping

Coprinus lagopus — Ink-cap, Bonfire

Coprinus micaceus — Ink-cap, Glistening

Coprinus picaceus — Fungus, Magpie

Coprinus plicatilis — Ink-cap, Furrowed

Copris lunaris — Beetle, English Scarab

Coprosma sp. — Coprosma

Copsychus luzoniensis — Shama

Copsychus malabaricus — Shama

Copsychus saularis — Magpie-robin

Copsychus seychallarum — Magpie-robin, Seychelles

Coracias bengalensis — Roller, Bengal or Indian

Coracias caudatus — Roller, Lilac-breasted

Coracias garrulus — Roller, Common or European

Coracina coerulescens — Cuckoo-shrike, Black

Coracina lineata — Cuckoo-shrike, Barred

Coracina novaehollandiae — Cuckoo-shrike, Black-faced or Large

Coracina papuensis — Cuckoo-shrike, Papuan

Coracina striata — Cuckoo-shrike, Bar-bellied

Coracina tenuirostris — Cicada-bird

Coracopsis nigra — Vasa

Coracopsis vasa — Vasa

Corallium nobile — Coral, Precious

Corallium rubrum — Coral, Precious

Corallorhiza trifida — Orchid, Coralroot

Corallus caninus — Boa, Green Tree

Coranus subapterus — Bug, Heath Assassin

Corbula gibba — Basket Shell

Corcorax melanorhamphus — Chough, White-winged

Cordulegaster sp. — Dragonfly

Cordulia sp. — Dragonfly, Downy Emerald

Cordyceps sp. — Fungus, Insect

Cordyline australis — Cabbage Tree

Cordyline indivisa — Cabbage Palm

Cordyline stricta — Palm Lily

Cordyline terminalis — Ti Plant

Cordylus cataphractus — Lizard, Armadillo

Cordylus giganteus — Sungazer

Coregonus albula — Pollan

Coregonus lavaretus oxyrhynchus — Houting

Coregonus sp. — Whitefish
Coreopsis sp. — Tickseed or Tickweed
Coreus marginatus — Bug, Squash
Coriandrum sativum — Coriander
Coriaria sp. — Coriaria
Coriolus versicolor — Fairy Stool
Coris angulata — Wrasse, Twin-spot
Coris julis — Wrasse, Rainbow
Corix punctata — Backswimmer, Lesser
Cornus alba — Dogwood, Tartarian or
    Siberian
Cornus capitata — Dogwood, Evergreen
Cornus kousa — Dogwood, Japanese or
    Oriental
Cornus mas — Cornelian Cherry or Cornel
Cornus sanguinea — Dogwood, Common
Cornus suecica — Cornel, Dwarf
Corokia sp. — Corokia
Coronella austriaca — Snake, Smooth
Coronilla emerus — Scorpion Senna
Coronilla glauca — Scorpion-vetch, Shrubby
Coronilla sp. — Scorpion-vetch
Coronilla varia — Crown-vetch
Coronopus didymus — Swine-cress, Lesser
Coronopus squamatus — Swine-cress,
    Common
Correa sp. — Fuchsia, Australian
Corrigiola litoralis — Strapwort
Cortaderia sp. — Grass, Pampas
Cortinarius alboviolaceus — Cortinarius,
    Violet
Cortinarius sp. — Cortinarius
Cortinarius violaceus — Cortinarius, Violet
Cortusa matthioli — Alpine Bells
Corucia zebrata — Skink, Prehensile-tailed
    Giant
Corvina nigra — Corb
Corvinella melanoleuca — Shrike, Magpie
Corvus albicollis — Raven, Cape or
    White-necked
Corvus albus — Crow, Pied
Corvus bennetti — Crow, Little
Corvus corax — Raven
Corvus corone cornix — Crow, Hooded
Corvus corone corone — Crow, Carrion
Corvus coronoides — Raven
Corvus enca — Crow, Little
Corvus frugilegus — Rook
Corvus macrorhynchos — Crow, Large-billed

Corvus monedula — Jackdaw
Corvus splendens — Crow, Indian House
Corybas sp. — Orchid
Corydalis cava — Corydalis, Hollow
Corydalis claviculata — Corydalis, Climbing
Corydalis lutea — Corydalis, Yellow
Corydalis solida — Corydalis, Purple
Corydalis sp. — Corydalis
Corydalis sp. — Fumewort
Corydon sumatranus — Broadbill, Dusky
Corydoras arcuatus — Corydoras, Arched
Corydoras julii — Corydoras, Leopard
Corydoras melanistius — Corydoras,
    Black-spotted
Corydoras myersi — Corydoras, Myer's
Corydoras paleatus — Corydoras, Peppered
Corydoras sp. — Catfish, Armored
Corylopsis pauciflora — Winter Hazel,
    Buttercup
Corylopsis sinensis — Winter Hazel, Chinese
Corylopsis spicata — Winter Hazel, Spike
Corylopsis veitchiana — Winter Hazel
Corylus avellana — Hazel, European or
    Common
Corylus avellana contorta — Hazel,
    Corkscrew or Curly
Corylus colurna — Hazel, Turkish
Corylus cornuta — Hazel, Beaked
Corylus maxima — Filbert
Corynactis viridis — Sea Anemone, Jewel
Coryne sarcoides — Fungus, Purple Knot
Corynephorus canescens — Hair-grass, Gray
Corynopoma riisei — Characin, Sword-tail
Coryphaena hippurus — Dolphin
Coryphaena equiselis — Dolphin
Coryphaenoides rupestris — Grenadier
Coryphantha sp. — Coryphantha
Coryphella sp. — Nudibranch
Coryphoblennius galerita — Blenny,
    Montagu's
Corythaixodes sp. — Go-away Bird
Corytheola cristata — Turaco, Great Blue
Corythophanes sp. — Iguana
Corythornis cristata — Kingfisher, Malachite
Coscoroba coscoroba — Swan, Coscoroba
Coscoroba olor — Swan, Coscoroba
Cosmopsarus regius — Starling, Golden-
    breasted
Cosmos bipinnatus — Cosmos

Cosmos sulphureus — Cosmos, Yellow
Cossus cossus — Moth, Goat
Cossypha caffra — Robin, Cape
Cossypha humeralis — Robin, White-throated
Costus sanguineus — Spiral Flag
Costus sp. — Costus
Cotinga sp. — Cotinga
Cotinus coggyria — Smoke Tree
Cotinus obovatus — Smoke Tree
Cotoneaster apiculatus — Cotoneaster, Cranberry
Cotoneaster conspicuus — Cotoneaster, Wintergreen
Cotoneaster dammeri — Cotoneaster, Bearberry
Cotoneaster frigida — Clusterberry
Cotoneaster horizontalis — Cotoneaster, Fishbone
Cotoneaster integerrimus — Cotoneaster, Common or Wild
Cotoneaster microphyllus — Rockspray
Cotoneaster multiflora — Cotoneaster, Many-flowered
Cotoneaster salicifolius — Cotoneaster, Willowleaf
Cotoneaster sp. — Cotoneaster
Cottus bubalis — Sea Scorpion
Cottus gobio — Bullhead
Cottus quadricornis — Bullhead, Four-horned
Cottus scorpius — Father Lasher
Cotula coronopifolia — Buttonweed
Cotula squalida — Brass Buttons, New Zealand
Coturnix chinensis — Quail, Blue-breasted or Painted
Coturnix coturnix — Quail, Common or Eurasian
Coturnix pectoralis — Quail, Stubble
Cotyledon sp. — Cotyledon
Cotyledon undulata — Silver Crown
Coua sp. — Coucal
Couroupita guianensis — Cannonball Tree
Crabro cribarius — Wasp, Slender-bodied Digger
Cracticus mentalis — Butcher-bird, Black-backed
Cracticus nigrogularis — Butcher-bird, Pied
Cracticus torquatus — Butcher-bird, Gray
Crambe cordifolia — Seakale

Crambe maritima — Seakale
Crangon vulgaris — Shrimp, Common
Craspedia uniflora — Woollyhead
Craspedophora magnifica — Rifle-bird, Magnificent
Crassula arborescens — Silver Dollar Plant
Crassula argentea — Jade Plant
Crassula falcata — Scarlet Paintbrush
Crassula lycopodioides — Crassula, Club Moss
Crassula perforata — String-of-buttons
Crassula sp. — Crassula
Crassula tillaea — Stonecrop, Mossy
Crataegus azarolus — Azarole
Crataegus crus-galli — Hawthorn, Cockspur
Crataegus laevigata — Hawthorn, Midland
Crataegus monogyna — Hawthorn, Common
Crataegus oxyacantha — Hawthorn, English
Crataegus oxyacanthoides — Hawthorn, Midland
Craterellus sp. — Horn of Plenty
Crateromys schadenbergi — Rat, Cloud
Crax sp. — Curassow
Creadion carunculatus — Saddleback
Creagus furcatus — Gull, Swallow-tailed
Creatophora cinerea — Starling, Wattled
Crenicara sp. — Cichlid
Crenicichla sp. — Cichlid, Pike
Crenilabrus sp. — Wrasse, Corkwing
Crenimugil labrosus — Mullet, Thick-lipped Gray
Creoleon lugdunense — Ant-lion
Crepidotus sp. — Crepidotus
Crepidula fornicata — Slipper Shell
Crepis aurea — Hawk's-beard, Golden
Crepis biennis — Hawk's-beard, Greater or Rough
Crepis capillaris — Hawk's-beard, Smooth
Crepis mollis — Hawk's-beard, Soft
Crepis paludosa — Hawk's-beard, Marsh
Crepis rubra and incana — Hawk's-beard, Pink
Crepis taraxacifolia — Hawk's-beard, Beaked
Crepis vesicaria — Hawk's-beard, Beaked
Crescentia cujete — Calabash Tree
Crex crex — Crake, Corn
Cricetomys emini — Rat, Giant Pouched
Cricetomys gambianus — Rat, Giant Pouched
Cricetus cricetus — Hamster, Common

Criniger bres — Bulbul, Gray-cheeked
Criniger flaveolus — Bulbul, White-throated
Crinodendrom hookerianum — Lantern Tree
Crinodonna corsii — Crinodonna
Crinum sp. — Lily, Crinum
Crioceris asparagi — Beetle, Asparagus
Crioceris liliae — Beetle, Lily
Cristatella sp. — Bryozoan, Freshwater
Crithmum maritimum — Samphire, Rock
Crocallis elinguaria — Moth, Scalloped Oak
Crocethia alba — Sanderling
Crocias langianus — Sibia, Langbian
Crocidura leucodon — Shrew, White-toothed
Crocidura russula — Shrew, White-toothed
Crocidura suaveolens — Shrew, Lesser
    White-toothed
Crocodylus acutus — Crocodile, American
Crocodylus cataphractus — Crocodile, African
    Long-nosed
Crocodylus johnstoni — Crocodile,
    Johnstone's
Crocodylus niloticus — Crocodile, Nile
Crocodylus novaeguineae — Crocodile, New
    Guinea
Crocodylus palustris — Crocodile, Mugger
Crocodylus porosus — Crocodile, Estuarine or
    Saltwater
Crocodylus siamensis — Crocodile, Siamese
Crocosmia aurea — Copper Tips
Crocosmia crocosmiflora — Montbretia,
    Garden
Crocosmia masonorum — Montbretia,
    Garden
Crocothemis nigrifrons — Dragonfly
Crocus albiflorus — Crocus, Purple
Crocus flavus — Crocus, Yellow
Crocus nudiflorus — Crocus, Autumn
Crocus purpureus — Crocus, Purple
Crocus sativus — Saffron
Crocus sp. — Crocus
Crocus tomasinianus — Saffron
Crocus vernus — Crocus, Spring
Crocuta crocuta — Hyena, Spotted
Croesus septentrionalis — Sawfly, Birch-leaf
Crossandra infundibuliformis — Firecracker
    Flower
Crossandra sp. — Crossandra
Crossaster papposus — Sunstar

Crossoptilon auritum — Pheasant, Blue
    Eared
Crossoptilon crossoptilon — Pheasant, Elwes'
    or Tibetan Eared
Crossoptilon mantchuricum — Pheasant,
    Brown Eared
Crotalaria sp. — Bird Flower
Crotalus durissus — Rattlesnake, South
    American
Croton sp. — Croton
Crotophaga ani — Ani, Smooth-billed
Crotophaga major — Ani, Greater
Crowea saligna — Crowea
Cruciata chersonensis — Crosswort
Cruciata laevipes — Crosswort
Crucibulum sp. — Fungus, Bird's-nest
Crypsirina cucullata — Treepie,
    Racket-tailed
Crypsirina temia — Treepie, Racket-tailed
Cryptanthus bivittatus — Earth Star
Cryptanthus fosterianus — Earth Star
Cryptanthus zonatus — Earth Star
Cryptoblepharus boutonii — Skink, Bouton's
    Snake-eyed
Cryptocheilus sp. — Wasp, Spider-hunting
Cryptogramma crispa — Fern, Parsley
Cryptomeria japonica — Cedar, Japanese
Cryptomys zechi — Mole-rat, Zech's
Cryptophagus saginatus — Beetle, Fungus
Cryptoprocta ferox — Fossa
Cryptostegia sp. — Rubber Vine
Crypturellus variegatus — Tinamou,
    Variegated
Crystallogobius linearis — Goby, Crystal
Crystallogobius nilssoni — Goby, Crystal
Ctenanthe sp. — Bamburanta
Ctenicera virens — Beetle, Click
Ctenocephalides canis — Flea, Dog
Ctenolabrus rupestris — Goldsinny
Ctenomys knighti — Tucotuco
Ctenomys mendocinus — Tucotuco
Ctenopoma acutirostre — Perch, Spotted
    Climbing
Ctenopoma ansorgei — Perch, Ansorge's
    Climbing
Ctenopoma oxyrhynchus — Perch,
    Sharp-nosed Climbing
Cucubalus baccifer — Catchfly, Berry

Cucujus sp. — Beetle, Bark
Cucullia campanulae — Moth, Shark
Cucullia umbratica — Moth, Shark
Cuculus canorus — Cuckoo, Common
Cuculus fugax — Cuckoo, Hawk
Cuculus pallidus — Cuckoo, Pallid
Cuculus saturatus — Cuckoo, Oriental
Cuculus solitarius — Cuckoo, Red-chested
Cuculus sparverioides — Cuckoo, Hawk
Cuculus vagans — Cuckoo, Hawk
Cucumaria sp. — Sea Cucumber
Cucumis melo — Cantaloupe or Muskmelon
Cucumis sativa — Cucumber
Cucurbita maxima — Squash
Cucurbita pepo — Pumpkin, Field
Cucurbita pepo ovifera — Gourd
Culex sp. — Mosquito
Culicicapa ceylonensis — Flycatcher,
    Gray-headed
Cuminum cyminum — Cumin
Cuniculus paca — Paca, Spotted
Cunninghamia lanceolata — Fir, Chinese
Cunonia capensis — Cunonia
Cuon alpinus — Dhole, Indian
Cupaniopsis anacardioides — Carrotwood or
    Tuckeroo
Cuphea sp. — Cigar Plant or Cigar Flower
Cupido minimus — Butterfly, Small Blue
Cupressocyparis leylandii — Cypress,
    Leyland
Cupressus arizonica and C. glabra — Cypress,
    Arizona
Cupressus macrocarpa — Cypress, Monterey
Cupressus sempervirens — Cypress, Italian or
    Mediterranean
Curculio sp. — Weevil, Nut
Cursorius coromandelicus — Courser, Indian
Cursorius cursor — Courser, Cream-colored
Cursorius temminckii — Courser,
    Temminck's
Cuscuta epithymum — Dodder, Common or
    Lesser
Cuscuta europaea — Dodder, Greater
Cuscuta sp. — Dodder
Cussonia sp. — Cabbage Tree, Spiked
Cutia nipalensis — Cutia
Cyanacompsa sp. — Grosbeak
Cyanea sp — Jellyfish
Cyanerpes caeruleus — Honeycreeper, Purple

Cyanerpes cyaneus — Honeycreeper, Red-
    legged or Yellow-winged
Cyanocorax sp. — Jay
Cyanolyca sp. — Jay
Cyanomitra verticalis — Sunbird
Cyanopica cyanus — Magpie, Azure-winged
Cyanoptila cyanomela — Flycatcher, Blue
Cyanoptila cyanomelana — Flycatcher,
    Japanese Blue
Cyanoramphus auriceps — Parakeet,
    Yellow-crowned
Cyanoramphus malherbi — Parakeet,
    Orange-fronted
Cyanoramphus novaezeelandiae — Parakeet,
    Red-crowned
Cyanosylvia svecica — Bluethroat
Cyanotis kewensis — Teddy-bear Plant
Cyanotis somaliensis — Pussy Ears
Cyathea sp. — Tree Fern
Cyathus sp. — Fungus, Bird's-nest
Cycas revoluta — Palm, Sago
Cyclamen coum — Cyclamen
Cyclamen europaeum — Cyclamen, European
Cyclamen hederifolium — Cyclamen
Cyclamen neapolitanum — Cyclamen,
    Neapolitan
Cyclamen persicum — Cyclamen, Florist's
Cyclamen purpurascens — Cyclamen,
    Common
Cyclamen repandum— Cyclamen, Repand
Cyclarhis gujanensis — Peppershrike,
    Rufous-browed
Cyclopes didactylus — Anteater, Two-toed or
    Dwarf
Cyclophorus lingua — Fern, Japanese Felt
Cyclopterus lumpus — Lumpsucker
Cyclorana sp. — Frog, Water-holding
Cyclothone sp. — Bristlemouth
Cycnoches chlorochilon — Orchid, Swan
Cydia pomonella — Moth, Codlin
Cydia saltitans — Moth, Codlin
Cydonia oblonga — Quince
Cydonia vulgaris — Quince
Cyema atrum — Snipe-eel
Cygnus atratus — Swan, Black
Cygnus bewickii — Swan, Bewick's
Cygnus columbianus — Swan, Bewick's
Cygnus cygnus — Swan, Whooper
Cygnus melanocoryphus — Swan,
    Black-necked

343

Cygnus olor — Swan, Mute
Cylindrophis rufus — Snake, Malayan Pipe
Cymatium sp. — Triton, Hairy
Cymbalaria muralis — Toadflax, Ivy-leaved
Cymbidium sp. — Orchid, Cymbidium
Cymbilaimus lineatus — Antshrike, Fasciated
Cymbirhynchus macrorhynchus — Broadbill, Black-and-Red
Cymbium sp. — Volute, West African
Cymothoe coccinata — Butterfly
Cymothoe herminia — Butterfly
Cynara cardunculus — Artichoke, Wild
Cynara scolymus — Artichoke, Globe
Cynictis penicillata — Meerkat, Red or Bushy-tailed
Cynocephalus variegatus — Colugo
Cynocephalus volans — Colugo
Cynodon dactylon — Grass, Bermuda
Cynogale bennetti — Civet, Otter
Cynoglossum creticum — Hound's-tongue
Cynoglossum germanicum — Hound's-tongue
Cynoglossum nervosum — Hound's-tongue
Cynoglossum officinale — Hound's-tongue
Cynolebias bellotti — Pearlfish, Argenine
Cynolebias nigripinnis — Pearlfish, Black-finned
Cynopithecus niger — Ape, Black or Celebes
Cynopoecilus ladigesi — Pearlfish
Cynops pyrrhogaster — Newt, Japanese
Cynopterus sp. — Bat, Indian Fruit
Cynosurus cristatus — Dog's-tail, Crested
Cynosurus echinatus — Dog's-tail, Rough
Cynthia cardui — Butterfly, Painted Lady
Cyornis sp. — Flycatcher, Blue
Cyperus alternifolius — Umbrella Plant
Cyperus diffusus — Umbrella Plant, Dwarf
Cyperus esculentus — Nutsedge or Nutgrass
Cyperus fuscus — Cyperus, Brown or Black
Cyperus longus — Galingale
Cyperus papyrus — Egyptian Paper Plant
Cyphomandra betacea — Tree Tomato
Cyphonia sp. — Treehopper
Cyphorinus thoracicus — Wren
Cyphus augustus — Weevil
Cypraea sp. — Cowry
Cypraecassis rufa — Helmet Shell, Bull-mouth
Cyprinus carpio — Carp, Common

Cypripedium acaule — Lady's-slipper, Pink
Cypripedium calceolus — Lady's-slipper
Cypripedium reginae — Lady's-slipper, Showy
Cypsiurus parvus — Swift, Palm
Cyrestis nivea — Butterfly, Map-wing
Cyrestis rusca — Butterfly, Map-wing
Cyrilla sp. — Leatherwood
Cyrtanthus sp. — Lily, Fire or Ifafa
Cyrtomium falcatum — Fern, Holly
Cyrtostomus frenatus — Sunbird, Yellow-breasted
Cystoderma sp. — Cystoderma
Cystopteris bulbifera — Fern, Bulblet Bladder
Cystopteris fragilis — Fern, Fragile or Brittle Bladder
Cystopteris montana — Fern, Mountain Bladder
Cytinus hypocistis — Cytinus
Cytinus sanguineus — Cytinus
Cytisus kewensis — Broom, Kew
Cytisus praecox — Broom, Warminster
Cytisus purpureus — Broom, Purple
Cytisus scoparius — Broom, Common or Scotch
Cytisus sp. — Broom

D

Daboecia cantabrica — Heath, Irish or St. Dabeoc's
Dacelo gigas — Kookaburra, Laughing
Dacelo leachii — Kookaburra, Blue-winged
Dacelo novaeguineae — Kookaburra, Laughing
Dacelo tyro — Kingfisher, Giant
Dacnis berlepschi — Dacnis
Dacrydium cupressinum — Pine, Red
Dacrymyces sp. — Dacrymyces
Dactylis glomerata — Cocksfoot
Dactylopsila picata — Phalanger, Striped
Dactylopsila trivirgata — Phalanger, Striped
Dactylopterus volitans — Gurnard, Flying
Dactylorchis sp. — Orchid
Dactylorhiza fuchsii — Orchid, Spotted
Dactylorhiza incarnata — Orchid, Early Marsh
Dactylorhiza maculata — Orchid, Moorland Spotted

Dactylorhiza majalis — Orchid, Broad-leaved Marsh

Dactylorhiza praetermissa — Orchid, Southern Marsh

Dactylorhiza purpurella — Orchid, Northern Marsh

Dactylorhiza sambucina — Orchid, Elder-flowered

Daedalea quercina — Fungus, Maze

Dais cotinifolia — Dais

Dalatias licha — Darkie Charlie

Daldinia concentrica — Cramp Balls

Dalechampia roezliana — Dalechampia

Dama dama — Deer, Fallow

Dama mesopotamica — Deer, Persian Fallow

Damaliscus albifrons — Blesbok

Damaliscus dorcas — Blesbok

Damaliscus hunteri — Hartebeest, Hunter's

Damaliscus korrigum — Topi

Damaliscus lunatus — Sassaby

Damaliscus lunatus topi — Topi

Damaliscus pygargus — Blesbok

Damasonium alisma — Star-fruit

Damon medius johnstoni — Whip-scorpion

Danae racemosa — Laurel, Alexandrian

Danaus sp. — Butterfly, Monarch

Danio malabaricus — Danio, Giant

Daphne cneorum — Garland Flower

Daphne gnidium — Mezereon, Mediterranean

Daphne laureola — Spurge-laurel

Daphne mezereum — Mezereon

Daphne odora — Daphne, Sweet or Winter

Daphnia sp. — Water-flea

Daphnis nerii — Hawkmoth, Oleander

Daption capensis — Petrel, Pintado

Darlingtonia california — Pitcher-plant, California

Darwinia sp. — Darwinia

Dascillus cervinus — Beetle

Dascyllus sp. — Damselfish

Dasya pedicellata — Chenille Weed

Dasyatis pastinaca — Stingray

Dasycercus cristicauda — Mouse, Marsupial

Dasychira pudibunda — Moth, Pale Tussock

Dasymutilla occidentalis — Wasp, Velvet-ant

Dasyornis brachypterus — Bristle-bird, Eastern

Dasyornis broadbenti — Bristle-bird, Rufous

Dasypeltis inornata — Snake, Egg-eating

Dasypeltis scaber — Snake, Egg-eating

Dasypoda hirtipes — Bee, Hairy-legged Mining

Dasypoda plumipes — Bee, Hairy-legged Mining

Dasyprocta aguti — Agouti, Orange-rumped

Dasypus novemcinctus — Armadillo, Nine-banded

Dasyurops maculatus — Tiger-cat

Dasyurus geoffroii — Dasyure, Common or Spotted

Dasyurus maculatus — Tiger-cat

Dasyurus quoll — Dasyure, Eastern

Dasyurus viverrinus — Dasyure, Common or Spotted

Datura arborea — Angel's Trumpet

Datura candida — Angel's Trumpet

Datura inoxia — Angel's Trumpet

Datura metel — Angel's Trumpet

Datura rosei — Datura, Scarlet

Datura sanguinea — Datura, Scarlet

Datura stramonium — Thorn-apple

Datura suaveolens — Angel's Trumpet

Daubentonia madagascariensis — Aye-aye

Daucus carota — Queen Anne's Lace

Daucus carota var. sativa — Carrot

Davallia fejeensis — Fern, Fiji or Rabbit's-foot

Davidia involucrata — Dove Tree

Decabelone grandiflora — Golden Horn

Decaisnea fargesii — Decaisnea

Decticus verrucivorus — Grasshopper, Wart-biter

Deilephila elpenor — Hawkmoth, Elephant

Deilephila euphorbiae — Hawkmoth, Elephant

Deilephila porcellus — Hawkmoth, Elephant

Delias sp. — Butterfly

Delichon dasypus — Martin, Asian House

Delichon urbica — Martin, House

Delma sp. — Lizard, Legless

Delonix regia — Poinciana, Royal

Delphinium elatum — Larkspur, Candle or Alpine

Delphinium sp. — Larkspur

Delphinus delphis — Dolphin, Common

Dendroaspis angusticips — Mamba, Green

Dendroaspis polylepis — Mamba, Black

Dendrobates sp. — Frog, Arrow-poison

Dendrobium nobile — Orchid, Dendrobium

Dendrobium sp. — Orchid, Dendrobium
Dendrocalamus giganteus — Bamboo, Giant
Dendrochilum sp. — Orchid, Chain
Dendrochirus sp. — Scorpionfish
Dendrocincla fulginosa — Woodcreeper, Plain Brown
Dendrocitta vagabunda — Treepie, Rufous
Dendrocolaptes certhia — Woodcreeper, Barred
Dendrocopus canicapillus — Woodpecker, Pygmy
Dendrocopus leucotus — Woodpecker, White-backed
Dendrocopus maculatus — Woodpecker, Pygmy
Dendrocopus major — Woodpecker, Great Spotted
Dendrocopus medius — Woodpecker, Middle Spotted
Dendrocopus minor — Woodpecker, Lesser Spotted
Dendrocopus syriacus — Woodpecker, Syrian
Dendrocygna arcuata — Tree-duck, Wandering
Dendrocygna bicolor — Tree-duck, Fulvous
Dendrocygna eytoni — Tree-duck, Plumed
Dendrocygna guttata — Tree-duck, Spotted
Dendrocygna viduata — Tree-duck, White-faced
Dendrohyrax arboreus — Hyrax, Tree
Dendrohyrax dorsalis — Hyrax, Tree
Dendrolagus sp. — Tree-kangaroo
Dendrolaphis punctulatus — Snake, Green Tree
Dendrolimus pini — Moth, Pine Lappet
Dendromecon rigida — Poppy, California Tree
Dendromus insignis — Mouse, African Climbing
Dendronanthus indicus — Wagtail, Forest or Tree
Dendronotus sp. — Nudibranch
Dendrophyllia sp. — Coral
Dendropicos fuscescens — Woodpecker
Denisonia sp. — Whipsnake, Little
Dennstaedtia punctilobula — Fern, Hay-scented
Dentalium sp. — Tusk Shell
Dentaria bulbifera — Coral-wort or Coralroot
Dentex sp. — Bream, Toothed

Deporaus betulae — Beetle, Birch-leaf Roller
Dermacentor sp. — Tick, Shield
Dermestes lardarius — Beetle, Larder
Dermestes maculatus — Beetle, Hide or Leather
Dermochelys coriacea — Turtle, Leatherback
Dermogenys pusillus — Halfbeak, Wrestling
Deroptyus accipitrinus — Parrot, Red-fan
Deschampsia cespitosa — Hair-grass, Tufted
Deschampsia flexuosa — Hair-grass, Wavy
Deschampsia sp. — Hair-grass
Descurainia sophia — Flixweed
Desfontainea hookeri — Desfontainea
Desfontainea spinosa — Desfontainea
Desmana moschata — Desman, Russian
Desmana pyrenaica — Desman, Pyrenean
Desmodus rotundus — Bat, Vampire
Deutzia gracilis — Deutzia, Slender
Deutzia lemoinei — Deutzia, Lemoine
Deutzia rosea — Deutzia
Deutzia scabra — Deutzia, Fuzzy
Deutzia sp. — Deutzia
Diacrisia sannio — Moth, Clouded Buff
Diacrium bicornutum — Orchid, Virgin
Dianella intermedia — Turutu (Blueberry)
Dianthus alpinus — Pink, Alpine
Dianthus arenarius — Pink, Sand
Dianthus armeria — Pink, Deptford
Dianthus barbatus — Sweet William
Dianthus carthusianorum — Pink, Carthusian
Dianthus caryophyllus — Carnation or Pink, Clove
Dianthus deltoides — Pink, Maiden
Dianthus gratianopolitanus — Pink, Cheddar
Dianthus monspessulanus — Pink, Fringed
Dianthus plumarius — Pink, Garden or Common or Chelsea
Dianthus superbus — Pink, Superb
Dianthus sylvestris — Pink, Wood
Diapensia lapponica — Diapensia
Dicaeum cruentatum — Flowerpecker, Scarlet-backed
Dicaeum hirundinaceum — Mistletoebird
Dicaeum ignipectus — Flowerpecker, Fire-breasted
Dicaeum trigonostigma — Flowerpecker, Orange-bellied
Dicentra cucullaria — Dutchman's Breeches

Dicentra spectabilis — Bleeding-heart, Common
Dicentrarchus labrax — Bass
Dicerorhinus sumatrensis — Rhinoceros, Sumatran
Diceros bicornis — Rhinoceros, Black
Diceros simus — Rhinoceros, Square-lipped or White
Dichondra repens — Dichondra
Dichorisandra reginae — Spiderwort, Queen's
Dicksonia antarctica — Fern, Tasmanian Tree
Dicranocephalus sp. — Beetle, Rose Chafer
Dicranum scoparium — Moss, Broom
Dicranura vinula — Moth, Puss
Dicrostonyx torquatus — Lemming
Dicrurus adsimilis — Drongo, Fork-tailed
Dicrurus annectans — Drongo, Crow-billed
Dicrurus bracteatus — Drongo, Spangled
Dicrurus hottentottus — Drongo, Spangled
Dicrurus leucophaeus — Drongo, Ashy or Gray
Dicrurus macrocercus — Drongo, Black
Dicrurus paradiseus — Drongo, Racket-tailed
Dicrurus remifer — Drongo, Racket-tailed
Dictamnus albus — Burning Bush
Dictyophora duplicata — Stinkhorn, Collared
Dictyophora europaea — Lanternfly, European
Dictyophora indusiata — Stinkhorn, Collared
Didelphis marsupialis — Opossum, Common
Didermocerus sumatrensis — Rhinoceros, Sumatran
Didierea madagascariensis — Didierea
Didunculus strigirostris — Pigeon, Tooth-billed
Dieffenbachia amoena — Dumb Cane
Dieffenbachia picta — Dumb Cane, Spotted or Variable
Dierama pulcherrima — Wand Flower, South African
Difflugia sp. — Amoeba
Digitalis ambigua — Foxglove, Yellow
Digitalis ferruginea — Foxglove, Rusty
Digitalis grandiflora — Foxglove, Yellow
Digitalis lutea — Foxglove, Small Yellow
Digitalis obscura — Floxglove, Rusty
Digitalis purpurea — Foxglove, Purple or Common

Digitaria ischaemum — Crab-grass, Smooth
Digitaria sanguinalis — Crab-grass
Diglossa sp. — Flowerpiercer
Dillenia indica — Hondapara
Dilophus febrilis — Fly, Fever
Dimorphotheca sp. — Marigold, Cape
Dinemellia dinemelli — Weaver, White-headed Buffalo
Dinocras cephalotes — Fly, Stone
Dinomys branickii — Pacarana
Dinopium javanense — Woodpecker, Three-toed
Diodon sp. — Porcupine-fish
Diodora apertura or D. italica — Limpet, Keyhole
Diomedea chlororhynchos — Albatross, Yellow-nosed
Diomedea chrysostoma — Albatross, Gray-headed
Diomedea epomophora — Albatross, Royal
Diomedea exulans — Albatross, Wandering
Diomedea immutabilis — Albatross, Laysan
Diomedea irrorata — Albatross, Waved
Diomedea melanophris — Albatross, Black-browed
Diomedea nigripes — Albatross, Black-footed
Dionaea muscipula — Venus' Fly-trap
Dionysia sp. — Dionysia
Dioscorea sp. — Yam
Diospyros kaki — Persimmon, Oriental
Diospyros lotus — Date-plum
Diospyros virginiana — Persimmon, Common
Dipelta floribunda — Dipelta
Diphyllodes magnificus — Bird of Paradise, Magnificent
Diphyllodes respublica — Bird of Paradise, Waigeu or Wilson's
Diphyscium foliosum — Moss, Nut
Dipladenia splendens — Dipladenia
Diplodactylus sp. — Gecko, Spiny-tailed
Diplodes sp. — Bream
Diploglossus sp. — Lizard
Diplolaemus darwinii — Lizard, Patagonian
Diplolepis rosae — Wasp, Mossy Rose Gall
Diplometopon zarudnyi — Worm-lizard, Arabian
Diplotaxis muralis — Rocket, Annual Wall
Diplotaxis tenuifolia — Rocket, Perennial Wall

Diporiphora bilineata — Dragon, Two-lined
Diprion pini — Sawfly, Pine
Dipsacus fullonum — Teasel, Common
Dipsacus pilosus — Teasel, Small
Dipsacus sylvestris — Teasel, Wild
Dipsas sp. — Snake, Snail-eating
Dipus sagitta — Jerboa
Disa sp. — Orchid
Disa uniflora — Pride of Table Mountain
Disanthus cercidifolius — Disanthus
Discosura longicauda — Hummingbird,
    Coquette
Dispersis sp. — Orchid
Dispholidus typus — Snake, Boomslang
Dissotis princeps — Dissotis
Distichodus rostratus — Distichodus
Distichodus sexfasciatus — Distichodus,
    Six-banded
Distoechurus pennatus — Phalanger
Dixippus sp. — Stick-insect
Dizygotheca elegantissima — Aralia, False
Dodecatheon sp. — Shooting Star
Dodonaea viscosa — Akeake
Dolichopus sp. — Fly, Long-headed or -legged
Dolichos lablab — Bean, Hyacinth
Dolichothele sp. — Dolichothele
Dolichotis patagonum — Mara
Dologale dybowskii — Mongoose, Pousargues'
Dolomedes fimbriatus — Spider, Pirate or
    Raft
Dombeya sp. — Wild Pear or Wild Plum
Domicella domicella — Lorikeet
Domicella garrula — Lorikeet
Donacia sp. — Beetle, Leaf or Reed
Dorcopsis sp. — Wallaby, Mountain
Dorcus parallelopipedus — Beetle, Lesser
    Stag
Doronicum clusii — Leopard's-bane, Clusius'
Doronicum grandiflora — Leopard's-bane,
    Large-flowered
Doronicum pardalianches — Leopard's-bane,
    Great
Doronicum sp. — Leopard's-bane
Dorstenia sp. — Dorstenia
Dorycnium sp. — Dorycnium
Doryopteris pedata palmata — Fern,
    Spear-leaved or Hand
Dosinia exoleta — Artemis, Rayed
Douglasia vitaliana — Douglasia

Doxanthus unguis-cati — Cat's Claw
Draba aizoides — Whitlow-grass, Yellow
Draba incana — Whitlow-grass, Hoary
Draba muralis — Whitlow-grass, Wall
Draba sp. — Whitlow-grass
Draco volans — Lizard, Common Flying
Dracocephalum sp. — Dragonhead
Dracophyllum sp. — Grass Tree
Dracunculus vulgaris — Arum, Dragon
Dreissena polymorpha — Mussel, Zebra
Drepana falcataria — Moth, Pebble Hook-tip
Drepanorhynchus reichenowi — Sunbird,
    Golden-winged
Drimys winteri — Winter's Bark
Dromaius novaehollandiae — Emu
Dromas ardeola — Plover, Crab
Dromiceius novaehollandiae — Emu
Dromicia nana — Phalanger
Drosanthemum floribunda — Redondo
    Creeper
Drosera anglica — Sundew, Great
Drosera binata — Sundew, Australian
Drosera capensis — Sundew, Cape Province
Drosera intermedia — Sundew, Long-leaved
    or Oblong-leaved
Drosera rotundifolia — Sundew, Common
Drosophila sp. — Fly, Fruit (Vinegar)
Drosophyllum lusitanicum — Sundew, Yellow
Drupa morum — Drupe, Purple
Dryandra sp. — Dryandra
Dryas octopetala — Avens, Mountain
Drymodes brunneopygia — Scrub-robin,
    Southern
Drymodes superciliaris — Scrub-robin,
    Northern
Drymophila sp. — Antbird
Dryocopus javensis — Woodpecker, Tristam's
Dryocopus martius — Woodpecker, Black
Dryomys nitedula — Dormouse, Forest or
    Russian
Dryophis nasutus — Snake, Long-nosed Tree
Dryops sp. — Beetle, Water
Dryopteris cristata — Buckler-fern, Crested
Dryopteris dilatata — Buckler-fern, Common
Dryopteris filix-mas — Fern, Male
Dryoscopus cubla — Shrike, Puff-backed
Dryotriorchis sp. — Eagle, Serpent
Dubusia taeniata — Mountain-tanager
Ducula aenea — Pigeon, Green Imperial

Ducula badia — Pigeon, Mountain Imperial

Ducula mindorensis — Pigeon, Mindoro Imperial

Ducula sp. — Pigeon, Imperial Fruit

Dugong dugong — Dugong

Dunckerocampus caulleryi — Pipefish

Dupetor flavicollis — Bittern, Black

Duranta repens — Pigeonberry

Durio zibethinus — Durian

Duvalia sp. — Duvalia

Dyckia sp. — Dyckia

Dynastes sp. — Beetle, Rhinoceros

Dynatosoma fuscicorne — Gnat, Fungus

Dysithamnus mentalis — Antvireo, Plain

Dytiscus sp. — Beetle, Predaceous Diving

E

Ecballium elaterium — Cucumber, Squirting

Eccremocarpus scaber — Glory Flower

Echeneis naucrates — Shark-sucker

Echeveria dactylifera — Plush Plant

Echeveria pulvinata — Plush Plant

Echeveria setosa — Firecracker Plant

Echeveria sp. — Echeveria

Echidna aculeata — Echidna

Echiniscus blumi — Water Bear

Echinocactus sp. — Cactus, Barrel

Echinocardium cordatum — Sea Urchin, Sand

Echinocereus engelmanii — Cactus, Strawberry Hedgehog

Echinocereus pectinatus — Cactus, Hedgehog or Rainbow

Echinocereus sp. — Cactus, Echinocereus

Echinochloa crus-galli — Grass, Barnyard or Cockspur

Echinococcus sp. — Tapeworm

Echinomyia sp. — Fly, Hedgehog or Parasitic

Echinops ritro — Globe-thistle

Echinops sp. — Globe-thistle

Echinops sphaerocephalus — Globe-thistle, Great

Echinops telfairi — Tenrec, Hedgehog

Echinopsis multiplex — Cactus, Urchin or Hedgehog

Echinorhynchus clavula — Worm, Fish

Echinosorex gymnurus — Moon-rat

Echinus esculentus — Sea Urchin, Edible or European

Echis carinatus — Viper, Saw-scaled

Echium bourgaeanum — Echium

Echium fastuosum — Echium

Echium lycopsis — Bugloss, Purple Viper's

Echium plantagineum — Bugloss, Purple Viper's

Echium vulgare — Bugloss, Viper's

Eciton sp. — Ant, Army

Eclectus roratus — Parrot

Ectobius lapponicus — Cockroach, Dusky

Edgeworthia papyrifera — Paper Bush

Edwardsia callimorpha — Sea Anemone, Burrowing

Egeria densa — Pondweed, Dense-leaved

Egernia sp. — Skink, Spiny-tailed

Egretta alba — Egret, Common or Great White

Egretta egretta — Egret, Little

Egretta eulophotes — Egret, Chinese

Egretta garzetta — Egret, Little

Egretta intermedia — Egret, Plumed or Lesser

Egretta sacra — Egret, Pacific Reef

Ehretia sp. — Ehretia

Eichhornia crassipes — Water-hyacinth

Eichhornia paniculata — Water-hyacinth

Eidoleon bistrigatus — Ant-lion

Eidolon helvum — Bat, Straw-colored Fruit

Eigenmannia virescens — Knifefish, Green

Eilema lurideola — Moth, Common Footman

Eira barbara — Tayra

Eisenia foetida or E. rosea — Earthworm, Brandling

Elaeagnus angustifolia — Oleaster

Elaeagnus multiflora — Gumi

Elaeagnus pungens — Oleaster

Elaeis guineensis — Palm, Oil

Elaeocarpus sp. — Elaeocarpus

Elagatis bipinnulatus — Rainbow-runner

Elanus caeruleus — Kite, Black-shouldered or -winged

Elanus notatus — Kite, Black-shouldered or -winged

Elanus scriptus — Kite, Letter-winged

Elaphe longissima — Snake, Aesculapian

Elaphe quatorlineata — Snake, Four-lined

Elaphe situla — Snake, Leopard

Elaphodus cephalophus cephalophus — Deer, Tufted

Elaphodus michianus — Deer, Tufted
Elaphomyces granulatus — Truffle, Deer
Elaphurus davidianus — Deer, Pere David's
Elaphus africanus — Elephant, African
Elaphus indicus — Elephant, Asian or Indian
Elaphus maximus — Elephant, Asian or
    Indian
Elasmucha grisea — Shield-bug, Parent
Elater sp. — Beetle, Click
Elatine sp. — Waterwort
Electrophorus electricus — Eel, Electric
Eledone cirrhosa — Octopus, Curled or Lesser
Elegia juncea — Elegia
Eleocharis acicularis — Spike-rush, Slender
Eleocharis multicaulis — Spike-rush,
    Many-stemmed
Eleocharis palustris — Spike-rush, Common
Eleocharis sp. — Spike-rush
Eleotris sp. — Goby, Sleeper
Elephantulus rozeti — Elephant-shrew,
    North African
Elettaria cardamomum — Cardamon
Eleusine indica — Goosegrass
Eliomys quercinus — Dormouse, Garden
Ellobius talpinus — Lemming
Elminius modestus — Barnacle
Elodea canadensis — Pondweed, Canadian
Elodea nuttallii — Waterweed, Esthwaite
Elop saurus — Ladyfish
Elsholtzia stauntonii — Elsholtzia
Elymus arenarius — Lyme-grass
Emarginula reticulata — Limpet, Slit
Ematurga atomaria — Moth, Common Heath
Emberiza aureola — Bunting, Yellow-
    breasted
Emberiza bruniceps — Bunting, Red-headed
Emberiza caesia — Bunting, Cretzschmar's
Emberiza calandra — Bunting, Corn
Emberiza cia — Bunting, Rock
Emberiza cineracea — Bunting, Cinereous
Emberiza cioides — Bunting, Siberian
    Meadow
Emberiza cirlus — Bunting, Cirl
Emberiza citrinella — Yellowhammer
Emberiza flaviventris — Bunting, Golden-
    breasted
Emberiza hortulana — Bunting, Ortolan
Emberiza leucocephala — Bunting, Pine
Emberiza melanocephala — Bunting, Black-
    headed

Emberiza pusilla — Bunting, Little
Emberiza rustica — Bunting, Rustic
Emberiza schoeniclus — Bunting, Reed
Emberiza spodocephala — Bunting, Black-faced
Emberiza striolata — Bunting, House or
    Striped
Emberiza tahapisi — Bunting, Rock
Embia sp. — Web-spinner
Emblema oculata — Firetail
Emblema picta — Finch, Painted
Embothrium coccineum — Fire Bush, Chilean
Emilia sp. — Cupid's Paintbrush
Empetrum nigrum — Crowberry
Empis tessellata — Fly, Dance or Empid
Empusa sp. — Mantis, Mediterranean
    Praying
Emydura sp. — Tortoise, Australian
    Snake-necked
Emys orbicularis — Tortoise, European Pond
Encephalartos sp. — Palm, Bread Fern
Endoliisoma tenuirostre — Cicada-bird
Endromis versicolora — Moth, Kentish Glory
Endymion hispanicus — Bluebell, Spanish
Endymion non-scriptus — Bluebell or
    Hyacinth, Wild
Engraulis encrasicholus — Anchovy
Enicurus leschenaulti — Forktail,
    Leschenault's or White-crowned
Enicurus maculatus guttatus — Forktail,
    Spotted
Enicurus schistaceus — Forktail, Slaty-backed
Enkianthus campanulatus — Enkianthus,
    Redvein
Enkianthus perulatus — Enkianthus, White
Ensifera ensifera — Hummingbird,
    Sword-billed
Ensis sp. — Clam, Razor
Entelurus aequoreus — Pipefish, Ocean or
    Snake
Entimus imperialis — Weevil
Entoloma sp. — Entoloma
Entomyzon cyanotis — Honey-eater,
    Blue-faced
Enyalius catenatus — Lizard, Brazilian Tree
Eolophus roseicapillus — Cockatoo, Roseate
Eophona sp. — Grosbeak
Eopsaltria australis — Robin, Southern
    Yellow
Eopsaltria capito — Robin, Pale Yellow

Eopsaltria chrysorrhoa — Robin, Northern Yellow

Eopsaltria griseogularis — Robin, Western Yellow

Eopsaltria leucops — Robin, White-faced

Eos squamata — Lorikeet

Epacris sp. — Heath

Ephedra sp. — Joint-pine or Joint-fir

Ephemera sp. — Fly, May

Ephestia kuhniella — Moth, Mediterranean Flour

Ephippiger sp. — Grasshopper, Steppe

Ephippiorynchus senegalensis — Stork, Saddle-billed

Ephthianura albifrons — Chat, White-faced

Ephthianura aurifrons — Chat, Orange

Ephthianura crocea — Chat, Yellow

Ephthianura tricolor — Chat, Crimson

Epiblemum scenicum — Spider, Jumping

Epicrates cenchris — Boa, Rainbow

Epidendrum atropurpureum — Orchid, Spice

Epidendrum ciliare — Orchid, Eyelash

Epidendrum cochleatum — Orchid, Clamshell

Epidendrum ibaguense — Orchid, Crucifix

Epidendrum sp. — Orchid, Epidendrum

Epigaea repens — Trailing Arbutus

Epilobium angustifolium — Fireweed

Epilobium hirsutum — Willow-herb, Hairy

Epilobium montanum — Willow-herb, Broad-leaved

Epilobium palustre — Willow-herb, Marsh

Epilobium sp. — Willow-herb

Epimachus meyeri — Bird of Paradise, Brown Sickle-billed

Epimedium macranthum — Barrenwort

Epimedium sp. — Barrenwort

Epinephelus sp. — Grouper

Epipactis atrorubens — Helleborine, Dark-red

Epipactis helleborine — Helleborine, Broad-leaved

Epipactis leptochila — Helleborine, Narrow-lipped

Epipactis palustris — Helleborine, Marsh

Epipactis sp. — Helleborine

Epiphile sp. — Butterfly

Epiphyllum sp. — Cactus, Orchid

Epiplatys sp. — Epiplatys

Epipogium aphyllum — Orchid, Ghost

Episcia cupreata — Flame-violet

Episcia dianthiflora — Lace Flower Vine

Epithelantha micromeris — Cactus, Button

Epitonium sp. — Wendletrap

Epixerus ebii — Squirrel, Palm

Epizoanthus arenaceus — Sea Anemone

Eptesicus pumilus — Bat, Serotine

Eptesicus serotinus — Bat, Serotine

Equisetum arvense — Horsetail, Common or Field

Equisetum fluviatile — Horsetail, Water

Equisetum hyemale — Horsetail, Rough

Equisetum palustre — Horsetail, Marsh

Equisetum scirpoides — Horsetail, Dwarf

Equisetum sylvaticum — Horsetail, Wood

Equisetum telmateia — Horsetail, Great

Equus africanus — Ass, African Wild

Equus asinus — Ass, African Wild

Equus burchelli — Zebra, Common

Equus caballus var. — Horse, Wild

Equus grevyi — Zebra, Grevy's

Equus hemionus var. — Ass, Asian or Indian Wild

Equus zebra — Zebra, Cape Mountain

Erannis defoliaria — Moth, Mottled Umber

Eranthemum nervosum — Sage, Blue

Eranthemum pulchellum — Sage, Blue

Eranthis hyemalis — Aconite, Winter

Eranthis tubergeniana — Aconite, Winter

Erasmia sanguiflua — Moth, Burnet

Erebia sp. — Butterfly, Ringlet

Eremialector quadricinctus — Sandgrouse, Four-banded

Eremiornis carteri — Spinifex-bird

Eremomela icteropygialis — Warbler, Yellow-bellied

Eremophila alpestris — Lark, Horned or Shore

Eremophila bilopha — Lark, Temminck's Horned

Eremophila sp. — Emu Bush

Eremopterix sp. — Finch-lark

Eremurus sp. — Lily, Foxtail

Eretmochelys imbricata — Turtle, Hawksbill

Ergates faber — Beetle, Carpenter's Longhorn

Erica arborea — Heath, Tree

Erica canaliculata — Heath, Christmas

Erica ciliaris — Heath, Dorset

Erica cinerea — Heather, Bell or Scotch
Erica darleyensis — Heath, Winter-flowering
Erica erigena — Heath, Irish
Erica herbacea — Heath, Spring or
  Flesh-colored
Erica lusitanica — Heath, Portugal or Spring
Erica mediterranea — Heath, Irish or Biscay
Erica tetralix — Heather, Bog
Erica vagans — Heath, Cornish
Erica veitchii — Heath, Tree
Ericulus setosus — Tenrec
Erigeron acer — Fleabane, Blue
Erigeron alpinus — Fleabane, Alpine
Erigeron annus — Fleabane, Annual
Erigeron borealis — Fleabane, Alpine
Erigeron sp. — Fleabane
Erinacea anthyllis — Broom, Hedgehog
Erinaceus europaeus — Hedgehog, Common
  or European
Erinaceus frontalis — Hedgehog, Southern
  African
Erinus alpinus — Erinus, Alpine
Eriobotrya japonica — Loquat
Eriocaulon aquaticum — Pipewort
Eriocaulon septangulare — Pipewort
Eriocheir japonica — Crab, Mitten
Eriocheir sinensis — Crab, Mitten
Eriocnemis luciana — Hummingbird, Puff-leg
Eriocnemis mirabilis — Hummingbird,
  Puff-leg
Eriogaster lanestris — Moth, Small Eggar
Eriophorum angustifolium — Cotton-grass,
  Common
Eriophorum scheucherzeri — Cotton-grass,
  Scheucherzer's
Eriophorum vaginatum — Cotton-grass,
  Hare's-tail
Eriosoma lanigerum — Aphid, Woolly Apple
Eriostemon sp. — Waxflower
Eristalis sp. — Fly, Drone
Erithacus calliope — Rubythroat, Siberian
Erithacus rubecula — Robin, European or
  Eurasian
Erithacus svecicus — Bluethroat
Eritrichium nanum — King of the Alps
Ero furcata — Spider, Pirate
Erodium cicutarium — Storksbill, Common
Erodium gruinum — Storksbill, Long-beaked
Erodium maritimum — Storksbill, Sea

Erodium sp. — Storksbill
Erolia alpina — Dunlin
Erolia ferruginea — Sandpiper, Curlew
Erolia ruficollis — Stint, Rufous-necked
Erophila verna — Whitlow-grass, Common
Erotylus varians — Beetle, Fungus
Erpobdella octoculata — Leech, Dog or Worm
Eruca sativa — Rocket Salad
Eruca vesicaria — Rocket Salad
Erucastrum gallicum — Rocket, Hairy
Ervatamia coronaria — Jasmine, Crape
Ervatamia divaricata — Jasmine, Crape
Eryngium alpinum — Eryngo, Alpine
Eryngium amethystinum — Eryngo, Blue
Eryngium campestre — Eryngo, Field
Eryngium giganteum — Sea Holly
Eryngium maritimum — Sea Holly
Eryngium sp. —Eryngo or Sea Holly
Erynnis icelus — Butterfly, Dingy Skipper
Erynnis tages — Butterfly, Dingy Skipper
Erysium sp. — Treacle-mustard
Erythea armata — Palm, Blue
Erythrina poeppigiana — Bucare Tree
Erythrina sp. — Coral Tree or Coral Bean
Erythrocebus patas — Monkey, Patas
Erythronium dens-canis — Dog's Tooth Violet
Erythropygia sp. — Scrub-robin
Erythrothlypis salmoni — Tanager, Scarlet-
  and-white
Erythrotriorchis radiatus — Goshawk, Red
Erythroxylon coca — Coca
Erythrura prasina — Parrot-finch, Pin-tailed
Erythrura psittacea — Parrot-finch
Erythrura trichroa — Parrot-finch,
  Blue-faced
Erythrura viridifacies — Parrot-finch
Eryx sp. — Boa, Sand
Esacus magnirostris recurvirostris —
  Thick-knee, Great
Escallonia sp. — Escallonia
Eschscholtzia californica — Poppy, California
Esomus danricus — Barb, Flying
Esomus malayensis — Barb, Malayan Flying
Esox lucius — Pike, Northern
Espeletia sp. — Espeletia
Espostoa lanata — Cactus, Peruvian Old-man
Espostoa melanostele — Cactus, Hedge
Estrilda amandava — Avadavat, Red
Estrilda astrild — Waxbill, Common or St.
  Helena

Estrilda melanotis — Waxbill, Swee or
Yellow-bellied
Estrilda nitidula — Twinspot, Green-backed
Estroplus maculatus — Chromide, Orange
Estroplus suratensis — Chromide, Green
Eubalaena glacialis — Whale, Right
Eublepharis macularius — Gecko, Panther
Eubucco bourcierii — Barbet, Red-headed
Eucalyptus ficifolia — Gum, Red-flowering
Eucalyptus globulus — Gum, Blue
Eucalyptus gunni — Gum, Cider
Eucalyptus leucoxylon — Gum, Yellow
Eucalyptus macrocarpa — Blue Bush or
Mottlecah
Eucalyptus sp. — Gum Tree
Eucera sp. — Bee, Long-horned or Solitary
Eucharis grandiflora — Lily, Amazon
Euchlornis riefferi — Fruit-eater,
Green-and-Black
Eucomis sp. — Pineapple Flower
Eucryphia sp. — Eucryphia
Eudocimus ruber — Ibis, Scarlet
Eudromecia caudata — Possum, Pygmy
Eudromia elegans — Tinamou, Crested
Eudromias morinellus — Dotterel
Eudynamis scolopacea — Koel
Eudynamis taitensis — Cuckoo, Long-tailed
Eudyptes chrysolophus — Penguin, Macaroni
Eudyptes crestatus — Penguin, Rockhopper
Eudyptes minor — Penguin, Little
Eudyptes pachyrhynchus — Penguin, Crested
Eudyptes schlegeli — Penguin, Royal
Eudyptes sclateri — Penguin, Crested
Eudyptula minor — Penguin, Little
Eugenia carophyllus — Clove
Eugenia sp. — Cherry, Brush or Scrub
Euglena sp. — Euglenoid
Eulabeornis castaneoventris — Rail,
Chestnut
Eulalia viridis — Worm
Eulophia quartiniana — Eulophia
Eumenes sp. — Wasp, Potter
Eumomota superciliosa — Motmot,
Turquoise-browed
Eumorphus marginatus — Beetle, Fungus
Eunectes murinus or E. notaeus — Anaconda
Eunereis sp. — Worm, Clam
Eunicella sp. — Sea Fan
Euonymus alatus — Spindle Tree, Winged

Euonymus europaeus — Spindle Tree,
Common
Euonymus fortunei — Winter Creeper
Euonymus japonica — Spindle Tree, Japanese
Euonymus sp. — Euonymus
Euoticus elegantulus — Galago,
Needle-clawed
Eupagurus bernhardus — Crab, Hermit
Eupatorium cannabinum — Agrimony, Hemp
Eupatorium coelestinum — Mist Flower
Eupatorium rugosum — Snakeroot, White
Eupatorium sordidum — Mist Flower
Eupetes macrocercus — Babbler, Rail
Eupetomena macroura — Hummingbird,
Swallow-tailed
Euphonia sp. — Euphonia
Euphorbia acanthothamnos — Spurge, Greek
Spiny
Euphorbia amygdaloides — Spurge, Wood
Euphorbia characias — Spurge, Large
Mediterranean or Hedge
Euphorbia cyparissias — Spurge, Cypress
Euphorbia dendroides — Spurge, Tree
Euphorbia epithymoides — Spurge, Cushion
Euphorbia esula — Spurge, Hungarian
Euphorbia exigua — Spurge, Dwarf
Euphorbia fulgens — Scarlet Plume
Euphorbia grandicornis — Corn's Horn
Euphorbia helioscopia — Spurge, Sun
Euphorbia hyberna — Spurge, Irish
Euphorbia ingens — Naboom
Euphorbia lactea — Euphorbia, Milk-striped
Euphorbia lathyris — Spurge, Caper
Euphorbia marginata — Snow-on-the-
mountain
Euphorbia milli — Crown of Thorns
Euphorbia myrsinites — Spurge, Broad-
leaved Glaucous
Euphorbia obesa — Euphorbia, Globe or
Basketball
Euphorbia palustris — Spurge, Marsh
Euphorbia paralias — Spurge, Sea
Euphorbia peplis — Spurge, Purple
Euphorbia peplus — Spurge, Petty
Euphorbia platyphyllos — Spurge, Broad-
leaved
Euphorbia portlandica — Spurge, Portland
Euphorbia pulcherrima — Poinsettia
Euphorbia robbiae — Mrs. Robb's Bonnet

Euphorbia sp. — Euphorbia
Euphorbia spinosa — Spurge, Spiny or Thorny
Euphorbia splendens — Crown of Thorns
Euphorbia tirucalli — Milkbush
Euphorbia wulfenii — Spurge
Euphractus sexcinctus — Armadillo, Six-banded
Euphractus villosus — Armadillo, Hairy
Euphrasia sp. — Eyebright
Euphydryas aurinia — Butterfly, Marsh Fritillary
Euphydryas cynthia — Butterfly, Marsh Fritillary
Eupithecia sp. — Moth, Pug
Euplectes afra — Weaver, Napoleon
Euplectes orix — Weaver, Grenadier
Euplectes progne — Widow, Long-tailed
Euploea sp. — Butterfly, Crow
Eupodotis afra — Korhaan, Black
Eupodotis australis — Bustard, Australian
Eupodotis bengalensis — Florican
Eupodotis senegalensis — Bustard, White-bellied
Euproctis similis — Moth, Gold- or Yellow-tail
Euproctus sp. — Newt
Eurocephalus anguitimens — Shrike, White-crowned
Eurostopodus guttatus — Nightjar, Spotted
Eurostopodus macrotis — Nightjar, Great-eared or Giant
Eurostopodus mystacalis — Nightjar, White-throated
Eurrhypara hortulata — Moth, Small Magpie
Eurylaimus javanicus — Broadbill, Banded
Eurylaimus ochromalus — Broadbill, Black-and-Yellow
Eurylaimus steerii — Broadbill, Banded
Euryops sp. — Marguerite Othonna
Eurypharaynx pelecanoides — Gulper-eel
Eurypyga helias — Bittern, Sun
Eurystomus orientalis — Dollarbird
Euschemon rafflesia — Butterfly, Regent Skipper
Euscorpius sp. — Scorpion
Euthynnus alletteratus — Albacore, False
Eutoxeres aquila — Sicklebill
Eutoxeres condamini — Sicklebill
Euxiphipops asfur — Angelfish
Euxiphipops xanthometopon — Angelfish
Evides elegans — Beetle, Wood-borer

Exaculum pusillum — Centaury, Guernsey
Exacum affine — Violet, Arabian or German
Excalfactoria chinensis — Quail, Chinese Painted or King
Exidia sp. — Witches' Butter
Exochorda sp. — Pearl Bush
Exocoetus volitans — Flying-fish, Tropical Two-wing
Extatosoma tiaratum — Leaf-insect

F

Fabiana imbricata — Fabiana
Fabriciana adippe — Butterfly, High Brown Fritillary
Facelina sp. — Nudibranch
Fagopyron esculentum — Buckwheat
Fagraea sp. — Tembusu
Fagus grandifolia — Beech, American
Fagus sylvatica — Beech, Common or European
Fagus sylvatica sp. — Beech, Hybrids
Falcaria vulgaris — Longleaf
Falco araea — Kestrel, Seychelles
Falco amurensis — Falcon, Amur
Falco berigora — Falcon, Brown
Falco biarmicus — Falcon, Lanner
Falco cenchroides — Kestrel, Nankeen
Falco cherrug — Falcon, Saker
Falco chiquera — Falcon, Red-headed or -necked
Falco columbarius — Merlin
Falco concolor — Falcon, Sooty
Falco dickinsoni — Kestrel, Dickinson's
Falco eleonorae — Falcon, Eleonora's
Falco hypoleucus — Falcon, Gray
Falco longipennis — Falcon, Little
Falco naumanni — Kestrel, Lesser
Falco novaeseelandiae — Falcon, New Zealand
Falco pelegrinoides — Falcon, Desert
Falco peregrinus — Falcon, Peregrine
Falco punctatus — Kestrel, Mauritius
Falco rusticolus — Gyrfalcon
Falco severus — Hobby, Oriental
Falco subbuteo — Hobby
Falco subniger — Falcon, Black
Falco tinnunculus — Kestrel, Common or Rock
Falco vespertinus — Falcon, Red-footed
Falculea palliata — Sicklebill

Falcunculus frontatus — Shrike-tit, Eastern
Falcunculus leucogaster — Shrike-tit, Western
Fannia canicularis — Fly, Little House
Fasciola hepatica — Fluke, Liver
Fatshedera lizei — Ivy, Japanese or Tree
Fatsia japonica — Fatsia, Japanese
Faucaria tigrina — Tiger's Jaws
Fedia cornucopiae — Valerian, African
Feijoa sellowiana — Guava, Pineapple
Felicia amelloides — Daisy, Blue
Felis aurata — Cat, Golden
Felis badia — Cat, Bay
Felis bengalensis — Cat, Leopard
Felis bieti — Cat, Chinese Desert
Felis caracal — Caracal
Felis cattus — Cat, Caffer Wild
Felis chaus — Cat, Jungle or Swamp
Felis colocolo — Cat, Pampas
Felis concolor — Cougar
Felis cougar — Cougar
Felis leo — Lion
Felis libyca — Cat, Caffer Wild
Felis lynx — Lynx, European or Northern
Felis lynx pardina — Lynx, Spanish
Felis manul — Cat, Pallas'
Felis margarita — Cat, Sand
Felis marmorata — Cat, Marbled
Felis nebulosa — Leopard, Clouded
Felis nigripes — Cat, Black-footed
Felis ocreata — Cat, Wild
Felis onca — Jaguar
Felis pardalis — Ocelot
Felis pardus — Leopard
Felis planiceps — Cat, Flat-headed
Felis rubiginosa — Cat, Rusty-spotted
Felis serval — Cat, Serval
Felis sylvestris — Cat, Wild
Felis temmincki — Cat, Golden
Felis tigrina — Cat, Margay
Felis tigris — Tiber, Bengal or Indian
Felis tigris altaica — Tiger, Siberian or Manchurian
Felis uncia — Leopard, Snow
Felis viverrina — Cat, Fishing
Felis wiedi — Cat, Margay
Felis yagouarondi — Cat, Jaguarundi or Otter
Fenestraria rhodalophylla — Fenestraria
Fennecus zerda — Fox, Fennec

Ferocactus latispinus — Devil's Tongue
Ferocactus sp. — Cactus, Fishhook
Ferula communis — Fennel, Giant
Festuca altissima — Fescue, Wood
Festuca arundinacea — Fescue, Tall
Festuca gigantea — Fescue, Giant
Festuca ovina — Fescue, Sheep's
Festuca ovina glauca — Fescue, Blue
Festuca pratensis — Fescue, Meadow
Festuca rubra — Fescue, Creeping or Red
Festuca sp. — Fescue
Festuca vivpara — Fescue, Viviparous
Ficedula albicollis — Flycatcher, Collared
Ficedula cyanomelana — Flycatcher, Japanese Blue
Ficedula hyperythra — Flycatcher, Snowy-browed or Thicket
Ficedula hypoleuca — Flycatcher, Pied
Ficedula narcissina — Flycatcher, Narcissus
Ficedula parva — Flycatcher, Red-breasted
Ficedula westermanni — Flycatcher, Little Pied
Ficus bengalensis — Banyan Tree
Ficus benjamina — Fig, Weeping
Ficus carica — Fig, Common
Ficus deltoides — Fig, Mistletoe
Ficus elastica — India Rubber Plant
Ficus lyrata — Fig, Banjo or Fiddle-leaved
Ficus pumila — Fig, Creeping or Climbing
Ficus religiosa — Peepul Tree or Bo Tree
Ficus sycomorus — Fig, Sycamore
Filago sp. — Cudweed
Filipendula sp. — Meadowsweet
Filipendula ulmaria — Queen-of-the-Meadow
Filipendula vulgaris — Dropwort
Finschia novaeseelandiae — Creeper, Brown
Firmiana simplex — Chinese Parasol Tree
Fissidentalium sp. — Tusk Shell
Fistulina hepatica — Fungus, Beefsteak
Fittonia sp. — Nerve Plant or Mosaic Plant
Flabelligera sp. — Worm
Flammulina velutipes — Mushroom, Winter
Flavicola pica — Tyrant, Pied Water
Florisuga mellivora — Hummingbird, White-necked Jacobin
Fodiator acutus — Flying-fish, Sharpchin
Foeniculum vulgare — Fennel, Common
Fomes fomentarius — Rot, Yellowish Sapwood
Fontinalis sp. — Moss, Water

Forcipiger longirostris — Butterflyfish, Long-nosed

Forficula auricularia — Earwig, Common

Formica rufa — Ant, Wood

Formicaleon tetragrammicus — Ant-lion

Formicarius analis — Ant-thrush

Formicarius rufipectus — Ant-thrush

Formicivora grisea — Antwren

Forpus sp. — Parrotlet

Forsythia sp. — Forsythia or Golden Bells

Fortunella japonica — Kumquat, Marumi

Fortunella margarita — Kumquat, Nagami

Fossa fossa — Civet, Malagasy

Fothergilla sp. — Witch-alder

Foudia madagascariensis — Fody

Foudia sechellarum — Fody

Francolinus coqui — Francolin, Coqui

Francolinus francolinus — Francolin, Black

Francolinus pintadeanus — Francolin, Chinese

Francolinus swainsoni — Francolin, Swainson's

Fragaria moschata — Strawberry, Hautbois

Fragaria vesca — Strawberry, Wild or Wood

Frailea sp. — Frailea

Frangula alnus — Buckthorn, Alder

Frankenia laevis — Sea Heath

Franklinia alatamaha — Franklinia

Fratercula arctica — Puffin, Atlantic or Common

Fraxinus excelsior — Ash, European or Common

Fraxinus excelsior pendula — Ash, Weeping

Fraxinus holotricha — Ash, Moraine

Fraxinus ornus — Ash, Flowering or Manna

Freesia sp. — Freesia

Fregata aquila — Frigatebird, Ascension

Fregata ariel — Frigatebird, Lesser

Fregata magnificens — Frigatebird, Magnificent

Fregata minor — Frigatebird, Greater

Fregetta grallaria — Petrel, White-bellied Storm

Fregetta tropica — Petrel, Black-bellied Storm

Fremontia californica — Flannel Bush

Freycinetia sp. — Kiekie

Fringilla coelebs — Chaffinch

Fringilla montifringilla — Brambling

Frithia pulchra — Frithia

Fritillaria acmopetala — Fritillary

Fritillaria imperialis — Crown Imperial

Fritillaria meleagris — Fritillary, Snake's Head

Fritillaria pallidiflora — Fritillary

Frullania sp. — Frullania

Fuchsia magellanica — Fuchsia, Magellan

Fuchsia sp. — Fuchsia

Fucus serratus — Wrack, Serrated

Fucus sp. — Wrack

Fulgora sp. — Lanternfly

Fulica atra — Coot, Black or Common

Fulica cristata — Coot, Crested or Red-knobbed

Fulmarus glacialis — Fulmar

Fulmarus glacialoides — Fulmar, Antarctic

Fumana procumbens — Heath-rose

Fumaria capreolata — Fumitory, Ramping

Fumaria densiflora — Fumitory, Dense-flowered

Fumaria muralis — Fumitory, Wall

Fumaria officinalis — Fumitory, Common

Fumaria parviflora — Fumitory, Small

Funambulus palmarum — Squirrel, Palm

Funaria hygrometrica — Moss, Cord

Funisciurus sp. — Squirrel, Striped

Furnarius leucopus — Hornero

Furnarius rufus — Hornero

Fusinus sp. — Spindle Shell

G

Gadus callarias — Cod, Atlantic

Gadus merlangus — Whiting

Gadus morhua — Cod, Atlantic

Gagea lutea — Yellow-star-of-Bethlehem

Gagea sp. — Yellow-star-of-Bethlehem

Gaillardia aristata — Blanket-flower

Gaillardia pulchella — Indian Blanket

Galactinia vesiculosa — Cup-fungus, Early

Galactites tomentosa — Galactites

Galago alleni — Galago, Allen's

Galago crassicaudatus — Galago, Thick- or Bush-tailed

Galago demidovii — Galago, Demidoff's or Dwarf

Galago elegantulus — Galago, Needle-clawed

Galago inustus — Galago, Needle-clawed

Galago senegalensis — Galago, Common or
  Senegal
Galanthus elwesii — Snowdrop, Giant
Galanthus nivalis — Snowdrop
Galathea sp. — Crab, Rock
Galax aphylla — Galax
Galbalcyrhynchus leucotis — Jacamar
Galbula dea — Jacamar, Paradise
Galbula ruficauda — Jacamar, Rufous-tailed
Galeatus maculatus — Bug, Lace
Galega officinalis — Goat's Rue
Galemys pyrenaicus — Desman, Pyrenean
Galeobdolon luteum — Archangel, Yellow
Galeocerdo cuvieri — Shark, Tiger
Galeodes sp. — Spider, Barrel
Galeopithecus volans — Colugo
Galeopsis angustifolia — Hemp-nettle,
  Narrow-leaved
Galeopsis segetum — Hemp-nettle, Downy
Galeopsis speciosa — Hemp-nettle,
  Large-flowered
Galeopsis tetrahit — Hemp-nettle, Common
Galeorhinus galeus — Tope
Galerida cristata — Lark, Crested
Galerida magnirostris — Lark, Thick-billed
Galerida modesta — Lark, Crested
Galerida theklae — Lark, Thekla
Galerina sp. — Galerina
Galeus melastomus — Dogfish, Black-
  mouthed
Galinsoga ciliata — Shaggy Soldier
Galinsoga parviflora — Gallant Soldier
Galium aparine — Cleavers
Galium boreale — Bedstraw, Northern
Galium cruciata — Crosswort
Galium mollugo — Bedstraw, Hedge or
  Smooth
Galium odoratum — Woodruff, Sweet
Galium palustre — Bedstraw, Marsh
Galium verum — Bedstraw, Lady's or Yellow
Galleria mellonella — Moth, Honeycomb
Gallicolumba luzonica — Pigeon, Bleeding
  Heart
Gallicrex cinerea — Watercock
Gallictis cuja — Grison, Lesser or Little
Galictis vittata — Grison
Gallinago gallinago — Snipe
Gallinago media — Snipe, Great
Gallinago megala — Snipe, Pintail

Gallinago nigripennis — Snipe, Ethiopian
Gallinago stenura — Snipe, Pintail
Gallinago stricklandii — Snipe
Gallinula chloropus — Gallinule, Common
Gallinula tenebrosa — Gallinule, Dusky
Gallirallus australis — Rail, Weka
Gallus gallus — Fowl, Red Jungle
Gallus sonneratii — Fowl, Sonnerat's Jungle
Gallus varius — Fowl, Green Jungle
Galtonia candicans — Hyacinth, Giant
  Summer
Gammarus locusta — Shrimp, Freshwater
Gammarus palex — Shrimp, Freshwater
Gamolepis sp. — Daisy, Paris or Sunshine
Gampsonyx swainsonii — Kite, Pearl
Gampsorhynchus rufulus — Babbler,
  White-hooded
Garcinia mangostana — Mangosteen
Gardenia jasminoides— Gardenia, Common
  or Cape Jasmine
Gardenia sp. — Gardenia
Garrodia nereis — Petrel, Gray-backed Storm
Garrulax sp. — Thrush, Laughing
Garrulous glandarius — Jay, Common or
  Eurasian
Garrya elliptica — Silk Tassel
Gasteracantha sp. — Spider
Gasteria liliputana — Gasteria, Lilliput
Gasteria maculata — Gasteria, Spotted
Gasteria sp. — Gasteria
Gasteria verrucosa — Ox-tongue
Gasteropelecus levis — Hatchetfish
Gasteropelecus sternicla — Hatchetfish
Gasterophilus intestinalis — Botfly, Horse
Gasterosteus aculeatus — Stickleback,
  Three-spined
Gastromyzon borneensis — Gastromyzon
Gastropacha quercifolia — Moth, Lappet
Gastrotheca sp. — Frog, Marsupial or Pouched
Gaterin sp. — Sweetlips
Gavia adamsii — Loon, White-billed
Gavia arctica — Loon, Black-throated or
  Arctic
Gavia immer — Loon, Common or Great
  Northern
Gavia stellata — Loon, Red-throated
Gavialis gangeticus — Gavial
Gazania sp. — Treasure Flower
Gazella arabica — Gazelle, Arabian

Gazella dama — Gazelle, Dama
Gazella dorcas — Gazelle, Dorcas
Gazella gazella — Gazelle, Mountain
Gazella granti — Gazelle, Grant's
Gazella gutturosa — Gazelle, Mongolian
Gazella leptoceros — Gazelle, Slender-horned
Gazella pelzelni — Gazelle, Pelzeln's
Gazella rufifrons — Gazelle, Red-fronted
Gazella soemmeringi — Gazelle,
    Soemmering's
Gazella spekei — Gazelle, Speke's
Gazella subgutturosa — Gazelle, Goitered or
    Persian
Gazella thomsoni — Gazelle, Thomson's
Geastrum sp. — Earth Star
Gekko gecko — Gecko, Great House
Gekko lineatus — Gecko, Great House
Gelochelidon nilotica — Tern, Gull-billed
Gelsemium sempervirens — Jessamine,
    Carolina Yellow
Genetta abyssinica — Genet, Abyssinian
Genetta genetta — Genet, Common or
    Small-spotted
Genetta servalina — Genet, Common or
    Small-spotted
Genetta tigrina — Genet, Large-spotted
Genetta victoriae — Genet, Giant
Genista aetnensis — Broom, Mt. Etna
Genista anglica — Petty Whin
Genista cinera — Woadwaxen, Ashy
Genista germanica — Greenweed, German
Genista hispanica — Gorse, Spanish
Genista lydia — Broom
Genista pilosa — Greenweed, Hairy
Genista radiata — Broom, Royal
Genista sagittalis — Broom, Winged
Genista tinctoria — Greenweed, Dyer's
Gennaeus nycthemerus — Pheasant, Silver
Gennaeus swinhoii — Pheasant, Swinhoe's
Gentiana acaulis — Gentian, Stemless or
    Trumpet
Gentiana andrewsi — Gentian, Closed or
    Bottle
Gentiana asclepiadea — Gentian, Willow
Gentiana clusii — Gentian, Stemless Trumpet
Gentiana corymbifera — Gentian, Snow
Gentiana crinata — Gentian, Fringed
Gentiana cruciata — Gentian, Cross
Gentiana kochiana — Gentian, Trumpet

Gentiana lutea — Gentian, Great Yellow
Gentiana nivalis — Gentian, Small Alpine
Gentiana pneumonanthe — Gentian, Marsh
Gentiana punctata — Gentian, Spotted
Gentiana purpurea — Gentian, Purple
Gentiana sino-ornata — Gentian (China)
Gentiana sp. — Gentian
Gentiana utriculosa — Gentian, Bladder
Gentiana verna — Gentian, Spring
Gentianella amarella — Gentian, Autumn
Gentianella campestris — Gentian, Field
Gentianella germanica — Gentian, Chiltern
Gentianella tenella — Gentian, Frined
Geocolaptes olivaceus — Woodpecker, Ground
Geoemyda spinosa — Terrapin, Spiny
Geogale aurita — Tenrec
Geogenanthus undatus — Seersucker Plant
Geoglossum sp. — Earth-tongue
Geometra papilionaria — Moth, Large
    Emerald
Geopelia cuneata — Dove, Diamond
Geopelia humeralis — Dove, Bar-shouldered
Geopelia striata — Dove, Barred Ground
Geophagus sp. — Cichlid
Geophaps scripta — Pigeon, Squatter
Geophaps smithi — Pigeon, Partridge
Geophilus sp. — Centipede, Luminous
Geopsittacus occidentalis — Parrot, Night
Georychus capensis — Mole-rat, Cape
Geospiza conirostris — Finch, Cactus Ground
Geospiza fortis — Finch, Medium Ground
Geospiza magnirostris — Finch, Large
    Ground
Geospiza scandens — Finch, Cactus Ground
Geotrupes sp. — Beetle, Forest Dung
Geotrygon montana — Dove, Quail
Geotrygon saphirina — Dove, Quail
Geotrygon veraguensis — Dove, Quail
Geranium columbianum — Cranesbill,
    Long-stalked
Geranium dissectum — Cranesbill,
    Cut-leaved
Geranium endressi — Cranesbill, French
Geranium lucidum — Cranesbill, Shining
Geranium macrorrhizum — Cranesbill,
    Italian or Rock
Geranium maculatum — Geranium, Spotted
    or Wild
Geranium molle — Cranesbill, Dove's-foot

Geranium nodosum — Cranesbill, Broad-leaved or Knotted

Geranium palustre — Cranesbill, Marsh

Geranium phaeum — Cranesbill, Dusky or Mourning Widow

Geranium pratense — Cranesbill, Meadow

Geranium purpureum — Herb Robert, Lesser or Little Robin

Geranium pusillum — Cranesbill, Small-flowered

Geranium pyrenaicum — Cranesbill, Mountain or Hedgerow

Geranium robertianum — Herb Robert

Geranium rotundifolium — Cranesbill, Round-leaved

Geranium sanguineum — Cranesbill, Bloody or Blood-red

Geranium sp. — Cranesbill

Geranium sylvaticum — Cranesbill, Wood

Geranium tuberosum — Cranesbill, Tuberous

Geranium versicolor — Cranesbill, Pencilled or Streaked

Geranoaetus melanoleucus — Buzzard, Gray Eagle

Geranospiza caerulescens — Hawk, Crane

Gerbera jamesonii — Daisy, Barberton or Transvaal

Gerbillus sp. — Gerbil

Geronticus calvus — Ibis, Bald or Hermit

Geronticus eremita — Ibis, Bald or Hermit

Gerrhosaurus validus — Lizard, Plated Rock

Gerris sp. — Water Strider

Gerygone chloronota — Warbler, Green-backed

Gerygone fusca — Warbler, Western or White-tailed

Gerygone igata — Warbler, Gray

Gerygone magnirostris — Warbler, Large-billed

Gerygone olivacea — Warbler, White-throated

Gerygone palpebrosa — Warbler, Black-throated

Gerygone sulphurea — Warbler, Yellow-breasted Wren

Gerygone tenebrosa — Warbler, Dusky

Gesneria sp. — Gesneria

Geum borsii — Avens

Geum chiloense — Avens, Scarlet

Geum coccineum — Avens, Scarlet

Geum montanum — Avens, Mountain

Geum reptans — Avens, Creeping

Geum rivale — Avens, Water

Geum urbanum — Avens, Common Wood or Herb Bennet

Gibbium sp. — Beetle, Spider

Gibbula sp. — Top Shell

Gigantorana goliath — Frog, Goliath

Gilia capitata or rubra — Gilia, Blue

Gingko bilboa — Maidenhair Tree

Ginglymostoma cirratum — Shark, Nurse

Giraffa camelopardalis — Giraffe

Giraffa camelopardalis reticulated — Giraffe, Reticulated

Giraffa camelopardalis tippelskirchi — Giraffe, Masai

Gladiolus byzantinus — Gladiolus, Eastern

Gladiolus segetum — Gladiolus, Field

Gladiolus sp. — Gladiolus

Glareola maldivarum — Praticole, Oriental

Glareola nordmanni — Pratincole, Black-winged

Glareola pratincola — Pratincole, European

Glaucidium brasilianum — Owl, Ferruginous Pygmy

Glaucidium brodiei — Owlet, Collared

Glaucidium cuculoides — Owlet, Asian Barred or Cuckoo

Glaucidium minutissimum — Owl, Pygmy

Glaucidium passerinum — Owl, Pygmy

Glaucidium perlatum — Owlet, Pearl-spotted

Glaucis hirsuta — Hummingbird, Hermit

Glaucium corniculatum — Horned-poppy, Red

Glaucium flavum — Horned-poppy, Yellow

Glaucus sp. — Nudibranch

Glaux maritima — Milkwort, Sea

Glechoma hederacea — Ground Ivy

Gleditsia sp. — Locust, Honey

Gleditsia triacanthos — Locust, Honey

Gleditsia triacanthos inermis — Locust, Thornless Honey

Gliciphila albifrons — Honey-eater, White-fronted

Gliciphila indistincta — Honey-eater, Brown

Gliciphila melanops — Honey-eater, Tawny-crowned

Glis glis — Dormouse, Edible or Fat

Globicephala melaena — Whale, Pilot

Globularia alypum — Globularia, Shrubby
Globularia cordifolia — Globe-daisy,
    Heart-leaved
Globularia elongata — Globe-daisy, Common
Globularia vulgaris — Globularia, Common
Gloeophyllum saepiarum — Gloeophyllum
Gloeoporus adustus — Gloeoporus
Gloeoporus fumosus — Gloeoporus
Glomeris sp. — Millipede, Pill
Glorisa rothschildiana — Lily, Glory
Glorisa superba — Lily, Glory or Climbing
Glossina sp. — Fly, Tsetse
Glossiphonia sp. — Leech, Snail
Glossobalanus sp. — Worm, Acorn
Glossodoris sp. — Nudibranch
Glossopsitta concinna — Lorikeet, Musk
Glossopsitta porphyrocephala — Lorikeet,
    Purple-crowned
Glossopsitta pusilla — Lorikeet, Little
Glottiphyllum sp. — Tongue-leaf
Gloxinera hybrids — Gloxinera
Gloxinia perennis — Gloxinia
Glyceria sp. — Sweet-grass
Glycine max — Soybean or Soya
Glycine soya — Soybean or Soya
Glycymeris sp. — Cockle, Dog
Glycyrrhiza glabra — Liquorice
Glyphorhynchus spirurus — Woodcreeper,
    Wedge-billed
Glyptocephalus cynocephalus — Witch
Glyptocephalus cynoglossus — Witch
Gmelina sp. — Snapdragon Tree
Gnaphalium luteoalbum — Cudweed, Jersey
Gnaphalium sp. — Cudweed
Gnaphalium supinum — Cudweed, Dwarf or
    Creeping
Gnaphalium sylvaticum — Cudweed, Wood or
    Heath
Gnaphalium uliginosum — Cudweed,
    Wayside or Marsh
Gnathodon speciosus — Jack, Yellow
Gnathonemus petersi — Mormyrid, Peter's or
    Ubangi
Gnathonemus sp. — Mormyrid, Elephant-
    snout
Gobio gobio — Gudgeon
Gobius minutus — Goby, Sand
Gobius niger — Goby, Black
Gobius paganellus — Goby, Black

Godetia grandiflora — Satin Flower
Goliathus sp. — Beetle, Goliath
Gomphidius roseus — Pink Nail
Gomphidius sp. — Gomphidius
Gompholobium sp. — Pea, Large Wedge
Gomphosus varius — Wrasse, Olive Club-
    nosed
Gomphus vulgatissimus — Dragonfly,
    Club-tailed
Gonepteryx cleopatra — Butterfly, Brimstone
Gonepteryx rhamni — Butterfly, Brimstone
Gonyocephalus sp. — Lizard, Angle-headed
Goodyera repens — Lady's Tresses, Creeping
Gordius sp. — Worm, Horsehair
Gordonia alatamaha — Franklinia
Gorgonocephalus sp. — Basket-star
Gorilla gorilla — Gorilla
Gorsachius goisagi — Bittern, Tiger
Gorsachius melanolophus — Bittern, Tiger
Gossypium herbaceum — Cotton, Levant
Gossypium hirsutum — Cotton, Upland
Gossypium sp. — Cotton
Goura cristata — Pigeon, Crowned
Goura victoria — Pigeon, Victoria Crowned
Gracula religiosa — Mynah, Hill or Talking
Graellsia isabellae — Moth, Silk
Grallaria perspicillata — Antpitta,
    Streak-chested
Grallaria squamigera — Antpitta,
    Streak-chested
Grallaricula ferrugineipectus — Antpitta
Grallaricula flavirostris — Antpitta
Grallaricula nana — Antpitta
Grallina cyanoleuca — Magpie-lark
Gramma loreto — Basslet, Fairy
Grammistes sexlineatus — Grouper, Six-lined
Grampus orca — Whale, Killer
Granatellus pelzelni — Chat, Rose-breasted
Grandala coelicolor — Grandala
Grantia sp. — Sponge, Purse
Grantiella picta — Honey-eater, Painted
Graphium sp. — Butterfly, Swallow-tail
Graphosoma italicum — Bug, Striped
Graphosoma lineatum — Bug, Striped
Graptopetalum sp. — Graptopetalum
Gratiola officinalis — Hedge-hyssop
Gravenhorstia sp. — Wasp
Greenovia aurea — Greenovia
Grevillea punicea — Red Spider Flower

Grevillea robusta — Silk-oak

Grevillea sp. — Grevillea

Grewia occidentalis — Starflower

Greyia sutherlandii — Bottlebrush

Grifolia frondosa — Hen of the Woods

Grimmia sp. — Grimmia

Grindelia robusta — Grindelia

Griselinia littoralis — Broadleaf

Groenlandia densa — Pondweed, Opposite-
leaved

Grus antigone — Crane, Sarus

Grus grus — Crane, Common

Grus japonensis — Crane, Japanese or
Manchurian

Grus leucogeranus — Crane, Siberian White

Grus monacha — Crane, Hooded

Grus rubicunda — Crane, Australia or Brolga

Grus vipio — Crane, White-necked

Gryllotalpa sp. — Cricket, Mole

Gryllus sp. — Cricket, Field

Guaiacum officinale — Lignum Vitae

Guara ruber — Ibis, Scarlet

Gubernatrix cristata — Cardinal, Yellow

Guira guira — Cuckoo, Guira

Gulo gulo — Wolverine

Gunnera sp. — Gunnera

Gustavia sp. — Membrillo

Guzmania sp. — Guzmania

Gygis alba — Tern, Fairy or White

Gymnadenia conopsea — Orchid, Fragrant

Gymnammodytes semisquamatus —
Sand-eel, Smooth

Gymnarchus niloticus — Mormyrid

Gymnobelideus leadbeateri — Possum,
Leadbeater's

Gymnocalycium bruchii — Cactus, Bruch's
Chin

Gymnocalycium mihanovichii — Cactus,
Plaid

Gymnocalycium quehlianum — Cactus, Plaid

Gymnocalycium sp. — Cactus

Gymnocephalus cernua — Ruffe

Gymnocephalus schraetzer — Ruffe

Gymnocichla nudiceps — Antbird

Gymnocorymbus ternetzi — Black Widow

Gymnodactylus sp. — Gecko, Barking

Gymnogenys typicus — Hawk, Harrier

Gymnomystax mexicanus — Blackbird,
Oriole

Gymnopilus sp. — Gymnopilus

Gymnopithys sp. — Antbird

Gymnorhina dorsalis — Magpie, Western

Gymnorhina hypoleuca — Magpie,
White-backed

Gymnorhina tibicen — Magpie, Black-backed

Gymnostinops montezuma — Oropendola,
Montezuma

Gymnotus carapo — Knifefish, Banded

Gynura aurantica — Velvet Plant, Purple

Gynura sarmentosa — Velvet Plant, Purple

Gypaetus barbatus — Vulture, Bearded

Gypohierax angolensis — Vulture, Palm-nut

Gyps africanus — Vulture, African
White-backed

Gyps bengalensis — Vulture, Indian
White-backed

Gyps coprotheres — Vulture, Cape

Gyps fulvous — Vulture, Griffon

Gyps ruppellii — Vulture, Ruppell's

Gypsophila elegans — Baby's-breath, Annual

Gypsophila fastigiata — Baby's-breath

Gypsophila muralis — Baby's-breath, Wall

Gypsophila paniculata — Baby's-breath

Gypsophila repens — Baby's-breath, Creeping

Gyrinocheilus aymonieri — Loach, Sucker

Gyrinus sp.— Beetle, Whirligig

Gyromitra esculenta — Morel, False

Gyroporus cyanescens — Boletus, Indigo

H

Haberlea sp. — Haberlea

Habia rubica — Ant-tanager, Red-crowned

Habrosyne pyritoides — Moth, Buff Arches

Hacquetica epipactis — Hacquetia

Haemanthus sp. — Lily, Blood

Haemaria discolor — Orchid, Jewel

Haematomma sp. — Lichen, Haematomma

Haematopinus suis — Louse, Hog

Haematopota pluvialis — Fly, Horse

Haematopus ater — Oystercatcher, Black

Haematopus fuliginosus — Oystercatcher,
Sooty

Haematopus moquini — Oystercatcher, Black

Haematopus ostralegus — Oystercatcher,
European or Pied

Haematopus unicolor — Oystercatcher, Sooty

Haematospiza sipahi — Finch, Scarlet

361

Haemopis sanguisuga — Leech, Horse
Hagedashia hagedash — Ibis, Hadeda
Hakea sp. — Cushionflower or Corkwood
Halcyon albiventris — Kingfisher, Brown-hooded
Halcyon chloris — Kingfisher, White-collared
Halcyon coromanda — Kingfisher, Ruddy
Halcyon fulgida — Kingfisher, Blue
Halcyon leucocephala — Kingfisher, Gray-headed
Halcyon macleayi — Kingfisher, Forest
Halcyon malimbica — Kingfisher, Blue
Halcyon pileata — Kingfisher, Black-capped
Halcyon pyrrhopygia — Kingfisher, Red-backed
Halcyon sancta — Kingfisher, Sacred
Halcyon senegalensis — Kingfisher, Angola or Woodland
Halcyon smyrnensis — Kingfisher, White-breasted or -throated
Haliaeetus albicilla — Eagle, White-tailed or Gray Sea
Haliaeetus leucogaster — Eagle, White-bellied Sea
Haliaeetus leucoryphus — Eagle, Pallas' Sea
Haliaeetus morphnoides — Eagle, Australian Little
Haliaeetus pelagicus — Eagle, Steller's Sea
Haliaeetus vocifer — Eagle, African Fish or Sea
Haliastur indus — Kite, Brahminy
Haliastur sphenurus— Kite, Whistling
Halichoerus grypus — Seal, Gray
Halichondria sp. — Sponge, Bread-crumb
Haliclystus sp. — Jellyfish, Stalked
Halicore australe — Dugong
Halictus sp. — Bee, Mining
Halimione pedunculata — Purslane, Sea
Halimione portulacoides — Purslane, Sea
Halimium sp. — Halimium
Haliotis sp. — Albalone
Halobaena caerulea — Petrel, Blue
Halobates sp. — Water Strider
Halocynthia papillosa — Sea Squirt, Red
Haloragis sp. — Toatoa
Hamamelis intermedia — Witch Hazel
Hamamelis japonica — Witch Hazel, Japanese
Hamamelis mollis — Witch Hazel, Chinese
Hamamelis virginiana — Witch Hazel

Hamatocactus setispinus — Cactus, Strawberry
Hamearis lucina — Butterfly, Duke of Burgundy Fritillary
Haminea sp. — Bubble Shell
Hamirostra melanosterna — Buzzard, Black-breasted
Hammarbya paludosa — Orchid, Bog
Hapale sp. — Marmoset
Hapalemur griseus — Lemur, Gray Gentle
Haplochromis sp. — Mouthbrooder
Harpa sp. — Harp Shell
Harpactes ardens — Trogon, Philippine
Harpactes fasciatus — Trogon, Indian or Malabar
Harpactes oreskios — Trogon, Orange-breasted
Harpagophytum sp. — Grapple Plant
Harpia harpyja — Eagle, Harpy
Harpyia bifida — Moth, Kitten
Harpyia furcula — Moth, Kitten
Harpyopsis novaeguineae — Eagle, New Guinea Harpy
Harrisia sp. — Harrisia
Hasarius adansoni — Spider, Jumping
Hatiora salicorniodes — Dancing Bones
Haworthia sp. — Haworthia
Hebe sp. — Hebe
Hebe speciosa — Hebe, Showy
Hebeloma crustuliniforme — Mushroom, Fairy Cake
Hedera canariensis — Ivy, Canary Island
Hedera colchica — Ivy, Persian
Hedera helix — Ivy, Common English or Evergreen
Hedycarya arborea — Pigeonwood
Hedychium coronarium — Ginger-lilly, White
Hedychium flavum or H. flavescens — Ginger-lily
Hedychium gardnerianum — Ginger-lily, Yellow
Hedychrum nobile — Wasp
Hedysarum coronarium — Honeysuckle, French
Hedysarum hedysaroides — Sainfoin, Alpine
Hedysarum sp. — Hedysarum
Helarctos malayanus — Bear, Sun
Helenium autumnale — Sneezeweed
Helenium sp. — Sneezeweed

Heliactin cornuta — Hummingbird, Sun Gem

Heliangelus amethysticollis — Hummingbird, Sun-angel

Heliangelus exortis — Hummingbird, Sun-angel

Helianthemum apenninum— Rockrose, White

Helianthemum canum — Rockrose, Hoary

Helianthemum chamaecistus — Rockrose, Common

Helianthemum nummularium — Rockrose, Common

Helianthemum sp. — Rockrose

Helianthus annus — Sunflower, Common

Helianthus decapetalus — Sunflower, Thin-leaved

Helianthus sp. — Sunflower

Helianthus tuberous — Artichoke, Jerusalem

Helicella sp. — Snail

Helichrysum arenarium — Everlasting, Yellow

Helichrysum bracteatum — Everlasting or Straw Flower

Helichrysum italicum — Everlasting, Italian

Helichrysum milfordiae — Everlasting

Heliconia bihai — Plantain, Wild

Heliconia collinsiana — Heliconia, Hanging

Heliconia humilis — Lobster Claw

Heliconia psittacorum — Parrot Flower

Heliconia wagneriana — Plantain, Wild

Heliconius sp. — Butterfly

Helictotrichon sp. — Oat-grass

Heliocopris sp. — Beetle, Dung

Heliodoxa sp. — Hummingbird, Brilliant

Heliomaster sp. — Hummingbird, Star-throat

Heliopais personata — Finfoot, Masked

Heliophila sp. — Heliophila

Heliopsis sp. — Sunflower, Orange

Heliornis fulica — Sungrebe

Heliosciurus sp. — Squirrel, Sun

Heliothryx aurita — Hummingbird, Fairy

Heliothryx barroti — Hummingbird, Fairy

Heliotropium arborescens — Heliotrope, Common

Heliotropium europaeum — Heliotrope

Heliotropium peruvianum — Heliotrope

Helipterum manglesii — Everlasting, Swan River

Helipterum roseum — Daisy, Everlasting

Helipterum sp. — Everlasting

Helix pomatia — Snail, Edible or Roman

Hellaetin cornuta — Hummingbird, Sun Gem

Helleborus cyclophyllus — Hellebore, Green or Bear's Foot

Helleborus foetidus — Hellebore, Stinking or Bear's Foot

Helleborus niger — Christmas Rose

Helleborus orientalis — Lenten Rose

Helleborus sp. — Hellebore

Helleborus viridis — Hellebore, Green

Helogale parvula — Mongoose, Dwarf

Helophilus sp. — Fly, Sun

Helostoma temmincki — Gourami, Kissing

Helxine soleirolii — Baby's Tears

Hemachatus haemachatus — Cobra, Ring-necked Spitting

Hemaris fuciformis — Hawkmoth, Bee

Hemaris tityus — Hawkmoth, Bee

Hemerobius sp. — Lacewing, Brown

Hemerocallis sp. — Day Lily

Hemerodromus africanus — Courser

Hemicentetes semispinosus — Tenrec, Striped

Hemichromis bimaculatus — Jewelfish

Hemicircus sp. — Woodpecker, Spotted

Hemidactylus turcicus — Gecko, Turkish

Hemiechinus auritus — Hedgehog, Long-eared Desert

Hemigalus derbyanus — Palm-civet

Hemigrammus caudovitatus — Tetra, Buenos Aires

Hemigrammus erythrozonus — Tetra, Glow-light

Hemigrammus ocellifer — Head-and-Tail-Light Fish

Hemigrammus pulcher — Tetra, Pretty

Hemigrammus rhodostomus — Tetra, Red

Hemiodus sp. — Characin

Hemiphaga novaeseelandiae — Pigeon, New Zealand

Hemiprocne comata — Swift, Tree

Hemiprocne longipennis — Swift, Tree

Hemiprocne mystacea — Swift, Tree

Hemirhamphus sp. — Halfbeak

Hemispingus atropileus — Hemispingus

Hemispingus verticalis — Hemispingus

Hemisus marmoratus — Frog, Pig-nosed

363

Hemitragus hylocrius — Tahr, Nilgiri
Hemitragus jemlahicus — Tahr, Himalayan
Heniochus sp. — Angelfish or Bullfish
Henricia sp. — Starfish, Scarlet
Heodes virgaureae — Butterfly, Scarce
  Copper
Hepatica nobilis — Hepatica, European
Hepsetus odoe — Characin, Pike
Heracleum mantegazzianum — Hogweed,
  Giant
Heracleum maxima — Cow-parsnip
Heracleum sphondylium — Hogweed
Herminium monorchis — Orchid, Musk
Hermodactylus tuberosus — Iris, Snake's
  Head
Herniaria sp. — Rupture-wort
Herpailurus yagouaroundi — Cat, Jaguar-
  undi or Otter
Herpestes edwardsii — Mongoose, Gray
Herpestes gracilis — Mongoose, Slender
Herpestes ichneumon — Mongoose, Egyptian
Herpestes pulverulentus — Mongoose, Gray
Herpestes sanguineus — Mongoose, Slender
Herpestes urva — Mongoose, Crab-eating
Herpeton tentaculatum — Snake, Fishing
Herpetotheres cachinnans — Falcon,
  Laughing
Herschelia sp. — Herschelia
Herse convolvuli — Hawkmoth, Convolvulus
Hesperis matronalis — Dame's-rocket or
  Dame's-violet
Heterocentrotus mammillatus — Sea Urchin
Heterocephalus sp. — Mole-rat, Naked
Heterodontus sp. — Shark, Horn
Heterohyrax brucei — Hyrax, Rock
Heterohyrax syriacus — Hyrax, Rock
Heteromunia pectoralis — Finch, Pictorella
Heteromyias cinereifrons — Robin,
  Gray-headed
Heteronetta atricapilla — Duck, Black-
  headed
Heteronotus sp. — Treehopper
Heteropelma sp. — Wasp, Ichneumon
Heterophasia melanoleuca — Sibia, Black-
  headed
Heteropneustes fossilis — Catfish, Stinging
Heterospiza meridionalis — Hawk, Savannah
Hexabranchus sp. — Nudibranch
Hexanchus griseus — Shark, Six-gilled

Hiatella arctica — Clam, Red-nose
Hibbertia scandens — Snake Vine
Hibiscus rosa-sinensis — Hibiscus
Hibiscus schizopetalus — Hibiscus, Fringed
Hibiscus sp. — Rose-mallow
Hibiscus syriacus — Rose of Sharon
Hibiscus trionum — Ketmia, Bladder or
  Flower-of-an-hour
Hieracium aurantiacum — Hawkweed,
  Orange
Hieracium brittannicum — Hawkweed
Hieracium brunneocroceum — Hawkweed,
  Brownish-orange
Hieracium pilosella — Hawkweed, Mouse-ear
Hieracium umbellatum — Hawkweed, Leafy
Hieracium villosum — Hawkweed, Woolly or
  Shaggy
Hieraetus fasciatus — Eagle, Bonelli's
Hieraetus kienerii — Eagle, Rufous-bellied
Hieraetus morphnoides — Eagle, Australian
  Little
Hieraetus pennatus — Eagle, Booted
Hierochloe odorata — Holy-grass
Hierophasis swinhoii — Pheasant, Swinhoe's
Hilara maura — Fly, Dance or Empid
Himantoglossum hircinum — Orchid, Lizard
Himantoglossum longibracteatum — Orchid,
  Giant
Himantopus himantopus — Stilt, Black-
  winged or Common
Himantopus leucocephalus — Stilt, Long-
  legged or Pied
Hipparchia sp. — Butterfly, Grayling or
  Wood-nymph
Hippeastrum equestre — Lily, Barbados
Hippeastrum hybrids — Amaryllis
Hippobosca equina — Fly, Forest
Hippocampus sp. — Seahorse
Hippocrepis comosa — Vetch, Horseshoe
Hippodamia sp. — Beetle, Ladybird
Hippoglossus hippoglossus — Halibut
Hippolais caligata — Warbler, Booted
Hippolais icterina — Warbler, Icterine
Hippolais olivetorum — Warbler, Olive-tree
Hippolais pallida — Warbler, Olivaceous
Hippolais polyglotta — Warbler, Melodius
Hippophae rhamnoides — Buckthorn, Sea
Hippopotamus amphibius — Hippopotamus
Hippopus sp. — Clam, Giant or Horse's Hoof

Hippotion celerio — Hawkmoth, Silver-
striped or Vine
Hippotragus equinus — Antelope, Roan
Hippotragus niger — Antelope, Sable
Hippuris vulgaris — Mare's-tail
Hirschfieldia incana — Mustard, Hoary
Hirudo medicinalis — Leech, Medicinal
Hirundapus sp. — Swift, Spine-tailed
Hirundo albigularis — Swallow, White-
throated
Hirundo abyssinica — Swallow, Striped
Hirundo concolor — Martin, Crag
Hirundo cucullata — Swallow, Striped
Hirundo daurica — Swallow, Red-rumped
Hirundo obsoleta — Martin, Crag
Hirundo rupestris — Martin, Crag
Hirundo rustica — Swallow, Barn
Hirundo senegalensis — Swallow, Mosque
Hirundo smithii — Swallow, Wire-tailed
Hirundo striolata — Swallow, Mosque
Hirundo tahitica — Swallow, Pacific
Hirundo tahitica neoxena — Swallow,
Welcome
Hister cadaverinus — Beetle, Hister
Histrio histrio — Sargassofish
Histrionicus histrionicus — Duck, Harlequin
Histriophaps histrionica — Pigeon, Flock
Hoheria sp. — Lacebark
Holacanthus sp. — Angelfish
Holcus lanatus — Yorkshire Fog
Holcus mollis — Soft-grass, Creeping
Holmskioldia sanguinea — Chinese Hat-plant
Holmskioldia tettensis — Cups-and-saucers
Holocentrus sp. — Squirrelfish or Soldierfish
Holosteum umbellatum — Chickweed, Jagged
or Umbellate
Holothuria sp. — Sea Cucumber
Homalopsis buccata — Snake, Broad-headed
Water
Homarus vulgaris — Lobster, European
Homogyne alpina — Coltsfoot, Alpine
Honkenya peploides — Sandwort, Sea
Hoodia bainii — Hoodia
Hoodia gordonii — Hoodia
Hookerella sp. — Gorgonian
Hoplia coerulea — Beetle, Cerulean Chafer
Hoplocampa testudinea — Sawfly, Apple
Hoplopterus armatus — Plover, Blacksmith
Hoplopterus spinosus — Plover, Spur-winged

Hoplosternum thoracatum — Cascadura
Hordelymus europaeus — Barley, Wood
Hordeum distichon — Barley, Two-rowed
Hordeum jubatum — Grass, Squirrel-tail
Hordeum marinum — Barley, Sea
Hordeum murinum — Barley, Wall
Hordeum secalinum — Barley, Meadow
Hordeum vulgare — Barley, Six-rowed
Horminum pyrenaicum — Dragonmouth
Hornungia petraea — Hutchinsia
Hosta sp. — Lily, Plantain
Hottonia palustris — Water Violet
Houstonia caerulea — Bluets
Houttuynia cordata — Houttuynia
Hovea longifolia — Mountain Beauty
Howea belmoreana — Palm, Curly
Howea forsteriana — Palm, Curly
Hoya bella — Wax Plant, Miniature
Hoya carnosa — Wax Plant or Wax Flower
Hucho hucho — Huchen
Huernia sp. — Huernia
Humea elegans — Incense Plant
Humulus lupulus — Hop, Common
Humulus scandens or H. japonica — Hop,
Japanese
Hunnemannia fumariaefolia — Poppy,
Mexican Tulip
Hutchinsia alpina — Hutchinsia, Alpine
Hyaena brunnea — Hyena, Brown
Hyaena crocuta — Hyena, Spotted
Hyaena hyaena — Hyena, Striped
Hydatina albocincta or H. physis — Bubble
Shell, Paper
Hydnora sp. — Hydnora
Hydrachna globosa — Mite, Round Water
Hydrangea macrophylla — Hydrangea, House
Hydrangea paniculata — Hydrangea, Panicle
or Plumed
Hydrangea petiolaris — Hydrangea,
Climbing
Hydrangea quercifolia — Hydrangea,
Oak-leaved
Hydrangea sp. — Hydrangea
Hydrangea villosa — Hydrangea
Hydrobates pelagicus — Petrel, Storm
Hydrocampa nympheata—Moth, Brown China
Mark
Hydrocharis morsus-ranae — Frog-bit
Hydrochoerus capybara — Capybara
Hydrochoerus hydrochoerus — Capybara

Hydrocleys nymphaeoides — Water Poppy

Hydrocotyle vulgaris — Pennywort, Marsh

Hydromantes geneii — Salamander, Brown Cave

Hydromantes gormani — Salamander, Gorman's

Hydromedusa tectifera — Terrapin, Cope's

Hydrometra stagnorum — Water-measurer

Hydrophasianus chirurgus — Jacana, Pheasant-tailed

Hydrophilus piceus — Beetle, Great Silver Water

Hydrophis sp. — Sea-snake, Banded

Hydropotes inermis — Deer, Water

Hydroprogne caspia — Tern, Caspian

Hydroprogne tschegrava — Tern, Caspian

Hydrosaurus amboinensis — Lizard, Fin-tailed

Hydrosaurus lesueuri — Lizard, LeSueur's Water

Hydrotragus leche — Waterbuck, Lechwe

Hydrous piceus — Beetle, Great Silver Water.

Hydrurga leptonyx — Seal, Leopard

Hyemoschus aquaticus — Chevrotain, Water

Hygrobia hermanni — Beetle, Screech or Squeak

Hygrobia tarda — Beetle, Screech or Squeak

Hygrocybe sp. — Hygrocybe

Hygrophoropsis aurantiaca — Chantarelle, False

Hygrophorus sp. — Hygrophorus

Hyla arborea — Treefrog, European

Hyla caerula — Treefrog, White's Green

Hyla faber — Treefrog, Blacksmith

Hyla infrafrenata — Treefrog, New Guinea Giant

Hyla meridionalis — Treefrog, Mediterranean

Hylacola pyrrhopygia — Heath-wren, Chestnut-tailed

Hylambates maculatus — Frog, Spotted

Hylarana albolaris — Frog, West African

Hylecoetus dermestoides — Beetle, Wood-borer

Hyles euphorbiae — Hawkmoth, Spurge

Hyles galii — Hawkmoth, Bedstraw

Hyles livornica — Hawkmoth, Striped

Hylobates agilis — Gibbon, Dark-handed

Hylobates cinereus — Gibbon, Gray or Silvery

Hylobates hoolock — Gibbon, Hoolock or White-browed

Hylobates lar — Gibbon, White-handed

Hylobates leuciscus — Gray or Silvery

Hylobates moloch — Gibbon, Gray or Silvery

Hylobates pileatus — Gibbon, Capped or Pileated

Hylobates syndactylus — Siamang

Hylobius abietis — Weevil, Large Pine

Hylochelidon ariel — Martin, Fairy

Hylochelidon nigricans — Martin, Tree

Hylochoerus meinertzhageni — Hog, Giant Forest

Hylocomium splendens — Moss, Mountain Fern

Hyloicus pinastri — Hawkmoth, Pine

Hylomis suillus — Moon-rat

Hylopezus perspicillatus — Antpitta, Streak-chested

Hylophilus decurtatus — Greenlet, Gray-headed

Hylophilus ochraceiceps — Greenlet, Tawny-crowned

Hylophylax naevioides — Antbird, Spotted

Hylophylax poecilonata — Antbird, Spotted

Hylotrupes bajulus — Beetle, House Longhorn

Hymenanthera sp. — Porcupine Plant

Hymeniacidon sp. — Sponge

Hymenocallis sp. — Spider-lily

Hymenolaimus malacorhynchos — Duck, Blue or Mountain

Hymenophyllum sp. — Filmy Fern, Tunbridge

Hymenosporum flavum — Sweetshade

Hyoscyamus albus — Henbane, White

Hyoscyamus aureus — Henbane, Golden or Yellow

Hyoscyamus niger — Henbane

Hypecoum sp. — Hypecoum

Hypericum androsaemum — Tutsan

Hypericum calycinum — Rose of Sharon

Hypericum elodes — St. John's Wort, Marsh

Hypericum hirsutum — St. John's Wort, Hairy

Hypericum humifusum — St. John's Wort, Trailing

Hypericum maculatum or H. dubium — St. John's Wort, Imperforate

Hypericum montanum — St. John's Wort, Mountain

Hypericum perforatum — St. John's Wort, Common

Hypericum pulchrum — St. John's Wort, Beautiful or Slender

Hypericum sp. — St. John's Wort

Hypericum tetrapterum — St. John's Wort, Square-stemmed

Hyperolius sp. — Frog, Reed or Sedge

Hyperoodon ampulatus — Whale, Bottle-nosed

Hyperoodon rostratus — Whale, Bottle-nosed

Hyperopisus bebe — Mormyrid

Hyperoplus lanceolatus — Sand-eel, Greater

Hyphaene thebaica — Palm, Doum

Hyphanturgus ocularis — Weaver, Spectacled

Hyphessobrycon callistus — Tetra, Jewel

Hyphessobrycon flammeus — Tetra, Flame or Red

Hyphessobrycon herbertaxelrodi — Tetra, Neon

Hyphessobrycon heterorhabdus — Tetra, Flag

Hyphessobrycon innesi — Tetra, Neon

Hyphessobrycon ornatus — Tetra, Ornate

Hyphessobrycon pulchripinnis — Tetra, Lemon

Hyphessobrycon rubostigma — Tetra

Hyphessobrycon serpae — Tetra, Serpae

Hypholoma fasciculare — Sulphur Tuft

Hypholoma sp. — Hypholoma

Hypholoma sublateritium — Brick Tuft

Hyphoraia aulica — Moth, Brown Tiger

Hypochera sp. — Widow-finch

Hypochoeris glabra — Cat's-ear, Smooth

Hypochoeris maculata — Cat's-ear, Spotted

Hypochoeris radicata — Cat's-ear, Common

Hypochoeris uniflora — Cat's-ear, Giant

Hypocolius ampelinus — Hypocolius, Gray

Hypoctonus rangunensis — Whip-scorpion

Hypocyrta nummularia — Goldfish Plant

Hypoderma bovis — Gadfly

Hypoestes sanguinolenta — Freckle Face

Hypoestes sp. — Baby's Tears

Hypolimnas dexithea — Butterfly, Diadem

Hypolimnas missippus — Butterfly, Diadem

Hypositta corallirostris — Nuthatch, Coral-billed or Madagascar

Hypotaenidia philippensis — Rail, Banded

Hypothymis azurea — Flycatcher, Black-naped Blue

Hypoxis sp. — Star-grass

Hypseleotris cyprinoides — Goby, Celebes Sleeper

Hypsignathus monstrosus — Bat, Hammer-head

Hypsipetes sp. — Bulbul

Hypsopygia costalis — Moth, Gold Fringe

Hyssop officinalis — Hyssop

Hystrix afrae-australis — Porcupine

Hystrix cristata — Porcupine, Crested

Hystrix indica — Porcupine, Indian Crested

Hystrix leucura — Porcupine

I

Iberis amara — Candytuft, Annual or Wild

Iberis saxatalis — Candytuft, Rock

Iberis sempervirens — Candytuft, Evergreen or Perennial

Iberis umbellata — Candytuft, Globe

Ibidorhyncha struthersii — Ibis-bill

Ibis ibis — Ibis, Wood or Yellow-billed

Ibis leucocephalus — Stork, Painted

Iboza riparia — Nutmeg Bush

Icerya purchasi — Scale, Cottony-cushion

Ichneumia albicauda — Mongoose, White-tailed

Ichneumon pisorius — Wasp, Ichneumon

Ichneumon suspiciosus — Wasp, Ichneumon

Ictailurus planiceps — Cat, Flat-headed

Ictalurus nebulosus — Bullhead, Brown

Icterus chrysater — Troupial

Icterus icterus — Troupial

Icthyophaga ichthyaetus — Eagle, Gray-headed Fishing

Ictinaetus malayensis — Eagle, Indian Black

Ictinia plumbea — Kite, Plumbeous

Ictonyx libyca — Weasel, Libyan Striped or Banded

Ictonyx striatus — Zorilla

Idesia polycarpa — Iigiri Tree

Idiurus zenkeri — Squirrel, Flying

Idotea sp. — Sowbug

Idria columnaris — Boojum Tree

Idus idus — Ide

Iguana iguana — Iguana, Common or Green

Ilex aquifolium — Holly, English
Ilex crenata — Holly, Japanese
Ilex cornuta — Holly, Chinese
Ilex opaca — Holly, American
Ilex paraguariensis — Mate
Ilex verticillata — Winterberry
Ilex vomitoria — Yaupon
Illecebrum verticillatum — Illecebrum
Illicium anisatum — Illicium, Japanese
Illicium floridanum — Polecat Tree
Illicium vernum — Star-anise
Ilyocoris cimicoides — Bug, Saucer
Impatiens balsamina — Balsam, Garden
Impatiens capensis — Jewel-weed
Impatiens glandulifera — Policeman's
   Helmet
Impatiens noli-tangere — Balsam, Touch-
   me-not
Impatiens parviflora — Balsam, Small
Impatiens petersiana — Busy Lizzie
Impatiens sp. — Balsam
Impatiens sultanii — Patience Plant or
   Patient Lucy
Impatiens walleriana — Busy Lizzie
Inachis io — Butterfly, Peacock
Incarvillea delavayi — Trumpet Flower,
   China
Incarvillea grandiflora — Trumpet Flowers
Indicator archipelagicus — Honey-guide
Indicator indicator — Honey-guide,
   Black-throated or Greater
Indicator minor — Honey-guide, Lesser
Indicator xanthonotus — Honey-guide
Indigofera sp. — Indigo
Indri indri — Indris
Inia geoffroyensis — Dolphin, Amazon
Inocybe sp. — Inocybe
Inonotus sp. — Inonotus
Inula britannica — Fleabane, British
Inula conyza — Ploughman's Spikenard
Inula crithmoides — Samphire, Golden
Inula helenium — Elecampane
Inula salicina — Fleabane, Irish
Inula sp. — Inula
Ipheion uniflorum — Star Flower, Spring
Iphiclides podalirius — Butterfly, Scarce
   Swallow-tail
Ipomoea batatas — Sweet Potato

Ipomoea hederacea — Morning-glory,
   Ivy-leaved
Ipomoea learii — Blue Dawn Flower
Ipomoea purpurea — Morning-glory, Common
   or Tall
Ipomoea sp. — Morning-glory
Ips sp. — Beetle, Bark
Irania gutturalis — Robin, White-throated
Irediparra gallinacea — Jacana, Comb-
   crested
Irena cyanogaster — Bluebird, Philippine
   Fairy
Irena puella — Bluebird, Blue-backed or
   Palawan
Iresine herbstii — Bloodleaf
Iridogorgia sp. — Gorgonian
Iridoprocne bicolor — Swallow, Tree
Iridosornis analis — Tanager, Golden-
   crowned
Iridosornis rufivertix — Tanager, Golden-
   crowned
Iris aphylla — Iris, Leafless
Iris bakeriana — Iris, Baker
Iris bucharica — Iris, Bokhara
Iris chamaeiris — Iris, Lesser
Iris cristata — Iris, Crested
Iris danfordiae — Iris, Danford
Iris florentina — Fleur-de-lis
Iris foetidissima — Iris, Stinking
Iris germanica — Iris, Common
Iris graminea — Iris, Grass-leaved
Iris histrio — Iris
Iris histrioides — Iris, Harput
Iris laevigata — Iris
Iris oratoria — Mantis
Iris pseudacorus — Flag, Yellow
Iris pumila — Iris
Iris reticulata — Iris, Netted
Iris sibirica — Iris, Siberian
Iris sisyrinchium — Barbary Nut
Iris spuria — Iris, Butterfly
Iris unguicularis — Iris, Algerian
Iris versicolor — Flag, Blue or Purple
Iris xiphioides — Iris, English
Iris xiphium — Iris, Spanish
Isatis tinctoria — Woad
Ischura elegans — Damselfly
Isoetes sp. — Quillwort

Isoloma sp. — Isoloma
Isoodon sp. — Bandicoot, Short-nosed
Isopyrum sp. — Isopyrum
Ispidina picta — Kingfisher, African Pygmy
Istiophorus sp. — Sailfish
Isurus oxyrhynchus — Shark, Mako
Itea ilicifolia — Sweetspire, Holly-leaf
Ithaginis cruentus — Pheasant, Blood
Ixia sp. — Lily, African Corn
Ixobrychus cinnamomeus — Bittern,
    Cinnamon or Little
Ixobrychus eurhythmus — Bittern,
    Schrenck's Least
Ixobrychus minutus — Bittern, Cinnamon or
    Little
Ixobrychus sinensis — Bittern, Yellow
Ixodes sp. — Tick, Caster Bean
Ixora coccinea — Flame of the Woods
Ixora sp. — Ixora

J

Jabiru mycteria — Jabiru
Jacamerops aurea — Jacamar
Jacana gallinacea — Jacana, Comb-crested
Jacaranda acutifolia or J. mimosifolia —
    Jacaranda
Jacobinia carnea — Plume Flower, Brazilian
Jacobinia sp. — Jacobinia
Jaculus jaculus — Jerboa, Lesser Egyptian
Jaculus orientalis — Jerboa, African
Japyx sp. — Double-tail
Jasione montana — Sheep's Bit
Jasione perennis — Sheep's Bit, Perennial
Jasminum mesnyi — Jasmine, Primrose
Jasminum nudiflorum — Jasmine, Winter
Jasminum officinale — Jasmine, Common
Jasminum officinale grandiflorum —
    Jasmine, Poet's
Jasminum polyanthum — Jasmine, Chinese
Jasminum revolutum — Jasmine, Yellow
Jasminum rex — Jasmine, King
Jasminum sambac — Jasmine, Arabian
Jasminum sp. — Jasmine
Jatropha sp. — Jatropha
Jeffersonia diphylla — Twin-leaf
Jeffersonia dubia — Twin-leaf
Jovibara sobolifera — Houseleek, Hen-and-
    chickens

Juanulloa sp. — Juanulloa
Jubula lettii — Owl, Crested or Maned
Juglans cinera — Butternut
Juglans nigra — Walnut, Black
Juglans regia — Walnut, English or Persian
Julodis sp. — Beetle, Wood-borer
Julus sp. — Millipede
Juncus articulatus — Rush, Jointed
Juncus bufonius — Rush, Toad
Juncus bulbosus — Rush, Bulbous
Juncus capitatus — Rush, Capitate or Dwarf
Juncus conglomeratus — Rush, Common or
    Compact
Juncus effusus — Rush, Soft
Juncus inflexus — Rush, Hard
Juncus maritimus — Rush, Sea
Juncus squarrosus — Rush, Heath
Juniperus chinensis — Juniper, Chinese
Juniperus communis — Juniper, Common
Juniperus horizontalis— Juniper, Creeping
Juniperus oxycedrus — Juniper, Prickly
Juniperus phoenicea — Juniper, Phoenician
Justicia sp. — Justicia
Jynx torquilla — Wryneck, Eurasian

K

Kachuga tecta tecta — Terrapin, Dura or
    Roofed
Kaempferia glanga — Lily, Spice
Kaempferia sp. — Peacock Plant
Kalanchoe beharensis — Velvet Leaf
Kalanchoe blossfeldiana — Kalanchoe
Kalanchoe fedtschenkoi — Kalanchoe,
    Rainbow
Kalanchoe sp. — Kalanchoe
Kalanchoe tomentosa — Panda Plant
Kallima inachus — Butterfly, Indian Leaf
Kalopanax pictus — Aralia, Castor
Kaloula pulchra — Bullfrog, Malayan
Kassina senegalensis — Frog, Senegal
Katsuwonus pelamis— Bonito, Oceanic
Kaupifalco monogrammicus — Buzzard,
    Lizard
Kenopia striata — Babbler, Wren
Kentia forsteriana — Palm, Curly
Kentranthus ruber — See Centranthus ruber
Kerria japonica — Globe Flower Bush
Ketupa blakistoni — Owl, Brown Fish

Ketupa ketupa — Owl, Malay Fish
Ketupa zeylonensis — Owl, Brown Fish
Kickxia elatine — Fluellen, Sharp-leaved
Kickxia spuria — Fluellen, Round-leaved
Kigelia africana or K. pinnata — Sausage
    Tree
Kinixys sp. — Tortoise, Hinged
Kirengeshoma palmata — Waxbell, Yellow
Kittacincla malabarica — Shama
Kleinia repens — Blue Chalk Sticks
Kleinia sp. — Kleinia
Knautia arvensis — Scabious, Field
Knautia sylvatica — Scabious, Mountain or
    Wood
Kniphofia sp. — Red Hot Poker
Kniphofia uvaria — Red Hot Poker
Kobresia simpliciuscula — False-sedge
Kobus ellipsiprymnus — Waterbuck,
    Common
Kobus defassa — Waterbuck, Defassa or
    Sing-sing
Kobbus kob — Kob
Kobbus leche — Waterbuck, Lechwe
Kobbus vardoni — Puku
Kochia lanciflora — Cypress, Summer
Kochia scoparia — Cypress, Summer
Koeleria sp. — Hairgrass
Koellikeria erinoides — Koellikeria
Koelreuteria paniculata — Golden Rain Tree
Koenigia islandica — Purslane, Iceland
Kohleria bogotenisis — Devil's Breeches
Kohleria sp. — Kohleria
Kohlrauschia prolifera — Pink, Childing
Kolkwitzia amabilis — Beauty-bush
Kryptopterus bicirrhis — Catfish, Indian
    Glass
Kunzea sp. — Tick Bush

L

Labeo sp. — Labeo
Labeotropheus sp. — Cichlid
Labia minor — Earwig
Labidura riparia — Earwig
Labroides dimidiatus — Wrasse, Cleaner
Laboides phthirophagus — Wrasse
Labrus bergylta — Wrasse, Ballan
Labrus mixtus — Wrasse, Cuckoo
Labrus ossifagus — Wrasse, Cuckoo

Laburnum anagyroides — Golden Rain or
    Golden Chain Tree
Laburnum watereri — Golden Rain or Golden
    Chain Tree
Laccaria amethystina — Fungus, Amethyst
Laccaria laccata — Fungus, Amethyst
Lacedo pulchella — Kingfisher, Banded
Lacerta agilis— Lizard, Sand
Lacerta lepida — Lizard, Eyed or Jeweled
Lacerta muralis — Lizard, Wall
Lacerta oxycephala – Lizard, Sharp-headed
Lacerta pityuensis — Lizard
Lacerta sicula — Lizard, Ruin
Lacerta viridis — Lizard, Green
Lacerta vivipara — Lizard, Common or
    Viviparous
Lachenalia aloides — Lily, Leopard
Lachesis muta — Bushmaster
Lachnostachys verbascifolia — Lamb's Tails
Lacrymaria velutina — Weeping Widow
Lactarius chrysorrheus — Milk-cap, Sulphur
Lactarius quietus — Milk-cap, Oak
Lactarius rufus — Milk-cap, Red
Lactarius torminosus — Milk-cap, Woolly or
    Shaggy
Lactarius volemus — Milk-cap, Orange-brown
Lactophrys sp. — Boxfish
Lactrodectus mactans tredecimguttatus —
    Spider, Black Widow
Lactuca perennis — Lettuce, Blue
Lactuca pulchella — Lettuce, Blue
Lactuca saligna — Lettuce, Least
Lactuca sativa — Lettuce, Garden or Common
Lactuca scariola — Lettuce, Wild or Prickly
Lactuca serriola — Lettuce, Prickly
Lactuca virosa — Lettuce, Acrid or Greater
    Prickly
Ladoga camilla — Butterfly, White Admiral
Ladoga populi — Butterfly, Large or Poplar
    Admiral
Laelia anceps — Orchid, Flor de San Miguel
Laelia sp. — Orchid, Laelia
Laeliocattleya — Orchid (Hybrid)
Lagenaria siceraria — Gourd, Bottle or
    Dipper
Lagerstroemia indica — Crape Myrtle
Lagerstroemia speciosa — Crape Myrtle
Lagidium peruanum — Chinchilla, Mountain
Lagidium viscacia — Chinchilla, Mountain

Lagonosticta rubricata — Firefinch, Blue-billed

Lagopus lagopus — Grouse, Red or Willow

Lagopus mutus — Ptarmigan, Rock

Lagorchestes sp. — Wallaby, Hare

Lagostomus maximus — Viscacha, Plains

Lagostrophus sp. — Wallaby, Hare

Lagothrix sp. — Monkey, Woolly

Lagurus lagurus — Lemming, Steppe

Lagurus ovatus — Hare's-tail Grass

Lalage leucomela — Triller, Varied

Lalage nigra — Triller, Pied

Lalage sueurii — Triller, White-winged

Lama glama — Llama

Lama glama huanacus — Guanaco

Lama guanacoe — Guanaco

Lama pacos — Alpaca

Lama peruana — Llama

Lama vicuna — Vicuna

Lambis sp. — Conch, Spider or Scorpion

Lamia textor — Beetle, Weaver Longhorn

Lamium album — Dead-nettle, White

Lamium amplexicaule — Dead-nettle, Henbit

Lamium galeobdolon — Archangel, Yellow

Lamium hybridum — Dead-nettle, Cut-leaved

Lamium maculatum — Dead-nettle, Spotted

Lamium orvala — Archangel, Balm-leaved

Lamium purpureum — Dead-nettle, Red or Purple

Lamna cornubica — Shark, Mackerel

Lamna nasus — Shark, Mackerel

Lampetra fluvialis — Lamprey, River

Lampetra planeri — Lamprey, Brook

Lampides boeticus — Butterfly, Long-tailed Blue

Lampra rutilans — Beetle, Wood-borer

Lampranthus sp. — Ice Plant

Lampranthus spectabilis — Ice Plant, Pink

Lampris sp. — Opah

Lamprocera litreillei — Firefly

Lamprocolius chalybeus — Starling, Glossy

Lamprologus sp. — Cichlid

Lamprotornis sp. — Starling, Glossy

Lampyris noctiluca — Glow-worm

Langaha intermedia — Snake, Madagascan Rear-fanged

Laniarius aethiopicus — Shrike, Bou-bou

Laniarius atrococcineus — Gonolek

Laniarius barbarus — Gonolek

Laniarius ferrugineus — Shrike, Bou-bou

Lanice conchilega — Worm, Sand-mason

Lanius cabanisi — Shrike, Fiscal

Lanius collaris — Shrike, Fiscal

Lanius collurio — Shrike, Red-backed

Lanius cristatus — Shrike, Brown

Lanius excubitor — Shrike, Northern

Lanius minor — Shrike, Lesser Gray

Lanius nubicus — Shrike, Masked

Lanius schach — Shrike, Long-tailed or Rufous-backed

Lanius senator — Shrike, Woodchat

Lanius tigrinus — Shrike, Tiger

Lantana camara — Lantana, Common

Lantana montevidensis or L. sellowiana — Lantana, Trailing

Lanthanotus borneensis — Monitor, Earless

Laothoe populi — Hawkmoth, Poplar

Lapageria rosea — Bellflower, Chilean

Lapeyrousia sp. — Lapeyrousia

Laphria sp. — Fly, Robber

Lapidaria margaretae — Karoo Rose

Laportea sp. — Laportea

Lappula myosotis — Forget-me-not, Bur

Lapsana communis — Nipplewort

Larix decidua — Larch, European

Larix eurolepis — Larch, Hybrid or Dunkeld

Larix kaempferi — Larch, Japanese

Larix laricina — Larch, American or Tamarack

Larosterna inca — Tern, Inca

Larus argentatus — Gull, Herring

Larus audouinii — Gull, Audouin's

Larus bulleri — Gull, Black-billed or Buller's

Larus canus — Gull, Common

Larus cirrhocephalus — Gull, Gray-headed

Larus dominicanus — Gull, Black-backed or Dominican

Larus fuliginosus —Gull,Lava

Larus furcatus — Gull, Swallow-tailed

Larus fuscus — Gull, Lesser Black-backed

Larus genei — Gull, Slender-billed

Larus glaucoides — Gull, Iceland or Kumlien's

Larus hyperboreus — Gull, Glaucous

Larus ichthyaetus — Gull, Great Black-headed

Larus marinus — Gull, Great Black-backed

Larus melanocephalus — Gull, Mediterranean

Larus minutus — Gull, Little
Larus novaehollandiae — Gull, Red-billed or Silver
Larus pacificus — Gull, Pacific
Larus ridibundus — Gull, Black-headed
Larus roseus — Gull, Ross'
Larus sabini — Gull, Sabine's
Larus scopulinus — Gull, Red-billed or Silver
Larus scoresbii — Gull, Dolphin
Larus tridactyla — Kittiwake, Black-legged
Lasiocampa quercus — Moth, Oak Eggar
Lasiommata megera — Butterfly, Wall Brown
Lasiorhinus latifrons — Wombat, Hairy-nosed or Soft-furred
Lasiorrhynchus barbicornis — Weevil, Giraffe
Lasiurus cinereous — Bat, Hoary
Lasius sp. — Ant, Meadow
Lathamus discolor—Parrot, Swift
Lathraea clandestina — Toothwort, Purple
Lathraea squamaria — Toothwort
Lathyrus aphaca — Vetchling, Yellow
Lathyrus clymenum — Pea, Sweet
Lathyrus hirsutus — Vetchling, Hairy
Lathyrus japonica — Pea, Beach or Sea
Lathyrus latifolius — Pea, Everlasting or Perennial
Lathyrus maritimus — Pea, Beach or Sea
Lathyrus montanus — Vetch, Bitter
Lathyrus niger — Pea, Black
Lathyrus nissolia — Vetchling, Grass
Lathyrus odoratus — Pea, Sweet
Lathyrus palustris — Pea, Marsh
Lathyrus pratensis — Vetchling, Meadow
Lathyrus sylvestris — Pea, Wild or Everlasting
Lathyrus tuberous — Pea, Earth-nut
Lathyrus vernus — Pea, Spring
Latiaxis sp. — Latiaxis Shell
Laticauda colubrina — Sea-snake, Banded
Laticauda laticauda — Krait, Black-banded
Latimeria chalumnae — Coelacanth
Laurus nobilis — Sweet Bay
Lavandula dentata —·Lavender, Fringed
Lavandula officinalis — Lavender, English
Lavandula spica — Lavender, English

Lavandula stoechas — Lavender, French or Spanish
Lavandula vera — Lavender, English
Lavatera arborea — Mallow, Tree
Lavatera cretica — Mallow, Lesser or Smaller Tree
Lavatera sp. — Mallow
Lavatera trimestris — Mallow, Annual
Lawsonia inermis — Mignonette Tree
Layia sp. — Tidy Tips
Leander serratus — Prawn, Common
Lebistes reticulatus — Guppy
Lecanora sp. — Lecanora
Lecidea sp. — Lecidea
Ledum groenlandicum — Labrador Tea
Ledum palustre — Ledum
Legousia speculum-veneris — Venus' Looking-glass
Leiocassis poecilopterus — Catfish
Leiocassis siamensis — Catfish
Leiopelma hochstetteri — Frog, Hochstetter's
Leiothrix argentaurus — Mesia
Leiothrix lutea — Robin, Pekin
Leipoa ocellata — Fowl, Mallee
Leistes militaris — Blackbird, Red-breasted
Lemmaphyllum microphyllum — Lemma-phyllum
Lemmus lemmus — Lemming, Norwegian
Lemna gibba — Duckweed, Gibbous or Fat
Lemna minor — Duckweed, Common or Lesser
Lemna polyrhiza — Duckweed, Great
Lemna triscula — Duckweed, Ivy
Lemur catta — Lemur, Ring-tailed
Lemur fulvus — Lemur, Brown
Lemur macaco — Lemur, Black
Lemur mongoz — Lemur, Mongoose
Lemur variegatus — Lemur, Ruffed
Lens culinaris — Lentil
Lentinellus cochleatus — Lentinellus
Lentinus sp. — Lentinus
Lenzites sp. — Lenzites
Leonitis leonurus — Lion's Ear or Wild Dagga
Leontideus rosalia — Marmoset, Golden Lion or Silky
Leontocebus rosalia — Marmoset, Golden Lion or Silky
Leontodon autumnalis — Hawkbit, Autumn

Leontodon hispidus — Hawkbit, Rough or Greater

Leontodon taraxacoides — Hawkbit, Lesser or Hairy

Leontopithecus rosalia — Marmoset, Golden Lion or Silky

Leontopodium alpinum — Edelweiss

Leonurus cardiaca — Motherwort

Leotia lubrica — Fungus, Green Slime

Lepadogaster sp. — Sucker

Lepas sp. — Barnacle, Goose

Lepidiota bimaculata — Beetle, Cockchafer

Lepidium campestre — Pepperwort, Field or Common

Lepidium heterophyllum — Pepperwort, Downy

Lepidium latifolia — Dittander

Lepidium ruderale — Pepperwort, Narrow-leaved

Lepidium sativum — Cress, Garden or Upland

Lepidochitona sp. — Chiton

Lepidocolaptes sp. — Woodcreeper

Lepidorhombus whiffiagonis — Megrim

Lepidosaphes ulmi — Scale, Mussel or Oyster-shell

Lepidosiren paradoxa — Lungfish, South American

Lepilemur mustelinus — Lemur, Sportive or Weasel

Lepilemur ruficaudatus — Lemur, Sportive or Weasel

Lepiota clypeolaria — Agaric, Fragrant

Lepiota cristata — Lepiota, Crested

Lepiota excoriata — Lepiota, Crested

Lepiota morgani — Mushroom, Green-lined Parasol

Lepiota procera — Mushroom, Parasol

Lepiota rhacodes — Mushroom, Shaggy Parasol

Lepisma sp. — Silverfish

Lepismachilis notata — Bristletail

Lepista sp. — Blewits

Leporinus affinis — Boga

Leporinus fasciatus — Leporinus, Banded

Leporinus sp. — Mieri

Leptailurus serval — Cat, Serval

Leptidea sinapis — Butterfly, Wood White

Leptodactylus sp. — Frog, Slender-fingered Bladder

Leptodirus hohenwarti — Beetle, Cave-dwelling Ground

Leptodon cayanensis — Kite, Cayenne or Gray-headed

Leptolophus hollandicus — Cockatiel

Leptonychotes weddelli — Seal, Weddell's

Leptopelis sp. — Frog, Leptopelid

Leptophis ahaetulla — Snake, Parrot

Leptophyes punctatissima — Bush-cricket, Speckled

Leptopterus madagascariensis — Vanga, Blue

Leptopterus madagascarinus — Vanga, Blue

Leptoptilos crumeniferus — Marabou, African

Leptoptilos dubius — Marabou, Greater

Leptoptilos javanicus — Marabou, Lesser

Leptosomus discolor — Cuckoo-roller

Leptospermum ericoides — Tea Tree, White

Leptospermum scoparium — Tea Tree, Manuka

Leptosynapta inhaerens — Sea Cucumber

Leptotes sp. — Orchid, Leptotes

Leptotyphlops sp. — Snake, Worm

Lepus capensis — Hare, Brown or European

Lepus europaeus — Hare, Brown or European

Lepus saxatilis — Hare, Scrub or Bush

Lepus timidus — Hare, Blue or Mountain

Lerwa lerwa — Partridge, Snow

Lesbia victoriae — Hummingbird, Black-throated Train-bearer

Leschenaultia biloba — Leschenaultia, Blue

Lestes sp. — Damselfly

Leto staceyi — Moth, Swift

Leto venus — Moth, Swift

Leucadendron argenteum — Silver Tree

Leucadendron discolor — Flame Gold Tips

Leucania impura — Moth, Wainscot

Leucania pallens — Moth, Wainscot

Leucaspius delineatus — Moderlieschen

Leuchtenbergia principis — Cactus, Agave or Prism

Leuciscus cephalus — Chub

Leuciscus idus — Ide

Leuciscus leuciscus — Dace

Leucocoryne ixiodes — Glory of the Sun

Leucodonta bicoloria — Moth, White Prominent

Leucogenes leontopodium — Edelweiss, North Island

Leucojum aestivum — Snowflake, Summer
Leucojum vernum — Snowflake, Spring
Leucopaxillus sp. — Leucopaxillus
Leucophaeus scoresbii — Gull, Dolphin
Leucopsar rothschildi — Mynah, Bali or
  Rothschild's
Leucopternis sp. — Hawk
Leucorchis albida — Orchid, Small White
Leucosarcia melanoleuca — Pigeon, Wonga
Leucospermum sp. — Pincushion
Leucotreron alligator — Pigeon,
  Yellow-bellied Fruit
Leucotreron cincta — Pigeon, Yellow-bellied
  Fruit
Leuctra fusca — Fly, Stone
Levisticum officinale — Lovage
Leycesteria formosa — Honeysuckle,
  Himalayan
Lialis burtonis — Lizard, Legless
Liasis sp. — Python
Libellula depressa — Dragonfly, Broad-
  bodied Libellula
Libythea celtis — Butterfly, Nettle-tree
Lichanura trivirgata — Boa, Rosy
Lichmera indistincta — Honey-eater, Brown
Ligia sp. — Sea Slater
Ligularia sp. — Ligularia
Ligusticum mutellina — Lovage, Alpine
Ligusticum scoticum — Lovage, Northern or
  Scots
Ligustrum japonicum — Privet, Wax-leaved
Ligustrum ovafolium — Privet, California
Ligustrum sinensis — Privet, Chinese
Ligustrum sp. — Privet
Ligustrum vicaryi — Privet, Vicary Golden
Ligustrum vulgare — Privet, Common
Lilium auratum — Lily, Golden Rayed
Lilium bulbiferum — Lily, Orange
Lilium candidum — Lily, Madonna
Lilium henryi — Lily, Henry's
Lilium longiflorum — Lily, Easter or
  Bermuda
Lilium margaton — Lily, Margaton or
  Turk's-cap
Lilium pomponium — Lily, Red
Lilium pyrenaicum — Lily, Yellow Turk's-cap
Lima sp. — Clam, File
Limacella sp. — Limacella
Limanda limanda — Dab

Limax sp. — Slug, Spotted
Limenitis camilla — Butterfly, White
  Admiral
Limenitis populi — Butterfly, Large or Poplar
  Admiral
Limicola falcinellus — Sandpiper, Broad-
  billed
Limnocorax flavirostra — Crake, Black
Limnodromus griseus — Dowitcher
Limnodromus semipalmatus — Dowitcher
Limnodynastes sp. — Bullfrog, Australian
Limodorum abortivum — Limodorum
Limonium bellidifolium — Sea-lavender,
  Matted
Limonium binervosum — Sea-lavender, Rock
Limonium latifolium — Sea-lavender,
  Wide-leaved
Limonium sinuatum — Sea-lavender, Winged
Limonium suworowii — Statice
Limonium vulgare — Sea-lavender, Common
Limosa lapponica — Godwit, Bar-tailed
Limosa limosa — Godwit, Black-tailed
Limosa melanuroides — Godwit, Black-tailed
Limosella sp. — Mudwort
Linanthus sp. — Linanthus
Linaria alpina — Toadflax, Alpine
Linaria arenaria — Toadflax, Sand or French
Linaria dalmatica — Toadflax, Dalmatian
Linaria maroccana — Toadflax, Annual
Linaria purpurea — Toadflax, Purple
Linaria repens — Toadflax, Pale
Linaria sp. — Toadflax
Linaria triphylla — Toadflax, Three-leaved
Linaria vulgaris — Toadflax, Common
Lindernia sp. — Lindernia
Lineus geniculatus — Worm, Bootlace
Lineus sp. — Worm, Bootlace or Ribbon
Linnaea borealis — Twinflower
Linophryne sp. — Angler, Deep-sea
Linum austriacum — Flax, Austrian
Linum bienne — Flax, Pale
Linum catharticum — Flax, Purging or Fairy
Linum elegans — Flax, Dwarf Flowering
Linum flavum — Flax, Yellow
Linum grandiflorum — Flax, Flowering or
  Scarlet
Linum perenne — Flax, Perennial
Linum sp. — Flax
Linum tenuifolium — Flax, Narrow-leaved

Linum usitatissimum — Flax, Cultivated
Linyphia sp. — Spider, Platform
Lipara lucens — Fly, Reed Gall
Lipaugus strephophorus — Piha
Lipaugus vociferans — Piha
Lipoptena cervi — Fly, Deer or Louse
Liposcelis divinatorius — Louse, Book
Lippia sp. — Lippia
Liquidambar styraciflua — Sweet Gum
Liriodendron tulipifera — Tulip Tree
Liriope muscari — Lily-turf, Blue
Liriope spicata — Lily-turf, Creeping
Lisianthus sp. — Funeral Flower
Lissotis melanogaster — Bustard, Black-
  bellied
Listera cordata — Twayblade, Lesser
Listera ovata — Twayblade, Common
Litchi chinensis — Litchi or Lychee
Lithobius sp. — Centipede, Brown
Lithocranius walleri — Gerenuk
Lithops sp. — Living Stones
Lithosia lurideola — Moth, Common Footman
Lithospermum arvense — Gromwell, Corn
Lithospermum diffusum — Gromwell,
  Scrambling
Lithospermum officinale — Gromwell,
  Common
Lithospermum purpurocaeruleum —
  Gromwell, Blue
Littorella uniflora — Shore-weed
Littorina sp. — Periwinkle
Livistona chinensis — Palm, Chinese Fan
Livistona sp. — Palm, Fan
Liza auratus — Mullet, Golden Gray
Lloydia serotina — Lily, Snowdon
Lo vulpinus — Rabbitfish
Lobelia cardinalis — Cardinal Flower
Lobelia deckenii — Lobelia
Lobelia dortmanna — Lobelia, Water
Lobelia erinus — Lobelia, Common
  (cultivated)
Lobelia fulgens — Cardinal Flower
Lobelia sp. — Lobelia
Lobelia urens — Lobelia, Blue
Lobibyx miles— Plover, Masked
Lobibyx novaehollandiae — Plover,
  Australian Spur-winged
Lobiophasis bulweri — Pheasant, Bulwer's
  Wattled

Lobipes lobatus— Phalarope, Northern or
  Red-necked
Lobivanellus indicus — Lapwing,
  Red-wattled
Lobivia famatimensis — Cactus, Cob
Lobivia hetrichiana — Cactus, Cob
Lobivia sp. — Cactus, Cob
Lobocraspis griseifusca — Moth, Owlet
Lobodon carcinophaga — Seal, Crab-eating
Locusta migratoria — Locust, Migratory
Locustella certhiola — Warbler, Pallas'
  Grasshopper
Locustella fasciolata — Warbler, Gray's
  Grasshopper
Locustella fluviatilis — Warbler, River
Locustella lanceolata — Warbler,
  Lanceolated Grasshopper
Locustella luscinioides — Warbler, Savi's
Locustella naevia — Warbler, Grasshopper
Loddigesia mirabilis — Hummingbird,
  Marvelous Spatule-tail
Lodoicea maldivica — Coconut, Double or
  "Coco de mer"
Loiseleuria procumbens — Azalea, Wild or
  Alpine or Creeping
Loligo forbesi — Squid, Common
Loligo vulgaris — Squid, Common
Lolium multiflorum — Rye-grass, Italian
Lolium perenne — Rye-grass, Perennial
Lomatogonium sp. — Lomatogonium
Lonchura atricapilla — Mannikin
Lonchura castaneothorax — Finch, Chestnut-
  breasted
Lonchura cucullata — Mannikin
Lonchura ferruginosa — Mannikin, Chestnut
Lonchura flaviprymna — Finch, Yellow-
  rumped
Lonchura fuscans — Mannikin
Lonchura malacca — Mannikin, Chestnut
Lonchura punctulata — Finch, Spice
Lonchura striata — Mannikin
Lonicera alpigena — Honeysuckle, Alpine
Lonicera americana — Honeysuckle
Lonicera caerulea — Honeysuckle, Blue
Lonicera caprifolium — Honeysuckle,
  Perfoliate
Lonicera etrusca — Honeysuckle
Lonicera fragrantissma — Honeysuckle,
  Winter or Fragrant

Lonicera fuchsiodes — Honeysuckle, Scarlet or Trumpet

Lonicera hildebrandiana — Honeysuckle, Giant

Lonicera japonica — Honeysuckle, Japanese

Lonicera maackii — Honeysuckle, Amur

Lonicera periclymenum — Honeysuckle or Woodbine

Lonicera sempervirens — Honeysuckle, Trumpet or Coral

Lonicera sp. — Honeysuckle

Lonicera tatarica — Honeysuckle, Tartarian

Lonicera xylosteum — Honeysuckle, Fly

Lophiomys imhausi — Rat, Crested or Maned

Lophius piscatorius — Angler, European

Lophoaetus occipitalis — Eagle, Long-crested

Lophoceros flavirostris — Hornbill, Yellow-billed

Lophocolea sp. — Lophocolea

Lophodiodon calori — Porcupine-fish

Lophoictinia isura — Kite, Square-tailed

Lopholaimus antarcticus — Pigeon, Top-knot

Lophonetta specularioides — Duck, Patagonian Crested

Lophophaps ferruginea — Pigeon, Plumed

Lophophaps plumifera — Pigeon, Plumed

Lophophora williamsii — Peyote

Lophophorus impejanus — Pheasant, Himalayan or Imperial Monal

Lophophorus lhuysi — Pheasant, Chinese Monal

Lophophorus sclateri — Pheasant, Sclater's Monal

Lophopteryx capucina — Moth, Coxcomb Prominent

Lophopus sp. — Bryozoan

Lophorina superba — Bird of Paradise, Superb

Lophornis sp. — Hummingbird, Coquette

Lophostrix cristata — Owl, Crested or Maned

Lophostrix lettii — Owl, Crested or Maned

Lophura bulweri — Pheasant, Bulwer's Wattled

Lophura diardi — Pheasant, Diard's or Siamese Fireback

Lophura edwardsi — Pheasant, Edward's

Lophura erythrophthalma — Pheasant, Malay Crestless Fireback

Lophura ignita — Pheasant, Crested Fireback

Lophura imperialis — Pheasant, Imperial

Lophura leucomelana — Pheasant, Kalij

Lophura nycthemera — Pheasant, Silver

Lophura swinhoii — Pheasant, Swinhoe's

Loranthus sp. — Mistletoe

Loricaria filamentosa — Catfish, Armored

Loricaria parva — Catfish, Armored

Loriculus gulgulus — Lorikeet, Malay

Loriculus philippensis — Lorikeet, Malay

Loris tardigradus — Loris, Slender

Lorius domicella — Lorikeet

Lorius pectoralis — Parrot, Wax-billed

Lorius roratus — Parrot, Wax-billed

Loropetalum chinense — Fringe Flower

Lota lota — Burbot

Lotus angustissimus — Trefoil, Least or Slender Bird's-foot

Lotus bertholetii — Coral Gem

Lotus corniculatus — Trefoil, Common Bird's-foot

Lotus creticus — Trefoil, Silvery or Southern Bird's-foot

Lotus hispidus — Trefoil, Hairy Bird's-foot

Lotus pedunculatus — Trefoil, Marsh Bird's-foot

Lotus sp. — Trefoil, Bird's-foot

Lotus uliginosus — Trefoil, Greater Bird's-foot

Loxia curvirostra — Crossbill, Common or Red

Loxia leucoptera — Crossbill, Two-barred or White-winged

Loxia pytyopsittacus — Crossbill, Parrot or Scottish

Loxodonta africana — Elephant, African

Lucanus cervus — Beetle, Stag

Lucilia sp. — Fly, Greenbottle

Luciocephalus pulcher — Pikehead

Luciola italica — Firefly

Lucioperca lucioperca — Pike-perch

Luculia sp. — Luculia

Ludwigia palustris — Purslane, Hampshire or Water

Luidia sp. — Starfish

Lullala arborea — Woodlark

Lumbriculus sp. — Worm

Lumbricus sp. — Earthworm

Lumpenus lampretaeformis — Blenny, Snake

Lunaria annua — Honesty

Lunaria biennis — Honesty

Lunaria rediviva — Honesty, Perennial
Lupinus angustifolius — Lupin,
  Narrow-leaved
Lupinus arboreus — Lupin, Tree
Lupinus hirsutus — Lupin, Blue
Lupinus hybrids — Lupin
Lupinus luteus — Lupin, Yellow
Lupinus nootkatensis — Lupin, Scottish
Lupinus sp. — Lupin
Luronium natans — Water-plantain, Floating
Luscinia calliope — Rubythroat, Siberian
Luscinia luscinia — Nightingale, Thrush
Luscinia megarhynchos — Nightingale
Luscinia svecica — Bluethroat
Lusciniola melanopogon — Warbler,
  Moustached
Lutjanus sebae — Snapper, Red or Emperor
Lutjanus sp. — Snapper
Lutra lutra — Otter, European or Common
  River
Lutra maculicollis — Otter, Spotted-necked
Lutraria sp. — Clam, Otter Shell
Lutreolina crassicaudata — Opossum,
  Thick-tailed
Lutrogale perspicillata — Otter, Indian
  Smooth-coated
Luzula arcuata — Woodrush, Curved or Arctic
Luzula campestris — Woodrush, Field
Luzula pilosa — Woodrush, Hairy
Luzula sp. — Woodrush
Luzula spicata — Woodrush, Spiked
Luzula sylvatica — Woodrush, Greater
Lybius leucomelas — Barbet, Pied
Lybius torquatus — Barbet, Collared
Lycaena dispar — Butterfly, Large Copper
Lycaena phlaeas — Butterfly, Small Copper
Lycaon pictus — Dog, Cape Hunting
Lycaste sp. — Orchid, Lycaste
Lychnis alba — Campion, White
Lychnis alpina — Catchfly, Alpine
Lychnis chalcedonica — Maltese Cross
Lychnis coeli-rosa — Rose-of-heaven
Lychnis coronaria — Campion, Rose
Lychnis flos-cuculi — Ragged Robin
Lychnis flos-jovis — Flower-of-Jove
Lychnis viscaria — Catchfly, Red German or
  Sticky
Lycia hirtaria — Moth, Brindled Beauty
Lycium chinense — Matrimony Vine

Lycium europaeum — Matrimony Vine
Lycium halimifolium — Matrimony Vine
Lycodonomorphus rufulus — Snake, Water
Lycodontis sp. — Eel, Moray
Lycogala epidendrum — Slime-mold, Wood
Lycoperdon echinatum — Puffball, Spiny
Lycoperdon gemmatum — Puffball, Gemmed
Lycoperdon giganteum — Puffball, Giant
Lycoperdon perlatum — Puffball, Common or
  Warted
Lycoperdon pyriforme — Puffball,
  Pear-shaped or Wood
Lycopersicum esculentum — Tomato
Lycopodium clavatum —Clubmoss, Stagshorn
  or Wolf's Claw
Lycopodium sp. — Clubmoss
Lycopsis arvensis — Bugloss, Small or Lesser
Lycopus europaeus — Gipsy-wort
Lycoris aurea — Lily, Golden Spider or
  Hurricane
Lycoris radiata — Lily, Red Spider
Lycoris squamigera — Amaryllis, Hardy or
  Garden
Lycosa amentata — Spider, Wolf
Lycosa narboensis — Spider, Wolf
Lycosa tarentula — Tarantula
Lycoteuthis diadema — Squid, Jeweled
Lyctus sp. — Beetle, Powder-post
Lygodium japonicum — Fern, Japanese
  Climbing
Lygodium palmatum — Fern, Hartford
Lygosoma sp. — Skink
Lymantria dispar — Moth, Gypsy
Lymantria monacha — Moth, Black Arches
Lymnaea sp. — Snail, Pond
Lymnocryptes minimus — Snipe, Jack
Lynchailurus pajeros braccatus — Cat,
  Pampas
Lynx lynx — Lynx, European or Northern
Lynx lynx pardina — Lynx, Spanish
Lyophyllum sp. — Lyophyllum
Lyrurus mlokosiewiczi — Grouse, Caucasian
  Black
Lyrurus tetrix — Grouse, Black
Lysandra corydon — Butterfly, Chalk-hill
  Blue
Lysimachia ciliata — Loosestrife, Fringed
Lysimachia nemorum — Pimpernel, Yellow or
  Wood

Lysimachia nummularia — Moneywort
Lysimachia punctata — Loosestrife, Large
  Yellow or Dotted
Lysimachia sp. — Loosestrife
Lysimachia thrysiflora — Loosestrife, Tufted
Lysimachia vulgaris — Loosestrife, Yellow
Lythrum hyssopifolia — Grass Poly
Lythrum portula — Purslane, Water
Lythrum salicaria — Loosestrife, Purple
Lythrum virgatum — Loosestrife, Slender
Lytta vesicatoria — Beetle, Blister

M

Mabuya sp. — Skink, Mabuya
Macaca arctoides — Macaque, Stump-tailed
Macaca fascicularis — Macaque, Crab-eating
  or Long-tailed
Macaca fuscata — Macaque, Japanese
Macaca irus — Macaque, Crab-eating or
  Long-tailed
Macaca maurus — Macaque, Moor
Macaca mulatta — Macaque, Rhesus
Macaca nemestrina — Macaque, Pig-tailed
Macaca radiata — Macaque, Bonnet
Macaca silenus — Macaque, Lion-tailed
Macaca speciosa — Macaque, Stump-tailed
Macaca sylvana — Ape, Barbara
Macadamia ternifolia— Queensland Nut
Machaerirhynchus flaviventris — Flycatcher,
  Boat-billed
Machaeropterus regulus — Manakin, Striped
Machilis sp. — Silverfish
Macleaya cordata — Poppy, Plume
Maclura pomifera — Orange, Osage
Macodes sp. — Orchid, Jewel
Macoma sp. — Tellin
Macrobiotus sp. — Water Bear
Macrocephalon maleo — Maleo
Macrocystidia cucumis — Cucumber Slice
Macroderma gigas — Bat, False Vampire
Macrodipteryx longipennis — Nightjar,
  Standard-wing
Macrodipteryx vexillarius — Nightjar,
  Pennant-wing
Macrodontia sp. — Beetle, Longhorn
Macroglossum stellatarum — Hawkmoth,
  Hummingbird
Macrognathus aculeatus — Eel, Spiny

Macronectes giganteus — Petrel, Giant
Macronous sp. — Babbler, Tit
Macronyx ameliae — Long-claw
Macronyx capensis — Long-claw
Macronyx croceus — Long-claw
Macropis labiata — Bee
Macropodia sp. — Crab, Spider
Macropodus sp. — Paradise-fish
Macropus agilis — Wallaby, Agile or Sandy
Macropus canguru — Kangaroo, Great Gray
Macropus giganteus — Kangaroo, Great Gray
Macropus major — Kangaroo, Great Gray
Macropus parma — Wallaby, Parma or
  White-Throated
Macropus robustus — Wallaroo, Gray
Macropus ruficollis — Wallaby, Bennett's or
  Red-necked
Macropus rufogriseus — Wallaby, Bennett's
  or Red-necked
Macropus rufus — Kangaroo, Red
Macropygia amboinensis — Cuckoo-dove
Macropygia phasianella — Cuckoo-dove
Macropygia ruficeps — Cuckoo-dove
Macropygia unchall — Cuckoo-dove
Macroramphosus sp. — Snipefish
Macroscelides proboscideus — Elephant-
  shrew, Short-eared
Macrothylacia rubi — Moth, Fox
Macrotis lagotis — Bandicoot, Rabbit
Mactra sp. — Trough Shell
Maculinea arion — Butterfly, Large Blue
Madoqua kirki — Dik-dik, Long-snouted
Madoqua saltiana — Dik-dik, Phillips'
Madrepora sp. — Coral, Stag's Horn
Magnolia denudata — Lily Tree
Magnolia grandiflora — Magnolia, Southern
Magnolia obovata — Magnolia, Whiteleaf
  Japanese
Magnolia sieboldii — Magnolia, Oyama
Magnolia soulangeana — Magnolia, Saucer
Magnolia stellata — Magnolia, Star
Mahonia aquifolium — Oregon Grape
Mahonia japonica — Mahonia
Mahonia lomariifolia — Mahonia
Mahonia repens — Mahonia
Maia squinado — Crab, Spiny Spider
Maianthemum bifolium — Lily, May
Makaira nigricans — Marlin, Black or Blue
Makaira sp. — Marlin, Striped

Malachius sp. — Beetle, Flower
Malacochersus tornieri — Tortoise, Pancake
Malaconotus sp. — Shrike, Bush
Malacopteron sp. — Babbler, Tree
Malacoptila fulvigularis — Puffbird
Malacoptila panamensis — Puffbird
Malacoptila striata — Puffbird
Malacorhynchos membranaceus — Duck, Pink-eared
Malacosoma neustria — Moth, Lackey
Malapterurus electricus — Catfish, Electric
Malaxis monophyllos — Orchid, One-leaved Bog
Malaxis paludosa — Orchid, Bog
Malcomia maritima — Stock, Virginia
Malleus malleus — Oyster, Hammerhead
Malope trifida — Malope
Malpighia coccigera — Singapore Holly
Malpighia glabra — Barbados Cherry
Malpolon moilensis — Snake, Mailed
Malpolon monspessulanus — Snake, Montpellier
Malurus amabilis — Wren, Lovely
Malurus assimilis — Wren, Purple-backed
Malurus callainus — Wren, Turquoise
Malurus coronatus — Wren, Lilac-crowned
Malurus cyaneus — Wren, Blue
Malurus cyanotus — Wren, Blue-and-white
Malurus elegans — Wren, Banded or Red-winged
Malurus lamberti — Wren, Variegated
Malurus leuconotus — Wren, Black-and-white
Malurus leucopterus — Wren, Black-and-white
Malurus melanocephalus — Wren, Red-backed
Malurus melanotus — Wren, Black-backed
Malurus pulcherrimus — Wren, Blue-breasted
Malurus splendens — Wren, Splendid
Malus baccata — Crabapple, Siberian
Malus floribunda — Crabapple, Japanese Flowering
Malus pumila — Apple, Common
Malus purpurea — Crabapple, Red Flowering
Malus purpurea lemoinei — Crabapple, Purple Flowering
Malus sp. — Apple

Malus sylvestris — Apple, Crab or Crabapple
Malva alcea — Mallow, Vervain
Malva moschata — Mallow, Musk
Malva neglecta — Mallow, Dwarf
Malva pusilla — Mallow, Small
Malva sylvestris — Mallow, Common
Mamestra brassicae — Moth, Cabbage
Manacus manacus — Manakin, White-bearded
Manacus vitellinus — Manakin, Golden-collared
Mandevilla splendens — Dipladenia
Mandevilla suaveolens — Jasmine, Chilean
Mandingoa nitidula — Twinspot, Green-backed
Mandragora officinarum — Mandrake
Mandrillus sphinx — Mandrill
Manettia bicolor — Firecracker Vine
Manettia inflata — Firecracker Vine
Mangifera indica — Mango
Manihot sp. — Cassava or Tapioca Plant
Maniola jurtina — Butterfly, Meadow Brown
Manis crassicaudata — Pangolin, Indian
Manis gigantea — Pangolin, Giant
Manis javanica — Pangolin, Malayan
Manis longicaudata — Pangolin, Long-tailed Tree
Manis temmincki — Pangolin, Temminck's or Cape Giant
Manis tetradactyla — Pangolin, Long-tailed Tree
Manis tricuspis — Pangolin, Small-scaled Tree
Manorina flavigula — Miner, Yellow-throated
Manorina melanocephala — Miner, Noisy
Manorina melanophrys — Miner, Bell
Manta birostris — Manta, Atlantic
Mantella auriantiaca — Frog, Golden
Mantichora sp. — Beetle, Tiger
Mantis religiosa — Mantis, Praying
Mantispa styriacus — Fly, Mantis
Maranta arundinacea — Arrowroot
Maranta bicolor — Prayer Plant
Maranta leuconeura — Prayer Plant
Marasimius androsaceus — Fungus, Horsehair
Marasimius oreades — Mushroom, Fairy Ring

Marasimius rotula — Marasimius, Little Wheel

Marasimius scorodonius — Marasimius, Garlic

Marcgravia sp. — Marcgravia

Marchantia polymorpha — Liverwort, Common

Margaritifera margaritifera — Mussel, Pearl

Margarornis squamiger — Tree-runner, Pearled

Marianthus candidus — Marianthus

Marjoram hortensis — Marjoram, Sweet

Marmosa murina — Opossum, Mouse or Murine

Marmota bobak — Marmot, Bobak or Steppe

Marmota campschatica — Marmot

Marmota caudata — Marmot

Marmota marmota — Marmot, Alpine

Marphysa sanguinea — Worm, Rock

Marrubium vulgare — Horehound

Marsilea quadrifolia or M. vestita — Water-clover

Martes flavigula — Marten, Yellow-throated

Martes foina — Marten, Beech or Stone

Martes martes — Marten, Pine

Martes zibellina — Sable, Eurasian

Marthasterias glacialis — Starfish, Green or Spiny

Marumba quercus — Hawkmoth, Oak

Masdevallia sp. — Orchid

Mastacembelus sp. — Eel, Spiny

Matricaria chamomilla — Chamomile, German

Matricaria inodora — Mayweed, Scentless

Matricaria maritima — Mayweed, Scentless

Matricaria matricaroides — Pineapple Weed

Matricaria recutita — Chamomile, Wild

Matteuccia struthiopteris — Fern, Ostrich

Matthiola incana — Stock, Common

Matthiola sinuata — Stock, Sea

Maurolicus muelleri — Pearl-side

Maxillaria sp. — Orchid

Mazama americana — Deer, Brocket

Mazama gouazoubira — Deer, Brocket

Mazus reptans — Mazus

Mecistomela sp. — Beetle, Leaf

Meconema thalassinum — Bush-cricket, Oak

Meconopsis cambrica — Poppy, Welsh

Medicago arabica — Medick, Spotted

Medicago falcata — Medick, Yellow or Sickle

Medicago hispida — Medick, Toothed or Hairy

Medicago lupulina — Medick, Black

Medicago marina — Medick, Sea

Medicago minima — Medick, Small

Medicago polymorpha — Medick, Hairy or Toothed

Medicago sativa — Alfalfa

Medicago sp. — Medick

Medinilla magnifica — Medinilla

Medinilla sp. — Medinilla

Mediolobivia sp. — Mediolobivia

Megacephalon maleo — Maleo

Megaceryle maxima — Kingfisher, Giant

Megachile sp. — Bee, Leaf-cutting

Megadyptes antipodes — Penguin, Yellow-eyed

Megalaima asiatica — Barbet, Blue-cheeked or -throated

Megalaima franklinii — Barbet, Golden-throated

Megalaima haemacephala — Barbet, Coppersmith or Crimson-breasted

Megalaima oorti— Barbet, Black-browed

Megalaima virens — Barbet, Great Hill

Megalamphodus megalopterus — Tetra, Black Phantom

Megalamphodus sweglesi — Tetra, Red Phantom

Megaleia rufa — Kangaroo, Red

Megalobatrachus japonicus — Salamander, Giant

Megalomma vesiculosum — Worm, Burrowing

Megaloprepia magnifica — Pigeon, Magnificent Fruit

Megalops cyprinoides — Herring, Ox-eye

Megalornis antigone — Crane, Sarus

Megalornis grus — Crane, Common

Megalornis leucogeranus — Crane, Siberian White

Megalotis zerda — Fox, Fennec

Megalurus gramineus — Grassbird

Megalurus palurus — Grassbird

Megalurus timoriensis — Grassbird

Megophrys nasuta — Frog, Asiatic Horned

Megapodius sp. — Fowl, Jungle or Scrub

Megaptera novaeangliae — Whale, Humpback

Megarhynchus pitangua — Flycatcher, Boat-billed

Megasoma elephas — Beetle, Elephant

Mehelya sp. — Snake, File

Melaenornis silens — Flycatcher, Fiscal

Melaleuca sp. — Cajeput Tree

Melampsalta sp. — Cicada

Melampyrum arvense — Cow-wheat, Field

Melampyrum cristatum — Cow-wheat, Crested

Melampyrum nemorosum — Cow-wheat, Blue-topped

Melampyrum pratense — Cow-wheat, Common

Melampyrum sylvaticum — Cow-wheat, Wood

Melanargia galathea — Butterfly, Marbled White

Melandrium album — Campion, White

Melandrium dioicum — Campion, Red

Melandrium rubrum — Campion, Red

Melanitta fusca — Scoter, Velvet

Melanitta nigra — Scoter, Black or Common

Melanitta perspicillata — Scoter, Surf

Melanocetus sp. — Angler, Deep-sea

Melanochlora sultanea — Tit, Sultan

Melanocorypha calandra — Lark, Calandra

Melanocorypha leucoptera — Lark, White-winged

Melanocorypha yeltoniensis — Lark, Black

Melanodera melanodera — Finch, Black-throated

Melanogrammus aeglefinus — Haddock

Melanoleuca sp. — Melanoleuca

Melanosuchus niger — Caiman, Black

Melanotaenia maccullochi — Rainbow-fish, Dwarf

Melanotaenia nigrans — Rainbow-fish, Australian Red-tailed

Melasoma populi — Beetle, Leaf

Meles meles — Badger, Eurasian

Melia azedarach — Chinaberry or Bead Tree

Melianthus sp. — Honey-flower

Melica ciliata — Melick, Hairy

Melica nutans — Melick, Mountain or Nodding

Melica uniflora — Melick, Wood

Melierax canorus — Goshawk, Chanting

Melierax gabar — Goshawk, Gabar

Melierax metabates — Goshawk, Chanting

Melierax musicus — Goshawk, Chanting

Melilotus alba — Melilot, White

Melilotus altissima — Melilot, Tall or Yellow

Melilotus officinalis — Melilot, Common

Meliornis niger — Honey-eater, White-cheeked

Meliornis novaehollandiae — Honey-eater, New Holland

Meliphaga albilineata — Honey-eater, White-lined

Meliphaga chrysops — Honey-eater, Yellow-faced

Meliphaga gracilis — Honey-eater, Graceful

Meliphaga lewinii — Honey-eater, Lewin

Meliphaga leucotis — Honey-eater, White-eared

Meliphaga melanops — Honey-eater, Yellow-tufted

Meliphaga ornata — Honey-eater, Yellow-plumed

Melissa officinalis — Balm, Lemon

Melitaea sp. — Butterfly, Heath Fritillary

Melithreptes albogularis — Honey-eater, White-throated

Melithreptes brevirostris — Honey-eater, Brown-headed

Melithreptes laetior — Honey-eater, Golden-backed

Melithreptes lunatus — Honey-eater, White-naped

Melitoda sp. — Gorgonian

Melittis melissophyllum — Balm, Bastard

Melittophagus bullockoides — Bee-eater, White-fronted

Melittophagus bulocki — Bee-eater, Red-throated

Melittophagus pusillus — Bee-eater, Little

Mellinus arvensis — Wasp, Digger

Mellivora capensis — Honey-badger

Melocactus sp. — Turk's Cap

Meloe sp. — Beetle, Oil or Blister

Melogale sp. — Badger, Ferret

Melolontha melolontha — Beetle, Cockchafer

Melophagus ovinus — Tick, Sheep

Melopsittacus undulatus — Budgerigar

Melursus ursinus — Bear, Sloth

Menispermum sp. — Moonseed

Mentha aquatica — Mint, Water

Mentha arvensis — Mint, Corn or Field

Mentha longifolia — Horsemint
Mentha piperita — Peppermint
Mentha pulegium — Penny-royal
Mentha requienii — Mint, Corsican or
    Creeping
Mentha rotundifolia — Mint, Round-leaved or
    Apple-scented
Mentha spicata — Spearmint
Menura alberti — Lyrebird, Prince Albert's
Menura novaehollandiae — Lyrebird, Superb
Menura superba — Lyrebird, Superb
Menyanthes trifoliata — Bogbean or
    Buckbean
Mercuralis annua — Mercury, Annual
Mercuralis perennis — Mercury, Dog's
Merganetta armata — Duck, Torrent
Mergus albellus — Smew
Mergus merganser — Merganser, Common
Mergus serrator — Merganser, Red-breasted
Merlangius merlangus — Whiting
Merluccius merluccius — Hake
Merodon equestris — Fly, Large Narcissus
Merops apiaster — Bee-eater, European
Merops bullockoides — Bee-eater,
    White-fronted
Merops bulocki — Bee-eater, Red-throated
Merops leschenaulti — Bee-eater,
    Chestnut-headed
Merops nubicoides — Bee-eater, Carmine
Merops nubicus — Bee-eater, Carmine
Merops orientalis — Bee-eater, Green
Merops ornatus — Bee-eater, Rainbow
Merops philippinus — Bee-eater, Blue-tailed
Merops pusillus — Bee-eater, Little
Merops superciliosus — Bee-eater,
    Blue-cheeked
Merops viridus — Bee-eater,
    Chestnut-headed
Mertensia maritima — Shorewort, Northern
Mertensia virginia — Bluebell, Virginia
Mertensiella sp. — Salamander
Merulius sp. — Fungus, Dry Rot
Mesoacidalia aglaia — Butterfly, Dark Green
    Fritillary
Mesocricetus auratus — Hamster, Golden
Mesoenas unicolor — Mesite, Brown
Mespilus germanica — Medlar
Messor barbarus — Ant, Harvester
Meta sp. — Spider, Cave Orb-weaver

Metachirus crassicaudatus — Opossum,
    Thick-tailed
Metachirus nudicaudatus — Opossum,
    Four-eyed
Metasequoia glyptostroboides — Redwood,
    Dawn
Metoecus paradoxus — Beetle, Wasp Fan
Metridium sp. — Sea Anemone, Plumose
Metrioptera brachyptera — Bush-cricket, Bog
Metrosideros sp. — Rata
Metroxylon sago — Palm, Sago
Metynnis sp. — Silver Dollar Fish
Meum athamanticum — Spignel
Mibora minima — Sand-grass, Early
Michauxia sp. — Michauxia
Michelia sp. — Banana Shrub
Mico argentata — Marmoset, Silvery
Micralestes interruptus — Tetra, Congo
Micrarchus sp. — Stick-insect
Micrastur ruficollis — Falcon, Barred Forest
Micrastur semitorquatus — Falcon, Collared
    Forest
Microcebus murinus — Lemur, Lesser Mouse
Microcephalophis gracilis — Sea-snake
Microchirus sp. — Solenette
Microeca fascinans — Flycatcher, Brown
Microeca flavigaster — Flycatcher,
    Lemon-breasted
Microeca leucophaea — Flycatcher, Brown
Microgale longicauda — Tenrec, Long-tailed
Microhierax caerulescens — Falconet,
    Red-legged
Microhierax erythrogonys — Falconet,
    Red-legged
Micromesistius sp. — Whiting, Blue
Micrommata sp. — Spider
Micromys minutus — Mouse, European
    Harvest
Micronisus gabar — Goshawk, Gabar
Micropotamogale lamottei — Otter-shrew
Micropteryx calthella — Moth, Primitive
Micropus erectus — Cudweed, Upright False
Microrhopias quixensis — Antwren,
    Dot-winged
Microstomus kitt — Sole, Lemon
Microtus agrestis — Vole, Field or
    Short-tailed
Microtus arvalis — Vole, Common
Microtus nivalis — Vole, Alpine or Snow

Microtus orcadensis — Vole, Common
Micrurus frontalis — Snake, Brazilian Coral
Milium effusum — Millet, Wood
Milium scabrum — Millet, Dwarf
Miltochrista miniata — Moth, Rosy Footman
Miltonia sp. — Orchid, Pansy
Milvago chimachima — Caracara, Yellow-headed
Milvus aegyptius — Kite, Yellow-billed
Milvus migrans — Kite, Black
Milvus milvus — Kite, Red
Mimas tiliae — Hawkmoth, Lime
Mimosa pudica — Sensitive Plant
Mimosa sensitiva — Sensitive Plant
Mimulus cupreus — Monkey-flower
Mimulus guttatus — Monkey-flower, Common
Mimulus sp. — Monkey-flower
Mimusops zeyheri — Moepel
Minla cyanouroptera — Siva, Blue-winged
Mino coronatus — Mynah, Golden-crested
Minuartia rubella — Sandwort, Mountain
Minuartia stricta — Sandwort, Bog
Minuartia tenuifolia — Sandwort, Fine-leaved
Minuartia verna — Sandwort, Spring or Vernal
Miopithecus talapoin — Guenon, Dwarf
Mirabilis jalapa — Marvel of Peru
Mirafra sp. — Lark, Bush
Mirounga leonina — Seal, Elephant (Southern)
Miscanthus sinensis — Eulalia
Misgurnus fossilis — Loach, Pond
Misopates orontium — Snapdragon, Lesser
Misumena vatia — Spider, Crab
Mitchella repens — Partridge-berry
Mitra sp. — Miter Shell
Mitu mitu — Curassow, Razor-billed
Mnium hornum — Moss, Forest-star
Modiolus barbatus — Mussel, Horse
Modiolus modiolus — Mussel, Horse
Moehringia muscosa — Sandwort, Mossy
Moehringia trinervia — Sandwort, Three-nerved or Three-veined
Moenchia erecta — Chickweed, Upright
Moenkhausia pittieri — Tetra
Moenkhausia sanctaefilomenae — Tetra

Mogurnda mogurnda—Gudgeon, Australian or Purple-striped
Mohoua albicilla — Whitehead
Mohoua ochrocephala — Yellowhead
Mola mola — Sunfish, Ocean
Molinia caerulea — Moor-grass, Purple
Mollugo verticillata — Carpetweed or Chickweed, Indian
Moloch horridus — Devil, Mountain or Thorny
Molothrus bonariensis — Cowbird, Glossy or Shining
Moltkia sp. — Moltkia
Molva molva — Ling
Momordica charantia — Balsam-pear
Momotus momota — Motmot, Blue-crowned
Monacanthus sp. — Filefish, Fringed
Monachus monachus — Seal, Mediterranean Monk
Monachus schauinslandi — Seal, Mediterranean Monk
Monarcha azurea — Flycatcher, Black-naped Blue
Monarcha frater — Flycatcher, Pearly
Monarcha leucotis — Flycatcher, White-eared
Monarcha melanopsis — Flycatcher, Black-faced
Monarcha trivirgata — Flycatcher, Spectacled
Moneses uniflora — Wintergreen, One-flowered
Monocentrus japonicus — Pine-cone Fish
Monocirrhus polycanthus — Leaf-fish, South American
Monodactylus argenteus — Fingerfish, Common
Monodactylus sebae — Silverfish
Monodon monoceros — Narwhal
Monodora myristica — Nutmeg, Calabash
Monotropa hypopitys — Bird's-nest, Yellow
Monstera deliciosa — Swiss Cheese Plant
Monstera pertusa — Swiss Cheese Plant
Montia fontana — Water Blinks
Montia perfoliata — Spring Beauty
Montia sibirica — Claytonia, Pink
Monticola sp. — Thrush, Rock
Montifringilla adamsi — Finch, Snow
Montifringilla nivalis — Finch, Snow
Monvillea sp. — Monvillea

Moraea pavonia — Iris, Peacock or Peacock Flower

Moraea sp. — Tulp, Yellow

Morelia argus or spilotes — Python, Carpet or Diamond

Moricandia arvensis — Cabbage, Purple or Violet

Morina longifolia — Himalayan Whorlflower of Nepal

Moringa oleifera — Horseradish Tree

Morisia monantha — Morisia

Mormolyce phyllodes — Beetle, Fiddler or Violin

Morpho sp. — Butterfly, Morpho

Morulius chrysophekadion — "Shark", Black

Morus alba — Mulberry, White

Morus bassanus — Gannet

Morus capensis — Gannet, Cape

Morus nigra — Mulberry, Black

Morus rubra — Mulberry, Red

Moschus moschiferus — Deer, Musk

Motacilla aguimp — Wagtail, African Pied

Motacilla alba — Wagtail, Pied or White

Motacilla capensis — Wagtail, Cape

Motacilla cinerea — Wagtail, Gray

Motacilla citreola — Wagtail, Citrine

Motacilla flava — Wagtail, Yellow

Motacilla flava flava — Wagtail, Blue-headed

Motacilla maderaspatensis — Wagtail, Pied or White

Mucuna sp. — Mucuna

Muehlenbeckia sp. — Pohuehue

Mugil sp. — Mullet

Muhlenbergia schreberi — Nimbleweed

Mulleripicus pulverulentus — Woodpecker, Great Slaty

Mullus surmuletus — Mullet, Red

Mungos mungo — Mongoose, Banded or Zebra

Munsa clypealis — Cicada

Muntiacus sp. — Muntjac

Muraena sp. — Eel, Moray

Murex sp. — Murex Shell

Murraya sp. — Jasmine, Orange

Mus musculus — Mouse, House

Musa paradisiaca — Banana

Musa sapientum — Banana

Musa sp. — Banana

Musca domestica — Fly, House

Muscardinus avellanarius — Dormouse, Common or Hazel

Muscari armeniacum — Grape Hyacinth, Armenian

Muscari atlanticum — Grape Hyacinth

Muscari botryoides — Grape Hyacinth, Common or Small

Muscari comosum — Grape Hyacinth, Tassel

Muscari sp. — Grape Hyacinth

Muscicapa latirostris — Flycatcher, Brown

Muscicapa parva — Flycatcher, Red-breasted

Muscicapa striata — Flycatcher, Spotted

Muscisaxicola sp. — Tyrant, Ground

Muscivora tyrannus — Flycatcher, Fork- or Swallow-tailed

Musculus discors or M. marmoratus — Mussel

Musophaga rossae — Turaco, Ross'

Mussaenda erythrophylla — Red-flag Bush

Mustela erminea — Weasel, Short-tailed

Mustela frenata — Weasel, Long-tailed

Mustela lutreola — Mink

Mustela nivalis — Weasel, European

Mustela putorius — Polecat, European

Mustela sibirica — Weasel, Siberian

Mustelus mustelus — Hound, Smooth

Mutilla sp. — Wasp, Velvet-ant

Mutinus caninus — Stinkhorn, Dog

Mutisia oligodon — Mutisia

Mutisia sp. — Mutisia

Mycelis muralis — Lettuce, Wall

Mycena epipterygia — Mycena, Yellow-stemmed

Mycena haematopus — Mycena, Bleeding

Mycena polygramma — Roof Nail

Mycena pura — Mycena, Amethyst or Lilac

Mycteria ibis — Ibis, Wood or Yellow-billed

Mycteria leucocephalus — Stork, Painted

Myctophum sp. — Lanternfish

Myiagra cyanoleuca — Flycatcher, Satin

Myiagra inquieta — Flycatcher, Restless

Myiagra rubecula — Flycatcher, Leaden

Myiagra ruficollis — Flycatcher, Broad-billed

Myioborus ornatus — Redstart, Golden-fronted

Myioborus torquatus — Redstart, Collared

Myiodynastes maculatus — Flycatcher, Streaked

Myiophoneus blighi — Thrush, Whistling

Myiophoneus caeruleus — Thrush, Whistling

Myiopsitta monachus — Parakeet, Green or Quaker

Myiornis auricularis — Tyrant, Pygmy

Myliobatis aquila — Ray, Eagle

Mylossoma argenteum — Silver Dollar Fish

Myocastor coypu — Nutria

Myoporum sp. — Ngaio

Myoprocta acouchi — Acouchi

Myoprocta pratti — Acouchi

Myopus schisticolor — Lemming

Myosciurus pumilio — Squirrel, Pygmy

Myosotidium hortensia — Forget-me-not, Chatham Island

Myosotis alpestris — Forget-me-not, Alpine

Myosotis arvensis — Forget-me-not, Common or Field

Myosotis discolor — Forget-me-not, Yellow

Myosotis palustris — Forget-me-not, Water

Myosotis secunda — Forget-me-not, Marsh

Myosotis scorpiodes — Forget-me-not, Water

Myosotis sp. — Forget-me-not

Myosotis sylvatica — Forget-me-not, Wood

Myosoton aquaticum — Chickweed, Water or Great

Myosurus minimus — Mouse-tail

Myotis bechsteini — Bat, Bechstein's

Myotis daubentoni — Bat, Daubenton's or Water

Myotis myotis — Bat, Mouse-eared

Myotis mystacinus — Bat, Whiskered

Myotis nattereri — Bat, Natterer's

Myrica cerifera — Bayberry

Myrica gale — Sweet Gale

Myrica pensylvanica — Bayberry

Myricaria germanica — Tamarisk, German

Myriophyllum sp. — Water-milfoil

Myriophyllum spicatum — Water-milfoil, Spiked

Myriophyllum verticillatum — Water-milfoil, Whorled

Myripristis murdjan — Soldierfish, Black-tipped

Myristica fragrans — Nutmeg and Mace

Myrmecia sp. — Ant

Myrmeciza exsul — Antbird, Chestnut-backed or -tailed

Myrmeciza hemimelaena — Antbird, Chestnut-backed or -tailed

Myrmecobius fasciatus — Anteater, Banded or Marsupial

Myrmecodia sp. — Myrmecodia

Myrmecophaga tridactyla or M. jubata— Anteater, Giant or Great

Myrmeleon sp. — Ant-lion

Myrmeleotettrix maculatus — Grasshopper, Mottled

Myrmotherula sp. — Antwren

Myrrhis odorata — Sweet Cicely

Myrsine sp. — Myrsine

Myrtillocactus geometrizans — Blue Flame

Myrtus communis — Myrtle, Common or True

Mystus vittatus — Catfish, Striped Dwarf

Mytilus edulis — Mussel, Blue or Edible

Myxilla sp. — Sponge

Myxine glutinosa — Hagfish

Myzantha flavigula — Miner, Yellow-throated

Myzantha melanocephala — Miner, Noisy

Myzomela sp. — Honey-eater

Myzostoma sp. — Worm, True

Myzus persicae — Aphid, Green Peach

N

Naemorhaedus goral — Goral, Gray or Short-tailed

Naja haje — Cobra, Egyptian

Naja naja — Cobra, Hooded or Indian

Naja nigricollis — Cobra, Black-necked Spitting

Naja nivea — Cobra, Cape

Najas sp. — Naiad

Nandayus nenday — Parrot, Black-headed

Nandina domestica — Bamboo, Heavenly or Sacred

Nandinia binotata — Palm-civet, African or Two-spotted

Nannopterum harrisi — Cormorant, Flightless

Nandus nandus or N. nebulosus — Leaf-fish

Nanacara anomala — Cichlid, Golden-eyed Dwarf

Nannaethiops tritaeniatus — Characin, Three-striped African

Nannaethiops unitaeniatus — Characin, One-striped African

Nannobrycon eques — Pencilfish, Tube-mouthed

Nannochromis sp. — Cichlid, African

Nannostomus sp. — Pencilfish

Napoleonaea imperialis — Napoleonaea

Napothera sp. — Babbler, Wren

Narcissus biflorus — Primrose Peerless

Narcissus bulbocodium — Daffodil, Hoop Petticoat

Narcissus cyclamineus — Narcissus, Cyclamen-flowered

Narcissus jonquilla — Jonquil

Narcissus poeticus — Narcissus, Pheasant's-eye

Narcissus pseudonarcissus — Daffodil, Wild

Narcissus tazetta — Narcissus, Polyanthus

Nardostachys jatamansi — Nardostachys

Nardus stricta — Mat-grass

Narthecium ossifragum — Asphodel, Bog

Nasalis larvatus — Monkey, Proboscis

Naso lituratus — Unicorn Fish, Striped-face

Nassarius sp. — Basket Shell

Nasturtium officinale — Watercress, Common

Nasua narica — Coati, White-nosed

Nasua nasua — Coati, Ring-tailed or Red

Natica sp. — Snail, Moon

Natrix maura — Snake, Viperine

Natrix sp. — Snake, Grass or Water

Natrix viperinus — Snake, Viperine

Nauclea latifolia — Peach, Guinea or African

Naucrates ductor — Pilotfish

Naumburgia thrysiflora — Loosestrife, Tufted

Nautilus sp. — Nautilus

Necrobia sp. — Beetle, Hide or Leather

Necrophorus sp. — Beetle, Sexton

Nectarinia afra — Sunbird, Double-collared

Nectarinia amethystina — Sunbird, Black

Nectarinia asiatica — Sunbird, Purple

Nectarinia calcostetha — Sunbird, Copper-throated

Nectarinia chalybea — Sunbird, Double-collared

Nectarinia famosa — Sunbird, Malachite

Nectarinia jugularis — Sunbird, Olive-backed

Nectarinia mediocris — Sunbird, Double-collared

Nectarinia reichenowi — Sunbird, Golden-winged

Nectarinia senegalensis — Sunbird, Scarlet-chested

Nectarinia sperata — Sunbird, Van Hasselt's

Nectarinia talatala — Sunbird, White-bellied

Nectarinia violacea — Sunbird, Orange-breasted

Nectarinia zeylonica — Sunbird, Purple

Neillia longiracemosa — Neillia

Nelumbo nucifera — Lotus, Sacred

Nemacheilus barbatulus — Loach, Stone

Nematanthus wettsteinii — Goldfish Plant

Nematobrycon palmeri — Tetra, Emperor

Nematocentrus maccullochi — Rainbow-fish, Dwarf

Nematus ribesii — Sawfly, Currant or Gooseberry

Nemichythys scolopaceus — Snipe-eel

Nemobius sylvestris — Cricket, Wood

Nemoptera sp. — Lacewing, Thread

Neoaratus sp. — Fly, Robber

Neobisium sp. — Scorpion, Book or False

Neoceratodus forsteri — Lungfish, Australian

Neochmia phaeton — Finch, Crimson

Neocrex erythrops — Crake, Paint-billed

Neodrepanis sp. — Sunbird, False

Neofelis nebulosa — Leopard, Clouded

Neofinetia falcata — Orchid, Samurai

Neolamprima sp. — Beetle, Stag

Neolebias ansorgei — Characin, Ansorge's

Neomarica caerulea — Marica

Neomorphus geoffroyi — Cuckoo, Ground

Neomys anomalus — Shrew, European Water

Neomys fodiens — Shrew, European Water

Neophema bourkii — Parrot, Bourke's

Neophema elegans — Parrot, Elegant

Neophema splendida — Parrot, Scarlet-chested

Neophoca cinerea — Sea Lion, Australian

Neophron percnopterus — Vulture, Egyptian

Neoporteria sp. — Neoporteria

Neoraimondia sp. — Neoraimondia

Neoregelia sp. — Neoregelia

Neositta chrysoptera — Sitella

Neositta pileata — Sitella

Neotermes sp. — Termite

Neotetracus sinensis — Hedgehog

Neotinea intacta — Orchid, Dense-flowered
Neotragus batesi — Antelope, Bates' Pygmy
Neotragus moschatus — Suni
Neotragus pygmaeus — Antelope, Royal
Neottia nidus-avis — Orchid, Bird's-nest
Nepa sp. — Water-scorpion
Nepenthes sp. — Pitcher Plant
Nepeta cataria — Catmint
Nepeta x faassenii — Catmint, Garden
Nepeta longibracteata — Catmint
Nepeta mussinii — Catmint
Nepeta nuda — Catmint, Hairless
Nephelium sp. — Lychee Nut
Nephila sp. — Spider, Golden Silk
Nephrolepis exaltata — Fern, Boston
Nephrops norvegicus — Lobster, Norway or
    King
Nephrurus laevis or N. asper — Gecko,
    Knob-tailed
Neptis sp. — Butterfly, Glider
Neptunea antiqua — Spindle Shell
Neptunus pelagicus — Crab, Swimming
Nereis sp. — Worm, Clam
Nerine bowdenii — Agapanthus, Pink or
    Nerine
Nerine sarniensis — Lily, Guernsey
Nerine sp. — Nerine
Nerium oleander — Oleander
Nerophis sp. — Pipefish
Nertera granadensis — Bead Plant
Nesasio solomonensis — Owl, Fearful
Neslia paniculata — Mustard, Ball
Nesomimus melanotis — Mockingbird,
    Galapagos
Nesomimus parvulus — Mockingbird,
    Galapagos
Nesomimus trifasciatus — Mockingbird,
    Galapagos
Nestor meridionalis — Kaka
Nestor notabilis — Kea
Netta erythrophthalma — Pochard, African
    and South American
Netta rufina — Pochard, Red-crested
Nettapus auritus — Goose, African Pygmy
Nettapus coromandelianus — Goose, Indian
    Pygmy
Nettapus pulchellus — Goose, Green Pygmy
Neurobasis chinensis — Dragonfly

Neuroterus quercusbaccarum — Wasp,
    Spangle Gall
Nicandra physaloides — Apple-of-Peru
Nicodemia diversiflora — Oak, Indoor
Nicolai sp. — Ginger, Torch
Nicotiana rustica — Tobacco, Small
Nicotiana sp. — Tobacco
Nicotiana tabacum — Tobacco, Large
Nigella arvensis — Fennel-flower
Nigella ciliata — Fennel-flower
Nigella damascena — Love-in-a-mist
Nigella sativa — Fennel-flower
Nigretella nigra — Orchid, Black Vanilla
Niltava grandis — Niltava, Large
Niltava sundara — Niltava, Rufous-bellied
Ninox boobook — Owl, Boobook
Ninox connivens — Owl, Barking
Ninox novaeseelandiae — Owl, Boobook
Ninox philippensis — Owl, Philippine
    Boobook
Ninox rufa — Owl, Rufous
Ninox scutulata — Owl, Brown or Oriental
    Hawk
Ninox strenua — Owl, Great Hawk
Nipponia nippon — Ibis, Japanese Crested
Noctua pronuba — Moth, Large Yellow
    Underwing
Nolana sp. — Bellflower, Chilean
Nolanea sp. — Nolanea
Nomada sp. — Bee, Cuckoo or Parasitic
Nomeus gronovi — Man-of-war Fish
Nonagria typhae — Moth, Bulrush Wainscot
Nonea pulla — Wrinklenut
Nonnula sp. — Nunlet
Nopalxochia phyllanthoides — Cactus,
    Empress
Notaden bennetti — Frog, Catholic
Notechis scutatus — Snake, Tiger
Notharchus macrorhynchus — Puffbird,
    Large-billed
Nothobranchius sp. — Nothobranchius
Nothofagus antarctica — Beech, Antarctic
Nothofagus obliqua — Beech, Roble
Nothofagus procera — Beech, Southern or
    Raoul
Nothofagus solandri — Beech, New Zealand
    Black
Nothoprocta peridcaria — Tinamou

Notiomystis cincta — Stitchbird
Notobasis syriaca — Thistle, Syrian
Notocactus leninghausii — Cactus, Golden Ball
Notocactus scopa — Cactus, Silver Ball
Notocactus sp. — Cactus, Ball
Notodonta dromedarius — Moth, Iron Prominent
Notodonta ziczac — Moth, Pebble Prominent
Notomys sp. — Mouse, Hopping or Kangaroo
Notonecta sp. — Backswimmer
Notophoyx novaehollandiae — Heron, White-faced
Notophoyx picata — Heron, Pied
Notopterus chitala — Featherback
Notornis hochstetteri — Takahe
Notornis mantelli — Takahe
Notoryctes typhlops — Mole, Marsupial or Pouched
Notospartium sp. — Broom, Pink
Nucella sp. — Dogwinkle
Nucifraga caryocatactes — Nutcracker
Nucula sp. — Clam, Nut
Numenius arquata — Curlew, Common or Eurasian
Numenius madagascariensis — Curlew, Long-billed
Numenius phaeopus — Whimbrel
Numenius tenuirostris — Curlew, Slender-billed
Numenius variegatus — Whimbrel
Numida meleagris — Guinea-fowl, Helmeted or Tufted
Nuphar lutea — Water-lily, Yellow
Nuphar sp. — Water-lily or Spatterdock
Nuytsia floribunda — Christmas Tree
Nyctalus noctula — Bat, Noctule or Nocturnal
Nyctea scandiaca — Owl, Snowy
Nyctereutes procyonoides — Dog, Raccoon-like
Nyctibius grandis — Potoo, Great
Nyctibius griseus — Potoo, Common
Nycticebus coucang — Loris, Slow
Nycticorax caledonicus — Heron, Nankeen or Rufous Night
Nycticorax nicticorax — Heron, Black-crowned Night
Nycticryphes semicollaris — Snipe, Painted

Nyctyornis amictus — Bee-eater, Red-bearded or -breasted
Nyctyornis athertoni — Bee-eater, Blue-bearded
Nymphaea alba — Water-lily, White
Nymphaea caerulea — Lotus, Blue
Nymphaea capensis — Water-lily, Cape Blue
Nymphaea rubra — Water-lily, Red
Nymphaea sp. — Water-lily
Nymphalis antiopa—Butterfly, Camberwell Beauty
Nymphalis io—Butterfly, Peacock
Nymphalis polychloros—Butterfly, Large Tortoise-shell
Nymphicus hollandicus — Cockatiel
Nymphoides peltata — Water-lily, Fringed
Nymphon sp. — Sea Spider
Nymphula nympheata — Moth, Brown China Mark
Nypa fruticans — Palm, Nipa
Nysson spinosus — Wasp, Cuckoo
Nystalus radiatus — Puffbird, Barred

O

Obregonia denegrii — Cactus, Artichoke
Oceanites gracilis — Petrel, White-vented Storm
Oceanites oceanicus — Petrel, Wilson's Storm
Oceanodroma castro — Petrel, Madeiran Storm
Oceanodroma leucorrhoa — Petrel, Leach's
Oceanodroma tethys — Petrel, Storm
Ochlodes venata — Butterfly, Large Skipper
Ochna atropurpurea — Carnival Bush
Ochna serrulata — Carnival Bush
Ochna sp. — Carnival Bush
Ochotona alpinus — Pika, Alpine or Siberian
Ochotona hyperborea — Pike, Northern or Japanese
Ochotona pusilla — Pika, Steppe
Ochotona roylei — Pika, Royle's
Ochthornis littoralis — Tyrant, Water
Ocimum basilicum — Basil, Sweet
Octocyon megalotis — Fox, Bat-eared
Octopus vulgaris — Octopus, Common
Ocyphaps lophotes — Pigeon, Crested
Ocypode sp. — Crab, Ghost

Ocypus olens — Coach-horse, Devil's
Odobenus rosmarus — Walrus
Odocoileus bezoarcticus — Deer, Pampas
Odocoileus dichotomus — Deer, Marsh
Odontites lutea — Bartsia, Yellow
Odontites verna — Bartsia, Red
Odontoglossum grande — Orchid, Tiger
Odontolabis sp. — Beetle, Stag
Odonus niger — Filefish, Red-toothed
Oecophylla sp. — Ant, Leaf-tying or Weaving
Oedemera sp. — Beetle, Flour
Oedipoda caerulescens — Grasshopper,
    Blue-winged Wasteland
Oedipomidas oedipus — Marmoset, Cotton-
    topped or Pinche
Oedura sp. — Gecko, Fat-tailed or Robust
Oena capensis — Dove, Cape or Manaqua
Oenanthe aquatica — Dropwort, Fine-leaved
    Water
Oenanthe crocata — Dropwort, Hemlock
    Water
Oenanthe deserti — Wheatear, Desert
Oenanthe fistulosa — Dropwort, Common
    Water
Oenanthe hispanica — Wheatear, Black-
    eared
Oenanthe isabellina — Wheatear, Isabelline
Oenanthe lachenalii — Dropwort, Parsley
    Water
Oenanthe leucopyga — Wheatear,
    White-crowned Black
Oenanthe leucura — Wheatear, Black
Oenanthe monacha — Wheatear, Hooded
Oenanthe monticola — Chat, Mountain
Oenanthe oenanthe — Wheatear
Oenanthe picata — Chat, Pied
Oenanthe pileata — Chat, Pied
Oenanthe pleschanka — Wheatear, Pied
Oenanthe xanthoprymna — Wheatear,
    Red-tailed
Oenothera biennis — Evening-primrose,
    Common
Oenothera erythrosepala — Evening-
    primrose, Large-flowered
Oenothera sp. — Evening-primrose
Oenothera stricta — Evening-primrose,
    Fragrant
Oestrus ovis — Botfly, Sheep
Okapia johnstoni — Okapi

Olea europaea — Olive
Olearia sp. — Daisy-bush
Olivancillaria sp. — Olive Shell
Olivia sp. — Olive Shell
Ommatophoca rossi — Seal, Ross
Omocestus sp. — Grasshopper, Green
Omphalina sp. — Omphalina
Omphalodes sp. — Navel-wort
Omphalodes verna — Blue-eyed-Mary
Ompok bimaculatus — Catfish, Glass or
    Ompok
Onchidoris sp. — Nudibranch
Onchiurus sp. — Springtail
Oncidium crispum — Orchid, Dancing Lady
Oncidium ornithorhynchum — Orchid,
    Dancing Lady
Oncidium papilio — Orchid, Butterfly
Oncidium sp. — Orchid, Oncidium
Oncidium varicosum — Orchid, Dancing Lady
Oncoba spinosa — Fried Egg Tree
Oneirodes carlsbergi — Angler, Deep-sea
Oniscus sp. — Sowbug
Onobrychis viciifolia — Sainfoin
Onoclea sensibilis — Fern, Sensitive or Bead
Ononis natrix — Restharrow, Large Yellow
Ononis reclinata — Restharrow, Small
Ononis repens — Restharrow, Common
Ononis sp. — Restharrow
Ononis spinosa — Restharrow, Prickly
Onopordum acanthium — Thistle, Scotch or
    Cotton
Onos mustela — Rockling, Five-bearded
Onosma sp. — Golden Drop
Onychogalea sp. — Wallaby, Nail-tail
Onychognathus morio — Starling,
    Red-winged
Onychorhynchus coronatus — Flycatcher,
    Royal
Onychorhynchus mexicanus — Flycatcher,
    Royal
Operophtera brumata — Moth, Winter
Ophicephalus sp. — Snakehead
Ophioderma longicauda — Brittle-star
Ophioglossum sp. — Fern, Adder's-tongue
Ophioglossum vulgatum — Fern, Adder's-
    tongue
Ophion sp. — Wasp, Ichneumon
Ophiophagus hannah — Cobra, King
Ophiopogon jaburan — Lily-turf

389

Ophiopogon japonicus — Lily-turf
Ophiothrix fragilis — Brittle-star, Common
Ophisaurus sp. — Lizard, Glass
Ophiura sp. — Brittle-star
Ophrys apifera — Orchid, Bee
Ophrys bertolonii — Orchid, Bertoloni's Bee
Ophrys bombyliflora — Orchid, Bumble Bee
Ophrys fuciflora — Orchid, Late Spider
Ophrys fusca — Orchid, Brown Bee
Ophrys insectifera — Orchid, Fly
Ophrys lutea — Orchid, Yellow Bee
Ophrys scolopax — Orchid, Woodcock
Ophrys speculum — Orchid, Mirror
Ophrys sphegodes — Orchid, Early Spider
Ophrys tenthredinifera — Orchid, Sawfly
Opilio parietinus — Harvestman
Opisthocomus hoazin — Hoatzin
Opisthograptis luteolata — Moth, Brimstone
Oplismenus hirtellus — Basket Grass or
  Ribbon Grass
Opopsitta coxeni — Parrot, Fig
Opopsitta diophthalma — Parrot, Fig
Opopsitta leadbeateri — Parrot, Fig
Opuntia ficus-indica — Fig, Barbary
Orchesella sp. — Springtail
Orchestria gammarella — Flea, Sand
Orchis coriophora — Orchid, Bug
Orchis fuchsii — Orchid, Spotted
Orchis laxiflora — Orchid, Jersey or
  Loose-flowered
Orchis mascula — Orchid, Early Purple
Orchis militaris — Orchid, Soldier or Military
Orchis morio — Orchid, Green-winged
Orchis palustris — Orchid, Marsh
Orchis papilionacea — Orchid, Pink Butterfly
Orchis purpurea — Orchid, Lady
Orchis quadipunctata — Orchid, Four-spotted
Orchis simia — Orchid, Monkey
Orchis strictifolia — Orchid, Early Marsh
Orchis tridentata — Orchid, Toothed
Orchis ustulata — Orchid, Dark-winged or
  Burnt
Orcinus orca — Whale, Killer
Orectogyrus bicostatus — Beetle, Whirligig
Orectolobus sp. — Wobbegong
Oreoica gutturalis — Bellbird, Crested
Oreoscopus gutteralis — Wren, Fern
Oreothraupis arremonops — Finch, Tanager
Oreotragus oreotragus — Klipspringer

Oreotrochilus estella — Hummingbird,
  Hillstar
Oreotrochilus melanogaster — Humming-
  bird, Hillstar
Orgyia antiqua — Moth, Vapourer
Orgyia recens — Moth, Vapourer
Origanum dictamnus — Dittany of Crete
Origanum vulgare — Marjoram, Pot or Wild
Origma solitaria — Warbler, Rock
Oriolus auratus — Oriole, Golden
Oriolus chinensis — Oriole, Black-naped
Oriolus flavocinctus — Oriole, Yellow
Oriolus larvatus — Oriole, Black-headed
Oriolus sagittatus — Oriole, Olive-backed
Oriolus traillii — Oriole, Maroon
Oriolus xanthonotus — Oriole, Dark-throated
Oriolus xanthornus — Oriole, Black-hooded
Orlaya grandiflora — Orlaya
Orneodes grammodactyla — Moth, Many-
  plumed
Ornithogalum nutans — Star of Bethlehem,
  Nodding
Ornithogalum pyrenaicum — Star of
  Bethlehem, Spiked
Ornithogalum sp. — Star of Bethlehem
Ornithogalum thyrsoides — Chincherinchee
Ornithogalum umbellatum — Star of
  Bethlehem, Common
Ornithoptera priamus — Butterfly, Birdwing
Ornithopus perpusillus — Birdsfoot, Common
  or Least
Ornithopus pinnatus — Birdsfoot, Orange
Ornithopus sativus — Serradella
Ornithorhynchus anatinus — Platypus
Orobanche caryophyllacea — Broomrape,
  Clove-scented
Orobanche elatior — Broomrape, Tall
Orobanche hederae — Broomrape, Ivy
Orobanche minor — Broomrape, Common or
  Lesser
Orobanche purpurea — Broomrape, Purple
Orobanche ramosa — Broomrape, Branched
Orobanche rapum-genistae — Broomrape,
  Greater
Orobanche sp. — Broomrape
Orontium aquaticum — Golden Club
Orthetrum sp. — Dragonfly
Orthilia secunda — Wintergreen, Nodding or
  Serrated or Toothed

Orthonyx spaldingi — Log-runner, Northern
Orthonyx temmincki — Log-runner, Southern
Orthotomus atrogularis — Tailorbird,
  Dark-necked
Orthotomus cuculatus — Tailorbird,
  Mountain
Orthotomus longicaudus — Tailorbird,
  Long-tailed
Orthotomus ruficeps — Tailorbird, Ashy
Orthotomus sepium — Tailorbird, Ashy
Orthotomus sericeus — Tailorbird,
  Rufous-tailed
Orthotomus sutorius — Tailorbird,
  Long-tailed
Ortygospiza atricollis — Quail-finch
Ortygospiza fuscocrissa — Quail-finch
Orycteropus afer — Aardvark
Oryctes nasicornis — Beetle, Rhinoceros
Oryctolagus cuniculus — Rabbit, European
  Wild
Oryruncus cristatus — Sharpbill, Crested
Oryx algazel — Oryx, Scimitar-horned or
  White
Oryx beisa — Oryz, Beisa
Oryx dammah — Oryx, Scimitar-horned or
  White
Oryx gazella — Gemsbok
Oryx leucoryx — Oryx, Arabian
Oryx tao — Oryx, Scimitar-horned or White
Oryza sativa — Rice
Oryzaephilus surinamensis — Beetle,
  Saw-toothed Grain
Osbornictis piscivora — Civet, Aquatic or
  Water
Oscinella frit — Fly, Frit
Oscularia deltoides — Oscularia
Osmanthus delavayi — Osmanthus
Osmanthus fragrans — Tea Olive
Osmanthus heterophyllus — Osmanthus,
  Holly
Osmanthus ilicifolius — Osmanthus, Holly
Osmerus eperlanus — Smelt, European
Osmia sp. — Bee
Osmunda regalis — Fern, Royal
Osmylus fulvicephalus — Lacewing, Giant
Osteoglossum bicirrhosum — Arawana
Osteolaemus tetraspis — Crocodile,
  Broad-fronted

Osteospermum fruticosum — Daisy, Trailing
  African
Ostichthys japonicus — Soldierfish,
  Deep-water
Ostracion lentiginosum — Box-fish, Blue
Ostracion tuberculatus — Boxfish
Ostrea sp. — Oyster, Native
Ostrowskia magnifica — Ostrowskia
Ostrya carpinifolia — Hornbean, European
  Hop
Otanthus maritimus — Cottonweed
Otaria byronia — Sea Lion, Patagonian or
  South American
Otidea onotica — Lemon Peel
Otidiphaps nobilis — Pigeon, Magnificent
  Ground
Otiorrhynchus clavipes — Weevil, Coal-black
Otiorrhynchus laevigatus — Weevil, Coal-
  black
Otis kori — Bustard, Kori
Otis tarda — Bustard, Great
Otis tetrax — Bustard, Little
Otocinclus sp. — Catfish
Otocolobus manul — Cat, Pallas'
Otus bakkamoena — Owl, Collared Scops
Otus choliba — Owl, Tropical Screech
Otus gurneyi — Owl, Giant Scops
Otus insularis — Owl, Seychelles
Otus leucotis — Owl, White-faced Scops
Otus rufescens — Owl, Rufous Scops
Otus sagittatus — Owl, White-fronted Scops
Otus scops — Owl, European Scops
Otus spilocephalus — Owl, Mountain or
  Spotted Scops
Oudemansiella mucida— Beech Tuft
Ourapteryx sambucaria — Moth, Swallow-
  tailed
Ouratea sp. — Ouratea
Ourebia ourebi — Oribi
Ourisia sp. — Ourisia
Ovis ammon ammon — Argali
Ovis ammon var. — Sheep, Wild
Ovis aries — Sheep, Wild
Ovis mouflon — Mouflon
Ovis orientalis — Sheep, Wild
Ovis musimon — Mouflon
Ovis vignel — Sheep, Wild
Oxalis acetosella — Wood-sorrel, Common
Oxalis adenophylla — Wood-sorrel

Oxalis articulata — Oxalis, Pink
Oxalis corniculata — Wood-sorrel,
Procumbent Yellow
Oxalis europaea — Sorrel, Upright Yellow
Oxalis floribunda — Oxalis, Pink
Oxalis pes caprae — Buttercup, Bermuda
Oxalis sp. — Oxalis
Oxidus gracilis — Millipede, Greenhouse
Oxymonacanthus longirostris — Triggerfish
Oxynotus centrina — Humantin
Oxypetalum caeruleum — Star-of-the-
Argentine
Oxypogon guerinii — Hummingbird, Bearded
Helmet-crest
Oxyria digyna — Sorrel, Mountain
Oxysternus maximus — Beetle, Hister
Oxytropis campestris — Milk-vetch, Yellow
Beaked
Oxytropis halleri — Milk-vetch, Purple
Mountain Beaked
Oxytropis pilosa — Milk-vetch, Hairy
Meadow Beaked
Oxytropis sp. — Milk-vetch
Oxyura australis — Duck, Blue-billed
Stiff-tail
Oxyura leucocephala — Duck, White-headed
Stiff-tail
Oxyuranus scutellatus — Taipan
Ozotoceros bezoarcticus — Deer, Pampas

P

Pachistima canbyi — Ratstripper
Pachycephala cinerea — Whistler, Mangrove
or White-bellied
Pachycephala pectoralis — Whistler, Golden
Pachycephala rufogularis — Whistler,
Red-throated
Pachycephala simplex — Whistler, Brown
Pachydactylus sp. — Gecko, African
Pachypanchax playfairi — Panchax,
Playfair's
Pachyphytum sp. — Moonstones or Thick
Plant
Pachypodium sp. — Pachypodium
Pachyptila belcheri — Prion, Thin-billed
Pachyptila desolata — Prion, Dove
Pachyptila turtur — Prion, Fairy
Pachyptila vittata — Prion, Broad-billed

Pachyrhamphus sp. — Becard
Pachyrrhynchus sp. — Weevil
Pachysandra terminalis — Spurge, Japanese
Pachystachys coccinea — Cardinal's Guard
Pachystachys lutea — Pachystachys
Pachyuromys duprasi — Gerbil
Padda oryzivora — Sparrow, Java
Paeonia lactiflora — Peony, Chinese
Paeonia lutea — Peony
Paeonia mascula — Peony, Wild
Paeonia officinalis — Peony, Common
Paeonia sp. — Peony
Paeonia suffruticosa — Peony, Tree
Paeonia tenuifolia — Peony, Fine-leaved
Pagellus erythrinus — Pandora
Pagellus sp. — Bream, Sea
Pagodroma nivea — Petrel, Snow
Pagophila eburnea — Gull, Ivory
Paguma larvata — Palm-civet, Masked
Paguristes sp. — Crab, Hermit
Pagurus bernhardus — Crab, Hermit or
Soldier
Pagurus japonicus — Crab, Hermit
Palaemon serratus — Prawn, Common
Paleosuchus palpebrosus — Caiman
Paleosuchus trigonatus — Caiman,
Smooth-fronted
Palingenia longicauda — Fly, May
Palinurus sp. — Lobster, Spiny
Paliurus spina-christi — Christ's-thorn
Pallenis spinosa — Pallenis
Palmatogecko rangei — Gecko, Web-footed
Palmipes placenta or P. membranaceus —
Starfish
Palomena prasina — Shield-bug, Green
Paltothyreus tarsatus — Ant
Pamianthe peruviana — Pamianthe
Pan paniscus — Chimpanzee, Pygmy
Pan satyrus — Chimpanzee
Pan troglodytes — Chimpanzee
Panacea prola — Butterfly
Panaeolus sp. — Panaeolus
Panaxia dominula — Moth, Scarlet Tiger
Pancratium illyricum — Lily, Corsican or
Spirit
Pancratium maritimum — Lily, Sea
Pandalus sp. — Prawn, Deep-sea
Pandanus sp. — Screwpine
Pandinus sp. — Scorpion, African

Pandion haliaetus — Osprey
Pandorea jasminoides — Bower-plant
Panellus stipticus — Fungus, Stiptic
Panicum miliaceum — Millet, Common or
  Broom-corn
Paniscus flavidella — Wasp, Ichneumon
Panonychus ulmi — Mite, Fruit Tree Spider
Panorpa sp. — Fly, Scorpion
Panthera leo — Lion
Panthera onca — Jaguar
Panthera pardalis — Ocelot
Panthera pardus — Leopard
Panthera tigris — Tiger, Bengal or Indian
Panthera tigris altaica — Tiger, Siberian or
  Manchurian
Panthera uncia — Leopard, Snow
Pantholops hodgsoni — Antelope, Tibetan
Pantodon buchholtzi — Butterflyfish
Panulirus sp. — Lobster, Spiny
Panurus biarmicus — Tit, Bearded
Papaver argemone — Poppy, Long
  Rough-headed
Papaver dubium — Poppy, Long-headed
Papaver hybridum — Poppy, Bristly
Papaver orientale — Poppy, Oriental
Papaver radicatum — Poppy, Arctic
Papaver rhoeas — Poppy, Corn
Papaver somniferum — Poppy, Opium
Papaver sp. — Poppy
Paphiopedilum sp. — Orchid
Papilio sp. — Butterfly, Swallow-tail
Papio anubis — Baboon, Dog-faced or Olive
Papio cynocephalus — Baboon, Yellow
Papio gelada — Baboon, Gelada
Papio hamadryas — Hamadryas
Papio leucophaeus — Mandrill
Papio mandrillus — Mandrill
Papio papio — Baboon, Western or Guinea
Papio richei — Baboon, Yellow
Papio sphinx — Mandrill
Papio ursinus — Baboon, Chacma
Paracanthurus sp. — Surgeonfish, Flagtail
Paracheirodon innesi — Tetra, Neon
Parachinus deserti — Hedgehog, Desert
Paracynictis selousi — Mongoose, Selous'
Paradisaea apoda — Bird of Paradise, Greater
Paradisaea minor — Bird of Paradise, Lesser
Paradisaea raggiana — Bird of Paradise,
  Count Raggi's

Paradisaea rudolphi — Bird of Paradise, Blue
  or Prince Rudolph's
Paradisea liliastrum — Lily, St. Bruno's
Paradisornis rudolphi — Bird of Paradise,
  Blue or Prince Rudolph's
Paradoxornis gularis — Parrotbill,
  Gray-headed
Paradoxornis nipalensis — Parrotbill,
  Black-throated
Paradoxornis webbiana — Parrotbill,
  Black-throated
Paradoxus sp. — Palm-civet
Paraechinus sp. — Hedgehog
Paramecium sp. — Paramecium
Paranemertes plana — Worm, Ribbon
Parapholis incurva — Hard-grass, Sea
Parapholis strigosa — Hard-grass, Sea
Pararge aegeria — Butterfly, Speckled Wood
Pararge megera — Butterfly, Wall Brown
Parasemia plantaginis — Moth, Wood Tiger
Paraxerus sp. — Squirrel, Bush
Pardalotus melanocephalus — Pardalote,
  Black-headed
Pardalotus ornatus — Pardalote, Striated
Pardalotus punctatus — Pardalote, Spotted
Pardalotus quadragintus — Pardalote,
  Forty-spotted
Pardalotus rubricatus — Pardalote,
  Red-browed
Pardalotus striatus — Pardalote,
  Yellow-tipped
Pardalotus substriatus — Pardalote, Striated
Pardalotus xanthopygus — Pardalote,
  Yellow-tailed
Pardofelis marmorata — Cat, Marbled
Pardosa sp. — Spider, Wolf
Parentucellia viscosa — Bartsia, Yellow
Parietaria diffusa — Pellitory-of-the-wall
Parietaria judaica — Pellitory-of-the-wall
Parietaria officinalis — Pellitory-of-the-wall.
Paris quadrifolia — Herb Paris
Parisoma subcaeruleum — Babbler, Common
  Tit
Parmelia sp. — Lichen
Parnassia palustris — Grass-of-Parnassus
Parnassius sp. — Butterfly, Apollo
Paroaria coronata — Cardinal, Red-crested
Paroaria cucullata — Cardinal, Red-crested
Paroaria gularis — Cardinal, Red-capped

393

Parodia sp. — Cactus, Parodia
Parrotia persica — Iron Tree
Parthenocissus henryana — Creeper,
    Silver-vein
Parthenocissus quinquefolia — Virginia
    Creeper
Parthenocissus tricuspidata — Ivy, Boston or
    Japanese
Parus afer — Tit, Gray
Parus ater — Tit, Coal
Parus caeruleus — Tit, Blue
Parus cinctus — Tit, Siberian
Parus cristatus — Tit, Crested
Parus cyanus — Tit, Azure
Parus dichrous — Tit, Crested
Parus lugubris — Tit, Somber
Parus major — Tit, Great
Parus montanus — Tit, Willow
Parus niger — Tit, Southern Black
Parus palustris — Tit, Marsh
Parus varius — Tit, Varied
Parvicardium exiguum — Cockle, Little
Passer domesticus — Sparrow, House or
    English
Passer flaveolus — Sparrow, Plain-backed
Passer hispaniolensis — Sparrow, Spanish
Passer melanurus — Sparrow, Cape
Passer moabiticus — Sparrow, Dead Sea
Passer montanus — Sparrow, Tree
Passer simplex — Sparrow, Desert
Pastinaca sativa — Parsnip
Patagona gigas — Hummingbird, Giant
Patella sp. — Limpet
Patina pellucida — Limpet, Blue-rayed
Patiniger polaris — Limpet
Paulownia imperialis — Empress Tree
Paulownia tomentosa — Empress Tree
Pavo cristatus — Peafowl, Blue or Indian
Pavo muticus — Peafowl, Green or Javan
Pavonia multiflora — Pavonia
Paxillus involutus — Fungus, Curled-edge
Paxistima canbyi — Ratstripper
Peachia sp. — Sea Anemone, Burrowing
Pecari tajacu — Peccary
Pecten jacobaeus — Scallop, Jacob's or
    Pilgrim's
Pecten maximus — Scallop, Great
Pedetes sp. — Springhare
Pedicularis foliosa — Lousewort, Leafy

Pedicularis oederi — Lousewort, Yellow
Pedicularis palustris — Red Rattle
Pedicularis recutita — Lousewort, Truncate
    or Alpine
Pedicularis sceptrum-carolinum — Moor-king
Pedicularis sp. — Lousewort
Pedicularis sylvatica — Lousewort, Common
Pediculus humanus var. — Louse, Head or
    Body
Pedilanthus tithymaloides — Redbird Cactus
Pedionomus torquatus — Plains Wanderer
Pegasus sp. — Dragonfish
Pelagodroma marina — Petrel, White-faced
    Storm
Pelamis platurus— Sea-snake, Black-and-
    yellow
Pelargopsis capensis — Kingfisher, Stork-
    billed
Pelea capreolus — Rhebuck, Gray or Vaal
Pelecanoides georgicus — Petrel, Magellan
    Diving
Pelecanoides magellani — Petrel, Magellan
    Diving
Pelecanoides urinatrix — Petrel, Common
    Diving
Pelecanus conspicillatus — Pelican,
    Australian
Pelecanus crispus — Pelican, Dalmatian
Pelecanus occidentalis — Pelican, Brown
Pelecanus onocrotalus — Pelican, White
Pelecanus rufescens — Pelican, Pink-backed
Pelecus cultratus — Ziege
Pelecyphora aselliformis — Hatchet-cactus
Pelidnota burmeisteri — Beetle
Pelidnota punctata — Beetle
Pelidnota sumptuosa — Beetle
Pellaea atropurpurea — Fern, Cliff-brake
Pellaea rotundifolia — Fern, Button
Pellaea viridis — Fern, Green Cliff-brake
Pellia sp. — Liverwort
Pellionia sp. — Pellionia
Pellorneum ruficeps — Babbler, Striped
    Jungle
Pelmatochromis sp. — Cichlid
Pelobates fuscus — Spadefoot, European
Peltigera canina — Lichen, Dog
Peltodoris atromaculata — Nudibranch,
    Spotted
Peltohyas australis — Dotterel, Australian

Pelvetia canaliculata — Wrack, Channelled
Penelope sp. — Guan
Peneoenanthe pulverulentus — Robin, Mangrove
Penicillium sp. — Penicillium
Penicillus sp. — Shaving Brush
Pennatula phosphorea — Sea Pen, Phosphorescent
Pentaglottis sempervirens — Alkanet, Green
Pentalagus furnessi — Rabbit, Ryukyu
Pentas lanceolata — Egyptian Star Cluster
Pentatoma rufipes — Bug, Forest
Penthimia nigra — Leafhopper
Peperomia sp. — Pepper-elder
Peplis protula — Purslane, Water
Perameles gunni — Bandicoot, Tasmanian Barred
Perameles nasuta — Bandicoot, Long-nosed
Perca fluviatilis — Perch, Common
Perdix barbata — Partridge, Bearded or Daurian
Perdix perdix — Partridge, Gray or Hungarian
Pereskia aculeata — Barbados Gooseberry
Pereskia grandiflora — Cactus, Rose
Pericrocotus cinnamomeus — Minivet, Small
Pericrocotus divaricatus — Minivet, Ashy
Pericrocotus ethologus — Minivet, Long-tailed
Pericrocotus flammeus — Minivet, Flame or Scarlet
Pericrocotus roseus — Minivet, Rosy
Pericrocotus solaris — Minivet, Gray-chinned or Mountain
Perilla frutescens — Beefsteak Plant
Periophthalmus sp. — Mudskipper
Periplaneta americana — Cockroach, American
Periploca graeca — Silk-vine
Perisama sp. — Butterfly
Perisoreus infaustus — Jay, Siberian
Perla sp. — Fly, Stone
Pernettya mucronata — Pernettya
Pernis apivorus — Buzzard, Honey
Perodicticus potto — Potto, Bosman's
Persea americana — Avocado Pear
Persea gratissima — Avocado Pear
Petasites albus — Butterbur, White
Petasites fragrans — Heliotrope, Winter

Petasites hybridus — Butterbur
Petaurista sp. — Squirrel, Flying
Petauroides volans — Phalanger, Greater Flying
Petaurus australis — Phalanger, Flying
Petaurus breviceps — Phalanger, Short-headed Flying
Petaurus norfolcensis — Phalanger, Short-headed Flying
Petaurus sciureau — Phalanger, Short-headed Flying
Peteronura brasiliensis — Otter, Brazilian or Giant
Petrobius sp. — Bristletail
Petrocallis pyrenaica — Rock-beauty, Pyrenean
Petrochelidon ariel — Martin, Fairy
Petrochelidon nigricans — Martin, Tree
Petrodomus sultani — Elephant-shrew, Forest
Petrodromus tetradactylus — Elephant-shrew, Four-toed
Petrogale penicillata — Rock-wallaby, Brush-tailed
Petrogale xanthopus — Rock-wallaby, Ring-tailed or Yellow-footed
Petroica cucullata — Robin, Hooded
Petroica goodenovii — Robin, Red-capped
Petroica macrocephala — Tomtit
Petroica multicolor — Robin, Scarlet
Petroica phoenicea — Robin, Flame
Petroica rodinogaster — Robin, Pink
Petroica rosea — Robin, Rose
Petroica vittata — Robin, Dusky
Petrolisthes sp. — Crab, Porcelain
Petromyzon fluvialis — Lamprey, River
Petromyzon marinus — Lamprey, Sea
Petronia brachydactyla — Sparrow, Rock
Petronia petronia — Sparrow, Rock
Petronia xanthocollis — Sparrow, Rock
Petrophassa albipennis — Pigeon, White-quilled Rock
Petrophassa rufipennis — Pigeon, Chestnut-quilled Rock
Petrophassa scripta — Pigeon, Squatter
Petrorhagia saxifraga — Tunic Flower
Petroscirtes sp. — Blenny
Petroselinum crispum — Parsley
Petroselinum segetum — Parsley, Corn

395

Peucedanum officinale — Hog's Fennel
Peucedanum oreoselinum — Parsley, Mountain
Peucedanum ostruthium — Masterwort
Peucedanum palustre — Parsley, Milk
Pezites militaris — Blackbird, Red-breasted
Pezoporus wallicus — Parrot, Ground
Phacellodomus rufifrons — Thornbill, Rufous-fronted
Phacochoerus aethiopicus — Warthog
Phaenicophaeus sp. — Malcoha or Malkoha
Phaenostictus mcleannani — Antbird, Ocellated
Phaeolepiota aurea — Phaeolepiota
Phaeolus schweinitzii — Fungus
Phaethon aethereus — Tropic Bird, Red-billed
Phaethon lepturus — Tropic Bird, White-tailed or Yellow-billed
Phaethon rubricauda — Tropic Bird, Red-tailed
Phaetornis sp. — Hummingbird, Hermit
Phaetusa simplex — Tern, Large-billed
Phalacrocorax africanus — Cormorant, Reed
Phalacrocorax albiventer — Cormorant, Kerguelen or King
Phalacrocorax aristotelis — Shag
Phalacrocorax atriceps — Cormorant, Blue-eyed or Antarctic
Phalacrocorax bougainvillii — Cormorant, Guanay or Peruvian
Phalacrocorax capensis — Cormorant, Cape
Phalacrocorax carbo — Cormorant, Common or European
Phalacrocorax carunculatus — Shag, New Zealand King
Phalacrocorax fuscescens — Cormorant, White-breasted
Phalacrocorax harrisi — Cormorant, Flightless
Phalacrocorax magellanicus — Cormorant, Rock
Phalacrocorax melanoleucus — Cormorant, Little Pied
Phalacrocorax punctatus — Shag, Spotted
Phalacrocorax pygmaeus — Cormorant, Pygmy
Phalacrocorax sulcirostris — Shag, Little Black
Phalacrocorax urile — Cormorant, Red-faced

Phalacrocorax varius — Shag, Pied
Phalaenopsis amabilis — Orchid, Moth
Phalaenopsis sp. — Orchid, Moth
Phalanger maculatus — Cuscus, Spotted
Phalanger nudicaudatus — Cuscus, Spotted
Phalanger orientalis — Cuscus, Gray
Phalangium opilio — Harvestman
Phalaris arundinacea — Canary-grass, Reed
Phalaris canariensis — Canary-grass
Phalaropus fulicarius — Phalarope, Red or Gray
Phalaropus lobatus — Phalarope, Northern or Red-necked
Phalaropus tricolor — Phalarope, Wilson's
Phalcoboenus sp. — Caracara
Phalera bucephala — Moth, Buff-tip
Phalichthys amates — Merry Widow
Phalium strigatum — Bonnet Shell
Phalloceros caudomaculatus — Caudo
Phallus impudicus — Stinkhorn, Common
Phallusia mammillata — Sea Squirt
Phaner furcifer — Lemur, Fork-marked Mouse
Phaneus sp. — Beetle, Dung
Phapitreron sp. — Dove, Fruit
Phaps chalcoptera — Pigeon, Forest Bronze-wing
Phaps elegans — Pigeon, Brush Bronze-wing
Phaps histrionica — Pigeon, Flock
Phascogale calura — Phascogale, Brush-tailed
Phascogale penicillata — Phascogale, Brush-tailed
Phascogale tapoatafa — Phascogale, Brush-tailed
Phascolarctos cinereus — Koala
Phascolomys mitchelli — Wombat
Phascolomys ursinus — Wombat
Phasianus colchicus — Pheasant, Ring-necked
Phasianus versicolor — Pheasant, Green or Japanese
Phebalium sp. — Phebalium
Phellinus igniarius — Willow Fomes
Phellodendron amurense — Cork Tree, Amur
Phelsuma madagascariensis — Gecko, Madagascar or Malagasy
Phelsuma sp. — Gecko, Madagascar or Malagasy

Phengodes plumosa — Beetle

Pheosia tremula — Moth, Swallow Prominent

Phibalura flavirostris — Cotinga,
Swallow-tailed

Philadelphus coronarius — Mock Orange,
Sweet

Philadelphus sp. — Mock Orange

Philaenus leucophthalmus — Froghopper,
Meadow

Philaenus spumarius — Froghopper,
Meadow

Philander opossum — Opossum, Four-eyed

Philanthus triangulum — Wasp, Bee-killer

Philantombia monticola — Duiker, Blue

Philemon argenticeps — Friarbird,
Silver-crowned

Philemon citreogularis — Friarbird, Little

Philemon corniculatus — Friarbird, Noisy

Philemon novaeguineae — Friarbird,
Helmeted

Philemon yorki — Friarbird, Helmeted

Philepitta castanea — Asity, Velvet

Philesia magellanica or P. buxifolia —
Philesia

Philesturnus carunculatus — Saddleback

Philetarius socius — Weaver, Sociable

Phillyrea decora — Mock-privet

Phillyrea latifolia — Mock-privet

Philodendron sp. — Philodendron

Philodromus sp. — Spider, Crab

Philomachus pugnax — Ruff

Philothamnus sp. — Snake, Tree or Water

Philudoria potatoria — Moth, Drinker

Phlebia radiata — Fungus, Orange-vein

Phlebodium aureum — Fern, Golden Polypody

Phlegopsis nigromaculata — Antbird,
Spectacled

Phleocryptes melanops — Rushbird,
Wren-like

Phleum alpinum — Cat's-tail, Alpine

Phleum arenarium — Cat's-tail, Sand

Phleum phleoides — Cat's-tail, Purple-stem
or Pointed

Phleum pratense — Grass, Timothy or
Cat's-tail

Phloemys sp. — Rat, Cloud

Phlomis fruticosa — Jerusalem Sage

Phlomis sp. — Jerusalem Sage

Phoca hispida — Seal, Ringed

Phoca vitulina — Seal, Hair or Harbor

Phocaena dioptrica — Porpoise, Spectacled

Phocaena phocaena — Porpoise, Common or
Harbor

Phodilus badius — Owl, Bay

Phodilus prigoginei — Owl, Bay

Phoebetria fusca — Albatross, Sooty

Phoebetria palpebrata — Albatross, Light-
mantled Sooty

Phoebis sp. — Butterfly, Sulphur

Phoenicircus nigricollis — Cotinga, Black-
necked Red

Phoeniconaias minor — Flamingo, Lesser

Phoenicoparrus andinus — Flamingo, Andean

Phoenicophaeus cummingi — Cuckoo, Scale-
feathered

Phoenicophaeus superciliosus — Cuckoo,
Rough-crested

Phoenicopterus antiquorum — Flamingo,
Greater

Phoenicopterus ruber — Flamingo, Greater

Phoenicopterus chilensis — Flamingo,
Chilean

Phoeniculus bollei — Hoopoe, Wood

Phoeniculus purpureus — Hoopoe, Wood

Phoenicurus erythrogaster — Redstart,
Guldenstadt's

Phoenicurus ochrurus — Redstart, Black

Phoenicurus phoenicurus — Redstart,
Common

Phoenix canariensis — Palm, Canary Islands

Phoenix dactylifera — Palm, Date

Phoenix reclinata — Palm, Senegal

Phoenix robellinii — Palm, Dwarf

Pholas dactylus — Piddock, Common

Pholcus phalangioides — Spider, Long-
bodied Cellar

Pholidoptera griseoaptera — Bush-cricket
Dark

Pholiota mutabilis — Fungus, Edible Stump

Pholiota squarrosa — Fungus, Scaly Cluster

Pholis gunnellus — Butterfish

Phonygammus keraudreni — Manucode

Phormium colensoi — Flax, Mountain

Phormium tenex — Flax, New Zealand

Photinia fraseri — Photinia, Fraser

Photinia glabra — Photinia, Oriental

Photinia serrulata — Hawthorn, Chinese

Photinia villosa — Photinia, Oriental

Photinus pyralis — Firefly

Photoblepharon palpebratus — Lantern-eye Fish

Phoxinus phoxinus — Minnow, European

Phragmipedium sp. — Orchid

Phragmites australis — Reed, Common

Phragmites communis — Reed, Common

Phrynocephalus sp. — Lizard, Agama

Phrynorhombus norvegicus — Topknot, Norwegian

Phthirus pubis — Louse, Crab

Phycis blennoides — Forkbeard, Greater

Phylidonyris albifrons — Honey-eater, White-fronted

Phylidonyris melanops — Honey-eater, Tawny-crowned

Phylidonyris niger — Honey-eater, White-cheeked

Phylidonyris novaehollandiae — Honey-eater, New Holland

Phylidonyris pyrrhoptera — Honey-eater, Crescent

Phyllanthus pulcher — Phyllanthus

Phyllastrephus sp. — Bulbul

Phyllitis scolopendrium — Fern, Hart's-tongue

Phyllium sp. — Leaf-insect

Phyllobates bicolor — Frog, Arrow-poison

Phyllobius sp. — Weevil, Tree

Phyllodactylus sp. — Gecko, European Leaf-fingered

Phyllodromia germanica — Cockroach, German

Phyllomedusa sp. — Frog, Leaf

Phyllomorpha laciniata — Bug, Plant-eating

Phyllopertha horticola — Beetle, Garden Chafer

Phyllopteryx eques — Seahorse, Australian

Phylloscopus bonelli — Warbler, Bonelli's

Phylloscopus borealis — Warbler, Arctic

Phylloscopus collybita — Chiffchaff

Phylloscopus fuscatus — Warbler, Dusky

Phylloscopus inornatus — Warbler, Yellow-browed

Phylloscopus nitidus — Warbler, Green

Phylloscopus olivaceus — Warbler, Leaf

Phylloscopus proregulus — Warbler, Pallas' Leaf

Phylloscopus schwarzi — Warbler, Radde's Willow

Phylloscopus sibilatrix — Warbler, Wood

Phylloscopus tenellipes — Warbler, Leaf

Phylloscopus trochiloides — Warbler, Greenish

Phylloscopus trochilus — Warbler, Willow

Phyllostachys sp. — Bamboo

Phyllotreta sp. — Beetle, Flea

Phyllurus sp. — Gecko, Leaf-tailed

Phymata sp. — Bug, Ambush

Phymatodes testaceus — Beetle

Physa fontinalis — Snail, Bladder

Physalia physalis — Portuguese Man-of-War

Physalis alkekengi — Cherry, Bladder or Winter or Ground

Physalis franchetti — Chinese Lantern

Physeter catodon — Whale, Sperm

Physeter macrocephalus — Whale, Sperm

Physignathus lesueuri — Lizard, LeSueur's Water

Physophora sp. — Jellyfish

Physospermum cornubiense — Bladder-seed

Phyteuma betonicifolium — Rampion, Blue Spiked

Phyteuma comosum — Devil's Claw or Rampion, Clustered

Phyteuma nigrum — Rampion, Black

Phyteuma orbiculare — Rampion, Round-headed

Phyteuma scheuchzeri — Rampion, Horned

Phyteuma spicatum — Rampion, Spiked

Phyteuma tenerum — Rampion, Round-headed

Phytolacca americana — Pokeweed or Pokeberry

Phytolacca sp. — Pokeweed or Pokeberry

Phytometra gamma — Moth, Silver-Y

Phytomyza illicis — Fly, Holly Leaf-miner

Phytophthora infestans — Mildew, Downy

Phytotoma rutila — Plant-cutter

Piaya cayana — Cuckoo, Squirrel

Pica pica — Magpie, Black-billed or Common

Picathartes gymnocephalus — Rockfowl

Picathartes oreas — Rockfowl

Picea abies — Spruce, Norway

Picea brachytyla — Spruce, Sargent

Picea likiangensis — Spruce, Likiang
Picea omorika — Spruce, Serbian
Picea orientalis — Spruce, Oriental
Picea polita — Spruce, Tiger-tail
Picea smithiana — Spruce, Morinda or West Himalayan
Picoides tridactylus — Woodpecker, Three-toed (Northern)
Picris echioides — Ox-tongue, Bristly
Picris hieracioides — Ox-tongue, Hawkweed
Picumnus sp. — Piculet
Picus canus — Woodpecker, Gray-headed
Picus flavinucha — Woodpecker, Yellow-naped
Picus puniceus — Woodpecker, Crimson
Picus rivolii — Woodpecker, Crimson
Picus viridus — Woodpecker, Green
Pieris brassicae — Butterfly, Cabbage White
Pieris forrestii — Pieris, Chinese
Pieris japonica — Andromeda, Japanese
Pieris napi — Butterfly, Green-veined White
Pieris rapae — Butterfly, Small White
Pieris taiwanensis — Pieris
Piezorhynchus alecto — Flycatcher, Shining
Pilea cadieri — Aluminum Plant
Pilea involucrata — Panamiga or Pan-american Friendship Plant
Pilea sp. — Panamiga
Pilosella auranticum — Hawkweed, Orange
Pilosella officinarium — Hawkweed, Mouse-ear
Pilularia globulifera — Pillwort
Pimelea sp. — Rice Flower
Pimelia sp. — Beetle
Pimelodella sp. — Catfish
Pimelodus sp. — Catfish
Pimenta dioica — Allspice or Pimento
Pimpinella anisum — Anise or Aniseed
Pimpinella major — Burnet-saxifrage, Greater
Pimpinella saxifraga — Burnet-saxifrage
Pinctada sp. — Oyster, Pearl
Pinguicula alpina — Butterwort, Alpine
Pinguicula grandiflora — Butterwort, Large-flowered
Pinguicula lusitanica — Butterwort, Pink or Pale
Pinguicula macrophylla — Butterwort
Pinguicula moranensis — Butterwort

Pinguicula vulgaris — Butterwort, Common
Pinicola enucleator — Grosbeak, Pine
Pinna sp. — Pen Shell, Mediterranean
Pinnotheres pisum — Crab, Oyster or Pea
Pinus bungeana — Pine, Lace Bark
Pinus cembra — Pine, Arolla
Pinus contorta — Pine, Beach or Lodge-pole
Pinus halepensis — Pine, Aleppo or Jerusalem
Pinus leucodermis — Pine, Bosnian
Pinus mugo — Pine, Mugo or Mountain
Pinus nigra — Black or Austrian
Pinus nigra var. maritima — Pine, Corsican
Pinus parviflora — Pine, Japanese White
Pinus peuce — Pine, Macedonian
Pinus pinaster — Pine, Maritime
Pinus pinea — Pine, Stone
Pinus radiata — Pine, Monterey
Pinus resinosa — Pine, Norway or Red
Pinus rigida — Pine, Pitch
Pinus strobus — Pine, Weymouth or White
Pinus sylvestris — Pine, Scots
Pinus thunbergii — Pine, Japanese Black
Pinus wallichiana — Pine, Bhutan
Pionites melanocephala — Parrot, Black-headed
Pionus sp. — Parrot
Pipa pipa — Toad, Surinam
Piper crocatum — Pepper, Saffron
Piper nigrum — Pepper, Black
Piper ornatum — Pepper, Celebes
Piper tiliifolium — Pepper
Pipistrellus pipistrellus — Bat, Pipistrelle
Pipra sp. — Manakin
Pipreola riefferii — Fruit-eater, Green-and-Black
Piptoporus betulinus — Fungus, Birch
Pisaura mirabilis — Spider, Wolf
Piscicola geometra — Leech, Fish
Pisidium sp. — Mussel, Pea
Pisolithus sp. — Puffball
Pisonia umbellifera — Birdcatcher Tree
Pistacia chinensis — Pistachio or Pistache, Chinese
Pistacia lentiscus — Mastic Tree
Pistacia terebinthus — Turpentine Tree
Pistacia vera — Pistachio (nut)
Pistia stratiotes — Water-lettuce
Pisum sativum — Pea, Garden
Pitcairnia sp. — Pitcairnia

Pithecia monachus — Saki, Hairy or Monk
Pithecia pithecia — Saki, White-faced
Pithecophaga jefferyi — Eagle, Monkey-
    eating
Pithys albifrons — Antbird, White-faced or
    -plumed
Pitta brachyura — Pitta, Indian
Pitta caerulea — Pitta, Great Blue or Giant
Pitta cyanea — Pitta, Blue
Pitta ellioti — Pitta, Elliot's
Pitta granatina — Pitta, Garnet or
    Red-headed
Pitta guajana — Pitta, Banded or Blue-tailed
Pitta iris — Pitta, Rainbow
Pitta mackloti — Pitta, Blue-breasted
Pitta moluccensis — Pitta, Blue-winged
Pitta sordida — Pitta, Hooded
Pitta steerii — Pitta, Steer's
Pitta versicolor — Pitta, Buff-breasted or
    Noise
Pittasoma sp. — Antpitta
Pittosporum crassifolium — Karo
Pittosporum sp. — Pittosporum
Pittosporum tenuifolium — Kohuhu
Pittosporum tobira — Pittosporum, Japanese
Pittosporum undulatum — Box, Victorian
Pityriasis gymnocephala — Wood-shrike,
    Bald-headed
Plagianthus betulinus — Manatu or Lacebark
Plagiodontia aedium — Hutia
Plagiothecium sp. — Moss, Slender
Planorbis sp. — Snail, Ram's Horn
Plantago coronopus — Plantain, Buck's-horn
Plantago indica — Plantain, Branched
Plantago lanceolata — Plantain, Ribwort or
    Narrow-leaved
Plantago major — Plantain, Common or Great
Plantago maritima — Plantain, Sea
Plantago media — Plantain, Hoary
Plantago rigida — Plantain, Alpine
Plantago sempervirens — Plantago, Shrubby
Platalea alba — Spoonbill, African
Platalea flavipes — Spoonbill, Yellow-billed
Platalea leucorodia — Spoonbill, European or
    White
Platalea minor — Spoonbill, Black-faced or
    Lesser
Platalea regia — Spoonbill, Royal

Platanista gangetica — Dolphin, Ganges
    River
Platanthera bifolia — Orchid, Lesser
    Butterfly
Platanthera chlorantha — Orchid, Greater
    Butterfly
Platanus acerifolia — Plane, London
Platanus hybrida or P. hispanica — Plane,
    London
Platanus orientalis — Plane, Oriental
Platax sp. — Batfish
Platichthys sp. — Flounder
Platycephalus indicus — Flathead
Platycercus adelaidae — Rosella, Adelaide
Platycercus adscitus — Rosella, Pale-headed
Platycercus caledonicus — Rosella, Green
Platycercus elegans — Rosella, Crimson
Platycercus eximius — Rosella, Eastern
Platycercus flaveolus — Rosella, Yellow
Platycercus icterotis — Rosella, Western
Platycercus venustus — Rosella, Northern
Platycerium alcicorne — Fern, Stag's-horn
Platycerium angolense — Fern, Angola
    Stag's-horn
Platycerium bifurcatum — Fern, Stag's-horn
Platycnemis pennipes — Damselfly,
    White-legged
Platycodon grandiflora — Balloon Flower
Platyptilia sp. — Moth, Plume
Platyspiza crassirostris — Finch, Vegetarian
    Tree
Platysternon megacephalum — Terrapin,
    Large-headed
Plautus alle — Dovekie
Plebejus argus — Butterfly, Silver-studded
    Blue
Plecotus auritus — Bat, Long-eared
Plecotus austriacus — Bat, Long-eared
Plectorhinchus sp. — Sweetlips
Plectranthus australis — Swedish Ivy
Plectranthus coleoides marginatus —
    Swedish Ivy
Plectranthus oertendahlii — Candle Plant
Plectranthus sp. — Plectranthus
Plectrophenax nivalis — Bunting, Snow
Plectropterus gambiensis — Goose,
    Spur-winged
Plectorhyncha lanceolata — Honey-eater,
    Striped

Plegadis falcinellus — Ibis, Glossy
Pleione sp. — Orchid, Pleione
Pleiospilos sp. — Mimicry Plant
Pleomele reflexa — Song of India
Pleurobrachia sp. — Comb-jelly
Pleurodeles sp. — Newt
Pleuronectes limanda — Dab
Pleuronectes platessa — Plaice
Pleurospermum austriacum — Rib-seed, Austrian
Pleurothallis sp. — Orchid
Pleurotomaria sp. — Slit Shell
Pleurotus ostreatus — Mushroom, Oyster
Plexippus paykulli — Spider, Jumping
Plocepasser mahali — Sparrow-weaver, White-browed
Ploceus capensis — Weaver, Cape
Ploceus cucullatus — Weaver, Spotted-backed
Ploceus intermedius — Weaver, Masked
Ploceus manyar — Weaver, Manyar or Streaked
Ploceus ocularius — Weaver, Spectacled
Ploceus philippinus — Weaver, Baya
Ploceus velatus — Weaver, Masked
Plodia interpunctella — Moth, Indian Meal
Plotosus anguillaris or P. lineatus — Catfish, Marine
Plumbago auriculata or P. capensis — Leadwort
Plumbago indica — Leadwort
Plumbago rosea — Leadwort
Plumeria sp. — Frangipani
Plusia gamma — Moth, Silver-Y
Plusia sp. — Moth, Burnished Brass
Pluteus cervinus — Mushroom, Deer
Pluvialis apricarius — Plover, Golden
Pluvialis dominica — Plover, Golden
Pluvialis squatarola — Plover, Gray
Pluvianus aegypticus — Plover, Egyptian
Poa alpinus — Meadow-grass, Alpine
Poa annua — Meadow-grass, Annual
Poa compressa — Meadow-grass, Flattened or Flat-stalked
Poa nemoralis — Meadow-grass, Wood or Woodland
Poa pratensis — Meadow-grass, Common or Smooth
Poa sp. — Meadow-grass

Poa trivialis — Meadow-grass, Rough
Podarcis muralis — Lizard, Wall
Podargus ocellatus — Frogmouth, Marbled
Podargus papuensis — Frogmouth, Papuan
Podargus plumiferus — Frogmouth, Plumed
Podargus strigoides — Frogmouth, Tawny
Podica senegalensis — Finfoot, African or Peter's
Podiceps auritus — Grebe, Slavonian
Podiceps caspicus — Grebe, Black-necked
Podiceps cristatus — Grebe, Great Crested
Podiceps grisegena — Grebe, Red-necked
Podiceps novaehollandiae — Grebe, Australian Little
Podiceps nigricollis — Grebe, Black-necked
Podiceps poliocephalus — Grebe, Hoary-headed
Podiceps ruficollis — Grebe, Little
Podocarpus dacrydoides — Pine, White
Podocarpus macrophyllus — Yew, Japanese
Podocarpus sp. — Podocarpus
Podoces sp. — Jay, Ground
Podocnemis sp. — Tortoise, Greaved
Podophyllum emodii — Mayflower, Himalayan
Poecilia reticulata — Guppy
Poecilictis libyca — Weasel, Libyan Striped or Banded
Poecilodryas cerviniventris — Robin, Buff-sided
Poecilogale albinucha — Weasel, Striped
Poelagus marjorita — Rabbit, African
Poephagus grunniens — Yak, Wild
Poephila acuticauda — Finch, Long-tailed
Poephila cincta — Finch, Black-throated
Poephila gouldiae — Finch, Gouldian
Poephila personata — Finch, Masked
Pogoniulus bilineatus — Tinkerbird, Golden-rumped
Pogoniulus chrysoconus — Tinkerbird, Yellow-fronted
Poiana richardsoni — Linsang, African
Poicephalus meyeri — Parrot, Meyer's
Poicephalus robustus — Parrot, Brown-necked or Cape
Poinciana gilliesii — Poinciana
Poinciana pulcherrima — Pride of Barbados
Polemaetus bellicosus — Eagle, Martial
Polemaetus coronatus — Eagle, Crowned

Polemonium caeruleum — Jacob's Ladder

Polihierax insignis — Falcon, Pygmy

Polihierax semitorquatus — Falcon, Pygmy

Poliocephalus ruficollis — Grebe, Little

Poliolimnus cinereus — Crake, White-browed

Polistes sp. — Wasp, Paper

Pollachius sp. — Pollack

Pollenia rudis — Fly, Cluster

Polyboroides radiatus — Hawk, Harrier

Polyboroides typus — Hawk, Harrier

Polycarpon tetraphyllum — Allseed, Four-leaved

Polycentropsis abbreviata — Leaf-fish, African

Polycentrus schomburgkii — Leaf-fish, Schomburgk's

Polycera quadrilineata — Nudibranch

Polycnemum majus — Polycnemum

Polydesmus sp. — Millipede

Polygala calcarea — Milkwort, Chalk

Polygala chamaebuxus — Milkwort, Box-leaved or Shrubby

Polygala myrtifolia — Sweet Pea Shrub

Polygala serpyllifolia — Milkwort, Heath

Polygala sp. — Milkwort

Polygala vulgaris — Milkwort, Common

Polygonatum canaliculatum — Solomon's Seal, Great

Polygonatum multiflorum — Solomon's Seal, Common

Polygonatum odoratum — Solomon's Seal, Sweet-scented

Polygonatum sp. — Solomon's Seal

Polygonatum verticillatum — Solomon's Seal, Whorled

Polygonia c-album — Butterfly, Comma

Polygonia egea — Butterfly, Comma

Polygonum affine — Fleece Flower, Himalayan

Polygonum amphibium — Bistort, Amphibious

Polygonum aubertii — Silver-lace Vine

Polygonum aviculare — Knotweed, Prostrate

Polygonum baldschuanicum — Russian Vine

Polygonum bistorta — Bistort, Snakeroot

Polygonum campanulatum — Knotweed, Himalayan

Polygonum convolvulus — Bindweed, Black

Polygonum cuspidatum — Knotweed, Japanese

Polygonum dumetorum — Bindweed, Copse

Polygonum hydropiper — Water-pepper

Polygonum lapathifolium — Smartweed, Pale

Polygonum minus — Water-peppr, Lease

Polygonum persicaria — Lady's Thumb or Redleg

Polygonum viviparum — Bistort, Alpine

Polynemus sp. — Threadfin

Polyommatus icarus — Butterfly, Common Blue

Polyphylla fullo — Beetle, Pine Chafer

Polyplectron bicalcaratum — Peacock-pheasant, Gray

Polyplectron emphanum — Peacock-pheasant, Palawan

Polyplectron germaini — Peacock-pheasant, Germain's

Polyplectron malacense — Peacock-pheasant, Malay

Polypodium aureum — Fern, Golden Polypody

Polypodium vulgare — Polypody, Common

Polypogon monspeliensis — Beard-grass, Annual

Polyporus squamosus — Dryad's Saddle

Polypterus ornatipinnis — Bichir, Ornate

Polypterus sp. — Bichir

Polyscias balfouriana — Aralia, Balfour

Polystachya sp. — Orchid

Polystichum sp. — Fern, Polystichum

Polysticta stelleri — Eider, Steller's

Polystictus sp. — Polypore

Polytelis alexandrae — Parrot, Princess

Polytelis anthopeplus — Parrot, Regent

Polytelis swainsonii — Parrot, Superb

Polytrichum commune — Moss, Hair-cap

Polyxenus lagurus — Millipede

Pomacanthodes sp. — Angelfish

Pomacanthus sp. — Angelfish

Pomacentrus sp. — Damselfish

Pomatomus saltatrix — Bluefish

Pomatorhinus erythrogenys — Babbler, Rusty-cheeked Scimitar

Pomatorhinus hypoleucos — Babbler, Large Scimitar

Pomatorhinus schisticeps — Babbler, White-browed

Pomatoschistus minutus — Goby, Sand

Pomatostomus halli — Babbler, Hall's

Pomatostomus ruficeps — Babbler, Chestnut-crowned

Pomatostomus superciliosus — Babbler, White-browed Scimitar

Pomatostomus temporalis — Babbler, Gray-crowned Scimitar

Poncirus trifoliata — Orange, Chinese or Hardy

Pongo pygmaeus — Orangutan

Pontia daplidice — Butterfly, Bath White

Poospiza sp. — Finch, Warbling

Popelairia conversii — Hummingbird, Thorntail

Popelairia popelarii — Hummingbird, Thorntail

Populus alba — Poplar, White

Populus berolinensis — Poplar, Berlin

Populus canescens — Poplar, Gray

Populus nigra — Poplar, Black

Populus tremula — Aspen, European

Porania sp. — Starfish, Cushion

Porcellana sp. — Crab, Porcelain

Porcellio sp. — Sowbug

Porichthys sp. — Midshipman

Porphyrio alleni — Gallinule, Allen's or Lesser

Porphyrio melanotus — Swamp-hen

Porphyrio poliocephalus bellus — Swamp-hen

Porphyrio porphyrio — Gallinule, Purple

Portulaca grandiflora — Rose-moss

Portulaca kermesina — Purslane

Portulaca oleracea — Purslane

Portulacaria afra — Elephant Bush

Portunus sp. — Crab, Swimming

Porzana cinerea — Crake, White-browed

Porzana fluminea — Crake, Spotted

Porzana fusca — Crake, Ruddy-breasted

Porzana parva — Crake, Little or Lesser

Porzana paykullii — Crake, Band-bellied

Porzana plumbea — Crake, Spotless

Porzana porzana — Crake, Spotted

Porzana pusilla — Crake, Baillon's or Lesser Spotted

Porzana tabuensis— Crake, Spotless

Posoqueria sp. — Needle-flower or Needle-bush

Potamilla reniformis — Worm

Potamochoerus koiropotamus — Bushpig

Potamochoerus porcus — Bushpig

Potamogale velox — Otter-shrew

Potamogeton acutifolius — Pondweed, Sharp-leaved

Potamogeton berchtoldii — Pondweed, Small or Lesser

Potamogeton coloratus — Pondweed, Fen

Potamogeton crispus — Pondweed, Curled or Crisp

Potamogeton gramineus — Pondweed, Various-leaved

Potamogeton lucens — Pondweed, Shining

Potamogeton natans — Pondweed, Broad-leaved or Floating

Potamogeton nodosus — Pondweed, Loddon

Potamogeton obtusifolius — Pondweed, Grassy or Blunt-leaved

Potamogeton pectinatus — Pondweed, Fennel-like

Potamogeton perfoliatus — Pondweed, Perfoliate

Potamogeton pusillus — Pondweed, Small or Lesser

Potamotrygon sp. — Stingray, Freshwater

Potentilla alba — Cinquefoil, White

Potentilla anglica — Cinquefoil, Procumbent

Potentilla anserina — Silverweed

Potentilla arbuscula — Cinquefoil, Bush

Potentilla argentea — Cinquefoil, Hoary or Silvery

Potentilla crantzii — Cinquefoil, Alpine

Potentilla erecta — Tormentil

Potentilla fruticosa — Cinquefoil, Shrubby

Potentilla norvegica — Cinquefoil, Norwegian

Potentilla palustris — Cinquefoil, Marsh

Potentilla recta — Cinquefoil, Sulphur or Upright

Potentilla reptans — Cinquefoil, Creeping

Potentilla rupestris — Cinquefoil, Rock or White

Potentilla sp. — Cinquefoil

Potentilla sterilis — Strawberry, Barren

Potentilla tabernaemontani — Cinquefoil, Spring

Poterium sanguisorba — Burnet, Salad

Potorous tridactylus — Rat-kangaroo, Long-nosed

Potos flavus — Kinkajou
Pratia sp. — Pratia
Prenanthes purpurea — Lettuce, Hare's or
  Purple
Prepona sp. — Butterfly
Presbytis entellus — Langur, Sacred
Presbytis pileatus — Langur, Capped
Priacanthus arenatus — Big-eye
Priacanthus macracantha — Big-eye
Primula auricula — Bear's-ear
Primula denticulata — Primrose, Himalayan
Primula elatior — Oxlip
Primula farinosa — Primrose, Bird's-eye
Primula halleri — Primrose, Long-flowered
Primula hirsuta — Primrose, Red Alpine
Primula integrifolia — Primrose, Entire-
  leaved
Primula japonica — Primrose, Japanese or
  Candelabra
Primula malacoides — Primrose, Fairy or
  Baby
Primula minima — Primrose, Least
Primula obsonica — Primrose, Poison
Primula polyantha — Primrose, Polyanthus
Primula rosea — Primrose, Bog
Primula scotia — Primrose, Scottish
Primula sieboldii — Primrose, Siebold's
Primula sinensis — Primrose, Chinese
Primula sp. — Primrose
Primula spectabilis — Primrose, Showy
Primula veris — Cowslip
Primula vulgaris — Primrose, English or
  Common
Pringlea antiscorbutica — Cabbage,
  Kerguelen
Prinia flavicans — Prinia, Black-chested
Prinia gracilis — Warbler, Wren
Prinia inornata — Warbler, Wren
Prinia subflava — Prinia, Tawny-flaned
Priodontes giganteus — Armadillo, Giant
Prionace glauca — Shark, Blue
Prionailurus bengalensis — Cat, Leopard
Prionailurus rubiginosus — Cat, Rusty-
  spotted
Prionailurus viverrinus — Cat, Fishing
Prioniturus sp. — Parrot, Racket-tailed
Prionochilus sp. — Flowerpecker
Prionodon linsang — Linsang, Banded
Prionodon pardicolor — Linsang, Spotted

Prionodura newtoniana — Bowerbird, Golden
  or Newton's
Prionops plumata — Shrike, White-helmeted
Prionus coriarius — Beetle, Tanner Longhorn
Pristella riddlei — X-ray Fish
Pristis pectinatus — Sawfish, Small-tooth
Probosciger aterrimus — Cockatoo, Palm
Procapra gutturosa — Gazelle, Mongolian
Procapra picticaudata — Gazelle, Tibetan
Procavia capensis — Hyrax, Cape or Rock
Procavia habessinica — Hyrax, Abyssinian
Procellaria aequinoctialis — Petrel, White-
  chinned
Procellaria cinerea — Petrel, Gray
Procellaria diomedea — Shearwater, Cory's
Procellaria puffinus — Shearwater, Manx
Procnias alba — Bellbird
Procnias averano — Bellbird
Procnias nudicollis — Bellbird, Bare-
  throated
Procnias tricarunculata — Bellbird, Three-
  wattled
Procolobus verus — Colobus, Olive
Procyon cancrivorus — Raccoon, Crab-eating
Prodotiscus regulus — Honey-guide
Profelis badia — Cat, Bay
Profelis temmincki — Cat, Golden
Progne modesta — Martin
Progne tapera — Martin
Promerops cafer — Sugarbird, Cape
Promicrops lanceolatus — Grouper, Giant
Pronolagus crassicaudatus — Hare, Red Rock
Pronolagus rupestris — Hare, Red Rock
Propithecus verreauxi — Sifaka, Verreaux's
Prosperpinus proserpina — Hawkmoth,
  Willow-herb
Prostanthera ovalifolia — Mintbush
Prosthemadera novaeseelandiae — Tui
Protea cynaroides — Protea, Giant or King
Protea mellifera — Honeysuckle, Cape
Protea sp. — Protea
Proteles cristatus — Aardwolf
Protemnodon elegans — Wallaby, Pretty-
  faced
Protemnodon rufogrisea — Wallaby,
  Bennett's or Red-necked
Proteus anguineus — Olm
Protopterus sp. — Lungfish, African
Protoreaster linckii — Starfish

Protoxerus stangeri — Squirrel, Giant Forest
Protula tubularia — Worm, Chalk-tube
Prunella collaris — Accentor, Alpine
Prunella grandiflora — Self-heal, Large
Prunella immaculata — Accentor, Alpine
Prunella laciniata — Self-heal, Cut-
leaved
Prunella modularis — Sparrow, Hedge
Prunella montanella — Accentor, Siberian
Prunella vulgaris — Self-heal
Prunella webbiana — Self-heal
Prunus avium — Cherry, Sweet or Wild
Prunus cerasifera — Cherry-plum
Prunus cerasus — Cherry, Sour
Prunus conradine — Cherry, Conradine's
Prunus domestica — Plum, Common or
European
Prunus dulcis — Almond
Prunus laurocerasus — Laurel, Cherry
Prunus lusitanica — Laurel, Portugal
Prunus maackii — Cherry, Manchurian
Prunus mahaleb — Cherry, St. Lucie's or
Mahaleb
Prunus padus — Bird-cherry
Prunus serrulata — Cherry, Flowering
Prunus spinosa — Blackthorn
Prunus triloba — Almond, Flowering
Prunus yedoensis — Cherry, Yoshino or Tokyo
Psalidoprogne sp. — Swallow, Rough-winged
Psalliota sp. — Mushroom, Psalliota
Psammechinus miliaris — Sea Urchin, Purple
Psammodromus sp. — Lizard, Sand
Psarisomus dalhousiae — Broadbill, Long-
tailed
Psarocolius wagleri — Oropendola, Wagler's
Psephotus chrysopterygius — Parrot, Golden-
shouldered
Psephotus haematodus — Parrot, Red-backed
Psephotus haematogaster — Parrot,
Blue-bonnet
Psephotus haematonotus — Parrot,
Red-backed
Psephotus narathae — Parrot, Blue-bonnet
Psephotus pulcherrimus — Parrot, Paradise
Psephotus varius — Parrot, Mulga
Pseudacraea boisduvalii — Butterfly, False
Acraea
Pseudaspis cana — Snake, Mole

Pseudechis porphyriacus — Snake,
Australian Black
Pseuderanthcmum reticulatum — Yellowvein
Bush
Pseuderanthemum sp. — Eranthemum, False
Pseudibis gigantea — Ibis, Giant
Pseudis paradoxa — Frog, Paradoxical
Pseudobranchus striatus — Siren, Broad-
tailed Dwarf
Pseudobufo subasper — Toad, False
Pseudocheirus sp. — Possum, Ringtail
Pseudochelidon sp. — Martin, River
Pseudogyps africanus — Vulture, African
White-backed
Pseudogyps bengalensis — Vulture, Indian
White-backed
Pseudois nayaur — Sheep, Blue
Pseudolarix amabilis — Larch, Golden
Pseudomugil signifer — Silversides
Pseudopanthera macularia — Moth, Speckled
Yellow
Pseudophryne corroboree — Toadlet,
Corroboree
Pseudoplatystoma fasciatum — Catfish,
Pimelodid or Tiger
Pseudopodoces sp. — Jay, Ground
Pseudoscarus guacamaia — Parrot-fish
Pseudotriccus sp. — Tyrant, Pygmy
Pseudotropheus auratus — Cichlid,
Turquoise-gold
Pseudotsuga menziesii — Fir, Douglas
Pseudupeneus sp. — Goatfish
Psidium guajava — Guava
Psila rosae — Fly, Carrot
Psilocybe semilanceata — Liberty Cap
Psilotum nudum or triquetrum — Fern,
Whiskbroom
Psithyrus sp. — Bee, Cuckoo or Parasitic
Psittacula alexandri — Parakeet,
Red-breasted
Psittacula cyanocephala rosa — Parakeet,
Blossom- or Red-headed
Psittacula derbyana — Parakeet, Chinese or
Derbian
Psittacula himalayana — Parakeet, Black-
headed
Psittacula krameri — Parakeet, Rose-ringed
Psittacula longicauda — Parakeet,
Long-tailed

Psittacula roseata — Parakeet, Blossom-
or Red-headed

Psittacus erithacus — Parrot, Gray

Psitteuteles goldiei — Lorikeet

Psitteuteles versicolor — Lorikeet

Psittinus cyanurus — Parrot

Psophia crepitans — Trumpeter

Psophia leucoptera — Trumpeter

Psophodes cristatus — Wedgebill

Psophodes nigrogularis — Whipbird, Western

Psophodes olivaceous — Whipbird, Eastern

Psoralea bituminosa — Trefoil, Pitch

Psyche sp. — Moth, Bag-worm

Psychoda sp. — Midge, Owl

Psylla mali — Sucker, Apple

Pteria sp. — Oyster, Wing

Pteridium aquilinum — Fern, Brake or
Bracken

Pteridophora alberti — Bird of Paradise, King
of Saxony

Pteris cretica — Fern, Cretan Brake

Pteris multifida — Fern, Spider Brake

Pterocles alchata — Sandgrouse, Pin-tailed

Pterocles exustus — Sandgrouse, Chestnut-
bellied

Pterocles indicus — Sandgrouse, Indian or
Painted

Pterocles lichtensteinii — Sandgrouse,
Lichtenstein's

Pterocles orientalis — Sandgrouse, Black-
bellied or Imperial

Pterocles senegallus — Sandgrouse, Spotted

Pterocnemia pennata — Rhea, Darwin's

Pterodiscus sp. — Pterodiscus

Pterodroma arminjoniana — Petrel, Herald or
Trinidade

Pterodroma brevirostris — Petrel, Kerguelen

Pterodroma lessonii — Petrel, White-headed

Pterodroma leucoptera — Petrel, White-
winged

Pterodroma macroptera — Petrel, Great-
winged

Pterodroma mollis — Petrel, Soft-plumaged

Pteroglossus sp. — Aracari

Pteroides griseum — Sea Pen, Gray

Pterois sp. — Lionfish

Pterolebias sp. — Longfin

Pteromalus puparus — Fly, Chalcid

Pteromys volans — Squirrel, Flying

Pterophorus pentadactyla — Moth, White
Plume

Pterophyllum sp. — Angelfish, Freshwater

Pteropodocys maxima — Cuckoo-shrike,
Ground

Pteropus giganteus — Bat, Greater Indian

Pteropus poliocephalus — Bat, Gray-headed
Flying Fox

Pteropus vampyrus — Bat, Malay Fruit

Pterostoma palpina — Moth, Pale Prominent

Pterostylis sp. — Orchid, Hooded

Pterostyrax hispida — Epaulette Tree

Pteruthius sp. — Babbler, Shrike

Ptilinopus jambu — Dove, Fruit

Ptilinopus luteovirens — Dove, Fruit

Ptilinopus melanospila — Dove, Fruit

Ptilinopus regina — Pigeon, Red-crowned

Ptilinopus superbus — Dove, Fruit

Ptilinus pectinicornis — Beetle, Death-
watch

Ptilocercus lowii — Tree-shrew

Ptilocichla sp. — Babbler, Ground

Ptilogyna ramicornis — Fly, Crane

Ptilolaemus tickelli — Hornbill, Brown or
Tickell's

Ptilonorhynchus maculatus — Bowerbird,
Satin

Ptilonorhynchus violaceus — Bowerbird,
Satin

Ptiloris magnificus — Rifle-bird, Magnificent

Ptiloris paradiseus — Rifle-bird, Paradise

Ptiloris victoriae — Rifle-bird, Victoria

Ptilotus sp. — Lamb's-tails

Ptinus sp. — Beetle, Spider

Ptyas sp. — Snake, Oriental Rat

Ptychiozoon kuhlii — Gecko, Kuhl's

Ptyodactylus hasselquisti — Gecko, Fan-
footed

Ptyonoprogne rupestris — Martin, Crag

Puccinellia distans — Salt-marsh Grass,
Reflexed

Puccinellia maritima — Salt-marsh Grass,
Common

Pucrasia macrolopha — Koklas

Pudu pudu — Deer, Pudu

Pueraria lobata — Kudzu Vine

Pueraria thunbergiana — Kudzu Vine

Puffinus assimilis — Shearwater, Little

Puffinus bulleri — Shearwater, Buller's

Puffinus carneipes — Shearwater, Flesh-footed
Puffinus gravis — Shearwater, Greater
Puffinus griseus — Shearwater, Sooty
Puffinus pacificus — Shearwater, Wedge-tailed
Puffinus puffinus — Shearwater, Manx
Pulchriphyllium sp. — Leaf-insect
Pulex irritans — Flea, Human
Pulicaria dysenterica — Fleabane, Common
Pulicaria vulgaris — Fleabane, Lesser or Small
Pulmonaria angustifolia — Lungwort, Narrow-leaved
Pulmonaria longifolia — Lungwort, Narrow-leaved
Pulmonaria montana — Lungwort, Mountain
Pulmonaria officinalis — Lungwort, Common
Pulmonaria sp. — Lungwort
Pulsatilla alpina — Anemone, Alpine
Pulsatilla halleri — Pasque Flower
Pulsatilla montana — Anemone, Mountain
Pulsatilla pratensis — Pasque Flower, Small
Pulsatilla vernalis — Anemone, Spring or Pale
Pulsatilla vulgaris — Pasque Flower
Pulsatrix perspicillata — Owl, Spectacled
Pungitius pungitius — Stickleback, Nine- or Ten-spined
Punica granatum — Pomegranate
Puntius sp. — Barb, Tiger
Purpureicephalus spurius — Parrot, Red-capped
Pusa hispida — Seal, Ringed
Pusa sibirica — Seal, Baikal
Puschkinia scilloides — Squill, Striped
Putoria calabrica — Putoria
Puya raimondii — Puya
Pycnogonum littorale — Sea Spider
Pycnonotus atriceps — Bulbul, Black-headed or Blue-eyed
Pycnonotus barbatus — Bulbul, Black-eyed or White-vented
Pycnonotus brunneus — Bulbul, Red-eyed
Pycnonotus cafer — Bulbul, Red-vented
Pycnonotus goiavier — Bulbul, Yellow-vented
Pycnonotus jocosus— Bulbul, Red-whiskered
Pycnonotus nigricans— Bulbul, Red-eyed

Pycnonotus plumosus — Bulbul, Olive-winged
Pycnonotus sinensis — Bulbul, Chinese or Light-vented
Pycnoptilus floccosus — Pilot-bird
Pycnostachys urticifolia — Blue Boys
Pygathrix nemaeus — Langur, Douc
Pygocentrus piraya — Piranha
Pygoplites diacanthus — Angelfish, Royal
Pygopus sp. — Slow-worm
Pygoscelis adeliae — Penguin, Adelie
Pygoscelis antarctica — Penguin, Bearded or Chinstrap
Pygoscelis papua — Penguin, Gentoo
Pygosteus pungitius — Stickleback, Nine- or Ten-spined
Pyracantha angustifolia — Firethorn, Orange
Pyracantha atlantiodes — Firethorn, Gibb's
Pyracantha coccinea — Firethorn, Scarlet
Pyracantha rogersiana — Firethorn, Yellow
Pyralis farinalis — Moth, Flour or Meal
Pyrgus sp. — Butterfly, Skipper
Pyrochroa coccinea — Beetle, Cardinal
Pyrochroa serraticornis — Beetle, Cardinal
Pyrola chlorantha — Wintergreen, Green-flowered or Yellow
Pyrola media — Wintergreen, Intermediate
Pyrola minor — Wintergreen, Common or Lesser
Pyrola rotundifolia — Wintergreen, Large or Round-leaved
Pyrola uniflora — Wintergreen, One-flowered
Pyronia sp. — Butterfly, Gatekeeper
Pyrope pyrope — Diucon, Fire-eyed
Pyrostegia ignea or P. venusta — Flame Vine
Pyrrhocorax graculus — Chough, Alpine
Pyrrhocorax pyrrhocorax — Chough
Pyrrhocoris apterus — Firebug, Common
Pyrrholaemus brunneus — Redthroat
Pyrrhosoma sp. — Dragonfly
Pyrrhula erythaca — Bullfinch
Pyrrhula leucogenys — Bullfinch
Pyrrhula nipalensis — Bullfinch
Pyrrhula pyrrhula — Bullfinch
Pyrrhulina sp. — Pyrrhulina
Pyrrhura sp. — Parakeet
Pyrus communis — Pear, Common
Pyrus salicifolia — Pear, Willow-leaved

Pyrus sp. — Pear
Python anchietae — Python, Angola or Dwarf
Python molurus — Python, Indian
Python regius — Python, Ball
Python reticulatus — Python, Reticulated
Python sebae — Python, African
Pytilia afra — Pytilia
Pytilia melba — Finch, Melba
Pytilia phoenicoptera — Pytilia
Pyxicephalus adspersus — Bullfrog, African

Q

Quassia amara — Bitterwood
Quelea quelea — Quelea, Red-billed
Quercus acuta — Oak, Japanese
Quercus castaneifolia — Oak, Caucasian
Quercus cerris — Oak, Turkey
Quercus coccifera — Oak, Kermes or Holly
Quercus frainetto — Oak, Hungarian
Quercus ilex — Oak, Holm or Holly
Quercus petraea — Oak, Sessile or Durmast
Quercus robur — Oak, Common or English
Quercus suber — Oak, Cork
Quercusia quercus — Butterfly, Purple
    Hairstreak
Quisqualis indica — Rangoon Creeper

R

Rachycentron canadus — Cobia
Radiola linoides — Allseed
Rafflesia sp. — Rafflesia
Raja clavata — Ray, Thornback
Raja radiata — Ray, Starry
Raja sp. — Ray
Rallina eurizonoides — Crake, Philippine
    Banded
Rallina fasciata — Crake, Malay Banded or
    Red-legged
Rallina tricolor — Rail, Red-necked
Rallus aquaticus — Rail, Water
Rallus mirificus — Rail, Banded
Rallus pectoralis — Rail, Water
Rallus philippensis — Rail, Banded
Rallus striatus — Rail, Slaty-breasted
Ramaria sp. — Fungus, Coral
Ramonda myconi — Ramonda
Ramphastos ariel — Toucan, Ariel

Ramphastos cuvieri — Toucan, Cuvier's
Ramphastos discolorus — Toucan,
    Red-breasted
Ramphastos sulfuratus — Toucan, Keel-billed
Ramphastos toco — Toucan, Toco
Ramphocaenas melanurus — Gnatwren,
    Long-billed
Ramphocelus bresilius — Tanager, Brazilian
Ramphocelus carbo — Tanager, Brazilian
Ramphocelus dimidiatus — Tanager,
    Scarlet-rumped
Ramphocelus icteronotus — Tanager,
    Yellow-rumped
Ramphocelus nigrogularis — Tanager,
    Masked Crimson
Ramphocelus passerinii — Tanager,
    Scarlet-rumped
Ramphodon naevius — Hummingbird, Hermit
Ramphomicron microrhynchum —
    Hummingbird, Purple-backed Thornbill
Ramphotyphlops sp. — Snake, Blind
Rana adspersa — Bullfrog, African
Rana arvalis — Frog, Field
Rana esculenta — Frog, Edible
Rana goliath — Frog, Goliath
Rana malabaricus — Frog, West African
Rana ridibunda — Frog, Marsh
Rana temporaria — Frog, European
Rana tigrinia — Bullfrog, Indian
Ranatra sp. — Water-scorpion or Water-
    stick Insect
Rangifer tarandus — Reindeer
Raniceps raninus — Forkbeard, Lesser
Ranunculus acontifolius — Buttercup, White
Ranunculus acris—Buttercup, Common
    Meadow
Ranunculus alpestris — Crowfoot, Alpine
Ranunculus aquatilis — Crowfoot, Common
    Water
Ranunculus arvensis — Buttercup, Corn
Ranunculus asiaticus — Buttercup, Persian
    or Turban
Ranunculus auricomus — Goldilocks
Ranunculus bulbosus — Buttercup, Bulbous
Ranunculus circinatus — Crowfoot,
    Rigid-leaved
Ranunculus ficaria — Celandine, Lesser
Ranunculus flammula — Spearwort, Lesser
Ranunculus glacialis — Crowfoot, Glacier

Ranunculus hederaceus — Crowfoot, Ivy-leaved

Ranunculus insignis — Buttercup, Hairy Alpine

Ranunculus lappaceus — Buttercup, Grassland

Ranunculus lingua — Spearwort, Great

Ranunculus lyallii — Buttercup, Giant

Ranunculus montanus — Buttercup, Mountain

Ranunculus parnassifolius — Crowfoot, Parnassus-grass

Ranunculus parviflorus — Buttercup, Small-flowered

Ranunculus pyrenaeus — Crowfoot, Pyrenean

Ranunculus repens — Buttercup, Creeping

Ranunculus sardous — Buttercup, Pale Hairy

Ranunculus sceleratus — Buttercup, Celery-leaved

Ranunculus thora — Buttercup, Thora

Raoulia sp. — Daisy, Mat

Raphanus raphanistrum — Radish, Wild

Raphicerus campestris — Steinbok

Raphicerus melanotis — Steinbok

Raphicerus sharpei — Steinbok

Raphidia notata — Fly, Snake

Raphiolepis indica — Hawthorn, Indian

Raphiolepis umbellata — Hawthorn, Yeddo

Rapistrum perenne — Cabbage, Bastard or Steppe

Rapistrum rugosum — Cabbage, Bastard or Steppe

Rasbora heteromorpha — Harlequin Fish

Rasbora sp. — Rasbora

Rattus norvegicus — Rat, Norway

Rattus rattus — Rat, Black

Rattus tunneyi — Rat, Black

Ratufa bicolor — Squirrel, Giant

Ratufa indica — Squirrel, Giant

Ravenela madagascariensis — Traveller's Tree

Rebutia sp. — Cactus, Rebutia

Rechsteineria cardinalis — Cardinal Flower

Rechsteineria leucotricha — Edelweiss, Brazilian

Recurvirostra avosetta — Avocet, European

Recurvirostra novaehollandiae — Avocet, Australian or Red-necked

Redunca arundinum — Reedbuck, Common or Southern

Redunca fulvorufula — Reedbuck, Mountain

Redunca redunca — Reedbuck, Bohor

Reduvius personatus — Bug, Masked Hunter

Regulus ignicapillus — Firecrest

Regulus regulus — Goldcrest

Rehmannia sp. — Rehmannia

Reinhardtius hippoglossoides — Halibut, Greenland

Reinwardtia sp. — Flax, Yellow

Remiz pendulinus — Tit, Penduline

Remora remora — Shark-sucker

Reseda alba — Mignonette, White or Upright

Reseda lutea — Mignonette, Wild

Reseda luteola — Greenweed, Dyer's

Reseda odorata — Mignonette

Reseda villosa — Mignonette

Restio sp. — Restio

Restrepia sp. — Orchid

Reynoutria japonica — Knotweed, Japanese

Rhabdomys pumilio — Mouse, Zebra

Rhabdornis inornatus — Creeper, Plain-headed

Rhabdornis mystacalis — Creeper, Stripe-headed

Rhabdothamus solandri — Taurepo

Rhacophorus sp. — Frog, Gliding

Rhagio scolopacea — Fly, Snipe

Rhagium sp. — Beetle, Longhorn

Rhagoletis cerasi — Fly, Fruit

Rhamnus alaternus — Buckthorn, Mediterranean

Rhamnus catharticus — Buckthorn, Common

Rhamnus sp. — Buckthorn

Rhamphocorys clotbey — Lark, Thick-billed

Rhapis excelsa — Palm, Lady

Rhea americana — Rhea

Rheinartia ocellata — Pheasant, Ocellated

Rheum palmatum — Rhubarb, Sorrel

Rheum rhaponticum — Rhubarb

Rheum sp. — Rhubarb

Rhina barbirostris — Weevil, Brazilian

Rhinanthus minor — Yellow-rattle

Rhinanthus serotinus — Yellow-rattle, Greater

Rhinanthus sp. — Yellow-rattle

Rhincodon typus — Shark, Whale

Rhinobatus sp. — Guitarfish

Rhinoceros sondaicus — Rhinoceros, Javan

Rhinoceros unicornis — Rhinoceros, Great
Indian

Rhinocoris iracundus — Bug, Red Assassin

Rhinocrypta lanceolata — Gallito, Crested or
Gray

Rhinoderma darwini — Frog, Darwin's Dwarf

Rhinolophus ferrumequinum — Bat, Greater
Horse-shoe

Rhinolophus hipposideros — Bat, Lesser
Horse-shoe

Rhinomyias sp. — Flycatcher, Jungle

Rhinopithecus roxellanae — Monkey, Snub-
nosed

Rhinoplax vigil — Hornbill, Helmeted

Rhinopoma microphyllus — Bat, Rat-tailed or
Tomb

Rhinoptilus sp. — Courser

Rhipicephalus sp. — Tick, Dog

Rhipidura fuliginosus — Fantail, Black or
Gray

Rhipidura javanica — Fantail, Pied or
Malaysian

Rhipidura leucophrys — Wagtail, Willie

Rhipidura phoenicura — Fantail, Rufous

Rhipidura rufifrons — Fantail, Rufous

Rhipidura rufiventris — Fantail, Northern

Rhipidura setosa — Fantail, Northern

Rhipsalidopsis gaertneri — Cactus, Easter

Rhipsalidopsis rosea — Cactus, Easter

Rhizina undulata — Fungus, Doughnut

Rhizocarpon geographicum — Lichen, Map

Rhizomys sp. — Rat, Bamboo or Root

Rhizophora sp. — Mangrove

Rhizopogon luteolus — Truffle, False

Rhizostoma sp. — Jellyfish

Rhodeus sericeus amarus — Bitterling

Rhodinocichla rosea — Thrush-tanager, Rose-
breasted

Rhodiola kirilowii — Roseroot

Rhodiola rosea — Roseroot

Rhododendron ferrugineum — Alpenrose

Rhododendron hirsutum — Alpenrose, Hairy

Rhododendron luteum — Azalea

Rhododendron ponticum — Rhododendron,
Wild

Rhododendron simsii — Azalea Indica

Rhodohypoxis sp. — Rhodohypoxis

Rhodomyrtus tomentosa — Myrtle, Downy or
Rose

Rhodopechys githaginea — Finch, Trumpeter

Rhodophyllus sp. — Rhodophyllus

Rhodospingus cruentus — Finch, Crimson

Rhodostethia rosea — Gull, Ross'

Rhodothamnus chamaecistus — Ground-
cistus

Rhodotypus kerroides or R. scandens —
Kerria, White or Jet Bead

Rhoeo sp. — Boat Lily

Rhogogaster viridis — Sawfly

Rhoicissus rhomboidea — Grapevine,
Evergreen

Rhomborrhina japonica — Beetle, Rose
Chafer

Rhopalostylis sapida — Palm, Nikau

Rhus cotinus — See Cotinus coggyria

Rhyacornis sp. — Redstart, Water

Rhynchites sp. — Weevil, Leaf-roller

Rhynchocyon cirnei — Elephant-shrew,
Checkered-back

Rhynchoelaps bertholdi — Bandy-bandy

Rhynchogale melleri — Mongoose, Meller's

Rhynchosinapis cheiranthos — Cabbage,
Wallflower

Rhynchosinapis monensis — Cabbage, Isle of
Man

Rhynchospora alba — Beak-sedge, White

Rhynchospora fusca — Beak-sedge, Brown

Rhynchotragus kirki — Dik-dik, Long-
snouted

Rhynchotragus guentheri — Dik-dik, Long-
snouted

Rhynchotus rufescens — Tinamou, Red-
winged

Rhynochetus jubatus — Kagu

Rhyssa persuasoria — Wasp, Ichneumon

Rhyticeros sp. — Hornbill

Rhytidiadelphus sp. — Moss, Shaggy

Ribes alpinum — Currant, Mountain or
Alpine

Ribes grossularia — Gooseberry, English

Ribes nigrum — Currant, Black

Ribes rubrum or R. sylvestre — Currant, Red

Ribes spicatum — Currant, Upright Red

Ribes uva crispa — Gooseberry

Richea scoparia — Kerosene Bush

Ricinus communis — Castor-oil Plant or Castor Bean

Riopa sp. — Skink

Riparia paludicola — Swallow, Plain Bank

Riparia riparia — Swallow, Bank

Rissa tridactyla — Kittiwake, Black-legged

Rivulus sp. — Rivulus

Robinia hispida — Acacia, Rose

Robinia pseudoacacia — Locust, Black (tree)

Robinia viscosa — Locust, Clammy (tree)

Rochea coccinea — Rochea

Rochea falcata — Scarlet Paintbrush

Rodgersia sp. — Rodgersia

Rodriguezia secunda — Orchid, Coral

Roemeria hybrida — Horned-poppy, Violet

Rollulus roulroul — Roulroul

Roloffia lineriense — Lyretail, Calabar

Romulea bulbocodium — Crocus

Romulea columnae — Crocus, Warren or Sand

Rondoletia amoena — Rondeletia

Rondoletia odorata — Rondeletia

Rooseveltiella nattereri — Piranha, Natterer's or Red

Rorippa amphibia — Yellow-cress, Great

Rorippa islandica — Yellow-cress, Marsh

Rorippa nasturtium — Watercress

Rorippa sylvestris — Yellow-cress, Creeping

Rosa arvensis — Rose, Field or Trailing

Rosa canina — Rose, Dog

Rosa chinensis — Rose, China

Rosa multiflora — Rose, Multiflora

Rosa pendulina — Rose, Alpine or Drooping

Rosa pimpinellifolia — Rose, Burnet

Rosa rubiginosa — Sweet Briar

Rosa rugosa — Rose, Ramanus

Rosa spinosissima — Rose, Burnet

Rosa tomentosa — Rose, Downy

Rosa villosa — Rose, Downy

Roscoea cautleoides — Roscoea

Roscoea humeana — Roscoea

Rosmarinus officinalis — Rosemary

Rostratula bengalensis — Snipe, Painted

Rotheschildia sp. — Moth, Silk

Rothmannia sp. — Gardenia, Tree

Rousettus aegyptiacus — Bat, Egyptian Fruit

Roystonea regia — Palm, Royal

Rozites caperata — Mushroom, Gypsy

Rubia peregrina — Madder, Wild

Rubus arcticus — Bramble, Arctic

Rubus caesius — Dewberry

Rubus chamaemorus — Cloudberry

Rubus fruticosus — Blackberry

Rubus idaeus — Raspberry, European

Rubus saxatilis — Bramble, Stone or Rock

Rubus sp. — Lawyer

Rubus ulmifolius — Blackberry

Rudbeckia laciniata — Coneflower

Ruellia macrantha — Velvet Plant, Trailing

Ruellia sp. — Velvet Plant, Trailing

Rumex acetosa — Sorrel, Common

Rumex acetosella — Sorrel, Sheep's

Rumex alpinus — Monk's Rhubarb

Rumex aquaticus — Dock, Water

Rumex conglomeratus — Dock, Sharp or Clustered

Rumex crispus — Dock, Curled

Rumex hydrolapathum — Dock, Great Water

Rumex maritima — Dock, Golden

Rumex obtusifolius — Dock, Broad-leaved

Rumex palustris — Dock, Marsh

Rumex pulcher — Dock, Fiddle

Rumex sanguineus — Dock, Red-veined or Wood

Rumex scutatus — Dock, Shield

Rupicapra rupicapra — Chamois

Rupicola peruviana — Cock-of-the-Rock, Andean or Peruvian

Rupicola rupicola — Cock-of-the-Rock, Golden or Guianan

Ruppia sp. — Pondweed, Tassel

Ruschia sp. — Ruschia

Ruscus aculeatus — Butcher's Broom

Ruscus hypoglossum — Butcher's Broom, Large

Russelia equisetiformis — Fountain Plant

Russula emetica — Russula, Pungent

Russula sp. — Russula

Ruta chalepensis — Rue, Fringed or Sicilian

Ruta graveolens — Rue, Common

Rutilus rutilus — Roach

Rynchops albicollis — Skimmer, Indian

Rynchops flavirostris — Skimmer, African

Rynchops nigra — Skimmer, Black

S

Sabal palmetto — Palmetto
Sabella pavonina — Worm, Peacock
Sabella penicillus — Worm, Peacock
Saccharum officinarum — Sugar Cane
Saccoglossus sp. — Worm, Acorn
Saccostomus sp. — Mouse, Marsupial
Sagartia sp. — Sea Anemone
Sagina apetala — Pearlwort, Annual or
 Common
Sagina maritima — Pearlwort, Sea
Sagina nodosa — Pearlwort, Knotted
Sagina procumbens — Pearlwort, Procumbent
 or Mossy
Sagina saginoides — Pearlwort, Alpine
Sagina sublata — Pearlwort, Heath or
 Corsican
Sagitta sp. — Worm, Arrow
Sagittaria sagittifolia — Arrowhead
Sagittaria sp. — Arrowhead
Sagittarius serpentarius — Secretary-bird
Sagra busqueti — Beetle, Longhorn
Saguinus argentata — Marmoset, Silvery
Saguinus imperator — Tamarin, Emperor
Saguinus mystax — Tamarin, Moustached
Saguinus oedipus — Marmoset, Cotton-
 topped or Pinche
Saiga tatarica — Saiga
Saimiri sp. — Monkey, Squirrel
Saintpaulia hybrids — Violet, African
Saintpaulia ionantha — Violet, African
Salamandra atra — Salamander, Alpine
Salamandra salamandra — Salamander, Fire
Salamandrina terdigitata — Newt,
 Spectacled
Saldula saltatoria — Bug, Shore
Salicornia europaea — Glasswort, Common
Salicornia sp. — Glasswort
Salix alba — Willow, White
Salix arbuscula — Willow, Mountain or Little
 Tree
Salix atrocinera — Sallow, Common
Salix aurita — Sallow, Eared
Salix babylonica — Willow, Weeping or Poet's
Salix caprea — Sallow, Great
Salix cinera — Sallow, Common
Salix daphnoides — Willow, Violet

Salix discolor — Willow, Pussy
Salix fragilis — Willow, Crack or Yellow
Salix herbacea — Willow, Least
Salix lanata — Willow, Woolly
Salix lapponum — Willow, Downy
Salix matsudana — Willow, Corkscrew
Salix myrsinites — Willow, Myrtle-leaved
Salix nigra — Willow, Black
Salix nigricans — Willow, Dark-leaved
Salix pentandra — Willow, Bay or Laurel
Salix phylicifolia — Willow, Tea-leaved
Salix purpurea — Willow, Purple
Salix repens — Willow, Creeping
Salix reticulata — Willow, Reticulate or
 Netted
Salix retusa — Willow, Blunt-leaved
Salix triandra — Willow, Almond
Salix viminalis — Osier, Common
Salmo gairdneri — Trout, Rainbow
Salmo salar — Salmon, Atlantic
Salmo trutta — Trout, Brown
Salpa sp. — Sea Squirt
Salpiglossis sinuata — Painted Tongue
Salpornis spilonotus — Creeper, Gray Spotted
Salsola kali — Saltwort
Saltator sp. — Saltator
Salticus scenicus — Spider, Jumping
Salvelinus alpinus — Char, Alpine
Salvelinus fontinalis — Trout, Brook
Salvia glutinosa — Jupiter's Staff
Salvia horminoides — Clary, Wild
Salvia horminum — Sage, Red-topped
Salvia patens — Sage, Gentian
Salvia pratensis — Clary, Meadow
Salvia sclarea — Clary
Salvia triloba — Sage, Three-lobed
Salvia verbenaca — Clary, Wild
Salvinia natans — Water-fern
Sambucus ebulus — Danewort
Sambucus nigra — Elder, European
Sambucus racemosa — Elder, Alpine or Red
Samia sp. — Moth, Ailanthus
Samolus valerandi — Brookweed
Sanchezia nobilis — Sanchezia
Sanguisorba minor — Burnet, Salad
Sanguisorba obtusa — Burnet, Japanese

Sanguisorba officinalis — Burnet, Great
Sanicula europaea — Sanicle, Wood
Sansevieria sp. — Mother-in-law's Tongue
Sansevieria trifasciata — Mother-in-law's Tongue
Santolina chamaecyparissus — Lavender Cotton
Santolina virens — Lavender Cotton
Sanzinia madagascariensis — Boa, Sanzinia
Saperda sp. — Beetle, Longhorn
Sapium sp. — Tallow-tree, Chinese
Saponaria ocymoides — Soapwort, Rock
Saponaria officinalis — Bouncing Bett or Soapwort
Saponaria sp. — Soapwort
Sappho sparganura — Hummingbird, Sappho Comet
Saprinus maculatus — Beetle, Hister
Sarciphorus tectus — Plover, Blackhead
Sarcocaulon sp. — Sarcocaulon
Sarcococca hookerana — Box, Sweet or Christmas
Sarcodon imbricatum — Fungus, Scaly Prickle
Sarcogyps calvus — Vulture, Indian Black
Sarcophaga sp. — Fly, Flesh
Sarcophilus harrisii — Tasmanian Devil
Sarcops calvus — Mynah, Coleto
Sarcoptes scabei — Mite, Itch
Sarcoramphus papa — Vulture, King
Sarda sarda — Bonito, Common
Sardina pilchardus — Sardine
Sarkidiornis melanotus — Duck, Comb
Sarothamnus scoparius — Broom
Sarothrura elegans — Crake, Buff-spotted
Sarpa salpa — Saupe
Sarracenia flava — Huntsman's Horn
Sarracenia purpurea — Pitcher Plant
Sasia sp. — Piculet
Satanellus hallucatus — Dasyure, Little Northern
Satureia hortensis or S. thymbra — Savory, Summer
Satureia montana — Savory, Winter
Saturnia pavonia — Moth, Emperor
Saturnia pyri — Moth, Great Peacock
Satyrium sp. — Satyrium
Saussurea alpina — Saw-wort, Alpine

Saussurea sp. — Saw-wort
Saxicola caprata — Stonechat, Pied
Saxicola jerdoni — Bushchat, Jerdon's
Saxicola leucura — Stonechat
Saxicola rubetra — Whinchat
Saxicola torquata — Stonechat
Saxifraga aizoides — Saxifrage, Yellow Mountain
Saxifraga aizoon — Rockfoil
Saxifraga androsacea — Saxifrage, Rock Jasmine
Saxifraga burseriana — Saxifrage, Burser's
Saxifraga caesia — Saxifrage, Bluish
Saxifraga cernua — Saxifrage, Lesser Bulbous or Drooping
Saxifraga fortunei — Saxifrage, Fortune
Saxifraga granulata — Saxifrage, Meadow or Bulbous
Saxifraga grisebachii — Saxifrage, Bell
Saxifraga hirculus — Saxifrage, Yellow Marsh
Saxifraga hirsuta — Saxifrage, Kidney
Saxifraga hypnoides — Saxifrage, Mossy
Saxifraga longifolia — Saxifrage, Pyrenean
Saxifraga moschata — Saxifrage, Musky
Saxifraga nivalis — Saxifrage, Arctic or Clustered Alpine
Saxifraga oppositifolia — Saxifrage, Purple
Saxifraga paniculata — Saxifrage, Livelong or Silver
Saxifraga rivularis — Saxifrage, Highland or Alpine
Saxifraga rosacea — Saxifrage, Irish
Saxifraga rotundifolia — Saxifrage, Round-leaved
Saxifraga spathularis — St. Patrick's Cabbage
Saxifraga stellaris — Saxifrage, Starry
Saxifraga stolonifera — Aaron's Beard
Saxifraga tridactylites — Saxifrage, Rue-leaved
Saxifraga umbrosa — St. Patrick's Cabbage
Scabiosa atropurpurea — Scabious, Sweet
Scabiosa canescens — Scabious, Hoary
Scabiosa caucasica — Scabious, Caucasian
Scabiosa columbaria — Scabious, Small
Scabiosa lucida — Scabious, Shining
Scabiosa ochroleuca — Scabious, Yellow
Scabiosa sp. — Scabious

Scalpellum sp. — Barnacle, Stalked
Scandix pecten-veneris — Shepherd's Needle
Scapania sp. — Scapania
Scarabaeus laticollis — Beetle, Dung or Sacred Scarab
Scarabaeus sacer — Beetle, Dung or Sacred Scarab
Scardinius erythrophthalmus — Rudd
Scarus sp. — Parrot-fish
Scatophaga stercoraria — Fly, Dung
Scatophagus argus — Argus-fish
Sceliphron sp. — Wasp, Mud-dauber
Sceloglaux albifacies — Owl, Laughing
Scenopinus fenestralis — Fly, Window
Scenopoeetes dentirostris — Bowerbird, Tooth-billed
Schefflera digitata — Schefflera
Schefflera polybotrya — Schefflera
Scheuchzeria palustris — Rush, Rannoch
Schinus molle — Pepper-tree
Schistes geoffroyi — Wedgebill, Geoffroy's
Schistocerca gregaria — Locust, Desert
Schistocerca peregrina — Locust, Desert
Schistometopum thomensis — Caecilian, Sao Thome
Schizaea pusilla — Fern, Curly-grass
Schizanthus sp. — Butterfly Flower
Schizocentron elegans — Spanish Shawl
Schizocodon soldanelloides — Schizocodon
Schizophyllum commune — Split Gill
Schizostylis coccinea — Lily, Kaffir or River
Schlumbergera bridesii — Cactus, Christmas
Schlumbergera buckleyi — Cactus, Christmas
Schlumbergera truncata — Cactus, Thanksgiving or Claw
Schlumbergera zygocactus — Cactus, Thanksgiving or Claw
Schoenus nigricans — Bog-rush, Black
Schoinobates volans — Phalanger, Greater Flying
Schotia brachypetala — Tree Fuchsia
Sciadopitys verticillata — Pine, Japanese Umbrella
Sciapteron tabaniformis — Clearwing
Sciara sp. — Gnat, Fungus
Scilla autumnalis — Squill, Autumn
Scilla bifolia — Squill, Alpine or Two-leaved
Scilla peruviana — Squill, Peruvian

Scilla sp. — Squill
Scilla verna — Squill, Spring
Scincus philbyi — Skink, Sand
Scincus scincus — Skink, Sand
Scindapsus aureus — Ivy, Devil's or Pothos
Scirpus americanus — Club-rush, Sharp
Scirpus cespitosus — Deer-grass
Scirpus fluitans — Club-rush, Floating
Scirpus holoschoenus — Club-rush, Round-headed
Scirpus lacustris — Bulrush
Scirpus maritimus — Club-rush, Sea
Scirpus setaceus — Club-rush, Bristle
Scirpus sylvaticus — Club-rush, Wood
Scirpus tabernaemontani — Club-rush, Grayish or Glaucous
Sciuropterus russicus — Squirrel, Flying
Sciurus vulgaris — Squirrel, Red
Scleranthus annus — Knawel, Annual
Scleranthus perennis — Knawel, Perennial
Scleroderma sp. — Earthball
Sclerurus albigularis — Leaf-scraper, Gray-throated
Scolia flavescens — Wasp, Digger
Scolia flavifrons — Wasp, Digger
Scoliopteryx libatrix — Moth, Herald or Scalloped
Scolopax rusticola — Woodcock, Eurasian
Scolopendra sp. — Centipede, Giant
Scolymus hispanicus or S. maculatus — Oyster Plant, Spanish
Scolytus sp. — Beetle, Bark
Scomber colias — Mackerel, Spanish
Scomber scombrus — Mackerel, Common
Scomberesox saurus — Saury
Scomberomorus maculatus — Mackerel, Spanish
Scophthalmus maximus — Turbot
Scophthalmus rhombus — Brill
Scopolia carniolica — Scopolia
Scops bakkamoena — Owl, Collared Scops
Scopus umbretta — Hammerhead or Hamerkop
Scorpaena sp. — Scorpionfish
Scorpaeniopsis sp. — Scorpionfish
Scorpiurus muricatus — Scorpiurus
Scorzonera humilis — Vipergrass, Common
Scorzonera purpurea — Vipergrass, Purple
Scotocerca inquieta — Warbler, Scrub

Scotopelia bouvieri — Owl, Fishing
Scotopelia peli — Owl, Fishing
Scotopelia ussheri — Owl, Fishing
Scrobicularia plana — Furrow Shell,
    Peppery
Scrophularia aquatica — Figwort, Water
Scrophularia nodosa — Figwort, Common
Scrophularia scorodonia — Figwort,
    Balm-leaved
Scrophularia sp. — Figwort
Scrophularia vernalis — Figwort, Yellow
Scutellaria alpina — Skullcap, Alpine
Scutellaria galericulata — Skullcap, Common
Scutellaria minor — Skullcap, Lesser
Scutellaria sp. — Skullcap
Scutigera coleoptrata — Centipede, House
Scutigerella sp. — Centipede, Garden
Scyliorhinus caniculus — Dogfish, Lesser
    Spotted
Scyliorhinus stellaris — Dogfish, Greater
    Spotted
Scytalopus unicolor — Tapaculo
Scythrops novaehollandiae — Cuckoo,
    Channel-billed
Scytodes sp. — Spider, Spitting
Sebastes marinus — Haddock, Norway
Sebastes viviparus — Haddock Norway
Secale cereale—Rye
Sechium edule — Chayote
Securidaca longipedunculata — Violet Tree
Sedum acre — Stonecrop, Mossy or Wall-
    pepper
Sedum adolphi — Sedum, Golden
Sedum album — Stonecrop, White or Worm-
    grass
Sedum alpestre — Stonecrop, Alpine
Sedum anglicum — Stonecrop, English
Sedum annum — Stonecrop, Annual
Sedum caeruleum — Stonecrop, Blue
Sedum dasyphyllum — Stonecrop, Thick-
    leaved
Sedum forsteranum — Stonecrop, Rock
Sedum kamtschaticum — Stonecrop
Sedum maximum — Stonecrop, Great
Sedum middendorfianum — Stonecrop
Sedum morganianum — Donkey's Tail
Sedum reflexum — Stonecrop, Rock or
    Reflexed
Sedum rosea — Roseroot

Sedum rubrotinctum — Christmas Cheer
Sedum sieboldii — October Plant or October
    Daphne
Sedum spectabile — Stonecrop, Showy
Sedum telephium — Orpine or Livelong
Sedum villosum — Stonecrop, Hairy or Pink
Segestria senoculata — Spider, Cellar
Sehirus bicolor — Shield-bug, Pied
Sehirus dubius — Shield-bug, Pied
Seicercus sp. — Warbler
Seisura inquieta — Flycatcher, Restless
Selaginella sp. — Clubmoss
Selenarctos thibetanus — Bear, Asiatic or
    Himalayan Black
Selenidera sp. — Toucanet
Selenis spinifex — Beetle, Tortoise
Seleucides ignotus — Bird of Paradise,
    Twelve-wired
Seleucides melanoleuca — Bird of Paradise,
    Twelve-wired
Selinum carvifolia — Parsley, Cambridge
Semeiophorus vexillarius — Nightjar,
    Pennant-wing
Semioptera wallacei — Standard-wing,
    Wallace's
Semnopithecus entellus — Langur, Sacred
Semnornis ramphastinus — Barbet, Toucan
Sempervivum arachnoideum — Houseleek,
    Cobweb
Sempervivum montanum — Houseleek,
    Mountain
Sempervivum tectorum — Houseleek,
    Common
Sempervivum wulfenii — Houseleek,
    Wulfen's
Senecio aquaticus — Ragwort, Marsh
Senecio cineraria — Ragwort, Silver
Senecio confusus — Flame-vine, Mexican
Senecio cruentus — Cineraria
Senecio doronicum — Ragwort, Chamois
Senecio erucifolius — Ragwort, Hoary
Senecio fluviatilis — Ragwort, Broad-
    leaved
Senecio integrifolius — Fleawort, Field
Senecio jacobaea — Ragwort, Common
Senecio incanus — Groundsel, Gray Alpine
Senecio mikanioides — German or Parlor Ivy
Senecio paludosus — Ragwort, Great Fen
Senecio palustris — Fleawort, Marsh

Senecio squalidus — Ragwort, Oxford
Senecio sylvaticus — Groundsel, Wood
Senecio tamoides — Canary Creeper
Senecio vernalis — Groundsel, Spring
Senecio viscosus — Groundsel, Sticky
Senecio vulgaris — Groundsel, Common
Sepia officinalis — Cuttlefish, Common
Sepiola atlantica — Cuttlefish, Little
Sepiola rondeleti — Cuttlefish, Little
Serapias cordigera — Serapias, Heart-
flowered
Serapias lingua — Orchid, Tongue
Serapias sp. — Serapias
Sericornis beccarii — Scrub-wren, Little
Sericornis frontalis— Scrub-wren, White-
browed
Sericornis lathami — Scrub-wren, Yellow-
throated
Sericornis maculatus — Scrub-wren, Spotted
Sericornis minimus — Scrub-wren, Little
Sericulus aureas — Bowerbird, Black-faced
Golden
Sericulus chrysocephalus — Bowerbird,
Regent
Serinus canaria — Canary, Yellow
Serinus canicollis — Canary, Cape
Serinus citrinella — Finch, Citril
Serinus flaviventris — Canary, Yellow
Serinus gularis — Canary, Yellow
Serinus mozambicus — Canary, Yellow-
fronted
Serinus pusillus — Serin, Gold- or Red-
fronted
Serinus serinus — Serin
Serinus syriacus — Serin
Serinus tristiatus — Serin
Serranus scriba — Sea-perch
Serrasalmus sp. — Piranha
Serratula shawii — Saw-wort
Serratula tinctoria — Saw-wort
Sesamum indicum or S. orientale — Sesame or
Benne
Sesia apiformis — Clearwing, Hornet
Sesleria caerulea — Moor-grass, Blue
Sesleria sp. — Moor-grass
Sesli sp. — Moon Carrot
Setaria italica — Millet, Foxtail or Italian
Setaria sp. — Bristle-grass
Setaria viridis — Bristle-grass, Green

Setcreasea purpurea — Purple Heart
Setifer setosus — Tenrec, Hedgehog
Setonix brachyurus — Quokka
Sherardia arvensis — Madder, Field
Sialis flavilatera — Fly, Alder
Sialis lutaria — Fly, Alder
Sibbaldia procumbens — Cinquefoil, Least
Sibthorpia europaea — Moneywort, Cornish
Sicalis flaveola — Finch, Saffron
Sicista betulina — Mouse, Birch
Sicista subtilis — Mouse, Birch
Sida spinosa — Sida
Sieglingia decumbens — Heath-grass
Siganus sp. — Rabbitfish
Sigelus silens — Flycatcher, Fiscal
Sika nippon — Sika
Silaum silaus — Saxifrage, Pepper
Silene acaulis — Campion, Moss
Silene alba — Campion, White
Silene armeria — Catchfly, Sweet-
william
Silene gallica or S. anglica — Catchfly,
Small-flowered
Silene colorata — Catchfly
Silene conica — Catchfly, Sand or Striated
Silene dichotoma — Catchfly, Forked
Silene dioica — Campion, Red
Silene italica — Catchfly, Italian
Silene maritima — Campion, Sea
Silene noctiflora — Catchfly, Night-
flowering or Night-scented
Silene nutans — Catchfly, Nodding or
Nottingham
Silene otites — Catchfly, Spanish or
Breckland
Silene pendula — Catchfly, Drooping
Silene rupestris — Catchfly, Rock
Silene vulgaris — Campion, Bladder
Silurus glanis — Wels, European
Silvicapra grimmia — Duiker, Gray or
Grimm's
Silybum marianum — Milk-thistle
Simethis planifolia — Lily, Kerry
Simia sylvanus — Ape, Barbary
Simulium sp. — Fly, Black
Sinapis alba — Mustard, White
Sinapis arvensis — Charlock
Singa sp. — Spider, Garden
Sinningia sp. — Gloxinia

Siphonops annulatus — Caecilian, Mikans

Siphonostoma sp. — Pipefish

Siren lacertina — Siren, Greater

Sison amomum — Parsley, Stone

Sisymbrium altissimum — Rocket, Tall

Sisymbrium irio — Rocket, London

Sisymbrium officinale — Mustard, Hedge

Sisymbrium orientale — Rocket, Eastern

Sisyra fuscata — Fly, Spongilla

Sisyrinchium bermudiana — Blue-eyed Grass

Sisyrinchium striatum — Spring Bell

Sitophilus granarius — Weevil, Granary

Sitta castanea — Nuthatch, Chestnut-bellied

Sitta europaea — Nuthatch, European

Sitta frontalis — Nuthatch, Velvet-fronted

Sitta krueperi — Nuthatch, Turkish

Sitta neumayer — Nuthatch, Rock

Sitta tephronota — Nuthatch, Rock

Sitta whiteheadi — Nuthatch, Corsican

Sittasomus griseicapillus — Woodcreeper, Olivaceous

Sium latifolium — Parsnip, Water

Skimmia japonica — Skimmia, Japanese

Smaragdolanius leucotis — Shrike-vireo, Slaty-capped

Smerinthus ocellata — Hawkmoth, Eyed

Smicrornis brevirostris — Weebill, Brown

Smilacina racemosa — Spikenard, False

Sminthopsis crassicaudata — Mouse, Fat-tailed Pouched

Sminthopsis murina — Mouse, Fat-tailed Pouched

Sminthurus sp. — Springtail

Smithiantha hybrid — Temple Bells

Smithornis capensis — Broadbill, Cape

Smyrnium olusatrum — Alexanders

Smyrnium perfoliatum — Alexanders

Solandra guttata — Chalice Vine

Solandra sp. — Chalice Vine

Solanum capsicastrum — Cherry, Christmas or Winter

Solanum crispum — Potato Tree, Chilean

Solanum dulcamara — Nightshade, Bittersweet or Woody

Solanum jasminoides — Potato Vine

Solanum macranthum — Potato Tree

Solanum melongena — Egg Plant or Aubergine

Solanum nigrum — Nightshade, Black, Common or Deadly

Solanum pseudo-capsicum — Jerusalem Cherry

Solaster sp. — Starfish, Sun

Soldanella alpina — Soldanella, Alpine

Soldanella montana — Soldanella, Mountain

Soldanella pusilla — Soldanella, Dwarf or Lesser

Solea fulvomarginata — Sole

Solea solea — Sole

Solen sp. — Clam, Razor

Solidago canadensis — Goldenrod, Canada

Solidago virgaurea — Goldenrod, European

Sollya fusiformis — Bluebell Creeper

Somateria fischeri — Eider, Spectacled

Somateria mollissima — Eider, Common

Somateria spectabilis — Eider, King

Somniosus microcephalus — Shark, Greenland

Sonchus arvensis — Sow-thistle, Field

Sonchus asper — Sow-thistle, Prickly

Sonchus oleraceus — Sow-thistle, Common or Smooth

Sonchus palustris — Sow-thistle, Marsh

Sonerila margaritacea — Sonerila

Sophora japonica — Pagoda Tree

Sophora microphylla — Kowhai, Yellow

Sophora secundiflora — Mescal Bean

Sophora tetraptera — Kowhai, Yellow

Sophronitis sp. — Orchid, Sophronitis

Sorbaria sorbifolia — Spirea, False

Sorbus alnifolia — Mountain-ash, Korean

Sorbus aria — Whitebeam

Sorbus aucuparia — Rowan

Sorbus chamaemespilus — Whitebeam, Dwarf

Sorbus domestica — Service Tree

Sorbus hupehensis — Rowan, Hupeh

Sorbus intermedia — Whitebeam, Cut-leaved

Sorbus latifolia — Whitebeam, Broad-leaved

Sorbus torminalis — Service Tree, Wild

Sorex alpinus — Shrew, Alpine

Sorex araneus — Shrew, Common European

Sorex caecutiens — Shrew, Laxman's

Sorex minutus — Shrew, Pygmy

Sorubim lima — Catfish, Shovel-nosed

Spalax microphthalmus — Mole-rat

Sparassis crispa — Fungus, Cauliflower
Sparaxis tricolor — Harlequin Flower
Sparganium angustifolium — Bur-reed,
  Floating
Sparganium emersum — Bur-reed,
  Unbranched
Sparganium erectum — Bur-reed, Branched
Sparganium minimum — Bur-reed, Small
Sparganium simplex — Bur-reed, Small
Sparganium sp. — Bur-reed
Sparmannia africana — Hemp, African
Spartina townsendii — Cord-grass
Spartium junceum — Broom, Spanish or
  Weaver's
Sparus auratus — Gilthead
Spathiphyllum sp. — Spathiphyllum
Spathiphyllum wallisii — White Sails
Spathodea campanulata — Tulip Tree,
  African
Spathularia flavida — Spathularia
Spatula clypeata — Shoveler
Spatula rhynchotis — Shoveler, Blue-
  winged
Speothos venaticus — Bushdog
Spergula arvensis — Spurrey, Corn
Spergula sp. — Spurrey
Spergularia marina — Spurrey, Lesser Sea
Spergularia media — Spurrey, Greater Sea
Spergularia rubra — Spurrey, Sand or Red
Spergularia rupicola — Spurrey, Cliff or Red
Sphaerechinus granulatis — Sea Urchin,
  Purple
Sphaerium corneum — Mussel, Honey Orb
Sphaeroides sp. — Pufferfish
Sphagnum sp. — Bog Moss
Sphecius speciosus — Wasp
Sphecodus sp. — Bee, Cuckoo or Parasitic
Sphecotheres flaviventris — Figbird, Yellow
Sphecotheres vieilloti — Figbird, Southern
Sphedamnocarpus pruriens — Sphedamno-
  carpus
Spheniscus demersus — Penguin, Jackass
Spheniscus humboldti — Penguin, Peruvian
Spheniscus magellanicus — Penguin,
  Magellanic
Spheniscus mendiculus — Penguin,
  Galapagos
Sphenodon punctatus — Tuatara
Sphenomorphus sp. — Skink, Water

Sphenostoma cristatus — Wedgebill
Sphex sp. — Wasp, Sand
Sphinx ligustri — Hawkmoth, Privet
Sphinx pinastri — Hawkmoth, Pine
Sphodromantis lineola — Mantis, African
  Praying
Sphodropsis ghilianii — Beetle, Cave-
  dwelling Ground
Sphyraena barracuda — Barracuda, Great
Sphyrna zygaena — Shark, Hammerhead
Spilornis sp. — Eagle, Serpent
Spilosoma sp. — Moth, Ermine
Spinachia spinachia — Stickleback,
  Fifteen-spined
Spinturnix sp. — Mite
Spiraea x bumalda — Spirea, Bumalda
Spiraea japonica — Spirea, Japanese
Spiraea prunifolia — Bridal Wreath
Spiraea salicifolia — Spirea, Willow
Spiraea vanhourrei — Bridal Wreath
Spiranthes aestivalis — Lady's Tresses,
  Summer
Spiranthes romanzoffiana — Lady's Tresses,
  American or Irish
Spiranthes spiralis — Lady's Tresses,
  Autumn
Spirobraphis spallanzanii — Worm, Sabellid
Spisula sp. — Trough Shell
Spizaetus cirrhatus — Eagle, Changeable
  Hawk
Spizaetus ornatus — Eagle, Ornate Hawk
Spiziapteryx circumcinctus — Falconet, Spot-
  winged
Spondyliosoma cantharus — Bream, Black
  Sea
Spondylus buprestoides — Beetle, Forest or
  Cylinder Longhorn
Spondylus sp. — Oyster, Spiny or Thorny
Sporopipes squamifrons — Weaver, Scaly
Sprattus sprattus — Sprat
Spreo superbus — Starling, Superb
Squalis acanthias — Dogfish, Piked or Spiny
Squalis cephalus — Chub
Squatina squatina — Monkfish
Squilla sp. — Squill
Stachyris sp. — Babbler, Tree
Stachys alpina — Woundwort, Alpine
Stachys arvensis — Woundwort, Field
Stachys germanica — Woundwort, Downy

Stachys lanata — Lamb's Ears
Stachys officinalis — Betony, Wood
Stachys palustris — Woundwort, Marsh
Stachys recta — Woundwort, Perennial
   Yellow
Stachys sylvatica — Woundwort, Wood
Stachyurus chinensis — Stachyurus
Stachyurus praecox — Stachyurus
Stanhopea sp. — Orchid
Stapelia gigantea — Carrion Flower
Stapelia nobilis — Carrion Flower
Stapelia sp. — Carrion Flower
Stapelia variegata — Carrion Flower
Staphylea colchica — Bladder-nut
Staphylea holocarpa — Bladder-nut
Staphylea pinnata — Bladder-nut
Staphylinus olens — Coach-horse, Devil's
Stauropus fagi — Moth, Lobster
Steatornis caripensis — Oilbird
Steganopus tricolor — Phalarope, Wilson's
Steganura paradisea — Whydah, Sharp-
   tailed Paradise
Stellaria alsine — Stitchwort, Bog
Stellaria graminea — Stitchwort, Lesser
Stellaria holostea — Stitchwort, Greater
Stellaria media — Chickweed, Common
Stellaria nemorum — Stitchwort, Wood
Stellaria palustris — Stitchwort, Marsh
Stenella sp. — Dolphin
Stenobothrus lineatus — Grasshopper,
   Stripe-winged
Stenocarpus sinuatus — Firewheel Tree
Stenodelphis blainvillei — Dolphin
Stenolobium stans — Yellow Elder
Stenorhynchus seticornis — Crab, Spider
Stenostira scita — Flycatcher, Fairy
Stenotaphrum secundatum — St. Augustine
   Grass
Stephanandra incisa — Stephanandra, Dwarf
   Cut-leaved
Stephanibyx coronatus — Plover, Crowned
Stephanoaetus coronatus — Eagle, Crowned
Stephanophorus diadematus — Tanager,
   Diademed
Stephanotis floribunda — Wax Flower
Stercorarius longicaudatus — Skua, Long-
   tailed
Stercorarius parasiticus — Skua, Arctic
Stercorarius pomarinus — Jaeger, Pomarine

Stercorarius skua — Skua, Great
Sterculia murex — Chestnut, Lowveld
Stereum sp. — Stereum
Steriphoma paradoxa — Steriphoma
Sterna albifrons — Tern, Little
Sterna anaetheta — Tern, Bridled
Sterna balaenarum — Tern, Damara
Sterna bengalensis — Tern, Lesser Crested
Sterna bergii — Tern, Crested or Swift
Sterna dougallii— Tern, Roseate
Sterna fuscata — Tern, Sooty
Sterna hirundo — Tern, Common
Sterna hirundinacea — Tern, South American
Sterna hybrida — Tern, Whiskered
Sterna leucoptera — Tern, White-
   winged Black
Sterna macrura — Tern, Arctic
Sterna maxima — Tern, Royal
Sterna nereis — Tern, Fairy or White
Sterna niger — Tern, Black
Sterna paradisea — Tern, Arctic
Sterna sandvicensis — Tern, Sandwich
Sterna striata — Tern, White-fronted
Sterna sumatrana — Tern, Black-naped
Sterna vittata — Tern, Antarctic
Sternoptyx diaphana — Hatchetfish, Deep-
   sea
Sterrhopteryx fusca — Moth, Bag-worm
Stewartia sp. — Stewartia
Sticta sp. — Lichen, Sticta
Stictocarbo punctatus — Shag, Spotted
Stictonetta naevosa — Duck, Freckled
Stigmaphyllon ciliatum — Golden Vine
Stigmatogobius hoeveni — Goby, Celebes
Stilbum sp. — Wasp, Cuckoo
Stiltia isabella — Pratincole, Australian
Stipa capillata — Feather-grass, Hair-like
Stipa pennata — Feather-grass, Common
Stipa sp. — Feather-grass
Stipiturus malachurus — Emu-wren,
   Southern
Stipiturus mallee — Emu-wren, Mallee
Stipiturus ruficeps — Emu-wren, Rufous-
   crowned
Stizoptera bichenovii — Finch, Double-
   barred
Stizostedion lucioperca — Pike-perch
Stolephorus heterolobus — Anchovy
Stomoxys calcitrans — Fly, Stable (Biting)

Stranvaesia davidiana — Stranvaesia, Chinese

Stratiomys sp. — Fly, Soldier

Stratiotes aloides — Water Soldier

Strelitzia reginae — Bird-of-paradise Flower

Strepera arguta — Currawong, Clinking

Strepera fulginosa — Currawong, Black

Strepera graculina — Currawong, Pied

Strepera versicolor — Currawong, Gray

Strepsiceros strepsiceros — Kudu, Greater

Streptocarpus hybrids — Primrose, Cape

Streptopelia bitorquata — Dove, Java or Philippine Turtle

Streptopelia capicola — Dove, Cape Turtle

Streptopelia chinensis — Dove, Spotted or Lace-necked

Streptopelia decaocto — Dove, Collared

Streptopelia orientalis — Dove, Oriental or Rufous Turtle

Streptopelia reichenowi — Dove, Turtle

Streptopelia risoria — Dove, Barbary

Streptopelia semitorquata — Dove, Red-eyed Turtle

Streptopelia senegalensis — Dove, Laughing or Palm

Streptopelia tranquebarica — Dove, Red Turtle

Streptopelia turtur — Dove, Turtle

Streptoprocne zonaris — Swift, White-collared

Streptopus amplexifolius — Twisted Stalk

Streptosolen jamesonii — Browallia, Orange

Striga elegans — Witchweed

Striga lutea — Witchweed

Strigops habroptilus — Kakapo

Strix aluco — Owl, Tawny

Strix butleri — Owl, Hume's Tawny

Strix leptogrammica — Owl, Brown Wood

Strix nebulosa — Owl, Great Gray or Lapland

Strix seloputo — Owl, Spotted Wood

Strix uralensis — Owl, Ural

Strobilanthes dyerianus — Persian Shield

Strobilomyces floccopus — Old Man of the Woods

Strobilomyces strobilaceus — Fungus, Pine-cone

Stromanthe sanguinea — Stromanthe

Strombus sp. — Conch, True

Strongylodon macrobotrys — Jade Vine

Strophanthus sp. — Strophanthus

Stropharia aeruginosa — Toadstool, Verdigris

Stropharia semiglobata — Dung Roundhead

Struthidea cinerea — Apostle-bird

Struthio camelus — Ostrich

Strychnos sp. — Strychnos

Strymonidia pruni — Butterfly, Black Hairstreak

Strymonidia w-album — Butterfly, White-letter Hairstreak

Stultitia sp. — Stultitia

Sturnus philippensis — Starling, Violet-backed

Sturnus roseus — Starling, Rose-colored or Rosy

Sturnus sinensis — Starling, Gray-backed

Sturnus tristis — Mynah, Common or Indian

Sturnus unicolor — Starling, Spotless

Sturnus vulgarus — Starling, Common or European

Styphelia sp. — Styphelia

Styrax japonica — Snowbell, Japanese

Styrax obassia — Snowbell, Fragrant

Styrax officinalis — Storax

Suaeda maritima — Sea-blite, Annual

Suaeda vera — Sea-blite, Shrubby

Subularia aquatica — Awlwort

Succinea putris — Snail, Amber

Succisa pratensis — Scabious, Devil's-bit

Suillus sp. — Boletus

Sula abbotti — Booby, Abbott's

Sula bassana — Gannet

Sula capensis — Gannet, Cape

Sula dactylatra — Booby, Masked

Sula leucogaster — Booby, Brown

Sula nebouxii — Booby, Blue-footed

Sula serrator — Gannet, Australian

Sula sula — Booby, Red-footed

Suncus etruscus — Shrew, Savi's Pygmy

Suricata suricatta — Meerkat, Gray or Suricate

Surnia ulula — Owl, Hawk

Surniculus lugubris — Cuckoo, Drongo

Sus cristata — Boar, Wild

Sus scrofa — Boar, Wild

Sutherlandia frutescens — Cancerbush

Swainsona galegifolia — Pea, Darling

Swertia perennis — Felwort, Marsh

Swietenia sp. — Mahogany
Sycon sp. — Sponge, Purse
Sylvia atricapilla — Blackcap
Sylvia borin—Warbler, Garden
Sylvia cantillans — Warbler, Subalpine
Sylvia communis — Whitethroat
Sylvia conspicillata — Warbler, Spectacled
Sylvia curruca — Whitethroat, Lesser
Sylvia hortensis — Warbler, Orphean
Sylvia melanocephala — Warbler, Sardinian
Sylvia nana — Warbler, Desert
Sylvia nisoria — Warbler, Barred
Sylvia ruppelli — Warbler, Ruppell's
Sylvia sarda — Warbler, Marmora's
Sylvia undata — Warbler, Dartford
Sylvietta rufescens — Crombec, Long-
   billed
Sylviorthorhynchus desmursii — Wire-tail,
   Des Mur's
Sympetrum sp. — Dragonfly, Sympetrum
Symphalangus syndactylus — Siamang
Symphoricarpos albus — Snowberry
Symphoricarpos orbiculatus — Coralberry
Symphoricarpos rivularis — Snowberry
Symphysodon sp. — Discus, Brown or Green
Symphytum officinale — Comfrey, Common
Symphytum orientale — Comfrey, Eastern
Symphytum tuberosum — Comfrey, Tuberous
Symplocos paniculata — Sapphireberry
Synallaxis sp. — Spinetail
Synanceja verrucosa — Stonefish
Synanthedon tipuliformis — Clearwing
Synbranchus afer — Eel, Swamp
Synbranchus marmoratus — Eel, Swamp
Syncerus caffer — Buffalo, African or Cape
Syngnathus acus — Pipefish, Greater
Syngnathus pelagicus — Pipefish
Syngnathus pulchellus — Pipefish, Lesser
   Freshwater
Syngnathus typhle — Pipefish, Broad-nosed
Syngonium sp. — Arrowhead Vine or Plant
Synodontis angelicus — Catfish
Synodontis nigriventris — Catfish, Upside-
   down
Syndotis sp. — Catfish
Synoicus australis — Quail, Brown
Sypheotides indica — Florican
Syrigma sibilatrix — Heron, Whistling

Syringa amurensis — Lilac, Amur or
   Japanese
Syringa chinensis — Lilac, Chinese or Rouen
Syringa microphylla — Lilac, Littleleaf
Syringa persica — Lilac, Persian
Syringa vulgaris — Lilac, Common
Syrmaticus ellioti — Pheasant, Elliott's
Syrmaticus humiae — Pheasant, Hume's
Syrmaticus mikado — Pheasant, Mikado
Syrmaticus reevesi — Pheasant, Reeves'
Syrmaticus soemmeringii — Pheasant,
   Soemmering's Copper
Syromastes marginatus — Bug, Squash
Syromastes rhombeus — Bug, Squash
Syrphus sp. — Fly, Flower or Hover
Syrrhaptes paradoxus — Sandgrouse, Pallas'
Systenocerus caraboides — Beetle, Stag
Syzygium aromaticum — Clove Tree

T

Tabanus sp. — Fly, Horse
Tabebui sp. — Trumpet Tree
Tacca sp. — Bat Flower
Tachybaptus ruficollis — Grebe, Little
Tachyeres sp. — Duck, Steamer
Tachyglossus aculeatus — Echidna
Tachyglossus setosus — Echidna
Tachyoryctes daemon — Mole-rat, African
Tachyoryctes splendens — Mole-rat, African
Tachysines asynamorus — Cricket,
   Greenhouse
Tadorna cana — Shelduck, South African
Tadorna ferruginea — Shelduck, Ruddy
Tadorna radjah — Shelduck, Rajah
Tadorna tadorna — Shelduck, Common
Tadorna tadornoides — Shelduck, Australian
Tadorna variegata — Shelduck, Paradise
Taenia sp. — Tapeworm
Taeniopygia castanotis — Finch, Zebra
Taeniopygia guttata — Finch, Zebra
Taeniura lymma — Stingray, Ribbon-tail
Tagetes erecta — Marigold, African
Tagetes patula — Marigold, French
Talpa caeca — Mole, European
Talpa europaea — Mole, European
Tamandua tetradactyla or T. tridactyla —
   Anteater, Lesser or Three-toed

421

Tamarindus indica — Tamarind
Tamarix gallica — Tamarisk, French
Tamarix pentandra — Tamarisk, Five-
stamened
Tamarix sp. — Tamarisk
Tambourissa sp. — Tambourissa
Tamus communis — Bryony, Black
Tanacetum parthenium — Feverfew
Tanacetum vulgare — Tansy
Tanagra sp. — Euphonia
Tangara chilensis — Tanager, Paradise
Tangara chrysophrys — Tanager, Silver-
throated
Tangara fastuosa — Tanager, Seven-colored
Tangara gyrola — Tanager, Bay-headed
Tangara icterocephala — Tanager, Silver-
throated
Tangara nigrocincta — Tanager, Masked
Tangara parzudakii — Tanager, Black-eared
Tanichthys albonubes — Minnow, White-
cloud Mountain
Tanygnathus sp. — Parrot
Tanysiptera sylvia — Kingfisher, White-
tailed
Tapirus indicus — Tapir, Malayan or Saddle-
backed
Tapirus pinchaque — Tapir, Mountain
Tapirus roulini — Tapir, Mountain
Tapirus terrestris — Tapir, Brazilian or South
American
Taraba major — Antshrike, Great
Taraxacum officinale — Dandelion
Tarentola mauritanica — Gecko, Moorish or
Wall
Tarentula inquilina — Tarantula
Tarpon atlanticus — Tarpon
Tarsiger cyanurus — Bluetail, Red-flanked
Tarsipes rostratus — Possum, Honey
Tarsipes spenserae — Possum, Honey
Tarsius sp. — Tarsier
Taterillus gracilis — Gerbil
Taurago corythaix — Turaco, Knysna
Tauraco leucotis — Turaco
Tauraco livingstonii — Turaco, Livingstone's
Tauraco persa — Turaco
Tauraco porphyreolophus — Turaco, Violet-
crested
Tauraco ruspollii — Turaco
Tauraco schuttii — Turaco

Taurotragus derbianus — Eland, Giant
Taurotragus euryceros — Bongo
Taurotragus oryx — Eland, Cape or Common
Taurulus lilljeborgi or T. bulbalis —
Bullhead, Norway
Taxodium distichum — Cypress, Bald or
Swamp
Taxus baccata — Yew, English
Taxus sp. — Yew
Tayassu albirostris — Peccary
Tayassu tajacu — Peccary
Tayra barbara — Tayra
Tchagra senegala — Shrike, Bush
Tchagra tchagra — Shrike, Bush
Tealia sp. — Sea Anemone, Dahlia
Tecomaria capensis — Honeysuckle, Cape
Tectarius pagodus — Periwinkle
Tectona grandis — Teak
Teesdalia nudicaulis — Cress, Shepherd's
Tegenaria domestica — Spider, European
House
Tegenaria saeva — Spider, European House
Teinopalpus imperialis — Butterfly, Kaiser-
I-Hind
Teleonema filicauda — Manakin, Wire-tailed
Telescopus sp. — Snake, Eastern Tiger
Tellina sp. — Tellin
Telmatherina ladigesi — Rainbow-fish,
Celebes
Telopea speciosissima — Waratah
Telophorus sp. — Shrike, Bush
Temenuchus pagadarum — Starling, Pagoda
Tenebrio molitor — Beetle, Mealworm
Tenrec ecaudatus — Tenrec, Tailless
Tenthredo mesomelas — Sawfly
Tephrodornis gularis — Wood-shrike
Tephrodornis pondicerianus — Wood-shrike
Terachrysis semicincta — Wasp, Cuckoo
Terathopius ecaudatus — Eagle, Bateleur
Terebra sp. — Auger Shell
Terebratula sp. — Lampshell
Teredo sp. — Shipworm
Terenura callinota — Antwren
Termes bellicosus — Termite
Terminalia sp. — Myrobalan
Terpsiphone atrocaudata — Flycatcher, Black
or Japanese Paradise
Terpsiphone corvina — Flycatcher, Seychelles
Paradise

Terpsiphone paradisi — Flycatcher, Asiatic Paradise

Terpsiphone viridis — Flycatcher, African Paradise

Tersina viridis — Swallow-tanager

Testudo elegans — Tortoise, Starred

Testudo graeca — Tortoise, Mediterranean or Greek

Testudo hermanni — Tortoise, Mediterranean

Testudo pardalis — Tortoise, Leopard

Testudo sp. — Tortoise, Galapagos Giant

Tethya sp. — Sponge

Tetracerus quadricornis — Antelope, Four-horned

Tetraenura fischeri — Whydah

Tetragonolobus maritimus — Dragon's Teeth

Tetragonolobus purpureus — Pea, Asparagus

Tetranychus sp. — Mite, Spider

Tetrao urogallus — Capercaillie

Tetraodon sp. — Pufferfish

Tetraogallus sp. — Snowcock

Tetraopes tetraophthalmus — Beetle, Four-eyed Milkweed

Tetrapanax papyriferum — Rice Paper Plant

Tetrastes bonasia — Grouse, Hazel

Tetrastigma voinierianum — Chestnut Vine

Tetrax tetrax — Bustard, Little

Tetrix sp. — Groundhopper

Tetrosomus gibbosus — Boxfish

Tettigonia viridissima — Grasshopper, Great Green

Teucrium botrys — Germander, Cut-leaved

Teucrium chamaedrys — Germander, Wall

Teucrium fruticans — Germander, Tree

Teucrium montanum — Germander, Mountain

Teucrium scordium — Germander, Water

Teucrium scorodonia — Sage, Wood

Teucrium sp. — Germander

Textor capensis — Weaver, Cape

Textor cucullatus — Weaver, Spotted-backed

Thalacomys lagotis — Bandicoot, Rabbit

Thalarctos maritimus — Bear, Polar

Thalassema sp. — Worm

Thalasseus bergii — Tern, Crested or Swift

Thalasseus maximus — Tern, Royal

Thalassoica antarctica — Petrel, Antarctica

Thalassoma lunare — Wrasse, Moon

Thalassophryne maculosa — Toadfish, Spotted

Thalictrum alpinum — Meadow-rue, Alpine

Thalictrum aquilegifolium — Meadow-rue, Great

Thalictrum flavum — Meadow-rue, Common

Thalictrum minus — Meadow-rue, Lesser

Thalurania furcata — Wood-nymph

Thamnolaea cinnamomeiventris — Chat, Mocking

Thamnophilus doliatus — Antshrike, Barred

Thamnophilus multistriatus — Antshrike, Barred

Thanasimus formicarius — Beetle, Ant

Thatcheria mirabilis — Thatcheria, Miraculous

Thaumastura cora — Hummingbird, Peruvian Sheartail

Thaumatibis gigantea — Ibis, Giant

Thaumetopoea processionea — Moth, Processionary

Thayeria sp. — Penguin-fish

Thecla betulae — Butterfly, Brown Hairstreak

Thecla quercus — Butterfly, Purple Hairstreak

Thelocactus bicolor — Glory-of-Texas

Thelotornis kirtlandi — Snake, Southern Bird

Thelycrania sanguinea — Dogwood or Cornel

Thelypteris palustris — Fern, Marsh

Thelypteris phegopteris — Fern, Beech

Themus generosus — Beetle, Soldier

Theobroma cacao — Cacao or Cocoa Tree

Therion circumflexum — Wasp, Ichneumon

Thermobia domestica — Firebrat

Theropithecus gelada — Baboon, Gelada

Thesium alpinum — Toadflax, Alpine Bastard

Thesium humifusum — Toadflax, Bastard

Thevetia peruviana — Oleander, Yellow

Thinocorus orbignyanus — Seedsnipe, D'Orbigny's or Gray-breasted

Thinocorus rumicivorus — Seedsnipe, Least or Patagonian

Thinornis novaeseelandiae — Plover, New Zealand or Shore

Thlaspi alpestre — Pennycress, Alpine

Thlaspi arvense — Pennycress, Field

Thlaspi montanum — Pennycress, Mountain

Thlaspi perfoliatum — Pennycress, Perfoliate

Thlaspi rotundifolium — Pennycress, Round-leaved

Thomisus sp. — Spider, Crab

Thos aureus — Jackal, Common or Golden

Thraupis episcopus — Tanager, Blue-gray

Thraupis virens — Tanager, Blue-gray

Threskiornis aethiopica — Ibis, Sacred or White

Threskiornis molucca — Ibis, Sacred or White

Threskiornis spinicollis — Ibis, Straw-necked

Thripadectes flammulatus — Treehunter

Thripadectes virgaticeps — Treehunter

Thripias namaquus — Woodpecker, Bearded

Thrips tabaci — Thrips, Onion

Thryonomys swinderianus — Cane-rat

Thuja orientalis — Arborvitae, Oriental

Thujopsis dolobrata — Arborvitae, False or Hiba

Thunbergia alata — Black-eyed Susan Vine

Thunbergia erecta — Bush Clockvine

Thunbergia grandiflora — Sky Flower or Vine

Thunnus alalunga — Albacore

Thunnus albacores — Tuna, Yellowfin

Thunnus germo — Albacore

Thunnus thynnus — Tuna, Bluefin

Thyatira batis — Moth, Peach-blossom

Thylacinus cynocephalus — Tasmanian Wolf

Thylogale billardierii — Pademelon, Red-bellied

Thymallus thymallus — Grayling, European

Thymelaea hirsuta — Thymelaea

Thymelaea passerina — Thymelaea

Thymelaea tartonraira — Thymelaea

Thymelicus sp. — Butterfly, Skipper

Thymus citriodorus — Thyme, Lemon

Thymus drucei — Thyme, Wild

Thymus pulegioides — Thyme, Large Wild

Thymus serphyllum — Thyme, Wild

Thymus sp. — Thyme

Thymus vulgaris — Thyme, Common

Thyrsites atun — Snoek

Thysania agrippina — Moth

Thysania zenobia — Moth

Tiaris olivacea — Grassquit, Yellow-faced

Tibouchina sp. — Glory Bush

Tichodroma muraria — Wall-creeper

Tigridia pavonia — Tiger Flower

Tigrisoma lineatum — Heron, Tiger

Tilapia sp. — Mouthbrooder

Tilia cordata — Linden, Small-leaved

Tilia euchlora — Lime, Caucasian

Tilia europaea — Lime, Common

Tilia petiolaris — Lime, Silver Pendant

Tilia platyphyllos — Linden, Large-leaved

Tilia tomentosa — Linden, Silver

Tilia vulgaris — Linden, Common

Tiliqua sp. — Skink, Blue-tongued

Tillandsia cyanea — Tillandsia

Tillandsia sp. — Tillandsia

Tillandsia usneoides — Spanish Moss

Timalia pileata — Babbler, Red-capped

Timarcha tenebricosa — Beetle, Bloody-nosed

Tinamotis pentlandii — Tinamou, Puna

Tinamus sp. — Tinamou

Tinca tinca — Tench

Tinea pellionella — Moth, Clothes (Case-making)

Tineola bisselliella — Moth, Clothes (Webbing)

Tingis sp. — Bug, Lace

Tinnea aethiopica — Tinnea

Tinnea galpinii — Tinnea

Tipula sp. — Fly, Crane

Tityra cayana — Tityra, Black-tailed

Tityra semifasciata — Tityra, Masked

Tockus alboterminatus — Hornbill, Crowned

Tockus camurus — Hornbill, Red-billed

Tockus deckeni — Hornbill, Van der Decken's

Tockus erythrorhynchus — Hornbill, Red-billed

Tockus flavirostris — Hornbill, Yellow-billed

Tockus nasutus — Hornbill, Gray

Todirostrum sp. — Flycatcher, Tody

Tofieldia calyculata — Asphodel, German

Tofieldia pusilla — Asphodel, Scottish

Tolmiea menziesii — Youth-on-age

Tolpis barbata — Tolpis

Tolypeutes matacus — Armadillo, Three-banded

Tolypeutes tricinctus — Armadillo, Three-banded

Tonna galea — Tun Shell, Giant

Topaza pella — Hummingbird, Topaz

Topaza pyra — Hummingbird, Topaz

Tordylium apulum — Hartwort

Tordylium maximum — Hartwort, Great

Torenia fournieri — Wishbone Flower

Torgos tracheliotus — Vulture, Lappet-faced
Torilis arvensis — Hedge-parsley, Spreading
Torilis japonica — Hedge-parsley, Upright
Torilis nodosa — Hedge-parsley, Knotted
Torpedo marmorata — Ray, Electric
Torreya californica — Nutmeg, California
Torreya sp. — Torreya
Tortella tortuosa — Moss, Twisted
Tortrix viridana — Moth, Oak-roller
Touit huetii — Parrotlet
Toxotes chatareus — Archerfish
Toxotes jaculator — Archerfish
Tozzia alpina — Tozzia, Alpine
Trachelium caeruleum — Throatwort
Trachelospermum jasminoides — Jasmine, Confederate Star
Trachinus draco — Weever, Greater
Trachinus vipera — Weever, Lesser
Trachurus trachurus — Scad
Trachycarpus fortunei — Palm, Chusan or Windmill
Trachydosaurus rugosus — Skink, Shingle-back
Trachymene caerulea — Lace-flower, Blue
Trachyphonus darnaudii — Barbet, D'Arnaud's
Trachyphonus erythrocephalus — Barbet, Red-and-Yellow
Trachyphonus vaillantii — Barbet, Crested
Trachystemon orientalis — Borage, Eastern
Tradescantia blossfeldiana — Inch Plant, Flowering
Tradescantia fluminensis — Wandering Jew
Tradescantia sillamontana — Wandering Jew, White Velvet
Tradescantia virginiana — Spiderwort, Common
Tragelaphus angasii — Nyala, Zululand
Tragelaphus buxtoni — Nyala, Mountain
Tragelaphus imberbis — Kudu, Lesser
Tragelaphus scriptus — Bushbuck
Tragelaphus spekei — Sitatunga
Tragelaphus strepsiceros — Kudu, Greater
Tragopan blythii blythii — Pheasant, Blyth's
Tragopan melanocephalus — Pheasant, Western
Tragopan satyra — Pheasant, Satyr

Tragopan temmincki — Pheasant, Temminck's
Tragopogon porrifolius — Salsify
Tragopogon pratensis — Goatsbeard, Meadow
Tragopogon sp. — Goatsbeard
Tragulus javanicus — Chevrotain, Malayan
Tragulus meminna — Chevrotain, Indian
Trametes sp. — Trametes
Trapa natans — Water-chestnut
Tremarctos ornatus — Bear, Spectacled
Trematomus bernacchi — Cod, Antarctic
Tremella foliacea — Fungus, Quivering or Trembling
Tremella mesenterica — Fungus, Quivering or Trembling
Tremex sp. — Wasp, Wood
Treron apicauda — Pigeon, Pin-tailed Green
Treron australis — Pigeon, Green
Treron bicincta — Pigeon, Orange-breasted Green
Treron capellei — Pigeon, Green
Treron curvirostra — Pigeon, Lesser Thick-billed Green
Treron formosae — Pigeon, Green
Treron phoenicoptera — Pigeon, Yellow-footed Green
Treron seimundi — Pigeon, Green
Treron sphenura — Pigeon, Wedge-tailed Green
Treron vernans — Pigeon, Pink-necked Green
Trialeurodes vaporariorum — Fly, White
Tribonyx mortierii — Native Hen, Tasmanian
Tribonyx ventralis — Native Hen, Black-tailed
Tribulus terrestris — Maltese Cross
Trichastoma sp. — Babbler, Ground
Trichechus senegalensis — Manatee, African
Trichilia sp. — Mahogany, Cape
Trichinium sp. — Mulla Mulla
Trichiosoma sp. — Sawfly, Hawthorne
Trichius fasciatus — Beetle, Bee Chafer or Banded Brush
Trichobatrachus robustus — Frog, Hairy
Trichocera sp. — Gnat, Winter
Trichodes sp. — Beetle, Checkered
Trichodectes canis — Louse, Dog
Trichodesma physaloides — Chocolate-creams
Trichogaster leeri — Gourami, Pearl

425

Trichogaster trichopterus — Gourami, Three-
spot
Trichoglossus chlorolepidotus — Lorikeet,
Scaly-Breasted
Trichoglossus haematodus — Lorikeet,
Rainbow
Trichoglossus moluccanus — Lorikeet,
Rainbow
Tricholoma flavovirens — Man on Horseback
Tricholoma gambosum or T. georgii—
Mushroom, St. George's
Tricholoma nudum — Blewit, Wood
Tricholoma pardinum — Tricholoma, Leopard
Tricholoma portentosum — Agaric, Dingy
Tricholoma rutilans — Agaric, Red-haired
Tricholoma saponaceum — Toadstool, Soap-
scented
Tricholoma sulphureum — Blewit, Narcissus
Tricholoma terreum — Agaric, Gray
Tricholompsis rutilans — Blewit, Purple
Trichomanes sp. — Fern, Killarney or Filmy
or Bristle
Trichophaga tapetzella — Moth, Tapestry
Trichopsis pumilus — Gourami, Dwarf
Trichopsis vittatus — Gourami, Croaking
Trichosanthes sp. — Gourd, Snake
Trichurus caninus — Phalanger, Brush-
tailed
Trichosurus vulpecula — Phalanger, Brush-
tailed
Triclaria malachitacea — Parrot
Tricolia pullus — Top Shell, Pheasant
Tricyrtis stolonifera — Lily, Toad
Tridaena sp. — Clam, Giant
Trientalis europaea — Wintergreen,
Chickweed
Trifolium alpestre — Clover, Oval-headed
Trifolium alpinum — Trefoil, Alpine
Trifolium arvense — Clover, Hare's-foot
Trifolium badium — Trefoil, Brown
Trifolium bocconei — Clover, Twin-flowered
Trifolium campestre — Hop-trefoil
Trifolium dubium — Clover, Lesser Yellow
Trifolium fragiferum — Clover, Strawberry
Trifolium glomeratum — Clover, Flat-headed
or Clustered
Trifolium hybridum — Clover, Alsike
Trifolium incarnatum — Clover, Crimson
Trifolium medium — Clover, Zigzag

Trifolium micranthum — Trefoil, Slender
Yellow
Trifolium montanum — Clover, Mountain
Trifolium ochroleucon — Clover, Sulphur
Trifolium ornithopodioides — Fenugreek
Trifolium pratense — Clover, Red or Purple
Trifolium purpureum — Clover, Purple
Trifolium repens — Clover, White or Dutch
Trifolium scabrum — Clover, Rough
Trifolium squamosum — Clover, Sea or
Teazel-headed
Trifolium stellatum — Clover, Star
Trifolium striatum — Clover, Soft or Knotted
Trifolium strictum — Clover, Upright
Trifolium subterraneum — Clover,
Subterranean
Trifolium suffocatum — Clover, Suffocated
Trigla gurnardus — Gurnard, Gray
Trigla lineata or T. lastovitza — Gurnard,
Streaked
Trigla lucerna — Gurnard, Tub or Yellow
Triglochin maritima — Arrow-grass, Sea
Triglochin palustris — Arrow-grass, Marsh
Trigonella foenum-graecum — Fenugreek
Trigonella sp. — Fenugreek
Trigonoceps occipitalis — Vulture, White-
headed
Trimeresurus sp. — Viper, Pit or Tree
Tringa brevipes — Tattler, Gray-tailed or
Siberian
Tringa cinereus — Sandpiper, Terek
Tringa erythropus — Redshank, Spotted
Tringa glareola — Sandpiper, Wood
Tringa guttifera — Greenshank, Nordmann's
or Spotted
Tringa hypoleucos — Sandpiper, Common
Tringa incana — Tattler, Wandering
Tringa nebularia — Greenshank
Tringa ochropus — Sandpiper, Green
Tringa stagnatilis — Sandpiper, Marsh
Tringa terek — Sandpiper, Terek
Tringa totanus — Redshank, Common
Trinia glauca — Honewort
Trionyx sp. — Turtle, Soft-shelled
Triphaena pronuba — Moth, Yellow
Underwing
Tripleurospermum inodorum — Mayweed,
Scentless

Tripleurospermum maritimum — Mayweed,
Scentless
Tripterygion sp. — Blenny
Trisetum flavescens — Oat-grass, Yellow
Trisopterus esmarkii — Pout, Norway
Trisopterus luscus — Pouting
Trisopterus minutus — Poor-cod
Triticum aestivum — Wheat
Triticum sp. — Wheat
Tritonalia rubeta — Frog Shell
Triturus alpestris — Newt, Alpine
Triturus cristatus — Newt, Crested
Triturus helveticus — Newt, Palmate
Triturus marmoratus — Newt, Marbled
Triturus vulgaris — Newt, Smooth
Triumfetta sp. — Triumfetta
Trivia sp. — Cowry
Trochilus polytmus — Hummingbird,
Streamer-tail
Trogium pulsatorium — Louse, Book
Troglodytes troglodytes — Wren, Winter
Troglodytes troglodytes hirtensis — Wren,
St. Kilda
Troglodytes troglodytes nipalensis — Wren,
Winter
Trogon rufus — Trogon, Black-throated
Trogonophis wiegmanni — Worm-lizard,
Wiegmanni's
Trogulus tricarinatus — Harvestman
Troides sp. — Butterfly, Birdwing
Trollius europaeus — Globe Flower, European
Trollius sp. — Globe Flower
Trombidium sp. — Mite, Velvet Spider
Tropaeolum majus — Nasturium
Tropaeolum peregrinum — Canary Creeper or
Canary Bird Flower
Tropaeolum polyphyllum — Nasturium,
Wreath
Tropaeolum speciosum — Flame Flower,
Scottish
Tropaeolum tuberosum — Nasturium,
Tuberous
Tropidurus grayi — Anole
Tryngites subruficollis — Sandpiper, Buff-
breasted
Tuber aestivum — Truffle, Summer
Tuber brumale — Truffle, Winter
Tuber magnatum — Truffle, White

Tuber melanosporum — Truffle, Black or
Perigord
Tuberaria guttata — Rockrose, Annual or
Spotted
Tubifex sp. — Worm, Tubifex
Tubulanus sp. — Worm, Ribbon
Tubularia larynx — Hydroid
Tulbaghia acutiloba — Tulbaghia
Tulbaghia fragrans — Tulbaghia, Fragrant
Tulbaghia violacea — Garlic, Society
Tulipa australis — Tulip, Southern
Tulipa sylvestris — Tulip, Wild
Tulostoma sp. — Puffball
Tunga penetrans — Flea, Chigoe
Tupaia glis — Tree-shrew
Tupaia tana — Tree-shrew
Tupinambis — Tegu, Common or Great
Turbo sp. — Turban Shell
Turdoides jardinei — Babbler, Arrow-
marked
Turdoides sp. — Babbler, Jungle
Turdus iliacus — Redwing
Turdus libonyanus — Thrush, Kurrichane
Turdus merula — Blackbird, European
Turdus naumanni — Thrush, Naumann's
Turdus obscurus — Thrush, Eye-browed
Turdus olivaceus — Thrush, Olive
Turdus philomelos — Thrush, Song
Turdus pilaris — Fieldfare
Turdus ruficollis — Thrush, Black-
throated
Turdus sibiricus — Thrush, Siberian
Turdus torquatus — Ouzel, Ring
Turdus viscivorus — Thrush, Mistle
Turnagra capensis — Thrush, New Zealand
Turnera sp. — Sage Rose
Turnix maculosa — Quail, Red-backed
Turnix melanogaster — Quail, Black-
breasted
Turnix nana — Quail, Button
Turnix ocellata — Quail, Button
Turnix suscitator — Quail, Barred Button
Turnix sylvatica — Quail, Kurrichane Button
Turnix varia — Quail, Painted Button
Turnix velox — Quail, Little
Turritella sp. — Shell, Tower or Turret
Turritis glabra — Mustard, Tower
Tursiops aduncus — Dolphin, Bottle-nosed

Tursiops truncatus — Dolphin, Bottle-nosed
Turtur abyssinicus — Dove, Wood
Turtur chalcospilos — Dove, Wood
Tussilago farfara — Coltsfoot
Tylopilus felleus — Boletus, Bitter
Typha angustifolia — Reedmace, Lesser or
 Narrow-leaved
Typha latifolia — Reedmace, Great
Typhaeus typhoeus — Beetle, Minotaur
Typhlops sp. — Snake, Blind
Tyto alba — Owl, Barn
Tyto capensis — Owl, Grass
Tyto castanops — Owl, Masked
Tyto longimembris — Owl, Grass
Tyto novaehollandiae — Owl, Masked
Tyto soumagnei — Owl, Grass
Tyto tenebricosa — Owl, Sooty

U

Uca sp. — Crab, Fiddler
Ulex europaeus — Gorse, Common
Ulex gallii — Gorse, Dwarf
Ulex minor — Gorse, Dwarf
Ullucus tuberosus — Ulluco
Ulmus campestris — Elm, English
Ulmus carpinifolia — Elm, Smooth-leaved
 or Wheatley
Ulmus glabra — Elm, Wych
Ulmus hollandica — Elm, Dutch
Ulmus laevis — Elm, Fluttering or European
 White
Ulmus minor — Elm, Smooth-leaved or
 Wheatley
Ulmus parvifolia — Elm, Chinese
Ulmus procera — Elm, English
Ulva lactuca — Sea Lettuce
Umbilicaria sp. — Lichen
Umbilicus spinosus — Pennywort, Wall
Umbonia sp. — Treehopper
Umbra krameri — Mudminnow, European
Uncarina sp. — Uncarina
Uncia uncia — Leopard, Snow
Unio margaritifera — Mussel, Pearl
Unio pictorum — Mussel, Painter's
Upupa africana — Hoopoe, African
Upupa epops — Hoopoe, Eurasian
Uraeginthus bengalus — Cordon Bleu, Red-
 cheeked

Uraeginthus granatinus — Waxbill, Violet-
 eared
Uragus sibiricus — Rosefinch, Long-billed
Urania leilus — Moth
Uranoscopus scaber — Stargazer, Common
Uratelornis chimaera — Roller, Long-tailed
 Ground
Urginea maritima — Squill, Sea
Uria aalge — Murre, Common
Uria lomvia — Murre, Thick-billed
Uroaetus audax — Eagle, Wedge-tailed
Urocerus sp. — Wasp, Wood
Urocissa erythrorhyncha — Magpie, Red-
 billed Blue
Urocissa ornata — Magpie, Ceylon Blue
Urocissa whiteheadi — Magpie, Whitehead's
 Blue
Urogale everetti — Tree-shrew
Uroglaux dimorpha — Owl, New Guinea
 Hawk
Urolestes melanoleucus — Shrike, Magpie
Uromastyx sp. — Lizard, Spiny-tailed
Uropeltis sp. — Snake, Shield-tail
Uroplatus fimbriatus — Gecko, Leaf-tailed
Urosalpinx cinerea — Drill, Oyster
Urotriorchis macrourus — Hawk, African
 Long-tailed
Ursinia sp. — Ursinia
Ursus arctos — Bear, Brown
Ursus arctos syriacus — Bear, Syrian Brown
Ursus maritimus — Bear, Polar
Urtica dioica — Nettle, Stinging
Urtica dubia — Nettle, Small
Urtica pilulifera — Nettle, Roman
Urtica sp. — Nettle
Urtica urens — Nettle, Small or Annual
Usnea sp. — Lichen
Ustilago maydis — Smut, Corn
Ustilago sp. — Smut
Utethesia pulchella — Moth, Crimson
 Speckled
Utricularia intermedia — Bladderwort,
 Intermediate or Irish
Utricularia minor — Bladderwort, Lesser or
 Small
Utricularia sp. — Bladderwort
Utricularia vulgaris — Bladderwort, Greater

V

Vaccaria pyramidata — Cow Basil

Vaccinium corymbosum — Blueberry, Highbush

Vaccinium myrtillus — Bilberry or Blaeberry or Whortleberry

Vaccinium oxycoccus — Cranberry

Vaccinium uliginosum — Whortleberry, Bog

Vaccinium vitis-idaea — Cowberry or Cranberry, Mountain

Valeriana dioica — Valerian, Lesser or Marsh

Valeriana montana — Valerian, Mountain

Valeriana officinalis — Valerian, Common

Valeriana tripteris — Valerian, Three-leaved or Three-winged

Valerianella dentata — Corn Salad, Smooth-fruited

Valerianella locusta — Lamb's Lettuce

Vallisneria spiralis — Tape-grass

Vallota speciosa — Lily, Scarborough

Valvata sp. — Snail, Valve

Vampyroteuthis infernalis — Squid, Vampire

Vampyrus spectrum — Bat, False Vampire

Vanda caerulea — Orchid, Blue Vanda

Vanda sp. — Orchid, Vanda

Vanellus armatus — Plover, Blacksmith

Vanellus cinereus — Lapwing, Gray-headed

Vanellus coronatus — Plover, Crowned

Vanellus duvaucelii — Lapwing

Vanellus gregarius — Plover, Sociable

Vanellus indicus — Lapwing, Red-wattled

Vanellus miles — Plover, Masked

Vanellus senegallus — Plover, Wattled

Vanellus spinosus — Plover, Spur-winged

Vanellus tectus — Plover, Blackhead

Vanellus tricolor — Plover, Banded

Vanellus vanellus — Lapwing

Vanessa atalanta — Butterfly, Red Admiral

Vanessa cardui — Butterfly, Painted Lady

Vanessa urticae — Butterfly, Small Tortoise-shell

Vanga curvirostris — Vanga

Vanilla imperialis — Orchid, Vanilla

Vanilla planifolia — Orchid, Vanilla

Varanus acanthurus — Monitor, Spiny-tailed

Varanus exanthematicus — Monitor, Bosc's

Varanus giganteus — Monitor, Perentie

Varanus gouldii — Monitor, Gould's

Varanus griseus — Monitor, Desert or Gray

Varanus komodoensis — Monitor, Komodo Dragon

Varanus niloticus — Monitor, Nile

Varanus salvadori — Monitor, New Guinea

Varanus salvator — Monitor, Two-banded

Varanus varius — Monitor, Lace or Variegated

Vasum sp. — Vase Shell

Velella sp. — By-the-wind-sailor

Velia caprai — Water-cricket

Velia saulii — Water-cricket

Vellozia retinervis — Lily, Black-stick

Veltheimia viridiflora — Veltheimia

Venerupis sp. — Carpet Shell

Venidium fastuosum — Monarch-of-the-Veldt

Venus sp. — Clam, Venus

Veratrum album — False-helleborine, White

Veratrum nigrum — False-helleborine, Black

Verbascum blattaria — Mullein, Moth

Verbascum lychnitis — Mullein, White

Verbascum nigrum — Mullein, Dark

Verbascum phoeniceum — Mullein, Purple

Verbascum phlomoides — Mullein, Orange

Verbascum pulverulentum — Mullein, Hoary

Verbascum sp. — Mullein

Verbascum thapsiforme — Mullein, Large-flowered

Verbascum thapsus — Mullein, Common or Great

Verbena officinalis — Vervain

Veretillum cynomorium — Sea Pen

Vermicella annulata — Bandy-bandy

Vermicella calonota — Snake, Black-striped

Veronica agrestis — Speedwell, Field

Veronica alpina — Speedwell, Alpine

Veronica anagallis-aquatica — Speedwell, Water

Veronica arvensis — Speedwell, Wall

Veronica beccabunga — Brooklime

Veronica catenata — Speedwell, Pink Water

Veronica chamaedrys — Speedwell, Germander or Bird's-eye

Veronica filiformis — Speedwell, Creeping

Veronica fruticans — Speedwell, Rock or
Shrubby
Veronica gentianoides — Speedwell, Veronica
Veronica hederifolia — Speedwell, Ivy-leaved
Veronica latifolia — Speedwell, Large
Veronica longifolia — Speedwell, Long-leaved
Veronica montana — Speedwell, Wood
Veronica officinalis — Speedwell, Common
Veronica persica — Speedwell, Buxbaum's
Veronica polita — Speedwell, Gray
Veronica prostrata — Speedwell, Prostrate
Veronica scutellata — Speedwell, Marsh
Veronica serpyllifolia — Speedwell, Thyme-
leaved
Veronica sp. — Speedwell
Veronica spicata — Speedwell, Spiked
Veronica teucrium — Speedwell, Saw-leaved
Veronica triphyllos — Speedwell, Fingered
Veronica verna — Speedwell, Spring
Verpa bohemica — Morel, Early
Verpa digitaliformis — Fungus, Thimble
Verruca stroemia — Barnacle, Acorn
Verrucaria sp. — Lichen
Vespa crabro — Hornet, European
Vespa silvestris — Wasp, Tree
Vespertilio sp. — Bat, Particolored
Vespula sp. — Wasp, Common
Vetilia sp. — Stick-insect
Viburnum lantana — Wayfaring Tree
Viburnum opulus — Guelder Rose
Viburnum plicatum — Snowball, Japanese
Viburnum rhytidophyllum — Viburnum,
Leatherleaf
Viburnum tinus — Laurustinus
Viburnum tomentosum — Snowball,
Japanese
Vicia angustifolia — Vetch, Narrow-leaved
Vicia cracca — Vetch, Tufted or Cow
Vicia hirsuta — Vetch, Hairy
Vicia lutea — Vetch, Yellow
Vicia orobus — Vetch, Bitter or Upright
Vicia sativa — Vetch, Common
Vicia sepium — Vetch, Bush
Vicia sylvatica — Vetch, Wood
Vicia villosa — Vetch, Sand or Fodder
Victoria amazonica — Water-lily, Royal
Victoria cruziana — Water-lily, Royal
Vicugna vicugna — Vicuna
Vidua macroura — Whydah, Pin-tailed

Vidua paradisea — Whydah
Vigna sp. — Cowpea
Vinca major — Periwinkle, Greater (plant)
Vinca minor — Periwinkle, Lesser or Common
(plant)
Vincetoxicum hirundinaria — Swallowwort
Viola arvensis — Pansy, Field
Viola biflora — Violet, Yellow Wood
Viola calcarata — Pansy, Long-spurred
Viola canina — Violet, Heath Dog
Viola cornuta — Pansy, Tufted
Viola elatior — Violet, Upright
Viola hederacea — Violet, Australian
Viola hirta — Violet, Hairy
Viola kitaibeliana — Pansy, Dwarf
Viola lactea — Violet, Pale Heath
Viola lutea — Pansy, Mountain
Viola odorata — Violet, Sweet
Viola palustris — Violet, Marsh or Bog
Viola reichenbachiana — Violet, Pale Wood
Viola riviniana — Violet, Common Dog
Viola rupestris — Violet, Teesdale
Viola sp. — Violet
Viola stagnina — Violet, Fen
Viola tricolor — Pansy, Wild
Vipera ammodytes — Viper, Long-nosed
Vipera aspis — Viper, Asp
Vipera berus — Viper, Northern
Vipera russelli — Viper, Russell's
Vipera ursinii — Viper, Orsini's
Vireolanius sp. — Shrike-vireo
Viscum album — Mistletoe, Eurasian
Viscum sp. — Mistletoe, Eurasian
Vitaliana primuliflora — Primrose, Gold
Vitex agnus-castus — Chaste Tree
Vitis coignetiae — Glory Vine
Vitis sp. — Grape Vine
Vittaria lineata — Fern, Shoestring
Viverra civetta — Civet, African
Viverra zibetha — Civet, Large Indian
Viverricula indica — Civet, Small Indian
Viviparus sp. — Snail, River
Volucella sp. — Fly, Flower or Hover
Volute sp. — Volute
Volvaria sp. — Volvaria
Volvox sp. — Alga, Green
Vormella peregusna — Polecat, Marbled
Vorticella sp. — Vorticella
Vriesia splendens — Flaming Sword

Vriesia sp. — Flaming Sword
Vulpes bengalensis — Fox, Bengal
Vulpes chama — Fox, Cape
Vulpes corsac — Fox, Corsac
Vulpes ferrilatus — Fox, Tibetan Sand
Vulpes pallidus — Fox, Pale or Sand
Vulpes vulpes — Fox, Red
Vulpia bromoides — Fescue, Squirrel-tail
Vulpia myuros — Fescue, Rat's-tail
Vultur gryphus — Condor, Andean

W

Wahlenbergia hederacea — Bellflower, Ivy-
leaved
Wahlenbergia sp. — Bluebell or Harebell
Wallabia agilis — Wallaby, Agile or Sandy
Wallabia bicolor — Wallaby, Black-tailed
or Swamp
Wallabia canguru — Wallaby, Pretty-faced
Wallabia elegans — Wallaby, Pretty-faced
Wallabia parma — Wallaby, Parma or White-
throated
Wallabia parryi — Wallaby, Pretty-faced
Washingtonia filifera — Palm, Petticoat or
Desert
Watsonia sp. — Lily, Bugle
Weigela florida — Weigela
Weigela sp. — Weigela
Weinmannia sp. — Weinmannia
Welwitschia mirabilis — Welwitschia
Wilcoxia poselgeri — Cactus, Lamb's-tail
Wilcoxia schmollii — Cactus, Lamb's-tail
Wisteria floribunda — Wisteria, Japanese
Wisteria sinensis — Wisteria, Chinese
Wolffia arrhiza — Duckweed, Least or
Rootless
Wombatus hirsutus — Wombat, Tasmanian
Wombatus ursinus — Wombat, Tasmanian
Woodsia ilvensis — Woodsia, Oblong
Wormskioldia sp. — Lion's-eye
Wyulda squamicaudata — Phalanger, Scaly-
tailed

X

Xanichthys ringens — Triggerfish, Pink-tail
Xanthium spinosum — Cocklebur, Spiny

Xanthium strumarium — Cocklebur,
Common
Xanthoceras sorbifolium — Yellow Horn
Xanthoria sp. — Lichen
Xanthornis viridis — Cacique
Xanthorrhoea sp. — Grass Tree
Xanthotis flaviventer — Honey-eater, Tawny-
breasted
Xema sabini — Gull, Sabine's
Xenicus gilviventris — Wren, Rock
Xenicus longipes — Wren, Bush
Xenocerus puncticollis — Weevil
Xenocerus semiluctuosus — Weevil
Xenodacnis parina — Dacnis
Xenomystis nigri — Knifefish, Black African
Xenopeltis unicolor — Snake, Rainbow
Xenopirostris sp. — Vanga
Xenops sp. — Xenops
Xenopus fraseri — Toad, Clawed
Xenopus laevis — Toad, Clawed
Xenorhynchus asiaticus — Jabiru
Xenus cinereus — Sandpiper, Terek
Xeranthemum annum — Pink Everlasting
Xerocomus sp. — Boletus, Velvet or Cracked
Xeronema callistemon — Lily, Poor Knight's
Xerus sp. — Squirrel, Ground
Xestobium rufovillosum — Beetle, Death-
watch
Ximenia caffra — Plum, Kaffir or Sour
Xiphias gladius — Swordfish
Xiphocolaptes promeropirhynchus —
Woodcreeper, Strong-billed
Xipholena punicea — Cotinga, Pompadour
Xiphorhynchus sp. — Woodcreeper
Xylaria sp. — Dead Man's Fingers (fungus)
Xylocopa violacea — Bee, Carpenter
Xylodrepa quadripunctata — Beetle, Four-
spot Carrion
Xylophaga dorsalis — Piddock, Wood
Xylosma senticosa — Xylosma
Xylotrupes australicus — Beetle, Elephant
Xysticus sp. — Spider, Crab

Y

Yponomeuta sp. — Moth, Ermine
Yuhina sp. — Yuhina

Z

Zabrus sp. — Beetle, Grain
Zaedus pichyi — Armadillo, Hairy
Zaglossus bruijni — Echidna, Long-beaked
Zaluzianskya sp. — Zaluzianskya
Zanclus canescens — Moorish Idol
Zanclus cornutus — Moorish Idol
Zannichellia palustris — Pondweed, Horned
Zantedeschia aethiopica — Calla Lily
Zantedeschia elliottiana — Calla Lily, Golden
Zantedeschia rehmannii — Calla Lily, Pink
Zanthomiza phrygia — Honey-eater, Regent
Zarhynchus wagleri — Oropendola, Wagler's
Zea mays — Corn
Zebrasoma sp. — Surgeonfish
Zebrina pendula — Wandering Jew
Zelkova serrata — Zelkova, Japanese
Zelkova sp. — Zelkova, Caucasian
Zenaida galapagoensis — Dove, Galapagos
Zerynthia hipsipyle — Butterfly, Birthwort
Zerynthia polyxena — Butterfly, Birthwort
Zeugopterus punctatus — Topknot, Common
Zeus faber — Dory, John
Zeuzera pyrina — Moth, Leopard
Zingel streber — Streber
Zingel zingel — Zingel
Zingiber officinale — Ginger Plant
Zingiber spectabile — Ginger Plant
Zirphaea crispata — Piddock, Oval
Zizyphus jujuba — Jujube
Zoarces viviparous — Eelpout
Zoneaeginthus bellus — Firetail

Zoneaeginthus oculatus — Firetail
Zonibyx modestus — Dotterel, Rufous-chested
Zonifer tricolor — Plover, Banded
Zonotrichia capensis — Sparrow, Rufous-collared
Zoothera citrina — Thrush, Orange-headed Ground
Zoothera dauma — Thrush, Golden or Scaly Ground
Zoothera interpres — Thrush, Chestnut-capped Ground
Zostera marina — Eel-grass
Zosterops albiventris — White-eye, Pale
Zosterops ceylonensis — White-eye, Mountain or Hill
Zosterops chloris — White-eye, Pale
Zosterops gouldi — White-eye, Western
Zosterops lateralis — White-eye, Gray-breasted
Zosterops lutea — White-eye, Yellow
Zosterops montana — White-eye, Mountain or Hill
Zosterops nigrorum — White-eye, Yellow
Zosterops pallidus — White-eye, Pale
Zosterops palpebrosa — White-eye, Indian or Oriental
Zoysia matrella — Zoysia Grass
Zygaena sp. — Moth, Burnet
Zygocactus truncatus — Cactus, Crab or Christmas
Zygopetalum intermedium and Z. mackaii — Orchid, Zygopetalum

Bibliography by Title

The ABC of Indoor Plants. Jocelyn Baines and Katherine Key. New York, Alfred A. Knopf, 1977. (BIP)

Africa. Leslie Brown. (The Continents We Live On series). New York, Random House, 1965. (CAF)

The Amazing World of Insects. Arend T. Bandsma and Robin T. Brandt. New York, Macmillan, 1963. (BIN)

Animal Atlas of the World. E. L. Jordan. Maplewood, N.J., Hammond, 1969 (JAW).

Animal Life of Europe. Jakob Graf. New York, Frederick Warne, 1968. (GAL)

Animals of Asia. Robert Wolff. Illustrated by Robert Dallet. New York, Lion Press, 1969. (DAA)

Animals of Europe. Robert Wolff. Illustrated by Robert Dallet. New York, Lion Press, 1969. (DAE)

Animals of the Antarctic. Bernard Stonehouse. New York, Holt, Rinehart and Winston, 1972. (SAR)

Animals of the World: Africa. Allan Cooper et al. London, Paul Hamlyn, 1968. (CAA)

Annuals. James Underwood Crockett. Illustrated by Allianora Rosse. (Time-Life Encyclopedia of Gardening.) New York, Time-Life Books, 1971. (EGA)

Asia. Pierre Pfeffer. (The Continents We Live On series.) New York, Random House, 1968. (CAS)

Australia and the Pacific Islands. Allen Keast. (The Continents We Live On series.) New York, Random House, 1966. (CAU)

Australian Birds. Robin Hill. New York, Funk and Wagnall, 1967. (HAB).

Australian Reptiles in Colour. Harold Cogger. Honolulu, East-West Center, 1967. (COG)

Australian Wildlife. Eric Worrell. London, Angus and Robertson, 1966. (WAW)

The Beauty of Birds. Cyril Newberry. New York, Hanover, 1958. (NBB)

The Beauty of the Wild Plant. H. Kleijn and P. Vermeulen. London, George C. Harrap, 1964. (KVP)

Beetles. Ewald Reitter. New York, G. P. Putnam's Sons, 1961. (RBT)

Birds of Asia. A. Rutgers. Illustrated by John Gould. New York, Taplinger, 1969. (RBI)

Birds of Australia. Michael Morcombe. New York, Scribner, 1971. (MBA)

Birds of Australia. A. Rutgers. Illustrated by John Gould. London, Methuen, 1967. (RBU)

The Birds of Britain and Europe. Richard Fitter. Illustrated by Herman Heinzel. Philadelphia, Lippincott, 1972. (FBE)

Birds of Europe. Bertel Bruun. Illustrated by Arthur Singer. New York, McGraw-Hill, 1970. (BBE)

Birds of Europe. A Rutgers. Illustrated by John Gould. London, Methuen, 1966. (RBE)

Birds of Prey of the World. Mary Louise Grossman and John Hamlet. Photos by Shelly Grossman. New York, Potter, 1964. (GBP)

Birds of South Vietnam. Philip Wildash. Rutland, Vt., Charles E. Tuttle, 1968. (WBV)

Birds of the Antarctic. Edward Wilson. Brian Roberts, editor. New York, Humanities Press, 1968. (WAN)

Birds of the Atlantic Ocean. Ted Stokes. Illustrated by Keith Shackleton. New York, Macmillan, 1968. (SAO)

The Birds of the Falkland Islands. Robin W. Woods. New York, Anthony Nelson, 1975. (WFI)

Birds of the Tropics. John A. Burton. New York, Crown, 1973. (BBT)

Birds of the World. Oliver L. Austin, Jr. Illustrated by Arthur Singer. New York, Golden Press, 1961. (ABW)

Book of British Birds. Richard Fitter, editor. London, Drive Publications for Reader's Digest Association, 1973. (BBB)

The Book of Reptiles. R.A. Lanworn. London, Hamlyn, 1972. (LBR)

Britian's Wildlife; Rarities and Introductions. Richard Fitter. Illustrated by John Leigh-Pemberton. London, Nicholas Kaye, 1966. (FBW)

Bulbs. James Underwood Crockett. Illustrated by Allianora Rosse. (Time-Life Encyclopedia of Gardening.) New York, Time-Life Books, 1971. (EGB)

Buller's Birds of New Zealand. Edited and revised by E.G. Turbott. Honolulu, East-West Center, 1967. (BNZ)

Cacti and Succulents. Philip Perl. (Time-Life Encyclopedia of Gardening.) Alexandria, Va., Time-Life Books, 1978. (ECS)

Cacti and Succulents; a Concise Guide in Colour. Rudolf Subik. London, Hamlyn, 1968. (SCS)

Collins Guide to Mushrooms and Toadstools. Morten Lange and F. Bayard Hora. London, Collins, 1963. (LHM)

Collins Guide to the Sea Fishes of Britain and Northwestern Europe. Bent J. Muus. Illustrated by Preben Dahlstrom. London, Collins, 1974. (CSF)

Collins Pocket Guide to the Seashore. John Barrett and C.M. Yonge. Illustrated by Elspeth Yonge et al. London, Collins, 1970. (CSS)

Collins Pocket Guide to Wild Flowers. David McClintock and R.S.R. Fitter. Illustrated by Dorothy Fitchew et al. London, Collins, 1971. (MFF)

Color Dictionary of Flowers and Plants for Home and Garden. Roy Hay and Patrick M. Synge. New York, Crown Publishers, 1969. (DFP)

The Color Dictionary of Shrubs. S. Millar Gault. New York, Crown, 1976. (GDS)

Color Treasury of Aquarium Fish. Elso Lodi. London, Crescent, 1972. (CTF)

Color Treasury of Butterflies and Moths. Introduction by Michael Tweedie. London, Crescent, 1972. (CTB)

Color Treasury of Herbs and Other Medicinal Plants. Introduction by Jerry Cowhig. London, Crescent, 1972 (CTH)

Color Treasury of Insects: World of Miniature Beauty. Umberto Parenti. London, Crescent, 1971. (CTI)

Color Treasury of Mushrooms and Toadstools. Introduction by Uberto Tosco and Annalaura Fanelli. London, Crescent, 1972. (CTM)

Color Treasury of Sea Shells. Sergio Angeletti. Introduction by Michael Tweedie. London, Crescent, 1973. (CTS)

Common Weeds of Canada. Gerald A. Mulligan. (no place), McClelland and Stewart, 1976. (MCW)

Common Weeds of the United States. United States Department of Agriculture. New York, Dover, 1971. (WUS)

Complete Book of Mushrooms. Augusto Rinaldi and Vassili Tyndale. New York Crown, 1974. (RTM)

Complete Guide to Plants and Flowers. Frances Perry, editor. New York, Simon and Schuster, 1974 (SPL)

The Concise British Flora in Colour. W. Keble Martin. New York, Holt, Rinehart and Winston, 1965. (MBF)

Decorating with Plants. Oliver E. Allen. (Time-Life Encyclopedia of Gardening.) Alexandria, Va., Time-Life Books, 1978. (EDP)

Dictionary of Birds in Color. Bruce Campbell. New York, Viking Press, 1974. (CDB)

Dictionary of Butterflies and Moths in Color. Allan Watson and Paul E.S. Whalley. New York. McGraw-Hill, 1975. (WDB)

Easy Gardens. Donald Wyman and Curtis Prendergast. (Time-Life Encyclopedia of Gardening.) Alexandria, Va., Time-Life Books, 1978. (EEG)

Encyclopedia of Mammals. Maurice and Robert Burton. London, Octopus Books, 1975. (BMA)

The Encyclopedia of Mushrooms. Colin Dickson and John Lucas. New York, G.P. Putnam's Sons, 1979 (DLM)

Encyclopedia of Reptiles, Amphibians and Other Cold-blooded Animals. Introduction by Maurice Burton. London, Octopus Books, 1975. (ERA)

Europe. Kai Curry-Lindahl. (The Continents We Live On series.) New York, Random House, 1964. (CEU)

Evergreens. James Underwood Crockett. Illustrated by R.A. Merrilees and John Murphy. (Time-Life Encyclopedia of Gardening.) New York, Time-Life Books, 1971. (EGE)

Exotic Mushrooms. Henri Romagnesi. New York, Sterling, 1971. (REM)

Exotic Plants. Julia F. Morton. Illustrated by Richard E. Younger. (Golden Nature Guide series.) New York, Golden Press, 1971. (MEP)

Ferns. Philip Perl. Illustrated by Richard Crist. (Time-Life Encyclopedia of Gardening.) Alexandria, Va., Time-Life Books, 1977. (EGF)

A Field Guide in Color to Plants. Jan Tykac and Vlastimil Vanek. London, Octopus Books, 1978. (TVG)

A Field Guide to Australian Birds (Non-passerines). Peter Slater. Wynnewood, Pa., Livingston, 1972. (SNB)

A Field Guide to Australian Birds (Passerines). Peter Slater. Wynnewood, Pa., Livingston, 1974. (SAB)

Field Guide to the Birds of Britain and Europe. Roger Tory Peterson. (Peterson Field Guide series.) Boston, Houghton Mifflin, 1967. (PBE)

A Field Guide to the Birds of East and Central Africa. J.G. Williams. Illustrated by R. Fennessy. Boston, Houghton Mifflin, 1964. (WBA)

A Field Guide to the Birds of Galapagos. Michael Harris. Illustrated by Barry Kent McKay. London, Collins, 1974. (HBG)

A Field Guide to the Birds of New Zealand and Outlying Islands. R.A. Falla et al. Illustrated by Chloe Talbot-Kelly. Boston, Houghton Mifflin, 1967. (FNZ)

Field Guide to the Birds of Southeast Asia. Ben F. King and Edward C. Dickinson. (International series.) Boston, Houghton Mifflin, 1975. (KBA)

A Field Guide to the Birds of Southern Africa. O.P.M. Prozesky. Illustrated by Dick Findlay. London, Collins, 1970. (PSA)

A Field Guide to the Ferns and Their Related Families. Boughton Cobb. Illustrated by Laura Louise Foster. (Peterson Field Guide series) Boston, Houghton Mifflin, 1963. (CGF)

A Field Guide to the Insects of Britian and Northern Europe. Michael Chinery. (International series.) Boston, Houghton Mifflin, 1974. (CIB)

A Field Guide to the Larger Mammals of Africa. Jean Dorst. Illustrated by Pierre Dendelot. Boston, Houghton Mifflin, 1970. (DDM)

A Field Guide to the Snakes of Southern Africa. V.F.M. FitzSimons. London, Collins, 1970. (FSA)

Field Guide to the Trees of Britain and Northern Europe. Alan Mitchell. Boston, Houghton Mifflin, 1974. (MTB)

Fieldbook of Natural History. E. Laurence Palmer and H. Seymour Fowler. 2nd ed. New York, McGraw-Hill, 1975. (PFF)

Fishes of the World: An Illustrated Dictionary. Illustrated by Alwyne Wheeler and Peter Stebbing. New York, Macmillan, 1975. (WFW)

Fishes of the World in Color. Hans Hvass. Illustrated by Wilhelm Eigener. New York, E.P. Dutton, 1965. (HFI)

Flowering House Plants. James Underwood Crockett. Illustrated by Allianora Rosse. (Time-Life Encyclopedia of Gardening.) New York, Time-Life Books, 1971. (FHP)

Flowering Plants of the World. V.H. Heywood, editor. Illustrated by Victoria Goaman et al. Oxford (England) University Press, 1978. (HPW)

Flowering Shrubs. James Underwood Crockett. Illustrated by Allianora Rosse. (Time-Life Encyclopedia of Gardening.) New York, Time-Life Books, 1972. (EFS)

Flowers in Color. W. Rytz and Herbert Edlin. Illustrated by Hans Schwarzenbach. New York, Viking Press, 1960. (EFC)

Flowers of Europe. Oleg Polunin. London, Oxford University Press, 1969. (PFE)

Flowers of the Islands in the Sun. Graham Gooding. Illustrated by Clarence E. Hall. New York, Barnes, 1966. (GFI)

Flowers of the Mediterranean. Oleg Polunin and Anthony Huxley. Boston, Houghton Mifflin, 1966. (PFM)

Flowers of the World. Frances Perry. Illustrated by Leslie Greenwood. New York, Crown, 1972. (PFW)

Flowers of the World (in Full Color). Robert S. Lemmon and Charles L. Sherman. Garden City N.Y., Hanover House, 1958. (LFW)

Foliage House Plants. James Underwood Crockett. (Time-Life Encyclopedia of Gardening.) New York, Time-Life Books, 1972. (EFP)

Freshwater Fishes. Juraj Holcik and Jozef Mihalik. Feltham, England, Hamlyn, 1968. (FWF)

Freshwater Fishes of the World. Gunther Sterba. Translated and revised by Denys W. Tucker. New York, Viking Press, 1963. (SFW)

Garden Birds of South Africa. Kenneth Newman. New York, Elsevier, 1968. (NBA)

Gardening Under Lights. Wendy B. Murphy. (Time-Life Encyclopedia of Gardening.) Alexandria, Va., Time-Life Books, 1978. (EGL)

The Glory of the Tree. B.K. Boom and H. Kleijn. Garden City, N.Y., Doubleday, 1966. (BKT)

The Great Book of Birds. John Gooders. Foreword by Roger Tory Peterson. New York, Dial Press, 1975. (GPB)

Greenhouse Gardening. James Underwood Crockett. (Time-Life Encyclopedia of Gardening.) Alexandria, Va., Time-Life Books, 1977. (EGG)

Grzimek's Animal Life Encyclopedia. Bernhard Grzimek, editor-in-chief. 13 vols. New York, Van Nostrand Reinhold, 1974. (ALE)

Guide to Garden Flowers. Norman Taylor. Illustrated by Eduardo Salgado. Boston, Houghton Mifflin, 1958. (TGF)

Guide to Garden Shrubs and Trees. Norman Taylor. Illustrated by Eduardo Salgado. Boston, Houghton Mifflin, 1965. (TGS)

A Guide to the Birds of South America. Rodolphe Meyer deSchauensee. Wynnewood, Pa., Livingston, 1970. (SSA)

A Guide to the Pheasants of the World. Philip Wayre. Illustrated by J.C. Harrison. London, Country Life, 1969. (WPW)

Guide to Trees. Stanley Schuler, editor. (Fireside series.) New York, Simon and Schuster, 1977. (SST)

Hallucinogenic Plants. Richard Evans Schultes. Illustrated by Elmer W. Smith. (Golden Guide series.) New York, Golden Press, 1976. (EHP)

The Hamlyn Encyclopedia of Plants. J. Triska. London, Hamlyn, 1975. (TEP)

Handbook of British Flowering Plants. Edward A. Melderis and E.B. Bangerter. New York, Abelard-Schuman, 1959. (BFB)

Health Plants of the World. Francesco Bianchini and Francesco Corbetta. Illustrated by Marilena Pistoia. New York, Newsweek Books, 1977. (AMP)

Herbs. James Underwood Crockett and Ogden Tanner. Illustrated by Richard Crist. (Time-Life Encyclopedia of Gardening.) Alexandria, Va., Time-Life Books, 1977. (EGH)

Hummingbirds. Walter Scheithauer. New York, Crowell, 1967. (SHU)

An Illustrated Encyclopedia of Australian Wildlife. Michael Morcombe. South Melbourne, Australia, Mac Millan, 1974. (MEA)

Illustrated Encyclopedia of Butterflies and Moths. V.J. Stanek. Edited by Brian Turner. London, Octopus Books. 1977. (SBM)

The Illustrated Encyclopedia of Trees. Herbert Edlin et al. New York, Harmony, 1978. (EET)

The Instant Guide to Successful House Plants. David Longman. New York, New York Times Books, 1979. (LHP)

Japanese Gardens. Wendy B. Murphy. (Time-Life Encyclopedia of Gardening.) Alexandria Va., Time-Life Books, 1979. (EJG)

Landscape Gardening. James Underwood Crockett. Illustrated by R.A. Merrilees and B. Wolff. (Time-Life Encyclopedia of Gardening.) Alexandria, Va., Time-Life Books, 1971. (ELG)

Larousse Encyclopedia of Animal Life. New York, McGraw-Hill, 1967. (LEA)

Larousse Encyclopedia of the Animal World. Introduction by Desmond Morris. New York, Larousse, 1975. (LAW)

The Last Pardises: On the Track of Rare Animals. Eugen Schuhmacher. Garden City, N.Y., Doubleday, 1967. (SLP)

Last Survivors. Noel Simon and Paul Geroudet. Illustrated by Helmut Diller and Paul Barruel. New York, World, 1970. (SLS)

Lawns and Ground Covers. James Underwood Crockett. Illustrated by Allianora Rosse. (Time-Life Encyclopedia of Gardening.) Alexandria, Va., Time-Life Books, 1971. (EGC)

Living Amphibians of the World. Doris M. Cochran. (World of Nature series.) Garden City, N.Y., Doubleday, 1961. (CAW)

Living Fishes of the World. Earl S. Herald. Garden City, N.Y., Doubleday, 1961. (HFW)

Living Insects of the World. Alexander B. and Elsie B. Klots. (World of Nature series.) Garden City, N.Y., Doubleday, 1965 (?). (KIW)

Living Mammals of the World. Ivan T. Sanderson. (World of Nature series.) Garden City, N.Y., Doubleday, 1961. (SMW)

Living Plants of the World. Lorus and Margery Milne. 2nd ed., rev. New York, Random House, 1975. (MLP)

Living Reptiles of the World. Karl P. Schmidt and Robert F. Inger. (World of Nature series.) Garden City, N.Y., Doubleday, 1957. (SIR)

Mammals. Donald F. Hoffmeister. (Golden Bookshelf of Natural History series.) New York, Golden Press, 1963. (HMA)

The Mammals: A Guide to the Living Species. Desmond Morris. New York, Harper and Row, 1965. (MLS)

The Mammals of Rhodesia, Zambia and Malawi. Reay H.N. Smithers. Illustrated by E.J. Bierly. London, Collins, 1966. (SMR)

Mammals of the World. Hans Hvass. Illustrated by Wilhelm Eigener. London, Methuen, 1961. (HMA)

Marine Game Fishes of the World. Francesca la Monte. Illustrated by Janet Roemhild. Garden City, N.Y., Doubleday, 1952. (LMF)

Miniatures and Bonsai. Philip Perl. (Time-Life Encyclopedia of Gardening.) Alexandria, Va., Time-Life Books, 1979. (EMB)

The Mushroom Hunter's Field Guide. Alexander H. Smith. Rev. ed. Ann Arbor, University of Michigan Press, 1963. (SMG)

Mushrooms. Albert Pilat. Illustrated by Otto Usak. London, Spring Books, (no date). (PMU)

Mushrooms and Fungi. Moira Savonius. New York, Crown, 1973. (SMF)

Mushrooms and Other Fungi. H. Kleijn. Garden City, N.Y., Doubleday, 1962. (KMF)

Mushrooms and Other Fungi. Albert Pilat. Illustrated by Otto Usak. London, Peter Nevill, 1961. (PUM)

Natural History of Europe. Harry Garms. Illustrated by Wilhelm Eigener. London, Paul Hamlyn, 1967. (NHE)

New Zealand Flowers and Plants in Colour. J. T. Salmon. Wellington, New Zealand, Reed, 1963. (SNZ)

Non-flowering Plants. Floyd S. Shuttleworth and Herbert S. Zim. (Golden Nature Guide series.) New York, Golden Press, 1967. (NFP)

Ocean Life in Color. Norman Marshall. Illustrated by Olga Marshall. (Macmillan Color series.) New York, Macmillan, 1971. (MOL)

Orchids. Alice Skelsey. (Time-Life Encyclopedia of Gardening.) Alexandria, Va., Time-Life Books, 1978. (EGO)

Owls of the World. John A. Burton, editor. Illustrated by John Rignall. New York, E.P. Dutton, 1973. (BOW)

Oxford Book of Birds. Bruce Campbell. Illustrated by Donald Watson. Oxford (England) University Press, 1964. (OBB)

Oxford Book of Flowerless Plants. Frank H. Brightman. Illustrated by B.E. Nicholson. Oxford (England) University Press, 1966. (ONP)

Oxford Book of Food Plants. S.G. Harrison et al. Illustrated by B.E. Nicholson. Oxford (England) University Press, 1969. (OFP)

Oxford Book of Garden Flowers. E. B. Anderson et al. Illustrated by B. E. Nicholson. Oxford (England) University Press, 1963. (OGF)

Oxford Book of Insects. John Burton et al. Illustrated by Joyce Bee et al. Oxford (England) University Press, 1968 (OBI)

Oxford Book of Invertebrates. David Nichols with John A. L. Cooke. Illustrated by Derek Whiteley. Oxford (England) University Press, 1971. (OIB)

Oxford Book of Trees. A.E. Clapham. Illustrated by B.E. Nicholson. Oxford (England) University Press, 1975. (OBT).

Oxford Book of Vertebrates. Marion Nixon. Illustrated by Derek Whiteley. Oxford (England) University Press, 1972. (OBV)

Oxford Book of Wild Flowers. S. Ary and M. Gregory. Illustrated by B.E. Nicholson. Oxford (England) University Press, 1960. (OWF)

Perennials. James Underwood Crockett. Illustrated by Allianora Rosse. (Time-Life Encyclopedia of Gardening.) New York, Time-Life Books, 1972. (EGP)

Philippine Birds. John Eleuthere duPont. Illustrated by George Sandstrom and John R. Peirce. Greenville, Del., Delaware Museum of Natural History, 1971. (DPB)

The Pictorial Encyclopedia of Fishes. S. Frank. London, Hamlyn, 1971. (FEF)

Pictorial Encyclopedia of Insects. V.J. Stanek. London, Hamlyn, 1969. (PEI)

Pictorial Encyclopedia of Plants and Flowers. F.A. Novak. New York, Crown, 1966. (EPF)

Plants of the World. H.C.D. deWit. 3 vols. New York, E.P. Dutton, 1966-1969. (DEW)

Pocket Encyclopedia of Cacti and Other Succulents in Color. Edgar and Brian Lamb. New York, Macmillan, 1970. (LCS)

Portraits of Tropical Birds. John S. Dunning. Wynnewood, Pa., Livingston, 1970. (DTB)

Pruning and Grafting. Oliver E. Allen.
(Time-Life Encyclopedia of Gardening.)
Alexandria, Va., Time-Life Books, 1978. (EPG)

Reptiles and Amphibians. Zdenek Vogel.
Illustrated by P. Pospisil and M. Rada. London,
Paul Hamlyn, 1966. (VRA)

Reptiles and Amphibians of the World. Hans
Hvass. Illustrated by Wilhelm Eigener.
London, Methuen, 1964. (HRA)

Rock and Water Gardens. Ogden Tanner.
(Time-Life Encyclopedia of Gardening.)
Alexandria, Va., Time-Life Books, 1979.
(ERW)

Sea Shells of the World. R. Tucker Abbott.
Illustrated by George and Marita Sandstrom.
(Golden Nature Guide series.) New York,
Golden Press, 1962. (GSS)

Shade Gardens. Oliver E. Allen. (Time-Life
Encyclopedia of Gardening.) Alexandria, Va.,
Time-Life Books, 1979. (ESG)

Shrubs in Colour. A.G.L. Hellyer. Garden City,
N.Y., Doubleday, 1966. (HSC)

Snakes of the World. John Stidworthy.
Illustrated by Dougal MacDougal. New York,
Bantam, 1972. (SSW)

South America and Central America. Jean
Dorst. (The Continents We Live On series.)
New York, Random House, 1967. (CSA)

Spiders and Their Kin. Herbert W. and Lorna
R. Levi. (Golden Nature Guide series.) New
York, Golden Press, 1968. (GSP)

Strange and Beautiful Birds. Josef Seget.
London, Spring Books, 1965. (SBB)

Strange Wonders of the Sea. H. Gwynne
Vevers. Garden City, N.Y., Hanover House,
1957. (VWS)

Success with House Plants. Anthony Huxley,
editor. Pleasantville, N.Y., Reader's Digest
Association, 1979. (RDP)

Trees. James Underwood Crockett. (Time-Life
Encyclopedia of Gardening.) New York,
Time-Life Books, 1972. (EGT)

Vanishing Species. Introduction by Romain
Gary. New York, Time-Life Books, 1974. (LVS)

Vanishing Wild Animals of the World.
Richard Fitter. Illustrated by John
Leigh-Pemberton. London, Midland Bank,
1968. (VWA)

Vegetables and Fruits. James Underwood
Crockett. Illustrated by Richard Crist.
(Time-Life Encyclopedia of Gardening.) New
York, Time-Life Books, 1972. (EVF)

Vines. Richard H. Cravens. (Time-Life
Encyclopedia of Gardening.) Alexandria, Va.,
Time-Life Books, 1979. (EGV)

What Flower is That? Stirling Macoboy. New
York, Crown, 1971. (MWF)

Wild Animals of the World. T.L.C. Tomkins.
Illustrated by Rein Stuurman. (no place),
Pitman, 1962. (TWA)

Wild Australia. Michael K. Morcombe. New
York, Taplinger, 1972. (MWA)

Wild Flowers of Britain. Roger Phillips. New
York, Quick Fox, 1977. (PWF)

**The Wild Flowers of Britain and Northern
Europe.** Richard and Alastair Fitter.
Illustrated by Marjorie Blamey. New York,
Scribner, 1974. (WFB)

Wild Flowers of the Transvaal. Cythna Letty.
Johannesburg, South Africa, Hortors, 1962.
(LFT)

Wild Flowers of the World. Brian D. Morley. Illustrated by Barbara Everard. New York, G.P. Putnam's Sons, 1970. (EWF)

The Wild Garden. Lys de Bray. London, Weidenfeld and Nicolson, 1978. (BWG)

Wildflower Gardening. James Underwood Crockett and Oliver E. Allen. Illustrated by Richard Crist. (Time-Life Encyclopedia of Gardening.) Alexandria, Va., Time-Life Books, 1977. (EWG)

Wildlife Crisis. H.R.H. Prince Philip, Duke of Edinburgh, and James Fisher. New York, Cowles, 1970. (PWC)

Winter Gardens. Oliver E. Allen. (Time-Life Encyclopedia of Gardening.) Alexandria, Va., Time-Life Books, 1979. (EGW)

World Atlas of Birds. London, Mitchell Beazley, 1974. (WAB)

The World Encyclopedia of Animals. Maurice Burton, editor. New York, World Publishing, 1972. (WEA)

World Guide to Mammals. Nicole Duplaix and Noel Simon. Illustrated by Peter Barrett. New York, Crown, 1976. (DSM)

The World of Amphibians and Reptiles. Robert Mertens. New York, McGraw-Hill, 1960. (MAR)

The World of Amphibians and Reptiles. Milli Ubertazzi Tanara. Translated by Simon Pleasance. New York, Abbeville Press, 1978. (ART)

The World of Butterflies and Moths. Alexander B. Klots. New York, McGraw-Hill, (no date). (KBM)

The World of Flowers. Herbert Reisigl, editor. New York, Viking Press, 1964. (RWF)

The World of Insects. Paul Pesson. New York, McGraw-Hill, 1959. (PWI).

The World of Insects. Adriano Zanetti. Translated by Catherine Atthill. New York, Abbeville Press, 1978. (ZWI)

The World of Reptiles and Amphibians. John Honders, editor. New York, Peebles Press, 1975. (HWR)

The World of Wildflowers and Trees. Uberto Tosco. New York, Crown, 1973. (WWT)

The World's Vanishing Animals: The Mammals. Cyril Littlewood. Illustrated by D.W. Ovenden. New York, Arco, 1970. (LVA)

The World's Vanishing Birds. Cyril Littlewood. Illustrated by D.W. Ovenden. New York, Arco, 1972. (LVB)